HANDBOOK OF FOREST RESOURCE ECONOMICS

It is increasingly recognized that the economic value of forests is not merely the production of timber. Forests provide other key ecosystem services, such as being sinks for greenhouse gases, hotspots of biodiversity and areas for tourism and recreation. They are also vitally important in preventing soil erosion and controlling water supplies, as well as providing nontimber forest products and supporting the livelihoods of many local people.

This handbook provides a detailed, comprehensive and broad coverage of forest economics, including traditional forest economics of timber production, economics of the environmental role of forests and recent developments in forest economics. The chapters are grouped into six parts: fundamental topics in forest resource economics; economics of forest ecosystems; economics of forests, climate change and bioenergy; economics of risk, uncertainty and natural disturbances; economics of forest property rights and certification; and emerging issues and developments. Written by leading environmental, forest and natural resource economists, the book represents a definitive reference volume for students of economics, environment, forestry and natural resource economics and management.

Shashi Kant is Professor of Forest Resource Economics and Management, Faculty of Forestry, University of Toronto, Canada. He is also leader of the Forest Resource Economics group of the International Union of Forest Research Organizations.

Janaki R. R. Alavalapati is Professor and Head, Department of Forest Resources and Environmental Conservation, Virginia Polytechnic and State University, Blacksburg, Virginia, USA.

HANDBOOK OF FOREST RESOURCE ECONOMICS

*Edited by Shashi Kant and
Janaki R. R. Alavalapati*

LONDON AND NEW YORK

First published 2014 by Routledge

2 Park Square, Milton Park, Abingdon, Oxfordshire OX14 4RN

711 Third Avenue, New York, NY 10017

Routledge is an imprint of the Taylor & Francis Group, an informa business

First issued in paperback 2017

British Library Cataloguing-in-Publication Data
A catalogue record for this book is available from the British Library

Library of Congress Cataloging-in-Publication Data
Handbook of forest resource economics / edited by Shashi Kant and Janaki R.R. Alavalapati.
    pages cm
    Includes bibliographical references and index.
  1. Forests and forestry—Economic aspects.   2. Forest ecology—Economic aspects.
  I. Kant, Shashi, 1957–.   II. Alavalapati, Janaki R. R.
  SD393.H35 2014
  333.75—dc23
  2013039214

ISBN: 978-0-415-62324-7 (hbk)
ISBN: 978-1-138-57318-5 (pbk)

Typeset in Bembo
by Apex CoVantage, LLC

CONTENTS

ABOUT THE CONTRIBUTORS

Janaki R. R. Alavalapati is Professor and Head of the Department of Forest Resources and Environmental Conservation at Virginia Tech. He is also a Senior Fellow of the Energy and Climate Partnership of the Americas with the US Department of State. His work focuses on market solutions to forest and natural resource problems at regional, national and global levels.

Gregory S. Amacher is the Julian N. Cheatham Professor of Natural Resource Economics in the Department of Forest Resources and Environmental Conservation at Virginia Tech. He has written more than 100 journal articles and two books focusing on topics such as forest land-owner decisions, incentives facing poor households in developing countries and the choice of policy instruments in cases where natural resources can be exploited.

Arild Angelsen is Professor at the School of Economics and Business, Norwegian University of Life Sciences, and a former Scientist and Associate at the Center for International Forestry Research (CIFOR), Indonesia. He has done extensive research on tropical deforestation, environmental income and forests and climate (REDD+).

Juliann Aukema is a Research Scientist at the National Center for Ecological Analysis and Synthesis at the University of California, Santa Barbara. Her research interests in ecology and conservation biology span plant-animal interactions and forest ecology, invasive pests, climate change and the interactions of people and ecosystems.

Onil Banerjee is an International Development Economist with the Commonwealth Scientific and Industrial Research Organization (CSIRO), Australia. His research focuses on developing the evidence base for improving land and water management policy and practice in South Asia and Latin America.

Tim Bogle is a veteran analyst and professional forester working in British Columbia. He uses his expertise in forest systems analysis, integrating ecological, economic and institutional components to advise public policy. His recent work explores the influence of tenure arrangement on public objectives during times of catastrophic natural disturbance.

Ole Martin Bollandsås has been a Researcher at the Norwegian University of Life Sciences since 2008. His fields of interest are mainly forest inventory and development of forest management models (biomass, volume, growth, recruitment, mortality, etc.).

Kevin J. Boyle is Professor in the Department of Agricultural and Applied Economics at Virginia Tech and Director of the undergraduate degree program in real estate. He has published extensively on the development of choice modeling to estimate economic values for natural resources, and many of the applications of his research have been on ecological aspects of forest ecosystems.

Richard J. Brazee is Associate Professor of Forestry and Natural Resource Economics in the Departments of Natural Resources and Environmental Sciences and Agricultural and Consumer Economics at the University of Illinois at Urbana-Champaign. His research focuses on developing forest economics models.

Gary Q. Bull is Professor of Forest Economics and Management at the University of British Columbia. He focuses on carbon/bioenergy economics, international trade in forest products, forest carbon financing and timber supply. He has worked as a management consultant, an economist for forest products companies and for the Food and Agriculture Organization.

Joseph Buongiorno is the Class of 1933 and John N. McGovern Professor Emeritus in the Department of Forest and Wildlife Ecology at the University of Wisconsin, Madison, and a foreign member of the French Academy of Agriculture.

Thomas Burkhardt graduated in physics and economics. Since 2001, he has held the Chair of Finance at Koblenz University in Germany. His primary research interests are financial decision making, asset management, derivatives, and long-term investment decisions, including forest investments. His preferred methods include mathematical modeling and optimization as well as economic experiments.

Sun Joseph Chang is Professor of Forest Management, Economics and Policy at Louisiana State University. His major research interests include the application of the generalized Faustmann formula in forest management, valuation, finance and taxation, as well as sawing optimization, strategies for reducing carbon dioxide emissions and the economics of tropical rainforest conservation.

Mindy S. Crandall is a Ph.D. Candidate in Applied Economics at Oregon State University. Her research includes use of forest products as rural development tools, outcomes in forest-based communities and rural poverty. She has degrees in Forest Management and Resource Economics.

Frederick W. Cubbage is Professor in the Department of Forestry and Environmental Resources at North Carolina State University. His current research includes analysis of global timber investments, natural resource policy, certification and agroforestry. He has authored or co-authored more than 400 publications and several books and teaches natural resource policy and forest economics.

Peter Deegen is Professor in the field of Forest Resource Economics at the Faculty of Environmental Sciences, Dresden University of Technology. His research areas are economics of silviculture, institutions and constitutional contracts of individuals concerning natural resources and the environment and demand and supply of forestry goods through politics.

Anna Ebers is a Doctoral Candidate in Energy and Environmental Economics and a Fulbright Fellow at the SUNY School of Environmental Science and Forestry in Syracuse, New York. Her research focuses on forest biomass policy analysis, investment and policy uncertainty in the renewable energy sector and economics of rural electrification in developing countries.

Jeffrey Englin is a Professor at Arizona State University. He received his B.S. in Economics from the University of Minnesota and his M.A. and Ph.D. in Economics from the University of Washington. His research has focused on applied econometrics with an emphasis on the valuation of the environment.

Gregory E. Frey is an Assistant Professor at Virginia State University, where he investigates forest systems that potentially produce public and private benefits. Interests include agroforestry, small-scale forestry and urban forestry. Previously, he worked at the World Bank, working with REDD+ Readiness strategies and at North Carolina State University.

Johannes Gerst holds Bachelor and Master of Science degrees from the Faculty of Forest Sciences and Forest Ecology at the University of Göttingen. Since 2010, he has been a member of the Department of Forest Economics and Forest Utilization at the Georg-August-Universität Göttingen.

Awang Noor Abd. Ghani is Professor in the Faculty of Forestry, Universiti Putra Malaysia, Serdang, Selangor, Malaysia. His research focuses on economic valuation of ecosystem services, economics of sustainable forest management, timber pricing, investment analysis of forestry projects, economic impacts of forest development and socioeconomics of forestry workers. He holds a B.S. in Forestry from Universiti Pertanian Malaysia and M.S. and Ph.D. degrees in Forest Economics from Michigan State University.

Terje Gobakken received his M.S. and Ph.D. degrees from the Agricultural University of Norway in 1995 and 2001, respectively. He has been working at the Norwegian National Forest Inventory and is now a Professor at the Norwegian University of Life Sciences.

Yazhen Gong is currently Assistant Professor in the School of Environment and Natural Resources, Renmin University, China. She received her Ph.D. in forest resource economics from the University of British Columbia. She is among the first in China to use field experimental methods to study resource users' behaviors in their resource management.

Haripriya Gundimeda is currently working as an Associate Professor in the Department of Humanities and Social Sciences at the Indian Institute of Technology in Bombay. She was the coordinator of the report *The Economics of Ecosystems and Biodiversity* for local and regional policy. She is actively involved in developing a framework for green national accounts in India.

Robert G. Haight is a Research Forester with the US Forest Service, Northern Research Station, in St. Paul, Minnesota. He studies the economics of public programs for invasive species management, wildfire management and metropolitan open space protection.

Espen Halvorsen holds an M.S. in Economics of Forest Industries from the Norwegian University of Life Sciences, where he was a Ph.D. candidate while conducting this study. He is currently an Emergency Medical Technician.

Ravi Hegde has a Ph.D. in Forest Resource Economics from the University of British Columbia, where his thesis investigated an agroforestry-based carbon sequestration project in Mozambique. He currently works as an Agriculture and Resource Economist at Horticulture Australia in Sydney.

Donald G. Hodges is the James R. Cox Professor of Forest Economics and Policy at the University of Tennessee. He also serves as Director of the University of Tennessee Natural Resource Policy Center. His research focuses on private forest management decisions, sustainable forest management, forest policy and ecosystem service markets.

Ole Hofstad is Professor of Forest Resource Economics at the Norwegian University of Life Sciences. His research interests are divided between Norwegian and tropical forest management. He has taught forest economics, forest policy, multiple-use forestry and nature-based tourism. He has 6 years of work experience in East Africa.

Thomas P. Holmes is a Research Forest Economist with the US Forest Service, Southern Research Station, Research Triangle Park, North Carolina. His research interests include the economic evaluation of forest protection programs, advancements in market and nonmarket valuation techniques, new methods for analyzing forest disturbance impacts and sustainable management options for tropical forests.

Martin Hostettler is Managing Partner of Cycad Inc. in Switzerland. He received a master's degree in Forest Engineering from the Swiss Federal Institute of Technology (ETH Zurich). He works in the fields of forest policy and environmental engineering. His research interests are forest economics, Austrian methodology, constitutional political economy and property rights. He is Chairman of the Editorial Board of the *Swiss Forestry Journal* and lectures at Dresden University of Technology.

Anwar Hussain earned a master's degree in Economics from the University of Peshawar, Pakistan, as well as a master's degree in Agricultural and Applied Economics and a Ph.D. in Forestry and Agricultural Economics from the University of Minnesota. His research interests include natural resource economics, regional economic analysis, applied econometrics and general equilibrium modeling.

William F. Hyde is an economist and policy analyst who has taught at Duke University and Virginia Tech. He has also served as a senior researcher with Resources for the Future and with the US government's Economic Research Service. He has participated on the editorial boards of seven international journals, including serving as the editor of *Forest Science*. His most recent book is *The Global Economics of Forestry* (RFF Press, 2012).

Jette Bredahl Jacobsen is Professor of Environmental Valuation and Management of Ecosystem Services at the University of Copenhagen. She has a background in forestry and has been working with various topics within environmental valuation, especially biodiversity, risk and uncertainty and forest management.

Craig Johnston is a Ph.D. student in the Department of Economics at the University of Victoria, Canada. His recent work uses advanced methods in mathematical programming to determine optimal forest management strategies when considering the global carbon cycle.

Shashi Kant is Professor of Forest Resource Economics and Management at the Faculty of Forestry, University of Toronto. In 2013, he published a book titled *Post-Faustmann Forest Resource Economics* and has guest edited a special issue of *Forest Policy and Economics* on New Frontiers of Forest Economics.

Rajendra Prasad Khajuria is an Indian Forest Service officer who holds a Ph.D. in Forest Economics from the University of Toronto. He has vast field experience in forestry and environment management. He has authored several reports and published papers on decision making in forestry under uncertainty.

G. Cornelis van Kooten is Professor and Canada Research Chair in the Department of Economics at the University of Victoria, Canada. He is also a Senior Research Fellow in the Agricultural Economics Institute at the Hague and a member of the Department of Social Sciences, Wageningen University, the Netherlands.

Kent Kovacs is an Assistant Professor in the Department of Agricultural Economics and Agribusiness at the University of Arkansas. His forest research includes the topics of invasive species and climate change. His economics expertise is in nonmarket valuation, optimal control simulation and resource forecast modeling.

Pradeep Kumar is a Ph.D. candidate in Forest Economics at the Faculty of Forestry, University of Toronto. His research interests include game-theoretic modeling, joint forest management and nonmarket valuation. He has about 13 years of experience in creating, managing and protecting forests in a joint forest management setup.

Pankaj Lal is Assistant Professor of Environmental Economics and Policy in the Earth and Environmental Studies Department at Montclair State University in New Jersey. His research focuses on bioenergy, climate change, modeling and analyses of forest and agricultural markets, sustainable forest management and collective resource management and conservation.

Larry A. Leefers is an Associate Professor in the Department of Forestry at Michigan State University. His research focuses on property value effects of natural resource amenities, timber supply, timber pricing and economic impacts of natural resource management and wood-based industries. He holds a B.S. in Forestry from Southern Illinois University and M.S. and Ph.D. degrees in Forest Economics from Michigan State University.

Brian Leung is Associate Professor of Biology in the School of the Environment at McGill University. His research interests include ecological forecasting of invasive species (which species will arrive, where they will occur and what effects will they have) and using decision theory and bioeconomics to structure their management.

M. K. (Marty) Luckert is Professor of Forest and Natural Resource Economics and Policy in the Department of Resource Economics and Environmental Sociology at the University of Alberta. He has been studying forest tenures in Canada for more than 30 years and has been involved in numerous policymaking processes across Canada and abroad.

Thomas Hedemark Lundhede is Associate Professor in Forest Economics at the University of Copenhagen. His research interests focus on the socioeconomic aspects of biodiversity and conservation, in particular revealing society's preferences for different species by means of non-market valuation techniques and econometric modelling.

D. Evan Mercer is a Research Economist at the US Forest Service Southern Research Station, Research Triangle Park, North Carolina, and an Adjunct Professor at Duke and North Carolina State Universities. Mercer's research focuses on the effects of government policies, market factors and societal values on the production of multiple benefits from public and private forests.

Bernhard Möhring studied forestry and obtained his Ph.D. and habilitation at Göttingen University. Later he was Forest Officer in Lower Saxony. Since 1997, he has held the Chair of Forest Economics at the University of Göttingen. His main research interests include forest decision making, management, accounting, valuation and taxation.

Claire A. Montgomery is a Professor in the Department of Forest Engineering, Resources and Management at Oregon State University. Her research includes economics of forest land management for wood products, biomass and biodiversity; economics of wildland fire; and housing markets. She teaches microeconomic theory, forest resource economics and investment analysis.

Susan E. Moore has been Extension Associate Professor and Director of the Forestry and Environmental Outreach Program in the Department of Forestry and Environmental Resources at North Carolina State University since 2000. Prior to that, she worked in forestry consulting and as the Forested Wetlands Extension Specialist at the University of Florida.

Ian A. Munn is Professor of Forestry at Mississippi State University, specializing in natural resource economics and forest management. He has a Ph.D. in Economics and Forestry from North Carolina State University. His research interests include nonindustrial private forestland owners, human dimensions of wildlife management and the forest products industry.

David Newman is Professor of Forest Resource Economics and Policy and Chair of the Department of Forest and Natural Resource Management at the SUNY School of Environmental Science and Forestry in Syracuse, New York. His primary research has been in the areas of returns to research in forestry, timber supply and the role of government policy in landowner decision making.

David Nogués-Bravo is Associate Professor of Global Change Biology at the University of Copenhagen. His research aims at unveiling the drivers of biological diversity, species extinctions integrating genomics, phylogeography and niche modeling.

Markku Ollikainen is Professor of Environmental and Resource Economics at the University of Helsinki. He has contributed to the topics of optimal forest taxation, biodiversity conservation, intergenerational use of forest resources and, recently, water quality and bioenergy issues in forestry. He is a member of the Finnish Scientific Climate Policy Panel.

John Perez-Garcia is a Professor in the School of Environmental and Forestry Sciences, College of the Environment, University of Washington. He manages the CINTRAFOR Global Trade Model (CGTM). With more than two decades of experience, he uses the model in a variety of trade policy, environmental and economic assessment studies.

Neelam C. Poudyal is an Assistant Professor of Natural Resource Policy and Human Dimensions, Department of Forestry, Wildlife and Fisheries, University of Tennessee, USA. His research focuses on economics, policy and human dimensions issues in forestry and natural resource management.

Colin Price holds degrees from Oxford University. His academic career was spent at Oxford Brookes, Oxford and Bangor Universities. He is now a freelance academic. His books include *Landscape Economics* (Macmillan, 1978), *The Theory and Application of Forest Economics* (Blackwell, 1989) and *Time, Discounting and Value* (Blackwell, 1993). He is the author of more than 200 academic papers, many concerning the economics of time.

Carsten Rahbek is Professor of Macroecology at the University of Copenhagen. He is leader of the Danish National Research Foundation Center of Macroecology, Evolution and Climate. His research interests are patterns of species distribution, species range sizes, species assemblages and species richness.

Alicia Robbins is a Postdoctoral Research Associate at the School of Environmental and Forest Sciences, College of the Environment, University of Washington. Her research addresses natural resource use and policy, primarily in China and the United States.

Yueqin Shen is Professor of Forest Economics and Policy and Dean of the College of Economics and Management, Zhejiang A&F University. She received her Ph.D. from Beijing Forestry University. She has been leading the group of Forest Economics studies at Zhejiang A&F University for more than a decade and has published extensively on forest economics and policy.

Erin Sills is Professor of Forest Economics and Coordinator of International Programs in the Department of Forestry and Environmental Resources at North Carolina State University. She has a Ph.D. in Natural Resource Economics from Duke University and a B.A. in Public Policy from Princeton University.

Brent Sohngen is Professor of Natural Resource and Environmental Economics in the Department of Agricultural, Environmental and Development Economics at Ohio State University. He has done extensive work in valuing environmental change, modeling land-use/land-cover change, timber market modeling and economics of nonpoint-source pollution.

Niels Strange is Professor of Forest and Management Planning at the University of Copenhagen. He has a background in forestry, and his research covers forest and nature management, conservation and spatial planning, environmental economics and decision making under uncertainty.

Bo Jellesmark Thorsen is Professor of Applied Economics of Forest and Natural Resources at the University of Copenhagen. His research covers a broad set of topics ranging from decision making under uncertainty over various regulation and policy issues to the valuation of externalities and management of ecosystem services.

Anne Toppinen has been a Professor at the University of Helsinki since 2008. Before that, she held research positions at EFI, Finnish Forest Research Institute and Oregon State University. She has authored about 70 scientific articles in such journals as *Applied Economics*, *Forest Science* and *Corporate Social Responsibility and Environmental Management* and more than 200 other publications.

Sen Wang is associated with the Canadian Forest Service. He specializes in economic analyses of issues concerning natural resources, with a focus on forest management. He holds a Ph.D. from the University of British Columbia. He is an Adjunct Professor at the University of Toronto and Beijing Forestry University.

David Wear has served as a Researcher with the US Forest Service since 1987 and now leads a new research center at North Carolina State University focused on applying forest science directly to policy-relevant questions. His methods focus on linking economics and ecology to address complex resource problems.

Yali Wen is Professor of Forest Economics and Policy and Associate Dean of the College of Economics and Management, Beijing Forestry University. He received his Ph.D. from Beijing Forestry University. He has been leader of the group of Forest Economics studies at Beijing Forestry University for more than a decade and has published extensively on forest economics and policy.

Yi Xie is Associate Professor of Forest Economics and Policy in the College of Economics and Management at Beijing Forestry University. He received his Ph.D. from Beijing Forestry University and has been a Visiting Researcher at the Swedish Agriculture University for 1 year. His recent research interests cover forest tenure reform, farmers' forest management behavior and the forest land-use rights market in China. He has published extensively on forest economics and policy.

Zhen Xu is a Ph.D. student in Geography at the University of Victoria, Canada. His research interests lie in forest economics with forest disturbances and climate change. He has conducted several research projects on the economic effects of wildfires and mountain pine beetle outbreaks on the forest industry and public policies in British Columbia.

Daowei Zhang is the Alumni and George Peake Jr. Professor of Forest Economics and Policy in the School of Forestry and Wildlife Sciences at Auburn University. He served on the Board of Directors of the Pinchot Institute for Conservation and is a recipient of the Society of American Foresters Award in Forest Science as well as multiple Auburn Author Awards.

Yaoqi Zhang is Professor of Forest Economics and Policy in the School of Forestry and Wildlife Sciences at Auburn University. He received his Ph.D. from the University of Helsinki in 2001. He has worked at the Chinese Academy of Forestry, University of Toronto and University of Alberta. He has published close to 100 journal articles and book chapters.

1

EVOLVING FOREST RESOURCE ECONOMIC THOUGHT

Shashi Kant[1] and Janaki R. R. Alavalapati[2]

[1]PROFESSOR, FOREST RESOURCE ECONOMICS AND MANAGEMENT, FACULTY OF FORESTRY, UNIVERSITY OF TORONTO, 33 WILLCOCKS STREET, TORONTO, ONTARIO, M5S 3B3, CANADA. SHASHI.KANT@UTORONTO.CA
[2]PROFESSOR AND HEAD, DEPARTMENT OF FOREST RESOURCES AND ENVIRONMENTAL CONSERVATION, COLLEGE OF NATURAL RESOURCES AND ENVIRONMENT, VIRGINIA TECH, BLACKSBURG, VA 24061, USA. JRRA@VT.EDU

Abstract

This introductory chapter provides an overview of the contents of this handbook. First, a historical perspective of dominant forest economic thought is provided, and six themes of forest resource economics are identified. These themes are discussed in the form of six interrelated parts of the handbook. For each part, an overview is provided, followed by short reviews of its chapters.

Keywords

Bioenergy, climate change, ecosystems, forest economics, forest certification, natural disturbances, property rights, risk and uncertainty

Introduction

The links between forests and human beings are as old as the existence of human life on this planet. The origin of forest economic thought can be traced back to the early phases of *Homo sapiens* because forests have always been a resource for human welfare. As humans evolved from being hunters and gatherers through the agrarian and industrial eras to the information and technology era, formal concepts, principles and theories relating to forest economics have also evolved. As such, economic thought that was relevant and rational in one era may be irrelevant and irrational in another. There are situations in which some concepts, theories and technologies become dominant due to path dependence and positive feedback effects, whereas others remain dormant (Arthur, 1994). In addition, many path-breaking ideas and concepts are never transmitted from one place to another or from one culture to another due to communication and cultural barriers. Hence, it is impossible to trace the origin and evolution of forest economic thought.

Evidence suggests that Kautilya (or Chanakya) discussed some economic aspects of forest resources in his famous book *Arthashastra* (economics) written during the fourth century BC in India (Basu, 2011). It is also believed that the first discussion of economic harvesting in Germany was held in the monasteries of Mauermunster during the 1100s (Amacher, Ollikainen and Koskela, 2009). During the 1700s, Denmark and England played a dominant role in developing basic concepts of forest economic thought (Amacher et al., 2009). Danish Count C.D.F. Reventflow proposed an economic theory of optimal forest rotation as early as 1801 (Helles and Linddal, 1997). Englishman William Marshall, in his writings in 1790 and 1809, suggested the need to include the opportunity cost of growing trees and the cost of occupying the land in the calculation of optimal forest rotation (Scorgie and Kennedy, 2000). Irrespective of these early writings, the origin of current dominant forest economic thought is largely attributed to Martin Faustmann's paper published in 1849.

In the first half of the 1800s, many foresters of Germany, such as Friedrich Pfeil, Gottlieb König and Johan Hundeshagen, published economic aspects of forest management in the first journal of forest science, *die Allgemeine Forst- und Jagt Zeitung*, which was started in 1824 (Amacher et al., 2009). However, it was the article by Edmund von Gehren on the determination of land value published in the same journal in 1849 that attracted the attention of Martin Faustmann, who published his critique and offered a different approach to calculate land value in the same year. In 1850, Pressler supported Faustmann's approach with a mathematical formulation (Pressler, 1860). In 1921, Bertin Ohlin, a Swedish economist, also presented a mathematical formulation of optimal forest rotation (Ohlin, 1921). Hence, Faustmann, Pressler and Ohlin are considered the founders of forest economic thought, which remained unnoticed by the English-speaking world for almost a century. The earliest reference to Faustmann's formulation in English was Gaffney (1957), followed by Bentley and Teeguarden (1965) and Pearse (1967). Faustmann's paper was translated into English in 1968. Samuelson (1976) gave the credit for current economic thought to Faustmann's formulation, and since then, Faustmann's formulation has become the cornerstone of forest economics (Newman, 2002).

Irrespective of the origin of current forest economic thought, two aspects – optimal forest rotation and the choice of discount rate – have dominated discussions in forest economics for the past 50 years. The ownership of forests and the trade of forest products are two other aspects that have been discussed heavily. The issue of ownership has multiple aspects. About 75% of global forests are publicly owned, whereas about 14% are privately owned (White and Martin, 2002). In the case of public forests, determining optimal timber prices is a challenging economic issue because of a large single ownership that does not satisfy the conditions of a competitive market. In the case of private forests, the challenge is to design economically optimal tax policies to advance societal goals. Another complexity arises when different forest owners have different forest management objectives. Similarly, forest products have been locally and internationally traded for centuries, and an understanding of trade issues is just as critical as understanding the local economic issues associated with ownership.

Although the foundations of forest economic thought laid by German foresters mainly focused on timber resources, the importance of nontimber resources in decision making started to emerge in the 1970s. In 1976, Hartman incorporated nontimber resources in determining optimal forest economics rotation (Hartman, 1976). Since then, efforts to advance nonmarket evaluation techniques to quantify the value of ecosystem services such as outdoor recreation, biodiversity, clean air and clean water have been intensified.

Climate change seems to be the greatest environmental challenge of the twenty-first century. Forest carbon sequestration and storage has been shown to play a critical role in mitigating climate change. For example, Bonan (2008) found that carbon sequestration in forest ecosystems was close to one-third of carbon emissions from the use of fossil fuels and land-use change. Approximately 75% of total terrestrial biomass carbon and more than 40% of soil organic carbon are stored in forest ecosystems (Jandl et al., 2007). Hence, the economics of climate change must be an integral part of forest management and conservation strategies.

The risks and uncertainties associated with markets and natural processes such as climate change, forest fires and biological invasion of species have stimulated many forest economists to incorporate them into the analysis.

The Faustmann formulation assumes that a forest owner operates under the conditions of a 'private property' that includes exclusive, perpetual, transferable and unfettered property rights. Forest ecosystems provide a web of goods and services that include private goods, public goods, common-pool goods and club goods; therefore, a simple concept of resource ownership may not be good enough for economic analysis of forest ecosystems (Kant, 2000). In fact, government's role in regulating and managing forests arises due to the existence of multiple types of goods and associated market failures (Kant, 2003a). Forest ecosystems are specifically susceptible to market failures because they are expected to contribute not only to the private goals of the forest owner, but also to social objectives, including the state of the environment. Most governments play an active role in designing forest property rights arrangements to achieve private as well as social goals. Hence, the economics of forest property rights has become a very important component of current forest economic thought.

Finally, there are many economic aspects of forests that cannot be dealt with in the boundaries of the Faustmann framework, and that leads to gaps between theoretical economic models and forestry practices. Kant (2003b, 2013) observed that the economics profession, as a whole, has been re-examining and challenging almost every basis of neoclassical economic thought, in order to reduce the gap between theoretical models and practices. Hence, it is imperative for forest economists to extend the boundaries of forest economics beyond Faustmann's economic thought. The forest economics profession seems to have taken up this challenge by drawing concepts from other streams of economics, such as new institutional economics and political economy.

Keeping these six themes of forest economics in perspective, we have divided this book into six parts. Each part contains chapters focusing on specific issues related to its theme. There is some continuity, including linkages, among the chapters of each part; however, each chapter stands alone. Given the importance of the fundamental topics that have been the main attraction of forest resource economics for 60 years or more, we start this book with Part 1, focusing on fundamental topics, and close it with Part 6, which focuses on emerging issues and developments.

Part 1: Fundamental topics in forest resource economics

The focus of Part 1 is on four topics – Faustmann's formulation, rate of discount, ownership and international trade of forest products. In Chapter 2, Deegen and Hostettler note that although the Faustmann model is a useful tool for making an economic decision, the underlying process of market mechanisms, known as catallactics, is also very critical. The authors discuss theoretical concepts and provide an overview of selected contributions of forestry to the inner processes of market functioning. In Chapter 3, Chang discusses the generalized Faustmann formula that allows stumpage prices, stand volumes, annual incomes, regeneration costs and interest rates to vary from timber crop to timber crop. As a result, optimal management and/or optimal rotation would also vary from timber crop to timber crop. Chang notes that this formulation represents a more realistic world relative to Faustmann's world, in which everything remains static forever.

Price, in Chapter 4, highlights various economic and ethical perspectives associated with different economic justifications for discounting, such as opportunity cost, time preference, diminishing marginal utility, declining discount rate and internal rate of return.

Next, three chapters are focused on economic issues associated with ownership. In Chapter 5, Wear presents US forest policy history and forest economics research related to timber supply by ownership groups. He raises many important issues in light of new models of private ownership, such as Timber Investment Management Organizations (TIMO) and Real Estate Investment Trusts (REIT). Leefers and Ghani, in Chapter 6, focus on various timber-pricing mechanisms such as administered charges, negotiated values and market-derived prices – the residual value method and transactions evidence method – used by governments. Ollikainen, in Chapter 7, reviews the results of forest taxation in the Faustmann and Hartman framework, discusses best and second-best forest tax policies, and relates the discussion to modern forest policies promoting ecosystem services such as biodiversity benefits, climate mitigation and nutrient loading. Finally, in Chapter 8, Perez-Garcia and Robbins provide an overview of global forest products trade, discuss economic theory and empirical models of trade and present economic assessments of selected forest products trade policies.

Part 2: Economics of forest ecosystems

Part 2 covers three topics – valuation methods for ecosystem services, economics of specific ecosystems and payment mechanisms for ecosystem services. In Chapter 9, Boyle and Holmes provide an overview of valuation methods and expand on choice experiments. The authors present the latest information on choice experiment methodologies and then discuss their applications to forest ecosystems. The next four chapters are focused on the economics of different forest ecosystems. In Chapter 10, Montgomery and Crandall place old-growth forests within the context of the Faustmann and Hotelling models and discuss old-growth forest values and methods of their measurement. Poudyal and Hodges, in Chapter 11, focus on the economics of open spaces (or green spaces) in urban environments. In particular, they review measures of open spaces, valuation methods (with an emphasis on hedonic price method) and recent studies in open space valuation. Chapter 12 focuses on forest ecosystems that are used to manage game and recreational hunting. Here, Munn and Hussain present the institutional context of these ecosystems in the United States, insights about hunting lease markets of the south-eastern United States and economy-wide implications of wildlife-associated recreation activities. Mercer, Frey and Cubbage, in Chapter 13, focus on the economics of agroforestry systems and review economic principles and approaches to assess agroforestry systems and demonstrate their application through a case study. The focus of the last chapter in Part 2, Chapter 14, is on the status of payment for ecosystem services schemes in developing countries. In particular, Gong, Hegde and Bull discuss schemes for watershed services, biodiversity conservation and forest carbon and present lessons learned and future challenges.

Part 3: Economics of forests, climate change and bioenergy

There are three very important aspects associated with climate change and forests. First, climate change will impact the productivity of forests and thus the forestry sector. Second, forests can be managed to sequester carbon, thereby moderating climate change. Third, carbon emissions can be reduced by using wood as a source of energy and by reducing forest degradation and deforestation. In this part, economic issues associated with the previous three aspects are discussed.

Part 3 begins with Chapter 15, in which Sohngen discusses the potential impacts of climate change on forest ecosystems and reviews studies that have analyzed the impact of climate change on the forest sector. In Chapter 16, van Kooten, Johnston and Xu discuss economic issues related to the creation of forest carbon offset credits through forest management strategies and the problems associated with additionality, leakage, duration or impermanence and governance. Buongiorno, Bollandsås, Halvorsen, Gobakken and Hofstad, in Chapter 17, focus on the economics of carbon storage through uneven-aged forest management strategies and present methods to derive a schedule of supply for carbon storage. Lal and Alavalapati, in Chapter 18, discuss economic aspects of forest biomass-based energy, including forest biomass supply, public preferences for woody bioenergy, competition with traditional forest industries, land-use change and greenhouse gas emissions. Part 3 concludes with Chapter 19, in which Angelsen focuses on the economics of REDD+ (Reducing Emissions from Deforestation and Forest Degradation) and presents four broad themes: REDD+ credits in international carbon markets, REDD+ as performance-based aid, national and local payment for ecosystem services and other national policy approaches to curb deforestation.

Part 4: Economics of risk, uncertainty and natural disturbances

Risk and uncertainty associated with natural phenomena, such as climate change, forest fires and biological invasions, and the growth process of forests and markets are important aspects of forest economics. In Chapter 20, Amacher and Brazee review the literature on risk and forest landowner decisions and elaborate on two themes – risk associated with future market parameters and risk associated with established forest stands being subject to natural or catastrophic events before harvest. Burkhardt, Möhring and Gerst, in Chapter 21, present a stochastic model that incorporates risk as a survival function to calculate land value and optimal rotation defined in terms of expectations suitable for a risk-neutral decision maker. In Chapter 22, Khajuria focuses on the applications of real options analysis to forest harvesting and conservation decisions. He discusses the literature that has modeled timber prices as the geometric Brownian motion, mean reversion, mean reversion with jumps and mean reversion with varying long-run marginal cost process. Strange et al., in Chapter 23, focus on economically optimal and biologically sound conservation decisions in an uncertain world and discuss theoretically consistent approaches that combine biodiversity and valuation modeling under uncertainty. Holmes et al., in Chapter 24, focus on the economic analysis of preinvasion and postinvasion management of biological invasions of forests under risk and uncertainty conditions and suggest new microeconomic and aggregate economic studies of damages caused by biological invasions across forest types and ownerships.

Part 5: Economics of forest property rights and certification

Some economic aspects associated with ownership are discussed in Part 1. However, the concept of property rights is so complex and issues are so diverse that it requires a separate part rather than combining it with other topics. In Part 5, four chapters are devoted to property rights issues – one chapter provides a broader and general perspective, and the other three provide national perspectives for Brazil, China and the United States. The last chapter deals with the economics of forest certification, which has some property rights implications.

Luckert, in Chapter 25, discusses various economic concepts relating to forest tenures, including rules as attenuations and subsidies, forest tenures and economic behavior, economic rent and market and government failures, and then explores the challenges in analyzing economic impacts of forest tenures. The focus of Chapter 26 is on the economics of the evolution of the Brazilian Amazon frontier. In this chapter, Sills discusses the historical drivers of deforestation, the Brazilian government policies that increased agricultural rents, new drivers of deforestation and current policy initiatives that seek to change the incentives by increasing tenure security for forest land, imposing penalties for illegal deforestation and creating new opportunities to earn revenue from standing forest. In Chapter 27, Zhang, Shen, Wen, Xie and Wang use changes in the bundle of rights to forests and forestland and the separation of use rights from ownership to examine the evolution of forest property rights in China. Ebers and Newman, in Chapter 28, focus on the economic analysis of conservation easements in the United States. They discuss landowner incentives for instituting conservation easements, methods for easement appraisal and ways to measure easement performance. In Chapter 29, Toppinen, Cubbage and Moore discuss the concepts, advantages and economic aspects of forest certification and corporate social responsibility and elaborate on the challenges of extending these approaches to smaller organizations and developing countries.

Part 6: Emerging issues and developments

The economics profession, as a whole, has been re-examining and challenging almost every basis of neoclassical thought in order to reduce the gap between theoretical models and practices or to increase the theory–evidence ratio. These efforts include the emergence of new streams of economics such as behavioral economics, evolutionary game theory and new institutional economics. Forest economists are also making similar efforts by incorporating these new streams of economics into forestry. The chapters of Part 6 are examples of such efforts. The first chapter in this part, Chapter 30, focuses on new institutional economics (NIE), and Wang, Bogle and van Kooten present an overview of the genesis, scope and main developments of NIE, with emphasis on property rights and contracting, transaction cost economics, moral hazard and information and principal–agent relationships. In Chapter 31, Zhang discusses various theories of political economy and their origin and reviews empirical studies of forestry in various countries. Kumar and Kant, in Chapter 32, provide an overview of game theory and review applications of game theoretic models to forestry issues such as people's participation in comanagement of forests, timber markets and interactions among stakeholders in the case of weak property rights. Gundimeda, in Chapter 33, emphasizes the need of expanded forest accounts and reviews two major approaches, namely, income as a return on wealth and income change as an indicator of welfare. Chapter 34 focuses on the applications of computable general equilibrium (CGE) modeling in forest economics. Banerjee and Alavalapati, in this chapter, present the application of a recursive dynamic CGE model to assess the regional economic impacts of Brazil's forest concessions policy in the Amazon.

We close the book with a chapter on twelve unanswered questions in forest economics. In this Chapter 35, Hyde observes that there are many situations in which Faustmann's formulation is either incomplete or inappropriate. The author identifies unresolved issues within the discipline of forest resource economics at the beginning of the twenty-first century and discusses two concerns – empirical assessment and incremental effects – for policy applications.

Conclusion

This is the first publication of a handbook of forest resource economics. We have tried to cover a wide range of issues associated with the subject, starting with fundamental topics and moving to recent emerging issues and developments. Each chapter provides a synthesis of the state of the topic covered and aims to be a comprehensive, up-to-date, authoritative source on the subject.

The current forest resource economic thought is more than 165 years old and is growing in many ways. The growth is largely coming because of an increased understanding of ecosystem services benefits for human welfare. Emerging global issues such as climate change, sustainable development and the green economy have provided further impetus to the growth and diversification of forest resource economics. The emergence of new streams in economics, such as agent-based computation economics, behavioral economics, complexity theory and economics, public choice theory and social choice theory, have also contributed to the growth of forest resource economics. Hence, it is impossible to cover all important topics in this volume, and we regret that.

References

Amacher, G. S., Ollikainen, M. and Koskela, E. (2009). *Economics of Forest Resources*, MIT Press, Cambridge, MA.

Arthur, W. B. (1994). *Increasing Returns and Path Dependence in the Economy,* University of Michigan Press, Ann Arbor, MI.

Basu, R. L. (2011). *Kautilya's Arthasastra (300 B.C.): Economic Ideas*, Smashwords Edition.

Bentley, W. R. and Teeguarden, D. E. (1965). 'Financial maturity: A theoretical review', *Forest Science*, vol. 11, no. 1, pp. 76–87.

Bonan, G. (2008). 'Forests and climate change: Forcings, feedbacks, and the climate benefits of forests', *Science*, vol. 320, pp. 1444–1449.

Faustmann, M. (1849). 'Berechnung des Wertes welchen Walboden sowie noch nicht haubare Holzbestände für die Waldwirschaft besitzen', *Allgemeine Forst und Jagd Zeitung*, vol. 15, pp. 441–455. Translated by W. Linnard (1968) as 'Calculation of the value which forest land and immature stands possess for forestry' in *Martin Faustmann and the Evolution of Discounted Cash Flow*, Inst. Pap. No. 42, Commonwealth Forestry Institute, Oxford, UK, pp. 27–55. (Reprinted in *Journal of Forest Economics*, vol. 1, no. 1, pp. 7–44)

Gaffney, M. M. (1957). *Concepts of Financial Maturity of Timber and Other Assets*, Agricultural Economics Information Series #62, North Carolina State University, Raleigh, NC.

Hartman, R. (1976). 'The harvesting decision when a standing forest has a value', *Economic Inquiry*, vol. 14, pp. 52–55.

Helles, F. and Linddal, M. (1997). 'Early Danish contributions to forest economics', *Journal of Forest Economics*, vol. 3, no. 1, pp. 87–103.

Jandl, R., Lindner, M., Vesterdal, L., Bauwens, B., Baritz, R., Hagedorn, F., . . . Bryne, K. A. (2007). 'How strongly can forest management influence soil carbon sequestration?' *Geoderma*, vol. 137, pp. 253–268.

Kant, S. (2000). 'A dynamic approach to forest regimes in developing economies', *Ecological Economics*, vol. 32, no. 2, pp. 287–300.

Kant, S. (2003a). 'Economic theory of emerging forest property rights', in *People and Forests in Harmony*, Proceedings of the XII World Forestry Congress, Quebec City, Canada.

Kant, S. (2003b). 'Extending the boundaries of forest economics', *Journal of Forest Policy and Economics*, vol. 5, pp. 39–58.

Kant, S. (Ed.) (2013). *Post-Faustmann Forest Resource Economics*, Springer, Dordrecht, The Netherlands.

Newman, D. H. (2002). 'Forestry's golden rule and the development of the optimal forest rotation literature', *Journal of Forest Economics*, vol. 8, no. 1, pp. 5–28.

Ohlin, B. (1921). 'Till frågan om skogarnas omloppstid', *Ekonomisk Tidskrift* 22. Translated by C. Hudson (1995) as 'Concerning the question of the rotation period in forestry', *Journal of Forest Economics*, vol. 1, no. 1, pp. 89–114.

Pearse, P. H. (1967). 'The optimum forest rotation', *Forestry Chronicle*, vol. 43, pp. 178–195.

Pressler, M. R. (1860). 'Aus der Holzzuwachlehre (zweiter Artikel)', *Allgemeine Forst und Jagd Zeitung*, vol. 36, pp. 173–191. Translated by W. Löwenstein and J. R. Wirkner (1995) as 'For the comprehension of net revenue silviculture and the management objectives derived thereof', *Journal of Forest Economics*, vol. 1, no. 1, pp. 45–87.

Samuelson, P. A. (1976). 'Economics of forestry in an evolving society', *Economic Inquiry*, vol. 14, pp. 466–492. (Reprinted in *Journal of Forest Economics*, vol. 1, no. 1, pp. 115–149)

Scorgie, M. and Kennedy, J. (2000). 'Who discovered the Faustmann condition?' *History of Political Economy*, vol. 28, pp. 77–80.

White, A. and Martin, A. (2002). *Who Owns the World's Forests? Forest Tenure and Public Forests in Transition*, Forest Trends and Centre for International Environmental Law, Washington, DC. Retrieved from www.forest-trends.org/documents/files/doc_159.pdf

PART 1

Fundamental topics in forest resource economics

2

THE FAUSTMANN APPROACH AND THE CATALLAXY IN FORESTRY

Peter Deegen[1] and Martin Hostettler[2]

[1]PROFESSORSHIP OF FOREST POLICY AND FOREST RESOURCE ECONOMICS, FACULTY OF
ENVIRONMENTAL SCIENCES, TECHNISCHE UNIVERSITÄT DRESDEN, 01735 THARANDT,
GERMANY. DEEGEN@FORST.TU-DRESDEN.DE
[2]CYCAD AG, 3011 BERN, SWITZERLAND. MARTIN.HOSTETTLER@CYCAD.CH

Abstract

There exist two different classes of market theories. One class, which is the well-known standard microeconomics, deals mainly with the results of the market process. The Faustmann model belongs here: The optimal rotation length is the very result of market exchanges. The other class of market theory focuses on the understanding of the underlying process or, in the slogan of Vernon Smith, of the intention 'to make the "invisible hand" "visible"'. This class of theory is called catallactics. The key problems of catallactics are how individuals coordinate their decentralized knowledge through exchange, how prices carry that knowledge from individual to individual and how individuals discover new answers for unanticipated changes via market competition. Those questions are of major interest for understanding the complexity of forestry, contemporary and in the long term. Because catallactics is not as well established in forestry economics as its microeconomic counterpart, each section of this chapter comprises two parts. One part presents a brief introduction into the theoretical concepts of the market process from the catallactic point of view. The other part refers to, summarizes and systematizes selected contributions of forestry to the understanding of the inner process of coordination through selling and buying.

Keywords

Competition, coordination, entrepreneur, exchange, Faustmann model, human action, knowledge, prices, unanticipated changes

Catallactics, the economics for understanding the market process

This section is about the coordination of human actions through selling and buying applied to the field of forestry. The beginning of the study of this kind of coordination can be traced back

11

directly to Adam Smith. He discovered that selling and buying leads to satisfactory results for any individual in the society. Moreover, in his two main books, *The Theory of Moral Sentiments* (Smith, 1759/1984, p. 184f) and *An Inquiry into the Nature and Causes of Wealth of Nations* (Smith 1776/1979, p. 456), he speaks of the market coordination as if guided by an 'invisible hand'.

The first study of the results of the 'invisible hand' with particular reference to forestry is *The Isolated State* by Thuenen (1826/1990). He analyzes land rents accruing from different land uses such as the production of vegetables, lumber and rye as a diminishing function of distance on an overall homogenous area surrounding a central town. At each distance, the landowner selects the product promising the highest rent. In consequence, the regular patterns of the cultivated landscape – the Thuenen rings – are the very result of market exchanges (cf. Niehans, 1998). The second important model for studying the effects of coordination through selling and buying in forestry is the Faustmann model (Faustmann, 1849), which is well known to every expert in forestry. A current survey of this type of analysis in forestry is provided by Amacher, Ollikainen and Koskela (2009).

Both the Thuenen and the Faustmann models allow studying the *results* of the 'invisible hand'. For understanding the *inner nature* of the 'invisible hand', which tries to make the 'invisible hand' 'visible',[1] there is another class of market theories.

The key questions of this class of theories include the following: How does the decentralized coordination of millions of human actions work without any central supervisor? How is the knowledge on the globe utilized, when it is not given to anyone in its totality but is separated among billions of individuals? How do individuals mutually adjust their individual plans of life in cases of unanticipated changes in the society? According to the suggestion of Whately (1832, p. 6), we name this class of theories catallactics.

Thus, there exist two different classes of market theories. One class deals mainly with the *results* of the market process, which is the well-known standard microeconomics. The other class focuses on the *understanding of the underlying process*, which we call catallactics. In this chapter, we do not deal with the standard microeconomic market theory, but, instead, we focus on catallactics, or the study of how the 'invisible hand' works.

Nevertheless, catallactics is not as well established in forestry economics as its microeconomics counterpart. Therefore, every section of this chapter comprises two parts. One part presents a brief introduction into the theoretical concepts of the market process from the catallactic point of view. The other part refers to, summarizes and systematizes selected contributions to the understanding of the inner process of coordination through selling and buying. One group of the selected papers is from the field of forestry economics, which investigates forestry-related problems of market coordination. The other group of papers is from other economic disciplines, which offer contributions for a better understanding of coordination through selling and buying inside forestry.

The two classes of market theories work differently, however, not because of the underlying assumptions and methodologies. They both understand market exchange as interactions of purposeful individuals, and both are based on the methodological individualism (Kohn, 2004, p. 308). Instead, the differences of the two classes of theories stem from their different intentions. While *result*-related theories produce explanations which are satisfactory in comparison to empirical data, catallactical theories are employed for understanding the inner nature of exchange.

Thus, the 'invisible hand' is essentially a wonderful metaphor for *result*-oriented thinking. *The Isolated State* by Thuenen and the Faustmann model apply these class theories equally. They study the *results* of the market process. These are a well-structured, cultivated landscape and an optimal rotation length as the very *results* of market exchange. Let us move now from the study of results to the study of the inner nature of exchange.

The coordination of decentralized knowledge through selling and buying

In his seminal paper 'The Use of Knowledge in Society', Hayek (1945) characterizes the economic problem of society as a coordination problem, but not as a problem of the allocation of scarce means among alternative ends. The coordination problem arises because the knowledge of a society is separated among millions, or nowadays billions, of individuals. Therefore, it exists only bitwise, incomplete, contradictory and changeable in the minds of those individuals. There is no central body in the world where the knowledge of the billions of individuals is collected.

The story *I, Pencil* by Read (2008) gives illustrative assistance by showing the complexity of coordination for the production of an ordinary pencil. Read (2008, p. 4) starts with the assertion that no single individual on this earth knows how an ordinary pencil would be produced.

Although the specialists in the pencil factory know how to assemble a pencil, they do not know how to produce all the essential inputs. Let us look at the wooden material of the pencil: It may have come from a Brazilian or an Indonesian forest or from a plantation in South Africa. A lot of knowledge and continuous management over many years are necessary to produce timber for an ordinary pencil. Which tree species are suitable? How many plants are necessary? What is the best stand density for trees to grow in the right quality and with enough timber volume? Or look at the 'loggers to fell the trees'. They 'depend on specialized, high-tech equipment, as well as coffee, meals, clothing, health care, and countless other goods and services to do their job adequately. The logging equipment is made, in part, from steel. So steelworkers had a hand in the making of pencils, too, whether they know it or not' (Heyne, Boettke and Prychitko, 2010, p. 100). The steel in turn is made from ore. Miners, maybe in Brazil, in the Ukraine, in Canada or anywhere may have mined it. Sailors and truckers have transported the ore and the steel and the pencil machine and the pencil. At last, all the different components which are necessary for the production of a pencil are the results of hundreds and thousands of specialists. All these foresters, miners, steel producers, pencil machine producers, color producers, sailors, truckers, and so forth, were involved in the production of the pencil (Deegen, Hostettler and Navarro, 2011, p. 358).

> None of the thousands of persons involved in producing the pencil performed their task because they wanted a pencil. Some among them have never even seen a pencil and would not know what it is for ... These people live in many lands, speak different languages, practice different religions, may even hate one another – yet none of these differences prevented them from cooperating to produce a pencil.
>
> *(Friedman and Friedman, 1990, p. 12f)*

For visualizing the market process, we prefer a graph in which a single bilateral exchange among two parties is embedded in and related to many other bilateral exchanges (Figure 2.1) (cf. Vanberg, 1995, p. 47ff). Clearly, such a network diagram is only a small window of the countless bilateral exchanges which we call 'market'. It illustrates that every change in a single bilateral exchange affects all the other bilateral exchanges, sometimes slightly and sometimes stronger. However, every single change will be absorbed by the system while the individuals adjust their exchange actions and balance them with the other bilateral exchanges. In this manner, the gigantic network of bilateral exchanges is always and continuously in a never-ending movement in which individuals coordinate their individual plans through selling and buying.

A recent paper by Buongiorno, Raunikar and Zhu (2011) may serve as an illustration of the complexity of the decentralized coordination through markets. Buongiorno et al. (2011) show the projection of consequences for the global forest sector of doubling the rate of growth of bioenergy demand relative to a base scenario by applying the Global Forest Products Model

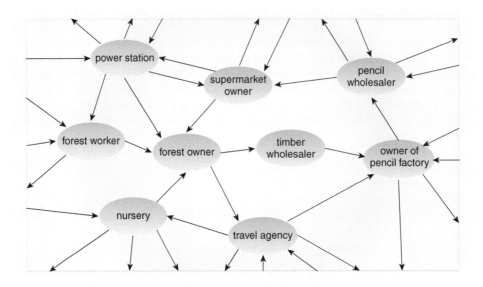

Figure 2.1 A network diagram for visualizing the coordination through market exchange.

(GFPM). They show, for instance, the prediction of the GFPM for the global forest stock change. Countries with the highest increases in fuelwood consumption, such as New Zealand with 364%, Germany with 334% and Canada with 329%, face only minor or even no reductions of their national forest stocks (i.e. 2%, 2% and 0%, respectively). As a consequence of the complex interdependencies of the global wood and bioenergy trade, forest stocks will decline significantly at completely different places of the world, such as in India by 50%, in Nigeria by 35%, in South Africa by 14% and in Indonesia by 10%. Although this study is far away from the complexity of the real world, it provides a little insight into the interweavements of global exchange.

Any concrete sale or purchase by an individual is embedded in and balanced with all other sales and purchases by the same individual (Figure 2.1). That means that each sale and each purchase unintentionally includes knowledge and preferences of all goods and services which the individual exchanges. To buy the ordinary pencil discussed previously is not only an action in the pencil market. Instead, it is an action that is simultaneously balanced with all other actions of the individual. Tullock (2005, p. 121) writes that the single individual makes something on the order of 15,000 to 20,000 buying decisions annually. This is a gigantic flow of information.

Illustrative forestry examples of the complexity of an ordinary individual action are also provided by studies of the determination of optimal rotation length of nonindustrial private forest owners when in situ preferences (Tahvonen and Salo, 1999), borrowing constraints (Tahvonen, Salo and Kuuluvainen, 2001) and nonforest income (Tahvonen and Salo, 1999; Tahvonen et al., 2001) are included. Because in those papers the same numerical example is applied, it can be used as an illustration for simultaneous balanced actions. The numerical examples in those papers show an optimal Faustmann rotation length of 83 years. However, considering the three components mentioned previously, the optimal rotation length ranges between 65 years and infinite, and it depends on the personal circumstances of the single forest owner.

Moreover, the single individual has to adjust his or her sales and purchases to the sales and purchases of the exchange partners. Therefore, buying the pencil is not only balanced with other actions of the single individual, but also adjusted to all sales and purchases of the store owner. The bilateral exchange is a result of various balancing actions of two involved parties.

Furthermore, the single sale or purchase is not only adjusted to the actions of the exchange partner, but strongly coupled with the other exchanges with the exchange partners of the exchange partner. For instance, the action of the customer of the pencil is not only coupled with the actions of the store owner; it also is coupled with the owner of the petrol station who sells petrol to the trucker who in turn transports the pencils from the wholesaler to the store.

Prices as carriers of knowledge in society

Prices

The complex, decentralized coordination of millions of individual actions through selling and buying takes place without any collection of all the knowledge in any single mind. It is not used in its totality in the contemporary society but is separated among millions of individuals. Usually, the single individual does not know all that much about the particular needs of her exchange partners. And the question arises, how can the single individual contribute to the satisfaction of the needs of which she does not know, and even satisfy those of individuals whom she does not know?

The carriers of this information are the prices, which are the results of previous and successful exchanges. The single individual can only become acquainted with those aspects of the many other unknown individuals which are reflected in these prices.

Let us imagine for a moment a well-working forest market, in which at every moment thousands of forest owners sell thousands of forests, and where most of these are immature. In this way, thousands of individuals become forest owners by buying forests.

Consider that the optimal rotation length is 50 years. Only the owners of the 50-year-old forest stand watch the prices for timber and for bare land. However, the sellers of the 49-year-old forests do not watch the prices for timber and bare land; instead, they watch the prices for 49-year-old forests. Only the buyers of these 49-year-old forests watch the timber and bare land prices and use this knowledge for their own asks in the market of 49-year-old forests. In the successful cases of selling and buying in the market of 49-year-old forests, the realized prices for the 49-year-old forests contain some information about the timber and bare land prices, which are necessary for the 50-year-old forest utilization.

In the same way, the sellers of the 48-year-old stands do not watch the prices for timber and bare land; they watch the prices for 48-year-old forests. The buyers of the 48-year-old forests also watch the prices of the 49-year-old forests and use this knowledge for their own asks in the market of 48-year-old forests. The realized prices for these 48-year-old forests contain some information about the prices of the 49-year-old forests, which again contain some information about the timber and bare land prices at the rotation length, and so forth.

Like a cascade, the forest prices carry stepwise the timber and bare land prices from the older to the younger forests and, finally, to the planting action through selling and buying of forests. From individual to individual, the prices of forests carry the knowledge 'which [enables] the sellers and the buyers to provide for needs of which he has no direct knowledge and by the use of means of the existence of which without it he would have no cognizance . . .' (Hayek, 1976, p. 115).

In the Faustmann model, the complex price cascade of the forests exchanges through markets is reduced to the beginning and the end point of the price cascade. It combines only the final timber and bare land price as the beginning of the price cascade and the planting cost as the end of the price cascade. As in every model, reductions in the Faustmann model are made for analytical reasons in order to find out the overall result of the market exchange but not to study the complex coordination through markets as a combination of many different sales and purchases.

However, the reduction of the price cascade to the beginning and the end point in the Faustmann model does not mean that the knowledge of timber prices at the end of the rotation is necessary at the moment of planting. With the help of prices, market exchange means exactly the opposite: to confine attention to the immediate circumstances of the individual actions.

The forest owner does not plant young trees because she knows that anybody will need wooden goods in 50 years. Instead, she plants trees because she expects that other individuals will buy her young immature forest stand when she sells the forest for various reasons, or as in the famous phrase by Samuelson (1976, p. 474): 'Even if my doctor assures me that I will die the year after next, I can confidently plant a long-lived olive tree, knowing that I can sell at a competitive profit the one-year-old sapling'.

For the same reason, an individual will buy an immature forest stand and conduct some pre-commercial thinnings, not because he knows which sorts of timber the demander at the time of the final rotation length will prefer. He conducts precommercial thinnings because he expects that another individual will buy the thinned forest stand for a satisfactory price (cf. Hayek, 1976, p. 115f).

Clearly, such a pure market process of many simultaneous exchanges of forests is a simplification because all these exchanges take place with some time lag: A forest owner plants trees not because he expects that other individuals will buy his young forest stand now and today, but, rather, he expects that other individuals will buy his forest stand someday in the future. As a consequence of unanticipated changes between the time of sale and the time of purchase, prices will change.

> It is these differences that bring about money profits and money losses . . . His (the entrepreneur's) success or failure depends on the correctness of his anticipation of uncertain events. If he fails in his understanding of things to come, he is doomed. The only source from which an entrepreneur's profit stems is his ability to anticipate better than other people the future demand of the consumers. If everybody is correct in anticipating the future state of the market . . . neither profit nor loss can emerge . . .
>
> *(Mises, 2007, p. 290)*

The adaptation of individuals to unanticipated changes by continuous price changes implies that the price cascade of forests is always in movement. Prices are not only the carriers of knowledge. Through selling and buying, the individuals substitute obsolete knowledge with new knowledge caused by the unanticipated changes. Thus, prices not only carry the knowledge, but also continually actualize the knowledge as well.

Nevertheless, the picture of thousands of simultaneous forest exchanges through markets illustrates how prices carry the information from exchange to exchange. When the forest owner sells an immature forest stand, it is neither possible nor necessary for him to have information on the future uses of this forest. Prices carry and actualize the whole complex of human knowledge and wants from individual to individual. When the individual considers the prices, he adjusts his individual actions with all the countless exchanges of all the other sellers and buyers. Nobody needs the information on the final needs, either for the present or for the future.

An illustrative case study for showing how individuals apply buying and selling for adjusting their living circumstances is the 'owner-consumer decisions on an amenity forest' by Christensen (1982). He describes the story of a New York businessman who bought a forest property with a number of different specific goals in view: He desired a rural retreat for his family as well as a secluded business place to bring associates for conferences together, and he anticipated horseback riding on the old logging roads. Time passed, his children grew up, other circumstances in his life changed and his aims shifted or deteriorated. The forest became more and more useless. Finally, after 12 years, he sold his forest property. In other words, he adjusted his asset endowments to his changing living circumstances in the long run by market exchange.

A careful step toward an understanding of how prices work as impersonal guides for individual actions is the generalized Faustmann model by Chang (1998), which is based on the Faustmann school of thought. In this model, a clear distinction between current and future prices with respect to the optimal rotation length is realized. Nobody knows or needs the prices of timber and production factors of future rotations. Instead, current land prices are used as the only available estimation of future land uses. This thinking is extended by price and product class watching during the time (Chang and Deegen, 2011).

Although exchanges through markets are independent of the ages of the sellers and buyers, they comprise intergenerational transfers of forest stocks. The buyer can be older or younger than the seller of the forests. It follows that some exchanges of forest stocks are exchanges among generations, and others are exchanges within the same generation. Every sale of forest stock from an older to a younger individual and vice versa is a smooth intergenerational transfer. This type of intergenerational exchange, however, is totally different from intergenerational transfer by bequest, which can be often observed in forestry and which is studied with overlapping generation models (cf. Amacher, Koskela and Ollikainen, 2002). These two types of intergenerational transfer should be clearly distinguished.

Learning by acting

Prices are the carriers of information and the transmitters of coordination, as we have demonstrated previously. Catallactics deals with the questions of how information comes into the prices and how the exchange through selling and buying utilizes information (cf. Smith, 2006, p. 2f).

For answering these questions, it is necessary to understand the learning process of individuals when they sell or buy. Market learning does not mean primarily reading, thinking and writing, as academics commonly do. In contrast, individuals in the market learn by acting, watching and listening. Literally in an endless feedback process, they realize the results of exchanges and repeat them in the same or an adapted manner. Experimental economics tries to make visible the learning process through selling and buying with the help of laboratory experiments (e.g. Smith, 1991). For the demonstration, an experiment inside the double auction institution is used (Figure 2.2).

> This trading institution, used throughout the world in financial, commodity and currency markets, is a two-sided multiple unit generalization of the ascending bid auction for unique items. Buyers submit bids to buy, while sellers submit offers or asks to sell, with a rich rule structure for defining priority based on price, quantity and arrival time . . . Notice that the demand crosses the supply at a range of market clearing prices, where demand = supply = 10 units, given by the interval (356, 360). Any whole number in this interval is a competitive equilibrium price. Only you and I know this, the subjects in this experiment know nothing of these facts . . . The subjects were inexperienced, meaning that none had previously been in a double auction experiment . . . The behavior shown in the right panel of Figure 1 is typical.
>
> *(Smith, 2006, pp. 4–5)*

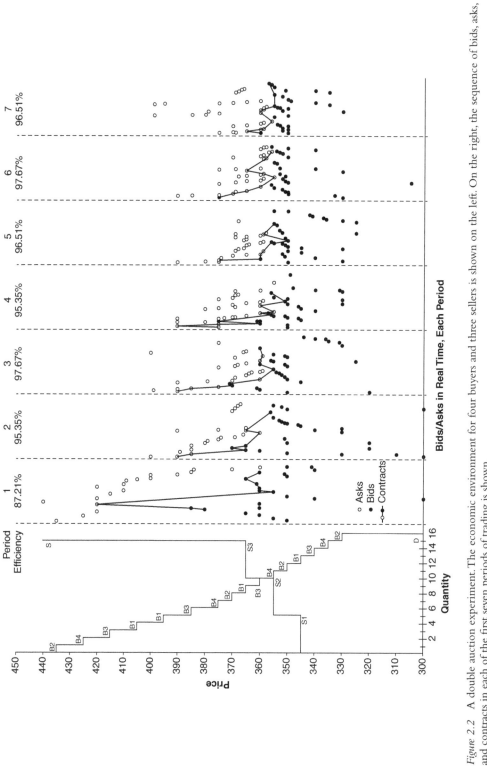

Figure 2.2 A double auction experiment. The economic environment for four buyers and three sellers is shown on the left. On the right, the sequence of bids, asks, and contracts in each of the first seven periods of trading is shown.

Reprint of Figure 1 in Smith (2006), with permission of John Wiley and Sons, Inc.

At the beginning, no participant has any idea about a 'realistic' bid or ask because there was not an auction before. Their offers come only from their individual wants and the initial expectations of the other participants. As a result, only a few bids find asks for an exchange contract. Most of them will be disappointed. Some of them do not adjust their individual expectations; others will change their asks or bids respectively. The successful traders learn as well. Their cognition involves that their expectations on the prices of other participants were not too bad. However, from period to period, the participants learn more and more about their own preferences and about the expectations of competitors and of trading partners via their own successful or unsuccessful trials of exchanges. During the periods, the participants learn more and more to coordinate their own actions with the actions of the other participants. During the periods, more and more bids and asks become successful. Or, in economic terms, the exchange process converges to the market equilibrium. The underlying way of learning is trial and error of acting, watching and listening, and of subsequent correcting or continuing.

During the bids and asks, individuals not only discover a little about how other individuals valuate goods. They often discover their own values that they give for the goods as well. After selling or buying, people are often astonished at how much they have paid for a good that they had valuated with a trifling amount of money at the beginning of the auction.

The evidence from approximately 150 to 200 individual economic experiments, conducted by many different researchers, studied for stationary, cyclical and irregular shifts in demand and supply in a wide spectrum of market institutions, such as posted-price, bilateral-bargaining games, continuous double auctions and others, shows that the participants converge with astonishing speed to the competitive equilibrium price and quantity (Smith, 1991, p. 226).

Thus, the microeconomic market theory is supported by the results from experimental economics: Market equilibrium is the consequence of the learning of individuals during the acts of selling and buying. From period to period, the participants watch their own success or failure to buy or to sell as well as the realized prices from the previous period, adapt their bids and asks in the present period to these observations and act again. Thus, the experiments visualize the learning process of the individual at the market.

Moreover, the many laboratory experiments with the wide variations of exchange rules show that no assumptions of price taking and of complete information are necessary for convergence to the competitive market equilibrium (Smith, 1991, p. 232). On the contrary, prices and quantity converge best to equilibrium under private incomplete information. Under complete information, the convergence process either fails or does less well (Smith, 1991, p. 803). Thus, the economic experiments support Hayek's (1945) hypothesis: 'The most significant fact about this (price) system is the economy of knowledge with which it operates, or how little the individual participants need to know in order to be able to take the right action . . .' (pp. 526–527). Or, in the words for testing the hypothesis at experimental markets, 'Strict privacy together with the trading rules of a market institution is sufficient to produce competitive market outcomes at or near 100% efficiency' (Smith, 1991, p. 223). These findings are also valid in the case of intertemporal competitive exchanges, which are typical in forestry (Miller, Plott and Smith, 1991).

In summary, selling and buying is a process in which individuals bring their own personal plans in accordance with the plans of the competitors and the exchange partners by learning stepwise with help of trial and error. Between the periods, prices carry and actualize the information of the exchange participants. The invisible hand of Adam Smith is nothing more than the learning process of humans by the trial and error of their actions.

As a consequence, the economic research on market exchange (microeconomics and catallactics) is on the right track. It shows that markets work in the way we think: Individuals coordinate their dispersed actions by selling and buying in a way that is self-regulating. Often enough, this coordination is much better than we expect from the standard market models (cf. Smith, 1991, p. 802).

Competition as a discovery procedure for finding answers to unanticipated changes

The existence of unanticipated changes is so extraordinarily prominent that Hayek wrote in his seminal paper, 'Competition as a Discovery Procedure': 'It is useful to recall at this point that *all* economic decisions are made necessary by unanticipated changes . . .' (Hayek, 2002, p. 17).[2] These unanticipated changes ask for adaptation of the individual plans as well as for readjustments of the individual plans with all other individual plans of the other individuals.

Prices are the carriers of information to show which of the changed circumstances ask for adaptation and adjustment and which do not. They show the single individual 'that what they have previously done, or can do now, has become more or less important . . .' (Hayek, 2002, p. 17) because the change of prices changes '. . . the compensation of the various services . . . without taking into account of the merits or defects of . . .' (Hayek, 2002, p. 17) the involved individuals. 'The most important function of prices, however, is that they tell us *what* we should accomplish, *not how much*' (Hayek, 2002, p. 17).

The seminal paper 'The View from John Sanderson's Farm: A Perspective for the Use of the Land' by Hugh M. Raup (1966) illustrates the land-use process as a result of unanticipated changes and their ensuing adaptations.

In 1740, the first settlers entered the virgin forest landscape of Petersham in central Massachusetts and started with subsistence agriculture in only small parcels. From 1791 to 1830, settlement continued, the regional road system in the landscape became a developed net, industrial towns grew and flourished continuously, regional markets evolved and agriculture changed from subsistence to a regional market economy. In other words, Petersham prospered. By 1850, the region was a full agricultural landscape with only a small amount of forest area.

In 1830, the opening of the Erie Canal changed the economic conditions: Settlers moved west. Railroads completed the traffic network, including changes from a system of isolated regional nets to a national network. Foodstuffs, in far greater quantity and produced more cheaply due to superior soil qualities in the west, were transported from western to eastern states. At the same time, these expansions attracted large sums of eastern capital for investments into mechanization and industrialization. As a result, Petersham's agriculture became uncompetitive; its economy collapsed. Over the decades, farmers emigrated. Agricultural use of the land was abandoned. Therefore, forests of nearly pure white pine came back by natural seeding. In 1900, Petersham was a full forest landscape again, yet without any value for the individuals who owned these former agricultural properties. However, some individuals discovered the value of the 'green gold'. As a consequence, the great logging and milling era between 1900 and 1920 arose in southern New England, with a new and a much higher prosperity than 100 years before.

The changes in prices as results of unanticipated changes do not lead only to a more or less unconscious balancing of the changing circumstances in everyday life. More importantly, the changes in prices offer incentives for discovering new solutions.

The fact that the white pine in Raup's (1966) paper becomes a raw material for containers, which were in high demand during the time of US industrialization, has nothing to do with the trees themselves. White pine had existed for a long time; it existed long before humans existed. Primarily, white pines were natural things, but not good for humans. Humans discover which of the billion different things in nature are goods. In the case of Raup's white pine, the pines came to maturity at the moment individuals demanded wood containers. Likewise, property owners from Petersham became aware that pines could be the raw material for those containers. Other people found niches in the price and wage structures of those days whereby the whole harvest process became economically feasible (Raup, 1966, p. 8).

> They all had first to be conceived in people's mind; then they had to be made attractive to investors so that capital would flow into them. A century earlier or even 50 years earlier, all that pine would have had very little value and most of it would, of necessity, have been cut down and burned to get it out of the way for farming.
>
> *(Raup, 1966, p. 8)*

In our economic analysis, we often reduce the adaptation to unanticipated changes to the rearrangement of the basket of the given goods according to the new price circumstances. But goods are not given. They are the result of human action (Hayek, 1948, p. 100f). Through market exchange, individuals do not make use of given knowledge. They discover, e.g. which natural things are goods, which technologies are most suitable for transforming things into goods, and so forth.

One great discovery in human history was the way to utilize ordinary trees as a raw material and as fuelwood because they existed at different places in the world in ancient and historical times in inconceivable dimensions in nature. Wooden raw material and fuelwood were not given as natural resources; instead, humans have discovered wood as material during history: Lips (1947) collected examples from the Stone Age and earlier of how humans discovered wood as common material.

Again, from century to century, individuals discovered more and more useful utilizations for this natural material (Perlin, 1997). When timber became scarce, humans were not troubled by this circumstance; instead, humans discovered substitutes and invented silviculture, the technology for producing 'natural' raw material. Kuester (1998, p. 69) remarks that the fast expansion of hazel after the Ice Age was a result of active 'silviculture' by humans during their resettlement of Central Europe. Koepf (1995/1996) notes that humans harvested forest trees in regular cutting cycles in the Modern Stone Age up to 4000 BC in southwest Germany as well as in Etruscan iron mining since 700 BC.

A recent example of discovering things as goods is the story of forest amenity evolution during the nineteenth and the twentieth centuries: Although forest scenic beauty has existed since time immemorial, the discovery of forest landscapes as a source of amenity services is a product of modern times (Mises, 2007, p. 645). Figures in Duerr (1993, p. 101), as well as in Anderson and Hill (1996, p. 516), give related illustrations of the increase in visitors to national parks during the twentieth century. Butler and Leatherberry (2004) show that the number of family forest owners in the United States has increased, and that the most common reason for these ownerships is enjoying beauty and scenery.

In the competitive market exchange, individuals also discover new technologies, new organizational solutions and new forms of cooperation as better answers to unanticipated changes. A typical example is silviculture, the forestry technology to reduce timber scarcity and boost forestland competitiveness. During the last 150 years, forestry practitioners have reduced the production time for timber (rotation length) from approximately 400 to 600 years (200 years ago) to nowadays 5 years in some forest plantations. According to Morozov (1928), forest practitioners first replaced succession with man-made forest regeneration. Secondly, they replaced slow-growing trees (oak and beech in Central Europe) with fast-growing trees (spruce and pine in Central Europe), and actually, they introduced biotechnology innovations (Sedjo, 1999,

p. 18f). That means forest practitioners have reduced interest costs for timber production of about 10^{13} euros/ha during the last 200 years, assuming a continuous interest rate of 5%.

An example of discovering new organizational solutions is the outsourcing of harvesters and forwarders. As an adaptation of vertical organization of forestry enterprises in Central Europe, they reorganized into specialized timber harvest companies. Before the introduction of harvesters and forwarders en masse, when harvest machines were mostly chainsaws, the timber harvest was typically part of forest ownership. After the introduction of harvesters and forwarders, both the capital cost and the cost of specialized knowledge and specialized organization increased and asked for adaptation. The adequate answer that forest enterprises found was the outsourcing of harvesters and forwarders and the foundation of specialized harvest companies.

An example of discovering new institutional arrangements as a reorganization of existing property rights is the story of conservation easements by forest trusts in the United States:

> [E]asements are based on the idea that property ownership is not a single indivisible right, but instead a collection of individual, often separable, rights. These individual rights include, for example, the right to erect structures, reside, grow crops and exclude other from property . . . The advantage of easements over ownership for land trusts is that they allow trusts to protect lands, not by acquiring the entire bundle of landowner rights, but by acquiring only those specific rights that are relevant to the trusts' conservation goals.
>
> *(Clark, Tankersley, Smith and Starnes, n.d., p. 2)*

The acting human: The maximizer and the entrepreneur

The underlying economic model of human action is the homo economicus: The individual maximizes her or his utility subject to constraints. This model is applied to the Faustmann model: The landowner maximizes the land expectation value with respect to the rotation length. Many different variations study various maximization and optimization problems, such as the optimal planting density (Chang, 1983) or the optimal choice between even- and uneven-aged forestry (Tahvonen, 2009).

The objective(s) is given, just as all involved products and production factors and their prices. The landowner in the Faustmann model knows every timber sort of her standing trees, knows every environmental service of her forest, which she can sell for known prices. Also, she knows everything about silvicultural and harvest technology. According to the underlying model structure, the economic choice of the forest owner is embedded in the objectives and their order, into the production factors and into the production functions which are all given. Choice means to find out the maximum or the optimal solution in the set of given factors and given objectives (Kirzner, 1979).

But the discovery procedure of competition needs the discoverer. As we pointed out in the fourth section of this chapter, the economic facts are not given but are the results of competition. Thus, although economic optimization is helpful for efficient allocation, it is only the second phase of human action. Before optimization can start, the identifying of objectives, products, production factors and production functions is necessary because these facts are not given. This part of discovery is called the phase of entrepreneur action (Kirzner, 1979).

Figure 2.3 illustrates the two phases of human action with the help of the structure of a Faustmann model: It shows the separation of human action into an entrepreneurial phase, in which the means and ends are discovered, and an economic phase, in which the means and ends are optimally allocated, where LEV is the land expectation value, P_j is the price of product class j, W_j is

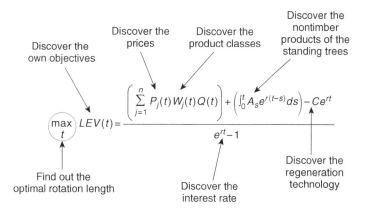

Figure 2.3 The distinction of human action into an entrepreneur phase and an economic phase, exemplified with the help of the structure of a Faustmann model.

the percentage of the product class j in the stand volume, Q is the total stand volume, A is the net revenue for the nontimber product of the standing trees, C is the regeneration cost, r is the interest rate, j is the number for a product class, t is the rotation length and s is the stand age with $s \leq t$.

The distinction of human action into an entrepreneur phase, in which the means and ends are discovered, and an economic phase, in which the means and ends are optimally allocated, is clearly an analytical tool. Every human is an entrepreneur and an economic person at the same time (Kirzner, 1979).

By studying the body of literature in the field of forestry economics with reference to market exchange, it is easy to see that the underlying model of human action focuses on the economic phase. Only a small amount of this literature deals with entrepreneurial aspects, such as Anderson and Leal (2001).

Conclusion

In this chapter, there is no presentation of catallactics as a unified, settled body of thought as the forest economist is accustomed to with the Faustmann school of thought. Instead, catallactics is more a progressive research program (Boettke, 2010, p. 159). Therefore, in this chapter, the main theoretical concepts of catallactics are combined with examples from the field of forestry-related research. This should be interpreted as an invitation to systematically inquire into the inner structure of the gigantic network of human exchanges. This comprises methodical challenges. One is the change in the point of view of what a theory of market coordination can explain because 'the predictive power of this theory is necessarily constrained to a prediction of the type of structure . . . that will result; it does not, however, extend to a prediction of particular events' (Hayek, 2002, p. 11). Another methodical job is the transformation of principally structural insight into operational theory, and lastly, to find ways for testing theorems empirically (Coyne, 2010, p. 26; Smith, 2006, p. 3; Boettke, 2010, p. 164f).

Acknowledgments

First of all, the writers would like to thank Renke Coordes very much for the extensive scientific and technical support during the recent months. The writers also wish to thank Shashi Kant and Janaki R. R. Alavalapati for their valuable comments and suggestions.

Notes

1   This slogan I noted at the Hayek lecture 'Hayek and Experimental Economics' by Vernon Smith in Freiburg, Germany, 27 June 2008.
2   The emphasis is found only in the German original of the paper (Hayek, 1968/2003, p. 142) but not in the English translation.

References

Amacher, G. S., Koskela, E. and Ollikainen, M. (2002). 'Optimal forest policies in an overlapping generations economy with timber and money bequests', *Journal of Environmental Economics and Management*, vol. 44, pp. 346–369.

Amacher, G. S., Ollikainen, M. and Koskela, E. (2009). *Economics of Forest Resources*, MIT Press, Cambridge, MA.

Anderson, T. L. and Hill, P. J. (1996). 'Appropriable rents from Yellowstone Park: A case of incomplete contracting', *Economic Inquiry*, vol. 34, pp. 506–518.

Anderson, T. L. and Leal, D. R. (2001). *Free Market Environmentalism* (Rev. ed.), Palgrave, New York.

Boettke, P. J. (2010). 'Back to the future: Austrian economics in the twenty-first century', in P. J. Boettke (Ed.), *Handbook on Contemporary Austrian Economics*, Edward Elgar, Cheltenham, UK, pp. 157–165.

Buongiorno, J., Raunikar, R. and Zhu, S. (2011). 'Consequences of increasing bioenergy demand on wood and forests: An application of the global forest products model', *Journal of Forest Economics*, vol. 17, pp. 214–229.

Butler, B. J. and Leatherberry, E. C. (2004). 'America's family forest owners', *Journal of Forestry*, October/November, pp. 4–13.

Chang, S. J. (1983). 'Rotation age, management intensity, and the economic factors of timber production: Do changes in stumpage price, interest rate, regeneration cost, and forest taxation matter?' *Forest Science*, vol. 29, no. 2, pp. 267–277.

Chang, S. J. (1998). 'A generalized Faustmann model for the determination of optimal harvest age', *Canadian Journal of Forest Research*, vol. 28, pp. 652–659.

Chang, S. J. and Deegen, P. (2011). 'Pressler's indicator rate formula as a guide for forest management', *Journal of Forest Economics*, vol. 17, no. 3, pp. 258–266.

Christensen, W. W. (1982). 'Owner-consumer decisions on an amenity forest', in W. A. Duerr, D. E. Teeguarden, N. B. Christiansen and S. Guttenberg (Eds.), *Forest Resource Management: Decision-Making Principles and Cases*, OSU Bookstores, Corvallis, OR, pp. 468–473.

Clark, C. D., Tankersley, L., Smith, G. F. and Starnes, D. (n.d.). *Farm and Forest Land Preservation with Conservation Easements*, Extension SP646, University of Tennessee, Knoxville, TN.

Coyne, C. J. (2010). 'Economics as the study of coordination and exchange', in P. J. Boettke (Ed.), *Handbook on Contemporary Austrian Economics*, Edward Elgar, Cheltenham, UK, pp. 14–29.

Deegen, P., Hostettler, M. and Navarro, G. A. (2011). 'The Faustmann model as a model for a forestry of prices', *European Journal of Forest research*, vol. 130, pp. 353–368.

Duerr, W. A. (1993). *Introduction to Forest Resource Economics*, McGraw-Hill, New York.

Faustmann, M. (1849). 'Berechnung des Werthes, welchen Waldboden, sowie noch nicht haubare Holzbestände für die Waldwirtschaft besitzen', *Allgemeine Forst- und Jagd-Zeitung*, December, pp. 441–455.

Friedman, M. and Friedman, R. (1990). *Free to Choose: A Personal Statement*, A Harvest Book, Harcourt, Inc., San Diego, CA.

Hayek, F. A. (1945). 'The use of knowledge in society', *American Economic Review*, vol. 35, no. 4, pp. 519–530.

Hayek, F. A. (1948). 'The meaning of competition', in F. A. Hayek, *Individualism and Economic Order*, University of Chicago Press, Chicago, pp. 92–106.

Hayek, F. A. (1968/2003): 'Der Wettbewerb als Entdeckungsverfahren' in Hayek, F. A., Rechtsordnung und Handelnsordnung. J.C.B. Mohr (Paul Siebeck) Tübingen, pp. 132–149.

Hayek, F. A. (1976). *Law, Legislation and Liberty: Vol. 2, The Mirage of Social Justice*, University of Chicago Press, Chicago.

Hayek, F. A. (2002). 'Competition as a discovery procedure', *The Quarterly Journal of Austrian Economics*, vol. 5, no. 3, pp. 9–23.

Heyne, P., Boettke, P. J. and Prychitko, D. L. (2010). *The Economic Way of Thinking*, Pearson Education, Upper Saddle River, NJ.

Kirzner, I. M. (1979). *Perception, Opportunity and Profit: Studies in the Theory of Entrepreneurship*, University of Chicago Press, Chicago.

Koepf, E. U. (1995/1996). 'Nachhaltigkeit: Prinzip der Waldwirtschaft – Hoffung der Menschheit?' *Scheidewege – Jahresschrift für skeptisches Denken, Special Edition*, pp. 307–317.

Kohn, M. (2004). 'Value and exchange', *Cato Journal*, vol. 24, no. 3, pp. 303–339.

Kuester, H. (1998). *Die Geschichte des Waldes. Von der Urzeit bis zur Gegenwart*, Verlag C. H. Beck, Muenchen, Germany.

Lips, J. E. (1947). *The Origin of Things: A Cultural History of Man*, A. A. Wyn, New York.

Miller, R. M., Plott, C. R. and Smith, V. L. (1991). 'Intertemporal competitive equilibrium: An empirical study of speculation', in V. L. Smith (Ed.), *Papers in Experimental Economics*, Cambridge University Press, Cambridge, UK, pp. 128–153.

Mises, L. V. (2007). *Human Action: A Treatise on Economics*, Liberty Fund, Indianapolis, IN.

Morozov, G. F. (1928). *Die Lehre vom Walde*, Neudamm, J. Neumann, Berlin.

Niehans, J. (1998). 'Thuenen, Johann Heinrich von (1783–1850)', in J. Eatwell, M. Milgate and P. Newman (Eds.), *The New Palgrave. A Dictionary of Economics* (Vol. 4), Stockton Press, New York, pp. 636–639.

Perlin, J. (1997). *A Forest Journey. The Role of Wood in the Development of Civilization*, Harvard University Press, Cambridge, MA.

Raup, H. M. (1966). 'The view from Sanderson's farm: A perspective for the use of the land', *Forest History*, vol. 10, pp. 2–11.

Read, L. E. (2008). *I, Pencil: My Family Tree as Told to Leonard E. Read*, Foundation for Economic Education, New York.

Samuelson, P. A. (1976). 'Economics of forestry in an evolving society', *Economic Inquiry*, vol. 14, pp. 466–492.

Sedjo, R. A. (1999). *Biotechnology and Planted Forests: Assessment of Potential and Possibilities*, Discussion Paper 00–06, Resources for the Future, Washington, DC.

Smith, A. (1979). *An Inquiry into the Nature and Causes of Wealth of Nations*, Liberty Press/Liberty Classics, Indianapolis, IN. (Original work published 1776)

Smith, A. (1984). *The Theory of Moral Sentiments*, Liberty Fund, Indianapolis, IN. (Original work published 1759)

Smith, V. L. (1991). *Papers in Experimental Economics*, Cambridge University Press, New York.

Smith, V. L. (2006). 'Markets, institutions and experiments', *Encyclopedia of Cognitive Science*. John Wiley and Sons, Inc, Hoboken, pp. 1–7.

Tahvonen, O. (2009). 'Optimal choice between even- and uneven-aged forestry', *Natural Resource Modeling*, vol. 22, no. 2, pp. 289–231.

Tahvonen, O. and Salo, S. (1999). 'Optimal forest rotation with in situ preferences', *Journal of Environmental Economics and Management*, vol. 37, pp. 106–128.

Tahvonen, O., Salo, S. and Kuuluvainen, J. (2001). 'Optimal forest rotation and land values under a borrowing constraint', *Journal of Economic Dynamics and Control*, vol. 25, pp. 1595–1627.

Thuenen, J. H. v. (1990). *Der isolierte Staat in Beziehung auf Landwirtschaft und Nationalökonomie*, Scientia Verlag, Aalen, Germany. (Original work published 1826)

Tullock, G. (2005). *The Economics of Politics*, Liberty Fund, Indianapolis, IN.

Vanberg, V. (1995). *Markt und Organisation: Individualistische Sozialtheorie und das problem korporativen Handelns.* J.C.B. Mohr (Paul Siebeck), Tübingen, Germany.

Whately, R. (1832). *Introductory Lectures on Political Economy*, B. Fellowes, London.

3

THE GENERALIZED FAUSTMANN FORMULA

Sun Joseph Chang

SCHOOL OF RENEWABLE NATURAL RESOURCES, LOUISIANA STATE UNIVERSITY AGRICULTURAL CENTER, BATON ROUGE, LA 70803 USA. TEL: 225-578-4167. XP2610@LSU.EDU

Abstract

This chapter examines the four core areas of the generalized Faustmann formula – the management of even-aged natural stands, even-aged plantations and uneven-aged stands, as well as the development of Pressler's indicator rate formula. Under the generalized formula, stumpage prices, stand volumes, annual incomes, regeneration costs and interest rates could vary from timber crop to timber crop. As a result, the optimal management of even-aged and uneven-aged stands also could vary from timber crop to timber crop. The optimal conditions for the decision variables are derived and their economic meanings explained. Although similar to those obtained under the classic Faustmann formula, the optimal conditions under the generalized Faustmann formula offer much broader and richer interpretations. The increment in stumpage value is shown to consist of price increment, quality increment and quantity increment. The results of comparative statics analysis showed that under the generalized Faustmann formula it is possible to untangle the impacts of changes in current and future production parameters and produce much sharper results. Pressler's indicator rate formula is also shown to maximize the land expectation value under the generalized Faustmann formula. The chapter closes with observations on ongoing efforts and future research opportunities.

Keywords

Generalized Faustmann formula, dynamic programming, even-aged management, uneven-aged management, Pressler's indicator rate formula, price increment, quality increment, quantity increment, comparative statics analysis

Introduction

For nearly 150 years the literature on the determination of optimal rotation age (see Newman, 2002, for a comprehensive compilation of the literature until that time) has relied on the classic Faustmann formula first advanced by Martin Faustmann (1849). In the economic literature, the optimal rotation problem is known as the tree-cutting problem or the wine-storage problem.

Over the years, it has attracted the attention of two Nobel laureates (Ohlin, 1921; Samuelson, 1976). Recognizing that stumpage prices, stand volume, regeneration cost and interest rate do not stay the same rotation after rotation, Chang (1998) developed the generalized Faustmann formula by allowing these factors to vary from harvest period to harvest period. In this chapter, four core areas of the generalized Faustmann formula – the management of (1) even-aged natural stands, (2) even-aged plantations and (3) uneven-aged stands, plus (4) the development of Pressler's indicator rate formula – will be addressed. As will be shown subsequently, these relaxations provide the generalized Faustmann formula with much greater flexibility and produce much richer analytical results.

Under the first topic, the question of optimal harvest age for even-aged natural stands will be examined. Given that about 93% of the world's forests are some type of natural stand (FAO, 2012), this topic is highly pertinent. The condition of reaching optimal harvest age will be examined along with a graphic analysis of the impact of changes in various production parameters. In addition, the total increment in stumpage value will be separated into price increment, quality increment and quantity increment. The relationship among the various formulas of optimal harvest age determination will also be discussed.

The second topic addresses the determination of optimal planting density and harvest age. With most of the industrial roundwood coming from plantations, its proper management is becoming ever more important and deserves careful examinations. The impact of changes in both current and future production parameters on the management decision variables will then be examined through comparative statics analysis.

Under the third topic, the generalized Faustmann formula for uneven-aged management will be developed. It will be shown that the formula resembles that of even-aged plantation management. With both management systems sharing the same theoretical foundation, further analyses are no longer needed. All of the analytical results for the management of even-aged plantations can be readily applicable to that of uneven-aged stands.

Under the fourth topic, Pressler's indicator rate formula will be shown to also represent the optimal condition for the generalized Faustmann formula. The chapter closes with observations on some current developments and future research opportunities.

The generalized Faustmann formula for even-aged natural stand management

Of the 4 billion hectares of forest in the world, 36% is primary forests and 57% is other naturally regenerated forests (FAO, 2012). Most of these forests are managed extensively as even-aged stands. After a clearcut, the stand is typically regenerated naturally, with or without incurring some expenses. The key management question thus revolves around how long one should wait before harvesting the new stand. As the simplest form of even-aged management, it will be discussed first.

Let

$$V_i(t_i) = \sum_{j=1}^{n} P_{ij}(t_i) W_{ij}(t_i) Q_i(t_i)$$

be the stumpage value of the ith timber crop at age t_i, with $\dfrac{\partial V_i(t_i)}{\partial t_i} > 0$ and $\dfrac{\partial^2 V_i(t_i)}{\partial t_i^2} < 0$. $P_{ij}(t_i)$ is the stumpage price of ith timber crop at age t_i for product class j. For example, in the US South, southern pine timber stands typically consist of pulpwood, chip-and-saw timber and sawtimber.

$W_{ij}(t_i)$ is the percentage of the product class *j* at age t_i of the *i*th stand volume,
$Q_i(t_i)$ is the total stand volume at age t_i and the volume of a particular product class,
$Q_{ij}(t_i) = W_{ij}(t_i)Q_i(t_i)$,
$A_i(s_i)$ is the net annual income for age s_i, $0 \leq s_i \leq t_i$ of the *i*th timber crop,
C_i is the regeneration cost for the *i*th timber crop,
r_i is the interest rate associated with the *i*th timber crop and
LEV_i is the land expectation value at the beginning of the *i*th timber crop.

To maximize the value of the land, we want to maximize the present value of profits from growing an infinite number of timber crops.

$$Max \, LEV_1 = \sum_{i=1}^{\infty} \left[V_i(t_i) + \sum_{s_i=1}^{t_i} A_i(s_i) \exp(r_i(t_i - s_i)) - C_i \exp(r_i t_i) \right] \exp\left(\sum_{j=1}^{i} -r_j t_j \right) \tag{1}$$

Note that as a special case, if all $V_i(t_i)$, $A_i(s_i)$, C_i and r_i remain the same for all timber crops, then equation (1) can be expressed as

$$LEV_1 = \left[V_1(t_1) + \sum_{i=1}^{t_1} A_1(s_1) \exp(r_1(t_1 - s_1)) - C_1 \exp(r_1 t_1) \right] \left[\exp(-r_1 t_1) + \exp(-2r_1 t_1) + \cdots \right]$$

$$= \left[V_1(t_1) + \sum_{i=1}^{t_1} A_1(s_1) \exp(r_1(t_1 - s_1)) - C_1 \exp(r_1 t_1) \right] \Big/ \left(\exp(r_1 t_1) - 1 \right) \tag{2}$$

and collapses to equation (2) as the classic Faustmann formula. Note also that equation (1) includes the Hartman (1976) formula as a special case. For easy comprehension, equation (1) can also be written as

$$Max \, LEV_1 = \left[V_1(t_1) + \sum_{s_1=1}^{t_1} A_1(s_1) \exp(r_1(t_1 - s_1)) - C_1 \exp(r_1 t_1) \right] \exp(-r_1 t_1)$$
$$+ LEV_2 \exp(-r_1 t_1) \tag{3}$$

In the previous equations the term 'timber crop' should be broadly interpreted. If future crops remain in forestry, they are naturally timber crops. If in the future, the land is switched to growing fruit trees, it would still be viewed as a timber crop. In this case, the income from annual fruit production becomes much more important, whereas that from the final harvest to replace the old fruit trees becomes far less important. Even in the case of conversion to annual crop production or real estate development, there are simply no timber crops in the future. Only the annual net incomes are involved. It should also be noted that over time, the timber crop species could change, for example, from southern pine to hardwood or from spruce to Douglas-fir. It could also change from timber production to fruit production or crop production and vice versa. The generalized Faustmann formula, therefore, could accommodate land-use changes by permitting different types of crops, may they be timber, fruit or grain, for different harvest periods. In the first case, the value of the timberland is determined endogenously, whereas in the latter cases, with land-use change under the generalized Faustmann formula, the value of the land in the future, as LEV_2 in equation (3), is determined exogenously as shown by Klemperer and Farkas (2001).

Equation (3) represents the famous recurrence relation of dynamic programming. In this equation, LEV_1 and LEV_2 represent the objective functions, and the expression

$$\left[V_1(t_1)+\sum_{s_1=1}^{t_1}A_1(s_1)\exp(r_1(t_1-s_1))-C_1\exp(r_1t_1)\right]\exp(-r_1t_1)$$ represents the payoff associated with

the decision variable t_1. Theoretically, equation (3) can be solved with the forward recursive solution method. However, such a solution would involve infinite numbers of stumpage prices, stand volumes, annual incomes or expenses, regeneration costs and interest rates, thus making it impractical. Fortunately, LEV_2 represents just a single value. It embodies all the optimal harvest age decisions for future timber crops that give rise to this specific value. Forest owners and/or managers need not know the details of these decisions, just that they give rise to the specific value. Therefore, solving for the optimal harvest age empirically would involve the insertion of a specific value of LEV_2 into equation (3) to solve for t_1. Such a value could be gleaned from various timberland transactions if there is an active timberland market. Or it could be chosen judiciously to determine the resulting harvest age for the first timber crop under various future values for the timberland.

On reaching the optimal harvest age

In addition to finding the optimal harvest age under equation (3), it is important to understand the economic meaning of reaching the optimal harvest age because it affords the opportunity to determine stepwise year by year the harvest decision by comparing the marginal benefit with the marginal cost of waiting. At the optimal t_1

$$\frac{\partial LEV_1}{\partial t_1}=\left[\frac{\partial V_1(t_1)}{\partial t_1}+r_1\sum_{s_1=1}^{t_1}A_1(s_1)\exp(r_1(t_1-s_1))+A_1(t_1)-C_1r_1\exp(r_1t_1)\right]\exp(-r_1t_1)$$

$$+\left[V_1(t_1)+\sum_{s_1=1}^{t_1}A_1(s_1)\exp(r_1(t_1-s_1))-C_1\exp(r_1t_1)\right](-r_1)\exp(-r_1t_1)$$

$$+LEV_2(-r_1)\exp(-r_1t_1)=0 \tag{4}$$

$$\frac{\partial V_1(t_1)}{\partial t_1}+A_1(t_1)=r_1V_1(t_1)+r_1LEV_2 \tag{5}$$

Equation (5) states that at the optimal harvest age, the extra amount of stumpage value earned by waiting one more year plus the extra annual income on the left-hand side of the equation must equal the cost of holding the trees plus the cost of holding the land on the right-hand side of the equation. When the left-hand side of equation (5) is greater than the right-hand side, one should wait another year. Conversely, the stand should be harvested. In the interest of brevity, no empirical examples for this topic will be presented. Readers interested in such examples are referred to Chang (1998).

The separation of the stumpage value increment

What is the benefit of waiting? Pressler (1860) pointed out that the stumpage value increment $\frac{\partial V_1(t_1)}{\partial t_1}$ consists of three types of increments when the harvest age is delayed one time period. They are the quantity increment (*Quantitätszuwachs*), the quality increment (*Qualitätszuwachs*) and, lastly, the price increment (*Teuerungszuwachs*). Over the years, these increments have been mentioned in various textbooks; however, it was Chang and Deegen (2011) who separated these satisfactorily both analytically and empirically. Given that

$$\frac{\partial V_1(t_1)}{\partial t_1} = \sum_{j=1}^{n} \left\{ \frac{\partial P_{1j}(t_1)}{\partial t_1} W_{1j}(t_1) Q_1(t_1) + P_{1j}(t_1) \frac{\partial W_{1j}(t_1)}{\partial t_1} Q_1(t_1) + P_{1j}(t_1) W_{1j}(t_1) \frac{\partial Q_1(t_1)}{\partial t_1} \right\} \quad (6)$$

$\sum_{j=1}^{n} P_{1j}(t_1) W_{1j}(t_1) Q'_1(t_1)$, the increase in stand value as a result of total stand volume increment, represents the quantity increment. The gain realized from changes in the composition of different product classes of the stand volume $\sum_{j=1}^{n} P_{1j}(t_1) W'_{1j}(t_1) Q_1(t_1)$ represents the quality increment. Finally, the gain realized from changes in prices of different product classes, $\sum_{j=1}^{n} P'_{1j}(t_1) W_{1j}(t_1) Q_1(t_1)$, represents the price increment. It should be noted that in some instances, the quality increment may not matter. For example, in the emerging biomass for energy market, sometimes no quality is recognized. In such a case, the quality increment simply falls out, and only price and quantity increments remain.

In practice, the growth in stumpage value over time can be determined by

$$V_1(t_1+1) - V_1(t_1) = \sum_{j=1}^{n} [P_{1j}(t_1+1) - P_{1j}(t_1)] W_{1j}(t_1+1) Q_1(t_1+1) + \sum_{j=1}^{n} P_{1j}(t_1) [W_{1j}(t_1+1)$$

$$-W_{1j}(t_1)] Q_1(t_1+1) + \sum_{j=1}^{n} P_{1j}(t_1) W_{1j}(t_1) [Q_1(t_1+1) - Q_1(t_1)] \quad (7)$$

Dividing $V_1'(t_1)$ by $V_1(t_1)$ results in

$$\frac{V_1'(t_1)}{V_1(t_1)} = \frac{\sum_{j=1}^{n} P'_{1j}(t_1) W_{1j}(t_1) Q_1(t_1)}{V_1(t_1)} + \frac{\sum_{j=1}^{n} P_{1j}(t_1) W'_{1j}(t_1) Q_1(t_1)}{V_1(t_1)} + \frac{\sum_{j=1}^{n} P_{1j}(t_1) W_{1j}(t_1) Q'_1(t_1)}{V_1(t_1)} \quad (8)$$

with the three terms on the right-hand side of equation (8) being the rates of price increment, quality increment and quantity increment, respectively. Among them, the last two increments in equation (8) are usually positive and under the control of a forester. Price increment or the rate of price increment, however, as Pressler warned, could be either positive or negative depending on the overall economy, specific technological developments or market conditions. For an example of separating these three increments empirically, the readers are referred to Chang and Deegen (2011).

Comparative statics analyses of the impact of changes in stumpage price levels, regeneration cost, annual income and regeneration cost

How will the current versus future changes in stumpage prices, annual income, regeneration cost and interest rate affect the optimal harvest age of the current timber crop? These analyses are important because they will show a priori how the optimal harvest age will be affected before any empirical analyses. Here the impact of these changes will be analyzed graphically. Mathematical analyses of the impact of these changes are available in Chang (1998). To analyze graphically the impact of changes in production factors both currently and in the future, first rewrite equation (5) as

$$\frac{V_1'(t_1) + A_1(t_1)}{V_1(t_1) + LEV_2} = r_1 \quad (9)$$

and name the left-hand side as the rate of marginal revenue growth (RMRG). As the timber stand ages, $V_1'(t_1)$ gradually declines. The numerator of the RMRG approaches $A_1(t_1)$, and the denominator increases and approaches the sum of the limit of $V_1(t_1)$ plus LEV_2. As shown in Figure 3.1, the RMRG curve gradually trends downward. On the other hand, the interest rate line is shown as a flat line. The point where these two curves cross is the optimal harvest age. With this graph, one can quickly see that a higher regeneration cost for the current timber crop, as a sunk cost, has no effect on the optimal harvest age of the current timber crop. On the other hand, a higher stumpage price level for the current timber crop would impact both the numerator and denominator of RMRG. When $V'(t)/V(t)$ is greater than r, higher stumpage prices would raise the current harvest age and vice versa. A higher annual income, on the other hand, would always move the RMRG curve up and raise the current harvest age. Finally, a higher current interest rate would simply move the interest rate line up and result in a lower harvest age for the current timber crop.

The impacts of all the changes in the production factors of future timber crops are reflected through LEV_2. For example, a higher stumpage price level for any of the future timber crops would result in a higher LEV_2 and consequently a smaller RMRG. A downward move of the RMRG curve will then lead to a lower harvest age for the current timber crop. The same is true for higher annual incomes for any of the future timber crops. On the other hand, a higher interest rate or a higher regeneration cost for any of the future timber crops would translate into a smaller LEV_2 and result in a bigger RMRG. As such, they will both lead to a higher harvest age for the current timber crop.

Table 3.1 summarizes the results of all of the comparative statics analyses and also compares these results with those under the classic Faustmann formula. Indeed, the generalized Faustmann formula yields much richer results. Under the classic Faustmann formula, a higher stumpage price level would always shorten the rotation. Yet under the generalized Faustmann formula, a higher current stumpage price level would either raise or lower the current harvest age, whereas a higher future stumpage price level would lower the current harvest age. Whereas

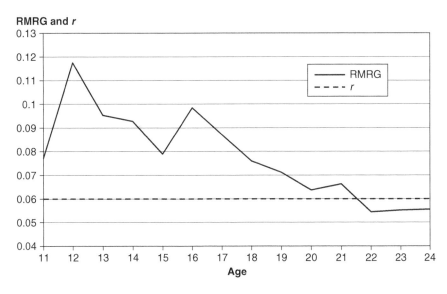

Figure 3.1 Rate of marginal revenue growth (RMRG) and interest rate r.

Table 3.1 The results of comparative statics analyses under the classic Faustmann formula and the generalized Faustmann formula.

Cause	Effect
	Under classic Faustmann formula
A one-time increase in	optimal rotation age t
C generation cost	increase
α in $\alpha P(t)$ stumpage price level	increase
β in $\beta A(s)$ annual income	increase if $A(s) < A(t)$ for all s no change if $A(s) = A(t)$ for all s decrease if $A(s) > A(t)$ for all s
r, interest rate	decrease
	Under generalized Faustmann formula
A one-time increase in	
Current timber crop	
C_k, regeneration cost	no change
α_k of $\alpha_k P_k(t_k)$, stumpage price level	
if $\dfrac{\partial V_1(t_1)}{\partial t_1} > r_1 V_1(t_1)$	increase
if $\dfrac{\partial V_1(t_1)}{\partial t_1} < r_1 V_1(t_1)$	decrease
β_k of $\beta_k A_k(s_k)$, annual income	increase
r_k, interest rate	decrease
Future timber crop	
C_n, regeneration cost	increase
α_n of $\alpha_n P(t_n)$, stumpage price level	decrease
β_n of $\beta_n A_n(s_n)$, annual income	decrease
r_n, interest rate	increase

a higher regeneration cost would raise the rotation age under the classic Faustmann formula, only a higher future regeneration cost would do so under the generalized Faustmann formula. Current regeneration cost under the generalized Faustmann formula, as a sunk cost, has no impact on the optimal harvest age. The impact of higher annual income levels under the classic Faustmann formula depends on whether $A_i(s_i)$ is an increasing or decreasing function of stand age. On the other hand, under the generalized Faustmann formula, the higher level of current annual income would raise the current harvest age, whereas higher levels of annual incomes in future timber crops would have the opposite effect. Lastly, a higher interest rate under the classic Faustmann formula lowers the optimal rotation age. Under the generalized Faustmann formula, a higher current interest rate lowers the optimal harvest age, whereas a higher future interest rate raises the optimal harvest age.

Other formulas of optimal harvest age determination and their relationship with the generalized Faustmann formula

Over the years, other formulas have been proposed to determine the optimal harvest age. Chief among them are the present net worth (PNW) formula, which maximizes the present value of the profit from growing just one crop of timber:

$$PNW = V_1(t_1)\exp(-r_1 t_1) - C_1$$

the forest rent (FR) formula of maximizing:

$$FR = [V_1(t_1) - C_1]/t_1$$

and the biological formula of maximizing the mean annual increment (MAI):

$$MAI = Q_1(t_1)/t_1$$

Regarding the relationship between LEV_1 and PNW, note that when all the annual incomes $A_i(s)$ of the current timber crop as well as LEV_2 – the present value of all incomes and expenses from future timber crops – are ignored, then LEV_1 becomes PNW. Given that the PNW formula ignores the cost of holding the land, it will lead to an optimal harvest age that is higher than that from the generalized Faustmann formula.

The relationship between the generalized Faustmann formula and the FR formula is examined through the land rent (R). Note that when all the annual incomes $A_i(s)$ are ignored,

$$R = r_1[V_1(t_1) - C_1\exp(r_1 t_1) + LEV_2]\exp(-r_1 t_1)$$

Applying L'Hopital's rule when r_1 approaches 0,

$$\lim_{r_1 \to 0} R = \frac{\lim_{r_1 \to 0} r_1\left[V_1(t_1) - C_1\exp(r_1 t_1) + LEV_2\right]}{\lim_{r_1 \to 0}\exp(r_1 t_1)}$$

$$= \frac{\lim_{r_1 \to 0}\left\{[V_1(t_1) - C_1\exp(r_1 t_1) + LEV_2] + r_1[-C_1 t_1\exp(r_1 t_1)]\right\}}{\lim_{r_1 \to 0} t_1\exp(r_1 t_1)}$$

$$= [V_1(t_1) - C_1]/t_1 = FR \text{ when } LEV_2 = 0$$

That is to say, R collapses to FR when all the annual incomes are ignored, $LEV_2 = 0$ and interest rate r_1 also equals 0. Given that when LEV_1 is maximized the land rent is also maximized, only when the previous conditions are satisfied will the FR formula result in the correct optimal harvest age.

For the biological formula of MAI maximization, note that when $P_1(t_1) = k$ and $C_1 = 0$, then

$$FR = \frac{P_1(t_1)Q_1(t_1) - C_1}{t_1} = \frac{kQ_1(t_1)}{t_1} = kMAI$$

That is to say, when all the annual incomes are ignored; LEV_2, interest rate r_1 and regeneration cost C_1 all equal to 0; and the stumpage prices of trees of different ages are all the same, implying that there is no premium for older and therefore larger diameter trees, then R collapses to MAI, and the MAI formula results in the correct optimal harvest age.

The generalized Faustmann formula for even-aged plantation management

Timber plantations now account for 7% of the forests in the world (FAO, 2012). Despite this relatively small percentage, in recent decades these plantations have been producing an ever-increasing amount of industrial roundwood supplies. Large acreages of pine plantations have been established in the US South, Brazil, Chile and New Zealand, as well as extensive Chinese fir plantations in China. Eucalyptus plantations have been established in Brazil, China, Australia and several Southeast Asian countries. Red pine and spruce plantations have been established widely in Europe. In the future, energy plantations could also emerge to play an important role in sequestering carbon dioxide emissions. More importantly, these plantations with their high productivity assure the possibility of conserving natural forests and ecosystems.

For even-aged plantations, both the harvest age and the initial planting density must be determined simultaneously. In this section, the notations defined earlier are expanded as follows:

$P_i(t_i, n_i)$ is the stumpage price for the jth product class of the ith plantation established with an initial planting density of n_i at age t_i.

$W_{ij}(t_i, n_i)$ is the percentage of the jth product class of the ith plantation established with an initial planting density of n_i at age t_i.

$Q_i(t_i, n_i)$ is the stand volume of the ith plantation established with an initial planting density of n_i at age t_i.

$V_i(t_i, n_i) = \sum_{j=1}^{n} P_{ij}(t_i, n_i) W_{ij}(t_i, n_i) Q_i(t_i, n_i)$ is the stumpage value of the ith plantation with an initial planting density of n_i at age t_i, with $\dfrac{\partial V_i(t_i, n_i)}{\partial t_i} > 0, \dfrac{\partial V_i(t_i, n_i)}{\partial n_i} > 0$ and

$\dfrac{\partial^2 V_i(t_i, n_i)}{\partial t_i^2} < 0, \dfrac{\partial^2 V_i(t_i, n_i)}{\partial n_i^2} < 0$.

Cs_i stands for the site preparation cost for the ith plantation.

Cp_i stands for the cost of planting per seedling, including the cost of both the labor and seedling.

All other variables are as defined previously.

Following equation (3), the generalized Faustmann land expectation value formula for plantation management can be expressed as

$$LEV_1 = \left[V_1(t_1, n_1) + \int_0^{t_1} A_1(s_1) \exp(r_1(t_1 - s_1)) ds_1 - (Cs_1 + Cp_1 n_1) \exp(r_1 t_1) \right] \exp(-r_1 t_1)$$
$$+ LEV_2 \exp(-r_1 t_1) \tag{10}$$

$$\frac{\partial LEV_1}{\partial t_1} = \left[\frac{\partial V_1(t_1, n_1)}{\partial t_1} + r_1 \int_0^{t_1} A_1(s_1) \exp(r_1(t_1 - s_1) ds_1 + A_1(t_1) - (Cs_1 + Cp_1 n_1) r_1 \exp(r_1 t_1) \right]$$
$$\exp(-r_1 t_1) + \left[V_1(t_1, n_1) + \int_0^{t_1} A_1(s_1) \exp(r_1(t_1 - s_1) ds_1 - (Cs_1 + Cp_1 n_1) \exp(r_1 t_1) \right]$$
$$(-r_1) \exp(-r_1 t_1) - r_1 \exp(-r_1 t_1) LEV_2 = 0 \tag{11}$$

$$\frac{\partial LEV_1}{\partial n_1} = \left[\frac{\partial V_1(t_1, n_1)}{\partial n_1} - Cp_1 \exp(r_1 t_1) \right] \exp(-r_1 t_1) = 0 \tag{12}$$

For notational simplicity, $\sum_{s_1=1}^{t_1} A_1(s_1)\exp(r_1(t_1-s_1))$ is replaced by $\int_0^{t_1} A_1(s_1)\exp(r_1(t_1-s_1))ds_1$. To maximize LEV_1, from equation (11):

$$\frac{\partial V_1(t_1,n_1)}{\partial t_1} + A_1(t_1) - r_1V_1(t_1,n_1) - r_1LEV_2 = 0 \qquad (13)$$

from equation (12):

$$\frac{\partial V_1(t_1,n_1)}{\partial n_1} - Cp_1\exp(r_1t_1) = 0 \qquad (14)$$

Equation (13) states that at optimal harvest age, the extra stumpage value plus the extra annual income earned by waiting one more year must equal the cost of holding the trees plus the cost of holding the land, similar to the case of even-aged natural stand management discussed earlier. Equation (14) suggests that at the optimal planting density, the extra stumpage value earned by planting an additional tree must equal the extra cost of planting the extra tree compounded to the end of the harvest period.

Table 3.2 presents an example of the simultaneous determination of optimal harvest age and planting density with an interest rate of 5.5% for the first harvest period, a site preparation cost of US$160 per acre and a planting cost of US$0.10 per tree, including the cost of the seedling and labor for planting, with no annual income and a future land value of US$800 per acre. Stumpage prices are US$80 per cord for chip-and-saw logs and US$28 per cord for pulpwood, with 76 cubic feet of solid wood per 128 cubic feet (4' × 8' × 8') of stacked volume. Given these parameters, the optimal planting density will be 700 trees per acre and optimal harvest age will be 26 years.

Comparative statics analysis of the generalized Faustmann formula for even-aged plantations

To carry out comparative statics analyses, the second-order conditions for the optimal combination of t_1 and n_1 must be established first.

$$\frac{\partial^2 LEV_1}{\partial t_1^2} = \left[\frac{\partial^2 V_1(t_1,n_1)}{\partial t_1^2} + \frac{\partial A_1(t_1)}{\partial t_1} - r_1\frac{\partial V_1(t_1,n_1)}{\partial t_1}\right]\exp(-r_1t_1)$$

$$+ \left[\frac{\partial V_1(t_1,n_1)}{\partial t_1} + A_1(t_1) - r_1V_1(t_1,n_1) - r_1\,LEV_2\right](-r_1)\exp(-r_1t_1)$$

$$= \left[\frac{\partial^2 V_1(t_1,n_1)}{\partial t_1^2} + \frac{\partial A_1(t_1)}{\partial t_1} - r_1\frac{\partial V_1(t_1,n_1)}{\partial t_1}\right]\exp(-r_1t_1) < 0 \qquad (15)$$

Equation (15) is less than 0 because the terms inside the bracket on the second line are the first-order condition for optimal t_1 and equal 0.

$$\frac{\partial^2 LEV_1}{\partial n_1^2} = \frac{\partial^2 V_1(t_1,n_1)}{\partial n_1^2}\exp(-r_1t_1) < 0 \qquad (16)$$

Table 3.2 The simultaneous determination of optimal planting density and harvest age.

Planting density (trees per acre)	Harvest age (years)							
	21	22	23	24	25	26	27	28
400								
Cords of C&S	15.46	17	18.72	20.92	22.63	24.2	26.14	27.35
Cords of pulpwood	12.6	12.61	12.99	12.17	12.08	12.68	12.28	12.29
Stumpage value/A	1,589.71	1,713.31	1,861.49	2,013.92	2,148.67	2,291.39	2,435	2,532.37
LEV/Acre	552.90	549.46	551.18	551.70	545.54	539.80	532.74	514.40
500								
Cords of C&S	14.92	16.44	18.36	21.01	22.65	24.28	26.27	28.42
Cords of pulpwood	14.94	14.94	15.3	14.43	14.69	14.79	14.83	14.53
Stumpage value/A	1,611.59	1,749.72	1,897.42	2,085.16	2,223.08	2,399.61	2,516.34	2,680.58
LEV/Acre	549.79	530.32	551.32	560.73	554.35	555.70	541.16	536.17
600								
Cords of C&S	14.46	16.13	18.61	20.72	22.75	24.46	26.27	28.28
Cords of pulpwood	16.97	17	17.11	16.45	16.67	16.74	16.87	16.53
Stumpage value/A	1,632.24	1,766.37	1,968.05	2,117.95	2,286.24	2,425.91	2,573.94	2,725.03
LEV/Acre	546.30	545.28	561.25	559.49	560.32	551.99	544.21	535.70
700								
Cords of C&S	13.87	15.69	18.5	20.48	21.94	24.68	26.59	28.21
Cords of pulpwood	18.93	19.37	18.45	18.14	18.85	19.1	18.36	18.12
Stumpage value/A	1,639.46	1,797.58	1,996.58	2,146.29	2,282.91	2,509	2,641.03	2,763.8
LEV/Acre	538.57	544.59	559.30	557.06	549.08	561.87	549.40	534.01
800								
Cords of C&S	13.39	15.16	18.19	20.02	22.04	24.14	26.29	28.53
Cords of pulpwood	20.68	21.11	20.24	20.49	20.7	20.66	20.16	19.91
Stumpage value/A	1,650.5	1,803.78	2,021.56	2,174.61	2,343.02	2,509.88	2,667.35	2,840.2
LEV/Acre	532.05	536.44	556.36	554.62	554.68	552.08	545.36	540.39

Stumpage price for chip and saw (C&S) = US$80/cord, pulpwood = US$28/cord, C_{s_1} = US$160/A, C_{p_1} = US$0.10/tree.
Interest rate = 5.5%, and LEV_2 = US$800/acre.

and

$$
\begin{aligned}
D &= \begin{vmatrix} \partial^2 LEV_1 / \partial t_1^2 & \partial^2 LEV_1 / \partial t_1 \partial n_1 \\ \partial^2 LEV_1 / \partial t_1 \partial n_1 & \partial^2 LEV_1 / \partial n_1^2 \end{vmatrix} \\
&= (\partial^2 LEV_1 / \partial t_1^2)(\partial^2 LEV_1 / \partial n_1^2) - (\partial^2 LEV_1 / \partial t_1 \partial n_1)^2 > 0
\end{aligned}
\tag{17}
$$

as part of the second-order conditions.

It should be noted that

$$
\begin{aligned}
\frac{\partial^2 LEV_1}{\partial t_1 \partial n_1} &= [\partial^2 V_1(t_1,n_1) / \partial t_1 \partial n_1 - Cp_1 r_1 \exp(r_1 t_1)]\exp(-r_1 t_1) + [\partial V_1(t_1,n_1) / \partial n_1 \\
&\quad - Cp_1 \exp(r_1 t_1)](-r_1)\exp(-r_1 t_1) \\
&= \partial^2 V_1(t_1,n_1) / \partial t_1 \partial n_1 \exp(-r_1 t_1) - Cp_1 r
\end{aligned}
\tag{18}
$$

because the terms of the second line are the first-order condition for the optimal n_1. Thus, although equation (17) must be true, a priori nothing is said about the sign of $\partial^2 V_1(t_1,n_1) / \partial t_1 \partial n_1$. Given that $\partial V_1(t_1,n_1) / \partial t_1$ represents the current annual increment in revenue, $\partial^2 V_1(t_1,n_1) / \partial t_1 \partial n_1 = \partial(\partial V_1(t_1,n_1) / \partial t_1) / \partial n_1$ represents changes in current annual increment in stumpage value as a result of changes in planting density. As Kent and Dress (1980) have shown, plantations of different initial planting densities eventually converge to the same random pattern. As such, given enough time, these stands of different initial planting densities will also converge to the same stand volume, and thus, value. Figure 3.2 shows two of the stumpage value curves and their corresponding current annual increments in stumpage value curves. For the stand with a higher planting density, its current annual increment (CAI) in stumpage value ascends faster, peaks at an earlier age and descends faster thereafter. For the stand with a lower planting density, its CAI ascends slower, peaks at a later age and descends slower thereafter. As shown in Figure 3.2, these two CAI curves will cross each other at an age T. Because the area below the CAI in stumpage value curve stands for the stumpage value, the vertically shaded area represents that period when the higher planting density stand outgrows the lower planting density stand in value. The horizontally shaded area, on the other hand, would represent the opposite case. At an age $\hat{T}$, these two shaded areas would be equal in size, and the two stands would end up with the same stumpage value thereafter. Once the optimal planting density is determined, the relevant CAI in stumpage value curve will be uniquely defined. The critical question, then, is the position of the optimal harvest age t_1 relative to T. If t_1 is less than T, $\partial^2 V_1(t_1,n_1) / \partial t_1 \partial n_1 > 0$. If t_1 is larger than T, $\partial^2 V_1(t_1,n_1) / \partial t_1 \partial n_1 < 0$. When t_1 and T coincide, $\partial^2 V_1(t_1,n_1) / \partial t_1 \partial n_1 = 0$. Thus, there are three possibilities.

Case 1, $\partial^2 V_1(t_1,n_1) / \partial t_1 \partial n_1 > 0$ and $[\partial^2 V_1(t_1,n_1) / \partial t_1 \partial n_1]\exp(-r_1 t_1) - Cp_1 r_1 < 0$

Case 2, $\partial^2 V_1(t_1,n_1) / \partial t_1 \partial n_1 > 0$ and $[\partial^2 V_1(t_1,n_1) / \partial t_1 \partial n_1]\exp(-r_1 t_1) - Cp_1 r_1 < 0$

Case 3, $\partial^2 V_1(t_1,n_1) / \partial t_1 \partial n_1 < 0$

As the subsequent analyses demonstrate, the sign of $\partial^2 V_1(t_1,n_1) / \partial t_1 \partial n_1$ plays an important role in discerning the impact of changes in site preparation cost, cost of planting, stumpage price and interest rate.

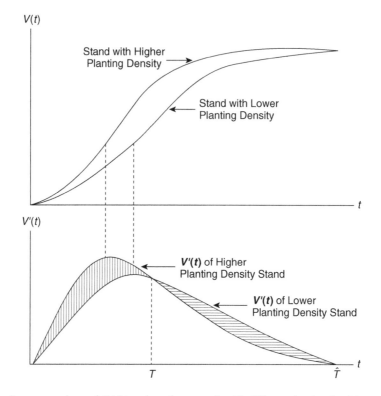

Figure 3.2 Stumpage value and CAI in value of two stands with different planting densities.

The impact of changes in current site preparation cost, Cs_1

As shown in Appendix A-1, $\dfrac{dt_1}{dCs_1} = 0$ and $\dfrac{dn_1}{dCs_1} = 0$, suggesting that a change in the current site preparation cost, as a sunk fixed cost, affects neither the harvest age nor the planting density of the current timber crop.

The impact of changes in current planting cost, Cp_1

As shown in Appendix A-2,

$$\frac{dt_1}{dCp_1} = \frac{-\partial^2 LEV_1 / \partial t_1 \partial n_1}{D} \tag{19}$$

and

$$\frac{dn_1}{dCp_1} = \frac{\partial^2 LEV_1 / \partial t_1^2}{D} < 0 \tag{20}$$

From equation (20), $dn_1/dCp_1 < 0$, suggesting that a higher current planting cost always leads to a lower planting density for the current stand.

The effect of a higher current planting cost on the optimal harvest age of the current stand, on the other hand, depends on the sign and the magnitude of $\partial^2 V_1(t_1,n_1)/\partial t_1 \partial n_1$.

Under case 1, when $\partial^2 V_1(t_1,n_1)/\partial t_1 \partial n_1 > 0$ and $[\partial^2 V_1(t_1,n_1)/\partial t_1 \partial n_1]\exp(-r_1 t_1) - Cp_1 r_1 > 0$, $dt_1/dCp_1 < 0$. Higher planting cost for the current timber crop lowers the optimal harvest age for the current timber crop.

Under case 2, when $\partial^2 V_1(t_1,n_1)/\partial t_1 \partial n_1 > 0$ but $[\partial^2 V_1(t_1,n_1)/\partial t_1 \partial n_1]\exp(-r_1 t_1) - Cp_1 r_1 < 0$, and under case 3, when $\partial^2 V_1(t_1,n_1)/\partial t_1 \partial n_1 < 0$, $dt_1/dCp_1 > 0$, higher planting cost for the current timber crop raises the optimal harvest age for the current timber crop. Whether the impact of higher planting cost on the optimal harvest age is case 1, 2 or 3 can only be determined empirically.

The impact of a higher current stumpage price level across the board, α_1

As shown in Appendix A-3,

$$\frac{dt_1}{d\alpha_1} = \left\{ -\left[\frac{\partial V_1(t_1,n_1)}{\partial t_1} - r_1 V_1(t_1,n_1) \right] \exp(-r_1 t_1) \frac{\partial^2 V_1(t_1,n_1)}{\partial n_1^2} \exp(-r_1 t_1) \right.$$
$$\left. - \frac{-\partial V_1(t_1,n_1)}{\partial n_1} \exp(-r_1 t_1) \left[\frac{\partial^2 V_1(t_1,n_1)}{\partial t_1 \partial n_1} \exp(-r_1 t_1) - Cp_1 r_1 \right] \right\} / D \tag{21}$$

$$\frac{dn_1}{d\alpha_1} = \left\{ (\partial^2 LEV_1 / \partial t_1^2) \left(-\left[\frac{\partial V_1(t_1,n_1)}{\partial t_1} - r_1 V_1(t_1,n_1) \right] \exp(-r_1 t_1) \right) \right.$$
$$\left. - \left[\frac{\partial^2 V_1(t_1,n_1)}{\partial t_1 \partial n_1} \exp(-r_1 t_1) - Cp_1 r_1 \right] \left(\frac{-\partial V_1(t_1,n_1)}{\partial n_1} \exp(-r_1 t_1) \right) \right\} / D \tag{22}$$

From equation (21) and (22) we reach the following conclusions.

If $\dfrac{\partial^2 V_1(t_1,n_1)}{\partial t_1 \partial n_1} \exp(-r_1 t_1) - Cp_1 r_1 > 0$, and $\dfrac{\partial V_1(t_1,n_1)}{\partial t_1} - r_1 V_1(t_1,n_1) \geq 0$, both $dt_1/d\alpha_1$ and $dn_1/d\alpha_1 > 0$. Higher stumpage price level across the board raises the harvest age and increases the initial planting density. Yet, when $\dfrac{\partial V_1(t_1,n_1)}{\partial t_1} - r_1 V_1(t_1,n_1) < 0$, both $dt_1/d\alpha_1$ and $dn_1/d\alpha_1$ are uncertain.

If $\dfrac{\partial^2 V_1(t_1,n_1)}{\partial t_1 \partial n_1} \exp(-r_1 t_1) - Cp_1 r_1 < 0$, and $\dfrac{\partial V_1(t_1,n_1)}{\partial t_1} - r_1 V_1(t_1,n_1) \geq 0$, both $dt_1/d\alpha_1$ and $dn_1/d\alpha_1$ are uncertain. When $\dfrac{\partial V_1(t_1,n_1)}{\partial t_1} - r_1 V_1(t_1,n_1) < 0$, both $dt_1/d\alpha_1$ and $dn_1/d\alpha_1 < 0$, meaning higher stumpage prices across the board lower the harvest age and lower the planting density.

The impact of higher annual income

As shown in Appendix A-4, with β_1 representing the level of annual income,

$$\frac{\partial^2 LEV_1}{\partial t_1 \partial \beta_1} = A_1(t_1) \text{ and } \frac{\partial^2 LEV_1}{\partial n_1 \partial \beta_1} = 0$$

$$\frac{dt_1}{d\beta_1} = -A_1(t_1)\left(\frac{\partial^2 LEV_1}{\partial n_1^2}\right)/D > 0 \text{ and}$$

$$\frac{dn_1}{d\beta_1} = -\left(\frac{\partial^2 LEV_1}{\partial t_1 \partial n_1}\right)(-A_1(t_1))/D.$$

That is to say, a higher level of annual income for the current timber crop always raises the harvest age for the current timber crop. Whether such annual income will increase or decrease the planting density depends on the sign of $\left(\dfrac{\partial^2 LEV_1}{\partial t_1 \partial n_1}\right)$.

The impact of higher interest rate for the current timber crop

As shown in Appendix A-5,

$$\frac{dt_1}{dr_1} = \left[(V_1(t_1,n_1) + LEV_2)\left(\frac{\partial^2 LEV_1}{\partial n_1^2}\right) - Cp_1 t_1 \exp(r_1 t_1)\left(\frac{\partial LEV_1}{\partial t_1 \partial n_1}\right)\right]/D$$

and

$$\frac{dn_1}{dr_1} = \left[\left(\frac{\partial^2 LEV_1}{\partial t_1^2}\right)(Cp_1 t_1 \exp(r_1 t_1)) - \left(\frac{\partial^2 LEV_1}{\partial t_1 \partial n_1}\right)(V_1(t_1,n_1) + LEV_2)\right]/D$$

Whether a higher interest rate for the current timber crop would lower the optimal harvest age depends on the sign of $\left(\dfrac{\partial^2 LEV_1}{\partial t_1 \partial n_1}\right)$. When it is greater than 0, the optimal harvest age will be lowered. Otherwise, the impact is uncertain and would depend on the magnitude of $\left(\dfrac{\partial^2 LEV_1}{\partial t_1 \partial n_1}\right)$. Similarly, the optimal planting density also depends on the sign of $\left(\dfrac{\partial^2 LEV_1}{\partial t_1 \partial n_1}\right)$. When it is greater than 0, a higher interest rate leads to a lower planting density. Otherwise, the impact is uncertain and would depend on the magnitude of $\left(\dfrac{\partial^2 LEV_1}{\partial t_1 \partial n_1}\right)$.

The impact of higher future land value

As shown in Appendix A-6,

$$\frac{dt_1}{dLEV_2} = r_1\left(\frac{\partial^2 LEV_1}{\partial n_1^2}\right)/D < 0 \text{ and}$$

$$\frac{dn_1}{dLEV_2} = -\left(\frac{\partial^2 LEV_1}{\partial t_1 \partial n_1}\right)r_1/D.$$

Higher future land value always lowers the optimal harvest age for the current timber crop. Its impact on the optimal planting density for the current timber crop depends on the sign of

$\left(\dfrac{\partial^2 LEV_1}{\partial t_1 \partial n_1}\right)$. When it is greater than 0, a higher future land value will decrease the planting density. Otherwise, it will increase the planting density.

The results of the previous comparative statics analyses are summarized in Table 3.3. A comparison of these results with those under the classic Faustmann formula (Chang, 1983) would indicate that the former produces much richer results regarding changes in the current parameters and clear-cut results regarding the future parameters unavailable under the classic Faustmann formula.

Table 3.3 Summary of comparative statics analyses of the impact of changes on the optimal t and n for even-aged plantation management under the generalized Faustmann formula.

Parameter	$\dfrac{\partial^2 V_1(t_1,n_1)}{\partial t_1 \partial n_1} < 0$	$\dfrac{\partial^2 V_1(t_1,n_1)}{\partial t_1 \partial n_1} > 0$ $\dfrac{\partial^2 V_1(t_1,n_1)}{\partial t_1 \partial n_1}\exp(-r_1 t_1)$ $-Cp_1 r_1 < 0$	$\dfrac{\partial^2 V_1(t_1,n_1)}{\partial t_1 \partial n_1} > 0$ $\dfrac{\partial^2 V_1(t_1,n_1)}{\partial t_1 \partial n_1}\exp(-r_1 t_1)$ $-Cp_1 r_1 > 0$
dt_1/dCs_1	$=0$	$=0$	$=0$
dn_1/dCs_1	$=0$	$=0$	$=0$
dt_1/dCp_1	>0	>0	<0
dn_1/dCp_1	<0	<0	<0
If $\dfrac{\partial V_1(t_1,n_1)}{\partial t_1} - r_1 V_1(t_1,n_1) \geq 0$			
$dt_1/d\alpha_1$	uncertain	uncertain	>0
$dn_1/d\alpha_1$	uncertain	uncertain	>0
If $\dfrac{\partial V_1(t_1,n_1)}{\partial t_1} - r_1 V_1(t_1,n_1) < 0$			
$dt_1/d\alpha_1$	<0	<0	uncertain
$dn_1/d\alpha_1$	<0	<0	uncertain
$dt_1/d\beta_1$	>0	>0	>0
$dn_1/d\beta_1$	<0	<0	>0
dt_1/dr_1	uncertain	uncertain	<0
dn_1/dr_1	uncertain	uncertain	<0
$dt_1/dLEV_2$	<0	<0	<0
$dn_1/dLEV_2$	>0	>0	<0

The generalized Faustmann formula under uneven-aged management

For various reasons, uneven-aged forest management has become the preferred method in many regions. For example, 'near natural' forest management favoring mixed stands of site-adapted tree species has become a dominant type of forest management in Europe (see, e.g. Pommerening, 2001; Detten, Wurz and Schraml, 2009). Given that uneven-aged management represents possibly the oldest form of forest management, extending the generalized Faustmann formula to uneven-aged management thus represents more than just an intellectual curiosity. Under uneven-aged management, instead of harvest age and planting density as in even-aged management, cutting cycle and residual growing stock level are the decision variables. As such, let

FV_0 be the value of the uneven-aged forest before any management activities.

S be the existing stand volume.

$Q_i(t_i, g_i), i = 1$ to ∞, be the volume of the ith uneven-aged stand with an initial residual growing stock of g_i and a cutting cycle of t_i years.

$V_0(S)$ be the stumpage value of the existing uneven-aged stand.

$v_i(g_i), i = 1$ to ∞, be the convex cost function for the residual growing stock value at the beginning of the ith cutting cycle. As such, $(\partial v_i(g_i)/\partial g_i) > 0$ and $(\partial^2 v_i(g_i)/\partial g_i^2) \geq 0$. These conditions imply that the value of the residual growing stock is always increasing as the level of residual growing stock increases. Further, the value is increasing either at an increasing rate or a constant rate as the level of residual growing stock increases.

$V_i(Q_i(t_i, g_i)), i = 1$ to ∞, be the quasi-concave stumpage value associated with the ith cutting cycle (timber harvest) before timber harvest. As such, over the relevant range $\partial V_i(Q_i(t_i, g_i))/\partial t_i > 0$ and $\partial^2 V_i(Q_i(t_i, g_i))/\partial t_i^2 < 0$, suggesting that the stumpage value of the uneven-aged stand increases with the elapsed time at a decreasing rate, and $\partial V_i(Q_i(t_i, g_i))/\partial g_i > 0$ and $\partial^2 V_i(Q_i(t_i, g_i))/\partial g_i^2 < 0$, indicating that the stumpage value of the uneven-aged stand also increases with the residual growing stock level at a decreasing rate.

$K_i, i = 1$ to ∞, be the fixed cost, for example the cost of obtaining a timber harvest permit, associated with the ith timber harvest.

$LEV_i, i = 1$ to ∞, be the land expectation value at the beginning of the ith cutting cycle under the generalized Faustmann formula.

Others are as defined earlier under even-aged management.

Under uneven-aged management, if t_0 years must elapse before the existing stand can be brought under management, then instead of $V_0(S)$ being the stumpage value for the existing stand, we need to wait t_0 years for the stand value to reach $V_0(Q_0(t_0, S))$. As such,

$$FV_0 = \left[V_0(t_0, S) - v_1(g_1) - K_1 + \int_0^{t_0} A_0(s_0) \exp(r_0(t_0 - s_0)) ds_0 \right.$$

$$+ \sum_{i=1}^{\infty} \left\{ V_i(Q_i(t_i, g_i)) - v_{i+1}(g_{i+1}) - K_{i+1} + \int_0^{t_i} A_i(s_i) \exp(r_i(t_i - s_i)) ds_i \right\}$$

$$\left. \exp\left(-\sum_{n=1}^{i} r_n t_n \right) \right] \exp(-r_0 t_0)$$

$$= \left[V_0(t_0, S) + \int_0^{t_0} A_0(s_0) \exp(r_0(t_0 - s_0)) ds_0 + LEV_1 \right] \exp(-r_0 t_0) \qquad (23)$$

Similar to earlier presentations,

$$LEV_1 = \left\{ V_1(Q_1(t_1, g_1) + \int_0^{t_1} A_1(s_1) \exp(r_1(t_1 - s_1)) \, ds_1 - [v_1(g_1) + K_1] \exp(r_1 t_1) \right\} \exp(-r_1 t_1)$$
$$+ LEV_2 \exp(-r_1 t_1) \tag{24}$$

Because the value of the forest consists of the value of the land and that of the trees, equation (23) is appropriately called the forest value.

If the existing stand can be brought under management immediately, $t_0 = 0$, then, as a special case, equation (23) can be expressed as

$$FV_0 = \{V_0(S) + LEV_1\} \tag{25}$$

Thus, uneven-aged management consists of two subproblems, that of determining the optimal cutting cycle for the existing stand and that of determining the cutting cycle and residual growing stock for future stands. Because the volume of the existing stand S is a given figure, the determination of its optimal cutting cycle becomes a simple problem once the value of LEV_1 is known, similar to the problem of even-aged natural stand management. The more interesting problem, therefore, is to solve for LEV_1 in equation (23). Note that equation (24) is similar to equation (10) for even-aged plantation management presented previously. For example, the fixed cost K_1 is the equivalent of the site preparation cost Cs_1, and the value of the residual growing stock $v_1(g_1)$ is the equivalent of the planting cost $Cp_1 n_1$. Given their similarity, expositions on solving for both t_1 and g_1, the meaning of their optimal conditions and the comparative statics analyses are no longer necessary. Readers interested in such topics should consult the relevant sections under even-aged plantation management and check out the article by Chang and Gadow (2010) for an empirical example.

The generalized Faustmann formula and Pressler's formula

In 1860 Max Robert Pressler published his famous indicator rate (Weiserprozent) formula, here shown with his then-used notations (Pressler, 1860, p. 190):

$$(a + b + c)[k/(k + 1)], \text{ with } k = h/g$$

where a is the rate of quantity increment (Quantitätszuwachs), b is the rate of quality increment (Qualitätszuwachs), c is the rate of price increment (Teuerungszuwachs) discussed previously in detail, h is the variable timber capital and g is the fixed land capital.

As Johansson and Löfgren (1985) pointed out, Pressler's indicator rate formula represents the earliest solution to maximizing the classic Faustmann land expectation value in its simplest form to determine the optimal rotation age.

$$\max_t LEV = \frac{[V(t) - C \exp(rt)]}{[\exp(rt) - 1]} \tag{26}$$

with all the variables as defined previously. At the optimal rotation age,

$$V'(t) = rV(t) + r\,LEV \tag{27}$$

where $V'(t) = dV(t)/dt$.

Equation (27) can also be written as Pressler's indicator rate (Weiserprozent) formula

$$\left[\frac{V'(t)}{V(t)}\right]\left(\frac{k}{k+1}\right)=r \tag{28}$$

where $k = V(t)/LEV$.

The more relevant question, therefore, for forest economics and management is the following: Can Pressler's indicator rate formula be used fruitfully under the generalized Faustmann formula?

Note that without $A_1(t_1)$, equation (5) as the first-order condition for the optimal harvest age t_1 can also be expressed as

$$\frac{V_1'(t_1)}{V_1(t_1)} = r_1[1 + LEV_2 / V_1(t_1)] \tag{29}$$

With $V_1(t_1)$ as the variable timber capital h, LEV_2 the fixed land capital g and $V_1(t_1)/LEV_2 = k$, equation (29) can be transformed into equation (28) as the famous Pressler's indicator rate formula. Thus, Pressler's indicator rate formula is also relevant under the generalized Faustmann formula. Moreover, in a recent article, Chang and Deegen (2011) showed how the price increment, quality increment and quantity increment can be combined with Pressler's indicator rate formula to determine the optimal harvest age in a dynamic world of constantly changing prices.

Conclusion

This chapter examines the four core areas of generalized Faustmann formula: the management of an even-aged natural stand, the management of a plantation, the management of an uneven-aged stand and the development of Pressler's indicator rate formula under the generalized Faustmann formula. Freed of the stringent assumptions about stumpage prices, stand volumes, regenerations costs and interest rates, harvest age and planting density or cutting cycle and residual growing stock level are allowed to vary under the generalized Faustmann formula. The ability to separate current and future production parameters under the generalized Faustmann formula makes it possible to untangle the impact of changes in these parameters.

Current efforts are being made to incorporate payment of carbon sequestration benefits and other ecological services as part of the $A_i(s_i)$. A manuscript on the generalized version of the van Kooten formula (van Kooten, Binkley and Delcourt, 1995) will soon be published (Susaeta, Chang, Carter and Lal, 2013). Its extension under uneven-aged management has recently been published by Parajuli and Chang (2012). Furthermore, a manuscript on extensions of the generalized Faustmann formula to incorporate catastrophic risk similar to the work of Reed (1984) and Reed and Errico (1986) is also under review (Susaeta, Carter, Chang and Adams 2013).

Incorporating various forms of forest taxation into the generalized Faustmann formula to examine their impact on the optimal management represents a promising line of research. Such an effort will also open the opportunity in forest valuation to determine the value of the forest, the timber stand and the land value.

It should be noted that the generalized Faustmann formula presented here addresses forest management under certainty. Despite recent progress in addressing optimal management under uncertainty (see, e.g. Alvarez and Koskela, 2006; Chaladná, 2007), this area remains fertile ground for additional research, particularly when it comes to the management of even-aged plantations and uneven-aged stands involving both harvest age and planting density or cutting cycle and residual growing stock as decision variables. Lastly, the relationship between the generalized Faustmann formula and the literature on the reservation price strategy (Brazee and Mendelsohn, 1988; Gong and Löfgren, 2007), as well as the real options theory, needs to be fully explored.

References

Alvarez, L.H.R. and Koskela, E. (2006). 'Does risk aversion accelerate optimal rotation under uncertainty', *Journal of Forest Economics*, vol. 12, no. 3, pp. 171–184.

Brazee, R. and Mendelsohn, R. (1988). 'Timber harvesting with fluctuating prices', *Forest Science*, vol. 34, no. 2, pp. 359–372.

Chaladná, Z. (2007). 'Determination of the optimal rotation period under stochastic wood and carbon prices', *Forest Policy and Economics*, vol. 9, no. 2007, pp. 1031–1045.

Chang, S. J. (1983). 'Rotation age, management intensity, and the economic factors of timber production: Do changes in stumpage price, interest rate, regeneration cost, and forest taxation matter?' *Forest Science*, vol. 29, no. 2, pp. 267–277.

Chang, S. J. (1998). 'A generalized Faustmann model for the determination of optimal harvest age', *Canadian Journal of Forest Research*, vol. 28, no. 5, pp. 652–659.

Chang, S. J. and Deegen, P. (2011). 'Pressler's indicator rate formula as a guide for forest management', *Journal of Forest Economics*, vol. 17, no. 3, pp. 255–266.

Chang, S. J. and Gadow, K. v. (2010). 'Application of the generalized Faustmann model to uneven-aged forest management', *Journal of Forest Economics*, vol. 16, no. 4, pp. 313–325.

Detten, R. v., Wurz, A. and Schraml, U. (2009). *Waldzukünfte – Herausforderungen für eine zukunftsfähige Waldpolitik in Deutschland*, Bundesministerium für Bildung und Forschung, Bonn, Germany.

Faustmann, M. (1849). 'Berechnung des Wertes welchen Walboden sowie noch nicht haubare Holzbestände für die Waldwirtschaft besitzen', *Allgemeine Forst und Jagd Zeitung*, vol. 15, pp. 441–455. Translated by W. Linnard (1968) as 'Calculation of the value which forest land and immature stands possess for forestry' in *Martin Faustmann and the Evolution of Discounted Cash Flow*, Inst. Pap. No. 42, Commonwealth Forestry Institute Oxford, U.K., pp. 27–55. (Reprinted in *Journal of Forest Economics*, vol. 1, no. 1, pp. 7–44)

Food and Agricultural Organization (FAO). (2012). *Global Forest Resource Assessment 2010: Key Findings*. Retrieved from http://foris.fao.org/static/data/fra2010/KeyFindings-en.pdf

Gong, P. and Löfgren, K.-G. (2007). 'Market and welfare implications of the reservation price strategy for the forest harvest decision', *Journal of Forest Economics*, vol. 13, no. 4, pp. 217–243.

Hartman, R. (1976). 'The harvesting decision when a standing forest has value', *Economic Inquiry*, vol. 14, no. 1, pp. 52–58.

Johansson, P.-O. and Löfgren, K. G. (1985). *The Economics of Forestry and Natural Resources*, Basil Blackwell, Oxford, UK.

Kent, B. M. and Dress, P. E. (1980). 'On the convergence of forest stand spatial pattern over time: The cases of regular and aggregated spatial patterns', *Forest Science*, vol. 26, no. 1, pp. 10–22.

Klemperer, W. D. and Farkas, D. R. (2001). 'Impacts on economically optimal timber rotations when future land use changes', *Forest Science*, vol. 47, no. 4, pp. 520–525.

Newman, D. H. (2002). 'Forestry's golden rule and the development of the optimal forest rotation literature', *Journal of Forest Economics*, vol. 8, no. 1, pp. 5–28.

Ohlin, B. (1921). 'Till frågan om skogarnas omloppstid', *Ekonomisk Tidskrift* 22. Translated by C. Hudson (1995) as 'Concerning the question of the rotation period in forestry', *Journal of Forest Economics*, vol. 1, no. 1, pp. 89–114.

Parajuli, R. and Chang, S. J. (2012). 'Carbon sequestration and uneven-aged management of loblolly pine stands in the southern USA: A joint optimization approach', *Forest Policy and Economics*, vol. 22, pp. 65–71.

Pommerening, A. (2001). 'Continuous cover forestry – chance and challenge for forest science', *Institute of Chartered Foresters (ICF) News 1/2001, Special Feature, Alternatives to Clearfelling*, pp. 8–10.

Pressler, M. R. (1860). 'Aus der Holzzuwachlehre (zweiter Artikel)', *Allgemeine Forst und Jagd Zeitung*, vol. 36, pp. 173–191. Translated by W. Löwenstein and J. R. Wirkner (1995) as 'For the comprehension of net revenue silviculture and the management objectives derived thereof', *Journal of Forest Economics*, vol. 1, no. 1, pp. 45–87.

Reed, W. (1984). 'The effects of the risk on the optimal rotation of a forest', *Journal of Environmental Economics and Management*, vol. 11, no. 2, pp. 180–190.

Reed, W. J. and Errico, D. (1986). 'Optimal harvest scheduling at the forest level in the presence of the risk of fire', *Canadian Journal of Forest Research*, vol. 16, no. 2, pp. 260–265.

Samuelson, P. A. (1976). 'Economics of forestry in an evolving society', *Economic Inquiry*, vol. 14, pp. 466–492. (Reprinted in *Journal of Forest Economics*, vol. 1, no. 1, pp. 115–149)

Susaeta, A., Carter, D., Chang, S. J. and Adams, D. (2013). 'The generalized Reed model and its application to determine the impact of hurricane risk on even-aged plantation management in southern United States', working paper, under review to in press, Journal of Forest Economics. http://dx.doi.org/10.1016/j.jfe.2013.08.001. available online 9 September 2013.

Susaeta, A., Chang, S. J., Carter, D. and Lal, P. (2013). 'Economics of carbon sequestration under fluctuating economic environment, forest management and technological changes: An application to forest stands in the southern United States,' working paper, under review.

van Kooten, G. C., Binkley, C. S. and Delcourt, G. (1995). 'Effect of carbon taxes and subsidies on optimal forest rotation age and supply of carbon services', *American Journal of Agricultural Economics*, vol. 77, no. 2, pp. 365–374.

Appendix A-1: The impact of a higher site preparation cost for the current timber crop

Applying the implicit function theorem,

$$d(\partial LEV_1 / \partial t_1) = (\partial^2 LEV_1 / \partial t_1^2) dt_1 + (\partial^2 LEV_1 / \partial t_1 \partial n_1) dn_1 + (\partial^2 LEV_1 / \partial t_1 \partial Cs_1) dCs_1 = 0$$

$$d(\partial LEV_1 / \partial n_1) = (\partial^2 LEV_1 / \partial t_1 \partial n_1) dt_1 + (\partial^2 LEV_1 / \partial n_1^2) dn_1 + (\partial^2 LEV_1 / \partial n_1 \partial Cs_1) dCs_1 = 0$$

As such,

$$\begin{bmatrix} \partial^2 LEV_1 / \partial t_1^2 & \partial^2 LEV_1 / \partial t_1 \partial n_1 \\ \partial^2 LEV_1 / \partial t_1 \partial n_1 & \partial^2 LEV_1 / \partial n_1^2 \end{bmatrix} \begin{pmatrix} dt_1 / dCs_1 \\ dn_1 / dCs_1 \end{pmatrix} = \begin{pmatrix} -\partial^2 LEV_1 / \partial t_1 \partial Cs_1 \\ -\partial^2 LEV_1 / \partial n_1 \partial Cs_1 \end{pmatrix}$$

Applying Kramer's rule,

$$dt_1 / dCs_1 = \begin{vmatrix} -\partial^2 LEV_1 / \partial t_1 \partial Cs_1 & \partial^2 LEV_1 / \partial t_1 \partial n_1 \\ -\partial^2 LEV_1 / \partial n_1 \partial Cs_1 & \partial^2 LEV_1 / \partial n_1^2 \end{vmatrix} / D = 0$$

$$dn_1 / dCs_1 = \begin{vmatrix} \partial^2 LEV_1 / \partial t_1^2 & -\partial^2 LEV_1 / \partial t_1 \partial Cs_1 \\ \partial^2 LEV_1 / \partial t_1 \partial n_1 & -\partial^2 LEV_1 / \partial n_1 \partial Cs_1 \end{vmatrix} / D = 0$$

because both $-\partial_2 LEV_1 / \partial t_1 \partial Cs_1$ and $-\partial^2 LEV_1 / \partial n_1 \partial Cs_1$ equal 0.

Appendix A-2: The impact of a higher planting cost per seedling for the current timber crop

$$dt_1 / dCp_1 = \begin{vmatrix} -\partial^2 LEV_1 / \partial t_1 \partial Cp_1 & \partial^2 LEV_1 / \partial t_1 \partial n_1 \\ -\partial^2 LEV_1 / \partial n_1 \partial Cp_1 & \partial^2 LEV_1 / \partial n_1^2 \end{vmatrix} / D$$

$$dn_1 / dCs_1 = \begin{vmatrix} \partial^2 LEV_1 / \partial t_1^2 & -\partial^2 LEV_1 / \partial t_1 \partial Cp_1 \\ \partial^2 LEV_1 / \partial t_1 \partial n_1 & -\partial^2 LEV_1 / \partial n_1 \partial Cp_1 \end{vmatrix} / D$$

Because $-\partial^2 LEV_1 / \partial t_1 \partial Cp_1 = 0$ and $-\partial^2 LEV_1 / \partial n_1 \partial Cp_1 = 1$,

$$\frac{dt_1}{dCp_1} = \begin{vmatrix} \partial^2 LEV_1 / \partial t_1 \partial Cp_1 & -\partial^2 LEV_1 / \partial t_1 \partial n_1 \\ \partial^2 LEV_1 / \partial n_1 \partial Cp_1 & -\partial^2 LEV_1 / \partial n_1^2 \end{vmatrix} / D = \partial^2 LEV_1 / \partial t_1 \partial n_1 / D$$

$$\frac{dn_1}{dCp_1} = \begin{vmatrix} \partial^2 LEV_1 / \partial t_1^2 & -\partial^2 LEV_1 / \partial t_1 \partial Cp_1 \\ \partial^2 LEV_1 / \partial t_1 \partial n_1 & -\partial^2 LEV_1 / \partial n_1 \partial Cp_1 \end{vmatrix} / D = \partial^2 LEV_1 / \partial t_1^2 / D < 0.$$

Appendix A-3: The impact of a higher stumpage price level for the current timber crop

The impact of a higher current stumpage price level for the current timber crop can be examined with the introduction of a price level variable α_1. As such, equations (13) and (14) will be rewritten as

$$\frac{\partial LEV_1}{\partial t_1} = \left[\frac{\partial \alpha_1 V_1(t_1, n_1)}{\partial t_1} + A_1(t_1) - r_1 \alpha_1 V_1(t_1, n_1) - r_1 LEV_2 \right] \exp(-r_1 t_1) = 0 \qquad (A\text{–}3\text{–}1)$$

$$\frac{\partial LEV_1}{\partial n_1} = \left[\frac{\partial \alpha_1 V_1(t_1, n_1)}{\partial n_1} - Cp_1 \exp(r_1 t_1) \right] \exp(-r_1 t_1) = 0 \qquad (A\text{–}3\text{–}2)$$

From equation (A-3-1)

$$\frac{\partial^2 LEV_1}{\partial t_1 \partial \alpha_1} = \left[\frac{\partial V_1(t_1, n_1)}{\partial t_1} - r_1 V_1(t_1, n_1) \right] \exp(-r_1 t_1) \qquad (A\text{–}3\text{–}3)$$

From equation (A-3-2)

$$\frac{\partial^2 LEV_1}{\partial n_1 \partial \alpha_1} = \frac{\partial V_1(t_1, n_1)}{\partial n_1} \exp(-r_1 t_1) > 0 \qquad (A\text{–}3\text{–}4)$$

Applying the implicit function theorem and the Kramer's rule,

$$\frac{dt_1}{d\alpha_1} = \begin{vmatrix} -\partial^2 LEV_1 / \partial t_1 \partial \alpha_1 & \partial^2 LEV_1 / \partial t_1 \partial n_1 \\ -\partial^2 LEV_1 / \partial n_1 \partial \alpha_1 & \partial^2 LEV_1 / \partial n_1^2 \end{vmatrix} / D$$

$$= \left\{ -\left[\frac{\partial V_1(t_1, n_1)}{\partial t_1} - r_1 V_1(t_1, n_1) \right] \exp(-r_1 t_1) \frac{\partial^2 V_1(t_1, n_1)}{\partial n_1^2} \exp(-r_1 t_1) \right.$$

$$\left. - \frac{-\partial V_1(t_1, n_1)}{\partial n_1} \exp(-r_1 t_1) \left[\frac{\partial^2 V_1(t_1, n_1)}{\partial t_1 \partial n_1} \exp(-r_1 t_1) - Cp_1 r_1 \right] \right\} / D \qquad (A\text{–}3\text{–}5)$$

$$\frac{dn_1}{d\alpha_1} = \begin{vmatrix} \partial^2 LEV_1 / \partial t_1^2 & -\partial^2 LEV_1 / \partial t_1 \partial \alpha_1 \\ \partial^2 LEV_1 / \partial t_1 \partial n_1 & -\partial^2 LEV_1 / \partial n_1 \partial \alpha_1 \end{vmatrix} / D$$

$$= \left\{ (\partial^2 LEV_1 / \partial t_1^2) \left(-\left[\frac{\partial V_1(t_1, n_1)}{\partial t_1} - r_1 V_1(t_1, n_1) \right] \exp(-r_1 t_1) \right) \right.$$

$$\left. - \left[\frac{-\partial^2 V_1(t_1, n_1)}{\partial t_1 \partial n_1} \exp(-r_1 t_1) - Cp_1 r_1 \right] \left(\frac{-\partial V_1(t_1, n_1)}{\partial n_1} \exp(-r_1 t_1) \right) \right\} / D \qquad (A\text{-}3\text{-}6)$$

Appendix A-4: The impact of a higher current annual income

The impact of a higher annual income for the current timber crop can be examined with the introduction of a price level variable β_1. As such, equations (13) and (14) will be rewritten as

$$\frac{\partial LEV_1}{\partial t_1} = \left[\frac{\partial V_1(t_1, n_1)}{\partial t_1} + \beta_1 A_1(t_1) - r_1 V_1(t_1, n_1) - r_1 LEV_2 \right] = 0 \qquad (A\text{-}4\text{-}1)$$

$$\frac{\partial LEV_1}{\partial n_1} = \left[\frac{\partial V_1(t_1, n_1)}{\partial n_1} - Cp_1 \exp(r_1 t_1) \right] = 0 \qquad (A\text{-}4\text{-}2)$$

From equations (A-4-1) and (A-4-2),

$$\frac{\partial^2 LEV_1}{\partial t_1 \partial \beta_1} = A_1(t_1) \quad \text{and} \quad \frac{\partial^2 LEV_1}{\partial n_1 \partial \beta_1} = 0$$

Take the total derivatives of equations (A-4-1) and (A-4-2), and applying Kramer's rule,

$$\frac{dt_1}{d\beta_1} = \begin{vmatrix} \partial^2 LEV_1 / \partial t_1 \beta_1 & -\partial^2 LEV_1 / \partial t_1 \partial n_1 \\ \partial^2 LEV_1 / \partial n_1 \partial \beta_1 & -\partial LEV_1 / \partial n_1^2 \end{vmatrix} / D = (-\partial^2 LEV_1 / \partial n_1^2)$$

$$(-A_1(t_1)) / D > 0$$

$$\frac{dn_1}{d\beta_1} = \begin{vmatrix} \partial^2 LEV_1 / \partial t_1^2 & -\partial^2 LEV_1 / \partial t_1 \partial \beta_1 \\ \partial^2 LEV_1 / \partial t_1 \partial n_1 & -\partial LEV_1 / \partial n_1 \partial \beta_1 \end{vmatrix} / D = (-\partial^2 LEV_1 / \partial t_1 \partial n_1)$$

$$(-A_1(t_1)) / D$$

Appendix A-5: The impact of a higher interest rate for the current timber crop

Applying the implicit function theorem to equations (13) and (14),

$$d\frac{\partial LEV_1}{\partial t_1} = \frac{\partial^2 LEV_1}{\partial t_1^2} dt_1 + \frac{\partial^2 LEV_1}{\partial t_1 n_1} dn_1 + \frac{\partial^2 LEV_1}{\partial t_1 \partial r_1} dr_1 = 0$$

$$d\frac{\partial LEV_1}{\partial n_1} = \frac{\partial^2 LEV_1}{\partial t_1 n_1} dt_1 + \frac{\partial^2 LEV_1}{\partial n_1^2} dn_1 + + \frac{\partial^2 LEV_1}{\partial n_1 \partial r_1} dr_1 = 0$$

Given that

$$\frac{\partial^2 LEV_1}{\partial t_1 \partial r_1} = -V_1(t_1, n_1) - LEV_2 \text{ and } \frac{\partial^2 LEV_1}{\partial n_1 \partial r_1} = -Cp_1 t_1 \exp(r_1 t_1), \text{ applying}$$

Kramer's rule

$$\frac{dt_1}{dr_1} = \begin{vmatrix} -\partial^2 LEV_1 / \partial t_1 \partial r_1 & \partial^2 LEV_1 / \partial t_1 \partial n_1 \\ -\partial^2 LEV_1 / \partial n_1 \partial r_1 & \partial^2 LEV_1 / \partial n_1^2 \end{vmatrix} / D$$

$$= [(V_1(t_1, n_1) + LEV_2)(\partial^2 LEV_1 / \partial n_1^2) - Cp_1 t_1 \exp(r_1 t_1)(\partial^2 LEV_1 / \partial t_1 \partial n_1)] / D$$

$$\frac{dn_1}{dr_1} = \begin{vmatrix} \partial^2 LEV_1 / \partial t_1^2 & -\partial^2 LEV_1 / \partial t_1 \partial r_1 \\ \partial^2 LEV_1 / \partial t_1 \partial n_1 & -\partial^2 LEV_1 / \partial n_1 \partial r_1 \end{vmatrix} / D$$

$$= [(\partial^2 LEV_1 / \partial t_1^2)(Cp_1 t_1 \exp(r_1 t_1) - (\partial^2 LEV_1 / \partial t_1 \partial n_1)(V_1(t_1 n_1) + LEV_2)] / D$$

Appendix A-6: The impact of a higher future land value

Applying the implicit function theorem to equations (13) and (14) and Kramer's rule,

$$\frac{\partial^2 LEV_1}{\partial t_1 \partial LEV_2} = -r_1 \text{ and } \frac{\partial^2 LEV_1}{\partial n_1 \partial LEV_2} = 0$$

Applying Kramer's rule,

$$\frac{dt_1}{dLEV_2} = \begin{vmatrix} -\partial^2 LEV_1 / \partial t_1 \partial LEV_2 & \partial^2 LEV_1 / \partial t_1 \partial n_1 \\ -\partial^2 LEV_1 / \partial n_1 \partial LEV_2 & \partial^2 LEV_1 / \partial n_1^2 \end{vmatrix} / D = r_1 \left(\frac{\partial LEV_1}{\partial n_1^2} \right) / D < 0$$

$$\frac{dn_1}{dLEV_2} = \begin{vmatrix} \partial^2 LEV_1 / \partial t_1^2 & -\partial^2 LEV_1 / \partial t_1 \partial LEV_2 \\ \partial^2 LEV_1 / \partial t_1 \partial n_1 & -\partial^2 LEV_1 / \partial n_1 \partial LEV_2 \end{vmatrix} / D = \left(\frac{-\partial^2 LEV_1}{\partial t_1 \partial n_1} \right) r_1 / D$$

4

TEMPORAL ASPECTS IN FOREST ECONOMICS

Colin Price

90 FARRAR ROAD, BANGOR, GWYNEDD LL57 2DU, UK. C.PRICE@BANGOR.AC.UK

Abstract

Time is crucial in many aspects of forest economics. It defines the units of resource use. It brings patterns of variation in costs and revenues. The long period between expenditure and ensuing benefit makes many forest investments inflexible and seemingly unprofitable under discounting protocols or less profitable than other land uses. Optimal forest rotation and thinning regime are strongly affected by the discount rate used, as are optimal intensity of silvicultural intervention and harvesting investment. Discounting favours maintenance of existing silvicultural systems, rather than adopting improved regimes. Long-term environmental values are made less important. Of the reasons given for discounting, the opportunity cost of investment funds argument assumes implausible reinvestment of revenues: opportunity cost can be dealt with more sophisticatedly. The time preference argument fatally misinterprets what people prefer, and neglects intergenerational justice. The diminishing marginal utility argument only applies to some classes of value. Risk is best treated by other protocols. The recently favoured declining discount rate protocol has questionable theoretical roots and creates major practical difficulties for forest economists. Apart from technical problems, internal rate of return as a choice criterion is ethically suspect. Foresters need to engage with these economic and ethical arguments.

Keywords

Time period, time lapse, discounting, investment, profitability, deforestation, silviculture, ecosystem services

Introduction: The importance of time in forestry

That time is an important factor – the most important factor – in forestry economics can hardly be doubted. Time gives the units – worker-hours, machine-hours, annual rentals of land – in which use of resources is expressed. Time within years or business cycles determines the availability and price of inputs and of products. Time gives the markers for the appropriateness and tractability of forest operations. But above all, the lapse of time between forest cost and forest benefit, is what has often called the economic viability of forestry into question, and has often led to the decline of forest extent and quality around the world. Moreover, the long-term consequences of deforestation, reforestation, afforestation and forest modification for such matters as atmospheric CO_2 and climate, for hydrological systems, for ecosystems and for landscapes are susceptible to whatever means are adopted to adjust for the lapse of time.

These long time periods until the maturity of investments – and the long-term consequences of actions – make seeing into the future, and accounting for projected events, of far greater significance than it is in other forms of industrial, personal and land-use investment, where payback periods of a few years are normal. Investments in forest roads, recreation facilities and harvesting equipment may have a shorter financial life-span and payback period, but even for these there may be long-term social and environmental consequences.

This chapter seeks to set forest decisions, many of which are evaluated in other chapters, in the context of passing time, and draws out some generalities about how time should be treated.

Time in the unit of resource

Unlike labour, capital and raw materials, which are required only for the duration of *operations*, land as a factor of production is occupied throughout the growth cycle of productive forests and is the permanent substrate for exercise of forest influences (Kittredge, 1948) – or, as would now be said, provision of ecosystem services. This is one of two quite separate influences of time in determination of optimal forest rotation.

The long duration of occupation raises the issue of nonadaptability. Whereas most production processes can be rearranged to produce, say, a different model of car, or food crop, within a timescale of months, and industrial sites may be restructured for entirely different modes and types of production within at most a few years, most forests are as they are, and apart from by catastrophic natural events or by clear felling, may not be changed to a very different form (species, age-class structure, etc.) in a time period less than decades. This makes functional flexibility, where a given physical crop can be adapted to a number of different purposes, an important feature of forestry in an era of multiple purposes and of rapid change in the emphasis given to each. For short-rotation woody crops (e.g. eucalypts and poplars) this is less of a problem, as is also the case for tropical plantations: but even for these the production cycle is generally substantially longer than that of agricultural crops.

The inertia of forest condition is also becoming an issue in relation to rapidly changing circumstances, as exemplified by global climate change, and by the associated migration of tree diseases, unprecedented in scale and rate (Price, 2010a). Species which are appropriate to climate and resistant to prevalent pests and diseases at the beginning of a rotation, may no longer be so by the end.

Seasonality

The season of the year may determine the feasibility and cost of forest operations. Ground conditions in the wet season may only allow access by use of prohibitively expensive modes of cross-country transport; or wet ground in the boreal zone may only be workable when frozen or blanketed by snow. Protection operations have a seasonality according to the flight time of insects or the advent of conditions (typically warm, moist ones) propitious for fungal infection.

In a mixed rural economy, where workers divide their time between forestry, agricultural, and tourist related activities, the opportunity cost of labour may also vary seasonally, according to the requirements of these other industries. Obversely, seasonality brings variation in revenues, from sale of forest recreation services or of seasonal vegetation such as Christmas trees and evergreen foliage.

Trends and cycles

Because of the long production period – for timber, and for high levels of environmental benefits such as creation of habitat, landscape and carbon stores – predicting future circumstances is unusually important for forest expenditures, and particularly for investment in afforestation or silvicultural improvements. Many changing factors – physical, biological, technological, shifts in the economic and political context – affect the expected quantity of timber and other benefits, and the actual or implicit price to be ascribed to them. Classical forest economics developed in a world of presumed constancy, where the consequences of any action were known with certainty, and where sustainable forest management meant repeating the same sequence of operations rotation after rotation (Faustmann, 1849). The reality always was that fires, storms, insects and funguses might bring rotations to an unexpected premature close; wars and the threat of wars came and receded, bringing fresh demands for timber; new political dispensations arose and were in their turn displaced; unimagined new technologies developed and replaced some traditional uses of wood; affluence and outdoor activities grew.

All these factors affect agriculture too, but the long forestry production cycle brings greatly increased prospects of disruptive changes' occurring within the production cycle.

In the meantime growth and development within the forest crop bring change in the benefits and costs delivered. Environmental influence increases with age, bringing some enhanced benefits such as climatic amelioration, but also some rising environmental costs, as with the increasing loss of water as the forest canopy expands laterally and pushes further into the zone of air turbulence. Again, such growth occurs over much longer periods than those typical in agriculture and encompasses a much greater cumulative change.

As well as these trends through the unfolding of earth time and developments with tree age, there are recurrent cycles. Seasonal cycles mediate the condition and effects, particularly of deciduous crops. Economic cycles bring variation in demand: in particular the construction industry, with its continuing dependence on timber as a material, drops into recession more deeply than other parts of the economy and likewise recovers more strongly.

In addition, there are less orderly short-term fluctuations depending on the aggregate of scarcely predictable climatic and political circumstances, such as those that underlie the world timber supply. But these are not just irritating noise with no overall effect on the forest economy: by intelligent response to high and low prices, forest owners and managers can fell – or indeed invest – at more advantageous times. Such price-responsive cutting leads not only to greater profitability (Lohmander, 1987) but increases the length of optimal rotation (so as to increase the possibility of its including a particularly high price peak). Khajuria's contribution to this volume (Chapter 22) reviews some relevant factors.

The effects of time discounting

However, the temporal aspect which has drawn universal attention – professional and lay, economic and political – is the great lapse of time between forest investment and the resultant accrual of enhanced revenues. The general perception is that money, goods, services and resources have a *time value*, its being supposed that early availability is more valuable than late availability, and early cost more burdensome than late.

The crudest manifestation of this is in a time horizon, the point in the future beyond which no further account is taken of events, costs and benefits. With rotations extending to decades or even centuries, the benefits of forestry often fall beyond any normal time horizon – which is thus not a popular concept among foresters. Many expedients have been adopted to avoid the problem, such as turning the focus away from individual stand investments, to considering instead the year-on-year flows of benefit and cost from a whole forest in which all age classes of stand or tree are represented (Markus, 1967). The illusion is thus created that expenditure and relevant revenue occur contemporaneously, not separated by a great remove of time. Clearly this is not the case: whatever it is that *results* from investment in regeneration of a stand, it can be neither the revenue from trees already felled and sold, nor the environmental benefits of mature trees that have heretofore accrued (Price, 1986).

A less severe form of censorship of the future is embodied in discounting: the process by which values are reduced progressively, the further into the future that they accrue.

Until recently, discounting was considered by popular agreement to take a form in which values were reduced by the same percentage, for each year further into the future that they accrued. This generates the following formulation of the discounted value, also known as present value:

$$PV = X_t / (1 + r)^t \tag{1}$$

where X_t is a benefit, cost or revenue expected t years after some reference date,
PV is the *present value* of X, discounted to that reference date,
r is the discount rate and
$1/(1 + r)^t$ is called the discount factor.

This formulation is often seen as the obverse of compound interest: instead of *multiplying* by 1 + [interest rate] once for each year in which present cash is invested *forwards* into the future, we *divide* by 1 + [discount rate] once for each year that values are brought *backwards* from the future. The concepts underlying discounting are, however, more complex and contentious, as we shall soon discuss.

As well as in this *quotient format*, discounting is also represented – particularly by professional economists – in the following *negative exponential format*.

$$PV = X_t \times e^{-\rho t} \tag{2}$$

where e is 2.718, the base of natural logarithms and ρ is the natural logarithm of $(1 + r)$.
Note that ρ is similar, but not equal, to r.

The two formulas deliver the same results, but negative exponential format is more flexible, for treatment of values accruing continuously through time and is more readily entered into algebraic arguments. Both formats are entirely general: they can be used with positive and negative and fractional time periods, and with positive and negative and fractional discount rates. Where values accrue continuously through time, or at regular discrete intervals, it is possible to use short-cut formulas to give a single summary discounted value.

The PV of £Y after 1 year, £Y after 2 years, after 3 years and after every subsequent year in perpetuity is

$$£Y/r \tag{3}$$

For a cash flow of £Y for every year *after* time B (i.e. at time $B + 1, B + 2, \ldots, \infty$), the PV is

$$\frac{£Y}{r} \times \frac{1}{(1+r)^B} \tag{4}$$

For a cash flow of $£Y$ every year up to and including time T (i.e. at time $1, 2, \ldots, T$), the PV is

$$\frac{£Y}{r} \times \left(1 - \frac{1}{(1+r)^{T}}\right) \tag{5}$$

For a cash flow of $£Y$ every year, starting at B until time T (i.e. at time $B, B+1, B+2, \ldots, T$), the PV is

$$\frac{£Y}{r} \times \left(\frac{1}{(1+r)^{B-1}} - \frac{1}{(1+r)^{T}}\right) \tag{6}$$

For a cash flow of $£Y$ which first occurs at time F and then recurs at F-year intervals thereafter, the PV is

$$\frac{£Y}{(1+r)^{F} - 1} \tag{7}$$

If $£Y$ also accrues at time 0, the formula for PV is

$$£Y \times \frac{(1+r)^{F}}{(1+r)^{F} - 1} \tag{8}$$

or

$$\frac{£Y}{1 - e^{-\rho F}} \tag{9}$$

Either formula can be used when $£Y$ is the summed PV of a series of cash flows lasting F years and then repeated in perpetuity.

If the cash flow is continuous at the rate $£Y$ per year from time B to time T, PV is

$$£Y/\rho \times (e^{-\rho B} - e^{-\rho T}) \tag{10}$$

If the continuous cash flow begins immediately, PV is

$$£Y/\rho \times (1 - e^{-\rho T}) \tag{11}$$

If it begins after B years and then lasts in perpetuity, PV is

$$£Y/\rho \times e^{-\rho B} \tag{12}$$

If it begins immediately and lasts in perpetuity, PV is

$$£Y/\rho \tag{13}$$

For detail of how such formulas are derived, see Price (1993, chapter 1).

It should be noted that in formulas like (3) and (7) the present value is finite, even though derived from an indefinitely continuing series of cash flows. This reduction of sustainable benefit to a limited, quite small sum, is one of the many results that makes environmentalists and foresters profoundly suspicious of discounting.

The effect of inflation

That discounting has nothing to do with inflation can hardly be said often enough. The purchasing power of money ordinarily decreases through time, sometimes at a spectacular rate, sufficient to cause political instability and draconian governmental measures to curb its influence. But discounting is about something more fundamental: the decline in value through time of individual goods and services purchased. Discounting is applied even in times and places of zero or trivial inflation.

Economists usually treat inflation by adjusting monetary sums according to a price index which shows the relative amounts of money required to purchase a given basket of goods at different times. Such adjustment is not usually needed in forest economics, or in investment appraisal generally. Because future prices cannot be observed, and because inflation rates fluctuate unpredictably, it is convenient to predict future values in terms of current (or real) prices, or in prices measured in the unit of the current value of money. To ensure consistency, the discount rate must itself be set in terms of current prices. Thus, a monetary interest rate does not represent the true rate of increase in purchasing power, which must be reduced by the margin of the inflation rate.

If money interest is to compensate fully for inflation, the appropriate rate is given by:

$$1 + [\text{money interest rate}] = (1 + [\text{real interest rate}]) \times (1 + [\text{inflation rate}]) \tag{14}$$

This precise formulation is not always recognised in the literature, and the difference from the approximate form

$$[\text{money interest rate}] = [\text{real interest rate}] + [\text{inflation rate}] \tag{15}$$

can become important in high-inflation economies.

In exponential format, however, the continuous real interest rate, ρ, may be summed with the continuous inflation rate, ι, to give the continuous monetary interest rate, μ.

The effects of time discounting in forest economics

The majority of writings on forest economics, perhaps the vast majority, make use of discounting, numerically or descriptively. In this section are considered some of the many decisions affected. More examples will be found in other chapters of this book.

Profitability: Absolute and relative

No one familiar with the power of compound interest will be surprised at the decisive – some would say catastrophic – effect of discounting on the long-term profitability of forestry, at anything other than a nominal discount rate. A rate of 6%–10% would normally be used in commercial decisions; the World Bank has advocated 10% (Adler, 1987); 10% was the rate mandated for public investments in the UK in the 1970s, including an application to afforestation (Treasury, 1972). Since then the UK's advised rate has ranged through 5%, 6% and 3.5%, but even the lowest of these rates has a profound influence on the economics of temperate forestry.

To take a simplified example: if €5,000 needs to be invested to establish 1 hectare of oak on the 200-year rotation normal in Europe, the required end-of-rotation revenue would be

$$€5,000 \times (1 + 0.10)^{200} = €950 \text{ billion, or around half of the GNP of France!}$$

To put it another way, the discounted value of an extremely high final revenue of €500,000 would be, at 10%

$$€500,000 / 1.10^{200} = €0.00263: \text{ that is, less than a single cent.}$$

Even the 4% discount rate used historically in Europe (Faustmann, 1849), the 3% commonly applied by foresters in Scandinavia, the 5% rate generally used in profitability analysis of non-industrial forestlands in the United States, all seriously compromise profitability. Wear's chapter in this volume (Chapter 5) explicitly acknowledges the time element as an obstacle to adequate private investment under discounting criteria.

On the other hand, discount rates have little adverse effect on the profitability of forest exploitation because the time lapse between investment (construction of forest roads and other infrastructure) and the last of the revenues is at most a few years. In fact, by giving little weight to the negative long-term consequences of removing tree cover, high discount rates can readily make exploitation seem *more* desirable from a societal perspective.

The chapter by Sills in this volume (Chapter 26) suggests that at high interest rates timber production is not a competitive land use, while that by Mercer, Frey and Cubbage (Chapter 13) notes that high discount rates favour agriculture over agroforestry. Angelsen's chapter in this volume (Chapter 19) mentions that even the lag between investment and 'payment for results' under REDD may be an obstacle to implementation.

The aggregate effect of all these influences is that high discount rates are hostile to creation and maintenance of tree cover in the world.

Optimal rotation and thinning

The question of the most profitable time lapse before harvesting should occur has long been, and remains, the most frequently discussed issue in the theory of forest economics (for a review, see Newman, 2002). On one hand, prolonging the rotation yields greater volume, generally a higher price per cubic metre, a lower harvesting cost per cubic metre and – where relevant – a better prospect for natural regeneration: on the other, as already discussed, it requires longer occupation of site, and thus an increasing opportunity cost from lost occupation of the site by successor crops. Crucially, it also entails an ever-heavier discounting penalty. Chapters in this volume by Chang (Chapter 3) and by Burkhardt, Möhring and Gerst (Chapter 21) include discounting as an explicit argument in rotation determination. The balance of factors, for a typical conifer crop in the UK, is shown in Figure 4.1.

NPVinf and NPVone are the net present values of respectively a perpetual series of rotations, and a single rotation. Where forestry is profitable at all, the value for a perpetual series culminates somewhat earlier, because of the opportunity cost, discussed previously, of the land for successor crops.

Even with moderate discount rates, the long rotations customarily adopted in European forestry can hardly be profitable, unless there is major early thinning revenue.

Removal of trees in the intermediate part of a rotation has been regarded both as a harvesting activity undertaken to derive early revenue and as a cultural operation intended to improve

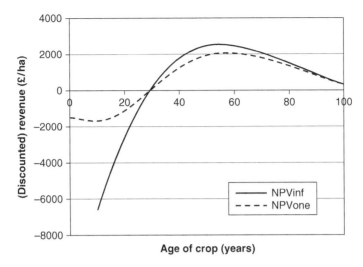

Figure 4.1 Profiles of net present value through time (own calculations).

future crop revenues. More generally, it is a means of allocating available increment, both between trees and through time. The rational balance among these should be heavily influenced by the discount rate. At high rates, emphasis will be given to early revenue, and a case can be made for removal of the best-quality, largest trees in the crop, once they can be harvested with reasonable profit: at low rates investment in future revenues may predominate, and low value trees should be removed, even perhaps at a loss, in order to allocate the increment of the remainder of the rotation to trees which may both be more photosynthetically efficient, and possess stem qualities required for a high final price per cubic metre (Price, 1987).

The discount rate also affects the intensity, frequency and pattern of thinning (Price, 1989, chapters 14 and 15).

Intensity of silviculture

Although silviculturists might be reluctant to characterise it in this way, silviculture is just a sequence of investments in a forest crop. Some, such as fertilising, are intended to increase the volume produced within a given time, or reduce the time taken to produce a given volume. Site amelioration, by activities like drainage, produces benefits over many future rotations and thus is particularly difficult to justify with a high discount rate. Other silvicultural investments, such as pruning, are intended to increase the quality of product as indexed by the price per cubic metre. Nowadays silvicultural investment may also be made in environmental improvements: pruning to improve aesthetic values; fertilising to accelerate carbon sequestration.

Crop protection activities also represent intensification of investment, to reduce the probability of *loss* of revenue.

Choice of hard-to-establish species or adoption of difficult-to-manage species mixtures are additional investments at crop initiation, often with a view of making the crop more amenable to a range of long-term environmental effects or more robust to long-term changes in climate or markets.

The net value of all these silvicultural activities is affected by discounting, and the optimal intensity of investment will always be less, the higher the discount rate.

In the context of urban forestry, high discount rates favour strategies that involve lower investment in tree establishment at the cost of higher lifetime maintenance. They also reduce the intensity of planting intended to achieve such benefits as air conditioning and shade creation.

The intensity of investment in forest road networks is similarly affected. The optimal density of the network is determined by the balance between early construction and maintenance costs of roads, especially in a new area of commercial exploitation, and late-in-rotation cross-country movement costs for timber. When in the 1970s the UK's public discount rate changed from 10% to 5%, the theoretically optimal forest road density increased by 40% overnight – though this did not seem to bring about a commensurate flurry of adjustment activity on the ground.

Discounting affects the cost of long-lived machinery and of facilities which generate income over a long time period. It thus affects optimal choice of technique and mix of output. High discount rates favour less capital-intensive harvesting systems, delaying the introduction of more technically advanced machinery.

Changing the system

Changing physical and economic conditions, changing availability of genetic material and forest machinery and new thinking (including that about discounting) indicate new 'ideal' forms of forestry. But, because of the inertia noted previously, change may only come about slowly. Thus, the benefits and/or cost savings of a changed system may be long delayed, compared with the lower but more immediate benefits of the present one. This has become a particularly relevant factor in relation to the adoption of continuous cover forestry (also called 'uneven-aged forestry'), which is now enshrined in at least one national forestry policy (National Assembly for Wales, 1999) not only as a target area, but also with a target timetable for transformation. Transformation could entail felling some trees before their optimal rotation, others after it, which would necessarily bring medium-term reduction of profit (Price and Price, 2006). Even if the system, after completion of transformation, was more profitable in perpetuity than had been the system which it displaced, the medium term penalties might outweigh this long-term enhancement.

Contrariwise, where continuous cover with multiple age classes exists already, economies in harvesting might be permanently achieved by an even-aged stand structure. But such advantage would also be brought about by felling trees in the short term at other than their optimal rotation, and the transformation in this direction might equally reduce the total of future discounted values.

The same argument applies to conversions between a single-aged forest and a forest with a normal series of age-classes, in either direction: whatever benefits are perceived to arise will be fully achieved only once the conversion is complete, while the costs occur in the short to medium term. Very often the conclusion will be to stay with whatever structure is in place, the more so at higher discount rates.

When changing circumstances, or changing perception of circumstances (as when a forest passes into new ownership or management), indicate that many different adjustments of the forest would improve the flow of long-term net benefit, the discount rate affects the cost of delaying each particular adjustment, influencing the sequence of adjustment, and possibly whether it is worth making any adjustments at all.

Discounting environmental values

Since it became feasible and fashionable to value environmental benefits and costs in monetary terms, these values have come within the purview of discounting. An example of such discounting appears in the chapter by Buongiorno, Bollandsås, Halvorsen, Gobakken and Hofstad (Chapter 17).

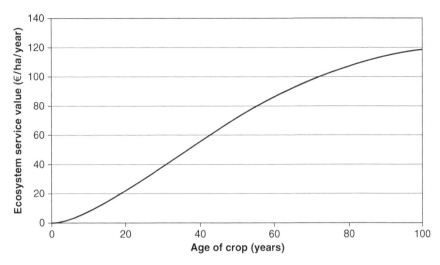

Figure 4.2 Conceptual profile for ecosystem services as forest matures (own calculations).

Table 4.1 The relative effects of reducing assigned value and increasing the discount rate.

Ecosystem service value	Value discounted at 1%	Discount rate	100% Ecosystem service value
100%	3,491	1%	3,491
33%	1,152	3%	1,153
10%	349	10%	107

Environmental values – for example carbon mitigation and gene conservation – generally accrue over a long period and usually achieve their greatest magnitude only with age of the forest. Thus, discounting may greatly affect their significance. Figure 4.2 shows a plausible profile of the effect.

Based on this profile, Table 4.1 shows that the effect of increasing the discount rate from a nominal 1% (favoured by the UK Treasury for very long-term discounting) by a factor of 10, to 10% (formerly used by the UK Treasury for all discounting) is more detrimental than reducing the monetised value of ecosystem services by a factor of 10. Appropriate discounting can be more important than accurate valuation.

On the other hand, when a mature forest is removed, environmental values are lost at a stage when they already have full value, so that discounting affects them less. Even so, over the same 100-year period as is embraced in Table 4.1's example, the effect of increasing the discount rate from 1% to 10% is to reduce discounted value by a factor of 6.3.

Discounting physical entities

Discounting for time is customarily applied to cash or benefit flows. But some applications will be found in which physical entities are discounted: for example hours worked periodically in costing machinery (Price, 1989, chapters 8 and 9); timber volume to be extracted in decisions about optimal road density (Price, 1989, chapter 10); tonnes of carbon transacted between atmosphere and forest in deriving a cost of climate change mitigation measures (Healey, Price and

Tay, 2000). No attempt should be made to envisage such discounted values in terms of reduced physical units: they are introduced as a computational convenience, to allow values to be derived algebraically.

The rationale of discounting

Discounting and interest payment have been in public discussion since the days of Moses, Plato and Aristotle. They continue to play a major part in debates about natural resource, environmental and particularly forestry economics. As well as thousands of academic papers discussing the reasons for discounting and the critical choice of a discount rate, there are a few entire books that explore the subject in more detail (Lind, 1982; Price, 1993; Portney and Weyant, 1999).

Given the crucial role of discounting, and some of the unpalatable conclusions that it brings about, and the fact that its foundations and validity are in endless dispute, it is appropriate to give a brief review of the justifications that have been put forward, and of the doubts that have been cast upon them.

Opportunity cost of investment funds

Investible funds and invested resources generate streams of income, or, in a societal context, of net benefit. The diversion of resources into forestry investments thus incurs an opportunity cost, measured by the real rate of return on alternative investments – particular ones, or ones having characteristics representative of the economy. Thus – it is argued – use of the real rate of return as a discount rate indicates whether the net benefits of the forestry investment are greater than the net benefits forgone elsewhere in the economy. This is perhaps the most frequently quoted argument for discounting, in private and public contexts alike.

But the validity of the argument depends on whether the 'opportunity cost rate of return' represents the reality of alternative investment. For it to do so, investment resources for forestry must be withdrawn *wholly* from alternative investments; the intermediate revenues from alternative investments must be *wholly* reinvested in investments earning a similar rate of return, until the time when all considered investments (including the forestry one) have come to a conclusion. Some benefits included in the social rate of return may, like landscape values, be inherently unmarketed, consumption-based, and thus incapable of direct reinvestment. Production processes designed to produce only capital goods which themselves produce only capital goods would rapidly lead to diminishing marginal product of that capital. Even if these difficulties did not exist, the temptation must always be present, for some intermediate revenues to be devoted to immediate consumption, whether by individuals impatient for the increase of their own well-being, or by organisations seeking to placate shareholders, or by governments under the exigency of dealing with short-term crises or perceived need to build up military capability. Despite fair words about 'the take-off into self-sustained growth' (Rostow, 1956, p. 26) or 'the role of the forest industries in the attack on economic under-development' (Westoby, 1962, p. 168), the revenues of tropical forest exploitation have all too frequently been mainly repatriated to rich countries, or devoted to self-indulgent prestige consumption devoid of development content. Similarly, revenues from North Sea Oil were not used (as had been announced) to regenerate the UK's manufacturing economy, but to give tax cuts enabling increased consumption, in a bid to seek political popularity and re-election of the incumbent government.

Revealingly, El Serafy (1989, p. 16) says: 'the setting aside of part of the proceeds [of exploitation] in reinvestment is only a metaphor.' But metaphorical reinvestment creates no flow of goods and services for the benefit of a real future population, or to compensate for environmental degradation. By contrast, d'Arge, Schulze and Brookshire (1982, p. 255) contend that 'Economists often use the notion of "hypothetical" compensation to justify discounting. In an ethical context such arguments play no role whatsoever. Rather, if no actual compensation occurs, the market rate of return has no relevance for discount rates.'

Pure time preference and its meaning (governmental and private)

In economies of every kind, for populations following every philosophy, every religion or none, impatience is a human characteristic. 'What do we want?' went the routine student chant of the 1960s UK. And, whatever the pertinent answer was, the follow-up question 'When do we want it?' universally evoked the response 'NOW!'.

This impatience has invariably been interpreted by economists as a preference for earlier over later consumption, and the political justification for reproducing this impatience has been that a democratic government is obliged to respect the preferences of its electorate.

Yet distinguished political economists and philosophers have been contemptuous of humanity's impatient propensities:

> There is no quality in human nature, which causes more fatal errors in our conduct, than that which leads us to prefer whatever is present to the distant and remote . . . (Hume, 1739, p. 538)

> It seems clear that the time at which a man exists cannot affect the value of his happiness from a universal point of view; and that the interests of posterity must concern a Utilitarian. (Sidgwick, 1874, p. 414)

> . . . [discounting is] ethically indefensible and arises merely from weakness of the imagination. (Ramsey, 1928, p. 543)

> . . . pure time preference is a polite expression for rapacity and the conquest of reason by passion. (Harrod, 1948, p. 40)

The universality of impatience has evolutionary roots, in conditions predating proprietary rights of use. In circumstances where a food resource or a mating opportunity forgone would yield, not its retention for future use, but the likelihood of its benefiting some other individual, there was naturally a selective advantage in 'seizing the moment'.

Proprietorship means that exclusive use can now be maintained over time. Ethical reflection may accord to consumption by others equal weight with consumption by oneself. Experience and imagination may be deployed to overrule the short-sighted instincts implanted by evolution, which are so often hostile to the individual's long-term well-being.

As for the political impropriety of overruling apparently continuing preference for early consumption, that itself may be set aside by reinterpretation of what time preference means. When do we want it – NOW! states the case precisely. It is consumption in the moment in which we exist that is given priority, not earlier consumption. And when time has passed and a later moment becomes NOW! we prefer consumption at that time over earlier consumption. Thus, life and psychological experiments are replete with examples of regret and remorse expressed, after indulgence of earlier preferences has led to a worse subsequent condition than might have been attained by a more thoughtful, even-handed appraisal. As Sen (1957, p. 746) says, 'We are interested in tomorrow's satisfaction as such, not in *today's assessment* of tomorrow's satisfaction.'

Even if it is argued that each generation is entitled to decide for itself how to allocate pains and pleasures through the span of its existence, discounting inevitably affects the balance between the interests of present and future generations. Future generations' well-being is only partly represented in the 'benevolent sentiments' of those presently living (Price, 2006). 'Sustainability' – that over-used and under-comprehended touchstone of recent political pronouncement – on the face of it places the interests of future generations at parity with those of the present. Yet future generations have no political voice, any more than they exert economic power: thus politicians endorse the practicalities of discounting, while proclaiming the theoretical virtue of sustainability. Overlapping generations models have vapidly argued that an intergenerational consensus on preferring to move forward a programme of rising consumption shows that intergenerational discounting is also ethically validated: such arguments achieve their conclusions by excluding questions on whether there is also intergenerational agreement on programmes which create net benefit to early generations at a net cost to later generations. A more impartial ethical construction is that of Rawls's 'original position' in which representatives of all stakeholders devise rules under a 'veil of ignorance', which prevents their knowing which stakeholders they will in fact become. Such a construct is wont to give rules weighting the well-being of all stakeholders equally, which in an intertemporal context means 'no discounting', with the exception which we shall meet in the next section.

How such rules, formed in detachment from self-interest, are to find their way into actual decision-making is more problematical. Stable societies may be more inclined to generate, for intertemporal issues, the equivalent of 'rules in use' for management of common property resources (Ostrom, 1990). 'The seventh generation' (Clarkson, Morrissette and Régallet, 1992) represents a beneficiary as though under the 'veil of ignorance', too distant in future to be personally known, and yet potentially a person to be treated with that equal respect which is the surest foundation for a just society. And, although sustainability as a word has been hijacked for political purposes, as a concept of good it nonetheless delivers a present 'warm glow' value to those who take the concept to their hearts, as well as paying it lip-service (Price, 2012).

Diminishing marginal utility

Rawls's means of constructing rules implies 'no discounting', except when there are due causes, such as: that future generations may have greater availability of the product or service or resource that is being discounted. This leads to what seems in ethical terms to be the most defensible case for discounting: diminishing marginal utility.

The progress of technology renders – or at least has so far rendered – greater production and greater satisfaction of wants through passing time. Yet this growth is not shared by all people and has not affected all kinds of consumption equally. The income of some sections of nations' populations, and that of some nations within the siblinghood of humanity – typically those already poorest – has stagnated, as has consumption of basic goods (Price, 2003). Forest-dwelling populations may voluntarily interpret 'sustainability' as living at an adequate level, because enough is enough, and more now may mean less for the future. For these, income is not subject to diminishing marginal utility, nor need the growth of mean gross national product per head be equivalent to a diminishing mean marginal utility for consumption, as it is spread over the whole population.

Nor are all goods and services equally susceptible to the power of technological advance (Kant, 2003). Those which are intensive in materials, or products of a perceived natural environment, may arise only in such quantity as their resource base allows.

Nor can transformation and substitution within the economy necessarily assure that the growing fulfilment of all needs will result from growth of some resources. Ever-growing electronic technology cannot be physically transformed into timber or painlessly deployed against climate change. Increasing computer visuals are not a complete substitute for lost trees in forest or urban landscape, and attempts to displace the experience of the natural world by ever-more-elaborate digital facsimiles are likely to be considered perverse and sinister, or an attempt to privatise what are traditionally regarded as birthrights.

And there are limitations imposed by the pressure of growing population on depleting resources. Unexpected breakthroughs are always a possibility, but there are unexpected setbacks too: for example widespread realisation of the limits that may be imposed by global climate change has emerged only in the past 20 years.

Unpredictability and time discounting

The earlier-mentioned effect of the long production cycle on vulnerability to many kinds of threats to profitability has sometimes been evaluated by means of a premium on the time discount rate. This expedient, discussed critically in the chapter by Amacher and Brazee (Chapter 20), is at best unsatisfactory and at worst dangerous as a generalisation, for the following reasons:

- The incidence of risk may vary systematically through the crop's life.
- The discount rate affects the relative value of successor rotations, as well as the value ascribed within a single, risk-laden production cycle.
- Discounting for risk gives less significance to far-future *costs*, such as those associated with global climate change, and may thus make a risky project or policy *more*, not less, likely to be accepted.
- Instead, a mean expected value should be calculated, using probabilities of outcomes as weights on each element of their range of possible net present values.

Risk of mortality for an individual too varies systematically through the individual's life cycle, so that discounting gives an inaccurate account of the probability of survival to enjoy benefits. Besides, except for those benefits that can only accrue to a particular individual, the lapse of time brings new life and new consumers for old: in a societal context the death of one individual does not eliminate the value of goods and services.

Less dramatically, the value to an individual or society of a particular product may be reduced by change in taste. But not all products are equally subject to the vagaries of fashion. Some have a basic value that is stable through time, and even fashionable premia may vanish and then return in cyclical fashion. A single discount rate cannot plausibly represent these very different profiles of change through time.

The extension of choice has also been proposed as a reason to prefer the early availability of resources which can be deployed in a number of ways. But early exercise of choice may deny that choice to later time periods, so need not extend the breadth of choice in the long term.

Declining discount rate

The conventional economic view of discounting has always presented it as negative exponential process, in which the same proportional decline in value is attributed to each successive time period. In recent times, however, it has been argued that the discount rate used should decline through time. This belief arises from a number of sources:

- the observed profile of ascribed value implied by consumers' choices;
- an ethical standpoint which permits discounting of an individual's own welfare, but not that of future generations, who come to play an increasingly prominent part in the beneficiary population as time rolls forwards;
- the effect of aggregating different components (goods, consumer groups, scenarios) to which different discount rates should apply.

These arguments are reviewed in Price (2005), which also provides a critique:

- consumers' time preferences should not be interpreted as any preference at all for earlier rather than later (as discussed previously);
- when different benefits or costs are accorded a different discount rate, discounting should be done separately, not at some mean or 'representative' rate;
- declining discount rates lead to 'dynamic inconsistency', in which a decision seen as optimal at one time is successively revised as time unfolds;
- evaluation of investments with a complex profile of benefits becomes excessively complex – for example successive rotation periods become longer (Price, 2011), and the effects of forests on climate change are no longer tractable to algebraic integration (Price, in review).

Nonetheless, both UK and French governments have mandated this form of discounting. It is to be doubted whether those who put the requirement in place were aware of the contrary theoretical arguments, nor of the practical difficulties that they were creating for working economists – not least, forest economists.

Internal rate of return

Because of the controversy surrounding choice of discount rate, it is sometimes suggested that choice of project, including of options within forestry, should favour the option with the highest internal rate of return (IRR). This criterion has, for example, been favoured by Boulding (1935) for the selection of optimal rotation. IRR is the rate of discount at which discounted cost equals discounted revenue, and hence the project breaks even. The argument is that adopting projects of highest IRR, then investing revenues in facsimile projects, allows the maximum growth rate of the economy.

Unfortunately, some projects have multiple IRRs, whereas others have no IRR at all. More seriously, perverse results arise when projects such as forest exploitation or business-as-usual responses to climate change have early benefits and later costs. Under these conditions, the lesser the benefits or the greater the costs, the higher the IRR.

The most fundamental argument against using IRR as a choice criterion is that it effectively pre-empts human choices, inevitably ones with an ethical element, between periods and particularly between generations. In no real circumstances are the proceeds of investment totally and endlessly reinvested in facsimile projects, so even the justification of maximising growth is illusory.

Due allowance for the opportunity cost of investment funds

All of the previous discussion casts doubt on the routine discounting of future consumption. Whatever the due causes are, for expecting that values will decline through time, they differ between products, income groups and predicted future scenarios. In many cases the appropriate rate will be zero, or at least much lower than the rates of return that have most commonly been used as discount rates.

This still does not deal with the opportunity cost of investment funds diverted into forestry. But sophisticated techniques for addressing this problem have been known since the time of Eckstein (1957), repeatedly espoused by economists specialising in investment appraisal, and applied particularly in appraisal of forestry development projects (Price, 1989, chapter 31; 1993, chapter 20). Combining these techniques with declining discount rate generates its own problems, but these are not in principle insoluble (Price, 2010b).

Conclusions

This paper has dealt with the various influences of time on value, from duration of resource use, through cycles, to lapses of time. The pervasive influence of time on the justification of forestry activity has been demonstrated, attention being drawn to some chapters by other contributors who have necessarily referred to the influence of time. The means of allowing for time's passage, and the controversies surrounding the justification of these means, have been laid out.

As for future research in this topic, it might be said that the theory has been worked over so often, over such a long period, that no substantially new concepts are likely to be discovered. What are required are a balanced attention to what has been discussed already, and an unbiased endeavour to craft workable criteria for the most desirable investments in forestry.

The nature of time has always been a mystery of the universe. Its treatment by economists remains, all too often, a mystery to foresters. Foresters should, however, engage with the problems, rather than turning their backs on them and going with their instincts. There is as much to question in economists' attitudes to future time, as there is in foresters' allegiance to the procedures and practices of time past. Time brings change, and foresters, conscious of the long-lasting effects of their decisions, should consider carefully how their own practice might change too.

References

Adler, H. A. (1987). *Economic Appraisal of Transport Projects*, Johns Hopkins University Press, Baltimore.
Boulding, K. E. (1935). 'The theory of a single investment', *Quarterly Journal of Economics*, vol. 49, pp. 475–494.
Clarkson, L., Morrissette, V. and Régallet, G. (1992). *Our Responsibility to the Seventh Generation: Indigenous Peoples and Sustainable Development*, International Institute for Sustainable Development, Winnipeg.
d'Arge, R. C., Schulze, W. D. and Brookshire, D. S. (1982). 'Carbon dioxide and intergenerational choice', *American Economic Review Papers and Proceedings*, vol. 72, pp. 251–256.
Eckstein, O. (1957). 'Investment criteria for economic development and the theory of intertemporal welfare economics', *Quarterly Journal of Economics*, vol. 71, pp. 56–84.
El Serafy, S. (1989). 'The proper calculation of income from depletable natural resources', in Y. J. Ahmad, S. El Serafy and E. Lutz (Eds), *Environmental Accounting for Sustainable Development*, The World Bank, Washington, DC.
Faustmann, M. (1849). 'On the determination of the value which forest land and immature stands possess for forestry', Oxford Forestry Institute Paper 42. Translated by W. Linnard (1968).
Harrod, R. F. (1948). *Towards a Dynamic Economics*, Macmillan, London.
Healey, J. R., Price, C. and Tay, J. (2000). 'The cost of carbon retention by reduced impact logging', *Forest Ecology and Management*, vol. 139, pp. 237–255.
Hume, D. (1739). *A Treatise of Human Nature*, Oxford University Press Oxford.
Kant, S. (2003). 'Choices of ecosystem capital without discounting and prices', *Environmental Monitoring and Assessment*, vol. 86, pp. 105–127.
Kittredge, J. (1948). *Forest Influences*, McGraw-Hill, New York.
Lind, R. C. (Ed.). (1982). *Discounting for Time and Risk in Energy Policy*, Johns Hopkins University Press, Baltimore.

Lohmander, P. (1987). *The Economics of Forest Management under Risk*, Swedish University of Agricultural Sciences, Umeå.

Markus, R. (1967). *Ostwald's Relative Forest Rent Theory*, Bayerischer Landwirtschaftsverlag, Munich.

National Assembly for Wales. (1999). *Woodlands for Wales*, Forestry Commission, Aberystwyth.

Newman, D. H. (2002). 'Forestry's golden rule and the development of the optimal forest rotation literature', *Journal of Forest Economics*, vol. 8, pp. 5–27.

Ostrom, E. (1990). *Governing the Commons*, Cambridge University Press, Cambridge.

Portney, P. R. and Weyant, J. P. (Eds). (1999). *Discounting and Intergenerational Equity*, Resources for the Future, Washington, DC.

Price, C. (1986). 'Will the real case against NDR please step forward?' *Quarterly Journal of Forestry*, vol. 80, pp. 159–162.

Price, C. (1987). 'Further reflections on the economic theory of thinning', *Quarterly Journal of Forestry*, 81, 85–102.

Price, C. (1989). *The Theory and Application of Forest Economics*, Blackwell, Oxford.

Price, C. (1993). *Time, Discounting and Value*, Blackwell, Oxford. Also available freely as an electronic text from the author: c.price@bangor.ac.uk

Price, C. (2003). 'Diminishing marginal utility: the respectable case for discounting?' *International Journal of Sustainable Development*, vol. 6, pp. 117–132.

Price, C. (2005). 'How sustainable is discounting?' in S. Kant and R. A. Berry, (Eds), *Economics, Natural Resources, and Sustainability: Economics of Sustainable Forest Management*, Kluwer, Dordrecht.

Price, C. (2006). 'Economics of sustainable development: Reconciling diverse intertemporal perspectives', in *Encyclopaedia of Life Support Systems*, EoLSS Publishers, Oxford.

Price, C. (2010a). 'Appraising the economic impact of tree diseases in Britain: several shots in the dark, and possibly also in the wrong ball-park?' *Scandinavian Forest Economics*, vol. 43, pp. 45–61.

Price, C. (2010b). 'Low discount rates and insignificant environmental values', *Ecological Economics*, vol. 69, pp. 1895–1903.

Price, C. (2011). 'Optimal rotation with declining discount rate', *Journal of Forest Economics*, vol. 17, pp. 307–318.

Price, C. (2012). 'Impatience, inconsistency, and institutions to counter their effects on sustainable forest management', in S. Kant (Ed.), *Post-Faustmann Forest Resource Economics*, Springer, Dordrecht.

Price, C. (in review). 'A nightmare waiting in the wings: the social cost of carbon, forest economics and declining discount rates'. MS available from the author: c.price@bangor.ac.uk

Price, M. and Price, C. (2006). 'Creaming the best, or creatively transforming? Might felling the biggest trees first be a win–win strategy?' *Forest Ecology and Management*, vol. 224, pp. 297–303.

Ramsey, F. P. (1928). 'A mathematical theory of saving', *Economic Journal*, vol. 38, pp. 543–559.

Rostow, W. W. (1956). 'The take-off into self-sustained growth', *Economic Journal*, vol. 66, pp. 26–48.

Sen, A. K. (1957). 'A note on Tinbergen on the optimum rate of saving', *Economic Journal*, vol. 67, pp. 745–748.

Sidgwick, H. (1874). *The Methods of Ethics*, Macmillan, London.

Treasury. (1972). *Forestry in Great Britain: An Interdepartmental Cost/Benefit Study*, HMSO, London.

Westoby, J. C. (1962). 'The role of forest industries in the attack on economic underdevelopment', *Unasylva*, vol. 16, pp. 168–201.

5

FOREST OWNERSHIP, POLICY AND ECONOMICS

David Wear

PROJECT LEADER, USDA FOREST SERVICE-SOUTHERN RESEARCH STATION, CENTER FOR INTEGRATED FOREST SCIENCE AND SYNTHESIS, NORTH CAROLINA STATE UNIVERSITY, COLLEGE OF NATURAL RESOURCES, 5227 JORDAN HALL, P.O. BOX 8008 RALEIGH, NC 27695, USA. DWEAR@FS.FED.US

Abstract

Ownership group has simplified and structured both forest policy rhetoric and forest economic analysis in the United States since the 1800s. Although questions and focus on particular ownership groups have evolved, analysis has distinguished between industrial private, nonindustrial private and public (generally national forest) owners' roles in determining timber supply and forest conditions. This chapter reviews this policy history and the research addressing differences in forest production behavior by ownership group and lays out new research challenges defined by recent and rapid changes in forest ownership.

Keywords

Forest policy, timber supply, conservation policy, industry landowners, national forests

Introduction

Institutional arrangements regarding forest ownership largely influence how forests are developed and managed around the globe. Four broad classes of landowners (forest industry, institutional investors, nonindustrial private and public) provide a common taxonomy and important distinctions of management motivation, capitalization and forest outcomes (see Hyde, 2012, chapters 7–10). This taxonomy provides a durable frame for organizing arguments regarding forest conservation policy and the institutional context for forest economics research. This chapter explores the use of ownership distinctions in policy rhetoric and applied forest economics using the United States, a nation with perhaps the greatest ownership diversity, as a case study.

'Ownership', or more precisely, 'ownership class', has served as an institutional frame for organizing and evaluating forest policy needs and forest economics in the United States. Forest ownership distinctions enter the language of forest policy and economics in two ways. Ownership distinctions have organized much of the rhetoric surrounding forest policy from the beginnings of the conservation movement in the mid-to-late 1800s. Not unrelated is the use of ownership distinctions as proxies for the institutional framework of forest management for empirical analysis of land allocation, harvest choice, management intensities and timber supply. Both uses emerge from a need to address a far greater complexity of owner motivations (and a general paucity of data) in either rhetorical or analytical settings, but the ease with which forest economists and policy analysts engage these broad distinctions might suggest unexamined elements.

The most substantial ownership distinction is between public and private. Public ownership implies resources managed by government entities to deliver goods and services for public consumption using decision processes guided primarily by administrative rules. Private ownership implies property rights held by individuals or firms whose decisions are organized to pursue their exclusive welfare, perhaps maximizing profit or utility. Of course, this strict dichotomy is false. Administrative rules may include directives to maximize revenues to the government, as is the case for state-owned forests managed to fund education in Montana and some other states (e.g. Jackson, 1987). Private owners are described by a variety of economic circumstances and regulatory milieus, implying varying constraints on resource allocation decisions. For example, the behaviors of private individuals and industry owners may fundamentally diverge (e.g. Newman and Wear, 1993). Empirical work confirms important distinctions in other management outcomes from these major types of owner classes (e.g. Arano and Munn, 2006) and owner surveys highlight a variety of owner objectives (Butler, 2008).

This chapter examines the concept of ownership as it organizes first forest policy rhetoric and then forest economics and policy analysis in the United States. It starts with a historical review of how forest policy has focused on ownership class distinctions and implicit hypotheses regarding differences in forest management approaches and outcomes. The second part of the chapter reviews forest economics research that accounts for ownership in understanding aggregate forest outcomes and in some cases directly addresses the hypotheses defined by the policy rhetoric. The chapter concludes with a discussion of unanswered questions about how ownership affects management, outcomes and policy, especially in light of recent changes in the structure of forest ownership in the United States. A concluding question asks whether ownership still provides a meaningful structure for contemporary forest policy analysis.

Ownership and forest policy rhetoric

Distinctions among forest ownership groups have organized the rhetoric of forest policy since the late nineteenth century. Ownership groupings reduce broad ranges of forest owners into a manageable set of classes that imply a meaningful homogeneity of management context and motivation within each class. Use of ownership class to organize rhetoric has simplified the focus of policy. The following discussion skims the historical context while highlighting some key ideas regarding the formation of forest and conservation policy.

The nineteenth-century forest conservation discourse starts with identifying the 'bad actors' profiting from unsustainable timber harvesting. George Perkins Marsh, in the book *Man and Nature* (1864/2003), associated overharvesting with a transient 'lumberman' ownership group distinct from the resident farmer ownership class. The wood products industry shifted from region to region, with wholesale land clearing often followed by land abandonment as shell companies failed to pay property taxes (see also Williams, 1989). Marsh enumerated several consequences of 'forest destruction' that ranged from the effects on water flows and disastrous accumulations of forest fuels to modification of local climate. Marsh saw indifferent management by these private landowners coupled with de facto treatment of public land as common property leading to widespread devastation. Although the farmer ownership class was not without blame – the first wave of deforestation was associated with conversion of forest to agriculture – Marsh saw rural farmers' long-term interests as a foundation for future conservation.

Marsh's ideas about the interconnectedness of forest conditions and human welfare formed the core of a speech delivered by Franklin Hough to the annual meeting of the American Association for the Advancement of Science in 1873. Hough's speech is important because he succeeded in motivating federal action that would eventually lead to establishing the US Forest Service and further framed the conservation discussion with an understanding of institutional and ownership context. Hough anticipated that ongoing and irreversible transition of public to private ownership of forests would limit the application of any central control over forest conditions in the United States: 'The title to the lands in our older (*eastern*) states (where the evils resulting from the loss of forests are liable to be first and most severely felt) has already passed into the hands of individuals, and from the theory of our system of government, the power that must regulate and remedy these evils must begin with the people, and not emanate from a central source.' As a result, informing private landowners (farmers) of the benefits that forest restoration and management would entail is one of the few potential remedies to forest losses. Hough saw as well an opportunity to protect resources by drawing some abandoned private lands back into the public domain through changes in state policy – that is, an opportunity to change the ownership distribution as a means to protect forests and benefit people.

Hough's speech and Marsh's treatise similarly implied that private landowners are subject to a market failure driven first by a lack of information regarding the long-term returns to forest management: 'It must come to be understood that a tree or a forest, planted, is an investment of capital, increasing annually in value as it grows, like money at interest, and worth at any time what it has cost – including the expense of planting, and the interest which this money would have earned at the given date' (Hough, 1873). Marsh more clearly sees the issue as a case where 'the value of its (the land's) timber will not return the capital expended and the interest accrued' (Hough, 1873, p. 278). Marsh appeals to a moral duty to restore forests for their benefits to subsequent generations in spite of challenging economics ('. . . we can repay our debt to our noble forefathers only by a like magnanimity'), whereas Hough appeals for government action that would clarify and operationalize reforestation goals on private lands, first through a careful assessment of natural resource conditions and the potential for government actions to encourage conservation.

As a result of Hough's call to action, Congress established the office of the Forester in US Department of Agriculture, initially filled by Hough and then by Gifford Pinchot, and the US Forest Service grew out of this chain of events. Pinchot's work as the Chief Forester and then the Chief of the Forest Service was fueled by an extension and refinement of this same rhetoric.

By associating lumbermen with big business, and with oligarch railroad barons, the conservation movement fit well within the progressive movement of the late nineteenth and early twentieth centuries and helped Pinchot and Theodore Roosevelt push Congress to establish national forests in the United States in 1905. In the process, a much more substantial public forestland ownership class was established. Policy rationale for establishing first the western forest reserves beginning in 1891 and then the initial national forests in 1905 focused on halting the disposal of the public domain and therefore the chain of harvesting and destruction by lumbermen. Pinchot argued not for setting these lands aside for preservation but for using them to develop a silviculture for the United States based on the continental European model. Public lands would help protect product and water flows for the nation but would also provide a place to demonstrate scientific forest management for private landowners. The tacit assumptions throughout the conservation dialogue of this period included (1) that farmer owners were fundamentally vested in good stewardship (their livelihoods depended on long-term stewardship though their resource constraints limited action); (2) that lumbermen would continue to ignore the care of land in making their harvest decisions; and (3) that public forest land could be used to develop a scientific sustained yield approach to management that might eventually be adopted across the broader landscape.

The Weeks Act of 1911 allowed for the establishment of Eastern national forests through purchase of lands deemed important to protect navigable streams – an idea advocated by Hough in 1873, among others. In many cases, these lands were available usually because they had been harvested and abandoned. Establishment of public forests in the West and then in the East failed to quell debates regarding the road to timber famine and the role of the forest industry. First the Forest Service and then the Society of American Foresters in 1919 called for the direct federal regulation of forest cutting on private lands as part of a national plan for forests (Graves, 1919; Committee for the Application of Forestry, Society of American Foresters, 1919; Steen, 1976). A protracted policy debate ensued and concluded in 1924 with the passage of the Clark-McNary Act (see especially Steen, 1976, pp. 173–195). The act did not provide for the regulation of private harvesting, but instead established cooperative programs with states to protect forests from fire, develop infrastructure for forest regeneration and provide technical assistance to small private landowners, and generally relied on landowner education and voluntary approaches to protect resource values on private lands.

Over the next 25 years, the US Forest Service and forest industry parried over forest regulation (see Steen, 1976) in the halls of Congress and in the realm of public opinion. Rather consistently, support for cooperative programs that leveraged state and private funding, especially for fire protection, developed as alternatives to any direct regulation of logging or logging practices. Industry urged Congress to vest regulatory authority in individual states. The US Forest Service advocacy for regulation from the US Forest Service essentially evaporated at the beginning of the Eisenhower administration as policy focus shifted away from forest protection per se and toward expanding the productivity of national forests and encouraging more efficient management of forests in small holdings; that is, those managed by farmers and other individuals.

Policy rhetoric had shifted rapidly from the close of World War II as housing demands boomed and timber scarcity loomed. Forest industry began to employ a growing cadre of college-educated professional foresters, and was increasingly viewed as practicing a sustained yield forest management. A shortfall in timber supply was seen to emerge, not from lands that the industry now held, but from other private forest lands held by a wide variety of owners with a broad range of objectives and management circumstances. The inelegant label that stuck to this group was 'nonindustrial private' forest (NIPF) landowners and it included farmers, commercial owners without processing facilities and all other private owners. During this period, forest policy concerns shifted from owners who were seen as overharvesting to those who were seen as underinvested in forest management. The primary policy issue was more squarely placed on mitigating timber scarcity by encouraging scientific management on NIPFs (see Boyd, 1984). Policy analysis also came to focus on national forests and their role in supporting the economic growth of the United States.

With national forests and forest industry producing at a maximum, further output growth required expanding productivity of the NIPF lands of the United States. The failure of NIPF owners to organize their forest management to produce more timber was seen as the key market failure behind timber scarcity. Forest policies from the 1950s to the 1990s addressed these failures – through the development of private landowner technical assistance especially (e.g. forestry extension programs), improved forest health and fire protection to protect private investments and, most notably, through subsidies (cost sharing) for tree planting, either for afforestation or reforestation.

This schema defined a constituent structure of much forest policy and forest economics for more than 50 years, from the late 1940s to the beginning of the 2000s. In 2000 the forest industry class controlled roughly 15% of timberland (20% in the productive southeastern United States). The NIPF owner had consistently been viewed and pursued as the target for enhancing the nation's forest productivity and improving resource conditions. Over a 40-year period, the Forest Service and forest industry had pioneered an intensive management style that provided the raw material foundation for a vast expansion in timber production in the United States and the interregional transfer of production from the West to the South. Then, at the start of the new century, these firms, with very few exceptions, rather suddenly sold their lands and left behind the business of growing timber. This change raises important questions, and not a little confusion, regarding the motivations behind this now-historical episode and its implications for the future of forests and forestry.

The divestiture of industry timberlands represents a substantial 'disintegration' of the vertically integrated wood products sector as timber growing was decoupled from wood products manufacturing. It also represents a decoupling of forest industry from land management and conservation policy issues. Most of these timberlands shifted to other forms of corporate ownership, largely in the form of fiduciary entities that package forest management assets for institutional investors (timberland investment management organizations, or TIMOs) or firms that directly invest in forest land portfolios with special tax treatments (timberland-focused real estate investment trusts, or REITs).

From the 1950s until the 1990s, forest policy had focused on encouraging effective management on nonindustrial forest lands as distinct from industrial forest lands. The tacit assumption had been that the former, with profit motives and ample resources, were predictable timber suppliers and were likely to pursue a sustained yield management as long as the economics of timber growing supported it. NIPFs were constrained in doing so by a lack of liquid capital and a lack of technical expertise. With disintegration, this dichotomy falls apart – if there are practically no industrial owners, then 'nonindustrial owners' applies to practically all private forests. A new set of policy and supply questions emerge.

Ownership in analysis of timber supply

Ownership defines the 'who' of forest management necessary to construct a compelling narrative for forest policy proposals. This forest policy dialogue has provided the primary motivation for forest economics research in the United States. Ownership distinctions in policy rhetoric define often-untested hypotheses regarding economic choices made by subgroups of forest landowners. Early economic analysis simply adopted many of these implicit hypotheses, whereas later studies attempted to quantify their implications and some directly tested the hypotheses. This literature indicates that ownership matters in the analysis of forest production and timber supply. Most fundamentally, the distinct behaviors among owner groups mean that timber production cannot be reduced to a representative producer model as the theoretical foundation for aggregate supply analysis.

Modern timber supply analysis has sought to build models to describe current production and predict future production possibilities from the nation's forests, by attaching an economic choice analysis to a description of forest conditions. Early economic models of timber supply (say, from the 1950s forward) apply optimal harvest choice defined by soil expectation values (discounted cash flow) to growth and yield models and sum solutions across a measured inventory. These models are inherently normative, in that they specify optimal solutions based on a 'correct' model of valuation and choice and allow for quite a bit of technical detail in the modeling of forest dynamics (e.g. through the growth and yield models). Still, they bring economic logic to supply analysis that had previously relied strictly on the analysis of forest inventories.

Vaux's (1954) examination of sugar pine production in California provides perhaps the most complete and best example of this general approach. A thorough assessment of sugar pine yield and management alternatives is used to construct optimal management choices across the range of the resource and, from this, to construct a long-term potential supply curve. By intersecting a statistical demand curve with this potential supply, Vaux identifies an economic production goal of 155 million board feet (mmbf)/year at a price of $50/thousand board feet (mbf) for the years 2010–2069.

Vaux's policy analysis examines how private forest landowners could be induced to select appropriate (optimal) management to attain the production goal. The premise of the analysis is clearly stated and similar in many ways to the language of Marsh and Hough:

> In the present state of our economy and institutions, almost all investments in any kind of timber growing are prejudiced in comparison with other investment alternatives.... Investments in timber growing are currently less attractive to liquid capital than many other classes of investment of no greater basic economic merit (Vaux, 1954, p. 46).

Vaux goes on to propose a set of specific policy actions that could overcome these obstacles and encourage management, including government matching funds for planning and management. In other words, the analysis takes the market failure (underinvestment by private landowners) as a given and constructs a policy response. The 'norm' of the model is that private landowners should be motivated to select management with the highest present net value for returns defined by the analyst (but would not be expected to pursue such management on their own). Although Vaux's analysis does not distinguish between types of owners, its rhetorical structure carries through to later treatments of NIPFs.

Shifts toward econometric models of markets beginning in the late 1960s and early 1970s place focus on allowing the data to reveal the responsiveness of private producers in forest and wood products markets, but findings are often interpreted within the frame of market failure (which may invalidate the assumptions of the estimated model). Early econometric analysis in the late 1960s and early 1970s focuses primarily on the potential role of the national forests to ameliorate increasing timber scarcity. Industrial owners (generally producers in the Pacific Northwest) were seen to be reaching peak production, and supply expansion could only derive from increasing harvests of national forest timber in the short run. The focus on structural lumber placed pressure especially on expanding harvests in the forest of Oregon, Washington and northern California.

William Bentley, writing in 1968, recognizes and explores this elemental shift in the focus of national forest management. After 40 years of essentially custodial management, the agency had been called upon to support national housing goals beginning in the 1950s; the Housing and Urban Development Act of 1968 formalized these goals at 26 million housing starts over the following decade. At the same time, econometric models of wood products markets, pioneered by William McKillop (1967), found their first policy application with a revived concern for timber scarcity. Attention turned away from private forests and almost exclusively toward national forests as the expedient option for keeping wood products, and therefore housing prices, low enough to accomplish the national housing goals. As the timber sale program grew to represent majority shares in many areas of the West, the market power of the US Forest Service defined substantial benefits but also costs to private producers of timber. Conflicts with other resource values eventually led to a backlash against this use of public lands to pursue market goals.

Adams and Blackwell (1973) apply the Wharton model of the United States economy to quantify the price escalation in wood products in the late 1960s. Robinson (1974) builds an econometric model with more of a resource economics foundation (in contrast to the macroeconomic/industrial process focus of Adams and Blackwell) to address the same questions. Neither explicitly models supply functions for public and private landowners, but both conclude that only expanding harvests from the national forests could turn back price growth for wood products. Robinson further discusses the use of log export restrictions and imports of lumber from Canada to service the economy's needs for timber. Still, the tacit assumption of both studies was that national forests were underutilized resources without an explicit analysis of supply potential. Neither study indicates how much would be enough to stem price increases or even how much production from national forests would be feasible in the short run or long run. The models are used as rhetorical devices for forwarding policy recommendations without directly modeling the policy mechanisms – i.e. supply functions linked to the production capacity of these owner groups.

As these two econometric studies were published, the push toward harvesting on public lands was encountering significant opposition, throwing into question the feasibility of using national forest timber to solve material scarcity problems. Writing in 1976, Adams, Darr and Haynes explore the potential impacts of a Fourth Circuit Court of Appeals ruling that targeted the authority for timber sales on the national forests – the so-called Monongahela decision. They explore the potential market implications of reducing timber harvesting from national forests across the country. Although not predicting the mechanisms that eventually led to harvest reductions on national forests – the National Forest Management Act broadly addressed the points litigated in the Monongahela case while the Endangered Species and National Environmental Policy Acts ultimately led to harvest reductions – the scenarios of Adams et al. (1976) essentially predicted the precipitous decline in public harvesting that occurred 15 years later and foreshadowed a shift in thinking about supply and ownership questions. Using Adams's (1977) model of interacting timber and wood products markets, they explored the price and production implications of redefining the role of national forests and shifted the policy question back toward private forest landowners.

Correctly predicting a strong shift in production from the PNW to the South, along with the expansion in softwood lumber imports from Canada and transitional increases in stumpage prices, Adams (1977) and Adams et al. (1976) identify two uncertainties at the foundation of their projections. One asks about the effects of capital fixity in limiting interregional shifts in production. Price impacts would be large if capital could not adjust rapidly. The other questions the ability of private landowners to adequately respond to increased prices with intensified management, especially in the southeastern United States.

The forest products market model used by Adams et al. (1976) and described in a subsequent paper by Adams (1977) advanced the quantitative analysis of timber supply in important ways and provided a template for forest product market analysis that persists to this day. Because the questions at hand focused on interactions between public and private producers, Adams distinguishes between public and private timber supplies in his model. He also accounts for potential cross-regional substitutions in timber production and defines supply regions consistent with homogenous product classes, an important element of policy debates. In addition, he incorporates substitutes for wood itself in his analysis, allowing for analysis of another linked environmental consequence of forest policy decisions.

Perhaps most important from the perspective of timber supply modeling, Adams's model augments the price relationship with an explicit link between forest inventories or stocks and timber supply. In so doing, Adams defines a coherent forecasting framework that links resource conditions to timber supply and harvesting and vice versa, thereby linking short-run decisions with long-run consequences – an essential element of forest policy analysis. By defining supply as a function of the standing forest inventory, mechanisms to adjust inventory over time needed to be incorporated in the models. Adopting a 'stand-table' projection method that would be refined over subsequent generations of models, Adams brings an explicit biological/forest management interface to forest product market modeling for the first time. This general frame of analysis has organized aggregate supply analysis to the present time.

Before the 1960s, timber market analysis derived from extrapolative inventory projections based on a high-resolution description of existing inventory and projections of growth and mortality coupled with extrapolated harvest projections, but failed to address underlying choice mechanisms. To be clear, economic sensibility to these analyses, in some cases quite sophisticated, undergirds these analyses, but they did not directly address the behavioral mechanisms of economic actors and were therein limited as policy analysis tools. The emergence of econometric models in the 1960s introduced a positive assessment of the behavior of economic actors, but only addressed the inventory, management and biological underpinnings of intertemporal supply in highly abstract ways. Adams's formulation brought together biophysical production relationships with economic choice mechanisms and defined market models consistent with the sensibility of forest economics in a way that allowed examining policy instruments targeting forest owners. The models also highlight the implications of uncertainty about how various owners adjusted their inventories, therein timber supply in the longer run.

In describing the logic for using the amount of forest inventory in the timber supply relationship, Adams appeals to an inverse correlation between inventory levels and harvesting costs. The logic derives from an old-growth harvesting schema where, as stocks are depleted, extraction costs rise for the remaining inventory. The interactions between supply and stock are less direct in the case of second growth, where harvesting infrastructure is in place and decisions to invest in forests are made with knowledge of future cost implications. Adams recognizes the issue especially as it relates to describing timber supply in the South and proffers alternative formulations (see also extensions by Adams and Haynes, 1980).

Nearly every empirical model of timber markets since 1980 adopts this specification of private timber supply with some aggregate measure of inventory (most commonly total growing stock) shifting private supply in the short run and, where relevant, long-run adjustments being driven by an inventory accounting model (e.g. Newman, 1987; Daniels and Hyde, 1986). Within such a model, public timber harvests can be set as an exogenous, administrative decision, and private supply follows the standard form:

$$\ln S = \alpha + \eta \ln P + \gamma \ln I \qquad (1)$$

where S is timber supply, P is timber price, I is timber inventory and α, η and γ are coefficients. Expressing the equation in logarithms provides for convenient derivations:

$$\frac{dS}{S} = \frac{\eta dP}{P} + \frac{\gamma dI}{I} \qquad (2)$$

defining η as the own price elasticity of supply and γ as the inventory elasticity, with both are expected to have a positive sign.

Within this general frame, ownership/policy questions have been approached in several ways. One application explored the development of markets with different assumptions about federal timber harvests. This framework accounts for private sector responses as a component of the net effects of federal supply policy and was the focus of Adams et al. (1976) and U.S. Forest Service Resources Planning Act (RPA) assessments published in 1980, 1990 and 2000. Policy instruments focused on private lands have also been addressed, though in indirect fashion, by adjusting the inventory trajectories of nonindustrial private landowners (the I in equation 1) to simulate policies that stimulate investments by these owners (see Adams et al., 1982).

Comparing production behavior across ownerships

Although most analysts have examined policy using ex ante projections to address proposed or hypothetical policy instruments, the same type of structural model has been used to estimate actual policy impacts using ex post counterfactual simulations. For example, Wear and Murray (2004) construct historical and counterfactual simulations to examine the substantial reduction in federal timber supply commencing in the early 1990s using an econometric softwood timber-lumber market model similar in structure to Adams's (1977) but without an inventory projection model. The counterfactual approach allows for estimation of standard errors on impacts and therefore a direct testing of hypotheses regarding policy effects. Wear and Murray's results quantify the types of impacts anticipated by Adams et al. in 1976, including the responses of private timber producers in both western and southeastern regions along with influence on lumber imports from Canada.

Another line of inquiry addresses some foundational hypotheses of forest policies: that supply behaviors of different owner groups are indeed different. Here the research has focused on differences in the production behaviors of industrial and nonindustrial forest owners. One approach compares own-price and inventory elasticities from short-run supply models for the two owner groups (using equation 2). Adams and Haynes (1980) find differences between NIPF and industrial supply, but only in the Southeast, and discuss the difficulties of disentangling the strategic behaviors of private producers in the West in the presence of sizable public harvest programs (further addressed by Adams, Binkley and Cardellichio, 1991).

Following a similar approach but accounting for the multiple product nature of supply in the South, Liao and Zhang (2008) find differences in supply from industrial and NIPF owners consistent with Adams and Haynes (1980) as well as subsequent studies (Brännlund, Johansson and Löfgren, 1985; Newman and Wear, 1993). Differences arise for the supply of both pulpwood and sawtimber-sized timber and generally show that price elasticities are higher for industrial owners, indicating a stronger responsiveness to market signals.

Few studies have gone beyond the ranking of elasticities to explicitly test for differences in production behavior across ownership classes. Newman and Wear (1993) test for differences in the supply and investment behaviors of industry and nonindustry owners in the southeastern United States. They test the hypothesis that both ownership groups' choices are consistent with profit-maximizing behavior using a restricted profit function with land and growing stock (standing timber) held as quasi-fixed factors of production. Newman and Wear find that profit maximization cannot be rejected for either group – i.e. that first and second order conditions for profit maximization hold for both ownerships – but they also reject the hypothesis that the two ownership groups have identical profit functions. Management on industry land was found to be more responsive to price signals (own and cross prices) than management on NIPF lands.

Significantly higher shadow prices for the growing stock held by NIPF owners are consistent with higher values placed on nontimber ecosystem services. This study rejects the assumption that NIPF production behaviors are unresponsive to price signals and challenges the broader notion that less intensive management by NIPF owners represents a market failure. Rather, lower management intensity may reflect multiple use objectives.

Wear and Newman also examine investment responsiveness to prices and regeneration costs. Industry investment is significantly influenced by pulpwood prices while NIPF investment is significantly influenced by sawtimber prices and regeneration costs. Using state-level data, Li and Zhang (2007) similarly find that forest industry planting is more responsive to pulpwood prices and harvest rates while NIPF planting is more responsive to sawtimber prices. Arano and Munn (2006) examine management intensity on private lands in Mississippi as measured by silvicultural expenditures per acre based on mail surveys of landowners. They also find NIPF management less intensive and less responsive than commercial management and find comparable investment levels on TIMO and industry lands.

The findings of Newman and Wear also indicate that optimal forest conditions should be distinct on NIPF and industry lands. That is, NIPF owners would be expected to hold higher volumes of growing stock on a per-unit area basis and to hold older forests. Accordingly, the flow of nontimber forest benefits would differ between ownership groups. This hypothesis of a distinct 'landscape signature' associated with different landowner classes was taken up by Wear and Flamm (1993) in a study of forest cover changes in the Southern Appalachian Mountains of North Carolina. With a spatial cross-sectional model, they test the propensity to harvest as a function of various spatially explicit cost factors interacted with ownership (in this case, national forest versus private). They reject the hypothesis that national forest and private owners had the same propensity to harvest in the 1980s. Over the period studied, they find much higher harvest rates for private lands, private harvest patterns largely organized by cost factors (e.g. distance to road, slope), but public harvest patterns seemingly indifferent to these factors. The result is a spreading out of management on public lands (consistent with multiple use objectives) and, perhaps surprisingly, higher harvest rates on national forest lands for the most remote areas during this period.

A common framework has organized much of the empirical work on timber supply and forest policy analysis since 1977: separate supply models for distinct ownership groupings (generally industrial and nonindustrial private along with public) using models that allow for short-run responses to price and long-run change through an aggregate inventory variable. This frame derives not only from general theory regarding material scarcity and prices, but also, we can assume, from data limitations. The use of an aggregate biophysical measure of forest capital (e.g. growing stock inventory) in the supply equation derives from an old-growth mining logic that may provide only limited insights into a second-growth timber economy (see Wear and Pattanayak, 2003). Aggregate inventory data are furthermore highly constructed data with observations recorded only every 5 to 15 years and representing multiple-year moving averages. As a result, the fit of supply models (and therefore the elasticity measures) may be strongly influenced by the technique used to interpolate and extrapolate the infrequent inventory data to fill a time series.

Cross-sectional models allow timber supply to be estimated without aggregating and interpolating inventory data and to interact with variables that describe the composition of the inventory. Newman and Wear (1993) hint at this with their inclusion of both growing stock and land in a county-level cross-sectional model and resulting indications that stock density as much as stock quantity influences supply possibilities. Prestemon and Wear (2000), followed by Polyakov et al. (2010), construct supply from an even finer cross-sectional grain by modeling harvest choice for inventory plots and aggregating results using the area frame structure of the inventory.

The former study finds differences between public and private supply, but more importantly, identifies important influences of various productivity and cost variables on harvest choice and aggregate supply. The implication is that the distribution of inventory across quality classes within an ownership grouping holds influence over supply, perhaps as much as ownership class per se.

Recent ownership dynamics

Ownership class has served as an organizing element of both forest policy and forest econom- ics analysis throughout the twentieth century. This frame has persisted in spite of changes in the nature and focus of the policy issues at hand – from overharvesting by industry to underinvest- ment by nonindustrial private landowners to the supply contributions of national forests – largely because of the relative fixity of the classes. Public forest area was largely fixed by mid-century (about a 1% difference between 1953 and 2002), and the area of industry forest land in 2002 was about 12% higher than observed in 1953 (Smith, Miles, Vissage and Pugh, 2003). In many ways, these ownership groupings helped organize and make manageable the policy dialogue.

The forest policy and ownership contexts in the United States have changed again over the past decade. Public timber harvesting, the focus of extensive environmental resource use debates for more than 25 years, seems no longer a central issue. Timber harvesting on western public lands fell by nearly 85% in the early 1990s in response to spotted owl endangerment but also in response to widespread challenges to timber harvesting on public lands. This harvest reduction has been sustained for about 20 years as domestic timber harvesting shifted strongly to private lands in the southeastern United States. If 'underinvestment' on nonindustrial lands remains problematic, it has not increased timber scarcity over the past 20 years – price patterns generally show a falling away from timber scarcity concerns (see Wear, Carter and Prestemon, 2007; Wear and Prestemon, 2004).

Most importantly, the perceived permanence of forest ownership fell away beginning at the turn of the current century. Nearly all land owned by forest industry companies changed owner- ship between 2000 and 2010, either through divestiture of the land or through reorganization of the company (as an illustration of the change, the 2007 Forest Service forest resource assessment by Smith, Miles, Perry and Pugh, 2009, dropped the forest industry classification from its report- ing). Where land was divested, most was acquired by REITs or TIMOs. Reorganization of the Weyerhaeuser Company, for example, transitioned some corporations from traditional vertically integrated wood products operations to real estate investment trusts. Both institutional arrange- ments (REIT and TIMO) suggest a continued focus on land and resources as investments but sig- nal possible changes in management outcomes. New business models would seem to suggest more rather than less market responsiveness in the long run (i.e. without intra-corporate constraints on production, timber supply and land-use switching would respond more directly to market signals), though timber sale agreements negotiated during land transactions might affect harvesting.

Conclusions

Forest economics research has challenged the notion that NIPF underinvest in forestry given their highly variable objectives for forest land ownership (Butler, 2008). The recent divestiture of industry forestlands or conversion of previously vertically integrated wood products firms to real estate investment trusts highlights the impermanence of ownership. In an era when landowners tend to allocate land to high-value uses irrespective of ownership, and ownership groups may shrink or swell in response to changing market conditions, can conservation and resource scarcity issues be usefully framed using the new ownership categories? Clearly the answer depends on what we mean by conservation issues. Timber scarcity seemingly no longer retains currency as a national or regional economic issue because broad trends toward scarcity have largely dissipated and the relative economic contribution of timber to economic prosperity has declined. Resource values associated with standing forests and biodiversity – i.e. services more consistent with public goods – may better describe this century's resource challenges.

Of course, alternative futures are a possibility, and policies encouraging bioenergy production using woody biomass along with carbon markets that value carbon sequestration through forestry could increase the relative scarcity of forests and their products in the future. Observed changes since the 1990s indicate a potential for rapid supply responses from the private sector through direct investment or changes in land ownership that would allow capital inflow and increased outputs. The question remains as to when growth in production, especially when combined with competing demands for rural land, impedes the capacity of forested ecosystems to deliver ecological services (Wear and Greis, 2012).

Whether conservation focus fixes on timber or on standing forests, economic analysis needs to focus on understanding the behavior of landowners managing forests, in the former case to forecast timber supply and in the latter case to project the consequences of management for standing forest conditions. For example, a price-responsive shift toward more planted pine forests in the highly productive southeastern Coastal Plain may hold implications for critically imperiled amphibian species, which in turn could influence future uses of these lands (Wear and Greis, 2012). This type of analysis remains central to policy debate by clarifying the potential magnitude of future problems (i.e. as a foundation for forest forecasting) and by examining the potential efficacy of proposed policy instruments. The literature on timber supply modeling indicates that historical ownership types helped explain production from and management of forests.

To make progress in our understanding of timber supply, foundational information needs to be developed regarding the responses of new institutional arrangements of timberland ownership in the United States. This includes TIMOs and REITs, which should be expected to make production and allocative decisions in response to both timber and land markets (consistent with stated business intent). Long-run timber supply analysis needs to effectively integrate forest management, land disposal and acquisition and land-use decisions. Analysts will also need to develop more refined insights into the implications of various allocative mechanisms such as closed-end and open-end timber funds and data on their prevalence. As is generally the case in modeling forest management decisions, data are limiting for this type of analysis and descriptive/case studies could be an important first step.

The quantity and quality of lands managed by TIMOs, REITs and all other ownerships may be considered exogenous in the short run but would be expected to be decision variables over time. It would seem that new timber supply/forest management models will need to address transitions between ownership types (and investment vehicles) as well as land-use switching in response to market signals. Recent history demonstrates that forest ownership categories can no longer be treated as fixed. Increasing scarcity of timber would likely lead to increased demand for timberland by investment organizations and shift some timberland from submarginal to inframarginal. Recent research makes clear that the area of planted forests, for example, communicates a strong signal of management intent perhaps more strongly than ownership category (see Polyakov et al., 2010). But it may be that ownership transition presages the transition to intensive management – that is, forest ownership and forest condition interact.

Although this discussion has focused on the history of policy and economics research in the United States, the questions raised here have currency throughout the developed and developing worlds. With the exception of some studies that have tested ownership differences, most forest economics implies a maintained hypothesis that ownership class defines a homogenous management context and response. These hypotheses demand rigorous testing to establish credible forecasts of forest uses and conditions in the future. Persistent questions regarding how private landowners respond to price signals (including expectations) in making harvest choice and investment decisions need reexamination in light of recent ownership changes and across the spectrum of institutional settings. The relative roles of public and various types of private ownership, including a clear understanding of the expected returns/outcomes associated with each, remain relevant questions around the globe.

References

Adams, D., Darr, D. R. and Haynes, R. W. (1976). 'Potential responses of softwood markets to the Monongahela decision', *Journal of Forestry*, vol. 74, pp. 668–670.

Adams, D. M. (1977). 'Effects of national forest timber harvest on softwood stumpage, lumber, and plywood markets: An econometric analysis', Research Bulletin No. 15, School of Forestry, Oregon State University, Corvallis, OR.

Adams, D. M., Binkley, C. S. and Cardellichio, P. A. (1991). 'Is the level of national forest timber harvest sensitive to price?' *Land Economics*, vol. 67, pp. 74–84.

Adams, D. M. and Haynes. R. W. (1980). 'The 1980 softwood timber assessment market model: Structure, projections and policy simulations', *Forest Science Monograph*, no. 22, Society of American Foresters, Washington, DC.

Adams, D. M., Haynes. R. W., Dutrow, G. F., Barber, R. L. and Vaseivich, J. M. (1982). 'Private investment in forest management and the long-term supply of timber', *American Journal of Agricultural Economics*, vol. 64, no. 2, pp. 232–241.

Adams, F. G. and Blackwell, J. (1973). 'An econometric model of the United States forest products industry', *Forest Science*, vol. 19, no. 1, pp. 82–96.

Arano, K. G. and Munn, I. A. (2006). 'Evaluating forest management intensity: A comparison among major forest landowner types', *Forest Policy and Economics*, vol. 9, pp. 237–248.

Bentley, W. R. (1968). 'Forest Service timber sales: A preliminary evaluation of policy alternatives', *Land Economics*, vol. 44, no. 2, pp. 205–218.

Boyd, R. (1984). 'Government support for nonindustrial production: The case of private forests', *Southern Journal of Economics*, vol. 51, pp. 89–107.

Brännlund, R., Johansson, P. O. and Löfgren, K. G. (1985). 'An economic analysis of aggregate sawtimber and pulpwood supply in Sweden', *Forest Science*, vol. 31, pp. 595–606.

Butler, B. J. (2008). *Family Forest Owners of the United States, 2006*, General Technical Report NRS–27, USDA Forest Service, Northern Research Station, Newtown Square, PA.

Committee for the Application of Forestry, Society of American Foresters. (1919). 'Forest devastation: A national danger and a plan to meet it', *Journal of Forestry*, vol. 17, no. 8, pp. 911–945.

Daniels, B. and Hyde, W. F. (1986). 'Estimation of supply and demand elasticities for North Carolina timber', *Forest Ecology and Management*, vol. 14, pp. 59–67.

Graves, H. S. (1919). 'A national lumber and forest policy', *Journal of Forestry*, vol. 17, no. 4, pp. 351–363.

Hough, F. B. (1878). *Report upon Forestry*, Prepared under the direction of the commissioner of Agriculture, in Pursuance of an act of Congress approved August 15, 1876, Government Printing Office, Washington, DC.

Hyde, W. F. (2012). *The Global Economics of Forestry*, Resources for the Future Press, New York.

Jackson, D. H. (1987). 'Why stumpage prices differ between ownerships: A statistical examination of state and Forest Service sales in Montana', *Forest Ecology and Management*, vol. 18, pp. 219–236.

Li, Y. and Zhang, Z. (2007). 'A spatial panel data analysis of tree planting in the US South', *Southern Journal of Applied Forestry*, vol. 31, no. 4, pp. 192–198.

Liao, X. and Zhang, Y. (2008). 'An econometric analysis of softwood production in the U.S. South: A comparison of industrial and nonindustrial forest ownerships', *Forest Products Journal*, vol. 58, no. 11, pp. 69–74.

Marsh, G. P. (2003). *Man and Nature*, D. Lowenthal (Ed.), University of Washington Press, Seattle, WA. (Original work published in 1864)

McKillop, W. L. (1967). 'Supply and demand for forest products: An econometric study', *Hilgardia*, vol. 38, pp. 1–132.

Newman, D. H. (1987). 'An econometric analysis of the southern softwood stumpage market: 1950–1980', *Forest Science*, vol. 33, pp. 932–945.

Newman, D. H. and Wear, D. N. (1993). 'The production economics of private forestry: A comparison of industrial and nonindustrial forest owners', *American Journal of Agricultural Economics*, vol. 75, pp. 674–684.

Poylakov, M., Wear, D. N. and Huggett, R. (2010). 'Harvest choice and timber supply models for forest forecasting', *Forest Science,* vol. 56, no. 4, pp. 344–355.

Prestemon, J. P. and Wear, D. N. (2000). 'Linking harvest choices to timber supply', *Forest Science*, vol. 46, no. 3, pp. 377–389.

Robinson, V. L. (1974). 'An econometric model of softwood lumber and stumpage markets 1947–1967', *Forest Science*, vol. 20, pp. 171–179.

Smith, W. B., Miles, P. D., Perry, C. H. and Pugh, S. A. (2009). *Forest Resources of the United States, 2007*, General Technical Report WO-78, USDA Forest Service, Washington, DC.

Smith, W. B., Miles, P. D., Vissage, J. S. and Pugh, S. A. (2003). *Forest Resources of the United States, 2002*, General Technical Report NC-243, USDA Forest Service, North Central Research Station, St. Paul, MN.

Steen, H. K. (1976). *The U.S. Forest Service: A History*, University of Washington Press, Seattle, WA.

Vaux, H. J. (1954). *Economics of Young Growth Sugar Pine Resources*, Bulletin No. 78, University of California-Berkeley, Division of Agricultural Sciences, Berkeley, CA.

Wear, D. N., Carter, D. and Prestemon, J. (2007). *The US South's Timber Sector in 2005: A Prospective Analysis of Recent Change*, General Technical Report SRS-99, USDA Forest Service, Southern Research Station, Asheville, NC.

Wear, D. N. and Flamm, R. (1993). 'Public and private disturbance regimes in the Southern Appalachians', *Natural Resource Modeling*, vol. 7, no. 4, pp. 379–397.

Wear, D. N. and Greis, J. G. (2012). *The Southern Forest Futures Project: Summary Report*, General Technical Report SRS-159, USDA Forest Service, Southern Research Station, Asheville, NC.

Wear, D. N. and Murray, B. C. (2004). 'Federal timber restrictions, interregional spillovers, and the impact on U.S. softwood markets', *Journal of Environmental Economics and Management*, vol. 47, no. 2, pp. 307–330.

Wear, D. N. and Pattanayak, S. K. (2003). 'Aggregate timber supply: From the forest to the market', in K. J. Abt and E. Sills (Eds.), *Forests in a Market Economy*, Kluwer Academic-Publishers, Dordrecht, The Netherlands.

Wear, D. N. and Prestemon, J. (2004). 'Timber market research, private forests, and policy rhetoric', in M. Rauscher and K. Johnsen (Eds.), *Southern Forest Science: Past, Present and Future*, USDA Forest Service, Southern Research Station, Asheville, NC.

Williams, M. (1989). *Americans and Their Forests*, Cambridge University Press, Cambridge, UK.

6

TIMBER PRICING

Larry A. Leefers[1] and Awang Noor Abd. Ghani[2]

[1]DEPARTMENT OF FORESTRY, MICHIGAN STATE UNIVERSITY, USA.
[2]FACULTY OF FORESTRY, UNIVERSITI PUTRA MALAYSIA.

Abstract

Forests are a dominant feature of the global landscape. Most of these forests are publicly owned and are the source of significant employment due to their timber resources. Timber owners need to receive compensation for the timber they sell, and revenues received by owners are based on a variety of approaches. For example, they may use administered charges or values, negotiated values using consultations, market-derived prices or other approaches. Administered values or charges are the most commonly used method for establishing timber value in developing countries, whereas the residual value method (RVM) and transactions evidence method (TEM) approaches have been used more extensively in North America. RVM and TEM are closely linked to markets. RVM uses the price of a timber-based wood product as the starting point for estimating timber value. It subtracts milling, logging, transportation and a profit-and-risk margin from the end-product price to derive the 'residual' timber price. TEM uses statistical analysis of recent timber sales and their characteristics to estimate timber price. Though economists often equate timber value with a market price, there is an absence of competitive markets in many regions. Hence, no approach is inherently superior to others.

Keywords

Timber pricing, administered prices, residual value method, transactions evidence method, hedonic pricing

Acknowledgments

The authors would like to thank Dr. Shashi Kant, Dr. Janaki R. R. Alavalapati and Dr. J. Michael Vasievich (Tessa Systems, LLC) for their helpful reviews of earlier versions of this chapter.

Introduction

Over 4 billion ha of the world, 31% of the land area, is covered by forests (FAO, 2010). Most of these forests (80%) are in public ownership. The dominance of public lands and lack of market competition in many areas has led to the development of several approaches for estimating timber value (or price). Though economists often equate value with market price, timber values may be set administratively, derived using different quantitative approaches or determined in the market. In this chapter, we use the term 'value' to mean, narrowly, the monetary amount resource owners receive for their timber. This definition excludes land values and values associated with nontimber and ecosystem services.

Today, more emphasis is being placed on the multitude of ecosystem services that forests provide and the value of those services (Nelson et al., 2009). Commodity production, however, remains a central focus for most forest-based companies, agencies, communities and individuals. The buying, selling and trading of timber supports wood-based employment and income around the world. For example, the Food and Agriculture Organization of the United Nations estimated that almost 11 million people were employed (full-time equivalents) in primary production of goods and in management of protected areas in 2005 (FAO, 2010).

The price of timber is of interest from local to international scales (Brown, Kilgore, Blinn and Coggins, 2012a; Smith, Markowski-Lindsay, Wagner and Kittredge, 2012). Quantitatively derived estimates of timber prices are used as a starting point for many market-based timber bidding procedures. Historical and projected timber prices are often inputs for investment analyses. And international trade disputes sometimes arise from disagreements regarding procedures for estimating timber prices (van Kooten, 2002). The 'bottom line': timber values matter a great deal, and foresters, economists and others need to understand how these values are derived in order to determine their suitability for different purposes.

In this chapter, we focus on timber products, administrative approaches to valuing timber, theoretical perspectives on timber-pricing approaches and applications of timber pricing in the United States (USA), Canada and elsewhere.

What is the product?

Stumpage, or 'standing timber as viewed by a commercial cutter', is usually the broad product of interest (Helms, 1998). This term is linked to markets so strongly that stumpage is also defined as 'the value of timber as it stands uncut' and is synonymous with royalty, or the 'payment to be made to the owner or lessor of a forest for the right of harvesting'. Standing timber can have many end uses, and as a result, the markets for these intended uses affect stumpage prices. For example, chipped or ground wood for biomass energy production has a lower price than trees going to veneer markets. Solid wood timber products include softwood and hardwood lumber, structural and nonstructural panels, engineered wood products, fuelwood and other industrial products (McKeever and Howard, 2011). Pulp and paper markets provide other outlets for stumpage. Most timber sales in North America differentiate between sawtimber and pulpwood for various species (e.g. white spruce) or species groups (e.g. spruce-fir), usually on the basis of size or quality. So timber and its pricing provide the starting point for a wide variety of value-added products.

Administrative approaches to timber pricing

Economists contend that timber values should approximate market prices. As Kant (2009) noted, however, a perfectly competitive market for timber will never exist. In many places, markets are poorly functioning or do not exist, and timber concessions of various forms are used for management and harvesting.

Consequently, a wide range of mechanisms for valuing timber are applied to set timber values in developed and developing countries. Value depends on the policy and forest revenue system of each country (FAO, 2001). Commonly used methods include administered charges or values, negotiated values via consultation and charges based on replacement cost. The most common method in developing countries is through administered values. The value is fixed by the government through setting up forest charges or royalties, either based on volume extracted (volume-based), area (area-based) or an area-volume combination (Amacher, Brazee and Witvliet, 2001; FAO, 2001; Kim, Phat, Koike and Hayashi, 2006; Crowe, 2008).

Administratively set values may be determined by the forestry agency or with consultation among multiple government agencies (FAO, 2001). Each situation is unique. For example, in Malaysia, administratively set timber fees or charges are based on (1) volume with a royalty, (2) area with a premium or (3) a percentage of the log value (Awang Noor and Mohd Shahwahid, 2003). The royalty, premium and log-value rates are set by the state and vary by state, species and other factors. Royalties tend to be higher for high-value species and lower for lesser-known species. The system is characterized by inflexible, undifferentiated and low rates. Under this approach, the government captures a low proportion of total potential stumpage values (Repetto and Gillis, 1988; Vincent, 1990; Awang Noor, Vincent and Yusuf, 1992; Awang Noor and Mohd Shahwahid, 2003).

Markets also foster use of administratively set prices. Some agencies have simple administrative procedures for using recent market evidence to set reservation or minimum-acceptable prices. The reservation price provides a starting point for competitive bidding.

Negotiated charges or prices for timber are also common. In some cases, stumpage rates are individually negotiated between governments and interested parties (FAO, 2001). In the absence of markets, historic precedent or price information from other regions may provide the basis for the charge. The negotiation approach is even applied in areas where timber markets exist, when there are no bids or when the reservation price is not met.

Another approach to timber pricing involves determining the current cost of replacement of timber removed, salvaged or damaged by producers (FAO, 2001). Establishment and management costs are compounded forward along with timber growth estimates to provide an estimate of value at the end of the rotation which could reproduce an income appraisal or net present value at an appropriate discount rate. At the end of the rotation, all costs accumulate as positive amounts because the seller wishes to recover costs with interest (Klemperer, 1996). This approach is appropriate for young timber stands that do not yet have sufficient market volume or value when site preparation and planting costs are known (Mayo and Straka, 2005).

Though administrative approaches to setting timber prices are widespread, two principal theoretically based approaches for timber pricing have evolved: the residual value method (RVM) and the transactions evidence method (TEM). Both have been used extensively in North America and applied to a lesser degree in other regions as well. In the absence of perfectly competitive markets, they represent 'second-best' approaches (Kant, 2009). The foundation of those approaches is presented in the next section.

Theoretical perspectives of timber-pricing approaches

The main focus in this section will be on the RVM and TEM approaches, with an emphasis on well-documented US Department of Agriculture Forest Service (USFS) efforts. The RVM approach is based on the sale of a market product (e.g. lumber) from which various manufacturing, transportation and harvesting costs are subtracted along with allowances for profit and other factors. The dollar amount left after subtraction is the residual value, an estimate of the value of the stumpage. The TEM or hedonic pricing approach relies on past timber sales and their characteristics as a basis for statistically estimating stumpage price. The RVM approach appraises the value for an operator of average efficiency, whereas the TEM approach estimates the highest bid (Niccolucci and Schuster, 1991).

The RVM approach was based on the derived demand economic principle that demand (and price) for factors of production are derived from the value of the final product (Weintraub, 1959; Schuster and Niccolucci, 1990). RVM is particularly useful when costs and productivity for processing factors are well known or can be derived from market sources. Hence, stumpage prices can be derived from lumber and other product prices. This provided the conceptual foundation for USFS stumpage pricing early in the agency's history.

The first significant effort by the USFS at standardization came in the 1914 timber appraisal manual, *Instructions for Appraising Stumpage on National Forests* (Wiener, 1982). This early work provided the foundation for the standard timber appraisal method used by the agency for most of the twentieth century. Following Wiener (1981), the simplest form of the RVM is:

$$SP - (MC + LC + P\&R) = S \tag{1}$$

where *SP* is the end product selling price, *MC* is the milling cost, *LC* is the logging and transport cost, *P&R* is the profit and risk margin and *S* is the residual stumpage price.

The early work was reinforced at the close of World War II by Julian Rothery (1945), the head appraisal specialist for the USFS. Though this was the accepted method, its application required flexibility to reflect local conditions. This approach was especially useful in the rural western USA where most markets were not competitive.

One major challenge with this approach was determining profit. The 1914 manual identified three methods for addressing profit (Wiener, 1982). The first was the investment method, which was based on a percentage return on money invested in the operations (e.g. average fixed investment and working capital). The second method was a percentage of the total unit cost of production, including depreciation of fixed investments, known as the 'overturn' method. The final method was to use a flat rate. The latter method was discouraged, and the overturn method was commonly used over time. Regardless of profit-deriving method, the RVM approach provided an estimate of the fair market value for stumpage. Weintraub (1959) was supportive of this approach but raised concerns especially about how to allocate the 'conversion return' (residual after costs were deducted from lumber selling price) between profit and stumpage price.

In recent decades, the RVM approach has been supplanted by the TEM approach within the USFS. Comparative or comparable sales have long provided a basis for timber valuation. Interestingly, over 20 years ago Schuster and Niccolucci (1990) compared the performance of six timber appraisal models using USFS sales in the central Rocky Mountains. They found that the residual value model, supplemented with an estimate for overbids, performed better than multiple-regression, TEM-based models. Nonetheless, the RVM approach was phased out soon afterwards in the early 1990s by the USFS, due in part to the high cost of developing and maintaining data on manufacturing and harvesting costs. Satisfactory adoption of TEM in the USA, especially in the Lake States, promoted this shift (Wiener, 1981). Both RVM and TEM approaches are used by various states across the USA (Brown, Kilgore, Blinn, Coggins and Pfender, 2010).

The TEM approach requires a statistical analysis of timber sales, relating selling price to various timber sale characteristics. In its simplest form, recent comparable sales transactions can be used as a basis for deriving reservation prices. Accordingly, some analysts are concerned with estimating sales price with little interest in details of the timber sale characteristics; whereas others are particularly interested in the role of influencing factors such as sale size, number of bids, contract length and so on. Buongiorno and Young (1984), for example, focused on predicting stumpage prices on a national forest in the upper Midwest of the USA; they started their analysis with 65 potential explanatory variables and narrowed the list to 14 which yielded the highest R^2. Others may be concerned about the effects of competition as reflected in number of bids and firm size and in the influence of these variables on sales price (Leefers and Potter-Witter, 2006).

Hedonic pricing theory underpins TEM modeling approaches. The theory, attributed to Rosen (1974) and others, posits that purchasers of a good are buying a bundle of attributes or characteristics of the good. Given market equilibrium in well-functioning competitive markets, the implicit price function is determined via the interaction of producers and consumers. Based on this theory, implicit price functions for differentiated goods can be assumed to exist. Therefore, researchers and others can proceed with estimating the functions. The marginal value of the sale attributes can be estimated in the form of regression coefficients with stumpage price as the dependent variable. Of course, statistical modeling must address selection of the appropriate model, detecting bias and misspecification, data outliers, autocorrelation, multicollinearity, heteroskedasticity and other considerations (Neter, Kutner, Nachtsheim and Wasserman, 1996; Seber and Lee, 2003). These statistical elements are central to the TEM approach, but beyond this scope of this chapter.

The general form of the hedonic model is:

$$\text{Price}_i = \beta_0 + \sum_j \beta_j \star x_j + \varepsilon_i \tag{2}$$

where Price_i is the stumpage price for sale i, β_0 is the intercept, $\sum_j \beta_j \star x_j$ is the sum of all coefficients multiplied by the quantity of their respective sale attributes for each attribute j and ε_i is the error for each timber sale i. Estimated model functional form may be either linear or transformed. Transformed models generally take a semi-log (i.e. logarithmic transformation of the dependent or independent variables) or log-log form (i.e. transformation of the dependent and independent variables). The Box-Cox transformation can be used, especially when testing hypotheses regarding the functional form.

Most published studies of timber pricing in the USA are based on transactions evidence. These studies are derived predominantly on public timber sales and reflect the need for governments to appraise the value of timber as part of their timber sales programs. Other publications center on the timber pricing and auction processes (Niccolucci and Schuster, 1994; Carter and Newman, 1998). The next section of this chapter provides a review of TEM and related studies.

Applications of timber pricing in the USA, Canada and elsewhere

In some countries, like Canada, Australia and New Zealand, the residual value method (RVM) is most common (Yang, Kant and Shahi, 2006; Kant, 2010). Many policy studies on timber pricing have advocated area-based fees – that is, fixed annual charges on the total area under contract. However, Boscolo and Vincent (2007) argued that the view of a neutral impact of area fees on decisions by timber concessionaires is incorrect. Instead, they suggested that area fees can induce concessionaires to accelerate timber harvests and to harvest more selectively.

Statistically derived pricing (via TEM) is more common in the USA; fewer TEM studies have been published on timber pricing in Canada and other countries. Little information was available on this pricing approach from developing countries. This reflects, in part, the TEM analysis requirement for good data sets with reliable measures of sale attributes.

Information on 25 stumpage-based TEM studies highlights the variability of timber-pricing research (Tables 6.1 and 6.2). The studies covered a wide expanse of the USA, from the

Table 6.1 Selected stumpage-based transaction evidence publications.

Authors (year)	Dependent variable	Model form	Solution algorithm	Data years	N	Location
Jackson and McQuillan (1979)	total sale price	semilog	ordinary least squares (OLS)	10 years	52	MT, USA
Buongiorno and Young (1984)	high bid	linear	stepwise	1976–80	101	WI, USA
Jackson (1987)	real bid price/mbf	semilog	ordinary least squares (OLS)	1979–84	39; 40	MT, USA
Puttock, Prescott and Meilke (1990)	lump sum price/ha	linear, semilog	Box-Cox model – logarithmic specification	1982–87	33	ON, Canada
Sendak (1992)	1n (real bid/cunit)	semilog	stepwise	1983–88	43; 34	VT, USA
Niccolucci and Schuster (1994)	real high bid + purchaser deposits	linear	weighted least squares	1984–91	~200–750	MT/ID, USA
Munn and Rucker (1995)	real price/ac	linear	ordinary least squares (OLS)	1987–91	220; 99	NC, USA
Nautiyal, Kant and Williams (1995a)	residual value/m^3	linear	ordinary least squares (OLS)	1986–91	5,000	ON, Canada
Nautiyal, Kant and Williams (1995b)	sale price/ha	linear	ordinary least squares (OLS)	1983–93	526	ON, Canada
MacKay and Baughman (1996)	total sale price	linear	ordinary least squares (OLS)	1991–92	779	MN, USA

Study	Dependent variable	Functional form	Estimation method	Sample size	Period	Location
Roos (1996)	price/ha	linear, semilog	ordinary least squares (OLS) and Box-Cox estimations	143	1992	Sweden
Munn and Palmquist (1997)	real price/ac	linear	stochastic frontier	298	1986–91	NC, USA
Paarsch (1997)	sales price + oral bonus bid	simulation	maximum likelihood and conditional maximum likelihood	129	1984–87	BC, Canada
Vasievich, Mills and Cherry (1997)	high bid, total costs, direct costs	linear	stepwise	445	1982–84	IN, USA
Bare and Smith (1999)	total sale price	linear	ordinary least squares (OLS)	306	1998	WA, USA
Li and Perrigne (2003)	bid price/m³	reduced form, linear	two-step nonparametric procedure	212	1993	France
Lynch, Huebschmann, Lewis, Tilley and Guldin (2004)	ln (real total bid)	log-log	weighted least squares	150	1992–98	AR/OK, USA
Dahal and Mehmood (2005)	bid price/ac	linear	ordinary least squares (OLS)	625	2002–03	AR, USA
Leefers and Potter-Witter (2006)	real price/m³	linear	ordinary least squares (OLS)	222; 205	2000–04; 2004–05	MI/MN/WI, USA
Niquidet and van Kooten (2006)	bid price/m³	truncated linear model	maximum likelihood and ordinary least squares (OLS)	639	1999–2002	BC, Canada

(Continued)

Table 6.1 (Continued)

Authors (year)	Dependent variable	Model form	Solution algorithm	Data years	N	Location
Préget and Waelbroeck (2006)	bid price/ha	seemingly unrelated regression (SUR)	Bayesian Metropolis–Gibbs Monte Carlo Markov Chain (MCMC) algorithm	2003	1,205	France
Rocha, Moreira, Reis and Carvalho (2006)	bid price/ha	mathematical modeling	Geometric Brownian Motion (GBM) and mean-reverting process (MRP)	1999–2002	–	Brazil
Yang and Kant (2008)	stumpage price/m^3	regression model (Parity Bounds Model [PBM])	simulated annealing (SA) algorithm (maximum likelihood estimation)	1995–2007	140	ON, Canada
Alzamora and Apiolaza (2010)	individual log recovery value or conversion return	linear	ordinary least squares (OLS) with Box-Cox estimation	–	156	Chile
Brown, Kilgore, Coggins and Blinn 2012b)	ln ($/cord equivalent)	semilog	ordinary least squares (OLS)	2001–06	4,395	MN, USA

OLS = ordinary least squares, GBM = geometric brownian motion, MRP = mean-reverting process

Table 6.2 Independent variables from selected stumpage-based transaction evidence publications (Table 6.1).

Timber or sale characteristics	Sale timing	Location	Management/policy	Other
Volume sold (or natural log)	Year	State region	Ownership type	Number of bidders
Cruise volume	Quarter	DNR areas	Auction method	Firm size
Proportion of volume of the species group	Season	National forest	Sale area	Wet weather
Volume by species and/or product (or natural log)	Month	Ranger districts	Contract length	Regional sale value/cunit
Volume/acre (or natural log)		Bid-up area	Operating restrictions	Pine sawtimber stumpage price
Maximum extraction log volume		Region	Logging method	Timber price as function of diameter
Volume (dummies)			Harvest method (% volume)	Truncated upset price
Other hardwood volume without crown			Selection harvests	Upset rates
Sawtimber volume (%)			Salvage	Reserve price or appraisal value
Average volume per tree			Reproduction method	Current sawnwood price export
Natural log of average sawtimber volume/tree			Road construction (mi)	Standard deviation of timber price
Average net cruise volume per tree			Road construction (cost)	Long-run equilibrium mean price
Number of trees			Haul distance or time	Purchaser credit/cunit
Number of species products			Paved haul distance	Lumber selling price or index
Species product (%)			Unpaved haul distance	Natural log of lumber price to sawlog price
Timber density			Yarding distance	Years of uncut timber under contract
Timber inventory growth with/without management			Yarding cost index	Total cost (US$/m³) sawnwood equivalent with/without forest management

(*Continued*)

Table 6.2 (Continued)

Timber or sale characteristics	Sale timing	Location	Management/policy	Other
Timber inventory density function			Lifetime of concession	Reversion speed/ half (yr)
Minimum timber inventory imposed by regulation				Development cost
Maximum timber inventory				Taxes (%)
Standard deviation of timber inventory				Discount rate
Species or timber quality class				Population per km² of forest land in the country
Average tree diameter or size				Concession area (ha)
Natural log of diameter				Site characteristics
Salvage by species (%)				
Conversion coefficient (%)				

West (Idaho, Montana and Washington), the Lake States (Indiana, Michigan, Minnesota and Wisconsin), New England (Vermont) and the South (Arkansas, North Carolina and Oklahoma). Other studies were in Canada, France, Sweden and South America (Brazil and Chile). Most early studies were done in the USA along with several Canadian studies in the 1990s.

The number of timber sales (N) included in the studies differed markedly. Earlier studies tended to have smaller sample sizes. The largest number of sales was analyzed by Nautiyal et al. (1995a) with 5,000 sales. Many of the studies used multiple datasets to compare ownerships (Jackson, 1987; Sendak, 1992; Munn and Rucker, 1995; Leefers and Potter-Witter, 2006) and different data subsets or time periods (Niccolucci and Schuster, 1994). Vasievich et al. (1997) separately estimated bid prices and sale costs. With one exception, researchers included multiple years to provide a sufficient number of sales for model estimation.

Most studies involved ordinary least squares (OLS) to solve multiple regression problems. We classified some studies as using OLS, though they were not specific regarding technique and might have only mentioned multiple regression analysis. Three papers used stepwise regression. Niccolucci and Schuster (1994) used an SAS feature for 'all possible' regressions; their research examined shortening data series and weighting recent sales more heavily to account for rapidly changing market prices. Munn and Palmquist (1997) explored information asymmetry between consultant and nonconsultant sales by using stochastic frontier analysis to estimate hedonic price functions.

The majority of researchers employed linear model forms, regardless of solution algorithm. The semi-log form was widely used as well; natural logarithms were associated occasionally with dependent variables and more often with independent variables (Tables 6.1 and 6.2). Model form was dictated in large part by the research question at hand. If elasticities were of interest, then log-log models may have been chosen. If ease of explanation of the effects of the independent variables were desired, a linear model may have been best. Several recent studies used more complex model forms (e.g. a truncated regression model, mathematical modeling, seemingly unrelated regression, etc.) and solution algorithms (e.g. Geometric Brownian Motion [GBM], mean-reverting process [MRP], simulated annealing [SA], etc.).

TEM models are composed of dependent and independent variables. Definitions for dependent variables all dealt with bid price, mostly high-bid price. Authors chose different language to describe this but were generally consistent in focusing on the winning bids or sale prices. Some studies expressed price in dollars per volumetric unit (e.g. mbf, cunit, m^3 or cord equivalent), whereas others adopted dollars per areal unit (e.g. ha). Total sales price was also used as the dependent variable in some cases. In others, authors referred to the dependent variable as bid price, but close review found it to be bid price per unit volume. Niccolucci and Schuster (1994) included the value of purchaser road credits as part of the dependent variable; this reflected the practice of having required road building, for which winning bidders paid part of the charges with their road credits.

Residual value was also included in one study (Nautiyal et al., 1995a); the residual value was calculated and then used as the dependent variable with supply-side and demand-side independent variables. Another variation of TEM was shown by Alzamora and Apiolaza (2010), who included the timber log recovery value as the dependent variable. They studied individual tree log transaction evidence of pine trees in Chile. They estimated the delivered log prices at the mill (conversion return) as a function of the log-based lumber value and log quality minus processing costs.

Most authors were careful to express the variables in real dollar terms, noting whatever index was applied. In other cases, authors did not indicate whether real or nominal dollars were used. Both transformed and untransformed dependent variables were employed.

A wide variety of independent variables has been used in studies (Table 6.2). They are classified for discussion into five categories: timber or sale characteristics, sale timing, location, management and policy and other. Researchers carefully selected their independent variables depending on the research question of interest, logical relationships among variables, and availability of data.

All studies included some measure of volume, whether it was total volume, volume per acre or a percentage of volume. In many cases, the volume was species and product specific. Transformed variables were sometimes chosen. The number of species and products was included to reflect the diversity of the sale. The majority of studies' attributes dealt with timber and sales characteristics. This is theoretically based; timber is priced on the basis of volume, species, grade, size, log quality, accessibility and so on that are likely to affect the yield and grade of lumber and other manufactured products (Jackson and McQuillan, 1979; Nautiyal et al., 1995a; Alzamora and Apiolaza, 2010). Also, different measures of tree quality and size may be contained in the models. If the dependent variable is total sale price, then sale volumes may be appropriate as explanatory variables. Alternatively, percentages of volumes could be used in conjunction with a variable for total sale area.

Given seasonality of timber markets and changes that occur over the years, sale year, quarter, month and season were sometimes selected as explanatory variables. Most often, these were treated as dummy variables. This allowed analysis of the effects of seasons or quarters within the annual sales cycle. Sale year provided a means to estimate effects of longer term trends (e.g. market slowdowns, inflation, etc.).

Location variables were commonly contained in the TEM models. Researchers and foresters recognized that location affects sale conditions, timber availability and many other factors. Examples of location variables from our selected studies included regions within a state, administrative areas used by a Department of Natural Resources (DNR), national forests, ranger districts and bid-up areas. Most of these represent some administrative unit; the bid-up area was associated with locations with higher sale bids. In hedonic pricing studies involving other natural resources (e.g. aesthetics, water access, etc.), proximity variables are usually included. For example, what is the distance to the nearest lake? A proximity variable of great interest for timber pricing is the 'distance to the mill'. Often, researchers do not have access to that information because multiple mills may receive timber from a given sale and there is sometimes a 'middleman' or logger between the sale and the mill. So, it is difficult to ascertain where harvested logs are going when many sales are involved.

Management and policy variables were typically represented in TEM models. Though the USFS can have fairly uniform procedures for timber pricing, state and local governments and private entities are more diverse in their approaches. As highlighted by Brown et al. (2012a), timber sale programs, policies and procedures are not uniform across the USA at the state level. This is the case around the world. USA states have received their policy and program direction from a variety of sources (e.g. constitutions, laws, administrative rules, etc.). Sale or auction methods, contract length, payment methods, competition and other elements are part of the mix that differentiates one state from another. All these factors can be explored within the TEM framework. As compiled in Table 6.2, ownership type, auction method, sale area, contract length and operating restrictions have been included in models. For example, Jackson (1987) examined pricing differences for state and federal ownership, and Schuster and Niccolucci (1994) used a two-step approach for estimating stumpage price to examine sealed bids versus oral-auction bids. Finally, harvest methods, road construction and haul and yarding distances are influenced by owners and reflected in TEM models.

There are a number of variables that are not easily classified, but nonetheless are important in model formulation. Number of bidders is used as a measure of competition, with increased competition leading to higher prices. Bidders can be treated as endogenous or exogenous variables (Carter and Newman, 1998; Leefers and Potter-Witter, 2006). Firm size, weather, end-product market variables and other variables are included in TEM models. In a limited number of studies, site characteristics were included. Each model is unique and reflects the research questions of interest and the researchers' approaches to addressing those questions.

Bare and Smith (1999) provided a novel TEM approach for estimating stumpage value for individual species and timber quality from lump-sum timber sales. Their model included species and quality variables (e.g. Douglas-fir poles, all classes of cedar, etc.) along with quarterly dummy variables. Regression results, in essence, dissected the lump-sum price per thousand board feet (mbf) into the prices for various species and timber qualities. Their study illustrated the flexibility of TEM, depending on the analysis questions pursued.

Concluding comments

The adoption and implementation of timber-pricing mechanisms depend on various factors – forest revenue system adopted by the government, policy and legal arrangements, administrative feasibility and practicality of the timber-pricing mechanism, market condition and the structure of the logging industry, among others. As a result, a wide range of timber-pricing mechanisms have been implemented in the developed and developing countries based on administrative pricing approaches, negotiated pricing, cost-based approaches, market evidence approaches and analytical approaches. In some cases, appraised value provides the starting point for stumpage bidding. In others, it provides the final price that will be paid.

In cases where appraisals provide a starting point for bidders, Combes, Niccolucci and Schuster (1989, p. 20) made an astute observation: 'When the timber is finally advertised and auctioned, differences between the various approaches to TEA [TEM] or between TEA and RV [RVM] appraisals become irrelevant. Appraisals end and actual markets take over.' This highlights two important points. First, many timber-pricing studies and models are intended to provide a starting point for market bidding (bringing together willing sellers and willing buyers). Appraised prices may be reduced by a standard percentage or to the lower bound of a confidence interval; this is intended to allow more bidders to participate in the process. Further, estimated prices may be adjusted due to hauling distance, stand conditions and so on. In competitive markets, the starting point may be far exceeded by the winning bid. In less competitive situations, the appraised price may be close to the winning bid. Second, many developers and implementers of these models may have no interest or need to publish their results. Hence, much of the work related to timber pricing is filed in agency or company records.

Published TEM studies come from agency and university researchers. They tend to be less concerned about implementation of their models and predicting the winning bid price and more interested in the research questions addressed by the models. In these cases, the independent variables are a source of information about competition, differences between ownerships, type of auctions and effects of management and policies (e.g. contract length).

At its foundation, one potential objective of timber pricing is to maximize timber sale revenue subject to various factors (forest characteristics, log quality, site condition, sale characteristics and management policy). Another objective might be to have a timber-pricing mechanism that would provide sufficient forest revenue to cover the current level of total expenditures of forestry operations while providing sustainable forest management for long-term socio-economic benefits, including forest industries' contributions to the national and regional economies. These objectives may conflict.

Given that timber pricing occurs in a social and political context, no system is inherently superior to others. In an idealized world, competitive markets (i.e. timber auctions) are often promoted. However, competitive markets do not exist in many areas, and other approaches are more salient. In some cases, competitive markets are illusory and collusion and corruption may distort prices. Thus, we are left with the age-old economist's question and answer: 'Which system is best?' 'It depends!'

Research on timber pricing has evolved over the past 50 years, and the current state of research is quite robust. This type of research was once the domain of forest economists and has now broadened to include environmental and resource economists along with more interdisciplinary efforts. More sophisticated models related to timber pricing have been developed, especially models incorporating supply and demand drivers. Studies of timber risks and their effects on pricing are more common, though some approaches, such as option pricing, have not taken hold. Timberland pricing studies are widely published in North America, with separation of timber and other values from land values. Studies highlight that timber pricing is also used for project evaluation, valuation of concessions for public offerings (e.g. valuation of the timber assets) and compensation.

Future research on timber pricing will address a number of problems. There will be less emphasis on North America and more emphasis on other developed and developing countries. Studies on the structure of timber markets (e.g. competitive vs. monopsony conditions) are expected because of their effects on timber pricing. In regions with little competition for stumpage, more studies, including comparative ones, will focus on administratively set prices. More efforts focused on the residual value method are likely. Much of this work will focus on how governments can capture more stumpage value from their concessions. This can lead to policy analyses related to impacts of various timber-pricing mechanisms on efficiency, resource utilization, income distribution and sustainable management of forest resources. Current foci of forest management will garner economists' attention: energy markets (e.g. pellets and other fuels), certified wood (e.g. price premiums), carbon sequestration (e.g. carbon markets) and environmental services (e.g. derivation and impacts of payments). So, there will be many opportunities for timber-pricing research in the future.

References

Alzamora, R. M. and Apiolaza, L. A. (2010). 'A hedonic approach to value *Pinus radiata* log traits for appearance-grade lumber production', *Forest Science*, vol. 56, no. 3, pp. 281–289.

Amacher, G. S., Brazee, R. J. and Witvliet, M. (2001). 'Royalty systems, government revenues, and forest condition: An application from Malaysia', *Land Economics*, vol. 77, no. 2, pp. 300–313.

Awang Noor, A. G. and Mohd Shahwahid, H. O. (2003). *Forest Pricing Policy in Malaysia*, Economy and Environment Program for Southeast Asia Research Report Series, Singapore.

Awang Noor, A. G., Vincent, J. R. and Yusuf, H. (1992). *Comparative Economic Analysis of Forest Revenue Systems in Peninsular Malaysia*, Final report submitted to Osborn Center Forestry Policy Grants Program, Washington, DC.

Bare, B. B. and Smith, R. L. (1999). 'Estimating stumpage values from transaction evidence using multiple regression', *Journal of Forestry*, vol. 97, no. 7, pp. 32–39.

Boscolo, M. and Vincent, J. R. (2007). 'Area fees and logging in tropical timber concessions', *Environment and Development Economics*, vol. 12, no. 4, pp. 505–520.

Brown, R. N., Kilgore, M., Blinn, C., Coggins, J. and Pfender, C. (2010). *Assessing State Timber Sale Policies, Programs and Stumpage Price Drivers*, College of Food, Agricultural and Natural Resource Sciences and the Agricultural Experiment Station, University of Minnesota, Minneapolis, MN.

Brown, R. N., Kilgore, M. A., Blinn, C. R. and Coggins, J. S. (2012a). 'State timber sale programs, policies, and procedures: A national assessment', *Journal of Forestry*, vol. 110, no. 5, pp. 239–248.

Brown, R. N., Kilgore, M. A., Coggins, J. S. and Blinn, C. R. (2012b). 'The impact of timber-sale tract, policy, and administrative characteristics on state stumpage prices: An econometric analysis', *Forest Policy and Economics*, vol. 21, pp. 71–80.

Buongiorno, J. and Young, T. (1984). 'Statistical appraisal of timber with an application to the Chequamegon National Forest', *Northern Journal of Applied Forestry*, vol. 1, no. 4, pp. 72–76.

Carter, D. R. and Newman, D. H. (1998). 'The impact of reserve prices in sealed bid federal timber sale auctions', *Forest Science*, vol. 44, no. 4, pp. 485–495.

Combes, J. A., Niccolucci, M. J. and Schuster, E. G. (1989). 'Stumpage appraisal: TEA and the northern region', *Forest Industries*, vol. 118, pp. 18–20.

Crowe, K. (2008). 'Modeling the effects of introducing timber sales into volume-based tenure agreements', *Forest Policy and Economics*, vol. 10, no. 3, pp. 174–182.

Dahal, P. and Mehmood, S. R. (2005). 'Determinants of timber bid prices in Arkansas', *Forest Products Journal*, vol. 55, no. 12, pp. 89–94.

Food and Agriculture Organization of the United Nations (FAO). (2001). *Reform Fiscal Policies in the Context of National Forest Programmes in Africa: Synthesis of Country Report*, Regional Forestry Meeting, Abuja, Nigeria.

Food and Agriculture Organization of the United Nations (FAO). (2010). *Global Forest Resources Assessment 2010: Main Report*, FAO, Rome, Italy.

Helms, J. A. (1998). *The Dictionary of Forestry*, CAB International and the Society of American Foresters, Bethesda, MD.

Jackson, D. H. (1987). 'Why stumpage prices differ between ownerships: A statistical examination of state and Forest Service sales in Montana', *Forest Ecology and Management*, vol. 18, no. 3, pp. 219–236.

Jackson, D. H. and McQuillan, A. G. (1979). 'A technique for estimating timber value based on tree size, management variables, and market conditions', *Forest Science*, vol. 25, no. 4, pp. 620–626.

Kant, S. (2009). *Global Trends in Ownership and Tenure of Forest Resources and Timber Pricing*, Report submitted to the Ontario Professional Foresters Association, Georgetown, ON, Canada.

Kant, S. (2010). 'Market, timber pricing, and forest management', *The Forestry Chronicle*, vol. 86, no. 5, pp. 580–588.

Kim, S., Phat, N. K., Koike, M. and Hayashi, H. (2006). 'Estimating actual and potential government revenues from timber harvesting in Cambodia', *Forest Policy and Economics*, vol. 8, no. 6, pp. 625–635.

Klemperer, W. D. (1996). *Forest Resource Economics and Finance*, McGraw-Hill, New York.

Leefers, L. A. and Potter-Witter, K. (2006). 'Timber sale characteristics and competition for public lands stumpage: A case study from the Lake States', *Forest Science*, vol. 52, no. 4, pp. 460–467.

Li, T. and Perrigne, I. (2003). 'Timber sale auctions with random reserve prices', *Review of Economics and Statistics*, vol. 85, no. 1, pp. 189–200.

Lynch, T. B., Huebschmann, M. M., Lewis, D. K., Tilley, D. S. and Guldin, J. M. (2004). 'A bid price equation for national forest timber sales in Western Arkansas and Southeastern Oklahoma', *Southern Journal of Applied Forestry*, vol. 28, no. 2, pp. 100–108.

MacKay, D. G. and Baughman, M. J. (1996). 'Multiple regression-based transactions evidence timber appraisal for Minnesota's state forests', *Northern Journal of Applied Forestry*, vol. 13, no. 3, pp. 129–134.

Mayo, J. H. and Straka, T. J. (2005). 'The holding value premium in standing timber valuation', *Appraisal Journal*, vol. 73, no. 1, pp. 98–106.

McKeever, D. B. and Howard, J. L. (2011). *Solid Wood Timber Products Consumption in Major End Uses in the United States, 1950–2009*, General Technical Report FPL-GTR-199, USDA Forest Service, Forest Products Laboratory, Madison, WI.

Munn, I. A. and Palmquist, R. B. (1997). 'Estimating hedonic price equations for a timber stumpage market using stochastic frontier estimation procedures', *Canadian Journal of Forest Research*, vol. 27, no. 8, pp. 1276–1280.

Munn, I. A. and Rucker, R. R. (1995). 'An economic analysis of the differences between bid prices on Forest Service and private timber sales', *Forest Science*, vol. 41, no. 4, pp. 823–840.

Nautiyal, J. C., Kant, S. and Williams, J. S. (1995a). 'A mechanism for tracking the value of standing timber in an imperfect market', *Canadian Journal of Forest Research*, vol. 25, no. 4, pp. 638–648

Nautiyal, J. C., Kant, S. and Williams, J. S. (1995b). 'A transaction evidence based estimate of the stumpage value of some southern Ontario forest species', *Canadian Journal of Forest Research*, vol. 25, no. 4, pp. 649–658.

Nelson, E., Mendoza, G., Regetz, J., Polasky, S., Tallis, H., Cameron, D., . . . Shaw, M. (2009). 'Modeling multiple ecosystem services, biodiversity conservation, commodity production, and tradeoffs at landscape scales'. *Frontiers in Ecology and the Environment*, vol. 7, no. 1, pp. 4–11.

Neter, J., Kutner, M. H., Nachtsheim, C. J. and Wasserman, W. (1996). *Applied Linear Statistical Models*, Irwin, Chicago, IL.

Niccolucci, M. J. and Schuster, E. G. (1991). 'Transactions evidence appraisal equations: Tests, developments, and problems', in M. A. Buford (Compiler), *Proceedings of the 1991 Symposium on Systems Analysis in Forest Resources*, General Technical Report SE-74, USDA Forest Service Southeastern Forest Experiment Station, Asheville, NC.

Niccolucci, M. J. and Schuster, E. G. (1994). 'Effect of database length and weights on transactions evidence timber appraisal models', *Western Journal of Applied Forestry*, vol. 9, no. 3, pp. 71–76.

Niquidet, K. and van Kooten, G. C. (2006). 'Transaction evidence appraisal: Competition in British Columbia's stumpage markets', *Forest Science*, vol. 52, no. 4, pp. 451–459.

Paarsch, H. J. (1997). 'Deriving an estimate of the optimal reserve price: An application to British Columbian timber sales', *Journal of Econometrics*, vol. 78, no. 1, pp. 333–357.

Préget, R. and Waelbroeck, P. (2006). *Hedonic Prices for Timber Auctions with Endogenous Participation*, Laboratoire d'Economie Forestiere, AgroParisTech-INRA, Nancy, France.

Puttock, G. D., Prescott, D. M. and Meilke, K. D. (1990). 'Stumpage prices in southwestern Ontario: A hedonic function approach', *Forest Science*, vol. 36, no. 4, pp. 1119–1132.

Repetto, R. and Gillis, M. (Eds.). (1988). *Public Policies and the Misuse of Forest Resources*, Cambridge University Press, Cambridge, MA.

Rocha, K., Moreira, A. R., Reis, E. J. and Carvalho, L. (2006). 'The market value of forest concessions in the Brazilian Amazon: A real option approach', *Forest Policy and Economics*, vol. 8, no. 2, pp. 149–160.

Roos, A. (1996). 'A hedonic price function for forest land in Sweden', *Canadian Journal of Forest Research*, vol. 26, no. 5, pp. 740–746.

Rosen, S. (1974). 'Hedonic prices and implicit markets: Product differentiation in pure competition', *The Journal of Political Economy*, vol. 82, no. 1, pp. 34–55.

Rothery, J. E. (1945). 'Some aspects of appraising standing timber', *Journal of Forestry*, vol. 43, no. 7, pp. 490–498.

Schuster, E. G. and Niccolucci, M. J. (1990). 'Comparative accuracy of six timber appraisal methods', *The Appraisal Journal*, vol. 58, no. 1, pp. 96–108.

Schuster, E. G. and Niccolucci, M. J. (1994). 'Sealed-bid versus oral-auction timber offerings: Implications of imperfect data', *Canadian Journal of Forest Research*, vol. 24, no. 1, pp. 87–91.

Seber, G. A. and Lee, A. J. (2003). *Linear Regression Analysis* (2nd ed.), John Wiley & Sons, Hoboken, NJ.

Sendak, P. E. (1992). 'State and federal timber stumpage prices in Vermont', *Northern Journal of Applied Forestry*, vol. 9, no. 3, pp. 97–101.

Smith, J. S., Markowski-Lindsay, M., Wagner, J. E. and Kittredge, D. B. (2012). 'Stumpage prices in southern New England (1978–2011): How do red oak, white pine, and hemlock prices vary over time?' *Northern Journal of Applied Forestry*, vol. 29, no. 2, pp. 97–101.

van Kooten, G. C. (2002). 'Economic analysis of the Canada-United States softwood lumber dispute: Playing the quota game', *Forest Science*, vol. 48, no. 4, pp. 712–721.

Vasievich, J. M., Mills, W. L., Jr. and Cherry, H. R. (1997). 'Hardwood timber sales on state forests in Indiana: Characteristics influencing costs and prices', in S. G. Pallardy, R. A. Cecich, H. G. Garrett and P. S. Johnson (Eds.), *Proceedings of the 11th Central Hardwood Forest Conference*, General Technical Report NC-188, USDA Forest Service North Central Forest Experiment Station, St. Paul, MN.

Vincent, J. R. (1990). 'Rent capture and the feasibility of tropical forest management', *Land Economics*, vol. 66, no. 2, pp. 212–223.

Weintraub, S. (1959). 'Price-making in Forest Service timber sales', *The American Economic Review*, vol. 49, no. 4, pp. 628–637.

Wiener, A. A. (1981). 'Appraising national forest timber values: A concept reexamined', *Journal of Forestry*, vol. 79, no. 6, pp. 372–376.

Wiener, A. A. (1982). *The Forest Service timber appraisal system: A historical perspective, 1891–1981*, USDA Forest Service, Washington DC.

Yang, F. E. and Kant, S. (2008). 'Rent capture analysis of Ontario's stumpage system using an enhanced parity bounds model', *Land Economics*, vol. 84, no. 4, pp. 667–688.

Yang, F. E., Kant, S. and Shahi, C. (2006). 'Market performance of the government-controlled but market-based stumpage system of Ontario', *Forest Science*, vol. 52, no. 4, pp. 367–380.

7

FOREST TAXATION

Markku Ollikainen

PROFESSOR OF ENVIRONMENTAL AND RESOURCE ECONOMICS, DEPARTMENT OF ECONOMICS AND
MANAGEMENT, UNIVERSITY OF HELSINKI, P.O. BOX 27, FI-00014, HELSINKI, FINLAND.

Abstract

This chapter surveys forest taxation literature from two angles. The traditional results of forest taxation in the Faustmann and Hartman models are reviewed and used to discuss first-best and second-best forest tax policies. Theoretical results are illustrated using numerical simulations. Forest tax literature is then related to modern forest policies promoting multiple ecosystem services. Using biodiversity benefits, climate mitigation and nutrient load policies as examples, it is shown that forest taxation plays an important role as a part of targeted policy packages promoting ecosystem services. The heterogeneity of landowners' preferences provides empirical and theoretical challenges for optimal tax policies.

Keywords

Optimal forest taxation, neutral and distortionary taxes, amenity benefits, climate policy, nutrient loads

Introduction

Forest taxation is a fascinating old subject of forest economics, which even inspired Martin Faustmann to write his celebrated contribution in 1849. One of Faustmann's motives was to define principles by which the value of the forest land could accurately be determined, to facilitate the efficient and fair taxation of forest landowners. Interestingly, forests have also indirectly contributed to general tax terminology. The roots of windfall profits and windfall taxes are found in forest utilization rights. Old British civil servants were allowed to freely use windfalls; forestlands were otherwise the property of the state, and timber utilization was subject to charges. Since those times, the windfall concept has meant extra revenues obtained through sheer luck and received either from nature or through policy, without a purposeful effort from the agent.

What makes forest taxation interesting is not so much the fact that governments raise funds for spending governmental programs. A more important issue is that taxes can be used to affect forest management and thereby the state of forests. Tax instruments are an appropriate way for countries to impact their timber supplies or the provisioning of amenity services. New instruments have recently been invented to encourage the provisioning of amenity benefits (payments for ecosystem services) or to control deforestation in concession areas (royalty payments), thus enlarging the role of forest taxation. (see for instance Amacher, Koskela and Ollikainen 2007 and 2012)

The set of possible forest taxes that governments have at their disposal is large, and policy-makers should choose the tax that best fits the forestry problem at hand. A typical tax problem is collecting a required amount of forest revenue for a government budget. Neutral taxes should be used in this case, as they do not distort private forest management. However, if impacting forest management is a goal, distortionary forest tax should be used. We will examine neutral and distortionary taxes and their proper uses.

Most forest tax literature examines the impacts which different forest taxes have on forest management. These comparative static results, or 'incentive impacts', are well established in both the Faustmann and the Hartman rotation frameworks (see Chang, 1982; Koskela and Ollikainen, 2001 and 2003a, respectively). We begin this chapter by reviewing their work and examining the ways in which forest taxes impact optimal rotation age, timber supply, amenities and forest rent. We complement our analytical discussion by insights from numerical simulations.

We next develop some basic features of optimal forest taxation under specified social welfare functions with the best for society as a target goal. Adding restrictions for tax rate determination leads to a second-best tax policy analysis. The issue of optimal taxation has a long history in public economics, beginning with the seminal paper written by Ramsey (1927). Its history in forestry, however, is surprisingly short-lived, dating back to Gamponia and Mendelsohn (1987), Englin and Klan (1990) and Koskela and Ollikainen (2003a).

As our third topic, we review new forest policy trends. Societies worldwide currently wish to jointly promote timber and amenity benefits from forests and alleviate negative externalities, for instance, nutrient loads entering waterways. Forest amenities comprise a large class of ecosystem services and are best promoted using differentiated and targeted tax and subsidy instruments. More often than not, a policy package is needed to promote ecosystem services. Forest taxes are often an integral part of these instrument packages.

We restrict our discussion to Faustmann and Hartman rotation frameworks and focus on three forest tax classes: harvest taxes, property taxes and profit taxes. Harvest taxes include the yield tax, levied on harvest revenue, and the unit tax (severance tax), levied on harvested timber volume. Property taxes are levied on land value. The site productivity tax is annually paid and is based on the yield potential of a given site, irrespective of the actual harvest or standing timber. Another lump-sum property tax is a proportional land value tax called the site value tax. Property tax may also be levied on tree value and is called the timber (stumpage) tax. Finally, a profit tax is assessed when aiming for net timber revenues from harvesting.[1]

Forest taxation in the Faustmann model

The comparative static results on forest taxation in the Faustmann model are well established by, e.g. Chang (1982, 1983) and Johansson and Löfgren (1985). We begin our discussion with harvest tax impacts. Assume that government levies a yield (τ) tax or unit tax (t) on harvesting. The after-tax net present harvest revenue values for a landowner in the presence these taxes are given by

$$V(\tau) = \left[p(1-\tau)f(T)e^{-rT} - c \right](1 - e^{-rT})^{-1} \qquad (1a)$$

$$V(t) = \left[(p-t)f(T)e^{-rT} - c \right](1 - e^{-rT})^{-1} \qquad (1b)$$

The difference between these two tax types is evident: Unit tax is independent of timber value; that is, stumpage price does not impact tax payments, but timber volume solely. Harvest taxes work similarly as a decrease in timber prices. We can thus immediately state that $\partial T/\partial \tau > 0$ and $\partial T/\partial t > 0$ (for proof, see Amacher, Koskela and Ollikainen, 2009). Hence, both yield and unit taxes will lengthen the rotation age.

In some countries landowners may be allowed to deduct management costs from the taxable harvest revenue. In this case, harvest tax impacts depend on whether tax deduction concerns all, or only part, of the costs. Under full yield tax deductibility equation (1a) reads as $V(\tau) = (1-\tau)\left[pf(T)e^{-rT} - c \right](1-e^{-rT})^{-1}$ and the tax has no impact on rotation age; it only represents a loss of net harvest revenue functioning as a neutral profit tax. Partial tax deductibility implies that the rotation age is lengthened, but this lengthening is less than in the absence of tax deduction in equation (1a).

Property taxes are considered next, beginning with the site value and site productivity taxes. Let β denote site value tax and $a(i)$ denote site productivity tax. The site productivity tax is based on a measure of site quality (denoted by i), such as site index. Equations (2a) and (2b) present the Faustmann model under these taxes:

$$V(\beta) = (1-\beta)\left[pf(T)e^{-rT} - c \right](1-e^{-rT})^{-1} \tag{2a}$$

$$V_i(a(i)) = \left[pf_i(T)e^{-rT} - c \right](1-e^{-rT})^{-1} - \frac{a(i)}{r} \tag{2b}$$

Choosing rotation age to maximize (2a) and (2b) shows that neither tax impacts the first-order condition, so that $T_\beta = T_{a(i)} = 0$. Thus, both taxes only cause a lump-sum reduction in the net present value of forest rents, but have no effect on rotation age.

When a timber property tax, α, is levied annually on the stumpage value of growing timber volume, the objective function of the landowner becomes

$$V(\alpha) = (1-e^{-rT})^{-1}\left[pf(T)e^{-rT} - c - \alpha \int_0^T pf(s)e^{-rs}ds \right] \tag{3}$$

The impact of the timber tax is more complicated, and it is useful to develop the first-order condition

$$V_T(\alpha) = pf'(T) - rf(T) - rV - \alpha(pf(T) - rU) = 0 \tag{4}$$

where $U = (1-e^{-rT})^{-1}\int_0^T pf(s)e^{-rs}ds$ denotes the present value of annual timber earnings, and V refers to the value of the forestland in the absence of the timber tax. The effect of the tax rate α on the optimal rotation age, $T_\alpha = -V_{T\alpha}/V_{TT}$, depends on the sign of $V_{T\alpha}(\alpha) = -(pf(T) - rU)$, which is negative (for original proof, see Koskela and Ollikainen, 2003a). Therefore, $T_\alpha^F < 0$ and a property tax levied on timber shortens the rotation age.

Table 7.1 presents a summary of forest tax impacts in terms of rotation age and timber supply in the short and long run. Long-term timber supply refers to average annual supply, s, over the steady-state rotation period: $s = f(T\star)/T\star$, where $T\star$ denotes the optimal rotation age and $f(T\star)$ the timber volume at optimal harvest time. We conventionally assume that forest volume, $f(T\star)$, is smaller than the Maximum Sustainable Yield (MSY) volume. Long-term timber supply is also impacted by changes in land allocation (Conrad, 2010; Amacher et al., 2009). We omit this

Table 7.1 Forest tax impacts on rotation age and timber supply: The Faustmann model.

Forest tax	Type of tax	Impact on rotation age	Impact on timber supply	
			In the short run	In the long run
yield tax	distortionary	lengthens	negative	positive
unit tax	distortionary	lengthens	negative	positive
site value tax	neutral	no impact	no impact	no impact
site productivity tax	neutral	no impact	no impact	no impact
timber tax	distortionary	shortens	positive	negative
profit tax	neutral	no impact	no impact	no impact

Table 7.2 Impacts of yield and timber taxes in the Faustmann model for Scandinavian boreal commercial forests.

	Rotation age (yrs)	Land value (€/ha)	Harvest revenue (€/ha)	Harvest volume (m³)	Average supply (m³)	Tax at harvest (m³)	PV of tax (€/ha)	Supply elasticity (%)
No tax	71.5	530	7,422	186	2.59	–	–	–
Yield tax rate (%)								
20	72.6	233	6,114	191	2.63	1,529	297	
30	73.4	84	5,457	195	2.66	2,339	444	0.03
40	74.4	−63	4,798	200	2.69	3,199	589	0.03
Timber tax (%)								
5	66.6	308	6,374	159	2.40	167	206	
10	61.9	116	5,518	138	2.23	282	358	−0.08
15	58.3	−52	4,814	120	2.06	361	471	−0.15
MSY	**113.8**			**364**	**3.20**			

impact by noting that all forest taxes, *ceteris paribus*, decrease the profitability of forestland relative to other land-use forms which reduce timber supplies.

From Table 7.1 we see that if society wishes to increase its timber supply in the long run, distortionary harvest taxes should be used. This is also the case if a government wishes to increase amenities associated with old stands. However, timber tax is convenient if the goal is to increase benefits associated with young stands. Neutral taxes are preferable if the goal is to only raise forest tax revenues without seeking other goals.

Table 7.2 numerically examines tax impacts when employing Fridth and Nilsson's growth function for Scandinavian boreal commercial forests. Yield and timber taxes are chosen as examples of distortionary taxes. We report their impacts on rotation age, land value, after-tax harvest revenue, harvested timber volume and average supply. The two latter figures are also reported at the MSY rotation age for the sake of comparison.

Table 7.2 confirms our theoretical findings: The yield tax lengthens and timber tax shortens rotation age; the former increases and the latter decreases long-term timber supply. More importantly, Table 7.2 complements our formal analysis in several ways which are seldom discussed in forest taxation literature. First, despite increasing yield tax rates, the private rotation ages always remain shorter than the MSY rotation age. Second, rotation age only changes slightly as a response to increases in the yield tax rate, but changes greatly in response to timber tax rate increases. This can be seen in the differences between timber supply elasticities. For yield tax, the tax elasticity of supply is roughly +3%, but for timber tax the (negative) elasticity of supply increases from −8% to −15%. Third, the detrimental effects of high tax rates on forestry become evident: Land values relatively quickly become negative. Heavy forest taxation may ruin forestry altogether.

Forest taxation in the Hartman model

Literature on forest taxation in the Hartman model (1976) is less plentiful than that in the Faustmann model. The basic contributions are by Gamponia and Mendelsohn (1987), Englin and Klan (1990) and Koskela and Ollikainen (2001) – for a compact presentation, see Amacher et al. (2009). Forest taxes impact harvesting in a complicated way. A landowner values amenity benefits and manages his forest so as to maximize the sum of the net present value of harvest revenue and amenity benefits. As forest taxes are levied on harvest revenue, they change the profitability of timber production relative to amenity production.

Starting with harvest taxes (yield tax τ and unit tax t), the Hartman model can be expressed as

$$W(\tau) = (1 - e^{-rT})^{-1} \left[p(1-\tau)f(T)e^{-rT} - c \right] + (1 - e^{-rT})^{-1} \int_0^T F(s)e^{-rs}ds \tag{5a}$$

$$W(t) = (1 - e^{-rT})^{-1} \left[(p-t)f(T)e^{-rT} - c \right] + (1 - e^{-rT})^{-1} \int_0^T F(s)e^{-rs}ds \tag{5b}$$

As with the Faustmann model, harvest taxes work identically as a decrease in timber price, but the impacts are more complicated. The tax impact is $\partial T/\partial\tau = -V_{T\tau}/V_{TT}$, where $\operatorname{sgn} V_{T\tau} = \operatorname{sgn}\left[rc(1 + e^{-rT})^{-1} + F'(T) \right]$. Thus, if a landowner values amenity benefits from old stands, or if the amenity benefits are independent of forest age ($F'(T) \geq 0$), tax impact is conventional, $\partial T/\partial\tau > 0$. However, if the landowner values amenity benefits from young stands and regeneration costs are 'small enough', $\partial T/\partial\tau < 0$ (for original proof, see Koskela and Ollikainen, 2001). Harvest taxes increase the relative profitability of amenity production, and the landowner adjusts the optimal rotation age in a direction dependent on the type of amenity valuation in question.

Next we focus on the site value and site productivity taxes. To keep the notation simple, let V denote the Faustmann part of the model and $E = (1 - e^{-rT})^{-1} \int_0^T F(s)e^{-rs}ds$ the amenity part. The site value tax and site productivity tax respectively show up as follows:

$$W(\beta) = (1 - \beta)V + E \tag{6a}$$

$$W(a(i)) = V - \frac{a(i)}{r} + E \tag{6b}$$

For the site value tax, the first-order condition is given by

$$W_T(\beta) = (1 - \beta)\left[pf'(T) - rpf(T) - rV \right] + F(T) - rE = 0 \tag{7}$$

101

Unlike in the Faustmann case, β does not vanish from the first-order condition (7); site value tax now becomes distortionary. Differentiating (7) with respect to β produces:

$$W_{T\beta}(\beta) = -(pf'(T) - rpf(T) - rV) \geq (<) \ 0 \ as \ F'(T) \geq (<) \ 0$$

Assuming that the second-order condition holds, we have the following result:

$$T_\beta^H \left\{ \begin{matrix} > \\ = \\ < \end{matrix} \right\} \ 0 \ as \ F'(T) \left\{ \begin{matrix} > \\ = \\ < \end{matrix} \right\} \ 0 \qquad (8)$$

From equation (8) we see that the site value tax does not affect rotation age (is neutral) if amenity valuation is site-specific ($F'(T) = 0$). If $F'(T) > (<) \ 0$, a rise in site value tax makes amenity production relatively more (less) profitable than timber production. Consequently, the landowner lengthens (shortens) the rotation age.

For the site productivity tax, the first-order condition is $W_T(a(i)) = V_T + F(T) - rE = 0$. The site productivity tax remains neutral because it does not affect the relative profitability of timber and amenity production. Finally, for the property tax on stumpage values of trees, α, the objective function of the landowner is given by

$$W(\alpha) = V - \alpha(1 - e^{-rT})^{-1} \int_0^T pf(s)e^{-rs}ds + E \qquad (9)$$

The first-order condition is $W_T(\alpha) = pf'(T) - rpf(T) - rV - \alpha(pf(T)-rU) + F(T) - rE = 0$. Just as in the Faustmann case, timber tax impacts only through term $(pf(T) - rU) > 0$. Hence, timber tax shortens the rotation age irrespective of the nature of amenity valuation.

Finally, we consider how profit tax affects rotation age in the presence of amenity benefits. We previously found this impact to be neutral in the basic Faustmann model. Profit tax is levied on net harvest revenue, so it resembles the site value tax, with its impacts depending on the amenity valuation of the landowner. Table 7.3 summarizes our results (short-term supply impacts are omitted).

The impacts of the site productivity and timber taxes are equivalent as in the Faustmann model. In contrast, the site value tax and profit tax become distortionary and lengthen or shorten rotation age depending on the type of amenity valuation. The same holds true for harvest taxes, which lengthen rotation age in most cases. Long-term supply impacts vary depending on the type of amenity valuation and also on how private rotation age relates to the MSY age for the valuation of old stands. Hence, finding a tax rate which well reflects the heterogeneous preferences of landowners becomes a demanding task and requires much empirical information.

Table 7.4 presents a numerical analysis using amenity valuation by Swallow et al. (1990): $b_0 TExp[-(1/b_1)T]$, where T is the rotation age, $b_0 = 0.735$ and $b_1 = 280$. These parameters imply that the landowner values amenity benefits received from older stands. We report two land values: Land value 1 is the sum of the harvest revenue and amenity benefits, and land value 2 is the harvest revenue only. Assuming that amenity valuation is not capitalized in commercial land values, land value 2 is closer to the Faustmann model.

Both taxes impact rotation ages exactly as the theory predicts. Average timber supply is increased only slightly by the yield tax. Tax elasticities of the timber supply increase with tax rate, and are practically equal to those in the Faustmann model. High forest taxation is detrimental to

Table 7.3 Forest tax impacts on rotation age and timber supply: Hartman model.

Forest tax	Type of tax	Impact on rotation age amenities valued from		Impact on timber supply from	
		Old stands	Young stands	Old stands	Young stands
yield tax	distortionary	lengthens	may shorten	increases	may decrease
unit tax	distortionary	lengthens	may shorten	increases	may decrease
site value tax	distortionary	lengthens	shortens	increases	decreases
site productivity tax	neutral	neutral	neutral	no impact	no impact
timber tax	distortionary	shortens	shortens	decreases	decreases
profit tax	distortionary	lengthens	shortens	increases	decreases

Table 7.4 Impacts of yield and timber taxes in the Hartman model for Scandinavian boreal commercial forests.

	Rotation age (yrs)	Land value 1 (€/ha)	Land value 2 (€/ha)	Harvest revenue (€/ha)	Harvest volume (m³/ha)	Average supply (m³/ha)	Tax at harvest (€/ha)	PV of tax (€/ha)	Supply elasticity (%)
No tax	77.6	1,386	512	8,618	215	2.8	–	–	–
Yield tax (%)									
20	80.3	1,099	211	7,308	228	2.8	1,827	284	
30	82.2	958	60	6,651	238	2.9	2,851	419	0.03
40	84.8	820	−92	5,993	250	2.9	3,996	545	0.06
Timber tax (%)									
5	71.6	1,128	291	7,429	186	2.6	200	240	
10	66.7	904	99	6,457	161	2.4	339	418	−0.07
15	62.6	708	−68	5,654	141	2.3	435	550	−0.14
MSY	**113.75**			**363.91**	**3.20**				

commercial forestry. However, land values show that the higher amenity benefits are valued by landowners, the more private forest management can sustain high taxation.

On optimal forest taxation

We now change our focus and ask: How should society use these taxes as a part of forest policy? We restrict our analysis to some representative cases (for original contributions, see Gamponia and Mendelsohn, 1987; Englin and Klan, 1990; Koskela and Ollikainen, 2003a). Throughout this section we assume that the forest landowner's goal is maximizing harvest revenue only. We defined society's objective function by assuming that it values either the net present value of harvest revenue solely, or alternatively, the sum of harvest revenue and amenity benefits. In the former case, the social welfare function is simply $SW = V$, and $SW = V + E$ in the latter case.

Optimal forest taxation when society solely values net harvest revenue

In this case, society and the forest landowner have identical preferences and social welfare is given by.

$$SW = V = \left[pf(T)e^{-rT} - c \right](1 - e^{-rT})^{-1} \tag{10}$$

As a warming-up problem, suppose that society wishes to find a tax rate that maximizes social welfare – what tax rate would accomplish this? The rotation age that maximizes the social welfare function is identical to the privately optimal rotation age (Johansson and Löfgren, 1985; Hellsten, 1988). Let $T^{\star}$ denote the optimal rotation age so that social welfare is $SW(T^{\star})$. Using a distortionary tax would not be optimal, as the optimal rotation age would be distorted. A neutral tax is not optimal either, as it would reduce the maximum value of SW. Hence, when society and the private landowner have identical preferences, there are no grounds for forest taxation. Only the necessity of collecting tax revenues would make forest taxation rational.

Next, assume that society must collect a given sum of tax revenue from forestry. The challenge is to raise the required sum with minimal distortion to forest management. The government should therefore only use neutral forest taxes. It is useful to establish this result formally. Define the social welfare function using the indirect target function of the landowner, $V^{\star}$ in the presence of forest taxes. Substituting the landowner's optimal rotation age choice, solved in the presence of taxes, $T = T(a(i), \tau, t, \beta, \theta, \alpha)$ into the objective function, produces

$$SW = V^{\star}(a(i), \tau, t, \beta, \theta, \alpha) \tag{11}$$

Assume that the government only responds to the discounted sum of tax revenue collected from forestry, given by

$$G = \left[(p\tau + t)f(T)e^{-rT} + \alpha e^{rT} \int_0^T pf(s)e^{-rs}ds \right](1 - e^{-rT})^{-1} + \frac{a(i)}{r} + \psi V \tag{12}$$

where $\psi = \theta, \beta$ and $V = (pf(T) - ce^{rT})(e^{rT} - 1)^{-1}$.

Let the site productivity tax be our benchmark neutral tax. The government maximizes the social welfare function (11) subject to the tax revenue target, $\overline{G}$, from (12). The Lagrangian for the problem is

$$\Omega = V^{\star} - \lambda(\overline{G} - G) \tag{13}$$

Choosing the size of the site productivity tax $a(i)$, and recalling that the landowner's response will not impact tax revenue because $T_a^F = 0$, gives

$$\Omega_{a(i)} = -\frac{1}{r} + \lambda \frac{1}{r} = 0 \tag{14}$$

The site productivity tax $a(i)$ is optimally chosen by equalizing the welfare loss suffered by landowners due to the tax with the marginal cost of public funds, represented by the shadow price of the revenue constraint, λ. This shadow price measures the marginal decrease in social welfare value for an incremental increase in government revenues due to a tax increase. In equation (14), λ is equal to 1, which is good news: The marginal cost of public funds is unity.

It is straightforward to show that the same result holds true for neutral site value and profit taxes. It is equally simple to demonstrate that no other taxes are required by the government to complement site productivity (site value or profit tax), as all required tax revenues can be collected using the neutral tax (for proof, see Amacher et al., 2009).

Suppose neutral taxes are not available and using distortionary taxes is necessary. This shifts us to the second-best real world. Harvest taxes are probably the most common form of forest taxes,

Table 7.5 Optimal second-best yield and timber tax rates under a binding tax revenue requirement for Scandinavian boreal commercial forests.

Tax revenue target (PV) (€/ha)	Tax rate (%)	Rotation age (yrs)	Land value (€/ha)	Harvest revenue (€/ha)	Taxes paid (€/ha)	Welfare loss (€/ha) in	
						Land value	Harvest revenue
No taxation	0	71.5	530.3	7,422	0	0	0
Yield tax							
200	13.4	72.2	330.1	6,543	1,017	200.3	878
300	20.2	72.6	229.6	6,101	1,545	300.7	1,321
400	27.0	73.2	128.9	5,655	2,091	401.4	1,767
Timber tax							
200	0.06	65.1	257.2	6,143	2,638	273.1	1,279
300	1.1	61.2	164.1	5,369	3,428	366.2	2,053
400	1.9	56.1	−27.4	4,389	3,781	557.7	3,033

so we focus on the yield tax. Choosing the optimal yield tax so as to maximize the Lagrangian (13) leads to the first-order condition:

$$\Omega_\tau = (\lambda - 1)pf(T^*)e^{-rT^*}(1 - e^{-rT^*})^{-1} - \lambda G_\tau T_\tau^* = 0 \tag{15}$$

In equation (15), λ is not equal to 1, indicating that taxation causes welfare losses: Collecting one unit of money for the government costs more than collecting one unit for landowners. The optimal yield tax rate depends on the direct cost (first term), the landowner's reactions to the yield tax (via T_τ^*) and its impact on the tax revenue. If alternative taxes are available, respective optimality formulas should be developed and the tax causing the smallest distortion chosen.

Table 7.5 provides numerical simulations on the second-best yield and timber property taxes under three alternative tax revenue targets. The welfare loss of the landowner is reported in terms of the reductions in land value and harvest revenue at the time of harvest. The first row shows the private optimum in the absence of forest taxation.

Both tax rates increase when the tax revenue target is increased, and landowners react to the taxes by adjusting their rotation ages. The yield tax rate increases with the tax revenue target from 13% to almost 30%. Timber tax rate increases from less than 1% to roughly 2%. The after-tax land value decreases with tax rates and more so with the timber tax. Harvest revenue also decreases more under the timber tax because by shortening the rotation age, the timber volume rapidly decreases. Prolonging timber tax payments until harvest time using a 2.5% interest rate shows that the tax burden is higher under timber tax, as is the loss in harvest revenue. Thus, yield tax causes a lower welfare loss than timber tax.

Forest taxation when forest amenities as public goods are accounted

Next suppose that society wishes to promote amenity production beyond the levels that private forest management produces. We express the social welfare function as

$$SW = V + (n - 1)E \tag{16}$$

where n refers to the number of citizens and $(n - 1)$ to the number of nonforest owners.

We start by defining the optimal first-best Pigouvian forest taxes. Their task is to induce the landowner to choose the socially optimal rotation age and thereby appropriate amenity benefits. The direction in which society wishes to shift the private rotation age depends on the society's amenity valuation. If society values amenities from old (young) stands, private rotation age will want to be lengthened (shortened). Formal analysis is needed to see how this can be achieved by taxation. The easiest method for solving the optimal tax rates is presented in the following. Define the first-order conditions characterizing the socially optimal rotation age and the privately optimal rotation age in the presence of forest taxes. Next, solve the tax rate that makes the two rotation ages identical (see Amacher et al., 2009). This approach produces the following optimal harvest tax rates:

$$\tau^* = -\frac{(n-1)\left[F(T) - rE\right]}{pf'(T) - rpf(T)(1 - e^{-rT})^{-1}} \leq (>) \, 0 \; as \; F'(T) \geq (<) \, 0 \tag{17a}$$

$$t^* = p\tau^* \leq (>) 0 \; as \; F'(T) \geq (<) \, 0 \tag{17b}$$

In (17a), both the denominator and numerator signs depend on society's valuation of amenities. First, if amenities are site-specific, $F'(T) = 0$ and $F(T) - rE = 0$, the optimal yield tax equals zero. Second, if society values amenities from old forests, we have $F(T) - rE > 0$, but $pf'(T) - rpf(T)(1 - e^{-rT})^{-1} < 0$, so that the optimal yield tax rate is positive. Finally, if amenities from young stands are valued, we have $F(T) - rE < 0$, the sign of $pf'(T) - rpf(T)(1 - e^{-rT})^{-1}$ is unknown, and the optimal yield tax can be either positive (tax), or negative (subsidy). A similar analysis applies to the unit tax in (17b).

Suppose that instead of harvest taxes, society wants to use the timber or site value taxes. Using the same procedure, the optimal first-best tax rates become:

$$\alpha^* = -\frac{(n-1)\left[F(T) - rE\right]}{(pf(T) - rU)} \leq (>) \, 0 \; as \; F'(T) \geq (<) \, 0 \tag{18a}$$

$$\beta^* = -\frac{(n-1)\left[F(T) - rE\right]}{pf'(T) - rpf(T) - rV} = \frac{(n-1)}{n} > 0 \tag{18b}$$

In equation (18a), the denominator is always positive, so the tax depends on amenity valuation. When amenities from older stands are valued, a timber subsidy is optimal, but if amenities from younger stands are valued, a positive timber tax becomes optimal. Equation (18b) defines the optimal site value tax. The second formula results from the observation that, due to the landowner's first-order conditions, we have $E_T^* = -V_T^*$ (numerator and denominator of 18b). Thus, the first-best site value tax is a classic Pigouvian tax: Tax rate choice depends on the value of amenities lost by harvesting – these in turn depend on the share of citizens in the economy.

Table 7.6 presents numerical simulations of the optimal forest tax design. We report both the social and private values of the forestland. The former value includes amenity benefits, while the latter is the private commercial land value. We also report the social welfare loss caused by reduced amenity services in the (too short) private rotation age and total tax outlays.

The socially optimal rotation age at the baseline is 77.59 years, which exceeds the private optimum by 6.07 years. The overly short private rotation age calls for corrective instruments, either a yield tax or a timber subsidy. Both instruments will establish the socially optimal

Table 7.6 Corrective Pigouvian forest taxes to promote forest amenities for Scandinavian boreal commercial forests.

Baseline	Rotation age (yrs)	Land value (€/ha)		Harvest revenue (€/ha)	Amenity loss (€/ha)
		Social	Commercial		
Social	77.59	1,386	512.3	8,618	0
Private	71.52	1,367	530.3	7,422	36.3
Optimal taxation					
	Tax rate	Land value (€/ha)		Harvest revenue (€/ha)	Tax outlays (€/ha)
		Social	Commercial		
Yield tax	0.5905	1,386	−342.0	3,529	5,089
Timber tax	−0.0047	1,386	770.2	8,618	−4,555

rotation age but their other impacts differ. A very high yield tax rate is needed to restore the social optimum, making forestry rents negative, which is a poor incentive for private forest management. The required timber subsidy is smaller (the subsidy strongly impacts rotation age), and it increases commercial land value.

The problem is that using taxation to establish the socially optimal rotation age in boreal forests entails heavy taxation or subsidization. We therefore ask if better ways of promoting forest amenity benefits exist. This question is examined next.

Forest taxes and other instruments for targeted forest policies

The set of public goods and negative externalities associated with forestry is large. The variety of forest ecosystem services ranges from climate and water services to game, recreation and cultural services. Much has been written on carbon sequestration and biodiversity conservation; recently bioenergy and water protection issues have also received attention. Our work suggests combining specific environmental instruments to be used by forest taxes in policy packages to promote given environmental targets. We present this discussion on water pollution and biodiversity conservation and climate policy issues.[2]

Nutrient loads from harvesting

Stand clearcutting causes nutrient loads because bare land is impacted by rain and is reinforced by soil preparation during the plantation of the new stand. Nutrient loads first increase and then gradually decrease toward the background state with the growth of the new stand. The key means of preventing nutrient loads caused by harvesting are prolonged rotation age and retaining unharvested buffer zones to fix nutrients (for original analysis, see Miettinen et al., 2012). We examine to what extent, if any, forest taxation can promote the reduction of nutrient load damages in waterways.

Let x denote the number of years that anthropogenic loads take place, and let m denote the buffer zone as the share of the stand. Nutrient load after clearcut, $g(s, m)$, is a function of time, s, and the size of the buffer zone, m, with $g_m < 0$, and $g_{mm} > 0$. Suppose the buffer zone is left permanently unharvested, so that the nutrient load on the subsequent stand is $z = (1 - m) g(s, m)$. Nutrient load damage, $D(z)$, is expressed as a function of the present value of the periodic loads, z, as follows: $D(z) = D(\int_0^x (1-m)g(s,m)e^{-rs}ds)$. Incorporating damages to the Faustmann model yields the following social welfare function:

$$SW = \{(1 - m)[pf(T)e^{-rT} - c] - e^{-rT} D(z)\}(1 - e^{-rT})^{-1} \tag{19}$$

The socially optimal choice for the rotation age is characterized by

$$SW_T: (1 - m)(pf'(T) - rpf(T)) + rD(z) - rSW = 0 \tag{20a}$$

$$SW_m: -(pf(T)e^{-rT} - c) - e^{-rT}D'(z)\left[\int_0^x z_m(s,m)e^{-rs}ds\right] = 0 \tag{20b}$$

From (20a) we see that nutrient loads impact the socially optimal rotation age via the damage term, tending to lengthen optimal rotation age. Equation (20b) defines the buffer zone size at the point where the net harvest revenue lost from this land equals the social benefits received from nutrient load reductions achieved by the buffer zone.

Assuming that nutrient loads are ignored by the private landowner, society needs instruments to lengthen the private rotation age and establish buffer strips. Pigouvian policy entails punishing harvesting and subsidizing buffer zones. Using the same approach as previously, we levy a yield tax and a buffer strip subsidy in the private choice and solve for the instrument levels that establish the social optimum:

$$\tau^* = rD(z)[(1 - e^{-rT})(1 - m)]^{-1} \tag{21a}$$

$$s^* = e^{-rT}D'(z)\left[\int_0^x z_m(s,m)e^{-rs}ds\right] \tag{21b}$$

The optimal yield tax rate reflects the interest cost of runoff damage, and the buffer zone subsidy reflects the present value of the reduction in marginal nutrient damages by the buffer zone. Hence, a distortionary forest tax for lengthening rotation age is a natural part of water policy in forestry. Taxation is incapable of performing the job alone, however; another instrument, the buffer strip subsidy, is needed to complement it. The Tinbergen rule for tax policy is witnessed here: Society should use at least as many instruments as it has goals.

We deepen our analysis with numerical simulations. Let nutrient loads vary from 1 kg/ha/yr to 10 kg/ha/yr, and let marginal damage be a constant 10 euros/kg. Table 7.7 presents the socially optimal rotation age and instruments.

The socially optimal rotation age is always longer than the private rotation age (71.5 years), and increases with increasing loads. The social land value is smaller than the private land value (€530.3/ha), and damages strongly increase with nutrient load increases. The optimal tax rate increases rapidly with loads and becomes extremely high. Establishing buffer zones is expensive as mature trees are left standing. Thus, it does not pay to establish a buffer zone if loads are 1–3 kg/ha/yr (the normal case in flat mineral soils); lengthening the rotation age with a yield tax suffices. For higher loads, however, buffer zones become desirable and we end up with the policy package suggested by the theory. Using buffer zones to fix nutrient levels reduces damages and

Table 7.7 Controlling the nutrient load from Scandinavian boreal commercial forests: The role of forest taxation.

Nutrient load (kg/ha)	Rotation age (yrs)	Buffer zone (share)	Land value (€/ha)	Runoff damage (€/ha)	Optimal tax rate (%)	Buffer zone subsidy (€/ha)
Buffer strips not policy option						
1	72.4	No option	499.1	30.8	16.4	No option
3	74.1	No option	439.1	87.8	37.1	No option
5	75.7	No option	381.9	139.5	49.6	No option
10	79.7	No option	250.2	248.8	66.3	No option
Buffer strips as a part of policy						
1	72.4	0	499.1	30.8	16.4	0
3	74.1	0	439.1	87.8	37.1	0
5	75.4	0.012	389.1	128.9	47.3	57.70
10	78.5	0.051	273.9	218.3	62.5	168.6

allows the reduction of the yield tax rate. All in all, forest taxation helps to reduce nutrient loads. For smaller nutrient loads, forest taxes alone can do the job, but for larger loads it must be complemented by the buffer strip subsidy.

Biodiversity conservation

It is tempting to think that lengthening rotation age with forest taxes takes care of biodiversity conservation, as biodiversity is high in pristine and old-growth forests. We now investigate whether this is true or not. We include two means to promote biodiversity: prolonged rotation age and the artificial increase of decaying and dead wood using retention trees (Koskela, Ollikainen and Pukkala, 2007a, 2007b).

In the steady state at the beginning of each rotation period, the planner has a given number of retention trees left from the previous harvest, denoted by $\bar{G}$. The planner decides on the amount of retention trees to be left during the final harvest. Thus, biodiversity benefits can be expressed as follows:

$$BB = \int_0^T F(x)e^{-rx}dx + \int_0^T B(\bar{G},x)e^{-rx}dx + e^{-rT}\int_0^{\bar{T}} B(x,G)e^{-rx}dx \qquad (22)$$

The first term, $a(T) = \int_0^T F(x)e^{-rx}dx$, with $F'(T) > 0$, indicates the valuation of biodiversity benefits from some old-growth species. The second term describes the biodiversity benefits from retention trees inherited from the previous harvest. The third term describes future benefits from retention trees to be left at the end of the current rotation age and accruing during the next rotation period (a bar above T indicates that it will be chosen during the next rotation period).

Retention trees from the previous period, $\bar{G}$, decrease the growth of the stand to be established, $f(T, \bar{G})$ and $f(\bar{T}, G)$. Let the regeneration costs be c, timber price p and real interest rate r. The planner chooses rotation age T and retention tree volume G so as to maximize

$$SW = \left[pe^{-rT} \left[f(T,\bar{G}) - G \right] - c + \int_0^T F(x)e^{-rx}dx + \int_0^T B(\bar{G},x)e^{-rx}dx \right] (1 - e^{-rT})^{-1}$$

$$+ e^{-rT} \left[\int_0^{\bar{T}} B(x,G)e^{-rx}dx + pf(\bar{T},G)e^{-r\bar{T}} \right] (1 - e^{-rT})^{-1} \tag{23}$$

The first-order conditions can be expressed as

$$SW_G = -p + \int_0^{\bar{T}} B_G(x,G)e^{-rx}dx + pf_G e^{-r\bar{T}} = 0 \tag{24a}$$

$$SW_T = pf_T - rp \left[f(T,\bar{G}) - G \right] + F(T) + B(\bar{G},T)$$

$$-r \left[\int_0^{\bar{T}} B(x,G)e^{-rx}dx + pf(\bar{T},G)e^{-r\bar{T}} \right] - rSW = 0 \tag{24b}$$

From equation (24a), the optimal retention tree volume is chosen so as to equate the present value of the sum of the marginal utility of retention trees over their whole decaying process with the sum of the marginal loss of the harvest revenue, or the value of decreased future growth. According to (24b), the optimal rotation age is chosen so that the marginal return of delaying the harvest by one unit of time equals the opportunity cost of delaying the harvest.

A combination of a tax and subsidy is required for promoting prolonged rotation ages and green-tree retention in commercial forests. Using the same approach as before, we exemplify the instrument combination using green-tree subsidy and timber tax:

$$s'(G)^* = \int_0^{\bar{T}} B_G(x,G)e^{-rx}dx \tag{25a}$$

$$\alpha^* = -\frac{\left[(F(T) - rE) + (B(T,\bar{G}) - rH) \right]}{pf(T,\bar{G}) - rU} \tag{25b}$$

where $U = (1 - e^{-rT})^{-1} \int_0^T pf(s,\bar{G})e^{-rs}ds$ and $(pf(T,\bar{G}) - rU) > 0$. The marginal subsidy for reten-

tion trees depends on the marginal biodiversity benefits of G. By (25b), the optimal timber subsidy depends on the present value of the retention tree subsidy and reflects the ratio of the net marginal biodiversity benefits (over their opportunity costs terms) and of the timber subsidy effects on timber production.

Table 7.8 exemplifies the policy package drawing on a more complex description of forest growth dynamics in Koskela et al. (2007a) and on their results of a landowner who may or may not value amenity benefits. The socially optimal volume of retention trees is roughly 8 m³ and optimal tax and subsidy instruments are chosen to produce it.

The yield tax rate is rather high (close to 30%–40%), whereas the timber subsidy and site value tax remain rather low. The retention tree subsidy is expressed for 10 m³. Note that while the timber tax rate is the same irrespective of landowner's motives, forest taxes and retention tree subsidy rates are lower under Hartman behavior.

Table 7.8 Promoting biodiversity conservation in Scandinavian boreal commercial forests: The role of forest taxation.

Instrument combination	Private landowner's harvesting follows			
Retention tree subsidy(ies) with	Faustmann model		Hartman model	
Yield tax (τ)	s = €1,000	τ = 40%	s = €750	τ = 30%
Timber tax (α)	s = €1,900	α = −1%	s = €1,900	α = −0.5%
Site value tax (β)	–	–	s = €1,700	β = 0.1%

Climate policies: Carbon sequestration

Among other ways, forests contribute to climate mitigation by acting as carbon sinks and as a source of energy wood to replace fossil fuels and emissions from electricity production. The role of forests as a carbon sink is recognized in international climate negotiations. The basic analysis of forests as a carbon sink is made by van Kooten, Binkley and Delcourt (1995).

Let $f(t)$ denote the growth function of trees, and α the carbon tons sequestered per cubic meter of timber biomass. Sequestered carbon is given by the stand growth rate, $\alpha f'(t)$. The present value of carbon uptake benefits $F(T)$ over a rotation period T is given by $F(T) = q\alpha \int_0^T f'(s)e^{-rs}ds$, where q denotes the marginal climate benefit. Integrating it by parts yields

$$F(T) = q\alpha \left[f(T)e^{-rT} + r\int_0^T f(s)e^{-rs}ds \right] \tag{26}$$

Let β denote the fraction of carbon that remains in long-term storage after harvesting, e.g. structures and landfills. The share $(1 - \beta)$ denotes the amount of carbon released into the air due to harvesting. The social net present value of timber harvested under zero regeneration cost is

$$\hat{W} = \left\{ [p - q\alpha(1 - \beta)]f(T)e^{-rT} + q\alpha\left[f(T)e^{-rT} + r\int_0^T f(s)e^{-rs}ds \right] \right\}(1 - e^{-rT})^{-1} \tag{27}$$

Differentiating (27) with respect to rotation age and arranging produces

$$(p + q\alpha\beta)f'(T) + rq\alpha f(T) = r\left[(p + q\alpha\beta)f(T) + q\alpha r\int_0^T f(s)e^{-rs}ds \right](1 - e^{-rT})^{-1} \tag{28}$$

In the absence of economic instruments, the private landowner omits climate benefits from carbon sequestration. As van Kooten et al. (1995) demonstrate, a subsidy/tax instrument is needed to promote carbon sequestration. The landowner receives unit payment h for each unit of carbon sequestered and pays a tax h for each unit released to the air by harvesting: $V = \left\{ [p - h\alpha(1 - \beta)]f(T)e^{-rT} + h\alpha\left[f(T)e^{-rT} + r\int_0^T f(s)e^{-rs}ds \right] \right\}(1 - e^{-rT})^{-1}$. The optimal tax/ subsidy rate is

$$h^\star = q \tag{29}$$

Thus, the optimal tax/subsidy is equal to the marginal climate benefit. This Pigouvian instrument is akin to timber tax/subsidy, but levied to promote climate mitigation, not to raise tax revenues. Thus, we have a modified timber tax for climate policies.[3]

Conclusions

We examined forest taxation as a means of guiding private forest management. Taxes can be used to impact timber supply, the provision of amenity benefits and the prevention of negative externalities. And of course, taxes can be used to collect budget revenues for the government. The possibility of using forest taxes in targeted environmental policy packages provides a new and innovative initiative.

It is surprising how seldom forest taxation has been systematically analyzed using simulation. Throughout this chapter we assessed tax rates and the burden caused by taxation in Scandinavian boreal forestry. We found that forest taxation has a surprisingly strong impact on the profitability of forestry. High taxes are detrimental for forest management. Lower tax rates enable other means of improving forest management in the provision of public goods and the reduction of negative externalities.

Finally, although we know many of the mechanisms through which forest taxation impacts, several questions posed by the literature remain open. An especially unstudied problem is how soil productivity differences and landowner-specific features impact landowners' reactions to forest taxes. The heterogeneity of forest plots and forest landowners' amenity valuation challenge simple forest tax designs. Thus, there is much theoretical and empirical work to be done.

Notes

1   This chapter omits many other forestry frameworks. To mention a few, Chang (1982) examines impacts of forest taxes on the rotation age and management effort. Mendelsohn (1993) and Koskela and Ollikainen (2003b) examine progressive forest taxation in rotation framework. Kovenock and Rothschild (1983), Koskela and Ollikainen (1997a) and Uusivuori (2000) examine forest taxation under the 'Austrian' sector problem. Taxation under pre-existing taxes is studied by Ovaskainen (1992). Barua, Kuuluvainen, Laturi and Uusivuori (2010) extend the tax analysis to thinning and lifetime uncertainty in the two-period framework. Alvarez and Koskela (2007) focus on forest taxation in the stochastic Faustmann framework, and Uusivuori and Kuuluvainen (2008) examine forest taxation in an age-class model.
2   Tropical forest concession policies are based on instrument combinations: the size of the concession, the royalty rate charged from harvesters and enforcement activities to reduce illegal logging and bribery (see Amacher, Koskela and Ollikainen, 2008, 2012; Barua, Uusivuori and Kuuluvainen, 2012).
3   Literature on taxing and subsidizing carbon sequestration is voluminous. An early treatment is by Tahvonen (1995). Pohjola and Valsta (2007) focus on thinning and Uusivuori and Laturi (2007) on management intensity in an age-class model.

References

Alvarez, L. and Koskela, E. (2007). 'Taxation and rotation age under stochastic forest stand value', *Journal of Environmental Economics and Management*, vol. 51, pp. 113–127.

Amacher, G. S., Koskela, E. and Ollikainen, M. (2007). 'Royalty reform and illegal reporting of harvest volumes under alternative penalty schemes', *Environmental and Resource Economics*, vol. 38, pp. 189–211.

Amacher, G. S., Koskela, E. and Ollikainen, M. (2008). 'Deforestation and land use under insecure property rights', *Environment and Development Economics*, vol. 14, pp. 281–303.

Amacher, G. S., Koskela, E. and Ollikainen, M. (2009). *Economics of Forest Resources*, MIT Press, Cambridge, MA.

Amacher, G. S., Koskela, E. and Ollikainen, M. (2012). 'Corruption and forest concessions', *Journal of Environmental Economics and Management*, vol. 63, pp. 92–104.

Barua, S., Kuuluvainen J., Laturi J. and Uusivuori, J. (2010). Effects of forest taxation and amenity preferences on nonindustrial private forest owners', *European Journal of Forest Research*, vol. 121, pp. 163–172.

Barua, S., Uusivuori, J. and Kuuluvainen J. (2012). 'Impacts of carbon-based policy instruments and taxes on tropical deforestation', *Ecological Economics*, vol. 73, pp. 211–219.

Chang, S. (1982). 'An economic analysis of forest taxation's impacts on optimal rotation age', *Land Economics*, vol. 58, pp. 310–323.

Chang, S. (1983). 'Rotation age, management intensity, and the economic factors of timber production: Do changes in stumpage price, interest rate, regeneration cost and forest taxation matter?' *Forest Science,* vol. 29, pp. 267–277.

Conrad, J. (2000). *Resource Economics*, Cambridge University Press, Cambridge, MA.

Conrad, J. (2010) Resource Economics, Second Edition, Cambridge University Press, Cambrigde MA.

Englin, J. E. and Klan, M. K. (1990). 'Optimal taxation: Timber and externalities', *Journal of Environmental Economics and Management*, vol. 18, pp. 263–275.

Gamponia, V. and Mendelsohn, R. (1987). 'The economic efficiency of forest taxes', *Forest Science*, vol. 33, pp. 367–378.

Hartman, R. (1976). 'The harvesting decision when a standing forest has value', *Economic Inquiry*, vol. 14, pp. 52–58.

Hellsten, M. (1988). 'Socially optimal forestry', *Journal of Environmental Economics and Management*, vol. 15, pp. 387–394.

Johansson, P.-O. and Löfgren, K.-G. (1985). *The Economics of Forestry and Natural Resources*, Basil Blackwell, Oxford, UK.

Koskela, E. and Ollikainen, M. (1997a). 'Optimal design of forest and capital income taxation in an economy with an Austrian sector', *Journal of Forest Economics*, vol. 3, pp. 107–132.

Koskela, E. and Ollikainen, M. (1997b). 'Optimal design of forest taxation with multiple-use characteristics of forest stands', *Environmental and Resource Economics*, vol. 10, pp. 41–62.

Koskela, E. and Ollikainen, M. (2001). 'Forest taxation and rotation age under private amenity valuation: New results', *Journal of Environmental Economics and Management*, vol. 42, pp. 374–384.

Koskela, E. and Ollikainen, M. (2003a). 'Optimal forest taxation under private and social amenity valuation', *Forest Science*, vol. 49, pp. 596–607.

Koskela, E. and Ollikainen, M. (2003b). 'A behavioural and welfare analysis of progressive forest taxation', *Canadian Journal of Forest Research*, vol. 33, pp. 2352–2361.

Koskela, E., Ollikainen, M. and Pukkala, T. (2007a). 'Biodiversity policies in commercial boreal forests: Optimal design of subsidy and tax combinations', *Forest Economics and Policy*, vol. 9, pp. 982–995.

Koskela, E., Ollikainen, M. and Pukkala, T. (2007b). 'Biodiversity conservation in boreal forests: Optimal rotation age and volume of retention trees', *Forest Science*, vol. 53, pp. 443–452.

Kovenock, D. and Rothschild, M. (1983). 'Capital gains taxation in an economy with an Austrian sector', *Journal of Public Economics*, vol. 21, pp. 215–256.

Mendelsohn, R. (1993). 'Nonlinear forest taxes. A note', *Journal of Environmental Economics and Management*, vol. 24, pp. 296–299.

Miettinen, J., Ollikainen, M., Finer, L., Koivusalo, H., Lauren A. and Valsta, L. (2012). 'Diffuse load management with biodiversity co-benefits: The optimal rotation age and buffer zone size', *Forest Science*, vol. 58, pp. 342–357.

Ovaskainen, V. 1992. Forest taxation: timber supply and economic efficiency. Acta Forestalia Fennica No. 233.

Pohjola, J. and Valsta, L. (2007). 'Carbon credits and management of Scots pine and Norway spruce stands in Finland', *Forest Policy and Economics*, vol. 9, pp. 789–798.

Ramsey, F. (1927). 'A contribution to the theory of taxation', *Economic Journal*, vol. 37, pp. 47–61.

Swallow, S., Park, S. and Wear, D. (1990). 'Policy-relevant nonconvexities in the production of multiple forest benefits', *Journal of Environmental Economics and Management*, vol. 19, pp. 264–280.

Tahvonen, O. (1995). 'Net national emissions, CO_2 taxation and the role of forestry', *Resource and Energy Economics*, vol. 17, pp. 307–315.

Uusivuori, J. (2000). 'Neutrality of forestry income taxation and inheritable tax exemptions for timber capital', *Forest Science*, vol. 46, pp. 219–228.

Uusivuori, J. and Kuuluvainen, J. (2008). 'Forest taxation in multiple-stand forestry with amenity preferences', *Canadian Journal of Forest Research*, vol. 38, pp. 806–820.

Uusivuori, J. and Laturi, J. (2007). 'Carbon rentals and silvicultural subsidies for private forests as climate policy instruments', *Canadian Journal of Forest Research*, vol. 37, pp. 2541–2551.

van Kooten, G. C., Binkley, C. and Delcourt, G. (1995). 'Effect of carbon taxes and subsidies on optimal forest rotation age and supply of carbon services', *American Journal of Agricultural Economics*, vol. 77, pp. 365–374.

8

INTERNATIONAL TRADE IN FOREST RESOURCES

John Perez-Garcia and Alicia Robbins

SCHOOL OF ENVIRONMENTAL AND FOREST SCIENCES, UNIVERSITY OF WASHINGTON,
BOX 352100, SEATTLE, WA 98915-2100, USA. PERJOHM@U.WASHINGTON.EDU;
ASTR@U.WASHINGTON.EDU

Abstract

Trade in forest products occurs at every level of the forest products supply chain. Imports of wood products allow an economy to consume more overall goods and services than in the absence of trade, while exports pay for purchases made abroad. Globalization occurs as barriers to trade are reduced and comparative advantages in production process across regions are realized. In 2010, global trade in forest products involved 170 countries, valued at US$224 billion. Even though there is evidence that the intensity of wood use in economies has declined over the past decade, world trade in logs, lumber and plywood has continued to grow, peaking in 2005. Forest products trading patterns change rapidly among countries in response to changing political and economic conditions and require analytical models to assess how the forest sector reacts to these changes. Trade models capture these changes in trading behavior using the principle of comparative advantage. There has been significant activity by forest trade modelers to measure trade policy impacts using equilibrium models over the past four decades. A variety of models are used to study political, economic and environmental policies affecting the forest sector.

Keywords

Forest sector, empirical trade models, comparative advantage, trade policy, tariffs, countervailing duties, spatial equilibrium, trade assessments, globalization, terms of trade

Global nature of wood products trade

Why does trade in forest products occur?

The globalization of forest products markets began in the 1960s and continues to expand today. Globalization occurs as barriers to trade are reduced and comparative advantages in production processes across regions are realized. Along with other commodities, trade in forest products can be distorted by both tariff and nontariff barriers, and international agencies along with national counterparts, industrial partners and nongovernmental organizations strive to reduce trade barriers and further the process of globalization.

An early example of forest products trade was the result of a strong windstorm in the US Pacific Northwest (PNW) that leveled large areas of forests in 1962. The storm literally created an excess supply of logs in the market overnight and led to an estimated 15 billion board feet of downed trees that could not be absorbed by the regional milling capacity. As a result, the sale of downed timber to mills in Japan was negotiated, thus opening a trading activity that continues today. Today, trade in forest products occurs at every level of the forest products supply chain. In 2010, global trade in forest products involved 170 countries, valued at US$224 billion (FAO, 2012).

Trade activity occurs as a response to meet consumptive needs in end-using sectors. The principle of comparative advantage is the underlying basis for trade activity. Imports of wood products allow an economy to consume more overall goods and services than in the absence of trade, while exports pay for purchases made abroad. What a region imports and exports depends on the region's comparative advantage. Comparative advantage exists because production conditions around the world differ. These differences in production conditions lead to one region being more efficient in the production of one good relative to another, with the potential to specialize in the production of that good, while importing others. In the forest sector, comparative advantage exists at every level of the supply chain from log production to finished products manufacturing.

A description of the global wood products trade begins with an examination of global wood consumption, the driver of trade activity. Following this overview, there is a brief description of global trade trends. An example of changes in trade patterns and their speculative causes is then presented for softwood logs and lumber, followed by a description of trade models and databases used to assess trade policies. A discussion of selected trade policy is presented next. The chapter concludes with remarks regarding some foreseeable challenges for trade sector research needs.

Global wood fiber consumption trends

Gross domestic product data is used to relate global wood fiber consumption levels. A change in consumption occurs with economic activity, e.g. when the global economy enters a period of recession and/or supply of forest products restrictions occur, or vice versa, when the global economy grows and/or when new areas of forest production enter the global supply stream. Figure 8.1 presents global consumption data for industrial roundwood against global gross domestic product data as a time series from 1970 to 2010. Apparent industrial roundwood consumption is calculated using Food and Agricultural Organization of the United Nations (FAO) statistics (FAO, 2012) on global production, import and export levels. Gross domestic product (GDP) data (World Bank, 2012) is used as a measure of income to illustrate how consumption patterns follow periods of either global economic recession or growth. The expectations are that as economies grow, so does demand for industrial roundwood, and when there is a contraction in the global economy, consumption of industrial roundwood falls globally.

Four points are evident in Figure 8.1. First, downturns in the global economy correspond to economic recessions beginning in 1973, 1979, 2000 and 2007. Although different factors are behind each of the recessions cited, there is little growth, no growth or a contraction in GDP beginning that year with a corresponding decline in consumption levels soon afterward. There are also two periods of substantial change in consumption associated with reductions in the amount of global wood fiber availability. The break beginning in 1990 is likely due to the collapse of the former Soviet Union, resulting in a sharp curtailment in the global availability of

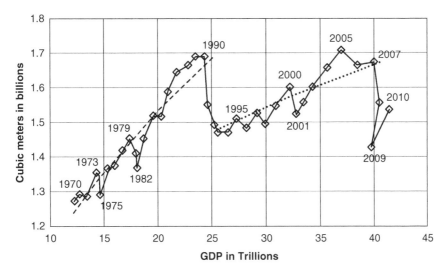

Figure 8.1 Consumption of industrial roundwood versus gross domestic product.
Data sources: FAO (2012) and World Bank (2012).

softwood fiber, and hence consumption, as well as restrictions in the level of harvests in the PNW due to protectionist efforts for the spotted owl (Sedjo, 1993; Perez-Garcia, 1995). Another observation is that the growth rate in consumption measured by income has changed. Prior to the collapse observed during the period 1990–1993, consumption of industrial roundwood measured as a function of income from 1970 to 1990 grew 1.2% per year. The growth trend is lower for the period 1992–2007. There are several plausible reasons for the reduction in wood use growth rate. One reason is that as a society we have lowered our intensity of use of roundwood as an industrial raw material by becoming more efficient in using it and by finding substitute materials that function in the same manner in respective end-use markets. Our changes in preferences can be both supply driven (i.e. it becomes more costly to produce, and manufacturers react to higher costs by reducing its use) and demand driven – our tastes for these raw materials are changing, reducing our desire to continue using them, i.e. we prefer plastic furniture. Because changes in economic activity occur disproportionately across the globe, relative prices change, so does comparative advantage in trading. A region may no longer be relatively more efficient in growing trees or milling roundwood than in providing services for the IT sector, so trade activity in forest products falls while trade in the service sector grows. A final observation regards the reduction in wood fiber consumption caused by the downturn in US housing demand starting in 2007. Because consumption drives trade behavior, the level of trade activity has also declined with the most recent recession.

Trends in global forest trade

World trade in logs, lumber and plywood has increased dramatically since the 1960s, reaching a peak in 2005 (Figure 8.2). Lumber trade fairly closely follows log trade, with some periods in which the potential substitution between these two products and their trade flows occurs. Trade in coniferous products has expanded more rapidly than in nonconiferous products since the early 1990s, driven mainly by the growth in consumption in the United States, Europe and

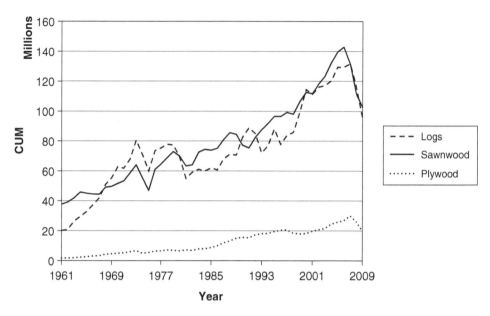

Figure 8.2 Global exports of logs sawnwood and plywood in cubic meters (CUM).
Data source: FAO (2012).

Japan. Although Europe would appear to be the largest force in the trade of many products, numbers are confounded by the fact that many exports and imports are to and from countries within the continent. In 2009, Russia was the largest exporter of coniferous logs, although exports have significantly decreased since the introduction of the log export tariff in April 2008, declining in 2009 by 55% over 2007. The tax, intended to spur the development of a domestic processing industry, has resulted in some increased lumber exports and only negligible plywood exports. China is the largest importer of both coniferous and nonconiferous logs, with imports having more than doubled since 2001, a result of both an expanding domestic process-ing industry and domestic logging restrictions imposed after 1998. Southeast Asia continues to be the largest source of nonconiferous logs, although total exports from the region are not much higher than those from West Africa or Russia. Canada has been the largest exporter of conifer-ous sawnwood, although its position has declined since 2007, while the United States remains the largest importer.

Trade activity response to changes in relative prices

Trade agents of forest products are responsive to changes in regional economic conditions that are captured by relative price measures. As an example, consider the changes observed in trade flows over the past two decades in softwood logs and lumber. Figure 8.3 contains trade flow patterns for four periods from pre-1990 to post-2008, beginning with the top-left panel and moving clockwise. Prior to the 1990s, softwood logs were a major export activity in the PNW. Major markets for US softwood logs were in Asia, including Japan, Korea and China. The early 1990s saw the introduction of legislation that curtailed harvest activity on public and private lands in the US PNW timber-supplying regions, raising absolute costs to US PNW log importers and US PNW domestic mill processors and costs relative to other regions and other

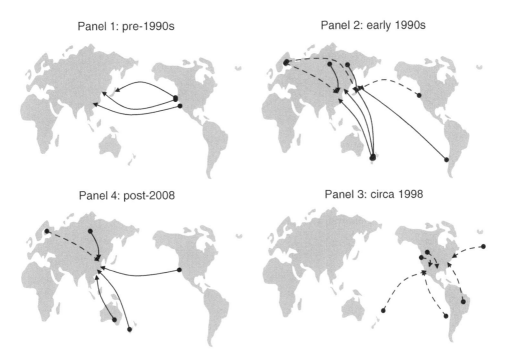

Figure 8.3 Changing trade patterns during the past three decades.

products. As a consequence, trade activity began to shift and replace US PNW log trade, not only by more processed wood products but by other nonforest products as well. The shift in trade activity included substitute softwood log exports from emerging forest sectors like Chile and New Zealand, increased log exports from Russia and increased softwood lumber exports from Canada and Scandinavia, all replacing US PNW log imports into Asia (Figure 8.3, Panel 2). The responses from producers were an indication of how the lack of trade barriers allowed producers in other regions to respond rapidly to changes in relative prices across the globe. However, these new flows were interrupted after the Asian economy faltered in 1998. Demand in Asia declined, while growth in housing in the United States accelerated (Figure 8.3, Panel 3). Trade activity again shifted in response to changes in consumption patterns and relative prices. Panel 3 indicates new lumber exports from New Zealand, Chile, Brazil and Europe entering the US market with existing Canadian exports increasing in response. The collapse of the housing sector in the United States, beginning in 2007, changed the relative prices of softwood forest products again. Panel 4 shows how logs, rather than lumber, in response to China's growing forest products manufacturing sector, were being exported from the US PNW region, New Zealand and Australia, as well as lumber from Scandinavia, while exports from Russia continued. Such rapid changes observed in the trade patterns of forest products require analytical models that assess how the forest sector reacts to changes in policies and economic conditions.

Trade assessments

The underlying economic theory behind forest products trade assessments is the concept of a spatial equilibrium. The derivation of spatial equilibria and the basis for spatial equilibrium

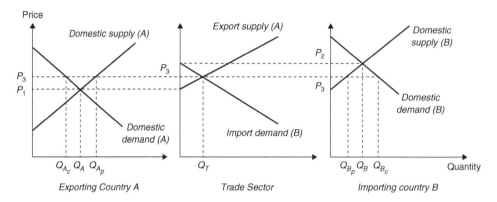

Figure 8.4 Spatial equilibrium with bilateral trade.

models is found in Samuelson's concept of the net social payoff realized through the trade of a single good (Samuelson, 1952). Figure 8.4 illustrates spatial equilibrium using a single bilateral trade flow. The diagram demonstrates the basic idea behind market equilibrium in the presence of trade activity. Suppose the world is made up of two countries, A and B. Both countries produce lumber. Without trade, the price of lumber in country A is P_1, and the price of lumber in country B is P_2. Because P_2 is higher than P_1, as trade between the two is introduced, P_1 rises to P_3 and P_2 falls to P_3. Country A produces more than its domestic market demand and begins exporting lumber to country B, and the equilibrium price P_3 emerges in both countries. At P_3 export supply meets import demand. Although Figure 8.4 depicts only a single bilateral trade flow with a single product, it contains the essential components needed to analyze trade policies and changes in economic conditions impacting the forest sector.

For instance, consider an increase in the transfer cost of getting lumber from country A to country B. Note that in Figure 8.4, the transfer cost of lumber from country A to country B is assumed to be zero. In reality, there are costs associated with insurance and freight that are added to the price of lumber in exporting country A. Or similarly, an export tariff by a government agency can be imposed on lumber exporters as a policy to reduce the volume of lumber leaving country A. Both act in the same fashion in the market model described and are easily incorporated in the diagram (see Figure 8.5). Either action will introduce a wedge (TC) between the trade equilibrium price observed in countries A and B. As a consequence, trade volume Q_T declines, and if costs or the tariff were high enough, trade activity would cease altogether.

In more realistic situations both log and lumber markets are examined simultaneously to capture feedback effects. Figure 8.6 illustrates both log and lumber trading agents and is used to trace the effects of higher costs associated with log production. These higher costs may be attributed to greater restrictions on harvesting activities due to environmental constraints or any other policy that raises costs to log producers in the exporting country. We use the example to illustrate the effects on product (lumber) and factor (log) markets known as the feedback effect (Wiseman and Sedjo, 1981). Let's say log producers in country A are faced with a permanent increase in logging costs. The cost increase is represented as a shift upward of the cost curve. It is labeled #1 in Figure 8.6 and depicted with a dashed line. Exporters also face the higher logging cost that leads to a shift upward in the export supply curve (#2) and less log volume traded (#3). As a consequence, the equilibrating log market price P_3 rises (#4) in both countries. The price increase in importing

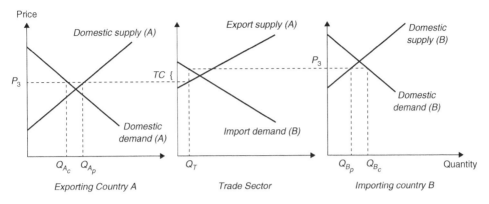

Figure 8.5 Spatial equilibrium in the presence of transfer costs.

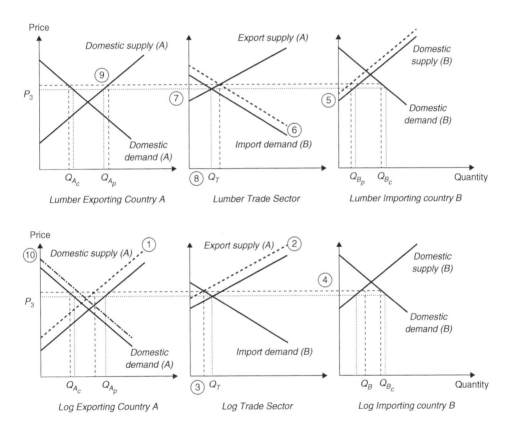

Figure 8.6 Spatial equilibrium for two sectors outlining the feedback effect.

country *B* (#4) leads to higher production costs for lumber producers in importing country *B* (#5). As a consequence, import demand for lumber from country *A* increases (#6). The chang-ing market conditions in the lumber trading sector result in higher prices for lumber in both countries (#7) and an increase in the quantity traded (#8). The quantity of lumber produced in the lumber exporting country *A* increases (#9), and as a consequence, the demand for logs shifts upward in log exporting country *A* market (#10). This triggers a new set of changes in the log markets, which, for the sake of clarity, are not shown here, but would continue to lead to market price changes until a stable equilibrium price is established in both log and lumber markets. One should note that the strength of this feedback effect depends on the slopes (elasticities) of supply and demand in both countries and the comparative advantages other regions have in log and lumber production around the globe.

One can achieve the same feedback effect with a policy targeting log exporters rather than all log producers. This exercise is left for the reader to complete, but the effect of a reduction in the volume of log exports in the exporting country is an increase in the demand for logs by its lumber mills by way of an increase in the demand for lumber exports. Proponents of log export bans use this argument to claim an increase in value-added processing with benefits to additional mill employees who would be hired. But this hiring occurs at the expense of port workers who would lose their jobs (Perez-Garcia, Lippke and Baker, 1997). Other environmental or policy changes that impact the forest sector can be modeled in this fashion. For instance, climate change impacts on productivity of timber supplying regions have been modeled by shifting the log supply functions in response to changes in productivity and observing the impacts across markets and regions around the globe (Adams, Alig, Callaway, McCarl and Winnett, 1996; Alig, Adams and McCarl, 2002; Kallio, Moiseyev and Solberg, 2006; Perez-Garcia, Joyce, Binkley and McGuire, 1997).

One final important component of trade policy regards exchange rates. In all the previous examples, the currency in exporting country *A* was on par with the currency in importing country *B*. Let's say the currency in importing country *B* strengthens relative to the currency in exporting country *A* via some form of government intervention. Such a situation is depicted in Figure 8.7 where the price axis for importing country *B* has shifted upward to represent the changes in values associated with the currency in country *B* relative to the currency in country *A*. This change is labeled as a change in the terms of trade ($\Delta T_{of} T$) in Figure 8.7. As a consequence, import demand for country *B* shifts upward as imports from country *A* become cheaper relative to country *B*.

Exchange rate intervention policies can lead to greater export levels (Q_T in Figure 8.7 increases) when currencies are manipulated differently through log and lumber markets. As an example, consider the following situation in which US furniture manufacturers have complained about Chinese counterparts manipulating their exchange rate in their favor. Consider China as a log importing country that allows its currency to appreciate relative to log exporting countries. This

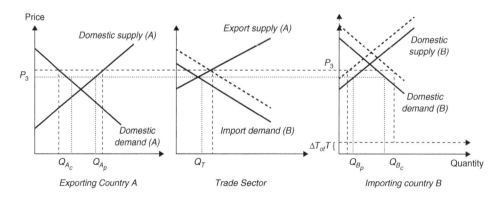

Figure 8.7 Spatial equilibrium with a change in the exchange rate.

position leads to a greater level of log imports. Also consider that products manufactured in China with log imports are then exported to the United States, and that the Chinese currency is held on par with the US currency. By allowing its currency to be tied directly to US dollar movements, while allowing the Chinese currency to float with log exporting country currencies, the policy causes greater volumes of logs to be imported by China than would otherwise have occurred. More log imports lower milling costs and allow Chinese finished product producers to expand their exports. That is, maintaining a country's currency on par through government intervention with a destination country's currency, rather than allowing it to appreciate in an open exchange market, as it does with log exporters, leads to lower cost through greater quantity of log available to the manufacturer of exported finished products. The intervention in exchange markets is considered a subsidy to Chinese exporters by US mill owners. It constitutes the source of a trade dispute based on a nontariff barrier due to exchange rate manipulations. Trade models can calculate the effect of the subsidy by measuring the terms of trade in exporting and importing countries in log and lumber sectors and measuring the differences observed in the trade volumes using counterfactual scenario analysis.

Empirical trade models

There has been a significant level of activity by forest trade modelers to measure trade policy impacts using equilibrium models over the past four decades. International trade models in forest products are an outgrowth of activity analysis techniques developed in the late 1970s and applied to the forest sector. The first global forest sector model developed was the International Institute for Applied Systems Analysis (IIASA) Global Trade Model (GTM) (Kallio, Dystra and Binkley, 1987). The IIASA project involved researchers from across the globe and developed a spatial equilibrium trade model of the type described previously. Prior to its development, market assessment models such as 'Papyrus' (Gilless and Buongiorno, 1987) and the Timber Assessment Market Model (TAMM) (Adams and Haynes, 1980) were used to analyze the US forest sector.

Most forest products trade models can be categorized as (1) spatial equilibrium/static simulation, (2) dynamic optimization/optimal control framework or (3) computable general equilibrium (CGE) models. These include the TAMM, the Center for International Trade in Forest Products (CINTRAFOR) Global Trade Model (CGTM) (Cardellichio, Youn, Adams, Joo and Chmelik, 1989), the European Forest Institute Global Trade Model (EFI-GTM) and the Global Forest Product Model (GFPM) (Buongiorno, Zhu, Zhang, Turner and Tomberlin, 2003). Both the CGTM and EFI-GTM are considered second- and third-generation extensions of the IIASA GTM. The Timber Supply Model (TSM) (Sedjo and Lyon, 1990) is an example of the dynamic optimization/optimal control model. The Global Trade Assessment Project (GTAP) (Hertel, 1997) is a computable general equilibrium model containing a forestry and forest products sector.

Static simulations differ from dynamic optimization methods in their treatment of model parameters over time. The static models update parameters period to period using updating functions for capital, industrial raw material inputs from the forest, and labor. Sector dynamics are independently achieved period by period in the static model. Dynamic optimization considers the effects conditions in future periods have on the present period. Current parameter values are a function of future conditions. Hence, the level of forest inventory in the future, for example, plays a role in determining the equilibrating price level today. In this way, the dynamic solution is optimized as in TSM, for example. These two approaches differ from CGE methods in that (1) they consider the forest sector only, while CGE models are representation of all sectors in the economy; and (2) CGE models are constrained to represent the structural economy for a specific period in time, i.e. underlying parameters are maintained throughout any projection of policy over time.

EFI-GTM focuses on Europe but contains a global perspective. It has been applied to analyze environmental policy such as the impact of increased forest conservation on European consumption of forest products (Kallio et al., 2006) as well as the impact of European measures aimed at curbing illegal logging (Moiseyev, Solberg, Michie and Kallio, 2010). CGTM has been applied to a number of global forest sector issues including the impacts of trade restrictions and climate change (Perez-Garcia, Joyce, Binkley and McGuire, 1997; Perez-Garcia, Lippke and Baker, 1997; Perez-Garcia, Joyce, McGuire and Xiao, 2002). TAMM was developed to support the Resource Planning Act (RPA) timber assessments conducted by the US Department of Agriculture Forest Service (Adams and Haynes, 1996) as well as many other forest sector issues. GFPM has its origins in the same roots as the IIASA GTM but evolved separately to incorporate a dynamic phase (Buongiorno et al., 2003). As with all of the previous models, GFPM has been applied to a number of topical areas including the effect of the imposition of a log export tariff (Turner, Buongiorno, Katz and Zhu, 2008) and the elimination of illegal logging (Li, Buongiorno, Turner, Zhu and Prestemon, 2008). More recently, GFPM was used to update the timber assessment in the United States (Buongiorno, Zhu, Raunikar and Prestemon, 2012).

TSM is the most widely referenced dynamic optimization model. TSM was developed to study timber supply from plantation regions across the globe, and their linkages to supply from natural forests to address the adequacy of long-term world timber supply (Sedjo and Lyon, 1990). In addition to the analysis of long-term timber supply, the model has been used to study the impacts of climate change (Sohngen, Mendelsohn and Sedjo, 2001).

CGE models that contain a forest sector are an outgrowth of advances made by World Bank economists in the development and application of CGE models for countries with large forest sectors. An important extension to this work is the GTAP. GTAP is a trade model designed more generally to project trade in all sectors but has been used to estimate the effects of non-tariff barriers to trade in forest products on economic growth, welfare and trade (Hertel, 1997).

Data on international trade in forest products

International public agencies

The development of global forest sector models capable of analyzing trade policies requires a comprehensive global database. Two global forestry data sets collected by the FAO and the International Tropical Timber Organization (ITTO) utilize the same standard survey instrument across all countries. The *FAO Yearbook of Forest Products* is the longest published source of information on the forest sector, publishing its 64th issue for 2010 activity in 2011. The data are gathered using a joint forest products questionnaire supported by the Forestry Department of FAO, the Economic Commission for Europe (ECE), the Statistical Office of the European Communities (EUROSTAT) and the ITTO. In cases where official statistics were not available, FAO produces an estimate or uses data from nonofficial sources.

The ITTO also produces an annual review and assessment of the world timber situation, focusing on tropical timber. Statistics for these publications are derived from members' responses to the Joint Forest Sector Questionnaire wherever possible. The ITTO is responsible for sending the questionnaire to all of its producer members and Japan, while responses from other consumer members are forwarded from partner agencies such as the ECE, EUROSTAT and FAO.

Discrepancies exist between available data sources in many categories even though there is a substantial effort by ITTO and FAO to ensure data consistency and accuracy. Each agency compiles data for presentation in their respective yearbooks after analysis and synthesis are completed by their respective staff, and consultations with member countries and other agencies.

The GTAP model uses collaborative efforts by individual countries to create a global database that is consistent with the CGE method and models employed by GTAP. Participation by countries is high and has led to a significant advancement in creating a global database with some forest sector details.

Private enterprises

In addition to public sources of data, commercial organizations collect trade data for forest products. The Global Trade Atlas® from Global Trade Information Services and Global Trade Information Services Inc. (GTI) are two commercially available data services. In addition to global datasets that can be used to calibrate a trade model, many countries maintain trade statistics that allow the construction of regional models. Many national, state and other data sources exist and can be found easily on the Internet.

Selected policy analysis

US-Canadian lumber dispute

One of the longest-standing trade disputes is between the United States and Canada concerning softwood lumber. The dispute began in 1982 when the US lumber industry formally complained that low Canadian stumpage rates, which were set by the government, constituted an unfair advantage for Canadian mill processors, who exported their lumber to the United States. US lumber producers claimed that stumpage rates were being manipulated by (1) the lack of market mechanisms to establish them and (2) a log export ban that maintained logs within the provinces for domestic processing. However, the US Commerce Department turned the claim down. As Canadian manufacturers captured an increasing share of the rebounding US market at a time when many mills in the United States had shut down as a result of a previous economic recession, a second claim was made in 1986, this time successfully. The US Commerce Department imposed a 15% duty on Canadian lumber but allowed the amount of alleged subsidy to be kept in Canada, becoming the largest self-imposed penalty in the history of world trade.

Eventually, the provinces of British Columbia and Québec declared that stumpage prices had increased sufficiently, and in 1991, Canada announced it was withdrawing from the memorandum, setting off another round of dispute. An agreement was not reached until 1996, this time using a quota system with export taxes becoming effective after quotas were exceeded. But when the agreement expired in 2001, tensions between mills in Canada and the United States resumed. In 2001 the US Department of Commerce and the International Trade Commission made preliminary determinations that softwood lumber exports were subsidized and these programs posed a threat of injury to the US industry. In 2006 the Canadian-US Softwood Lumber Agreement (SLA) was implemented. Again, the agreement used an export tax or export tax plus volume quota as the basis to control Canadian lumber trade.

At least three of the models described previously, TAMM, CGTM and GTAP, have been used to study the dispute, and many more specific models have been developed to measure the size of the subsidy and injury to mill owners (Adams, 2003; Boyd and Krutilla, 1987; DeRosa, Hufbauer and Perez-Garcia, 2000; van Kooten, 2002). Figure 8.8 characterizes the 1996 SLA using export supply and import demand functions and the spatial equilibrium approach described earlier. The baseline export supply and import demand functions are depicted in the top left panel. Moving

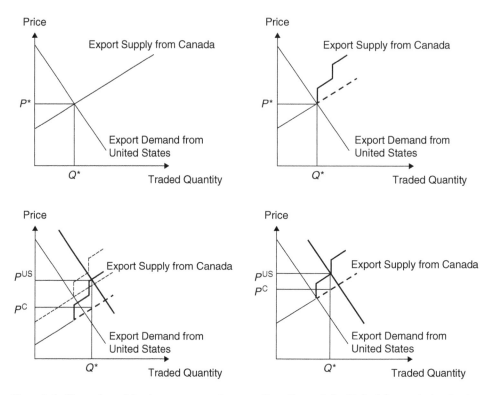

Figure 8.8 The softwood lumber agreement between Canadian and the United States, depicted using export supply and import demand curves and spatial equilibrium.

clockwise, the imposition of a combined tariff and quota system on export supply from Canada is represented by the step function in Panel 2. With no growth in the demand in the United States, i.e. import demand is within the volume of baseline trade condition, the amount of lumber exported from Canada is less than the quota and no tariff is assessed. Lumber prices in the United States and Canada are the same (with exception of transportation costs, which are not shown in the figure). During the 1990s, growth in the US housing sector expanded, increasing the import demand and allowing Canadian exports of lumber to the United States to expand, but only with an added tariff. The tariff value is the difference between P^{US} and P^{C}. Following the collapse of the Asian economies in 1998, the decline in the demand for Canadian exports in Asia led to a downward shift in the export supply function between the United States and Canada (lower left panel in Figure 8.7). This downward shift in the graphical representation of bilateral trade flows represents the reduction in demand recorded in the domestic market due to lower exported volumes to Asian consumers, and consequently, the downward shift in export supply to US markets. Excess lumber production after the collapse in demand in Asia could not be placed in the United States without paying a higher tariff. As a consequence, the spread between US and Canadian prices increased and the price of lumber in Canada declined further. The loss of the Asian markets for Canadian lumber and the quota on Canadian exports to the United States led mill owners in Canada to reject any extension of the 1996 agreement.

From the previous discussion, one notes that there are winners and losers when changes in trade volumes occur. US mills lose profits when Canadians mills increase lumber trade volumes. US consumers enjoy lower prices and Canadian mills garner higher profits with greater trade volumes. Earlier disputes involved Canadian and US mills as stakeholders interested in lumber trade. More recently, US consumers, represented by big box retailers and home associations, have entered the fray (Cashore, 1998; Brink, Groombridge and Lougani, 2000; Zhang, 2007). Left out to date has been the Canadian consumer. Their small size relative to the US consumer is likely to inhibit any real interest in these disputes. Of more interest has been the perspective presented by environmental groups during the last dispute. Environmental goals of preserving forests by reducing harvest levels in Canadian provincial forests can be achieved through maintaining trade restrictions on Canadian lumber.

Why has the competitive market outcome been so elusive? One factor is ownership, or more importantly, the public policies reflecting ownership views. In Canada, the vast majority of forestlands are provincially owned. Timber harvesting rights are leased to private firms with stumpage fees set by the government. Consideration is given to various costs associated with harvesting timber, but this system of price setting led to low prices and was the basis for the softwood lumber dispute. In the United States, federal timber, an important source for mills in the West during the first dispute, was offered through a competitive auction. The role of federal timber in supplying the needs of mills in the West has declined dramatically since the 1990s, however. The majority of logs harvested in the United States today do not use any administratively set pricing mechanism. The competitive nature of the stumpage market ensures that lumber markets are not distorted by timber price-setting policies. Incorporating competitiveness into Canadian stumpage markets would help alleviate the lumber trade disputes. Interestingly, one result of the continued bantering between the two countries has been the improvement in the cost structure of the mills in Canada, making them more competitive (Nagubadi, Thompson and Zhang, 2009).

Accelerated tariff liberalization policy

Notwithstanding the use of quotas and tariffs to manage the impacts of forest sector policies on the volume and value of trade between the United States and Canada, there exist ongoing attempts to liberalize trade by lowering tariffs on forest products and reducing nontariff barriers to trade. The need to understand the effect tariffs have on the forest sector and their central role in policy discussion has grown with the globalization of the wood products industry and calls for greater trade liberalization. The introduction or removal of tariffs has important consequences for the allocation of resources across sectors and ultimately on wood and its use in the forest products sector. Understanding the changes in the production, consumption and trade activity in the forest sector for wood inputs and intermediate products of lumber and plywood is an important first step in realizing any tariff reduction in the forest sector. Secondary impacts on the labor, capital and environmental sectors may be extracted from primary forest sector effects.

Whether a reduction in tariffs leads to an increase in consumption globally is a testable hypothesis. At least one analysis that simulated a tariff elimination policy resulted in greater trade activity with a decline in softwood production and consumption globally (Brooks, 2003). The study suggested that tariffs have restricted markets mostly for North America and European producers. Because the North American, particularly the US market, and Europeans are also the major consumers of softwood lumber, an import tariff elimination scenario resulted in greater *international* demand for their products raising *domestic* prices and lowering *domestic* consumption. This reduction in North American and European domestic consumption outweighs consumption gains in Japan, Australia and Mexico, regions where import tariff elimination would occur. The previous conclusion assumes that real exchange rates remained constant throughout the simulation, however, and effective changes with an adjustment in the terms of trade were not measured.

Export tariffs

The most recent example of trade intervention has been the implementation of export tariffs on Russian timber. Early work on Russia's forest sector and policies to help develop it were aimed at attracting investments to improve the collapsing milling infrastructure. A suggested tariff on roundwood exports from Russia was proposed (Backman and Perez-Garcia, 1995). Recently, implementation of the tariff has been studied by Moiseyev et al. (2010), Turner et al. (2008) and Solberg, Moiseyev, Kallio and Toppinen (2010). The effects of the tariff on development of Russian milling infrastructure can be measured by new domestic and foreign investments directed toward this sector, subject to a lagged time period for the effects to be felt.

Challenges for trade in forest products

Trade strategies

Early trade strategies promoted by nongovernmental groups to achieve a reduction in consumption and protect resources associated with forests resulted in unintended consequences. The first attempts to ban log exports, for example, led to substitution, which had little, if not detrimental, effect on the forest sector in the log-banning nation (Vincent, 1992; Perez-Garcia, Lippke and Baker, 1997). Yet managing the outflow of forest resources used to produce commodity products continues at the same time that international efforts exist to lower trade barriers, both tariff and nontariff ones.

World Trade Organization (WTO) members continue to struggle with how to deal with nontariff barriers (NTBs), those barriers that do not explicitly use monetary means to reduce trade volume, but act as one. There are many policies, regulations or other measures that countries consider implementing that constitute NTBs. Several of these policies are measures that promote environmental, social and developmental national objectives. Although WTO members note that many of the NTBs are legitimate public policy measures, it is still unclear how consistent they are with already agreed-upon trade regulations. It seems that tariffs and NTBs will continue to exist in global forest products trade, particularly given the added importance forest resources play in economic and social development plans and environmental quality. A clear understanding of trade policy impacts and their potential distorting effects is needed to minimize unintended consequences of trade and other legislation that may be considered NTBs. Participation of developing countries is seen as a must because the reduction of trade barriers is viewed as important to attract foreign direct investments and increase their trade portfolio.

New research areas

Interest in environmental services from forests that are not valued in market places has grown recently. As an example, carbon sequestered and stored in forests and forest products is viewed as a part of the solution to global climate change (Perez-Garcia, Lippke, Comnick and Manriquez, 2005; van Kooten and Sohngen, 2007). Trading schemes such as cap and trade are being investigated to determine their effectiveness. The complexity of monitoring and measuring the fate of carbon as it moves from being sequestered in trees to its final use is likely to limit forests' and their products' economic role in the trade of carbon credits, however (Lippke and Perez-Garcia, 2008).

There is also interest in the use of forest residuals and lower-valued timber in the production of bioenergy. The area has received a great deal of attention, particularly since the implementation of the renewable energy standard requirement. These new uses of forest products add new values to economic activity that did not exist a decade ago. Their study under trading regimes is an area ripe for investigation.

Finally, there is a need to improve upon trade modeling efforts on two fronts. First is the never-ending task of improving data collection and validation for use in forest sector assessments, particularly in developing countries with large forest resources and nascent forest industries. Primarily there is a need to invest in training of forest specialists who are responsible for monitoring and measuring forest sector activities for study. Improvements in data quality occur with investments in personnel used to collect the data. Second, there is a need to extend forest sector models beyond the traditional forest products to include environmental services and lifecycle assessments. Traditional economic input-output models have been extended to include the environmental services associated with economic activity (Shmelev, 2012). These environmentally enhanced input-output methods can be applied to forest sector analysis. The extended forest sector models would be useful to study the value of environmental goods and services produced by the forest sector to the global economy and how trade can contribute to achievement of higher environmental studies.

References

Adams, D. M. (2003). 'Market and resource impacts of a Canadian lumber tariff', *Journal of Forestry*, vol. 101, no. 2, pp. 48–52.

Adams, D. M., Alig, R. J., Callaway, J. M., McCarl, B. A. and Winnett, S. M. (1996). *The Forest and Agricultural Sector Optimization Model (FASOM): Model Structure and Policy Applications*, Research Paper – US Department of Agriculture, Forest Service, PNW-RP-495. Portland, OR

Adams, D. M. and Haynes, R. W. (1980). 'The 1980 softwood timber assessment market model: Structure, projections, and policy simulations', *Forest Science*, Monograph 22, pp. 54.

Alig, R. J., Adams, D. M. and McCarl, B. A. (2002). 'Projecting impacts of global climate change on the US forest and agriculture sectors and carbon budgets', *Forest Ecology and Management*, vol. 169, no. 1–2, pp. 3–14.

Backman, C. and Perez-Garcia, J. (1995). *The Former USSR Forest Sector and Capital Constraints*, International Institute for Applied Systems Analysis Collaborative Paper CP95-1, Laxenburg, Austria.

Boyd, R. A. and Krutilla, K. (1987). 'The welfare impacts of the US trade restrictions against the Canadian softwood lumber industry: A spatial equilibrium analysis', *The Canadian Journal of Economics*, vol. 20, no. 1, pp. 17–35.

Brink, L., Groombridge, M. A. and Lougani, P. (2000). *Nailing the Homeowner: The Economic Impact of Trade Protection of the Softwood Lumber Industry*, The Cato Institute, Washington, DC.

Brooks, D. J. (2003). 'Analysis of environmental effects of prospective trade agreements: The forest products ATL as a case study in the science-policy interface', *Forest Policy and Economics*, vol. 5, no. 4, pp. 385–394.

Buongiorno, J., Zhu, S., Raunikar, R. and Prestemon, J. (2012). *Outlook to 2060 for World Forests and Forest Industries: A Technical Document Supporting the Forest Service 2010 RPA Assessment*, General Technical Report SRS-151, US Department of Agriculture – Forest Service, Southern Research Station, Asheville, NC.

Buongiorno, J., Zhu, S., Zhang, D., Turner, J. and Tomberlin, D. (2003). *The Global Forest Products Model: Structure, Estimation and Applications*, Academic Publishers, San Diego, CA.

Cardellichio, P. A., Youn, Y. C., Adams, D. M., Joo, R. W. and Chmelik, J. T. (1989). *A Preliminary Analysis of Timber and Timber Products Production, Consumption, Trade, and Prices in the Pacific Rim until 2000*, CINTRAFOR Working Paper 22. College of Forest Resources, Seattle.

Cashore, B. (1998). *Flights of the Phoenix: Explaining the Durability of the Canada-U.S. Softwood Lumber Dispute*, Canadian-American Public Policy No. 32, The Canadian-American Center, University of Maine.

DeRosa, D., Hufbauer, G. and Perez-Garcia, J. (2000). *Lifting the U.S.-Canada Softwood Lumber Agreement: A Partial Equilibrium Analysis of the Economic Impact*, Peterson Institute for International Economics, Washington, DC.

Food and Agricultural Organization of the United Nations (FAO). (2012). 'Food and Agricultural Organization database on forest statistics'. Retrieved from www.fao.org/forestry/statistics/en/

Gilless, J. K. and Buongiorno, J. (1987). 'PAPYRUS: A Model of the North American Pulp and Paper Industry', *Forest Science*, Monograph 28, pp. 37.

Hertel, T. W. (Ed.). (1997). *Global Trade Analysis: Modeling and Applications*, Cambridge University Press, Cambridge, MA.

Kallio, A.M.I., Moiseyev, A. and Solberg, B. (2006). 'Economic impacts of increased forest conservation in Europe: A forest sector model analysis', *Environmental Science and Policy*, vol. 9, no. 5, pp. 457–465.

Kallio, M., Dystra, D. and Binkley, C. (1987). *The Global Forest Sector: A Global Perspective*, John Wiley and Sons. New York.

Li, R., Buongiorno, J., Turner, J.A., Zhu, S. and Prestemon, J. (2008). 'Long-term effects of eliminating illegal logging on the world forest industries, trade and inventory', Forest Policy and Economics, vol. 10, pp. 480–490.

Lippke, B. and Perez-Garcia, J. (2008). 'Will either cap and trade or a carbon emissions tax be effective in monetizing carbon as an ecosystem service?' *Forest Ecology and Management*, vol. 256, pp. 2160–2165.

Moiseyev, A., Solberg, B., Michie, B. and Kallio, A.M.I. (2010). 'Modeling the impacts of policy measures to prevent import of illegal wood and wood products', Forest Policy and Economics, vol. 12, pp. 24–30.

Nagubadi, R.V., Thompson, H. and Zhang, D. (2009). 'Productivity and trade during the softwood lumber dispute', *International Trade Journal*, vol. 23, no. 3, pp. 301–329.

Perez-Garcia, J. M. (1995). 'Global forest land use consequences of North American timber land withdrawals', *Journal of Forestry*, vol. 93, no. 7, pp. 35–38.

Perez-Garcia, J. and Lippke, B. (1997). 'The CINTRAFOR Global Trade Model and the forest assessment process', in R. Sedjo and A. Goetzl (Eds.), *Wood Fiber Market Models to Assist in the Development of a National Fiber Supply Strategy for the 21st Century*, Resources for the Future, Washington, DC.

Perez-Garcia, J., Lippke, B., Comnick, J. and Manriquez, C. (2005). 'Energy displacement and wood products market substitution using life cycle analysis', *Wood Fiber Science*, vol. 37, no. 5, pp. 140–148.

Perez-Garcia, J. M., Joyce, L., Binkley, C. S. and McGuire, A. D. (1997). 'Economic impacts of climate change on the global forest sector', *Critical Reviews in Environmental Technology*, vol. 27, pp. S123–S138.

Perez-Garcia, J. M., Joyce, L. A., McGuire, A. D. and Xiao, X. (2002). 'Impacts of climate change on the global forest sector', *Climatic Change*, vol. 54, pp. 439–461.

Perez-Garcia, J. M., Lippke, B. R. and Baker, J. (1997). 'Who wins and who loses: rade barriers in the PNW forest sector', *Contemporary Economic Policy*, vol. XV, no. 1, pp. 87–103.

Samuelson, P. A. (1952). 'Spatial price equilibrium and linear programming', *American Economic Review*, vol. 42, pp. 283–303.

Sedjo, R.A. (1993). Global consequences of American environmental policies', Journal of Forestry, vol. 91. no. 4. pp. 19–21.

Sedjo, R. A. and Lyon, K. S. (1990). *The Long-term Adequacy of World Timber Supply*, Resources for the Future, Washington, DC.

Shmelev, S. (2012). *Ecological Economics: Sustainability in Practice*, Springer, New York.

Sohngen, B., Mendelsohn, R. and Sedjo, R. (2001). 'A global model of climate change impacts on timber markets', *Journal of Agricultural and Resource Economics*, vol. 26, no. 2, pp. 326–343.

Solberg, B., Moiseyev, A., Kallio, A.M.I. and Toppinen, A. (2010). 'Forest sector market impacts of changed roundwood export tariffs and investment climate in Russia', *Forest Policy and Economics*, vol. 12, no. 1, pp. 17–23.

Turner, J. A., Buongiorno, J., Katz, A. and Zhu, S. (2008). 'Implications of the Russian roundwood export tax for the Russian and global wood products sectors', *Scandinavian Journal of Forest Research*, vol. 23, pp. 154–166.

Turner, J. A., Buongiorno, J. and Zhu, S. (2005). 'Effects of the free trade area of the Americas on forest resources', *Agricultural and Resource Economics Review*, vol. 34, no. 1, pp. 104–118.

van Kooten, G. C. (2002). 'Economic analysis of the Canada-United States softwood lumber dispute: Playing the quota game', *Forest Science*, vol. 48, pp. 712–721.

van Kooten, G. C. and Sohngen, B. (2007). 'Economics of forest ecosystem carbon sinks: A review', *International Review of Environmental and Resource Economics*, vol. 1, pp. 237–269.

Vincent, J. R. (1992). 'The tropical timber trade and sustainable development', *Science*, vol. 256, pp. 1651–1655.

Wiseman, A.C., and Sedjo, R.A. (1981). 'Effects of an export embargo on related goods: Logs and lumber', American journal of Agricultural Economics, vol. 65, pp 113–116.

World Bank. (2012). 'World Bank statistics database', Retrieved from http://databank.worldbank.org/Data/Views/VariableSelection/SelectVariables.aspx?source=World%20Development%20Indicators

Zhang, D. (2007). *The Softwood Lumber War: Politics, Economics, and the Long U.S.-Canadian Trade Dispute*, Resources for the Future Press, Washington, DC.

PART 2

Economics of forest ecosystems

9

CHOICE EXPERIMENTS AND VALUING FOREST ATTRIBUTES

Kevin J. Boyle[1] and Thomas P. Holmes[2]

[1]DIRECTOR, PROGRAM IN REAL ESTATE (0156), VIRGINIA TECH - THE NAME ENTERED IS NO LONGER USED BY THE UNIVERSITY AND WE ARE SUPPOSED TO USE VIRGINIA TECH ON ALL CITATIONS OF THE UNIVERSITY, 430A BISHOP-FAVRAO HALL, BLACKSBURG, VA 24061, USA.
[2]RESEARCH FOREST ECONOMIST, USDA FOREST SERVICE, SOUTHERN RESEARCH STATION, P.O. BOX 12254, RESEARCH TRIANGLE PARK, NC 27701, USA.

Abstract

This chapter considers the use of choice experiments to value services of forest ecosystems that are not traded in markets and conditions that might be marketed but have not been experienced in the market. A choice experiment is a survey-based valuation method in which the unique focus is the estimation of marginal values for individual elements of forest ecosystems. We begin with an explanation of the state of the art in choice experiments and then discuss common types of forest ecosystem applications from around the globe. The application of choice experiments to forestry issues has grown rapidly in recent years, and we anticipate continued growth. Concurrent with this growth in applications, there have been substantial enhancements in the design of choice questions and the econometric analysis of the choice data. Future forestry applications should search for the best procedures in study design and data analysis when implementing studies. Overall, choice modeling has provided forest decision makers in both the public and private sectors with richer information on economic values to enhance the quality and sustainability of forests at the local, national and global levels.

Keywords

Choice modeling, choice experiments, nonmarket valuation, use value, nonuse use value, experimental design, attributes, conditional logit model, random parameter logit model

Introduction

Around the globe, forests provide a suite of ecosystem services that are valued by people. Over time, as economies develop and household incomes increase, the relative value of specific ecosystem services can change from an emphasis on resource extraction (such as timber or fuelwood harvesting) toward a greater emphasis on nonmarket goods and services (Cubbage, Harou and Sills, 2007). In the United States, for example, this transition is evidenced by legislation requiring that national forests be managed for timber, range, watershed, recreation and fish and wildlife resources (Multiple Use-Sustained Yield Act of 1960, P.L. 86–517) and that long-range plans be established to balance the multiple uses of the nation's national forests (Forests and Rangeland Renewable Resources Planning Act of 1974, P.L. 93–378; National Forest Management Act of 1976, P.L. 94–5888). The emphasis on the provision of multiple goods and services from national forests provided a rationale for developing estimates of nonmarket values of forests that could be used for planning purposes (e.g. Loomis, 2005).

The broadening of forest management objectives to include the provision of nonmarket values is not unique to the United States. The European Union (EU) Forest Action Plan, for example, recently called for new research and the development of databases on the nonmarket value of forest resources to support forest-related initiatives (Stenger, Harou and Navrud, 2009). We also note that, over the past decade, the number of nonmarket valuation studies focused on forest ecosystems has grown rapidly, not only in the United States and Europe, but also in Asia and South and Central America.

People who visit public forests for recreation, hunting or fishing typically do not pay a fee, so there is no market price to assess the marginal values of these experiences. In addition, people who do not visit public forests can hold nonuse or passive use values for forest resources that are not expressed through market prices. These conditions require economists working on forest plans to consider nonmarket valuation methods such as travel-cost models and contingent-valuation surveys to estimate values for current and future forest conditions (Champ, Boyle and Brown, 2003). Travel-cost models are used to estimate recreation use value based on people's actual recreation experiences (Parsons, 2003; see Zanderson and Tol, 2009, for examples of travel-cost applications in Europe). Contingent-valuation surveys provide more flexibility, as this method can be used to estimate use, nonuse use and total values (Boyle, 2003; see Barrio and Loureiro, 2010, for contingent-valuation applications around the globe). Travel-cost models are restricted to conditions that people have actual experience, whereas contingent valuation does not have this restriction. Contingent-valuation scenarios can be designed to estimate the loss in recreation use values when a forest is damaged by disease or a pest infestation. If this degraded condition had not been previously experienced, there would not be observed use behavior to estimate a travel-cost model.

Nonmarket-valuation approaches naturally evolved to include choice experiments (Holmes and Adamowicz, 2003), and a number of the early applications valued forest characteristics (Adamowicz, Boxall, Williams and Louviere, 1998; Hanley, Wright and Adamowicz, 1998). A choice experiment, like contingent valuation, is a survey-based nonmarket valuation method. The unique feature of a choice experiment is that the change to be valued is described via a number of attributes, and marginal values can be estimated for each of these attributes. A contingent-valuation scenario provides a single value, whereas a choice-experiment scenario allows the estimation of multiple values. Thus, forest managers can learn what attributes provide the highest value and how values change with the levels of each attribute. For example, in a study of changes in forest management practices in the state of Maine in the United States, we used a choice experiment to investigate different logging practices that would reduce the ecological impacts of timber harvesting (Boyle, Holmes, Teisl and Roe, 2001). Attributes included in the design of the choice experiment included the density of logging roads, dead or dying trees left in harvest openings, live trees left in harvest openings, size of harvest openings, percentage of land available for timber harvesting, size of riparian protection zones and slash disposal. This choice experiment with these attributes allowed forest managers to learn about the values that the public placed on reducing different aspects of timber harvesting and customize forest regulations and harvesting guidance to minimize environmental concerns and maximize public satisfaction, while considering the productivity of forest operations. .

Thus, the complex nature of forest management and the multiple services provided by forests are the stimulating factors for considering choice experiments to value changes in forest conditions to support forest planning and management. It can be logically argued that considerations of multiple uses of public and private forests is a stimulating factor in the expansion of choice experiment applications to support forest ecosystems decision making around the globe (Bengston, 1994; Jensen and Bourgeron, 1994; Kennedy, 1985).

Although some current applications of choice experiments to value forest attributes still consider recreational use values for forest-based recreation, many current applications typically estimate total values for changes in forest ecosystems that include both use and nonuse values. Although the outcomes of choice experiments have broad management appeal, the rigor of doing a high-quality study has advanced considerably in recent years in terms of the design of choice experiments and the econometric analysis of the resultant data. In this chapter we provide an overview of the design of choice experiments and data analysis, as well as a summary of the types of empirical applications in the literature.

Choice experiments

Choice experiments are a survey-research approach to collecting data to estimate values people place on items that are not traded in markets, or for items traded in markets for which the conditions to be valued have not been experienced (Hanley et al., 1998; Holmes and Adamowicz, 2003). An example of the latter would be a change in forest management that created conditions that people recreating had not experienced previously, such as cessation of timber harvesting or other change that would impact forest recreation experiences. The choice experiment is applied in a survey that is typically administered through mail, in-person interview or Internet modes.

The choice experiment portion of the survey proposes alternative profiles that are defined in terms of attributes. Respondents typically are asked to choose between two or more alternatives where one alternative is the current or status quo condition (no change) and the other alternatives represent improvements or decrements in forest conditions. The alternatives are typically described in three or more attributes where one attribute is a monetary cost. If there were only two attributes, one forest attribute and the cost, the question would be equivalent to a dichotomous-choice, contingent-valuation question. Although it is possible to design a choice experiment that does not include a monetary attribute, the exclusion of the monetary attribute would preclude the possibility of developing monetary values for each forest attribute to use in decision making.

A sample choice question is presented in Table 9.1, which is taken from a study whose goal was to estimate household willingness to pay for public and private programs that would reduce homeowners' risk of economic damages from forest fires (Holmes, González-Caban, Loomis and Sánchez, 2013). To convey different risk levels, which are very small, the authors included graphic displays in the survey (similar to graphics used to display changes in health risks in health valuation experiments).

The key elements in designing a choice experiment include six steps: (1) selecting the attributes, (2) choosing levels for the attributes, (3) deciding the number of alternatives for each choice question, (4) picking the number of choice questions each survey subject will be asked, (5) developing the experimental design to select the combinations of attribute levels to be

Table 9.1 Example choice question.

	Alternative #1: Public fire prevention	Alternative #2: Private fire prevention	Alternative #3: Do nothing additional
Risk of your house being damaged in next 10 years	40 in 1,000 (4%)	10 in 1,000 (1%)	50 in 1,000 (5%)
Damage to property	$40,000	$80,000	$100,000
Expected 10-year loss = Risk × damage	$1,600 during 10 years	$800 during 10 years	$5,000 during 10 years
One-time cost to you for the 10-year program	$300	$100	$0
I would choose: Please check one box	□	□	□

presented in each choice question and (6) analyzing the response data. These are the elements that are unique to choice experiments. Other elements of the application of choice experiments would be common to the design of contingent valuation surveys and other types of economic surveys.

Experimental design

The foundation of any choice experiment is completing steps 1 through 5 identified previously. Steps 1 through 4 do not need to be followed in the order we present here, but they do need to be considered before completing the fifth step. Most choice experiments have a decision-making goal, whether it is a primary or secondary goal, and attributes and attribute levels are chosen to support the anticipated decisions. Thus, the attributes are selected to represent features that will be affected by (change due to) the anticipated decisions.

For example, in the Boyle et al. (2001) study cited previously, forest policy decision making could involve reducing the environmental impacts of timber harvesting, reducing the density of logging roads, leaving more dead or dying trees in harvest openings, leaving more live trees in harvest openings, reducing the size of harvest openings, limiting the percentage of land available for timber harvesting, increasing the size of riparian protection zones and changing slash disposal practices. Likewise, the Gelo and Kock (2012) study considered changes in forest type, increasing land productivity and differing levels of wood biomass harvesting by households. Thus, attribute and attribute level selection are not creations of the investigator, but reasoned choices based on three considerations: needs of decision makers, concerns of survey respondents and practical design features of the choice experiment.

The first element is developed through discussion between investigators and decision makers to select attributes and attribute levels, and the second step involves pretesting of the survey instrument. The discussions and pretesting can reduce or expand the number of attributes and possibly change attribute levels. If economic analyses are warranted, one of the attributes will always be a cost or payment where the estimated preference parameter on this attribute is interpreted as the marginal utility of money and is used to compute implicit values for each of the nonmonetary attributes. More will be said about the role of this attribute when we turn to the econometric analysis discussion subsequently.

A reduction in the number of attributes might occur if pretesting shows that an attribute of interest to decision makers has no relevance to respondents or is too difficult to convey to respondents. An example might be some element of the ecological functioning of a forest that does not easily translate into use or nonuse services of the forest that survey respondents' value. An added attribute might arise because survey respondents have a concern that is not fundamental to decision making, but if omitted from the study it could potentially bias the estimated effects of the included attributes. An example of this occurred in a study we are doing for the National Park Service to value improvements in visibility from reductions in anthropogenic haze (Paterson et al., 2013). The initial choice question did not have an attribute for the effects on forest flora and fauna from changes in haze. To avoid confounding the haze attribute with ecosystem effects, we found that it was necessary to include a forest ecosystem impact attribute in the design.

In addition to the number of attributes in a choice question, the design complexity of a choice experiment can be affected by the number of alternatives respondents are asked to consider in a choice question and the number of questions that are posed to each respondent. At a minimum, choice questions should have two alternatives, a status quo and a proposed change from the status quo. The status quo alternative is needed because the statistical outcome of a choice experiment is an unanchored utility index. If the results are to be used for economic analyses, particularly in cost-benefit analyses, then the status quo is crucial for measuring changes in value from baseline conditions. Beyond the status quo, a choice question may have one or more proposed changes that respondents are asked to choose from. For example, Rolfe and Bennett (2009) considered choice questions with two and three alternatives where one alternative in each design was the status quo defined as current conditions. Day et al. (2012) considered ordering effects when subjects are presented with multiple choice questions in a single choice survey. These and other studies have shown that increasing the number of alternatives and number of choice questions can have desirable and undesirable effects on estimated preference parameters; when designing a choice study, the analyst must carefully consider the information needs and the insights provided in these and other studies in the peer-reviewed literature.

A fundamental advancement in the design of choice questions in recent years has been the development of optimal design strategies for assigning combinations of attributes to choice questions (Kanninen, 2002; Rose, Bliemer, Hensher and Collins, 2008; Scarpa and Rose, 2008). Software for design includes Ngene (www.choice-metrics.com/features.html) and SAS (http://support.sas.com/techsup/technote/ts723.html).

The fundamental feature of choice experiments that enhances the appeal of this valuation method to researchers and decision makers is that it is possible to increase the amount of valuation information that can be gleaned from a fixed sampling budget. However, investigators must balance this desire for more information with caution that they do not create a design that is too complex for survey respondents, is not statistically efficient for estimating preference parameters and does not induce undesirable experimental effects such as anchoring or order effects.

Conceptual framework for welfare evaluation and data analysis

Analyses of choice experiment data are based on the standard random utility framework, but because survey respondents are asked to answer multiple choice questions in a single choice survey, panel data sets are collected. Thus, there are a variety of econometric approaches that have been employed to analyze these data. We will review two of these approaches here: the conditional logit (CL) model that does not consider the correlation between panel data observations, and the random parameters logit (RPL) model that does allow for this correlation. In addition, the CL model assumes that preference parameters are fixed, whereas the RPL model allows for heterogeneity in preference parameters across survey respondents.

The random utility framework is based on the assumption that survey respondents can make choices over alternatives, but from an econometric perspective, the analyst only observes a systematic component of respondents' utility/preferences (v_{ij}), and there is a random component that is not observable (ε_{ij}). Assuming that utility is a linear function of preference parameters over the attributes included in the choice question design for a particular study, the systematic component of utility is written as:

$$v_{ij}(Z_{jk}, \ y - p_j) + \varepsilon_{ij} = \sum_{k=1}^{K} \beta_{ik} Z_{jk} + \lambda_i \ y - p_j + \varepsilon_{ij} \tag{1}$$

where v_{ij} is the true but unobservable utility for alternative j for individual i, Z_{jk} is a vector of k attributes associated with alternative j excluding the cost attribute, p_j, y is income, β is a vector of preference parameters for the Z_{jk} attributes, λ is the marginal utility of money and ε_{ij} is a random error term with zero mean. The probability that individual i will choose alternative j is:

$$P_{ij} = Pr\left[\{v_{ij}(Z_{jK}, \ y - p_j) + \varepsilon_{ij} > v_{il}(Z_{lK}, \ y - p_l) + \varepsilon_{il}\} \ \forall j, l \in C, j \neq l \right] \tag{2}$$

where C is a choice set containing all of the alternatives presented in the choice question. Equation (2), with a little algebra, can be reorganized as the difference between utility of alternatives ($v_{ij} - v_{il}$). Therefore, any variable that remains constant between alternatives, such as the characteristics of the respondent, will drop out of the estimated model. Characteristics that are constant across alternatives are only included in the estimated model as interaction terms with variables that do change across alternatives.

A general specification of v_{ij} is employed where β and λ are index by i, which allows them to vary over people. This is generalized notation which supports preference heterogeneity estimated in the RPL model. In the more restrictive CL model, the i is suppressed and a single preference parameter is estimated for each attribute.

In economic applications of choice questions, the goal is typically to estimate changes in economic welfare for benefit-cost analyses. Willingness to pay (WTP), or compensating variation, is the amount of money that a person would give up to obtain a change in forest attributes that would keep them just as well off after the change as they were before the change. The value definition is specified as:

$$CV = \frac{1}{\hat{\lambda}}\{\hat{V}^j - \hat{V}^l\} \tag{3}$$

where j denotes a new forest condition and l denotes status quo, V denotes the nonprice attribute components of utility, λ is as defined previously and the hats denote parameters estimates based on the choice-experiment data. In equation (3), j and l can differ in the levels of one or more attributes. For example, if we consider a case with two forest attributes, z_1 and z_2, and wish to compute willingness to pay for changes in both attributes, WTP is computed as:

$$WTP = -\frac{\hat{\beta}_1 \Delta z_1 + \hat{\beta}_2 \Delta z_2}{\hat{\lambda}} \tag{4}$$

This assumes a linear specification of utility, which is common in analyses of choice data. It is also typical to compute the implicit price or marginal willingness to pay (MWTP) for a change in individual attributes:

$$MWTP_k = -\frac{\hat{\beta}_k \Delta z_k}{\hat{\lambda}} \; \forall k \in K \tag{5}$$

This is simply the parameter estimate on the attribute divided by the negative of the parameter estimate on the price variable.

Conditional logit model

It is often assumed that the random component of utility (ε) is independently and identically distributed (IID) as extreme value type I distribution where the difference ($\varepsilon_j - \varepsilon_l$) is distributed logistically. This assumption leads to the multinomial logit (also known as conditional logit) model (McFadden, 1974) and the probability that individual i chooses alternative j has the following expression:

$$P_j = \frac{\exp(\mu v_j)}{\sum_{j=1}^{N} \exp(\mu v_j)} \tag{6}$$

where μ is a scale factor that captures the variance of the unobserved component of utility and is typically normalized to one.

The CL logit model is restrictive in that it depends on the independence of irrelevant alternatives (IIA) property, which means that the ratio of choice probabilities between two alternatives in a choice set is unaffected by other alternatives that are available in the choice set. This restriction may not hold, for example, if two alternatives in a choice set are similar, but very different than a third alternative. The IIA property is easily tested using a procedure suggested by Hausman (Hensher, Rose and Greene, 2005) that examines how parameter estimates change when one alternative is dropped from the estimation. If the test indicates that IIA is violated, other models that do not rely on IIA should be used (such as the RPL model discussed subsequently). We note that CL model estimates were reported in about half of the forestry applications we summarize subsequently.

It is common practice to include a binary variable to identify alternatives in the choice set that describe the status quo or changes from current conditions; these are known as alternative specific constants (ASCs). These binary variables capture effects of respondents choosing or not choosing the status quo versus the changes independent of the attributes presented in choice questions. For example, a status quo option in analyses of choice data for woodland caribou protection indicated that there were reasons that respondents favored the status quo that were not explained by the attributes in the question design (Adamowicz et al., 1998). The authors postulated that this effect might reflect a status-quo bias, mistrust of resource managers to carry out the protection programs, uncertainty regarding complex trade-offs, or a form of protest response. Failure to include the ASC would have led to misspecification and biased parameter estimates on the attributes of interest.

Random parameter logit model

In equations (1) and (2), heterogeneous preferences for attributes are allowed by indexing preference parameters by i to designate each individual. The RPL model allows preferences to randomly vary over a continuous range of values (Train, 2003; Hensher et al., 2005). Further, the IIA assumption is relaxed by introducing additional stochastic components to the utility function through the preference parameters:

$$\beta_{ik} = \beta_k + \Gamma\omega_{ik} \tag{7}$$

where β_k is the mean value for the kth preference parameter, ω_{ik} is a random variable with mean zero and variance equal to one and Γ is a lower triangular matrix that allows free variances and correlations of parameters. The analyst specifies which parameters are random and the distribution of the random parameters (e.g. normal, triangular or log normal). It is typically the case that λ (marginal utility of money) is not randomly distributed to facilitate computation of WTP.

Probabilities in a RPL model are weighted averages of the standard logit formula evaluated at different values for β. The weights are determined by a density function $f(\beta|\theta)$, where θ represents the underlying parameters of the density function (such as the mean and covariance). Letting P_{ij} represent the probability that individual i chooses alternative j, the weighted probability function is:

$$P_{ij} = \int L_{ij} (\beta Z_j) f(\beta|\theta) d\beta \tag{8}$$

where L is the logit function shown in equation (6). The density function $f(\beta|\theta)$ can be simulated using a large number of draws of ω_{ik} using the functional form specified by the analyst (Train, 2003; Hensher et al., 2005). With this approach, the estimated $\hat{\beta}$'s are the average effects of the attributes, and the estimated standard deviations portray unobserved heterogeneity in the sampled population.

The RPL model is becoming the most commonly used format to analyze choice-experiment data, and this holds for forestry applications. Recent forestry applications have commonly assumed a normal distribution for the random parameters (Wang, Bennett, Xie, Zhang and Liang, 2007; Qin, Carlsson and Xu, 2011; Holmes et al., 2012), although the triangular distribution has also been used (Brey, Riera and Mogas, 2007; Farreras and Mavsar, 2012). A triangular distribution might be used when an analyst has good reason to restrict a parameter estimate to be only positive or only negative. All forestry studies we reviewed treated the price variable as fixed. Statistically significant estimates of the attribute parameters were reported in these studies and, in some cases, it was found that some of the sampled population had the unanticipated sign on their preferences. When part of the estimated distribution indicates that some respondents have the wrong preference-parameter sign, caution is warranted in evaluating the statistical results, and the analyst must ask if there were problems in their prior expectations of parameter signs or in implementing the choice experiment, or if an inappropriate statistical model is being used to analyze the data.

When computing WTP and MWTP using estimates from a random parameters model, it is recommended that MWTP include the standard deviation of each distributed parameter. In particular, mean WTP is computed as β_k / λ with standard deviation σ_k / λ (Hensher et al., 2005).

Forestry applications of choice experiments

Next, we discuss some peer-reviewed publications that dealt explicitly with a forestry application. We exclude studies that included a forestry attribute, but for which the parameter estimate was not significantly different than zero. The included studies were conducted in Europe (*n* = 13), North America (*n* = 6), Asia (*n* = 4), South or Central America (*n* = 2), Australia (*n* = 1) and Africa (*n* = 1). The first study we identified was published in 1998, and more than half of the studies have been published since 2007.

Although the literature review is not exhaustive, it will provide the reader with a flavor for the types of forestry applications where choice experiments are useful in informing decision making. We have categorized studies according to three major themes: (1) forest ecosystem services and nonuse values, (2) forestry contracts and (3) forest risk analysis.

Forest ecosystem services and nonuse values

Natural systems are increasingly viewed as critical capital assets that provide a broad suite of ecosystem services valued by people (Mäler, Aniyar and Jansson, 2008). The Millennium Ecosystem Assessment (2003) listed four categories of ecosystem services: provisioning (such as food, water and timber), regulating (such as carbon sequestration), cultural (such as recreation) and supporting (such as nutrient cycling). None of the forestry applications we reviewed were explicitly concerned with supporting services, and only a few forestry studies focused on trade-offs that included provisioning services, such as timber (Boyle et al., 2001; Holmes and Boyle, 2005) and nontimber forest products (Riera and Mogas, 2004; Mogas, Riera and Bennett, 2005; Mogas, Riera and Brey, 2009). Several studies included attributes related to regulating services, such as carbon sequestration (Riera and Mogas, 2004; Mogas et al., 2005, 2009; Brey et al., 2007; Balderas Torres, MacMillan, Skutsch and Lovett, 2012) and soil retention (Boyle et al., 2001; Riera and Mogas, 2004; Holmes and Boyle, 2005; Brey et al., 2007; Wang et al., 2007; Mogas et al., 2009). Studies focused on trade-offs involving cultural services have also been popular, such as recreation and ecotourism (Riera and Mogas, 2004; Mogas et al., 2005, 2009; Naidoo and Adamowicz, 2005; Horne, Boxall and Adamowicz, 2005; Brey et al., 2007; Christie, Hanley and Hynes, 2007; Elsasser, Englert and Hamilton, 2010).

Although biological diversity is not an ecosystem service per se, but rather is a structural feature of ecosystems that influences ecological outcomes, forestry studies investigating the value of biodiversity conservation have been popular. One of the primary challenges of including biological diversity in a choice experiment is the determination of ecosystem endpoints that enter the utility functions of respondents which also have relevance to managers and policymakers. One approach has been to link indices of biological diversity (such as avian species richness) with the likelihood of viewing such species during visits to conservation areas (Naidoo and Adamowicz, 2005). However, it has been more common to treat biological diversity as a nonuse value in forestry studies, where diversity has been characterized by measures of species richness (Rolfe, Bennett and Louviere, 2000; Lehtonen, Kuuluvainen, Pouta, Rekola and Li, 2003; Horne et al., 2005; Ohdoko and Yoshida, 2012) or by protecting natural processes in conservation areas (Bienabe and Hearne, 2006; Horne, 2006; Czajkowski, Busko-Briggs and Hanley, 2009).

Nonuse values estimated in forestry applications of choice experiments have demonstrated that nonuse values are a critical component of the total value of forest protection and conservation programs. In some cases, nonuse values have been found to exceed use values for attributes such as scenic beauty (Bienabe and Hearne, 2006), although trade-offs between scenic beauty and nonuse values have also been shown to be spatially dependent (Horne et al., 2005). Failure to recognize nonuse values for forests in decision making is a type of market failure that leads to misallocation of resources (Rolfe et al., 2000). Further, the identification of nonuse values can help forest managers decide where to implement protection and conservation activities, as nonuse values may be associated with remote locations that are seldom or never visited by recreationists or other users of forest ecosystems (Moore, Holmes and Bell, 2011).

Forestry contracts

Private forest land produces both private and public goods. The supply of public goods from private forests is underprovided when private forest owners do not account for the public benefits they produce or receive from other forest owners. Because the production of public goods (such as bio-diversity) from private forests may require diminished production of private goods (such as timber), one approach to maximizing social welfare from private forests is to establish voluntary contracts that compensate forest owners for decreases in commercial value. Voluntary contracts may be more socially acceptable than establishing new rules or regulations for the provision of public goods, and an emerging forestry literature has used choice experiments to investigate the preferences of private forest owners (or those who gain property rights) for attributes of forestry contracts.

Concern with biological diversity and endangered forest organisms in Finland led, in 2002, to a pilot program to enhance the conservation of forest biodiversity. Nearly three-fourths of the forest land located where biodiversity concerns are greatest is owned by private nonindustrial owners, and a choice experiment was designed to evaluate private forest owners' preferences for the attributes of a voluntary incentive program (Horne, 2006). The results indicated that forest owners would accept compensation for restrictions on the use of small patches of their land and the development of a nature management plan, although they disliked restrictions on silvicultural activities. Forest landowners also preferred shorter contracts to longer ones, thereby maintaining autonomy over long-run decisions.

Forest protection programs are also public goods in that the protective actions taken by one landowner convey protection benefits to neighboring landowners. The southern pine beetle (SPB) is the most damaging insect in southern U.S. forests, and forest management activities, such as thin-ning or planting resistant tree species, can reduce the risk of timber damage that benefit owners of forest on neighboring land. To understand the preferences of forest landowners for selected forest management treatments offered under a hypothetical cost-share program, a choice experiment was designed and implemented (Rossi, Carter, Alavalapati and Nowak, 2011). The authors found that replanting was the most preferred management option, whereas prescribed burning significantly reduced landowner utility, and landowners were indifferent regarding thinning activities – perhaps because they would occur many years in the future. Landowners were also indifferent regarding an attribute describing what percentage of their neighbors participated in the program.

Economists generally argue that clear property rights are necessary for investment and eco-nomic growth. Experiments with property rights are currently occurring in China, and a recent choice experiment was undertaken to investigate the preferences of Chinese farmers for the property right attributes of forest contracts (Qin et al., 2011). The authors concluded that farm-ers are very concerned with the types of rights provided by a contract and favor contracts that reduce the risk of contract termination and offer the priority to renew the contract at expiration.

Forest risk analysis

During the past decade, public concern has grown rapidly in many regions around the world regarding the threat of wildfires. Wildfires are a natural element in fire-prone forest ecosystems, and fire management involves a complex set of relationships between homeowners living in those ecosystems and fire managers faced with the responsibility of protecting life, homes and other resource values. A relatively new and emerging forestry literature has applied the choice experiment method to two aspects of the wildfire problem. One line of research has focused on decision making by fire managers faced with making fire suppression trade-offs. Within the United States, there is concern that past fire suppression paradigms have led to inefficient firefighting strategies and tactics under current conditions. To investigate the trade-offs that fire managers are willing to make among multiple objectives, a two-tiered choice experiment was designed and implemented with fire managers who were asked to provide expected responses (based on agency and political expectations) and their preferred responses (ignoring agency and political expectations) (Calkin, Venn, Wibbenmeyer and Thompson, 2012). The authors found that, for the expected response model, fire managers preferred more expensive options, while the opposite result was found for the preferred response model, and concluded that fire managers currently treat firefighting budgets as a free good.

A standard economic assumption is that, when faced with risky outcomes, decision makers attempt to maximize expected utility. This hypothesis was tested using a choice experiment on fire managers in the United States in which the probability of a successful outcome was varied across fire suppression alternatives (Wibbenmeyer, Hand, Calkin, Venn and Thompson, 2012). Study results led to the conclusion that fire managers' decision making is inconsistent with the expected utility model, and that managers may overspend firefighting resources when the likelihood or potential magnitude of damages is low.

In the United States, homeowners living in fire-prone landscapes are being asked to engage in vegetation management and home construction activities that reduce the risk of wildfire damages. To evaluate how homeowners make trade-offs between wildfire risks, potential loss and cost of risk-reducing measures, a choice experiment was developed and implemented with homeowners living predominantly in medium- and high-risk areas of Florida (Holmes et al., 2013). Similar to the study results described previously (Wibbenmeyer et al., 2012), the authors concluded that homeowner choices were generally inconsistent with expected utility theory. Rather, they suggested that their results were more consistent with prospect theory, under which people place greater weight on a certain loss (the cost of risk reduction) rather than on a probabilistic loss.

Wildfire risk is also a concern in many Mediterranean forests. In order to understand the trade-offs that citizens in Catalonia, Spain, would make between fire management programs which resulted in varying levels of fire intensity (trees per unit area killed) and total area burned, a choice experiment was designed and implemented (Farreras and Mavsar, 2012). The results showed that respondents are concerned about wildfire risks and that they preferred programs that focus on the reduction of the area burned.

Conclusion

The application of choice experiments to issues of concern in forestry has grown rapidly in recent years and we anticipate continued growth in applications for the foreseeable future. Concurrent with this growth in applications have been substantial enhancements in the design of choice experiments and the econometric analysis of the choice data. Future forestry applications should search for the best procedures in study design and data analysis when implementing studies.

As such, we emphasize the importance of developing meaningful survey instruments through the rigorous use of focus groups and pretests. Choice experiments need to be consistent with current ecological knowledge, and have forest management and policy relevance. Discussions with natural scientists and forestry decision makers during the survey development phase are prerequisite to the development of a meaningful survey instrument. This recommendation is particularly germane to the issue of identifying the endpoints of forest ecosystem attributes and attribute levels that are of interest to decision makers and enter the utility function of respondents, especially regarding nonuse values.

This qualitative design must be followed with rigorous experimental designs that assign attributes to choice alternatives, and to the econometric analysis of the response data. The literature on the validity of choice experiments is expanding rapidly in the transportation, environmental and health economics literatures, and forest economics researchers are encouraged to stay current with these developments. Our review of the choice-experiment literature suggests that forestry applications have been mainly empirical applications to support decision making and are not always consistent with the frontiers in the methodology of choice experiments.

A relatively new innovation in forestry applications of choice experiments is to use this methodology to investigate trade-offs that agency decision makers are willing to make when outcomes are uncertain. This type of application appears poised to provide substantial insights into how public expenditures are weighted vis-à-vis other agency objectives when making public choices. A related innovation is the use of choice experiments to investigate decision making under conditions of risk. The few forestry applications of this topic have revealed that decision makers tend not to use the expected utility framework, but rely on other modes of decision making. It seems that there is much that can be learned from the application of choice experiments under conditions of risk and uncertainty, especially as applied to natural disturbances such as wildfires and forest pest outbreaks, and in the protection of forest ecosystems. Overall, choice modeling has provided forest decision makers in both the public and private sectors with richer information on economic values to enhance the quality and sustainability of forests at the local national and global levels. Empirical applications have only begun to consider the many applications where choice experiments can help to improve forest decision making.

References

Adamowicz, W., Boxall, P., Williams, M. and Louviere, J. (1998). 'Stated preference approaches for measuring passive use values: Choice experiments and contingent valuation', *American Journal of Agricultural Economics*, vol. 80, no. 1, pp. 64–75.

Balderas Torres, A., MacMillan, D., Skutsch M., and Lovett, J. (2012). 'The valuation of forest carbon services by Mexican citizens: The case of Guadalajara city and La Primavera biosphere reserve', *Regional Environmental Change*, vol. 13, no. 3, pp. 661–680.

Barrio, M. and Loureiro, M. L. (2010). 'A meta-analysis of contingent valuation forest studies', *Ecological Economics*, vol. 69, no. 5, pp. 1023–1030.

Bengston, D. N. (1994). 'Changing forest values and ecosystem management', *Society and Natural Resources: An International Journal*, vol. 7, no. 6, pp. 515–533.

Bienabe, E. and Hearne, R. R. (2006). 'Public preferences for biodiversity conservation and scenic beauty within a framework of environmental service payments', *Forest Policy and Economics*, vol. 9, no. 4, pp. 335–348.

Boyle, K. J. (2003). 'Contingent valuation in practice', in P. Champ. K. J. Boyle and T. C. Brown (Eds.), *A Primer on Nonmarket Valuation*, Springer.

Boyle, K. J., Holmes, T. P., Teisl, M. F. and Roe, B. (2001). 'A comparison of conjoint analysis response formats', *American Journal of Agricultural Economics*, vol. 83, no. 2, pp. 441–454.

Brey, R., Riera, P. and Mogas, J. (2007). 'Estimation of forest values using choice modeling: An application to Spanish forests', *Ecological Economics*, vol. 64, no. 2, pp. 305–312.

Calkin, D., Venn, T., Wibbenmeyer, M. and Thompson, M. P. (2012). 'Estimating US wildland fire managers' preferences towards competing strategic suppression objectives', *International Journal of Wildland Fire*, vol. 22, no. 2, pp. 212–222.

Champ. P., Boyle, K. J. and Brown, T. C. (2003). *A Primer on Nonmarket Valuation*, Springer.

Christie, M., Hanley, N. and Hynes, S. (2007). 'Valuing enhancements to forest recreation using choice experiment and contingent behavior methods', *Journal of Forest Economics*, vol. 13, no. 2–3, pp. 75–102.

Cubbage, F., Harou, P. and Sills, E. (2007). 'Policy instruments to enhance multi-functional forest management', *Forest Policy and Economics*, vol. 9, pp. 833–851.

Czajkowski, M., Busko-Briggs, M. and Hanley, N. (2009). 'Valuing changes in forest biodiversity', *Ecological Economics*, vol. 68, pp. 2910–2917.

Day, B., Bateman, I. J., Carson, R. T., Dupont, D., Louviere, J. J., Morimoto, S., . . . Wang, P. (2012). 'Ordering effects and choice set awareness in repeat-response stated preference studies', *Journal of Environmental Economics and Management*, vol. 63, no. 1, pp. 73–91.

Elsasser, P., Englert, H., and Hamilton, J. (2010). 'Landscape benefits of a forest conversion programme in North East Germany: Results of a choice experiment', *Annals of Forest Research*, vol. 53, no. 1, pp. 37–50.

Farreras, V. and Mavsar, R. (2012). 'Burned forest area or dead trees? A discrete choice experiment for Catalan citizens', *Economia Agraria y Recursos Naturales*, vol. 12, no. 2, pp. 137–153.

Gelo, D. and Koch, S.F. (2012). 'Does one size fit all? Heterogeneity in the valuation of community forest programs', Ecological Economics, vol. 74, pp. 85–94.

Hanley, N., Wright, R. E. and Adamowicz, V. (1998). 'Using choice experiments to value the environment', *Environmental and Resource Economics*, vol. 11, no. 3–4, pp. 413–428.

Hensher, D. A., Rose, J. M. and Greene, W. H. (2005). *Applied Choice Analysis: A Primer*, Cambridge University Press, New York.

Holmes, T. P. and Adamowicz, W. L. (2003). 'Attribute-based methods', in P. Champ, K. J. Boyle and T. C. Brown (Eds.), *A Primer on Nonmarket Valuation*, Springer.

Holmes, T. P. and Boyle, K. J. (2005). 'Dynamic learning and context-dependence in sequential, attribute-based, stated preference valuation questions', *Land Economics,* vol. 81, no. 1, pp. 114–126.

Holmes, T. P., González-Caban, A., Loomis, J. and Sánchez, J. (2012). 'The effects of personal experience on choice-based preferences for wildfire protection programs', *International Journal of Wildland Fire*, vol. 22, no. 2, pp. 234–245.

Horne, P. (2006). 'Forest owners' acceptance of incentive-based policy instruments in forest biodiversity conservation – a choice experiment-based approach', *Silva Fennica*, vol. 40, no. 1, pp. 169–178.

Horne, P., Boxall, P. C. and Adamowicz, W. L. (2005). 'Multiple-use management of forest recreation sites: A spatially explicit choice experiment', *Forest Ecology and Management*, vol. 207, no. 1–2, pp. 189–199.

Jensen, M. E. and Bourgeron, P. S. (1994). *Volume II: Ecosystem Management: Principles and Applications*, USDA Forest Service General Technical Report PNW-GTR-318.

Kanninen, B. (2002). 'Optimal design for multinomial choice experiments', *Journal of Marketing Research*, vol. 39, no. 2, pp. 214–227.

Kennedy, J. J. (1985). 'Conceiving forest management as providing for current and future social value', *Forest Ecology and Management*, vol. 13, no. 1–2, pp. 121–132.

Lehtonen, E., Kuuluvainen, J., Pouta, E., Rekola, M. and Li, C.-Z. (2003). 'Non-market benefits of forest conservation in southern Finland', *Environmental Science & Policy*, vol. 6, no. 3, pp. 195–204.

Loomis, J. (2005). *Updated Outdoor Recreation Use Values on National Forests and Other Public Lands*, USDA Forest Service General Technical Report PNW-GTR-658.

Mäler, K-G., Aniyar, S. and Jansson, Å. (2008). 'Accounting for ecosystem services as a way to understand the requirements for sustainable development', *Proceedings of the National Academy of Sciences*, vol. 105, no. 28, pp. 9501–9506.

McFadden, D. (1974). 'Conditional logit analysis of qualitative choice behavior', in P. Zarembka (Ed.), *Frontiers in Econometrics*, Academic Press, New York.

Millennium Ecosystem Assessment. (2003). *Ecosystems and Human Well-Being: A Framework for Assessment*, Island Press, Washington, DC.

Mogas, J., Riera, P. and Bennett, J. (2005). 'Accounting for afforestation externalities: A comparison of contingent valuation and choice modeling', *European Environment*, vol. 15, no. 1, pp. 44–58.

Mogas, J., Riera, P. and Brey, R. (2009). 'Combining contingent valuation and choice experiments: A forestry application in Spain', *Environmental and Resource Economics*, vol. 43, no. 4, pp. 535–551.

Moore, C. C., Holmes, T. P. and Bell, K. P. (2011). 'An attribute-based approach to contingent valuation of forest protection programs', *Journal of Forest Economics*, vol. 17, no. 1, pp. 35–52.

Naidoo, R. and Adamowicz, W. L. (2005). 'Economic benefits of biodiversity exceed costs of conservation at an African rainforest reserve', *Proceedings of the National Academy of Science*, vol. 102, no. 46, pp. 16712–16716.

Ohdoko, T. and Yoshida, K. (2012). 'Public preferences for forest ecosystem management in Japan with emphasis on species diversity', *Environmental Economics and Policy Studies*, vol. 14, no. 2, pp. 147–169.

Parsons, G. R. (2003). 'The travel cost model', in P. Champ, K. J. Boyle and T. C. Brown (Eds.), *A Primer on Nonmarket Valuation*, Springer.

Paterson, R., Boyle, K. J., Carson, R., Kanninen, B., Leggett, C. and Molenar, J. (2013). *National Park Service Visibility Valuation Study: Pilot Survey Results*, Report to National Park Service, Air Resources Division.

Qin, P., Carlsson, F. and Xu, J. (2011). 'Forest tenure reform in China: A choice experiment on farmers' property rights preferences', *Land Economics*, vol. 87, no. 3, pp. 473–487.

Riera, P. and Mogas, J. (2004). 'Finding the social value of forests through stated preference methods: A Mediterranean forest valuation exercise', *Silva Lusitana*, vol. 12, no. especial, pp. 17–34.

Rolfe, J. and Bennett, J. (2009). 'The impact of offering two versus three alternatives in choice modeling experiments', *Ecological Economics*, vol. 68, pp. 1140–1148.

Rolfe, J., Bennett, J. and Louviere, J. (2000). 'Choice modeling and its potential application to tropical rainforest preservation', *Ecological Economics,* vol. 35, pp. 289–302.

Rose, J. M., Bliemer, M.C.J., Hensher, D. A. and Collins, A. T. (2008). 'Design efficient stated choice experiments in the presence of reference alternatives', *Transportation Research Part B. Methodological*, vol. 42, no. 4, pp. 395–406.

Rossi, F. J., Carter, D. R., Alavalapati, J.R.R. and Nowak, J. T. (2011). 'Assessing landowner preferences for forest management practices to prevent the southern pine beetle: An attribute-based choice experiment approach', *Forest Policy and Economics*, vol. 13, no. 4, pp. 234–241.

Scarpa, R. and Rose, J. M. (2008). 'Design efficiency for non-market valuation with choice modeling: How to measure it, what to report and why', *Agricultural and Resource Economics*, vol. 52, no. 3, pp. 253–282.

Stenger, A., Harou, P. and Navrud, S. (2009). 'Valuing environmental goods and services derived from forests', *Journal of Forest Economics*, vol. 15, pp. 1–14.

Train, K. E. (2003). *Discrete Choice Methods with Simulation*, Cambridge University Press, New York.

Wang, X., Bennett, J., Xie, C., Zhang, Z. and Liang, D. (2007). 'Estimating non-market environmental benefits of the conversion of cropland to forest and grassland program: A choice modeling approach', *Ecological Economics*, vol. 63, pp. 114–125.

Wibbenmeyer, M. J., Hand, M. S., Calkin, D. E., Venn, T. J. and Thompson, M. P. (2012) 'Risk preferences in strategic wildfire decision-making: A choice experiment with U.S. wildfire managers', *Risk Analysis*, vol. 33, no. 6, pp. 1021–1037.

Zanderson, M. and Tol, R.S.J. (2009). 'A meta-analysis of forest recreation values in Europe', *Journal of Forest Economics*, vol. 15, no. 1–2, pp. 109–130.

10

THE ECONOMICS OF OLD-GROWTH FORESTS

Claire A. Montgomery and Mindy S. Crandall

DEPARTMENT OF FOREST ENGINEERING, RESOURCES, AND MANAGEMENT, 280 PEAVY HALL,
OREGON STATE UNIVERSITY, CORVALLIS, OR 97331, USA. CLAIRE.MONTGOMERY@OREGONSTATE.EDU;
MINDY.CRANDALL@OREGONSTATE.EDU

Abstract

Issues surrounding old-growth forest incorporate many of the classic problems in resource and environmental economics, such as optimal use of nonrenewable resources and valuation of nonmarket and public goods, and many newer problems, such as spatial dependencies and institutional arrangements. Because old-growth forests are a temporal condition rather than an ecosystem type, management must also contend with disturbance and uncertainty in developing these forests. This chapter traces out a brief history of the development of old-growth forests as an issue, places them within the context of the classic Faustmann and Hotelling resource optimization problems, discusses potential values of these forests and means of measuring them and concludes with thoughts for future economic analysis.

Keywords

Old-growth forest economics, optimal resource stocking, optimal forest rotation age

Introduction: What is old-growth forest?

Few issues have brought as much public visibility and conflict to forestry as that of old-growth forest. Old-growth forest management and valuation are now critical topics within forest economics, forest management and policy analysis. But the issue of old-growth forest did not arise in a vacuum. It came about in a particular place and time – the Pacific Northwest of the United States in the 1970s and 1980s – and this context has influenced the perception of and policy surrounding old-growth forest to this day.

Scarcity of a resource increases its marginal value, both in the marketplace and in society. As the conversion of forest proceeded westward across the United States, there were few calls for the protection, preservation or veneration of old forest stands. Conventional forestry marked the oldest stands for harvest first, to capture both the volume of standing timber and also to convert these perceived 'negative growth' stands to young fast-growing forest. Timber harvest reached the Pacific Northwest in earnest in the mid-to-late nineteenth century, with lumber concerns abandoning the cut-over land of New England and the Great Lakes states for the valuable stands of huge Douglas-fir and Ponderosa pine of the western states.

Concurrent with the rise of the lumber industry in the Pacific Northwest, fear over the destructive power of forest fires, the need for clean water supplies and a possible dearth of harvestable timber for industry drove the nation to begin creating 'forest reserves' (now national forests) in 1891. These national forests, located predominantly in the west where there was remaining public land to withdraw, were initially harvested little. Harvest of the original stands of timber proceeded apace on private lands in the Pacific Northwest until the scarcity of harvestable stands encountered the great demand for timber of World War II and the economic boom following it. Harvest on public lands in the west began to play a larger role in timber supply at that point in time, and the Pacific Northwest's wet temperate forests, dominated by Douglas-fir, were recognized as some of the last large amounts of old-growth in the country. This decline in old-growth forest area was noticeable enough that, in the early 1980s, foresters became concerned that continued harvest of public old-growth forest would be controversial. The issue of dwindling area of old-growth Douglas-fir forest came to the forefront of federal forest policy in the United States when, in 1990, an old-growth-dependent bird species, the northern spotted owl (*Strix occidentalis caurina*), was listed as threatened throughout its range in the old-growth forest of the Pacific Northwest under the Endangered Species Act. The listing of the northern spotted owl led to an injunction against any timber harvest on federal land in the historical range of the owl. And because of the owl's dependence on old-growth forest structure for nesting habitat, the issue of old-growth forest exploded on the public consciousness. In fact, many viewed the listing of the northern spotted owl as a proxy for listing old-growth forest itself.

Because the issue of old-growth forest arose from the particular circumstances of the Pacific Northwest, the visual idea and ecological definitions of old-growth were also tied to the particular species life characteristics of the dominant trees in the region, particularly Douglas-fir (Spies, 2004; Hilbert and Wiensczyk, 2007). Douglas-fir is a large, long-lived species that is relatively fire tolerant and shade intolerant. Old-growth forests composed of Douglas-fir are thus both very old, dominated by large-diameter trees, and likely to persist until a large disturbance, such as infrequent, high severity fire, allows for regeneration of a new forest.

Although the issue of old-growth came to prominence in the context of the forests of the Pacific Northwest, old-growth forest exists the world over. A quick scan of the forestry literature reveals interest in old-growth forest in Israel, Australia, Scandinavia, Japan and elsewhere. Significant area of intact forest remains in the boreal forest in Asia, the tropical rainforest in South America and the mixed forest in Australia. In places where little old-growth forest remains, there is a growing interest in restoration of forest with old-growth characteristics. Each of these forest types is likely to have a very different type of old-growth forest, as determined by the life growth characteristics of the dominant tree species, the time necessary for characteristics to develop and the disturbance regime (Spies, 2004).

Old-growth is not a forest type, but rather a temporal condition that may arise in any forest type; it is not a distinct type of forest ecosystem, but an ecosystem condition (Kimmins, 2003). In fact, old-growth forest has unknown beginnings; it is not evident what the starting conditions were that produced the old-growth that we observe today. This is clear even in the Pacific Northwest, where the origin of the stands that have become the 'type' definition of old-growth likely originated from wide-scale regional fires dating from ca. 1400 to ca. 1650 (Weisberg and Swanson, 2003). Difficulties in managing for old-growth are confounded by uncertain definition of old-growth and how it differs by forest type, past stand dynamics and the temporal nature of old-growth.

Although there are economic, social and cultural dimensions to any definition of old-growth, it rests fundamentally on ecological science and understanding. The most basic definition of old-growth is simply a forest with old trees. This is related in the public's mind to large trees, ecosystem attributes (such as multilayered canopies, snags and decaying downed wood) and absence of evidence of human-caused disturbances (Kimmins, 2003; Hilbert and Wiensczyk, 2007). However, the relationship of size to age varies widely by species and growth conditions. In fact, the age at which a tree becomes 'old' is somewhat arbitrary and it varies by species. Although management and policy require a clear definition of old-growth forest in order to establish management goals, hundreds of definitions exist. They differ in their emphases on spatial scale (tree, stand or forest level), stand characteristics (e.g. age, size of trees), ecological processes (e.g. time since disturbance) and evidence of human activity. An agreed-upon definition of old-growth forest is important even in areas with insignificant amounts of old forest remaining, particularly if there is interest in reestablishing forest with old-growth attributes.

In this chapter, we begin with the question of old-growth valuation: Why do we care? We then describe how old-growth values have been incorporated into economic models of optimal forest management at the stand level and at the forest level. We then discuss old-growth-related issues in the context of a working landscape and in the face of disturbance. We close with some thoughts about how the economics of old-growth forest will develop in the future.

Valuation of old-growth forest: Why do we care?

How much value should be placed on increasingly scarce old-growth forest, relative to other forest-related values? From an economics perspective, we are ultimately concerned with contributions of old-growth forest to overall utility of society that can arise from commodity value, aesthetic value, values associated with conserving biodiversity and other ecosystem services, existence value and option value. Moyer, Owen and Duinker (2008) developed a typology of forest-based value based on interviews with Canadian citizen groups and forestry leaders. Material values include economic goods and services (such as timber, food and shelter) and life-support services (such as air and water quality, carbon sequestration and biodiversity). Nonmaterial values include socio-cultural, ethical, spiritual and aesthetic experiences. Many of these values occur in young, as well as old, forests, but they differ in quality, intensity of experience and uniqueness. Nonmaterial values that arise from absence of evidence of human impact and agedness are unique to old-growth forest.

Financial incentives to hold old-growth forest or to hold timber to older ages do exist. Old-growth Douglas-fir, long known for its high-quality, knot-free timber, once dominated the supply of logs in the Pacific Northwest; because mills were designed to process large-diameter logs, wood processors were willing to pay more to capture the improved recovery associated with large-diameter and high-quality wood. Lower costs to handle and load fewer larger trees at logging sites leads to estimated harvest cost equations that are often a decreasing function of average stand diameter and volume per unit area. Nonetheless, the price premium associated with old-growth timber has largely disappeared in the last few decades. The combined effect of declining supply of large-diameter logs, increasing supply of second-growth and technological innovation that allows substitution of engineered wood products for large timbers has led to a retooling of almost all mills in the region to focus on consistent, smaller-diameter supply (Haynes, 2009), so that it is now common for large logs to require additional processing at the mill. This evolving trend toward producing and processing uniform-size, second-growth trees has left little financial incentive for timber landowners to grow large trees that provide some old-growth forest benefits while growing.

Standing timber values include some marketable services and some private amenity values. In some areas, ecotourism based on the presence of old-growth forest may generate revenue. People may rely on old-growth forest for an array of products that have direct use or market value such as berries, mushrooms or medicines. Option value arises from the potential for unique old-growth forest to yield new knowledge and discoveries that have direct use value. For example, the bark of Pacific yew, which grows in the old-growth Douglas-fir forest of the Pacific Northwest, contains a chemical that was discovered in 1967 to have treatment benefits in cancer chemotherapy. It was subsequently produced synthetically and marketed under the trademark Taxol. In economic models of timber supply, amenity benefits of standing timber appear in the utility functions of nonindustrial private forest landowners; landowners trade off the amenity value of the old forest with income from timber harvest (Kuuluvainen, Karppinen and Ovaskainen, 1996). Where carbon markets function well, forest landowners may capture some of the value of standing timber by selling carbon offsets. To the extent that individuals or communities hold property rights of ownership or use, these values will be capitalized into the market value of old-growth forested land.

However, much of the controversy over old-growth forest arises because so many of the benefits of old-growth forest involve ecosystem services, the values of which are inadequately captured in markets, or they are public goods for which there are no functioning markets at all. It is these spiritual, ethical and cultural values that stimulate such passion in the debate over how much and how to protect old-growth forest where it remains. Although this question will ultimately be resolved in a political arena, economists have tried to understand the trade-offs associated with old-growth so as to be able to inform the policy debate. Valuations of benefits use an array of methods. For example, Hagen, Vincent and Welle (1992) used contingent valuation to measure willingness to pay to preserve old-growth forest as habitat for the northern spotted owl. Englin and Mendelsohn (1991) used a hedonic travel cost model to estimate willingness to pay for old-growth forest along hiking trails in wilderness areas in the state of Washington. Measurement of costs typically use an opportunity cost approach; what is the value of the stream of marketable commodities foregone when old-growth forest is preserved (e.g. Van Kooten, 1995; Montgomery, Brown and Adams, 1994)?

Economics of old-growth forest at the stand level

Conventional forest economics has been dominated by the Faustmann model, which identifies the harvest age, A, of an even-aged forest stand that maximizes the net present value of a perpetual series of timber harvests that occur at that age, or bare land value, $F(A)$:

$$\max_A F(A) = \frac{PQ(A)}{e^{rA} - 1} \tag{1}$$

where the (constant) price of wood is P, the volume of wood harvested at age, A, is $Q(A)$, the discount rate is r and there are no interim costs or revenues. Harvestable volume increases at a decreasing rate over the relevant range: $Q_A(A) > 0$ and $Q_{AA}(A) < 0$. The marginal condition for optimal timber harvest age is:

$$PQ_A(A) = r[PQ(A) + F(A)] = \frac{rPQ(A)}{1 - e^{-rA}} \tag{2}$$

The left-hand side, $PQ_A(A)$, is the value of the marginal growth (the marginal benefit of holding timber), and the right-hand side is foregone interest earnings on postponed timber harvest revenue and bare land value (the marginal cost of holding timber). Wealth-maximizing land-owners will hold timber as long as the marginal benefit of doing so exceeds the marginal cost, and they will harvest at the age at which they are equal. This basic Faustmann model is discussed in detail in Chapter 3.

A main limitation of the basic Faustmann model is that it incorporates only benefits realized through timber harvest. In the latter part of the twentieth century, however, forest economists began to explore the effects on optimal rotation age when all the possible benefits provided by standing forests are valued, which may include recreation, watershed protection and wildlife forage. Richard Hartman, an economics professor at the University of Washington, proposed a specific extension of the Faustmann model applicable 'when a standing forest has value' (Hartman, 1976).

Hartman extended this basic model by including a function to represent the annual value of the standing timber, $G(A)$, so that equation (1) becomes:

$$F(A) = \frac{PQ(A)}{e^{rA} - 1} + \frac{e^{rA} \int_{a=0}^{A} [G(a)e^{-ra}]da}{e^{rA} - 1} \qquad (3)$$

The second term on the right-hand side capitalizes the stream of annual standing timber values into the bare land value. The numerator is the accumulation of standing timber values for one rotation compounded to the end of the rotation. The marginal condition for optimal timber harvest age with standing timber values is:

$$PQ_A(A) + \left[G(A) - \frac{\int_{a=0}^{A} [G(a)e^{-ra}]da}{1 - e^{-rA}} \right] = \frac{rPQ(A)}{1 - e^{-rA}} \qquad (4)$$

The term in brackets is the net marginal value of holding timber accruing from standing timber values. $G(A)$ can take many forms depending on the nature of value arising from the standing timber. In Figure 10.1, two possible forms for $G(A)$ are illustrated: one in which the standing timber values are high early in the life of the stand when deer forage or habitat for a species like the porcupine, which prefers young forest, is abundant, and they drop off as the stand ages, and one in which standing timber values increase as the stand ages and is able to provide habitat for old-growth-dependent species. Because Hartman was primarily interested in old-growth issues and, in particular, the conditions under which it might be optimal to never harvest a forest, he considered only the latter case, in which standing timber value increases with age of the stand: $G_A(A) \geq 0$.

Three possible solutions to equation (4) are illustrated in Figure 10.2. The solid black line is the right-hand side of equation (4) and represents the foregone interest earnings on postponed timber harvest revenue and bare land value (the marginal cost of postponing harvest). The dashed lines represent the left-hand side of equation (4) in three cases. In case (1), the dashed line represents only the value of the marginal growth of harvestable volume; there is no standing timber value, $G(A) = 0$. It is optimal to harvest at age A^*. This is the solution to the basic Faustmann model in equation (1). In case (2), standing timber value is positive, $G(A) > 0$, and exceeds the opportunity cost arising from foregone interest earnings on the standing timber value portion of the bare land value; that is, the bracketed term in equation (4), is also positive. This increases the marginal benefit of holding timber and the optimal harvest age, A^{**}, is greater than in the basic Faustmann model, $A^{**} > A^*$. In case (3), standing timber values are large

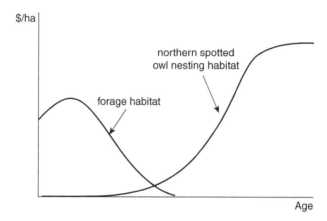

Figure 10.1 Standing timber value functions, $G(A)$, for deer forage habitat in young forest stands and northern spotted owl nesting habitat in old forest stands.

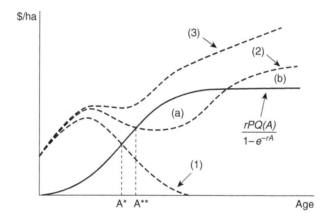

Figure 10.2 First-order conditions for optimal rotation age when standing timber has value, equation (4) for three cases: (1) standing timber has zero value; (2) net marginal value of standing timber is positive and increases with age so that optimal rotation age is longer; (3) net marginal value of standing timber is positive and big enough so that it is optimal to never harvest.

enough, and increase in age sufficiently, so that it is optimal to never harvest this stand. Extensions of the Hartman model have noted that in case (2), it is not enough to look at marginal conditions (Strang, 1983). Total value must be considered as well. The area labeled (a), where marginal cost exceeds marginal benefit of postponing harvest, is a loss. But the area labeled (b), where marginal benefit again exceeds marginal cost, is positive and may be large enough to offset loss (a). If not, so that area (a) > area (b), it is optimal to harvest at age A^{**}; otherwise, it is optimal to never harvest.

Standing timber value need not rise or fall uniformly with age. For example, standing timber value may be high in a young stand as it provides forage, decline in the middle years and rise again as the stand ages to provide habitat for old-growth-dependent species, merging the two standing timber value functions depicted in Figure 10.1. In such cases, there may be multiple local optima so that it is necessary to consider total values (Calish, Fight and Teeguarden, 1978); indeed, changes in economic conditions may cause optimal harvest age to shift between local optima.

The optimal stock of old-growth forest

Although trees are a renewable resource, with rotation ages that are species and end-use dependent, some aspects of forests function more as non-renewable resources. Old-growth structure, as a temporal condition, could return to any one stand on the landscape. However, the long time frame under which it regenerates, coupled with the incentives to harvest long before that, indicates that it may be relevant to consider old-growth an exhaustible (nonrenewable) resource.

In the standard Hotelling model of resource extraction, it is optimal to mine an exhaustible resource at a rate at which increasing scarcity drives real price for the resource to increase at the real rate of return on the next best alternative (e.g. the market interest rate). When there is a backstop technology that can be used to produce a substitute for the exhaustible resource, the initial resource price is set so that the resource is completely exhausted when its price is just equal to the marginal production cost of the substitute. In the case of old-growth forest, plantation forestry is a backstop technology for producing wood. Indeed, real softwood lumber prices in the United States and Douglas-fir stumpage prices in the United States have grown at an average annual rate of 1.5 percent since 1880 and began to slow in 1945; Douglas-fir stumpage prices have grown at an average annual rate of 3.8 percent since 1910 and began to slow in 1945 (Cleveland and Stern, 1993). Real wood prices, although volatile, appear to have leveled off and have been roughly constant since the 1970s. Thus, it appears that plantation forestry began functioning as a backstop technology for old-growth timber soon after World War II and was well-established in the 1970s. Why did this switch not occur when old-growth forest reserves were fully exhausted, as the Hotelling model predicts, but rather while some old-growth forest remains?

Although plantation forest combined with technological innovation in wood processing may provide a close substitute for old-growth forest as a source of wood products, it is a poor substitute for old-growth as a standing forest that provides ecosystem services and the nonmaterial values described in the previous section. It is conceivable that growing public concern over the dwindling area of old-growth forest and public demand for its protection are a reflection of the increasing marginal value of the benefits it provides as it becomes more scarce. In that case, the policy question becomes not one of optimal harvest age, but one of optimal stocking: How much old-growth forest is enough?

Conrad and Ludwig (1994) addressed this question by formulating a dynamic land allocation model to identify an optimal stopping rule for conversion of old-growth forest to plantation forest with a regular Faustmann harvest cycle. Their model differs from the Hotelling model in that plantation forestry occurs at the same time as ongoing mining of old-growth forest, even though wood price is assumed to be constant. In their model, scarcity-driven wood price growth is replaced with an old-growth benefit function. Letting $X(t)$ be the area of old-growth remaining at time t, there is a concave benefit function, $B(X(t))$, for which $B_X(X(t)) > 0$ and $B_{XX}(X(t)) < 0$, so that the marginal benefit of old-growth forest increases as the stock is depleted (as $X(t)$ decreases). It is optimal to harvest old-growth (convert old-growth to plantation forest) as long as net harvest revenue plus bare land value exceeds the marginal benefit of standing old-growth forest stock:

$$PQ(A_X) + \frac{PQ(A_F)}{e^{rA_F} - 1} > B_x(X(t)) \tag{5}$$

where P is stumpage price and $Q(A)$ is harvestable volume at age A as before. A_X represents the age of old-growth at which net volume growth becomes zero and A_F is the Faustmann rotation age. The first term on the left-hand side is the net revenue from harvesting the marginal unit area of old-growth forest and the second term is bare land value on a Faustmann rotation. The right-hand side is the marginal value of a unit area of old-growth forest left standing, given the remaining stock, $X(t)$. Harvest of old-growth forest stops at $X(t) = X^*$ when:

$$PQ(A_X) + \frac{PQ(A_F)}{e^{tA_F} - 1} = B_x(X^*) \tag{6}$$

By itself, this would lead to instantaneous conversion of all old-growth in excess of X^*. A gradual conversion is enforced in this model by constraining the area that can be converted to planta-tion in any period, effectively regulating the forest using area control. An alternative approach to avoiding a gap in supply would be to allow endogenous price determination in the log market while the harvested area of old-growth reaches Faustmann age.

In any case, the ultimate outcome of this model is a static landscape where equation (6) allo-cates fixed areas to industrial wood production and old-growth forest. Old-growth forest is set aside in reserves. This reserve mentality dominated conservation policy in the latter half of the twentieth century. Biodiversity conservation emphasized nature reserve design. Wilderness was set aside to preserve areas 'where the earth and its community of life are untrammeled by man, where man himself is a visitor who does not remain' (Wilderness Act of 1964). The Northwest Forest Plan of 1994 effectively implements an old-growth forest reserve strategy on federal forest land in the US Pacific Northwest; timber harvest is allowed on 22 percent of the federal land within the plan area, with the remainder in some form of reserve, including late successional reserve. Industrial wood production occurs on private land outside the federal reserve system. The result is a static forested landscape allocated between industrial forest production on private land and old-growth forest reserve on federal land.

Tahvonen (2004) generalized the optimal old-growth stocking model in two ways. First, wood price is determined endogenously by specifying a strictly increasing economic surplus function to represent wood products markets. This smooths harvest during a conversion period and leads to a dynamic equilibrium. Second, and more importantly, the standing forest benefit function, $B(X(t))$ depends on the entire age class structure of the forest so that:

$$X(t) = \sum_{s=1}^{N} \alpha_s x_{st} \tag{7}$$

where x_{st} is the area in age class s at time t, α_s represents the contribution of area in age class $s = 1, \ldots N$ to standing timber values and N is the age class above which the forest is considered to be old-growth. Area accumulates in x_{Nt} once it reaches N unless it is harvested. Tahvonen solved this model over a 100-year time horizon for three different specifications for α. In case (1), $\alpha_s = 0$ for $s < N$ and $\alpha_N = 1$; that is, only old-growth forest has standing timber value. The forest is allocated between industrial wood production on a Faustmann rotation and old-growth forest reserve. This case is the same as the Conrad and Ludwig solution with the exception that endogenous wood price smooths harvest during the conversion period. In case (2), $0 < y < N, \alpha_s = 0$ for $s < y$ and $\alpha_s = 1$ for $s = y, \ldots, N$. In this case, standing timber has value prior to becoming old-growth. Existing old-growth is depleted and forest is allocated between wood production on two harvest cycles of Faustmann rotation age and an extended rotation age. In case (3), $\alpha_s = 0$ for $s < y$ and α_s increases linearly from 0 to 1 as the forest ages from age class y to N. Case (3) merges Hartman's stand level model of optimal rotation age when standing timber has value and Conrad and Ludwig's model of optimal stocking. In this case, the forest is allocated between old-growth forest reserve and wood production on an extended rotation age.

Working forest landscape

In practice, there are several difficulties with a reserve approach to protecting old-growth forest. First, the most highly productive, accessible forest land is already allocated to production forestry. The areas that remain as candidates for old-growth forest reserves are typically poor site quality and on steep terrain. They do not represent the full range of variation in forested ecosystems. Thus, to the extent that preserving a natural heritage is an objective, reserves only partially fulfill the purpose. Second, a landscape split between production forests and old-growth reserves will have young forest and old forest and nothing in between. Healthy forest ecosystems are dynamic systems in which forest stands progress through the full range of age classes; the pattern of young and old forest shifts across the entire landscape over time as fire and other disturbance (including timber harvest) reset the successional clock. Finally, a reserve system carries with it risk. As noted, the lack of forest stands in successional stages between plantation forestry and old-growth means that, should severe disturbance strike many old-growth areas, there are no ready replacements waiting with some old-growth structure and characteristics.

Tahvonen's model provides an analytical framework for investigating alternatives to a fixed reserve system. A similar framework has been used to solve numerically for optimal forest management on actual landscapes when standing timber has value. Case study applications incorporated spatial and dynamic complexities into the standing timber benefit function using simulation models for ecosystem services. Examples include Calkin, Montgomery, Schumaker, Arthur and Nalle (2002) and Nalle, Montgomery, Arthur, Schumaker and Polasky (2004). In both of these studies, an actual landscape is managed jointly for wood production and wildlife population viability. Each potential harvest unit is assigned a standing timber value (the α's in Tahvonen's model) in each time period as a function of its vegetation and the vegetation in surrounding units. The benefit function for standing timber is based on wildlife population viability which, in turn, depends on the trajectory of attributes of the vegetation and its spatial configuration, as represented by the α's, over a 100-year time horizon. The benefit function was estimated using output from a spatially and temporally explicit simulation model of wildlife population dynamics on the landscape, given the trajectory of vegetation. Calkin et al. modeled the flying squirrel (*Glaucomys sabrinus*), a species that favors old-growth forest and does not disperse very far as juveniles seek breeding habitat, so that proximity of suitable habitat is important. The α's for each unit increase as the timber in the unit ages and as timber in nearby units age. This is an application of Tahvonen's case (3) extended to include spatial dependencies. Nalle et al. modeled two species with different habitat preferences and different dispersal patterns; the great horned owl (*Bubo virginianus*) prefers old-growth forest and disperses long distances, and the common porcupine (*Erethizon dorsatum*) prefers young forest and disperses short distances. This is an example of a nonconvexity in Hartman's amenity function in which standing timber value is high for young forest, drops off in the middle age classes and increases again as the stand ages. Wood price is endogenously determined in a regional log market. The solution to Nalle et al.'s application is a shifting mosaic of young and old forest on a 1.7 million-ha landscape over a 100-year time horizon. Wildlife populations and timber harvest levels stabilize about halfway through the time horizon. When compared with a 'status quo' simulation of industrial production forestry on private land and forest reserve on federal land, the model solutions are superior, in the sense that wildlife populations could be increased over the status quo without decreasing wood production and vice versa.

Where a reserve approach has been implemented, it is to preserve what little remains of intact old-growth forest. Very little, if any, exists outside of reserved areas. To restore old-growth forest to areas where it has been eliminated is, by definition, a lengthy process. In the Douglas-fir region, it takes more than 200 years to develop old-growth forest by natural processes. In many places, current forest conditions are very different from those that produced the old-growth forests of the past (even in reserves). In particular, stocking is denser and fire has been excluded. It may be that the attributes that we associate with old-growth forest will not develop without active management. Silvicultural methods to promote or maintain old-growth structure are currently being developed (Bauhaus, Puettmann and Messier, 2009). These methods typically involve restoration thinning, long rotations, snag creation (killing trees) and retention of standing dead trees.

What would such active management involve financially and would private forest landowners be willing to do it? In the state of Oregon, the Oregon Department of Forestry (ODF) has set a target of 20 percent of state forest land to be managed for old forest structure, which they take to mean widely spaced large trees, multiple canopy layers and standing and downed large dead wood. Latta and Montgomery (2004) used approximate methods to identify cost-effective management regimes that would achieve ODF criteria in approximately 1,000 measured Douglas-fir forest inventory plots on private land in western Oregon for at least 30 years prior to regeneration harvest. The management regimes they identified involved repeated high-volume commercial thinning that removed 50 to 60 percent of the standing volume; the first occurred at Faustmann rotation age, and subsequent thinnings occurred at 20-year intervals; regeneration harvest occurred at approximately 150 years of age. They found that the opportunity cost of active management for old-growth forest structure was approximately 30 percent of land and timber value for net present value maximizing industrial forest management regimes, on average. That was true across site classes, ecoregions and stocking levels for the existing stands. They also found, in a subsequent study (Montgomery, Latta and Adams, 2006), that a target of 20 percent area with old-growth forest structural attributes within 95 years had an opportunity cost of 254 million 1992 US dollars or $6 per adult Oregonian per year (discounting at 6 percent). This compares to an opportunity cost of 4.4 billion 1992 US dollars of achieving the same objective using reserves and no active management. Interestingly, the optimal allocation of land frequently assigned high-site land to old-growth structure rather than just timber production; high-quality land was able to contribute to the overall old-forest structure goal faster, even though the opportunity costs of foregone timber harvest were higher. One avenue to explore to incentivize such a management strategy could be to make a combined carbon/old-growth payment to forest landowners who subscribed.

Disturbance

In many places, the threat to old-growth forest arises not as much from timber harvest as it does from disruption of fire disturbance cycles. Old-growth forests are dynamic ecosystems and are adapted to cycles of disturbance, of both natural and human origin, that occur at multiple temporal and spatial scales.

The old-growth Ponderosa pine forests of the dry regions of the western United States are a case in point. These forests have adapted over many millennia to frequent, light surface fire that serves a fuel reduction and thinning function. Fires were regularly ignited both naturally by lightning storms and purposefully by indigenous people for an array of reasons. Like the old-growth Douglas-fir forests of the wet temperate zone west of the Cascade Mountains, these forests have dwindled in area dramatically since European settlement. The cause is twofold. First, selective timber harvest of high-value, large Ponderosa pine removes overstory vegetation, primarily Ponderosa pine, and changes the structure and species composition of the forest. But, more importantly for the future, a policy of aggressive fire suppression over the past century has led to forest fire fuel accumulation and the development of understories of ladder fuels that carry fire into the canopies of these forests, where it kills trees. The consequence is that fire, when it does occur, is far more likely to be catastrophic and stand destroying. This is a poignant example of C. S. Holling's (1995) 'brittle ecosystem' in which people, seeking to dampen disturbance cycles in the interest of protecting the resource, have, instead, created conditions that make the forest ecosystem vulnerable to collapse and put the resource at increased risk.

Fire, likewise, has played an important role in the wet temperate forests of the Pacific Northwest, where widespread catastrophic fire served a regenerative function, releasing nutrients into the soil and adding large dead woody material to stream systems where it provides the basis for complex habitat that support salmon populations that are now threatened or endangered throughout the Pacific Northwest. It is highly unlikely that these widespread catastrophic fires will ever be allowed to burn unsuppressed on the densely populated forest landscapes of today.

As a consequence, protection of old-growth forest will require more than set-asides in reserves, especially in the fire-adapted forested ecosystems where fire exclusion has created conditions that put remaining old-growth forest at risk. It will require active management to restore these dry-site fire-adapted forests to conditions in which fire disturbance can be reintroduced as a healthy process. It is important to note that people have been an integral part of every forested landscape for millennia. And fire has been used by people to manipulate ecosystems wherever they go (Pyne, 2001). Although there is an existing literature dealing with cost-effective strategies for suppressing wildfire and optimal placement of fuel treatment to block or slow wildfire, restoration forestry (including restoration thinning, fuel reduction, purposeful prescribed fire and reintroduction of wildfire into forested systems) is a new frontier for forest economists. Most of the economics of restoration forestry has emphasized cost estimation for various treatments, such as prescribed burning. However, Calkin, Hummel and Agee (2005) set the stage for trade-off analysis; they used simulation to quantify trade-offs between restoration treatments to reduce fire threat and late serial forest structure in a late successional reserve mixed fir forest in Washington State.

Concluding remarks

The specific economics of old-growth forest were developed in the context of the dwindling areas of old-growth forest in the wet temperate regions of the Pacific Northwest. It was triggered by social and political events surrounding the ending of a period of mining of old-growth forest in those places. Most economic analyses have dealt with questions of valuation of old-growth and of preserving old-growth in reserves. When is it optimal to never harvest? How much old-growth is enough? What is the opportunity cost of preserving old-growth? How can the benefits of old-growth be quantified?

In the future, the economics of old-growth forest will be more broadly defined. It will be informed by the understanding that old-growth forests are complex dynamic systems rather than static monuments of the past to be preserved in museum-like reserves. The question of old-growth forest will be approached in the context of healthy forested landscapes, the parts of which are interrelated.

Attention is already turning from preservation to restoration of old-growth forest. However, as forest restoration treatments do not typically yield high-quality wood products that can generate revenue, the economics of cost-effective active management for old-growth forest attributes will become even more relevant than in the past. How to incentivize active management for old-growth structure will become a central research and policy question. Some answers may come in the form of payments for carbon offsets. Markets for the by-products of restoration thinning and fuel treatment may emerge as the biofuels industry develops technologies for processing small woody biomass. Experiments in markets for ecosystem services may lead to viable market mechanisms that allow people to bid on forest management strategies that maintain or promote the attributes of old-growth forest that hold cultural or aesthetic value to them.

It is increasingly apparent that disturbance has played a formative role in the development of all old-growth forests and the attributes of old-growth forest that we treasure. In the future, economics of old-growth will have to account for that function. In places where fire hazard is high, placing existing old-growth forest at risk, research questions regarding management of that risk will arise. Where reintroduction of wildfire is infeasible, because of the growing wildland urban interface, we will be looking for cost-effective means of mimicking disturbance on the forested landscape.

Although the interest in the economics of old-growth forests is relatively new, the recognition that these forests provide essential ecological, cultural and social benefits to humans has made them a major issue in forestry and forest economics. The preservation of existing stands, the management for future stands and the value that these forests hold have become critical policy questions debated among ecologists, politicians and the public. Forest economists can play a vital role in providing information about the costs and benefits of old-growth and the trade-offs inherent in these forest management decisions. They can also provide guidance for reforming institutional arrangements and incentive structures for preserving or restoring old-growth forest – or what Bauhaus et al. (2009) labeled 'old-growthness'. Issues surrounding old-growth forest incorporate many of the classic problems in resource and environmental economics, such as optimal use of nonrenewable resources and valuation of nonmarket and public goods, and many newer problems, such as spatial dependencies and institutional arrangements. The evolution of defining, valuing and managing for old-growth forest will continue to drive new ideas, models and solution techniques in forest economics.

References

Bauhaus, J., Puettmann, K. and Messier, C. (2009). 'Silviculture for old-growth attributes', *Forest Ecology and Management*, vol. 258, pp. 525–537.

Calish, S., Fight, R. D. and Teeguarden, D. E. (1978). 'How do nontimber values affect Douglas-fir rotations?' *Journal of Forestry*, vol. 4, pp. 217–221.

Calkin, D. E., Hummel, S. S. and Agee, J. K. (2005). 'Modeling trade-offs between fire threat reduction and late-seral forest structure', *Canadian Journal of Forest Research*, vol. 35, pp. 2562–2574.

Calkin, D. E., Montgomery, C. A., Schumaker, N. H., Arthur, J. L. and Nalle, D. J. (2002). 'Developing a production possibility set of wildlife species persistence and timber harvest value using simulated annealing', *Canadian Journal of Forest Research*, vol. 32, no. 8, pp. 1329–1342.

Cleveland, C. J. and Stern, D. I. (1993). 'Productive and exchange scarcity: An empirical analysis of the U.S. forest products industry', *Canadian Journal of Forest Research*, vol. 23, pp. 1537–1549.

Conrad, J. and Ludwig, D. (1994). 'Forest land policy: The optimal stock of old-growth forest', *Natural Resource Modeling*, vol. 8, no. 1, pp. 27–45.

Englin, J. P. and Mendelsohn, R. O. (1991). 'A hedonic travel cost analysis for valuation of multiple components of site quality: The recreation value of forest management', *Journal of Environmental Economics and Management*, vol. 21, pp. 275–290.

Hagen, D. A., Vincent, J. W. and Welle, P. G. (1992). 'Benefits of preserving old-growth forests and the spotted owl', *Contemporary Policy Issues*, vol. 10, pp. 13–25.

Hartman, R. (1976). 'Harvesting decision when a standing forest has value', *Economic Inquiry*, vol. 14, no. 1, pp. 52–58.

Haynes, R. W. (2009). 'Contribution of old-growth timber to regional economics in the Pacific Northwest', in T. A. Spies and S. Duncan (Eds.), *Old-growth in a New World: A Pacific Northwest Icon Reexamined*, Island Press, Washington, DC.

Hilbert, J. and Wiensczyk, A. (2007). 'Old-growth definitions and management: A literature review', *BC Journal of Ecosystems and Management*, vol. 8, no. 1, pp. 15–31.

Holling, C. S. (1995). 'What barriers? What bridges?', in L. H. Gunderson, C. S. Holling and S. S. Light (Eds.), *Barriers and Bridges to the Renewal of Ecosystems and Institutions*, Columbia University Press, New York.

Kimmins, J. P. (2003). 'Old-growth forest: An ancient and stable sylvan equilibrium, or a relatively transitory ecosystem condition that offers people a visual and emotional feast?' *The Forestry Chronicle*, vol. 79, no. 3, pp. 429–440.

Kuuluvainen, J., Karppinen, H. and Ovaskainen, V. (1996). 'Landowner objectives and nonindustrial private timber supply', *Forest Science*, vol. 42, no. 3, pp. 300–309.

Latta, G. and Montgomery, C. A. (2004). 'Minimizing the cost of stand level management for older forest structure in western Oregon', *Western Journal of Applied Forestry*, vol. 19, no. 4, pp. 221–231.

Montgomery, C. A., Brown, G. M., Jr. and Adams, D. M. (1994). 'The marginal cost of species preservation: The northern spotted owl', *Journal of Environmental Economics and Management*, vol. 26, pp. 111–128.

Montgomery, C. A., Latta, G. S. and Adams, D. M. (2006). 'The cost of achieving old-growth forest structure', *Land Economics*, vol. 82, no. 2, pp. 240–256.

Moyer, J. M., Owen, R. J. and Duinker, P. N. (2008). 'Forest values: A framework for old-growth forest with implications for other forest conditions', *The Open Forest Science Journal*, vol. 1, pp. 27–36.

Nalle, D. J., Montgomery, C. A., Arthur, J. L., Schumaker, H. H. and Polasky, S. (2004). 'Modeling joint production of wildlife and timber in forests', *Journal of Environmental Economics and Mangement*, vol. 48, no. 3, pp. 997–1017.

Pyne, S. J. (2001). *Fire: A Brief History*, University of Washington Press, Seattle.

Spies, T. A. (2004). 'Ecological concepts and diversity of old-growth forests', *Journal of Forestry*, vol. 102, no. 1, pp. 14–20.

Strang, W. J. (1983). 'On the optimal forest harvesting decision', *Economic Inquiry*, vol. 21, no. 4, pp. 576–583.

Tahvonen, O. (2004). 'Timber production versus old-growth preservation with endognous prices and forest age-classes', *Canadian Journal of Forest Research*, vol. 34, pp. 1296–1310.

Van Kooten, G. C. (1995). 'Economics of protecting wilderness areas and old-growth timber in British-Columbia', *The Forestry Chronicle*, vol. 71, no. 1, pp. 52–58.

Weisberg, P. J. and Swanson, F. J. (2003). 'Regional synchroneity in fire regimes of western Oregon and Washington, USA', *Forest Ecology and Management*, vol. 172, no. 1, pp. 17–28.

Wilderness Act of 1964, Pub. L. 88–577. Retrieved from http://wilderness.nps.gov/document/Wilderness Act.pdf

11

VALUATION OF OPEN SPACES IN URBAN ENVIRONMENTS

Neelam C. Poudyal[1] and Donald G. Hodges[2]

[1]ASSISTANT PROFESSOR, NATURAL RESOURCE POLICY AND HUMAN DIMENSIONS, DEPARTMENT
OF FORESTRY, WILDLIFE AND FISHERIES, UNIVERSITY OF TENNESSEE, 274 ELLINGTON PLANT
SCIENCE BLDG., KNOXVILLE, TN 37996-4563, UGA
[2]JAMES R. COX PROFESSOR OF FOREST ECONOMICS AND POLICY, DEPARTMENT OF FORESTRY,
WILDLIFE AND FISHERIES, UNIVERSITY OF TENNESSEE, 273 ELLINGTON PLANT
SCIENCES BLDG., KNOXVILLE, TN 37996-4563, USA.

Abstract

As green spaces in urban areas face increasing pressure for development, sustaining the public good benefit of these resources will depend on land managers' abilities to prove that the welfare associated with open space policies outweighs the cost of acquisition and preservation. This chapter introduces and discusses basic premises and practices of the hedonic method of housing price and reviews some of the recent economic studies applying this method in open space valuation. It also introduces readers to some of the issues and recent advancement in the hedonic method, which has been the model of open space valuation thus far. The chapter concludes with key points summarizing the state of the art of open space valuation research and offering some directions for future research.

Keywords

Aesthetic value, housing value, implicit prices, land-use diversity, nonmarket valuation, product attribute, public good, recreation land, spatial pattern, willingness to pay

Introduction

Open spaces provide urban residents with a variety of environmental, recreational and aesthetic benefits. Despite these benefits, open spaces in urban environments are shrinking due to market pressure for development. However, rapidly growing populations in metropolitan areas have increased the demand for open spaces such as recreation parks, whereas such public resources are relatively constant in supply, resulting in declining open space per capita (Kline, 2006). Increasing congestion of the urban environment necessitates acquisition of additional open space and enforcement of new public policies.

In response to this demand, government and nongovernmental organizations have developed initiatives in open space protection and management. The US Department of Agriculture

162

Forest Service, for example, developed the Open Space Conservation Strategy, which emphasizes active engagement of communities and landowners to identify and protect priority open spaces (USDA Forest Service, 2007). Similarly, the Trust for Public Land generated $34 billion in public funds, which have been invested in creating and helping to pass 425 conservation ballot measures (TPL, 2012). In fact, 80% of those ballots were passed in favor of open space. These observations are evidence of overwhelming support for open space policies among citizens, and leadership and institutional commitment among land management agencies.

Open space policies, especially investments in the acquisition of new open space or easements, often face the question of whether the benefits outweigh the costs. In fact, one of the six major strategic goals of *National Research Plan for Urban Forestry 2005–2015* is to understand the economic benefits and real estate value added by open space resources like urban forests (Clarke, Kruidenier and Wolf, 2007). Estimating the value that residents place on open space may help derive expected economic benefits of open space protection policies in urban environments, where vacant lots usually have a very high opportunity cost.

The remainder of this chapter will introduce nonmarket approaches in valuing public goods and then present a description of hedonic valuation, the most commonly used method in open space valuation. It will discuss some of the economic as well as econometric issues surrounding hedonic valuation and alternative methods. Finally, the chapter will conclude with a review of the recent advances in open space valuation. This chapter draws mainly from open space literature in the United States, but the principles and practices discussed here are equally applicable globally.

Valuation of environmental goods

Economists have developed a variety of nonmarket approaches to estimate the economic benefits associated with such resources. When development expands in the urban environment, natural spaces become scarce (e.g. short supply) and developed properties receive a positive externality from existing or newly acquired open space. The value of this externality is estimated by the premium residents might be willing to pay for its consumption. Two approaches, i.e. stated preference and revealed preference, are commonly used in estimating the value of open space. Stated preference methods directly ask beneficiaries (i.e. households) their willingness to pay (WTP) for protecting open space or willingness to accept (WTA) compensation for forgoing the open space benefit they currently enjoy. One of the methods of stated preference is contingent valuation (CV), which describes the good (i.e. open space) to be provided, and then asks a randomly selected sample of residents to state their maximum willingness to pay in order to protect open space for a given period of time (Boyle, 2003). Such questionnaires can also ask residents to accept or reject a suggested payment amount. Carefully designed CV questions explicitly explain to residents the exact payment to be made, duration of payment and vehicle of payment (e.g. tax, donation, use fee, utility cost, etc.).

The revealed preference approach, on the other hand, employs a rather indirect method of estimating value for open space resources. A consumer's preference for a nonmarket good is indirectly analyzed by examining his or her behavior on purchasing and consuming market goods that are somehow related to the nonmarket good of interest. In the case of an urban environment, the house typically is a market good that is directly related to the open space nearby, and therefore the hedonic model of housing price is a commonly used technique (Taylor, 2003). The basic premise behind this proposition is that when a homebuyer decides to pay a certain price for a house, he or she considers not only the physical attributes of the house itself but also the amenities nearby (e.g. proximity to work, neighborhood quality, air quality, open space and greenery, etc.).

Hedonic valuation is widely applied in open space valuation as researchers analyze market data to trace the amenity effect as well as the implicit price of the open space good based on a regression analysis, which is free from bias issues prevalent in the CV method (e.g. starting point bias, social desirability bias, hypothetical bias). Hence, this chapter focuses on the hedonic valuation method, as it has been the primary workhorse in open space valuation. The next section will present a simple illustration of hedonic valuation with specific reference to a household's utility and externality of environmental goods like open spaces.

Homeowner's utility and hedonic method

The basic premise of hedonic theory as Rosen (1974) originally proposed is that the value of a product is defined by its attributes (Lancaster, 1966). This theory posits that goods are aggregates of difference attributes and not all of those attributes can be bought or sold separately. Hence, the buyer of the good pays a price for all those attributes. In the case of the hedonic model of housing price, for example, a homebuyer maximizes his or her own utility while purchasing a bundle of attributes associated with the property. The property comes with both market goods such as the number of bedrooms, swimming pool and so forth, as well as nonmarket goods like open space. The price that a buyer agrees to pay for a house reflects the value placed on the property itself and a number of other items (e.g. view of open space) that are not separately sold to the homeowner but become available to him or her with the purchase of the house.

Of note, even though an individual has no control over the quantity or quality of open space near a property, he or she can be strategic in finding a property located close to an open space of desired quantity and quality, and the premium associated with that reveals the value he or she places on open space. According to the theory of marginal rate of substitution, a homebuyer's utility is maximized when he or she is ready to give up one good in exchange for another while maintaining the same level of utility. By the same token, a price-taking buyer, facing a range of choices in the housing market, maximizes his or her household utility by choosing the quality and quantity of open space up to the point where the marginal benefit (of living within a certain distance of open space, or having an open space of certain size in an adjacent lot) equals marginal cost (extra price of the house). Observing many transactions with considerable variation in the level of open space amenity consumed across transactions allows economists to estimate the marginal implicit price of this nonmarket good. Typically, the utility and income constraints are expressed in a set of equations that can be solved with the first partial derivative of the price equation to estimate the willingness to pay for open space resource (Taylor, 2003). Willingness to pay estimated for an average household can be aggregated among all targeted beneficiaries to compute the total value of open space for the entire community. The next section will discuss several measures of open space resources.

Measuring open space amenity

Although a variety of approaches have been adopted for measuring open space amenities, the selection of the appropriate measure depends on the policy question (i.e. what aspect of open space is of value concern). A review of previous research on open space valuation reveals that measures of open space can be broadly classified into two types, quantity and quality.

Open space in quantity

Measuring the actual quantity of open space in terms of area is a simple and logical approach. Size of the nearest open space is measured in the most appropriate areal units (e.g. square footage, acreage). Typical sources of data in these applications are square footage of the open space parcel from the county or city tax assessor's parcel database or satellite imagery of land cover and/or land-use data processed with remote sensing software.

Researchers have used area measures in valuing a variety of open space types, including forest (Cho, Poudyal and Roberts, 2008; Poudyal, Hodges, Fenderson and Tarkinson, 2010), wetland (Mahan, Polasky and Adams, 2000), conservation land (Irwin, 2002), public park (Poudyal, Hodges and Merrett, 2009), private backyard (Peiser and Schwann, 1993) and general open space (Smith, Poulos and Kim, 2002). Further, two recent publications (Waltert and Shlapfer, 2010; McConnell and Walls, 2005) have summarized most of the empirical research on open space valuation. Although some open space types such as forests, greenbelts, lakes and nature preserves are consistently found to have a positive amenity effect on housing prices (i.e. increase in size of nearby open space increases the house price), other types such as neighborhood parks and forested wetlands have provided mixed results. This is not surprising, as open spaces like forests hold higher aesthetic appeal and sense of naturalness, whereas human-modified open spaces like croplands and neighborhood parks are often linked with negative externalities, potentially due to a range of factors like farm odor (Johnston et al., 2001), human presence and congestion (McConnell and Walls, 2005).

Open space in quality

Recent assessments of open space valuation have adopted a variety of techniques for measuring the quality of open space. An issue with the *size* measure of open space as discussed in the previous section is that it considers open space as a homogenous commodity and does not account for differences in attributes (e.g. proximity to house, public access for recreation use, land cover characteristics). Such an approach is not very consistent with the theory of product attributes (Lancaster, 1966).

Nevertheless, the quality approach of measuring open space benefit is not free of criticism. The primary challenge is translating quality-based measures to a quantifiable policy in open space protection. In other words, quality measures are often nonquantifiable for implementation and evaluation purposes. However, economists have used a variety of interdisciplinary tools, such as landscape ecology and geospatial technology, that are useful in appropriately capturing the quality aspect of open space.

Proximity

Proximity has been a widely applied measure in open space valuation. This is considered an important attribute as it is linked to accessibility. For example, all things being equal, a green lawn that is in an adjacent parcel could yield a higher positive externality (i.e. amenity benefit) to a household than other similar lawns located a block away. Public open spaces (e.g. parks, sports fields) that are at closer in distance involve less cost in accessing them (walking time) and come with more local benefits (e.g. cooling effect, scenic value). For this reason, open space benefits are considered local, and any policy protecting open space should take the targeted service area into consideration.

Alternative measures of proximity have been used in terms of dummy variables, i.e. whether there is an open space within a certain distance. This allows one to compare the difference in price of two otherwise identical houses, with and without immediate access to open space, and attribute that difference in price to open space. Lutzenhiser and Netusil (2001), for example, employed this technique and found the marginal value associated with having a house within 1,500 feet of natural areas to be $10,648. Equivalent values for specialty parks facilities and urban parks were found to be $5,657 and $1,214, respectively (McConnell and Walls, 2005). Another study by

Thorsnes (2002) estimated a $5,800 to $8,400 premium for houses that were located next to a forest preserve. Comparatively higher premiums for adjacency to a forest preserve are noteworthy and perhaps justifiable because forest preserves are permanently protected open spaces with little likelihood of development and provide some assurance of a long-term benefit to homebuyers.

For many environmental goods (or bads), distance measures seem to be the appropriate measures of externalities. If urban foresters or city planners know the marginal benefit associated with the distance to public open space, for example, they can estimate the utility gain of an individual household from an open space acquisition program that aims to decrease the average distance between houses and public space. This information could be of significant value in strategically locating new open space plots within a city.

Beyond 'proximity'

A distance measure is not always appropriate for measuring open space benefits; therefore, alternative measures of open space quality have been used in recent years (Geoghegan, Wainger and Bockstael, 1997; Cho et al., 2008). These measures are more advanced and comprehensive in representing the amenity benefits associated with open space resources at the neighborhood level but not represented by a distance measure (Acharya and Bennett, 2001). In particular, such measures can capture the compositional diversity in vegetation, viewshed, spatial pattern and configuration of open space patches in the urban land-use matrix. Various theories and methodological frameworks borrowed from environmental psychology (prospect-refuge theory) and landscape ecology (e.g. theory of island biogeography, interior area, edge effect) have been incorporated into typical hedonic models of nonmarket valuation to quantify and evaluate these quality measures of open space (Geoghegan et al., 1997).

Variation in premium

Open space is also valued differently depending on location. For example, an open space plot located in a less developed neighborhood with abundant natural areas nearby could be valued significantly less by nearby residents, compared to a similar plot located in an urban neighborhood with few natural areas nearby. A simple approach to reveal variation in premium would be to split the sample into rural and urban and estimate the amenity benefit from two separate hedonic models. However, this would not be justifiable from an econometric standpoint unless the hedonic function significantly differs between these two regions. Geoghegan et al. (1997) and Cho et al. (2008) employed two versions of varying parameter models, which allowed regression coefficients associated with open space variables to vary across the city area. Geoghegan et al. (1997) used a spatial expansion method, and Cho et al. (2008) used geographically weighted regression; both noticed a significant variation in value of open space amenity depending on the location within a city (e.g. urban, rural, rural-urban interface). Anderson and West (2006) also noted that the proximity to space was valued higher in high-density neighborhoods than elsewhere.

Variety matters

Size and proximity measures of open space as widely used in earlier hedonic studies fell short of including the other aspect of space quality: variety in cover. One would expect that open space of heterogeneous land cover (i.e. multiple cover types) may be more appealing from an aesthetic standpoint than that of homogenous land cover (i.e. one cover type only). According to theories in environmental psychology, land cover heterogeneity in open space adds to the variety of attraction and aesthetic beauty, appreciated more by humans and therefore resulting in higher values (Appleton, 1975). With the increasing availability of user-friendly GIS interfaces, economists have been able to quantify the diversity aspects of open space. For example, Geoghegan et al. (1997) and Acharya and Bennett (2001) evaluated land-use diversity index, originally used in landscape ecology literature (Turner, 1990) in a hedonic model. Geoghegan et al. (1997) noted that the amenity value depended partially on the location of the neighborhood in relation to the city center, whereas Acharya and Bennett (2001) concluded that land-use diversity and richness were not desirable regardless of location. This mixed result is partly attributable to the fact that they included developed land as well as open spaces in creating the diversity index (McConnell and Walls, 2005). Although this is a good measure of land-use diversity, it is not the same as open space diversity, which was later used by Poudyal, Hodges, Tonn and Cho (2009) in creating a separate index for open space diversity and developed land diversity. Poudyal, Hodges, Tonn, et al. (2009) found that, compared to their homogeneous counterparts, heterogeneous open space plots (i.e. with a variety of land cover types such as pine trees, hardwood and mixed species, water, grassland, farmland) significantly increased the housing price within the neighborhood.

Few bigger versus many smaller

Urban foresters and landscape designers involved in open space acquisition often face the questions of number and location of new plots. The challenge of resource allocation often involves a trade-off between efficiency and equity, i.e. whether to allocate the available budget to acquire a large plot in one single location or rather split that budget to acquire several smaller plots in many locations. From the environmental justice perspective, the latter option is justified, but from an efficiency standpoint, the former seems preferable. This is because providing open space of some size to as many residents as possible would be more equitable than providing a big open space in close proximity to a few residents. Conversely, larger open space can have higher social and physical carrying capacity, conserve ecosystem values and functions, and accommodate a variety of recreation activities for the residents. In addition, open spaces of significant size are considered landmarks of high cultural and social significance as well as community identity (e.g. Central Park in New York, Humboldt Park in Chicago).

Hedonic models using the size of open space have usually revealed a positive effect on housing price, suggesting that larger open spaces are valued higher. However, this still does not answer the question on the trade-off between size and number. According to a study by Poudyal, Hodges, Tonn, et al. (2009) that used a fragmentation index (McGarigal and Marks, 1995) to measure how a given acreage of open space is disaggregated among plots (few vs. many), a house located in proximity to a large open space would be valued more than a similar house with many smaller plots of open spaces nearby. That means residents place a higher value on a few large plots of open space than on many small plots. This supports investing more open space budget monies in acquiring a few larger plots than many smaller ones. Consistent with this notion, local governments in recent years have seen rising public demand for bigger parks rather than smaller pocket parks. For example, based on the expressed preferences of its residents, the Athens-Clark County government in Georgia incorporated this design consideration as the acquisition policy in its new strategic plan (Aued, 2012).

What about shape?

Spatial configuration of an open space plot may affect its aesthetic value substantially, as well as its ability to offer a recreation spectrum. For example, a linear narrow strip of greenbelt along a stream may be spacious enough for jogging or pet walking and may add significant greenery, but it may not be suitable for other recreational activities requiring wider space. Urban forest-ers and landscape planners involved in land acquisitions for open space protection face a choice set of available parcels that vary in shape (e.g. linear, square, rectangular, triangular, trapezoidal). Managing parcels with smooth edges and rectangular or square shapes could be cheaper as far as landscape design and maintenance are concerned. Understanding whether and how shapes or boundary structure may influence the value of open space plots can be useful in preserving human values and aesthetic beauty, and adding real estate tax revenue.

Recent open space studies have demonstrated that plots of square or rectangular shapes with straight edges are more likely than those with irregular shapes to have a positive impact on hous-ing prices (Nelson, Kramer, Dorfman and Bumback, 2004; Poudyal, Hodges, Tonn, et al., 2009). According to Poudyal, Hodges, Tonn, et al. (2009), all things being equal, a house located in a neighborhood containing open spaces of square or rectangular plots with linear and straight edges would sell for a significantly higher price than an identical house located in a neighbor-hood containing plots of irregular shapes with convoluted and rough boundaries. Analogous to habitat ecology, open space plots in irregular and more linear shapes have a higher edge effect (i.e. less core area) that could arguably often translate to more noise and less natural-looking space.

View

Another important attribute of an open space is its view, for which open space literature still shows a variety of results. Insignificant relationship between visibility of open space from a house and the price of that house was reported by Kask and Maani (1992) and Beron, Murdoch and Thayer (2001). Those who did find a statistically significant relationship noticed a different kind of externality, i.e. positive (Tyrvainen and Miettienen, 2000; Bond, Seiler and Seiler, 2002; Bourassa, Hoesli and Sun, 2004) and negative (Paterson and Boyle, 2002). Criticism exists on the measurement validity and reliability of sample in those studies (Bourassa et al., 2004; Sander and Polasky, 2009). Most recent attempts in valuing views have adopted a GIS-based program to compute the total area covered by open space within a house's viewshed (i.e. total area visible from the location of a given house). A study by Paterson and Boyle (2002) in rural Connecticut used open space of various types as a percentage of the extent of view within 1 km and found that the visibility of forestland was negatively related to housing price, whereas a Minneapolis study by Sander and Polasky (2009) noted a positive effect of viewshed area in general but found an insignificant effect of the proportion of forest within the viewshed. According to this study, the estimated premium associated with a 100-m^2 increase in viewshed area regardless of land-use type was \$386. Yet another study in Nashville, Tennessee, by Poudyal, et al. (2010) estimated a premium of \$30 for each acre of forestland visible from a house. Although the value of viewshed in general and view of forests are not directly comparable, discrepancy in the premium between these two studies could be attributed to the differences in the level of urbanization and, more importantly, the abundance of open space between Minneapolis and Nashville. The implicit price (i.e. marginal willingness to pay) for a view of open space could be relatively high in a city where natural areas are relatively scarce.

Open space valuation: A two-step process

Almost all studies in hedonic valuation of open space estimate the implicit price of open space by regressing house price against an open space variable, in addition to variables representing structural characteristics of house and neighborhood features. Basically, the regression coefficient gives the partial derivative, ceteris paribus effect, of the open space variable on the housing price, which is interpreted as the implicit price (or marginal WTP) of open space amenity. This process is commonly known as a first-stage hedonic model. Even though this is sufficient for evaluating marginal prices, policies of open space protection are often nonmarginal in nature. In such cases, the implicit price estimated from the first-stage regression does not accurately reflect the benefit ex ante (Taylor, 2003) and therefore requires developing an uncompensated demand function that shows a relationship between the implicit price and actual quantity of open space being consumed. Evaluating the area under the demand curve between two quantity levels allows one to estimate the consumer surplus, which is the monetary measure of the benefit a household enjoys from the consumption of additional open space amenity. This stage of estimating the demand function is called the second stage in hedonic valuation.

The second stage of hedonic valuation is rarely estimated, as it is difficult to identify the demand with the same set of data for a given housing market. That means deriving a demand curve requiring multiple implicit prices, whereas results from a single hedonic equation in the first stage yields only one implicit price. For this reason, economists use housing transaction data from multiple markets (i.e. geographical areas such as cities and towns) or multiple market periods (i.e. time periods such as 1980s, 1990s, 2000s). Once those markets or periods are proved to be distinct from each other (i.e. hedonic function is significantly different), implicit prices from those markets could be used in estimating the demand for open space (Freeman, 1993). Typically, the demand function models the quantity of open space amenity (observed) as a function of the implicit price of open space itself (estimated from the first-stage regression), the implicit price of complement or substitute goods (e.g. implicit price of living area), and exogenous socio-demographic characteristics of the consumer. Integration of the inverse demand function at two levels of open space amenity is calculated and interpreted as the consumer surplus that an average household receives from the change in open space quantity.

Existing hedonic studies involving second-stage demand modeling have mostly used data from multiple cities or metropolitan areas (Brasington and Hite, 2005). However, it is difficult to get data from multiple markets while dealing with valuation of an open space amenity that is unique to a region or a city. The implicit price of open space must be estimated for each submarket (e.g. distinct regions of housing market within a city) and a demand curve derived. A few recent applications of second-stage demand models in open space valuation are worthy of discussion. In an effort to value wetlands, Mahan et al. (2000) divided the city of Portland, Oregon, into four quadrants and estimated the implicit price for a wetland area, based on a relatively robust first-stage model. However, the estimated demand curve showed a positive relationship between quantity and price, which is inconsistent with the economic theory of demand for normal goods. This raised a question on the assumption of distinct submarkets within a city, even though the study also suffered from two conceptual and methodological issues, replying on ad hoc submarkets, and not controlling for the substitutes or complements in inverse demand function.

Emerging literature in housing economics suggests that a metropolitan area may constitute multiple submarkets, which can be identified by multivariate cluster analysis of housing data (Bourassa, Hoesli and Peng, 2003). The clustering approach segments a market based on similarities and differences in housing properties, locational characteristics, and other features. The resulting segments (or submarkets) possess maximum similarity among houses within a submarket, but minimum similarity between houses of different submarkets. Based on this notion, a recent study by Poudyal, Hodges and Merrett (2009) used a multivariate statistical clustering method to identify submarkets within the city of Roanoke, Virginia, and then successfully estimated a second-stage demand function for public park acreage. The estimated demand function was consistent with the economic theory of demand, and the welfare value associated with open space policy is on par with results from comparable studies. Similarly, Day, Bateman and Lake (2007) have also successfully estimated a second-stage demand model with the identification of submarkets based on statistical clustering of hedonic variables. Poudyal, Hodges and Merrett (2009) reported that expanding public open space within the city limit by 20% from its current level would bring an estimated $160 in consumer surplus per household, which was aggregated over 40,000 properties to estimate the total surplus of $6.5 million associated with the policy.

Economic and econometric issues with hedonic methods

Hedonic analyses of open space valuation encounter a number of econometric issues, such as endogeneity of open space variables, spatial autocorrelation of the residual term, nonstationarity of the relationship between housing price and open space, reliability of property price data and selection of substitute or complement variables in second-stage demand model. The following paragraphs introduce these issues and briefly elaborate upon the ways economists have dealt with them.

Endogeneity of open space variables in hedonic equations was reported by Irwin and Bockstael (2001) and Irwin (2002). They acknowledged that the value of a house might be affected by the presence and quantity as well as quality of open space nearby, but at the same time, the state of open space plots (e.g. presence, size, quality) could be affected by the average housing value in the neighborhood. Open space plots that are privately owned or not under any protected status (e.g. conservation easement, public park) could be in high demand for real estate development. Further, new homebuyers may consider the risk of their potential future development. This could result in biased estimates of implicit prices, and for this reason, recent efforts in open space valuation have instrumented open space variables with exogenous instrumental variables (e.g. Cho et al., 2008; Poudyal, Hodges, Tonn, et al., 2009).

Spatial autocorrelation is the correlation of error term that results from the omission of variables that are believed to influence housing price. Because houses in close proximity share similar neighborhood features (e.g. amenities, crime, school quality), failure to take account of all possible factors can result in a biased implicit price. Irwin (2002) took a subsample of houses that were not close to each other to minimize the correlation in error term, whereas Anderson and West (2006) used local fixed effect models to account for missing spatial variables. Recent developments in spatial econometrics (Anselin and Bera, 1998) have also developed a number of spatial autoregressive models (SAR) that can correct for spatial dependence in housing data, and appropriately adjust the premium estimates.

Spatial nonstationarity in the relationship between open space and property price has been discussed in several studies (Geoghegan et al., 1997; Cho et al., 2008; Anderson and West, 2006). This issue arises when the marginal contribution of the open space amenity on housing price does not stay constant across the urban area of interest (Fotheringham, Brunsdon and Charlton, 2002). Therefore, using the average benefit estimated for the entire city can be inaccurate. As discussed earlier, a number of varying parameter models have been applied, e.g. spatial expansion (Geoghegan et al., 1997) and geographically weighted regression (Cho et al., 2008), to compute a range of premiums over a city area. Local variation in coefficients often indicates that site-specific open space programs might be needed, but the policies are homogenously designed and implemented over a political unit (e.g. city, township, etc.) and consistency could be a challenge.

The hedonic method can yield biased estimates of willingness to pay for open space if the housing market is not in *equilibrium*. This happens when the homebuyers are not aware of the environmental benefit (perceived vs. real risk/benefit), and are not knowledgeable of the cost of mobility (Pope, 2008; Rosen, 1974). In a disequilibrium market, the price differential between houses with different levels of open space would not correspond to the marginal willingness of the homebuyer to pay for the open space. Similarly, sales transactions observed in a relatively short period of time may not fully capture the market in equilibrium. Data from a disequilibrium market (e.g. economic recession or temporary surge in housing demand) can also result in biased welfare estimates of open space policy. Further, there are some other assumptions regarding the housing market that must be met for the hedonic valuation method to be valid. For example, a market should be defined by a particular geographical area, and each household in the market is assumed to demand one house. It is assumed that the buyers are free to choose among all properties in the market. Making sure that the housing market under study meets all these criteria can be a real issue.

Selection of appropriate *substitutes and complements* in the second-stage demand equation still remains a challenge. Per demand theory, the quantity of open space demanded should be explained by its own price, price of substitutes and complements, and socio-demographic factors. The implicit price of open space itself can be computed from the first-stage hedonic regression, but little is known about what should be considered reliable substitutes or complements for open space. Studies have used implicit price of living area (square footage of house) as substitute for different environmental amenities (e.g. Boyle, Poor and Taylor, 1999, for water quality of nearby lake; Poudyal, Hodges and Merrett, 2009, for nearby public open space). Poudyal, Hodges and Merrett (2009) concluded that public open spaces are substitutes for household lot size, suggesting that providing more public open spaces could be an effective policy in fostering high-density development and reducing sprawl.

Aggregating benefits among beneficiaries remains another economic question of concern in weighing the costs and benefits of an open space policy. For example, most of the hedonic studies use transactions of single family residential properties only, and the estimated benefit from such analysis does not reflect the value for apartment renters who also enjoy the benefit. This is perhaps justifiable, as the homeowners are the long-term residents as well as consumers of these amenities and are more influential in local policies through their voting. Nevertheless, a complete analysis should also incorporate how the total benefits of open space polices are transferred to the public through increased premiums on rented properties. Another problem in aggregation is to identify the population of beneficiaries. Once the consumer surplus is estimated on a per-household basis, it is not clear over how many households within the urban area the welfare should be aggregated. All these questions are very important if the objective of valuation is to establish an efficient market-based mechanism for payment for ecosystem services.

Other methods of open space valuation

In addition to hedonic valuation, a number of alternative approaches have been applied in recent years. Analyzing community voting and the cost of bonds related to open space protection (Bates and Santerre, 2001) is one example. Total funds raised by a community through some sort of referendum or the bonds issued by a municipality to implement an open space project are some of the monetary measures applicable in this approach. Another example is a replacement cost method, which relies on the estimation of total cost involved in providing all kinds of public goods by alternative means, if the open spaces were not present (Polasky, 2012). This method has been popular recently with the emerging markets for ecosystem services and the need to estimate the value of such services.

Other methods of valuation utilize a production function approach to assess the amenity or disamenity of land-use variables. For example, Frey, Luechinger and Stutzer (2009) used a life satisfaction approach, which uses a typical multiple regression framework to explain an empirical measure of human well-being (usually measured at a community level) as a function of environmental amenities like open space, in addition to many other community characteristics. Similarly, Poudyal, Hodges, Bowker and Cordell (2009) used a life expectancy production function, where a ceteris paribus effect of open spaces amenities of different types on average expectancy of community residents was analyzed. The appropriate method depends on the availability of relevant data and expertise and, more importantly, the policy question to be answered.

Conclusion

Nonmarket valuation techniques have been widely used in estimating the value of open spaces in urban environments. Even though some stated preference approaches such as contingent valuation have been used to directly solicit the dollar value beneficiaries attach to the open space amenity of concern, the revealed preference-based hedonic valuation of property price remains the most popular method. The estimated value can be used in quantifying the welfare effect of a public policy protecting open space for public use, or in estimating welfare compensation residents may deserve in exchange for a development project. Findings from recent valuation studies that exclusively focused on the quality of open space (e.g. diversity, shape, pattern) have provided remarkable insights into understanding public preference for landscape designs, in allocating limited tax dollars to acquire parcels of desirable features in desirable locations, and in justifying municipal investment in landscape engineering.

Open space valuation has made substantial progress over the last few decades in terms of minimizing measurement error and accurately measuring the amenities of interest and, more importantly, tracing their effects on property price in robust econometric models. Recent developments in remote sensing and GIS technology have facilitated the analysis of land-use data in precisely measuring open space plots of any size and shape. Furthermore, resource economists have used interdisciplinary tools in quantifying the qualitative aspect of open space and evaluating them in advanced econometric models that correct for a number of bias and inefficiency issues prevalent in earlier efforts.

Available methods may not agree on the estimated benefit, but research has shown that carefully designed nonmarket valuation studies come fairly close (e.g. convergent validity) to confirming the positive externality value of open space in an urban environment. Thus, not estimating or not accounting for the value of open space in the benefit-cost equation of land management policies is like attaching zero value to these important resources, which will result in an inefficient market. Open spaces come in a variety of forms, and each has a unique set of ecosystem service and cultural values to offer to urban residents. Therefore, estimating the benefit for a single household and aggregating over the population of potential beneficiaries merits thoughtful discussion to clearly understand the policy question on hand and estimate the value from a household utility and welfare economics standpoint. Future research on open space valuation could focus on examining the sensitivity of welfare estimates from hedonic methods to model assumptions, comparing the marginal willingness to pay estimated from various valuation methods, and assessment of economic efficiency of open space policies that are designed to meet local needs within an urban area.

References

Acharya, G. and Bennett, L. L. (2001). 'Valuing open space and land-use patterns in urban watersheds', *Journal of Real Estate Finance and Economics*, vol. 22, pp. 221–237.
Anderson, S. T. and West, S. E. (2006). 'Open space, residential property values, and spatial context', *Regional Science and Urban Economics*, vol. 36, pp. 773–789.

Anselin, L. and Bera, A. (1998). 'Spatial dependence in linear regression models with an introduction to spatial econometrics', in A. Ullah and D. Giles (Eds.), *Handbook of Applied Economics Statistics*, Marcel Dekker, New York.

Appleton, J. (1975). *The Experience of Landscape*, John Wiley and Sons, Chichester, UK.

Aued, B. (2012). 'Athens–Clarke study says don't build neighborhood parks'. Retrieved from www.online athens.com/local-news/2012-03-14/athens-clarke-study-says-dont-build-neighborhood-parks

Bates, L. J. and Santerre, R. E. (2001). 'The public demand for open space: The case of Connecticut communities', *Journal of Urban Economics*, vol. 50, pp. 97–111.

Beron, K., Murdoch, J. and Thayer, M. (2001). 'The benefits of visibility improvement: New evidence from the Los Angeles metropolitan area', *Journal of Real Estate Finance and Economics*, vol. 22, pp. 319–337.

Bond, M. T., Seiler, V. L. and Seiler, M. J. (2002). 'Residential real estate prices: A room with a view', *Journal of Real Estate Research*, vol. 23, pp. 129–137.

Bourossa, S. C., Hoesli, M. and Peng, V. S. (2003). 'Do housing submarkets really matter?' *Journal of Housing Economics*, vol. 12, pp. 12–28.

Bourassa, S. C., Hoesli, M. and Sun, J. (2004). 'What's in view?' *Environment and Planning A*, vol. 36, no. 8, pp. 1427–1450.

Boyle, K. (2003). 'Contingent valuation in practice', in P. Champ, K. Boyle and T. Brown (Eds.), *A Primer on Non-Market Valuation*, Kluwer Academic Publishers, Boston, MA.

Boyle, K. J., Poor, P. J. and Taylor, L. O. (1999). 'Estimating the demand for protecting water lakes from eutrophication', *American Journal of Agriculture Economics*, pp. 1118–1122.

Brasington, D. M. and Hite, D. (2005). 'Demand for environmental quality: A spatial hedonic analysis', *Regional Science and Urban Economics*, vol. 35, pp. 57–82.

Cho, S., Poudyal, N. C. and Roberts, R. K. (2008). 'Spatial analysis of the amenity value of green open space', *Ecological Economics*, vol. 66, no. 2, pp. 403–416.

Clarke, J., Kruidenier, W. and Wolf, J. (2007). *A National Research Plan for Urban Forestry 2005–2015*, National Urban and Community Forestry Advisory Council, USDA Forest Service, Washington, DC.

Day, B., Bateman, I. and Lake, I. (2007). 'Beyond implicit prices: Recovering theoretically consistent and transferable values for noise avoidance from a hedonic price model', *Environmental and Resource Economics*, vol. 37, no. 1, pp. 211–232.

Fotheringham, A. S., Brunsdon, C. and Charlton, M. E. (2002). *Geographically Weighted Regression: The Analysis of Spatially Varying Relationships*, John Wiley and Sons, West Sussex, UK.

Freenman, A. M., III. (1993). *The Measurement of Environmental and Resource Values, Theory and Methods*, Resources for the Future, Washington, DC.

Frey, B. S., Luechinger, S. and Stutzer, A. (2009). *The Life Satisfaction Approach to Environmental Valuation*, IZA Discussion Paper No. 4478, Institute for the Study of Labor, Bonn, Germany.

Geoghegan, J., Wainger, L. A. and Bockstael, N. E. (1997). 'Spatial landscape indices in a hedonic framework: An ecological economics analysis using GIS', *Ecological Economics*, vol. 23, pp. 251–264.

Irwin, E. G. (2002). 'The effects of open space on residential property values', *Land Economics*, vol. 78, no. 4, pp. 465–480.

Irwin, E. G. and Bockstael, N. E. (2001). 'The problem of identifying land use spillovers: Measuring the effects of open space on residential property values', *American Journal of Agricultural Economics*, vol. 83, no. 3, pp. 698–704.

Johnston, R. J., Opaluch, J. J., Grigalunas, T. A. and Mazzotta, M. J. (2001). 'Estimating amenity benefits of coastal farmland'. vol. 32 (summer 2001), pp. 305–325.

Kask, S. B. and Maani, S. A. (1992). 'Uncertainty, information, and hedonic pricing', *Land Economics*, vol. 68, pp. 170–184.

Kline, J. D. (2006). 'Public demand for preserving local open space', *Society and Natural Resources*, vol. 19, pp. 645–659.

Lancaster, K. J. (1966). 'A new approach to consumer theory', *Journal of Political Economy*, vol. 74, pp. 132–157.

Lutzenhiser, M., and Netusil, N. R. (2001). 'The effect of open spaces on a home's sale price', *Contemporary Economics Policy*, vol. 19, no. 3, pp. 291–298.

Mahan, B. L., Polasky, S. and Adams, R. M. (2000). 'Valuing urban wetlands: A property price approach', *Land Economics*, vol. 76, no. 1, pp. 100–113.

McConnell, V. and Walls, M. (2005). *The Value of Open Space: Evidence from Studies of Non-Market Benefits*, Resources for the Future, Washington, DC.

McGarigal, K. and Marks, B. J. (1995). *FRAGSTAT: Spatial Pattern Analysis Program for Quantifying Landscape Structure*, General Technical Report PNW-GTR-351, USDA Forest Service, Pacific Northwest Research Station, Portland, OR.

Nelson, N., Kramer, E., Dorfman, J. and Bumback, B. (2004). *Estimating the Economic Benefit of Landscape Pattern: A Hedonic Analysis of Spatial Landscape Indices*, Institute of Ecology, Department of Agricultural and Applied Economics, The University of Georgia, Athens, GA.

Paterson, R. W. and Boyle, K. J. (2002). 'Out of sight, out of mind? Using GIS to incorporate visibility in hedonic property value models', *Land Economics*, vol. 78, no. 3, pp. 417–425.

Peiser, R. and Schwann, G. (1993). 'The private value of public open space within subdivisions', *Journal of Architectural and Planning Research*, vol. 10, no. 2, pp. 91–104.

Polasky, S. (2012). 'Ecosystem valuation', in A. Hastings and L. J. Gross (Eds.), *Encyclopedia of Theoretical Ecology*, University of California Press, Berkeley, CA.

Pope, J. C. (2008). 'Does seller disclosure affect property values? Buyer information and the hedonic model', *Land Economics*, vol. 84, pp. 551–572.

Poudyal, N. C., Hodges, D. G., Bowker, J. M. and Cordell, H. K. (2009). 'Evaluating natural resource amenities in a human life expectancy production function', *Forest Policy and Economics*, vol. 11, pp. 253–259.

Poudyal, N. C., Hodges, D. G., Fenderson, J. and Tarkinson, H. (2010). 'Realizing the value of forest landscape in a viewshed', *Southern Journal of Applied Forestry*, vol. 34, no. 2, pp. 72–78.

Poudyal, N. C., Hodges, D. G. and Merrett, C. D. (2009). 'A hedonic analysis of demand for and benefit of urban recreation parks', *Land Use Policy*, vol. 26, pp. 975–983.

Poudyal, N. C., Hodges, D. G., Tonn, B. E. and Cho, S. (2009). 'Valuing the diversity, spatial pattern, and configuration of open space plots in urban neighborhood', *Forest Policy and Economics*, vol. 11, pp. 194–201.

Rosen, S. (1974). 'Hedonic prices and implicit prices: Product differentiation in pure competition', *The Journal of Political Economy*, vol. 82, no. 1, pp. 34–35.

Sander, H. A. and Polasky, S. (2009). 'The value of views and open space: Estimates from a hedonic pricing model for Ramsey County, Minnesota, USA', *Land Use Policy*, vol. 26, pp. 837–845.

Smith, V. K., Poulos, C. and Kim, H. (2002). 'Treating open space as an urban amenity', *Resource and Energy Economics*, vol. 24, no. 1, pp. 107–129.

Taylor, L. (2003). 'The hedonic method', in P. Champ, K Boyle and T. Brown (Eds.), *A Primer on Non-Market Valuation*, Kluwer Academic Publishers, Boston, MA.

Thorsnes, P. (2002). 'The value of a suburban forest preserve: Estimates from sales of vacant residential building lots', *Land Economics*, vol. 78, no. 3, pp. 426–441.

The Trust for Public Land (TPL). (2012, Fall/Winter). *The Trust for Public Land: Annual Report Issue*, TPL, San Francisco, CA.

Turner, M. G. (1990). 'Spatial and temporal analysis of landscape patterns', *Landscape Ecology*, vol. 4, no. 1, pp. 21–30.

Tyrvainen, L. and Miettinen, A. (2000). 'Property prices and urban forest amenities', *Journal of Environmental Economics and Management*, vol. 39, pp. 205–223.

USDA Forest Service. (2007). *Forest Service Open Space Conservation Strategy*, USDA Forest Service, Washington, DC.

Waltert, F. and Shlapfer, F. (2010). 'Landscape amenities and local development: A review of migration, regional economic and hedonic pricing studies', *Ecological Economics*, vol. 70, no. 2, pp. 141–152.

12

HUNTING LEASES

Markets and economic implications

Ian A. Munn[1] *and Anwar Hussain*[2]

[1]PROFESSOR, FOREST ECONOMICS AND MANAGEMENT, COLLEGE OF FOREST RESOURCES, FORESTRY DEPARTMENT BOX 9680 MISSISSIPPI STATE, MS 39762, USA.
ECONOMIST, CONSERVATION ECONOMICS INSTITUTE, P.O. BOX 755, BOISE, ID 83701, USA

Abstract

Wildlife-associated recreation activities and the related hunting lease markets in the United States have assumed special significance. They provide landowners the opportunity to supplement their incomes, impact regional economies, increase rural land values, generate revenues for wildlife management agencies through hunting license sales and serve as game management and ecosystem conservation instruments. Although demand for hunting access on private lands has induced a market for hunting leases, it is still thin and fragmented, as only 3% to 15% of private landowners across the nation allow fee access, primarily due to high transaction costs, concerns related to recreational liability and tax implications of managing a fee-access business. Concurrently, academic research conceptualizing the workings of the budding hunting lease market and its economic implications for local and regional economies has intensified. This chapter reviews some of this research that has focused on landowner willingness to allow fee access, hunters' preferences and willingness to pay for fee access, factors underlying regional differences in per-acre hunting lease prices, capitalization of lease income into rural land values and regional economic impact in the context of the southeastern United States. Suggestions for further research on multiple hunting leases by individual hunters, data generation and modeling are also presented.

Keywords

Big game hunting, cultural ecosystem services, deer, regional economic impact, fee-access hunting, hunting leases, land values, rural development, southeastern United States, turkey, waterfowl, wildlife-associated recreation outfitter.

Introduction

The significance of hunting varies around the globe. In developing countries where agriculture and extractive industries dominate the economic structure, hunting is essentially a cultural and subsistence activity. As economies transform and incomes rise, leisure hunting takes over, but its significance still differs across developed countries depending on acres of forestland, regional differences in body size of game animals and people's attitudes toward hunting. Moreover, it takes diverse forms with differing sets of regulations and norms governing where, when and how hunting takes place (Katwata, 2011).

In North America, hunting is more than a subsistence and cultural ecosystem service (Wallace, 2007). It provides landowners the opportunity to supplement their incomes (Loomis and Fitzhugh 1989; Boxall and MacNab, 2000), induces private landowners to provide and improve wildlife habitat (Jones, Munn, Grado and Jones, 2001), promotes rural economic development and generates revenues for wildlife management agencies (Mehmood, Zhang and Armstrong, 2003) and serves as a means to manage game (Schwabe and Schuhmann, 2002; Hussain et al., 2007). Within the United States, there are significant regional differences in the ability of wildlife-associated activities to compete with other land uses. For instance, in the US Southwest (e.g. Texas), hunting lease income often exceeds agricultural income and recreational use is the highest and best use of the land (Baen, 1997; Little and Berrens, 2008), whereas in the US upper Midwest, landowners require twice as much as landowners in the US Southeast to allow fee access (Gray, 1998; Kilgore, Snyder, Schertz and Taff, 2008).

In the remainder of this chapter, we first describe the US state and federal regulations that define the parameters within which wildlife-associated recreation activities (e.g. fishing, hunting) are allowed to occur. Then, we present insights learned about the hunting lease market in the southeastern United States. Specifically, we identify factors underlying landowner willingness to allow fee access (supply), hunters' willingness to pay for a single and multiple leases (demand), and local differences in per-acre lease prices. Next, we highlight economy-wide economic implications of wildlife-associated recreation activities for forestland values, employment and value added (e.g. wages, capital earnings, taxes), followed by concluding remarks.

Institutional setting

In the United States, hunting is a major recreational and economic activity (Straka, 2011; Hussain, Munn, Holland, Armstrong and Spurlock, 2012). In order to pursue their sport legally, hunters must meet three criteria. First, they must have a valid state hunting license. Provided the hunter meets minimum criteria, such as hunter safety training, age, no criminal record and/or game violations, licenses can be purchased from the relevant state agency. Second, they must have the appropriate approval to hunt their targeted species. For example, migratory game species such as waterfowl are controlled at the federal level, and additional permits or stamps are required. In other instances, for example where game populations cannot withstand heavy hunting pressure, permits that limit the number of animals harvested are often required by state or federal agencies. Permits are typically issued on a first-come, first-served basis or by some form of lottery. In rare instances, permits are auctioned off. For most game species, however, additional permits or licenses are not required, e.g. deer, wild turkey and small game. Finally, hunters must have a place to hunt. It is with this regard that hunting markets have developed.

Places for the general public to hunt for free are in relatively short supply compared to the demand. Even though hunting access to public lands is relatively unrestricted, requiring at most a nominal fee or an easily acquired permit, availability varies dramatically across regions and can be scarce in local markets. Furthermore, because access is unrestricted, overcrowding often results. Nonindustrial private forest (NIPF) lands are generally closed to public hunting (Jones et al., 2001). Due to liability concerns, generally landowners do not allow unrestricted public hunting (Wright, Kaiser and Nicholls, 2002; Sun, Pokharel, Jones, Grado and Grebner, 2007).

Historically the percentage of NIPF landowners that allowed fee access was small, less than 3% nationally, and less than 8% in the South, where the practice is most prevalent (Cordell, Bergstrom, Teasely and Maetzold, 1998). More recent research suggests that the percentage of NIPF landowners providing hunting access for a fee is increasing. For example, Jones et al. (2001) reported that in 1998 up to 14% of Mississippi NIPF landowners provided hunting access to their lands for a fee. By 2003, that percentage had increased to 16% (Munn et al., 2006), suggesting that markets for access to hunting lands are continuing to develop.

Fee access to private lands such as NIPF, forest industry and timberland investment organizations (TIMOs) can take various forms (Thomas, Adam and Thigpen, 1990; Straka, 2011). The most common form of access is by hunting lease, whereby a landowner grants the lessee the right to hunt any or all game species present for the duration of the lease. One year is the typical lease duration but longer periods are not uncommon, particularly where lease or habitat improvements are involved. In some cases, the lease duration is less than a year and usually mirrors the hunting season for one or more species. A second method is a permit system, whereby the landowner issues a permit to a hunter for access to the property. Typically, the permit is for a specified period of time and species. As an example, landowners offering dove hunts most often use some sort of permit system. The landowner prepares a field in advance and sells a limited number of permits to hunters wishing to hunt the field on a given day. A third method is guided hunts. In this case, the landowner also provides guide services along with the hunt, which may include lodging, meals, hunt setup (e.g. setting out decoys), game retrieval, transportation, calling the game (e.g. waterfowl or turkey) and access to pre-established blinds, to name a few. When offering guided hunts, landowners may provide the guide service directly, or by arrangement with an outfitter. In the latter case, the outfitter may pay the landowner a flat fee for access or may have a share agreement with the landowner. Leases are by far the most common form of fee access. According to Jones et al. (2001), 92% of the landowners who sell hunting access used leases to convey hunting access to hunters, 14% used permit systems and only 2% offered guided hunts. Some landowners utilized more than one method. As the dominant form of payment for fee-hunting access, leases have attracted the most attention in the literature.

Hunting leases are more prevalent on industry lands than NIPF lands. They provide forest industry an excellent source of annual revenue, valuable public relations and a land security program (Guynn and Marsinko, 2003; Cook, 2007). Hence, during the period 1984–1999, the percentage of total forest industry landholdings involved in hunting lease programs on forest industry lands increased from about 66% to nearly 84%. Although TIMOs also engage in leasing, the managers maintain hunting lease fees at levels lower than the market would bear, because unlike forest industry, TIMOs have minimal on-site supervision of their forestlands, making access control and public relations vital functions to prevent unwanted intrusions on their holdings. They recognize they are giving up significant incremental hunting lease revenue, but the increased tract security and public relations more than pay for the difference (Straka, 2011).

The hunting lease market

Supply-side considerations

Leasing has positive and negative aspects, and it is the landowners' perceptions of the relative magnitudes of these aspects that ultimately determine whether they engage in fee hunting. Considering the positive aspects, leasing provides protection to the land as hunt clubs and their members provide oversight, limiting trespass, poaching, vandalism, timber theft and crop damage, in addition to the regular flow of nontimber income (Morrison, Marsinko and Guynn, 2001). Markets for hunting leases can provide landowners incentive to improve habitat (Noonan and Zagata, 1982), and for large landowners, especially the forest industry and TIMOs, leasing can generate public goodwill if the lands are leased to local hunters (Morrison et al., 2001).

Landowner decision to allow hunting access

According to Wright et al. (2002), a landowner decision to allow hunting access is a matter of the degree to which the public is allowed or restricted. Five types of access may be distinguished: *prohibition* (precluding recreation), *exclusion* (reserving hunting opportunities to personal enjoyment), *restriction* (expanding access to include friends, acquaintances), *open* access (allowing everyone) and *fee access* (charging a fee for access). The decision to allow fee access depends on three broad sets of factors: resource attributes, landowner socioeconomic characteristics and user characteristics (Gray, 1998). Of these, user attributes relate to landowner perceptions of access-related problems induced by hunter behavior.

Resource attributes that influence landowner willingness to lease include size of ownership, type of habitat, game species present, habitat improvements and current use. The size of ownership is particularly important and influences landowner willingness to lease in several ways. First, ownership objectives for larger landholders are more likely to include revenue generation. Second, larger landowners are more likely to have surplus acreage above that needed for personal hunting. Third, large ownership sizes reduce per-acre transaction costs, resulting in some economies of scale (Baen, 1997). Finally, landowners engaged in wildlife habitat improvement, either on their own or with the assistance of government programs such as the Conservation Reserve Program (CRP), are more likely to be aware of the value their land has to hunters and thus be more likely to lease (Messonnier and Luzar, 1990).

Landowner characteristics could be important in fee access. For instance, a landowner may not find it attractive to allow fee access because potential benefits (tangible and intangible such as stewardship) may not be sufficient to offset the opportunity costs of time, material resources and foregone returns attributed to conflicts with other land uses (Zhang, Hussain and Armstrong, 2006). In general, at least some subset of socio-demographic characteristics does influence landowner willingness to lease, although the results for specific characteristics have not been consistent across studies. Age, employment status and income have been shown to influence landowner willingness to lease to varying degrees. Messonier and Luzar (1990) found that landowner willingness to lease increased with income, while Zhang et al. (2006) reported that landowners employed full-time were less likely to lease than those in employment categories typically associated with lower income. Hussain et al. (2007) reported that those 40 to 50 years old were less likely to lease than those younger and presumably with lower incomes or those older and presumably with higher incomes. Moreover, they found that landowners who were currently hunting on their own property were less likely to lease their land than landowners who were not.

User characteristics or hunter behavior-related factors include concerns that adversely influence the decision to allow access. Concerns over loss of privacy, safety, loss of control as to who is using the land, accident liability and damage to property are among the more prominent. Landowners who had had a bad experience with hunters or had become aware of hunter problems through neighbors and acquaintances were less likely to allow access (Guynn and Schmidt, 1984; Wright et al., 2002). Media attention to problems with fee hunting may also contribute to a hesitation to actively become involved in this enterprise. Concerns about loss of privacy and increased liability resulting from leasing negatively impact landowner willingness to lease (Zhang et al., 2006;

Munn, Hussain, West, Grado and Jones, 2007). Concerns about safety have a similar effect. Jones et al. (2001) compared the perceived severity of problems associated with leases by landowners who did not lease to the actual severity of problems experienced by landowners who did lease for an extensive suite of potential problems associated with leasing. Across the board, the severity of problems actually experienced by landowners engaged in leasing was substantially lower than the perceived severity of those problems by landowners not engaged in fee hunting. Although these concerns are very real to individual landowners, evidence suggests that these concerns are unwarranted because the number of successful lawsuits against landowners has been virtually nil (Sun et al., 2007).

Demand-side considerations

Purchasing access has become increasingly necessary given that public hunting lands are perceived to be crowded or low quality. From the hunter's perspective, each of the three primary forms of fee-access hunting (viz. purchase hunt, permit hunt and leases) offers distinct advantages. Purchasing hunts through an outfitter usually involves numerous additional benefits such as meals and lodging, guide services, game processing and at least the promise of a superior hunting experience. Outfitted hunts, however, tend to be very expensive and not an option for many hunters. Permit hunts allow hunters to focus their efforts on specific species at specific times. Hunting conditions are generally good, as hunter crowding and overhunting are limited by the permit system. Costs vary depending on the provider and the species targeted. Leases allow season-long access and typically include multiple game species. Leases can also provide hunters opportunities to regulate their own harvest levels subject to game laws, manipulate and improve the habitat in order to increase the number or quality of game and tailor the level of amenities such as blinds, food plots and facilities to their own needs. Hunters often form hunt clubs whose members share the cost of the lease, thereby reducing the per-person costs. Hunt clubs limit hunter crowding by restricting membership and by establishing club rules (e.g. limiting harvest levels). In general, leases offer the lowest cost per hunt of the three fee-access vehicles. Leases in prime locations with high-quality game can, however, be very expensive.

Hunter decision to purchase a lease and willingness to pay for hunting access

A *hunter decision to purchase a lease* is influenced by hunter income and age, hunter perception of crowding on public lands relative to private lands, hunter hunting avidity and alternative hunting access options (Munn, Hussain, Hudson and West, 2011). Hunters with other hunting access options are less likely to purchase a lease. All else equal, households with higher incomes are more likely to purchase a lease; likewise, older hunters are more likely to purchase a lease than are younger hunters. Compared to the population as a whole, lease hunters are disproportionately Caucasian males, as this demographic group also accounts for a disproportionately large share of those who engage in leisure hunting (USFWS 2007).

Determinants of hunter WTP for hunting leases can be grouped into three broad categories: lease characteristics (e.g. size in acres, onsite access, location relative to hunter residence, duration, price per acre), game characteristics (e.g. abundance, quality, diversity) and hunter characteristics (e.g. income, age). Regarding *lease characteristics*, Hussain, Munn, Hudson and West (2010) reported that hunters preferred lease sizes in the 500- to 1,000-acre range over smaller tracts; tracts over 1,000 acres did not command any additional premium. Onsite access consisting of all-weather roads positively impacted hunter choice. Leases located closer to a hunter's residence are favored over leases located farther away; all else being equal, hunters pay less for a lease located farther away than an otherwise similar lease located closer to the hunter's residence. Lease price per acre impacts hunter choice of a lease alternative negatively.

Among *game characteristics*, game quality is an important attribute that influences hunters' valuation of a hunting lease (Hussain, Zhang and Armstrong, 2004). In a study of hunting leases on private lands in California, Loomis and Fitzhugh (1989) found that hunters were willing to pay $106 more per hunter for a 10% increase in trophy quality deer in the total deer harvest. Likewise, Standiford and Howitt (1993) observed a positive correlation between lease price and trophy size of deer in California hardwood rangeland. Both studies classified game quality as the number or percent of trophy deer; trophy size was not defined and was subjective. Average local Boone and Crockett scores provide an objective and quantifiable measure of game quality. Rhyne, Munn and Hussain (2009) demonstrated that lease prices were positively linked to average Boone and Crockett scores and could account for up to 17% difference in lease prices. Game diversity also positively impacts hunter WTP; hunting leases that have multiple species such as deer, turkey and waterfowl to hunt are favored over leases that have only deer as a game species (Hussain et al., 2010).

Hunter decision to purchase multiple leases

Many hunters purchase multiple hunting leases in any given year. However, virtually all studies on hunting lease markets ignore this complexity of the hunting lease market. Munn et al. (2011) reported that 21% of Mississippi's lease hunters purchased more than one lease annually and found that factors influencing hunter willingness to participate in the hunting lease market and factors influencing the number of leases purchased were not the same. The former were sensitive to factors bearing on budget allocations across broad commodity groups, whereas the latter were sensitive to intra-group influences such as the ability of a given lease to satisfy a hunter need for a given species. Munn et al. (2011) addressed these complexities of the hunting lease market by estimating a two-equation system, with the first equation modeling the decision to participate in the hunting lease market and the second equation modeling the number of leases purchased conditional on the hunter first opting to participate in the hunting lease market.

Analysis of multiple leases has important implications for survey design and implementation of contingent valuation experiments. Specifically, to more completely specify factors that influence WTP for hunting access, one needs to make a decision about whether to invoke WTP for all leases or for a specific lease (e.g. a lease for a certain game such as deer, turkey or waterfowl; or a particular period such as a season or year), and how bid price would be asked in each lease. Munn et al. (2011) considered it appropriate to confine the focus of the research to the most expensive lease while framing the contingent valuation question. The decision to focus on the most expensive lease was warranted because the existence of a most expensive lease implies that the lessee was indeed willing to pay over and above the price paid for the less expensive leases. In contrast, invoking WTP for all leases would have been tedious and time-consuming for lessee hunters and could have compromised information quality and/or resulted in a lower response rate. Nonetheless, it is important to emphasize that their results pertain only to the incremental WTP for the most expensive leases and can be extrapolated to leases in general only with caution.

Two important findings emerged. First, the initial decision to participate in the hunting lease market was much more complex than the subsequent decision about how many leases to purchase. The decision to participate was impacted by a wide range of factors, but the decision to purchase multiple leases was influenced by only a few factors, specifically hunters' access to alternative hunting sites and perception of congestion on public lands. Second, having alternative access options negatively influenced the decision to lease and positively influenced the number of leases purchased, suggesting that there are some who have free hunting options, yet actively engage in leasing and gain utility from having a range of hunting options.

Hunting lease prices

Factors influencing hunting lease prices

Hunting lease prices depend on intrinsic site attributes, attributes under the landowner's control, provision of services, landowner skills in managing and marketing a lease operation, lease size and region-specific factors. Site attributes such as game abundance, diversity and quality positively relate to hunter willingness to pay for access and consequently higher lease revenues for the landowner. However, these indicators are often difficult to ascertain a priori. Thus, hunters form their expectations of game quality on the basis of observed site attributes (e.g. tract size, relative proportions of agricultural, forest and pasture land, forest cover type, forest stand structure and understory conditions). As this information is at the same time available to landowners, habitat differentiation based on site attributes facilitates understanding between landowners and hunters about lease transactions. Forest cover type has proved to be particularly important; for instance, the richness and high productivity of bottomland hardwoods translate into a comparable richness of wildlife (Shrestha and Janaki 2004; Meilby, Strange, Thorsen and Helles 2006; Hussain et al. 2007).

Evidence regarding the role of services is mixed. Pope and Stoll (1985), Loomis and Fitzhugh (1989) and Messonnier and Luzar (1990) did not find evidence to support the claim that hunters paid more for landowner-provided services, whereas Zhang et al. (2006) did. These conflicting results probably reflected differences in services and data quality. Differences in landowner management skills are important; landowners who acquire skills through extension programs earn higher lease revenue than otherwise similar landowners. Hussain et al. (2007) noted that technical expertise in managing a fee-access operation in terms of tax and other legal ramifications, hunter relations and wildlife habitat management can be important. Pope and Stoll (1985) and Messonnier and Luzar (1990) noted that in the Texas and Louisiana hunting lease markets, large landowners were able to earn higher earnings per acre leased. This is understandable because hunters place a premium on larger parcels, whereas smaller tracts with otherwise good site attributes pose many issues such as potential safety problems with a high likelihood of conflict with hunters in neighboring lands.

Differences in local hunting lease rates

Estimates of lease rates (per-acre prices) that ignore location or region-specific differences are likely to be suspect. Although a number of studies have noted differences in local hunting lease rates (e.g. Pope and Stoll, 1985; Rhyne et al., 2009), analyses of the determinants of local lease rate differences are lacking. In general, lease rates vary between locales either because the characteristics of leases differ or characteristics are valued differently between areas. Using the Oaxaca-Blinder decomposition technique (Oaxaca, 1973), Munn and Hussain (2010) identified resource and economic factors underlying differences in hunting lease rates for two regional markets in Mississippi. The data were generated from leases sold by sealed-bid auction. Differences in the characteristics of leases accounted for 71% of the price differential, whereas differences in valuation accounted for only 29% of the differential in lease rates. Although these results may not be typical for all regional markets, they do demonstrate the importance of understanding the relative mix of differences due to characteristics versus differences due to valuation. Generally speaking, differences in valuation are due to market forces and, as such, cannot be influenced by the actions of individual stakeholders. The levels of characteristics, however, can be influenced by landowners. Efforts by landowners to increase lease revenues by manipulating the attributes under their control (e.g. land allocation to various forest types, lease duration and lease size) will be more effective where lease rate differentials are due primarily to differences in characteristics.

Valuation issues in hunting lease markets

Numerous issues continue to complicate a clear understanding of hunting lease markets (Mozumder, Starbuck, Berrens and Alexander, 2007; Taff, 1991). First, most leases issued on private lands result in no public record of the market transactions. To gather information, researchers have relied on nonmarket valuation methods to infer lease values from landowners or hunters. Hypothetical data generated via nonmarket valuation methods, however, do not represent actual market conditions and there are concerns about their reliability and accuracy. Even market data are not ideal for valuating lease characteristics if the market is not competitive. Evidence suggests that NIPF hunting leases are not necessarily issued in a fully competitive manner. Lease prices are typically negotiated. Few landowners advertise to the public, either on the Internet or in print media. Most rely on word-of-mouth or family and friends to develop their customer base. Munn et al. (2006) and Hussain et al. (2007) found that landowners experienced in the hunting lease market generated higher lease prices than their less experienced, less knowledgeable counterparts. Given that less than 12% of NIPF landowners in the study were classified as experienced, a large majority of landowners were not capturing full market value. Studies of auctioned leases have been limited (Bishop and Heberlein, 1979; Rhyne et al., 2009) but provide valuable insights into the actual value hunters place on leases. Rhyne et al. (2009) used hunting lease data from market transactions of hunting lease auctions in Mississippi. Lease rates in this study averaged 34% higher than self-reported, negotiated lease rates reported by Munn et al. (2006) for NIPF leases in Mississippi even after adjusting for inflation.

Second, hunting lease markets tend to be thin and involve significant transactions costs (Kaiser and Wright, 1985; Sun et al., 2007). The small minority of landowners who provide fee access do so because net revenues and hunter-provided benefits (e.g. protection against trespass and vandalism) more than offset the opportunity cost of leasing inclusive of transaction costs. Besides serving as a barrier to landowners' willingness to allow fee access, these features of the hunting lease market call for nonrandom surveys which in turn determine estimation procedures that need to be used to quantify (a) landowners' decisions to provide fee access and factors influencing lease price per acre, and (b) hunters' decisions to purchase hunting lease(s) and factors influencing willingness to pay per acre.

Third, the majority of studies on hunting leases have either been focused on willingness to pay for deer hunting (e.g. Mackenzie, 1990; Knoche and Lupi, 2007) or waterfowl economic impact (e.g. Grado, Kaminski, Munn and Tullos, 2001). Topics such as (a) hunter willingness to pay for hunting access and landowner willingness to allow fee access for turkey and waterfowl, (b) hunters pursuing multiple leases, (c) hunting operations by outfitters and (d) impact of hunting lease income on forestland values have not received enough attention. Species-specific analyses are particularly needed because deer and turkey require large tracts, whereas waterfowl do not. And given that tract size is a predominant consideration in a hunting lease, lumping waterfowl with deer and turkey could be a major source of bias in estimation. Moreover, turkey and waterfowl hunters are a distinct group of hunters that occurs with relatively low probability, whereas deer hunters are ubiquitous, which calls for nonrandom survey sampling procedures and associated estimation methods.

Fourth, future research needs to recognize the phenomenon of multiple leases and how it can be appropriately analyzed. The purchase of multiple leases is probably motivated by the inability of a given lease to satisfy the lessee's demand for alternative hunting experiences (e.g. some hunting sites may be good for deer hunting, whereas others may be good for turkey and waterfowl hunting). Given the complexity entailed by invoking incremental WTP for all leases simultaneously, choice modeling using choice experiments may be more amenable to dealing with these complexities.

Local and regional economic implications

Capitalization of hunting lease income into local forestland values

Expenditures on wildlife-associated recreation activities by US households are impacting rural economies in important ways (Hussain, Munn, Grado and Henderson, 2008; Munn, Hussain, Spurlock and Henderson, 2010). In the southeastern United States, where more than 60% of the land is privately owned, recreation-related revenues (e.g. hunting leases, wildlife viewing) are being capitalized into land values. Based on data from Texas, Baen (1997) found that recreation income could enhance land values to the point that recreation becomes the highest and best use of rural land. Henderson and Moore (2006) argued that farmland values in Texas were higher in locations with more developed markets for wildlife recreation activities. They estimated that a dollar increase in a county's per-acre recreation income was associated with a 1.3% increase in per-acre land value, and a dollar increase in per-acre hunting lease rates was associated with a 6% increase in per-acre land value. Using rural land sales data from Mississippi, Jones et al. (2006) reported that real estate appraisers familiar with land sales ascribed 36% of the sale values to recreational opportunities the tracts provided. According to Straka (2011), the average net contribution of hunting lease revenue to land expectation value for forest industry and TIMO investments in the South was $184 per acre (in 2010 dollars). Although the aforementioned studies highlight the significance of wildlife recreation-related income streams for rural land values, the wide disparity in their definitions of recreation income clearly warranted more research. Hussain, Munn, Brashier, Jones and Henderson (2013) attempted to fill this gap in knowledge, addressing how hunting lease income was capitalized into forestland values in north Mississippi. The study employed individual land sales and hunting lease income data across sub-state ecoregions rather than county averages or expert opinion to determine the capitalization rate for hunting lease income. Per-acre forestland value was regressed on per-acre lease price, transaction features, county characteristics and ecosystem characteristics. A dollar increase in per-acre annual lease price was found to increase per-acre forestland value by 0.80%. Evaluated at the mean per-acre forestland value of US$1,598, this translated into an implicit price of US$13.24 per acre and a capitalization rate of 7.55%. This capitalization rate indicated that $286.36 of the total per-acre forestland value (approximately 18%) could be attributed to hunting leases.

Economy-wide local and regional economic impacts

Wildlife-associated recreation activities have assumed a significant role in the US economy. According to the 2006 USFWS survey, 87.5 million people aged 16 and above participated in wildlife-associated recreation activities, spending $122.4 billion on trips and equipment. Note, however, these dollar estimates of direct expenditures constitute only a component of the total impact associated with wildlife-associated activities. In addition to these direct expenditures, indirect and induced expenses also arise when industries supply wildlife recreation-related goods and services. Collectively, these impacts could be significant depending on a region's economic and natural resource base (Reeder and Brown, 2005); moreover, economic development based on wildlife-associated recreation activities may be environment-friendly (English and Bergstrom, 1994). Despite the contribution of wildlife-associated recreation to regional economies, research on the subject is limited. In particular, some studies have been less realistic as they ignored linkages of recreation activities with the rest of the economy, whereas others using input-output methods assumed factor supplies to be perfectly elastic, thus assuming away changes in input prices and potential product market responses.

Addressing the aforementioned limitations, Hussain et al. (2012) used a regional general equilibrium model. Exogenous demand shocks to the regional economy used estimates of expenditures incurred by wildlife recreationists on hunting, fishing and wildlife-watching activities. Making alternative assumptions about labor and capital mobility and their supply, counterfactual simulation results suggested that without wildlife-associated recreation expenditures, regional employment would have been smaller by 396,000 to 783,000 jobs, and value added would have been $22 billion to $48 billion less. These findings underscored the use of general equilibrium analysis and suggested that although the economic impacts induced by wildlife-associated recreation activities can be significant, there was a need to be cognizant of their highly variable nature and dependence on factor mobility and factor supply elasticity.

Conclusions

Hunting lease markets impact local economies, rural land values and wildlife conservation. This chapter summarized research focused on the hunting lease market in the southeastern United States. In this region, private landownership dominates and there is excess demand for hunting access. Although the percentage of landowners who allow fee access has been increasing over the last 15 years, less than 20% currently allow fee access. Underlying reasons why landowners are hesitant to allow fee access include lack of knowledge about managing a fee-access business, tax implications of running a fee-access business and hunter liability concerns. Because there is great potential for improved outcomes for all concerned (i.e. landowners, hunters, local governments, regional economies and wildlife conservation), the functioning of hunting lease markets in the southeastern United States can be characterized as sub-optimal.

Landowners seeking to maximize hunting lease revenues should be aware of hunters' preferences and cater to those preferences with appropriate land management activities and lease provisions. Public land managers should take into account the preferences and experiences of hunters when developing management goals and constructing leases and better target educational campaigns. Knowing the social and economic characteristics of hunters as well as their hunting preferences and experiences can guide private and public land managers toward cultivating management goals that maximize lease revenue and/or help meet hunter expectations.

The sub-optimal state of the hunting lease market notwithstanding, it impacts local and regional economies in significant ways. For instance, rural lands that offer quality hunting earn significantly more lease income than otherwise similar lands and this, in turn, translates to higher capitalized land values. To an extent, landowners are able to enhance game abundance and quality on their lands by investing in wildlife habitat improvements whereby state and federal government conservation programs defray part of the cost. At the broad regional level, hunting leases have the potential to generate gainful seasonal employment and induce higher value added. Wildlife habitat enhancement and promotion of hunting lease opportunities are viable rural development strategies for local governments. In particular, education programs addressing landowner informational needs and legislative reforms that reduce landowner liability would be steps forward in facilitating the functioning of hunting lease markets.

Our understanding of hunting lease markets is still incomplete. Definitely, more research is needed. Fortunately, as hunting markets continue to mature in the southeastern US and to develop in other regions, both within the US and around the world, opportunities to address these research needs and to improve our understanding of hunting markets will be plentiful.

Acknowledgments

Anwar Hussain is grateful to James B. Armstrong and Daowei Zhang, School of Forestry & Wildlife Sciences, Auburn University, for stimulating his interest in wildlife-associated recreation economics and providing financial support for several related research projects.

References

Baen, J. S. (1997). 'The growing importance and value implications of recreational hunting leases to agricultural land investors', *Journal of Real Estate Research*, vol. 14, no. 3, pp. 399–414.

Bishop, R. C. and Heberlein, T. (1979). 'Measuring values of extramarket goods: Are indirect measures biased?' *American Journal of Agricultural Economics*, vol. 61, no. 5, pp. 926–930.

Boxall, P. C. and MacNab, B. (2000). 'Exploring the preferences of wildlife recreationists for features of boreal forest management: A choice experiment approach', *Canadian Journal of Forest Research*, vol. 30, no. 12, pp. 1931–1941.

Cook, F. C. (2007). 'An empirical analysis of hunting leases by timber firms', MS thesis, Montana State University, Bozeman, MT.

Cordell, H. K., Bergstrom, J. C., Teasely, R. J. and Maetzold, J. A. (1998). 'Trends in outdoor recreation and implications for private land management in the East', in J. S. Key, G. R. Goff, P. J. Smallidge, W. N. Grafton and J. A. Parkhurst (Eds.), *Proceedings: Natural Resources Income Opportunities for Private Lands*, University of Maryland, Cooperative Extension Service, College Park, MD, pp. 4–10.

English, B. K. and Bergstrom, J. C. (1994). 'The conceptual links between recreation and site development and regional economic impacts', *Journal of Regional Science*, vol. 34, no. 4, pp. 599–611.

Grado, S. C., Kaminski, R. M., Munn, I. A. and Tullos, T. A. (2001). 'Economic impacts of waterfowl hunting on public lands and at private lodges in the Mississippi Delta', *Wildlife Society Bulletin*, vol. 29, no. 3, pp. 846–855.

Gray, C. D. (1998). 'An economic analysis of private land leases for outdoor recreation', MS thesis, University of Georgia, Atlanta, GA.

Guynn, D. C., Jr. and Marsinko, A. P. (2003). 'Trends in hunt leases on forest industry lands in the Southeastern United States', in *Proceedings of the First National Symposium on Sustainable Natural Resource-Based Alternative Enterprises*, May 28–31, 2003, Starkville, MS, pp. 68–74.

Guynn, D. E. and Schmidt, J. L. (1984). 'Managing deer hunters on private lands in Colorado', *Wildlife Society Bulletin*, vol. 12, no. 1, pp. 12–19.

Henderson, J. and Moore, S. (2006). 'The capitalization of wildlife recreation income into farmland values', *The Journal of Agriculture and Applied Economics*, vol. 3, no. 3, pp. 597–610.

Hussain, A., Armstrong, J. B., Brown, D. B. and Hogland, J. (2007). 'Land-use pattern, urbanization and deer-vehicle collisions in Alabama', *Human-Wildlife Conflicts*, vol. 1, no. 1, pp. 89–96.

Hussain, A., Munn, I. A., Brashier, J., Jones, W. D. and Henderson, J. E. (2013). 'Capitalization of hunting lease income into northern Mississippi forestland values', *Land Economics*, vol. 89, no. 1, pp. 137–153.

Hussain, A., Munn, I. A., Grado, S. C. and Henderson, J. E. (2008). 'Economic impact of wildlife-associated outfitters and their clientele', *Human Dimensions of Wildlife*, vol. 13, no. 4, pp. 243–251.

Hussain, A., Munn, I. A., Grado, S. C., West, B. C., Jones, W. D. and Jones, J. C. (2007). 'Hedonic analysis of hunting lease revenue and landowner willingness to allow hunting access', *Forest Science*, vol. 53, no. 4, pp. 493–506.

Hussain, A., Munn, I. A., Holland, D. J., Armstrong, J. B. and Spurlock, S. (2012). 'Economic impact of wildlife-associated recreation expenditures in the southeast United States: A general equilibrium analysis', *Journal of Agriculture and Applied Economics*, vol. 44, no. 1, pp. 63–82.

Hussain, A., Munn, I. A., Hudson, D. M. and West, B. C. (2010). 'Attribute-based analysis of hunters' lease preferences', *Journal of Environmental Management*, vol. 91, no. 12, pp. 2565–2571.

Hussain, A., Zhang, D. and Armstrong, J. B. (2004). 'Willingness to pay for hunting leases in Alabama', *Southern Journal of Applied Forestry*, vol. 28, no. 1, pp. 21–27.

Jones, W. D., Munn, I. A., Grado, S. C. and Jones, J. C. (2001). *Hunting: An Income Source for Mississippi's Nonindustrial Private Landowners*, Forest and Wildlife Research Center Research Bulletin FO-164, Mississippi State University, MS.

Jones, W. D., Ring, J. K., Jones, J. C., Watson, K., Parvin, D. W. and Munn, I. A. (2006). 'Land valuation increases from recreational opportunity: A study of Mississippi rural land sales', *Proceedings of the Annual Southeastern Association of Fish and Wildlife Agencies*, vol. 60, pp. 49–53.

Kaiser, R. A. and Wright, B. A. (1985). 'Recreational access to private land – beyond the liability hurdle', *Journal of Soil and Water Conservation*, vol. 40, no. 6, pp. 478–481.

Katwata, Y. (2011). 'Economic growth and trend changes in wildlife hunting', *Acta Agriculture Slovenica*, vol. 97, no. 2, pp. 115–123.

Kilgore, M. A., Snyder, S. A., Schertz, J. M and Taff, S. J. (2008). 'The cost of acquiring public hunting access on family forests lands', *Human Dimensions of Wildlife*, vol. 13, no. 3, pp. 175–186.

Knoche, S. D. and Lupi, F. (2007). 'Valuing deer hunting services from farm landscapes', *Ecological Economics*, vol. 64, no. 2, pp. 313–320.

Little, J. M. and Berrens, R. P. (2008). 'The southwestern market for big-game hunting permits and services: A hedonic pricing analysis', *Human Dimensions of Wildlife*, vol. 13, no. 3, pp. 143–157.

Loomis, J. B. and Fitzhugh, L. (1989). 'Financial returns to California landowners for providing hunting access: Analysis and determinants of returns and implications to wildlife management', in *Transactions of the North American Wildlife Natural Resource Conference*, Wildlife Management Institute, University of California-Davis, CA, pp. 197–201.

Mackenzie, J. (1990). 'Conjoint analysis of deer hunting', *Northeastern Journal of Agricultural and Resource Economics*, vol. 19, no. 1, pp. 109–117.

Mehmood, S. R., Zhang, D. and Armstrong, J. B. (2003). 'Factors associated with declining hunting license sales in Alabama', *Human Dimensions of Wildlife*, vol. 8, pp. 243–262.

Meilby, H., Strange, N., Thorsen, B. J., and Helles, F. (2006). 'Determinants of hunting rental prices: A hedonic analysis', *Scandinavian Journal of Forest Research*, vol. 21, no. 1, pp. 63–72.

Messonnier, M. L. and Luzar, E. J. (1990). 'A hedonic analysis of private hunting land attributes using an alternative functional form', *Southern Journal Agricultural Economics*, vol. 22, no. 2, pp. 129–135.

Morrison, H. S., IV, Marsinko, P. C. and Guynn, D. C., Jr. (2001). 'Forest industry hunt-lease programs in the southern United States: 1999', *Proceedings of the Annual Conference of the Southeastern Association of Fish & Wildlife Agencies*, vol. 55, pp. 567–574.

Mozumder, P., Starbuck, M., Berrens, R. P. and Alexander, S. (2007). 'Lease and fee hunting on private lands in the U.S.: A review of the economic and legal issues', *Human Dimensions of Wildlife*, vol. 12, no. 1, pp. 1–14.

Munn, I. A., and Hussain, A. (2010). 'Factors explaining differences in local hunting lease revenues: Insights from Oaxaca decomposition', *Land Economics*, vol. 86, no. 1, pp. 66–78.

Munn, I. A., Hussain, A., Grado, S. C., Jones, W. D., West, B. C., Miller, J. E. and Loden, E. K. (2006). *Wildlife and Fisheries Economic Enterprises in Mississippi*, Forest and Wildlife Research Center Research Bulletin FO164, Forest and Wildlife Research Center, Mississippi State University, MS.

Munn, I. A., Hussain, A., Hudson, D. and West, B. C. (2011). 'Hunter preferences and willingness to pay for hunting leases', *Forest Science*, vol. 57, no. 3, pp. 189–200.

Munn, I. A., Hussain, A., Spurlock, S. and Henderson, J. E. (2010). 'Economic impact of fishing, hunting, and wildlife-associated recreation expenditures on the Southeast U.S. regional economy: An input-output analysis', *Human Dimensions of Wildlife*, vol. 15, no. 6, pp. 433–449.

Munn, I. A., Hussain, A., West, B. C., Grado, S. C. and Jones, W. D. (2007). 'Analyzing landowner demand for wildlife and forest management information', *Journal of Agricultural and Applied Economics*, vol. 39, no. 3, pp. 557–569.

Noonan, P. F., and Zagata, M. D. (1982). 'Wildlife in the market place: Using the profit motive to maintain wildlife habitat', *Wildlife Society Bulletin*, vol. 10, no. 1, pp. 46–49.

Oaxaca, R. L. (1973). 'Male-female wage differentials in the urban labor markets', *International Economic Review*, vol. 14, no. 3, pp. 693–709.

Pope, C. A., III and Stoll, J. R. (1985). 'The market value of ingress rights for white-tailed deer hunting in Texas', *Southern Journal Agricultural Economics*, vol. 17, no. 1, pp. 177–182.

Reeder, J. R. and Brown, D. M. (2005). *Recreation, Tourism and Rural Well-being*, ERR-7, Economic Research Service/USDA, Washington, DC.

Rhyne, J. D., Munn, I. A. and Hussain, A. (2009). 'Hedonic analysis of auctioned hunting leases: A case study of Mississippi sixteenth section lands', *Human Dimensions of Wildlife*, vol. 14, no. 4, pp. 227–239.

Schwabe, A. K. and Schuhmann, P. W. (2002). 'Deer-vehicle collisions and deer value: An analysis of competing literature', *Wildlife Society Bulletin*, vol. 30, no. 2, pp. 609–615.

Shrestha, R. K. and Janaki, R. A. (2004). 'Effect of ranchland attributes on recreational hunting in Florida: A hedonic price analysis', *Journal of Agricultural and Applied Economics*, vol. 36, no. 3, pp. 763–772.

Standiford, R. B. and Howitt, R. E. (1993). 'Multiple use management of California's hardwood rangelands', *Journal of Range Management*, vol. 46, no. 2, pp. 176–182.

Straka, T. J. (2011). 'Contribution of wildlife to the value of U.S. southern forestlands', *Journal of the American Society of Farm Managers and Rural Appraisers*, vol. 74, no. 1, pp. 23–32.

Sun, C., Pokharel, S., Jones, W. D., Grado, S. C. and Grebner, D. L. (2007). 'Extent of recreational incidents and determinants of liability insurance coverage for hunters and anglers in Mississippi', *Southern Journal of Applied Forestry*, vol. 31, no. 3, pp. 151–158.

Taff, S. J. (1991). *Perspectives on the Sale of Hunting Access Rights to Private Lands in Minnesota*, Staff Paper # P91-39, Department of Agricultural and Applied Economics, University of Minnesota, Saint Paul, MN.

Thomas, J. K., Adam, C. E. and Thigpen, J. (1990). *Texas Hunting Leases: Statewide and Regional Summary*, Texas Agricultural Extension Service, The Texas A&M University System, College Station, TX.

US Fish and Wildlife Service (USFWS). (2007). '2006 National Survey of Fishing, Hunting, and Wildlife-Associated Recreation.' Retrieved from www.census.gov/prod/2008pubs/fhw06-nat.pdf

Wallace, K. J. (2007). 'Classification of ecosystem services: Problems and solutions', *Biological Conservation*, vol. 139, no. 3–4, pp. 235–246.

Wright, B. A., Kaiser, R. A. and Nicholls, S. (2002). 'Rural landowner liability for recreational injuries: Myths, perceptions, and realities', *Journal of Soil and Water Conservation*, vol. 57, no. 3, pp. 183–191.

Zhang, D., Hussain, A. and Armstrong, J. B. (2006). 'Supply of hunting leases from nonindustrial private forestlands in Alabama', *Human Dimensions of Wildlife*, vol. 11, no. 1, pp. 1–14.

13

ECONOMICS OF AGROFORESTRY

D. Evan Mercer,[1] Gregory E. Frey[2] and Frederick W. Cubbage[3]

[1]RESEARCH ECONOMIST, USDA FOREST SERVICE, USA.
[2]ASSISTANT PROFESSOR AND EXTENSION SPECIALIST FOR FORESTRY,
VIRGINIA STATE UNIVERSITY, USA.
[3]PROFESSOR, INTERNATIONAL FORESTRY, NATURAL RESOURCE POLICY, ECONOMICS,
CERTIFICATION AND AGROFORESTRY, NC STATE UNIVERSITY, USA.

Abstract

This chapter provides principles, literature and a case study about the economics of agroforestry. We examine necessary conditions for achieving efficiency in agroforestry system design and economic analysis tools for assessing efficiency and adoptability of agroforestry. The tools presented here (capital budgeting, linear programming, production frontier analysis and risk analysis) can help determine when agroforestry is a feasible option and provide arguments for cases when agroforestry systems are economically, socially and environmentally appropriate, fostering improved sustainable development for landowners, farmers and communities. The chapter closes with a case study applying the capital budgeting and real options analysis to evaluate the potential for agroforestry to augment efforts to restore bottomland hardwood forests in the Lower Mississippi Alluvial Valley. Agroforestry systems provide multiple outputs, potentially reducing risk and increasing income while also purportedly producing more ecosystem services than conventional agriculture. Our review and case study, however, provide cautionary tales about the limits of agroforestry and the need for rigorous economic research and analysis to design efficient and productive agroforestry systems and to optimize private and public investments in agroforestry.

Keywords

Capital budgeting, linear programming, real options, silvopasture, production frontier, agroforestry

Introduction

Agroforestry is 'a land-use system that involves deliberate retention, introduction, or mixture of trees or other woody perennials in crop or animal production systems to take advantage of economic or ecological interactions among the components' (SAF, 2012). Examples include: intermixed crops and trees on small farms (most often in developing countries), where the trees provide shade, fuel or fodder; silvopasture (mixed grazing and trees); shade-grown coffee or cocoa; and windbreaks, shelterbelts and riparian buffers. Potential advantages include reducing financial and biophysical risks, improving crop yields or quality, reducing fertilizer or other chemical inputs, improving livestock health, adapting to climate change through more resilient production systems, retaining more land at least partially forested, reducing soil erosion and increasing biodiversity.

Small-scale agroforestry, common in the tropics, provides multiple products for small farmers and good mixes of low-cost inputs. Medium-scale agroforestry may involve larger crop systems and focus on two or three simple tree and crop or grazing systems. Large-scale agroforestry remains uncommon, with silvopasture perhaps the most promising (Cubbage et al., 2012). No matter how efficient and eco-friendly they are, agroforestry systems can contribute to sustainable land use only if they are adopted and maintained over long time periods (Mercer, 2004). Adoption of agroforestry is considerably more complex than most agricultural innovations, because it usually requires establishing new input-output mixes of annuals, perennials and other components, combined with new conservation techniques such as contour hedgerows, alley cropping and enriched fallows (Rafiq, Amacher and Hyde, 2000). The multicomponent, multiproduct nature of agroforestry may limit adoption due to the complex management requirements and long periods of testing, experimenting and modification. For example, most agroforestry systems take 3 to 6 years before benefits begin to be fully realized compared to the few months needed to evaluate a new annual crop (Franzel and Scherr, 2002). The additional uncertainties in adopting new agroforestry input-output mixes suggests that agroforestry projects will require longer time periods to become self-sustaining and self-diffusing than earlier Green Revolution innovations.

Efficiency in agroforestry design

The efficiency objective of agroforestry is to optimize the use of all available resources to enhance the sustainable economic development of farms and communities. Meeting the efficiency objective requires the social marginal benefits from agroforestry (e.g. increased wood and food production, reduced soil erosion, carbon sequestration, etc.) to exceed the social marginal costs (e.g. production inputs, externalities). The net benefits from agroforestry must also equal or exceed the net benefits from identical investments in alternative land uses. Given efficient local markets, sustainable self-initiated agroforestry systems meet these requirements or they would not continue to exist. However, nonmarket (external) benefits and costs of land use are often ignored in private land-use decision making. Therefore, projects and policies initiated by governments or donor agencies require more formal efficiency analysis and explicit consideration of positive and negative externalities with alternative land-use scenarios.

The production possibility frontier (PPF) shows efficient combinations of the annual and woody perennial crops that can be produced with a given level of inputs. Agroforestry systems typically exhibit a composite of three possible production relationships (Figure 13.1). In the extreme areas (*ab* and *de*), combinations of trees and annual crops are *complementary*. This could occur when adding trees to agriculture reduces weeding, increases available nitrogen, improves microclimate or reduces erosion control costs (as in area *ab*). At the other extreme (area *de*), intercropping annual crops during tree plantation establishment may also reduce weeding costs and increase tree production. However, as either more trees are added to crop production or more crops to tree production, competition for nutrients and light dominate to produce the *competitive* region *bd*. *Supplementary* relationships occur at points *b* and *d*.

Data on the value of the outputs, usually the market price, are used to construct the iso-revenue line in Figure 13.1. The slope of the iso-revenue line is the rate at which the two goods can be exchanged (i.e. market prices). The landowner's goal is to reach the highest possible

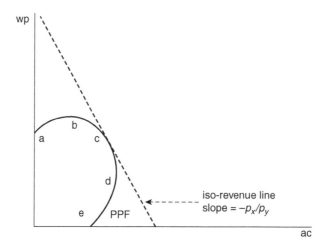

Figure 13.1 Concave production possibility frontier (PPF) resulting in agroforestry as optimal choice wp = woody perennial output, ac = annual crop output, c = optimal combination of wp and ac (Mercer and Hyde, 1992).

iso-revenue line, where income is maximized; this occurs at the tangency of the iso-revenue line and the PPF, point *c* in Figure 13.1. A PPF concave to the origin is a necessary condition for agroforestry to be feasible. Specialization can also be optimal with a concave PPF when the iso-revenue line is either very flat or steep, i.e. when one product is much more highly valued than the other. With convex PPFs, the optimal solution is to specialize in annual crops or forestry, depending on the price ratio, but not agroforestry. This occurs when the trees and crops are strictly competitive or when economies of scale favor monocultures over mixed tree and crop regimes.

Payments for ecosystem services (carbon sequestration, soil erosion control, water quality, biodiversity) will often be crucial for the widespread adoption of agroforestry systems (Frey, Mercer, Cubbage and Abt, 2010). Optimal production decisions from the landowner's perspective are determined by the PPF and the relative private value (e.g. market prices) of all alternatives. From society's viewpoint, however, market prices rarely reflect the social value of the ecosystem services associated with alternate land uses. For example, nonmarket benefits (e.g. erosion control and water quality protection) provided by the trees may increase their value to society. Likewise, negative externalities associated with the annual crops may reduce their social value. Therefore, when determining optimal production for society, the values (shadow prices) of the outputs should be adjusted to include external costs and benefits.

Figure 13.2 illustrates how optimal production decisions can vary between landowners and society. The private iso-revenue line reflects the landowner's relative valuation of the outputs based on their market prices, p_x and p_y; the social iso-value line reflects the social value of all outputs based on their shadow prices, sp_x and sp_y. Here, society values woody perennial production more than is reflected in the market price because of the positive externalities associated with trees. As a result, $sp_y > p_y$ and the slope of the social iso-value line is lower than the private iso-revenue line ($-sp_x/sp_y < -p_x/p_y$). Optimal production from the landowner's perspective occurs at *A*, specialization in annual crop production. From society's viewpoint, however, optimal production occurs at point *B*, an agroforestry combination. Therefore, it may benefit society to provide incentives to encourage adoption of agroforestry to move closer to point *B*.

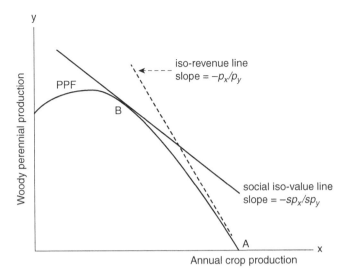

Figure 13.2 Social versus private optimal production decisions (p_x and p_y = market prices of *x* and *y*, sp_x and sp_y = shadow prices of *x* and *y*; Mercer and Hyde, 1992).

Economic approaches for assessing agroforestry

Capital budgeting

Capital budgeting (cash flow or cost-benefit analysis) is a simple, powerful tool for comparing the efficiency (profitability) of alternative land uses (Klemperer, 1996). Although the previous description of production possibilities is useful for identifying the optimal combinations of inputs and outputs within a continuum of land uses, capital budgeting allows comparisons of alternatives that utilize different inputs and produce different outputs. The most common capital budgeting tool is Net Present Value (NPV), the sum of the discounted periodic net revenues per unit of land over a given time horizon. If the NPV is higher for agroforestry than for all feasible alternatives, it is potentially adoptable. The Soil Expectation Value (SEV) is more appropriate when the time horizons of alternatives vary. SEV calculates the net return per hectare assuming the regime will be repeated in perpetuity. Frey et al. (2010) showed how SEV can be altered for regimes that do not involve fixed rotations. Multiplying SEV by the interest rate, *r*, gives the Annual Equivalent Value (AEV). AEV is useful when comparing forestry and agroforestry to systems such as agriculture, where yearly returns are the norm.

NPV, SEV and AEV are appropriate when land is the most limiting factor of production. In many common agroforestry situations, however, capital, labor or time will be the most limiting factor (Franzel, 2004). Table 13.1 provides a scenario for each production input and relevant capital budgeting criteria that maximizes returns to the most limited factor. The Benefit-Cost Ratio (BCR) compares discounted benefits to costs as a unitless proportion rather than a difference as in NPV. Potential benefits and costs can be expressed per unit of land or as a total for the project because the units cancel. Internal Rate of Return (IRR) is the discount rate that makes the NPV equal zero. It is often used in practice, even though it is not as theoretically appropriate as NPV for most producers with limited land and relatively high levels of access to capital. IRR has intuitive appeal and is appropriate when a producer does not have a set discount rate.

Table 13.1 Limiting factors of production, potential scenarios and appropriate capital budgeting criteria.

Limiting factor	Scenario	Capital budgeting criteria
Land	• Landowner with access to credit and labor • High transaction costs for acquiring land • Larger family forest landowner in developed country	NPV, SEV, AEV
Capital	• Fixed level of investment capital • No constraints to land acquisition • Timber Investment Management Organization • Limited-resource farmer with sufficient land and family labor but no/limited access to credit	BCR, IRR
Labor	• Limited-resource or small family farmer with sufficient land but thin or nonexistent labor markets	DRW
Time	• Limited-resource farmer with access to capital at high interest rates and/or needs for quick returns for subsistence	Payback period, IRR

Conceptually, constraints to time are similar to constraints to capital (both indicate a high discount rate), so IRR is often a good criteria for both. Discounted Returns per Workday (DRW), the ratio of the discounted net revenues to discounted wages, expressed in dollars per workday, can be used when labor is the limiting asset.

Linear programming

When the objective is maximizing long-term profits from the entire farm under multiple constraints, linear programming (LP) is often the tool of choice. LP models differ from capital budgeting in two important ways. LP models the entire farm, not just the activity of interest, and accounts for diversity among farms. Each farm is modeled separately and aggregated to evaluate potential adoptability in a particular region. Mudhara and Hildebrand (2004) use LP to assess the impact of adopting improved fallows on household welfare and discretionary income in Zimbabwe. Thangata and Hildebrand (2012) used ethnographic linear programming (ELP) to examine the potential for agroforestry to reduce carbon emissions in sub-Saharan Africa. ELP provides insights into the complexity and diversity of smallholder farm systems by accounting for three important aspects in agroforestry decision making: (1) farmers' resource endowments (land, labor, capital), (2) farmers' multiple objectives (profits, subsistence needs, education, etc.) and (3) market conditions (prices, access, etc.). Dhakal, Bigsby and Cullen (2012) used LP to model the effects of government forest policies on households using community forests in Nepal. Their model captures the economic impacts of forest policy changes on landowners and the supply of forest products from private and community forests.

Production frontier analysis

Figure 13.3 depicts a production frontier, the maximum output that can be produced for any given level of input. Points *a*, *b* and *c* represent three farms or 'decision-making units' (DMUs).

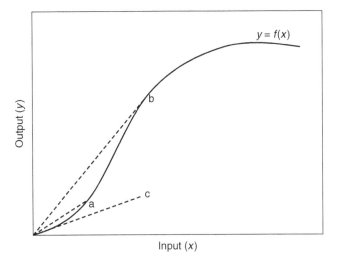

Figure 13.3 Production frontier: Slope of dashed lines represents technical efficiency of decision-making units *a*, *b* and *c*.

Total technical efficiency (TE)[1] is measured as the slope of the line segment from the origin to that DMU's point on the production frontier. In Figure 13.3, DMU *b* has the highest TE, *a* the middle and *c* the lowest. TE can be decomposed into two components: scale efficiency and pure TE. Both DMU *a* and *b* are operating at 100% pure TE because they lie on the production frontier. However, DMU *a* has a lower scale efficiency than DMU *b* because it could increase its total TE by expanding. DMU *c*, on the other hand, has lower pure TE than both *a* and *b* because it produces the same output with more inputs than *a* and less output with the same level of inputs as *b*.

In one input/one output cases, calculating TE is simple. The most efficient DMU produces the maximum output per unit of input (the slope of the line through the origin to the DMU's point). With multiple inputs and outputs, however, measuring relative efficiency becomes more complicated. One possible measure is the ratio of the weighted inputs to weighted outputs. If all outputs and inputs have market values, the prices are the weights and TE is equivalent to the benefit-cost ratio. However, in many situations (particularly in the developing world), markets are thin, prices may not exist and/or farmers may lack access to markets. In this case, benefit-cost ratios are not comparable between farms, but two methods (parametric and nonparametric) are available that account for the curvature of the production frontier (Bravo-Ureta and Pinheiro, 1993).

Parametric methods assume a specific functional form (e.g. Cobb-Douglas) and typically use corrected ordinary least squares or maximum likelihood to estimate parameter coefficients. Bright (2004) provided examples of production possibilities frontiers for agroforestry systems and multiple monocultures within a single farm. Lindara, Johnsen and Gunatilake (2006) applied stochastic frontier analysis using a Cobb-Douglas production function to evaluate factors affecting the TE of spice-based agroforestry systems in Sri Lanka.

Data envelopment analysis (DEA) uses LP to determine the weights that maximize TE without specifying a functional form and assuming that no DMU or linear combinations of the DMUs are 100% efficient. DEA is suited for comparing the efficiency of DMUs faced with multiple inputs and outputs, some of which may have no market value, a common situation in developing regions. Essentially, DEA picks weights (relative shadow prices) for each input and output to maximize TE.

Allowing the weights to vary is useful in at least three ways. First, using weights rather than market prices is critical when markets are thin or nonexistent. Second, prices for inputs or outputs often vary between regions, so that choosing a price from a single region or using a mean price can affect the efficiency measure. Third, individual farmers may value inputs or outputs differently than the market price due to government subsidies, individual preferences, subsistence and so forth. DEA can reduce the effects of the resulting distortions.

Figure 13.4 displays DMU_0 as a linear combination of the other DMUs. If any linear combination produces at least as much of each output and uses less input than DMU_0, then DMU_0 is inefficient. In other words, efficient farms and linear combinations of efficient farms form an envelope, which represents the production frontier. Inefficient levels are calculated as the relative distance from the efficient envelope. In Figure 13.4, DMUs *a* and *b*, located on the empirical efficient frontier (the 'envelope'), are 100% efficient. DMU *c* produces the same output using more inputs than a linear combination of *a* and *b*, located at point *x*, and thus, *c* is inefficient.

Frey et al. (2012) applied DEA to compare the relative efficiency among silvopasture, conventional pasture and plantation forestry in Argentina. Then, they applied nonparametric statistical analysis to compare the systems within farms. Silvopasture was found to be more efficient than conventional cattle ranching, but results were inconclusive for conventional forestry. Pascual (2005) utilized both parametric (stochastic production function analysis) and nonparametric data envelope analysis to examine the potential for reducing deforestation by improving the efficiency of traditional slash-and-burn milpa systems in Mexico. They found that deforestation would be reduced by 24% if households operated on the production frontier.

Risk and uncertainty

The expected utility paradigm is the theoretical foundation for most analyses of investment under uncertainty (Hildebrandt and Knoke, 2011). Rather than analyzing risk and return separately, expected utility theory examines the entire distribution of returns simultaneously. The

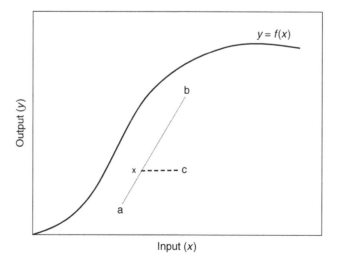

Figure 13.4 Envelope formulation: Decision Making Unit *c* is inefficient relative to *a* and *b* because *x*, a linear combination of *a* and *b*, produces the same output as *c* with fewer inputs.

decision maker chooses between uncertain prospects by comparing their expected utility values, i.e. the weighted sums obtained by adding the utility values of outcomes multiplied by their respective probabilities.

Mean variance analysis

The objective of risk research based on Markowitz's (1959) mean-variance hypothesis is to find the subset of 'efficient' portfolios that minimize risk for any given level of returns or, conversely, maximize returns for any given level of risk. Mean-variance (usually denoted E-V for 'expected value/variance') analysis can be very powerful, but the underlying assumptions limit it to a fairly restricted set of situations. E-V assumes that either the agent has a quadratic utility function, assets have normally distributed returns or both (Feldstein, 1969). These assumptions can cause large deviations from expected utility maximization, depending on the form of the utility function. Nevertheless, E-V is quite robust in approximating efficient expected utility maximization under a wide variety of utility functions and common levels of risk-aversion (Kroll, Levy and Markowitz, 1984).

E-V is often evaluated using quadratic programming (Steinbach, 2001) in which variance for given levels of expected returns is minimized by allowing the investments in each asset to vary. Repeating the quadratic program for a range of expected returns produces the frontier of efficient portfolios. Lilieholm and Reeves (1991) and Babu and Rajasekaran (1991) used E-V to analyze the efficient allocation of agroforestry within the whole farm and showed that adopting agroforestry can be optimal for certain levels of risk aversion. Ramirez et al. (2001) compared the financial returns, stability and risk of six cacao-laurel-plantain systems, and Ramirez and Sosa (2000) assessed the financial risk and return tradeoffs for coffee agroforestry systems in Costa Rica. Both studies evaluated expected returns and financial risk based on E-V analyses of estimated cumulative distribution functions of the NPVs and demonstrated the need to allow for the possibility of nonnormality of the variables in NPV analyses.

Stochastic dominance

Stochastic dominance (SD) encompasses the entire probability distribution of outcomes, does not require normality for the utility functions and requires only minimal assumptions about preferences (Hadar and Russell, 1969). Due to less restrictive assumptions and use of partial information, SD results are less deterministic and typically only provide a partial ranking of efficient and inefficient alternatives. Therefore, SD is commonly used for initial screenings of alternatives to provide a partial ordering based on partial information (Hildebrandt and Knoke, 2011). Castro, Calvas, Hildebrandt and Knoke (2013) applied SD to analyze the uncertainties associated with using conservation payments (CP) to preserve shade coffee in Ecuador. They investigated the effects of land-use diversification on CP by allowing different combinations of coffee agroforestry and monoculture maize production on farms. CP were two to three times higher when calculated with SD compared to maximizing a concave utility function, and Castro et al. concluded that the assumptions underlying SD are inappropriate for risk-averse farmers.

Real options

Land-use practices vary widely in their flexibility, and the best land managers include the value of the option to change or postpone actions in their decision calculus. Although deterministic models can incorporate changing future conditions and optimize decisions that adapt to these circumstances, they are inappropriate under risky or uncertain conditions because they assume perfect foresight. Stochastic analyses using real options (RO) techniques can estimate the value of flexibility given uncertain future conditions. The key difference between RO and capital budgeting is the recursive nature of the RO decision-making process. RO assumes that decisions made in the current year can be put off until the future. For example, a land manager can put off timber harvest and reforestation decisions, based on current conditions. Utilizing both stochastic and deterministic models can provide important insights about financial decisions (Frey, Mercer, Cubbage and Abt, 2013).

RO analyses are based on the Bellman equation, which assumes that a decision maker chooses a management regime to maximize the sum of current and discounted expected future rewards (profit, utility, etc.):

$$V_t(s) = \max_{x \in X(s)} \left\{ f(s,x) + \delta \cdot E_\varepsilon \left[V_{t+1} \left(g(s,x,\varepsilon) \right) \right] \right\},$$

$$s \in S,$$

$$t = 1, 2, \ldots, T \tag{1}$$

$V_t(s)$ denotes the total land value at time t in state s, $f(s,x)$ are the gains from choosing x under state s, $\delta = 1/(1 + \rho)$ is the discount factor, ρ is the discount rate and $E[\cdot]$ is the expectation operator. T is the time horizon, and $g(\cdot)$ is the transition function from states s, actions x and shocks ε (variability, risk).

Most forest harvesting RO models have used Markov-chain, Monte-Carlo techniques to solve the Bellman equation. Recently, however, partial differential methods are usually preferred due to improved precision (Miranda and Fackler, 2002). In a partial differential, infinite-horizon model, all points in time become equivalent and the Bellman equation simplifies to:

$$V(s) = \max_{x \in X} \left\{ f(s,x) + \delta \cdot E_\varepsilon \left[V \left(g(s,x,\varepsilon) \right) \right] \right\} \tag{2}$$

Partial differential collocation methods are used to solve equation (2) and determine the optimal regime for each state, $x(s)$.

Behan, McQuinn and Roche (2006) used RO to show that it is optimal for Irish farmers to wait longer to reforest or afforest than suggested by standard discounted cash-flow analyses because of establishment costs and the relative irreversibility of switching to forestry. Rahim, van Ierland and Wesseler (2007) used RO to analyze economic incentives to abandon or expand gum agroforestry in Sudan. They found that a 315% increase in gum Arabic prices would be needed to induce a shift in land use from agricultural production to gum agroforestry. Mithofer and Waibel (2003) used RO to analyze investment decisions for tree planting in Zimbabwe. They found that indigenous fruit tree planting is affected by tree growth rates and costs of collecting fruits from communal forests. Isik and Yang (2004) applied RO to examine participation in the Conservation Reserve Program (CRP) in Illinois. Although option values, land attributes and farmer characteristics significantly influenced participation, uncertainties in crop prices and program payments and irreversibilities associated with fixed contract periods were also crucial.

Next, we provide a case study (from Frey et al., 2010, 2013) applying capital budgeting and RO analysis to examine the potential for agroforestry to solve land-use problems in the Lower Mississippi Alluvial Valley (LMAV).

Case study: Agroforestry potential in the LMAV

The LMAV, the floodplain of the Mississippi River below the Ohio River (Figure 13.5), once contained the largest contiguous area of bottomland hardwood forest (BLH) in the United States. Beginning in the 1800s, converting BLH to agriculture has had a long history in the LMAV. For example, between 1950 and 1976, approximately one-third of the LMAV's bottomland forests were converted. Now, only a quarter of the original BLH survives, and what remains is degraded by fragmentation, altered hydrology, sedimentation, water pollution, invasive exotic plants and timber harvesting (Twedt and Loesch, 1999).

BLH forests provide critical ecosystem services, including wildlife habitat, clean water, flood mitigation and groundwater recharge, biogeochemical processes such as nutrient uptake and sediment deposition and carbon sequestration (Walbridge, 1993). However, the existing forest base has been reduced to the point where it can no longer meet society's demand for these services (Dosskey, Bentrup and Schoeneberger, 2012).

Beginning in the 1970s, a number of initiatives were introduced promoting BLH restoration in the LMAV; foremost are the CRP and the Wetlands Reserve Program (WRP). Although a significant amount of reforestation has occurred, most BLH remain characterized by continued deforestation and degradation. Agroforestry has been suggested as a means to augment BLH restoration by restoring trees on agricultural lands and producing at least some of the ecosystem services of natural BLH such as wildlife habitat and improved water quality (Dosskey et al., 2012).

In this case study, we illustrate the use of capital budgeting and RO to evaluate the potential adoptability of agroforestry in the LMAV. First, we use capital budgeting to compare profitability of agroforestry, production forestry and annual cropping with and without government incentives. Then, we apply RO analysis to examine how risk, uncertainty and flexibility affect the adoption decision.

Methods

Capital budgeting

Although agricultural and forestry management activities can take place year-round, we approximated them with discrete, yearly costs and benefits, as is common with forestry financial estimations. SEVs were used to compare expected returns from alternative investments in the LMAV (Klemperer, 1996). First, we calculated the NPV of the inputs required to produce a mature forest stand. Then, for even-aged management regimes, we estimated the financial returns from a clearcut, repeated in perpetuity to find SEV using:

$$SEV = \sum_{t=0}^{T} \frac{B_t - C_t}{(1+r)^t} \left[1 + \frac{1}{(1+r)^T - 1} \right] \tag{3}$$

where B_t and C_t are benefits (e.g. revenues from timber harvest or hunting lease) and costs (e.g. site preparation and maintenance) per hectare accrued in year t, T is the total number of time periods and r is the annual discount rate.

For uneven-aged regimes, we approximated the periodic sustainable harvest as a yearly harvest exactly equal to the mean annual increment and calculated the SEV for the annual sustainable return as:

$$SEV_{sust} = \left[\sum_{t=0}^{T-1} \frac{B_t - C_t}{(1+r)^t} \right] + \left[\frac{B_T - C_T}{r} \cdot \frac{1}{(1+r)^T} \right] \tag{4}$$

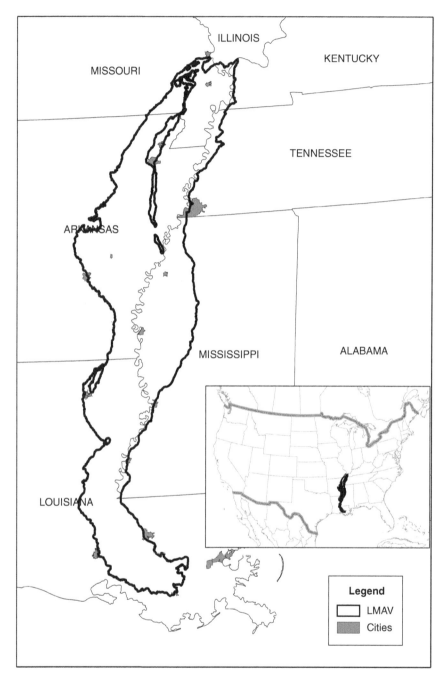

Figure 13.5 Geographic extent of the Lower Mississippi Alluvial Valley (LMAV) (Lower Mississippi Valley Joint Venture, 2002).

A base case SEV was calculated for each scenario assuming no policy interventions and two land capability classes. We also estimated the impacts of government incentive payments, such as fixed direct payments (FDPs) for agricultural crops, forestry and agroforestry systems, WRP and CRP and potential payments for carbon sequestration services.

Real options

In order to solve the partial differential collocation problem for the agriculture versus forestry (or agroforestry) optimal switching problem, we utilized a discrete-time dynamic program (Miranda and Fackler, 2002).[2] The method utilizes n nodes to generate a system of n linear equations to approximate the value function (equation 2) within the pre-defined state space for each possible action. The action producing the highest value function is preferred. For a relatively simple forestry management regime, such as cultivation of cottonwood for pulpwood with no intermediate thinning, the value function is:

$$f(s,x) = \begin{cases} s^{AG} \mid s^{SA} = 0 \\ sprep \mid s^{SA} = 1 \\ cc \mid s^{SA} = 2\,or\,3 \\ GY(s^{SA}) \star s^{TIMB} \mid x = 2 \\ GY(s^{SA}) \star s^{TIMB} + lclear \mid s^{SA} \neq 0 \,\&\, x = 0 \\ 0 \mid otherwise \end{cases} \tag{5}$$

where *sprep* is the cost of site preparation, *cc* is the cost of competition control in years 2 and 3, $GY(s^{SA})$ is the growth and yield function, and *lclear* is the cost of land clearing for agriculture (stump removal). The parameters utilized in the RO model are listed in Table 13.2.

Data

To compensate for lack of data on agroforestry in the LMAV and to validate existing information on forestry and agriculture, we organized three Delphi panels of forestry, agriculture and agroforestry experts to estimate key factors such as yields, costs, prices and management regimes. Additional data were obtained from the NRCS Soil Survey, USDA Agriculture and Resource Management Survey, state crop budget worksheets and Louisiana Quarterly Timber Price Reports. The details of the Delphi methodology and other data sources are described in Frey et al. (2010).

Results

Capital budgeting

BASE CASE

Table 13.3 presents the base case (no incentive payments) results for two Land Capability Classification (LCC) types, LCC3 and LCC5, with the highest potential for agroforestry in the LMAV. LCC3 lands are rarely flooded with poor drainage and severe limitations for agricultural production. LCC5 lands are frequently flooded, very poorly drained soils limited mainly

Table 13.2 Parameters used in the RO models (Frey et al., 2013).

Description	Source*	Units	Value**	
			LCC3	LCC5
Agricultural returns				
Equilibrium returns to agriculture	3, 4	$/ha/yr	382	110
Standard deviation of returns to agriculture	3, 4	$/ha/yr	253	238
Agricultural returns mean reversion rate	2, 4	unitless	0.35	0.35
Timber growth/yield and output prices				
Growth rate of cottonwood in pure plantation	1, 5	ton/yr	19.5	21.9
Growth rate of short-rotation woody crop species	1, 5	ton/yr	21.0	23.2
Growth rate of bottomland oak species in pure plantation	1, 5	ton/yr	7.9	7.9
Equilibrium of mixed hardwood pulpwood price	2, 6	$/ton	5.90	
Standard deviation of mixed hardwood pulpwood price	2, 6	$/ton	1.01	
Ratio of mixed hardwood sawtimber to pulpwood price	2, 6	unitless	5.67	
Ratio of low value to mixed hardwood sawtimber price	1, 6	unitless	0.8	
Ratio of oak to mixed hardwood sawtimber price	1, 6	unitless	1.15	
Timber (pulpwood) price mean reversion rate	2, 6	unitless	0.50	
Other forestry parameters				
Cost of site preparation and planting	1, 7	$/ha	−699	
Cost of competition control	1, 7	$/ha	−32	
Cost of clearing forested land	1	$/ha	−1,356 or −500	
Cost of coppicing cottonwood	1	$/ha	−148	
Yearly administration cost	1, 7	$/ha/yr	−20	
Value of hunting lease in mixed hardwood stand	8	$/ha/yr	15	
Value of hunting lease in cottonwood stand	1	$/ha/yr	7.5	
Relative yield of cottonwood in a cottonwood-oak intercropping system	1, 9		0.90	
Relative yield of oak in a cottonwood-oak intercropping system	1, 9		0.45	
Pecan yield and output prices				
Maximum yield of pecan in orchard (achieved years 19–50)	10	lbs/ha	2,371	
Proportion of maximum yield produced in years 1–7	1	unitless	0	
Proportion of maximum yield produced in years 8–9	1	unitless	0.5	
Proportion of maximum yield produced in year 10	1	unitless	0.63	
Proportion of maximum yield produced in year 11	1	unitless	0.65	
Proportion of maximum yield produced in years 12–16	1	unitless	0.83	
Proportion of maximum yield produced in years 17–18	1	unitless	0.92	
Proportion of maximum yield produced in years 19–50	1	unitless	1	
Equilibrium of pecan nut price	2, 11	$/lb	0.88	
Standard deviation of pecan nut price	2, 11	$/lb	0.32	
Pecan nut price mean reversion rate	2	unitless	0.90	
Other pecan parameters				
Cost of site preparation and planting for pecan	10	$/ha	−1,467	
Yearly fixed costs for pecan management	10	$/ha/yr	−611	
Variable costs for pecan management (mult. by yield rate)	10	$/ha/yr	−982	

Description	Source*	Units	Value**	
			LCC3	LCC5
Agroforestry parameters				
Cost of pruning	1, 7	$/ha	−148	
Relative yield of trees in an alley cropping system	1		0.58	
Ratio of planted acres in an alley cropping system	1	unitless	0.67	
Relative yield of agricultural crop per planted acre in a cottonwood alley cropping system	1	unitless	[0.75, 0.7, 0.65, 0.6, 0.55, 0.5, 0.5, 0.5, 0.5]	
Relative yield of agricultural crop per planted acre in a hard hardwood alley cropping system	1	unitless	[0.8, 0.75, 0.7, 0.065, 0.6, 0.55, 0.55, 0.55, 0.55]	
Relative yield of agricultural crop per planted acre in a pecan alley cropping system in year 2	1	unitless	0.67	
Same, year 3	1	unitless	0.63	
Same, year 4	1	unitless	0.60	
Same, year 5	1	unitless	0.57	
Same, year 6	1	unitless	0.53	
Same, years 7–9	1	unitless	0.50	
Same, years 10–18	1	unitless	0.47	
Same, years 19–50	1	unitless	0.43	
Other model parameters				
Discount rate		unitless	0.05	
Minimum agricultural returns in model state space		$/ha	−800	
Maximum agricultural returns in model state space		$/ha	800	
Minimum mixed hardwood pulpwood price in model state space		$/ton	0	
Maximum mixed hardwood pulpwood price in model state space		$/ton	20	
Minimum pecan price in model state space		$/lb	0	
Maximum pecan price in model state space		$/lb	3	
Covariance			0	
			[0, 0]	

*Number indicates source of the parameter estimate: 1 = Delphi assessment; 2 = mean reversion model; 3 = Monte-Carlo crop switching model; 4 = ERS (2009); 5 = NRCS (2008); 6 = LA DAF (2008); 7 = Smidt et al. (2005); 8 = Hussain et al. (2007); 9 = Gardiner et al. (2004); 10 = Ares et al. (2006); 11 = NASS (2008).[9]
**LCC, Land Capability Classification.

to pasture, range, forestry or wildlife. On LCC3 soils, none of the agroforestry or production forestry systems were competitive with agriculture at any discount rate. However, the SEVs for most of these systems, particularly the agroforestry systems, were substantially higher than on LCC5 soils. In particular, the alley cropping systems had SEVs over $2,000 per hectare at the lowest discount rate (5%).

On the most marginal land (LCC5), at the lowest discount rate (5%), three agroforestry practices and production forestry have higher SEVs than agriculture (soybeans), assuming no policy interventions. At discount rates of 7%–10%, soybean crops dominate all agroforestry and forestry systems on LCC5 lands. The only systems with positive SEVs on LCC5 sites with a 7% discount

Table 13.3 Soil expectation values (SEVs, 2008$ per hectare) for production systems with no policy interventions and varying discount rates, on Land Capability Classes (LCC) 3 and 5 in the Lower Mississippi Alluvial Valley (LMAV) (Frey et al., 2010).

Discount rate (%)	LCC3			LCC5		
	5	7	10	5	7	10
Soybeans	5,150	3,679	2,575	925	661	463
Rice	7,771	5,551	3,886	−768	−548	−384
Cottonwood for pulpwood	−257	−499	−689	−338	−625	−844
Cottonwood for sawtimber	1,180	275	−347	1,210	205	−479
Short-rotation woody crop	−2,217	−1,839	−1,565	−2,253	−1,941	−1,713
Hard hardwoods (clearcut)	52	−495	−758	−129	−667	−922
Hard hardwoods (sustainable)	−179	−613	−794	−357	−783	−957
Cottonwood and oak interplanting (clearcut)	158	−495	−885	18	−649	−1,048
Cottonwood and oak interplanting (sustainable)	−12	−589	−915	−158	−743	−1,077
Pecan silvopasture	1,020	−918	−2,255	−28	−1,864	−3,106
Hard hardwoods silvopasture	811	190	−122	321	−246	−513
Pine silvopasture	2,512	951	−12	1,861	404	−477
Hard hardwoods riparian buffer	−333	−652	−784	−510	−822	−947
Cottonwood and oak riparian buffer	−590	−956	−1,138	−769	−1,135	−1,317
Pecan alley crop	2,355	7	−1,640	−235	−2,000	−3,191
Hard hardwoods alley crop	843	275	−13	−8	−467	−656
Cottonwood alley crop	2,144	1,076	362	1,367	393	−234

rate are pine silvopasture, cottonwood alley cropping and cottonwood for sawtimber; at a 10% discount rate, SEVs are negative for all agroforestry and forestry systems.

In the absence of incentive payments, landowners are more likely to adopt agroforestry than conventional forestry on moderately marginal land (LCC3), while on the most marginal land (LCC5) the returns for agroforestry and forestry are similar. Still, the low SEVs for agroforestry compared to agriculture predict little success for agroforestry or forestry in the LMAV. Our estimates are less favorable for forestry than earlier studies (e.g. Anderson and Parkhurst, 2004) because we include tree seedling mortality and recent increase in crop prices.

GOVERNMENT INCENTIVES CASE

Table 13.4 provides SEV results that include government incentive payments (Average Crop Revenue Election [ACRE] and FDP agricultural subsidies and enrollment in WRP and CRP programs). The ACRE and FDP program increase the value of agriculture by 15% on LCC3 and 60% on LCC5 lands. WRP and CRP enrollment is competitive with agriculture on LCC5 land with a slightly lower return from CRP at a 5% discount rate. Higher discount rates make CRP and agriculture less competitive because the WRP easement is paid up front, whereas agriculture and CRP receive annual payments. On LCC3 lands, WRP is less competitive, because of the $2,223 per hectare rate cap, while CRP payments are based on the typical land rental rate, which is higher for LCC3 soils. Therefore, CRP is somewhat more competitive than WRP on moderate soils.

Next, we examine the impacts of a market for carbon sequestration credits. Table 13.5 shows the CO_2 net price per ton that equalizes SEVs for forestry, agroforestry and agriculture, including

Table 13.4 Soil expectation values (SEVs, 2008 $ per hectare, 5% discount rate) for production systems under existing incentive policies: soybeans with Average Crop Revenue Election and Fixed Direct Payment (ACRE and FDP) programs, hard hardwoods with Wetlands Reserve Program (WRP) and hard hardwoods riparian buffer with Conservation Reserve Program Conservation Practice 22 (CRP CP22), on Land Capability Classes (LCC) 3 and 5 in the Lower Mississippi Alluvial Valley (LMAV) (Frey et al., 2010).

	System	*No policy*	*ACRE & FDP*	*WRP*	*CRP*
LCC3	Soybeans	5,150	5,950		
	Hard hardwoods	52		2,233	
	Hard hardwoods riparian buffer	−333			3,696
LCC5	Soybeans	925	1,478		
	Hard hardwoods	−129		2,233	
	Hard hardwoods riparian buffer	−510			2,184

Table 13.5 Break-even net revenue per metric ton CO_2 (2008 $) in various forestry and agroforestry systems compared to soybeans with Average Crop Revenue Election (ACRE) and Fixed Direct Payment (FDP) payments, on Land Capability Classes (LCC) 3 and 5 in the Lower Mississippi Alluvial Valley (LMAV) (Frey et al., 2010).

System	*LCC3*	*LCC5*
Cottonwood for pulpwood	59.58	15.90
Cottonwood for sawtimber	32.47	1.66
Short-rotation woody crop	254.60	102.36
Hard hardwoods (clearcut)	26.59	7.24
Hard hardwoods (sustainable harvest)	15.15	4.54
Cottonwood and oak interplanting (clearcut)	30.87	7.62
Cottonwood and oak interplanting (sustainable)	17.39	4.77
Pecan silvopasture	40.35	12.32
Hard hardwoods silvopasture	29.37	6.61
Pine silvopasture (optimistic returns per head)	35.39	0.00
Hard hardwoods riparian buffer	31.78	10.05
Cottonwood and oak riparian buffer	39.19	13.46
Pecan alley crop	29.42	14.02
Hardwood alley crop	31.55	9.18
Cottonwood alley crop	32.64	0.87

ACRE and FDP payments. At any higher price for CO_2, the respective forestry/agroforestry system becomes more profitable than soybeans. Additional costs/barriers to selling carbon credits from forestry/agroforestry systems, however, may limit participation in CO_2 markets. These include the costs of verifying and registering carbon credits and demonstrating additionality (i.e. proof that the reforestation would not have taken place without the carbon payment).

Real Options

The RO model allowed landowners to adopt the most profitable land use and then convert to other land uses based on knowledge of past returns and expectations of future returns. RO provided a powerful and realistic reflection of the actual decisions that landowners make and extended previous analyses of farm, forest and agroforestry decision making. We found that the decision to switch is driven almost entirely by agricultural returns, given the mean-reversion assumption and the long waiting period between agroforestry establishment and the final timber harvest. For example, if the pulpwood price in the current year was $10/ton and the agricultural returns in the current year were $100 per hectare, continuing in agriculture would be optimal. Switching to alley cropping is only optimal when agriculture loses $800 or more per hectare.

ADOPTION THRESHOLDS

The point at which agroforestry becomes more desirable than agriculture is the 'adoption threshold'. The adoption thresholds are summarized in Table 13.6 for LCC3 and Table 13.7 for LCC5 land. The 'RO value' is the estimate of the value function, $V(s)$, assuming forestry/agroforestry at the year of site planting and equilibrium prices. This is comparable to the SEV in some cases but allows for increased value from flexibility, including the option to switch back to agriculture. In many cases on recently planted forestry or agroforestry LCC3 land, at equilibrium prices, the optimal decision is to switch back to agriculture immediately.

AEV in Tables 13.6 and 13.7 can be viewed as the 'SEV adoption threshold'. The AEV does not account for the value of being able to wait to convert agricultural land to forestry or select the optimal timber rotation given dynamic timber prices. In most cases, the greater flexibility associated with annual cropping results in lower probabilities of adopting forestry or agroforestry than the simple AEV analysis suggests. Systems with the RO value closest to SEV are the least flexible; most notably, the WRP, which essentially has no flexibility. The RO threshold is lower than the SEV threshold for WRP because we assumed that enrollment in WRP is irreversible and no timber harvest is permitted. The only income allowed after the easement payment is from hunting leases.

Table 13.6 RO and SEV adoption thresholds in terms of agricultural returns per hectare, and RO value and SEV for production forestry and agroforestry systems on land capability class (LCC) 3 land ($/ha/yr) (Frey et al., 2013).

	RO adoption threshold ($/ha/yr) Land clearing cost:		Prob. of crossing threshold* (%)	AEV (SEV adoption threshold)	RO value, at land clearing cost $1,356/ha	SEV**
	$1,356/ha	$500/ha				
Wetlands Reserve Program	−1,000	−1,000	<0.1	112	2,236	2,233
Cottonwood	−1,000	−1,000	<0.1	59	5,581★★★	1,180
Short rotation woody crop	−980	−980	<0.1	−111	6,678★★★	−2,217
Hard hardwoods	−1,000	−1,000	<0.1	3	5,544★★★	52
Cottonwood-oak intercrop	−1,000	−1,000	<0.1	8	5,544★★★	158
Pecan alley crop	−1,000	−1,000	<0.1	118	5,406★★★	2,355
Hard hardwoods alley crop	−1,000	−900	<0.1	42	6,632★★★	843
Cottonwood alley crop	−1,000	−1,000	<0.1	107	6,259★★★	2,144

*At land clearing cost $1,356/ha.
**SEV from Frey et al. (2010).
***The optimal decision at the equilibrium agricultural return value and timber price for a recently planted forestry/agroforestry plot is to return immediately to agriculture.

Table 13.7 RO and SEV adoption thresholds in terms of agricultural returns per hectare, and RO value and SEV for production forestry and agroforestry systems on land capability class (LCC) 5 land ($/ha/yr) (Frey et al., 2013).

	RO adoption threshold ($/ha/yr) Land clearing cost:		Prob. of crossing threshold* (%)	AEV (SEV adoption threshold)	RO value, at land clearing cost $1,356/ha	SEV**
	$1,356/ha	$500/ha				
WRP	−240	−240	11	112	2,236	2,233
Cottonwood	140	140	39	61	3,770	1,210
Short rotation woody crop	−550	−550	0.9	−113	1,548	−2,253
Hard hardwoods	−730	−690	<0.1	−6	955	−129
Cottonwood-oak intercrop	−510	−420	1	1	1,469	18
Pecan alley crop	−450	−450	2	−12	1,834	−235
Hard hardwoods alley crop	−830	−600	<0.1	0	1,346	−8
Cottonwood alley crop	270	270	43	68	3,471	1,367

*At land clearing cost $1,356/ha.
**SEV from Frey et al. (2010).

Nevertheless, the returns to WRP enrollment on LCC5 land make it more attractive in the RO model than many forestry and agroforestry systems. For example, on LCC5 sites, the RO adoption threshold was significantly more negative (i.e. more difficult to reach) than the SEV adoption threshold (the AEV) for WRP enrollment, cottonwood timber plantation, short-rotation woody crops, hard hardwood timber plantation, cottonwood-oak intercrop plantation, pecan alley cropping and hardwood alley cropping. On LCC3 land, all RO adoption thresholds were lower than the SEV adoption thresholds.

At first glance, agricultural returns must become negative for it to be optimal to switch to forestry or agroforestry. However, agricultural returns need only turn negative for 1 year for switching to be attractive, and it is certainly feasible that net agricultural returns on these marginal lands will occasionally be negative. On LCC5 sites, three forestry and agroforestry systems have a greater than 10% chance of being adopted on any given plot in any given year. On LCC3 sites, however, no system had a greater than one in a thousand chance of being an optimal choice in any given year.

Approximately 40%–50% of the LMAV is classified as LCC3, and any large-scale effort at reforestation would need to include these soils. To examine the impact of market changes on adoption on LCC3 land, we calculated ROs and AEVs under three scenarios: (1) timber prices double, (2) timber prices double and volatility declines 50% and (3) timber prices double and volatility of agricultural returns increases 50%. We compared adoption thresholds at age 10 and maximum stand age using the equilibrium timber price.

Reducing timber price volatility had little effect on the outcomes in scenarios 1 and 2, and increasing the timber price did not significantly affect adoption thresholds on LCC3 land. All were still below a 0.1% probability of crossing the threshold in any given year.

Under scenario 3 (timber prices double and volatility of agricultural returns increases 50%), changes in adoption thresholds were similar but smaller in magnitude than in Scenarios 1 and 2, suggesting that increased volatility in agricultural returns actually favors agriculture. This is likely due to the assumption of risk neutrality.

Disadoption thresholds for scenarios 1, 2 and 3 were affected more strongly than adoption thresholds for all systems, particularly at older stand ages. In all cases, all three scenarios increased, or kept the same disadoption threshold, meaning forestry and agroforestry would be less likely to be disadopted.

The base case did not include Farm Bill agricultural payments, so a scenario (which we did not model) similar to the present-day scenario which includes Farm Bill agricultural payments but no payments for ecosystem services, would favor agriculture more strongly than the base case. However, when payments for ecosystem services are allowed, forests are more strongly favored than the base case, indicating that these payments can more than counteract farm bill agricultural payments. In fact, these payments have a stronger effect relative to the base case than doubling the timber price in scenario 2 (Frey et al., 2013).

Conclusions

This chapter provides principles, literature and a case study about the economics of agroforestry. The tools presented here can help determine when agroforestry is a feasible option and provide arguments for cases where agroforestry systems are economically, socially and environmentally appropriate, fostering improved sustainable development for landowners, farmers and communities. Agroforestry systems provide multiple outputs, potentially reducing risk and increasing income while also purportedly producing more ecosystem services than conventional agriculture. Our review and case study, however, provide cautionary tales about the limits of agroforestry.

In a few cases where complementary production relationships occur, agroforestry is obviously superior to tree, crop or pasture monocultures. There still may be some resistance to adoption of agroforestry in these cases due to the management challenges with complex systems, but at least the economics may lead to adoption in the long run. In the more common case of competitive production relationships, finding the right mix of inputs and products requires more economic analyses, cautious generalizations about the merits of the cases examined and more extension efforts to encourage farm adoption where agroforestry appears most warranted. The principles we posit here and the literature cited provide a basis for such reviews and recommendations.

The analyses do suggest that to reach its promise, even in cases where the research and economics indicate clear benefits, substantial outreach efforts must occur. The decision to adopt agroforestry systems involves judgments about which systems generate the highest short- and long-run returns, are easiest to manage, readily marketable and fit in with cultural traditions. These factors are not all economic, and farmers may err in evaluating the financials. In some cases, government action and support may be needed to create proper markets and institutions. In almost all cases, better knowledge of inputs, outputs, costs and markets will be required.

The economics and adoption of agroforestry systems will also be determined by the scale of the specific operations. Small-scale subsistence farms in developing countries have higher likelihoods of producing clear, net benefits from agroforestry, based on the need for multiple livelihood products, the availability of on-farm hand and animal labor that can be used with few adverse effects on multiple crops and the benefits of fertilization from trees and/or livestock. These small-scale farm systems can extend beyond the subsistence level with moderate ease in many countries, through production of fruits, nuts, bananas and similar outputs that can be sold in local markets. In these cases, one could say that the findings of economic studies and the adoption of agroforestry systems are often congruent. Farmers have developed these promising systems, economic analyses often support the merits of the systems and outreach programs help extend these systems to a broader range of producers. The environmental benefits of these systems are probably undervalued by the farmer, however, and some type of payments for ecosystem services may be necessary, in most cases, to increase adoption to socially desirable levels.

The discrepancy between the purported income diversification, risk reduction, environmental benefits and the limited farm adoption of agroforestry systems seems to be much wider for medium-sized farms. Numerous studies have found biological and economic benefits from agroforestry systems, which at least have returns greater than monoculture forestry and at times greater than agriculture on poor crop or pasture lands. Yet, adoption rates are often low. These discrepancies suggest either that our science and economic models are faulty, that farmers are irrational, that nonmarket benefits need monetization or that tradition and ease of management trump purely economically rational decision making. A common adage says that if your economic models suggest that farmers are making bad decisions, it is probably the models that are in error.

The case study in the LMAV suggests that agroforestry adoption may be even more difficult than cash-flow analyses alone indicate. For example, the higher opportunity cost to convert back from forests to agriculture reduces agroforestry's desirability. Research on production systems and economics for more mechanized agroforestry systems is still inconclusive about overall merits, so sound economic analyses need to be conducted on individual cases being considered. Our review provides the tools to do so. In situations where large-scale, highly mechanized pure monoculture systems dominate the landscape (as in the LMAV), it will be difficult for mixed agroforestry systems to reverse this situation, except at the windbreak, stream buffer or ornamental level. But those benefits have been well documented, and the economics can be analyzed and promoted.

Additional investments in economics research will be required, however, for agroforestry to achieve its full potential (Mercer and Alavalapati, 2004). Economic analyses need to move beyond enterprise-specific foci and focus on whole-farm analyses as well as move beyond strictly financial analyses to also include the impacts of policy constraints, market failures, farmer preferences and the impact of cultural taboos. Additional dynamic optimization research is needed that includes impacts of stochastic prices, yields and weather variables. Developing time series or panel data sets for econometric analyses is crucial to advancing agroforestry economics research. A large hole in agroforestry economics is studies examining economy-wide impacts of agroforestry adoption using applied general equilibrium analyses such as input-output models, computable general equilibrium models (CGE) and social accounting matrices (SAM). Finally, decisions to adopt agroforestry are complicated by the multiple biophysical, social and economic objectives involved, many of which are difficult to value monetarily. In addition to more research with traditional approaches to nonmarket valuation (contingent valuation, conjoint analysis, travel cost and hedonic approaches), studies using alternative approaches such as the analytical hierarchy process (AHP) should also be expanded (Shrestha, Alavalapati and Kalmbacher, 2004).

Notes

1 Technical efficiency is the purely biophysical effectiveness of production inputs.
2 Because the stand age varies depending on when the landowner switches to forestry, solving the Bellman equation (equation 1) with a Markov-chain dynamic program was not possible.

References

Anderson, J. D. and Parkhurst, G. M. (2004). 'Economic comparison of commodity and conservation program benefits: An example from the Mississippi Delta', *Journal of Agricultural and Applied Economics*, vol. 36, no. 2, p. 415.

Ares, A., Reid, W. and Brauer, D. (2006). 'Production and economics of native pecan silvopastures in central United States', *Agroforestry Systems*, vol. 66, no. 3, pp. 205–215.

Babu, S. C. and Rajasekaran, B. (1991). 'Agroforestry, attitude towards risk and nutrient availability – a case-study of South Indian farming systems', *Agroforestry Systems*, vol. 15, no. 1, pp. 1–15.

Behan, J., McQuinn, K. and Roche, M. J. (2006). 'Rural land use: Traditional agriculture or forestry?' *Land Economics*, vol. 82, no. 1, pp. 112–123.

Bravo-Ureta, B. C. and Pinheiro, A. E. (1993). 'Efficiency analysis of developing country agriculture: A review of the frontier function literature', *Agricultural and Resource Economics Review*, vol. 22, no. 1, pp. 88–101.

Bright, G. (2004). 'Exploring the economics of agroforestry systems using a production function approach', in J.R.R. Alavalapati and D. E. Mercer (Eds.), *Valuing Agroforestry Systems*, Kluwer Academic Publishers, Dordrecht, The Netherlands, pp. 79–93.

Castro, L. M., Calvas, B., Hildebrandt, P. and Knoke, T. (2013). 'Avoiding the loss of shade coffee plantations: How to derive conservation payments for risk-averse land-users', *Agroforestry Systems*, vol. 87, no. 2, pp. 331–347. Retrieved from http://link.springer.com/article/10.1007/s10457-012-9554-0/fulltext. html#Sec3

Cubbage, F., Balmelli, G., Bussoni, A., Noellemeyer, E., Pachas, A. N., Fassola, H., . . . Hubbard, W. (2012). 'Comparing silvopastoral systems and prospects in eight regions of the world', *Agroforestry Systems*, vol. 86, pp. 303–314.

Dhakal, B., Bigsby, H. and Cullen, R. (2012). 'Socioeconomic impacts of public forest polices on heterogenous agricultural households', *Environmental and Resource Economics*, vol. 53, pp. 73–95.

Dosskey, M. G., Bentrup, G. and Schoeneberger, M. (2012). 'A role for agroforestry in forest restoration in the Lower Mississippi Alluvial Valley', *Journal of Forestry*, vol. 110, pp. 48–55.

ERS (Economics Research Service). (2009). Agricultural Resource Management Survey (ARMS). USDA Economic Research Service, Washington, DC. Available from http://www.ers.usda.gov/Briefing/ARMS/ (accessed 1/15 2009).

Feldstein, M. S. (1969). 'Mean-variance analysis in theory of liquidity preference and portfolio selection', *Review of Economic Studies*, vol. 36, no. 1, pp. 5–12.

Franzel, S. (2004). 'Financial analysis of agroforestry practices: Fodder shrubs in Kenya, woodlots in Tanzania, and improved fallows in Zambia', in J.R.R. Alavalapati and D. E. Mercer (Eds.), *Valuing Agroforestry Systems*, Kluwer Academic Publishers, Dordrecht, The Netherlands, pp. 9–37.

Franzel, S. and Scherr, S. J. (2002). 'Introduction', in S. Franzel and S. J. Scherr (Eds.), *Trees on the Farm: Assessing the Adoption Potential of Agroforestry Practices in Africa*, CABI, Wallingford, pp. 1–11.

Frey, G. E., Fassola, H. E., Pachas, A. N., Colcombet, L., Lacorte, S. M., Renkow, M., . . . Cubbage, F. W. (2012). 'A within-farm efficiency comparison of silvopasture systems with conventional pasture and forestry in northeast Argentina', *Land Economics*, vol. 88, no. 4, pp. 639–657.

Frey, G. E., Mercer, D. E., Cubbage, F. W. and Abt, R. C. (2010). 'Economic potential of agroforestry and forestry in the Lower Mississippi Alluvial Valley with incentive programs and carbon payments', *Southern Journal of Applied Forestry*, vol. 34, no. 4, pp. 176–185.

Frey, G., Mercer, E., Cubbage, F. and Abt, R. (2013). 'A real options model to assess the role of flexibility in forestry and agroforestry adoption and disadoption in the Lower Mississippi Alluvial Valley', *Agricultural Economics*, vol. 44, pp. 73–91.

Gardiner, E. S., Stanturf, J.A. and Schweitzer, C.J. (2004). 'An afforestation system for restoring bottomland hardwood forests: Biomass accumulation of Nuttall oak seedlings interplanted beneath eastern cottonwood', *Restoration Ecology*, vol. 12, no. 4, pp. 525–532.

Hadar, Josef, and William R. Russell. (1969). 'Rules for ordering uncertain prospects', *The American Economic Review*, vol. 59, no. 1, pp. 25–34.

Hildebrandt, P. and Knoke, T. (2011). 'Investment decisions under uncertainty – A methodological review on forest science studies', *Forest Policy and Economics*, vol. 13, pp. 1–15.

Hussain, A., Munn I.A., Grado S. C., West B. C., Jones W. D., and Jones, J. (2007). 'Hedonic analysis of hunting lease revenue and landowner willingness to provide fee-access hunting', *Forest Science*, vol. 53, no. 4, pp. 493–506.

Isik, M. and Yang, W. (2004). 'An analysis of the effects of uncertainty and irreversibility on farmer participation in the conservation reserve program', *Journal of Agricultural Resource Economics*, vol. 29, no. 2, pp. 242–259.

Klemperer, W. D. (1996). *Forest Resource Economics and Finance*, McGraw Hill, New York.

Kroll, Y., Levy, H. and Markowitz, H. M. (1984). 'Mean-variance versus direct utility maximization', *Journal of Finance*, vol. 39, no. 1, pp. 47–61.

LA DAF (Louisiana Department of Agriculture and Forestry). (2008). Louisiana Quarterly Report of Forest Products. Louisiana Department of Agriculture and Forestry, Baton Rouge, LA. Available from http://www.ldaf.state.la.us/portal/Offices/Forestry/ForestryReports/QuarterlyReportofForestProducts/tabid/451/Default.aspx (accessed 1/15 2009).

Lilieholm, R. J. and Reeves, L. H. (1991). 'Incorporating economic risk aversion in agroforestry planning', *Agroforestry Systems*, vol. 13, no. 1, p. 63.

Lindara, L.M.J.K., Johnsen, F. H. and Gunatilake, H. M. (2006). 'Technical efficiency in the spice based agroforestry sector in Matale district, Sri Lanka', *Agroforestry Systems*, vol. 68, pp. 221–230.

Lower Mississippi Valley Joint Venture (LMVJV). (2002). *Conservation Planning Atlas, Volume 1*, Lower Mississippi Valley Joint Venture, Vicksburg, MS. Retrieved from www.lmvjv.org/cpa_volume1.htm

Markowitz, H. (1959). *Portfolio Selection; Efficient Diversification of Investments*, Wiley, New York.

Mercer, D. E. (2004). 'Adoption of agroforestry innovations in the tropics: A review', *Agroforestry Systems*, vol. 20441, pp. 311–328.

Mercer, D. E. and Alavalapati, J.R.R. (2004). 'Summary and future directions', in J.R.R. Alavalapati and D. E. Mercer (Eds.), *Valuing Agroforestry Systems: Methods and Applications*, Kluwer Academic Publishers, Dordrecht, The Netherlands, pp. 303–310.

Mercer, D. E. and Hyde, W. F. (1992). 'Economics of Agroforestry', in W. Burch and K. Parker (Eds.), *Social Science Applications in Asian Agroforestry*, Winrock International/South Asia Books, pp. 111–144.

Miranda, M. J. and Fackler, P. L. (2002). *Applied Computational Economics and Finance*, MIT Press, Cambridge, MA.

Mithofer, D. and Waibel, H. (2003). 'Income and labour productivity of collection of indigenous fruit tree products in Zimbabwe', *Agroforestry Systems*, vol. 59, pp. 295–305.

Mudhara, M. and Hildebrand, P. E. (2004). 'Assessment of constraints to the adoption of improved fallows in Zimbabwe using Linear Programming Models', in J.R.R. Alavalapati and D. E. Mercer (Eds.), *Valuing Agroforestry Systems: Methods and Applications*, Kluwer Academic Publishers, Dordrecht, The Netherlands, pp. 201–218.

NASS (National Agricultural Statistics Service). (2008). Noncitrus Fruits and Nuts 2007 Summary. Report No. Fr Nt 1-3 (08) a. USDA National Agricultural Statistics Service, Washington, DC.

NRCS. (2008). Soil Survey Geographic (SSURGO) database. USDA Natural Resources Conservation Service, Ft. Worth, TX.

Pascual, U. (2005). 'Land use intensification potential in slash-and-burn farming through improvements in technical efficiency', *Ecological Economics*, vol. 52, pp. 497–511.

Rafiq, M., Amacher, G. S. and Hyde, W. F. (2000). 'Innovation and adoption in Pakistan's Northwest Frontier Province', in W. F. Hyde and G. S. Amacher (Eds.), *Economics of Forestry and Rural Development: An Empirical Introduction from Asia*, University of Michigan Press, Ann Arbor, MI, pp. 87–100.

Rahim, A., van Ierland, E. C. and Wesseler, J. (2007). 'Economic incentives for abandoning or expanding gum arabic production in Sudan', *Forest Policy and Economics*, vol. 10, pp. 36–47.

Ramirez, O. A. and Sosa, R. (2000). 'Assessing the financial risks of diversified coffee production systems: An alternative nonnormal CDF estimation approach', *Journal of Agricultural and Resource Economics*, vol. 25, no. 1, pp. 267–285.

Ramirez, O. A., et al. (2001). 'Financial returns, stability and risk of cacao-plantain-timber agroforestry systems in Central America', *Agroforestry Systems*, vol. 51, no. 2, pp. 141–154.

Shresta, R. K., Alavalapati, J.R.R. and Kalmbacher, R. S. (2004). 'Exploring the potential for silvopasture adoption in south-central Florida: An application of SWOT-AHP method', *Agricultural Systems*, vol. 81, no. 3, pp. 185–199.

Smidt, M., Dubois M. R., and Folegatti, B. S. (2005). 'Costs and cost trends for forestry practices in the South', *Forest Landowners*, vol. 65, pp. 25–31.

Society of American Foresters (SAF). (2012). 'Agroforestry definition', Society of American Foresters. Retrieved from http://dictionaryofforestry.org/dict/term/agroforestry

Steinbach, M. C. (2001). 'Markowitz revisited: Mean-variance models in financial portfolio analysis', *SIAM Review*, vol. 43, no. 1, pp. 31–85.

Thangata, P. H. and Hildebrand, P. E. (2012). 'Carbon stock and sequestration potential of agroforestry systems in smallholder agroecosystems of sub-Saharan Africa: Mechanisms for "reducing emissions from deforestation and forest degradation" (REDD+)', *Agriculture, Ecosystems and Environment*, vol. 158, pp. 172–183.

Twedt, D. J. and Loesch, C. R. (1999). 'Forest area and distribution in the Mississippi Alluvial Valley: Implications for breeding bird conservation', *Journal of Biogeography*, vol. 26, no. 6, p. 1215.

Walbridge, M. R. (1993). 'Functions and values of forested wetlands in the Southern United States', *Journal of Forestry*, vol. 91, no. 5, pp. 15–19.

14

PAYMENT FOR ECOSYSTEM SERVICES

Lessons from developing countries

Yazhen Gong,[1] Ravi Hegde[2] and Gary Q. Bull[3]

[1]ASSISTANT PROFESSOR, RESOURCE ECONOMICS, RENMIN UNIVERSITY OF CHINA.
[2]ECONOMIST, HORTICULTURE AUSTRALIA LTD.
[3]PROFESSOR, FOREST POLICY AND ECONOMICS, UNIVERSITY OF BRITISH COLUMBIA.

Abstract

This chapter synthesizes the status of payment for ecosystem services (PES) schemes implemented in developing countries in recent years. It provides an overview on PES schemes for watershed services, biodiversity conservation and forest carbon. It uses a conceptual framework that recognizes the roles of socio-economic, environmental and institutional factors in determining outcomes of PES schemes, in particular PES providers' participation in the schemes. It also reviews key impacts of PES programs at the participant household level and presents main challenges for PES in the future. It provides the following main lessons learned from PES practiced in developing countries: (1) the transaction costs of PES programs are often high, and this can affect the cost-effectiveness of the programs; (2) local communities' social capital and their ability to enforce contracts are critical for PES programs to successfully generate ecosystem services (ES), improve social welfare and facilitate equity; (3) the complexity of PES programs with multiple objectives could easily doom the program to failure.

Keywords

PES, watershed, carbon, biodiversity, participation, impact, transaction cost, social capital

Introduction

The degradation of ecosystem services (ES) is a serious environmental problem that may ultimately hinder human society from achieving a sustainable future. ES are broadly defined as 'the benefits people obtain from ecosystems' (MA, 2003), and they can be divided into four categories: (1) provisioning services (e.g. food, water, timber and fiber), (2) regulating services (e.g. climate regulation, flood control and water purification), (3) supporting services (e.g. soil formation and nutrient cycling) and (4) cultural services (e.g. recreational, religious and other nonmaterial benefits) (MA, 2003). However, as the human population continues to grow rapidly and demand for ES intensifies, these vital services are used unsustainably. In fact, approximately 60% of ES declined or were used unsustainably in the second half of the twentieth century and without acting proactively, ES will continue to decline (MA, 2005). Declining ES have become a barrier for human society to reducing global poverty and achieving the Millennium Development Goals (MA, 2005).

To reverse the degradation of ES while still meeting increasing human demands, there must be significant changes made to policies, institutions and market practices (MA, 2005). One response is the recent creation of a conservation initiative, known as payments for environmental services (PES). It has been proffered as a means to deal with ES degradation (Landell-Mills and Porras, 2002; Wunder, 2005; Pigiola et al., 2007; Engel, Pagiola and Wunder, 2008). PES uses market-based instruments to arrange transactions between ES users and providers,[1] thus directly internalizing what would otherwise be an externality (Pigiola et al., 2007; Engel et al., 2008); in other words, PES provide a direct and tangible incentive to the ES providers to enhance ES provisions and manage ecosystems more sustainably (Pattanayak, Wunder and Ferraro, 2010).

This chapter seeks to synthesize the status of PES schemes implemented in recent years. It focuses on PES schemes in developing countries where the rural poor are in greatest need of assistance, paying particular attention to payments for water services, biodiversity conservation and forest carbon. The lessons learned from PES practiced in developing countries can shed light on future design and implementation of PES to seek a potential win-win solution for ecosystem conservation and poverty alleviation.

The conceptualization of PES

Although the conceptualization of PES remains a topic of debate amongst scholars (Engel et al., 2008; Muradian, Corbera, Pascual, Kosoy and May, 2010; Tacconi, 2012),[2] the Coasian theory is the foundational framework used. From an environmental economics perspective, declining ES represent a classic 'market failure' problem arising from externalities. Many ES (e.g. species preservation, carbon sequestration and soil erosion control) are public goods and the market often fails to fully account for their benefits (Kinzig et al., 2011; Engel et al., 2008). From the PES perspective, this is occurring because these nonmarket benefits lack a 'price tag'. Without monetary incentive, ES providers – acting on their own self-interest – have little incentive to provide a socially optimal level of ES, leading to a situation where ES are underprovided.

Conceptually, PES can be viewed as an attempt to strike a Coasian bargain between ES users and providers to achieve optimal economic efficiency (Whittington and Pigiola, 2012).[3] In Coasian theory, despite the presence of externalities, economic efficiency can still possibly be attained through individual bargaining, provided that transaction costs are sufficiently low and the property rights are well-defined (Coase, 1960). In line with the Coase theory, Wunder (2005) coined the currently well-accepted definition for PES: 'PES is a voluntary transaction, where a well-defined environmental service (or a land use likely to secure that service) is being "bought" by at least one service buyer from at least one service provider, if and only if the service provider secures service provision' (p. 3). Therefore, PES represents the commitment of ES providers and users (or governments acting on their behalf) to form contracts that incentivize conservation of the ES (Whittington and Pigiola, 2012). In essence, PES is an incentive-based mechanism particularly focusing on addressing the market failure problem resulting from the externalities (Engel et al., 2008; Pagiola and Platais, 2005).[4]

In practice, the PES design is based on two major principles: (1) 'users pay', which means that users (such as downstream users of clean water) who benefit from the ES should pay for the services; and (2) 'producers get paid', which means that the producers (such as upstream land users) who provide the ES should get paid for supplying the services (Wunder, 2005; Pagiola and Platais, 2007; Engel et al., 2008). Key guidelines for the design of PES schemes to achieve potential economic efficiency include: defining property rights, designing proper bargaining processes and reducing transaction costs (Tacconi, 2012; Engel et al., 2008; Gong, Bull and Baylis, 2010).

PES schemes may take a variety of forms, but almost all have the following common elements identified: types of ES to be provided, ES seller(s) (providers), buyers (ES users), the recipient(s) (individuals vs. communities) of the payments, payment mechanism and eligibility rules for participation (Engel et al., 2008; Jack, Kousky and Sims, 2008). PES schemes vary in terms of operationalization of these elements. For example, with regard to buyers, PES schemes can utilize one of two basic types: (1) user-financed programs, in which the buyers are the actual ES users; and (2) government-financed programs, in which a third party (usually the national government) acting on behalf of ES users, provides payment to ES providers (Pagiola and Platais, 2007; Engel et al., 2008).[5] Both the user-financed and government-financed PES projects have been implemented to ES provision, notably in the developing world.

Status of PES in the developing world

Currently, PES schemes are targeting three main types of ES in developing countries: (1) watershed protection, (2) biodiversity conservation and (3) carbon sequestration. Among the existing PES schemes, some are nationwide programs financed by the national governments, while most are regional- or local-scale projects financed by the private sector.[6] For some PES programs/ projects, multiple ES are bundled into a single arrangement.[7] In the following sections, we present overviews of PES programs and projects designed to generate one of the three types of ES mentioned previously.

Payments for watershed services

Payments for watershed services (PWS) have been the focus of most implemented PES schemes (Kosoy, Martinez-Tuna, Muradian and Martinez-Alier, 2007; Muñoz-Piña, Guevara, Torres and Braña, 2008). By 2011, 205 active PWS programs had been implemented in 29 countries, accounting for a total dollar value of US$74.17 billion and covering 312 million ha of watershed. Of the total number of 205 active PWS programs, China and the United States accounted for 30% (61 programs) and 33% (67 programs), respectively. In 2011, China became a global leader in funding watershed protection, replacing North and Latin America (Bennett et al., 2013).

In order to calculate the appropriate compensation for ES, existing PWS programs and projects have mainly followed one of three criteria: (1) the opportunity cost of land and labor (e.g. China), which differs greatly across countries and regions; (2) willingness to pay from downstream users (e.g. Bolivia); and (3) the cost of alternative land management practices, such as protection versus reforestation or restoration (e.g. Costa Rica) (Stanton, Echavarria, Hamilton and Ott, 2010).

In developing countries, most PWS programs are user-financed and implemented on a regional or local scale (Pagiola and Platais, 2007), while Costa Rica, China and Mexico have taken a lead in implementing national-scale programs.[8] The user-financed projects often involve downstream water users and upstream landholders, with the former paying the latter to protect forests in critical watersheds to provide watershed services. Water Trust Funds (WTF) established in Ecuador, Colombia and Peru are typical examples of linking urban water users to landowners (Stanton et al., 2010).

For the national-scale PWS programs, Costa Rica pioneered the use of a formal PES mechanism in 1997 by establishing a countrywide PWS program called Pago por Servicios Ambientale (PSA), which remains the best-known PES example in a developing region. The PSA program is jointly financed by the national government, international donors (such as the World Bank and Overseas Development Assistance) and environmental service buyers (Pattanayak et al., 2010). The payment levels made by the PSA program were based on the cost of alternative land management practices. Specifically, approximately $41/ha/year was paid for natural regeneration, $64/ha/year for forest preservation and $980/ha/year over a 5-year period for new forest plantations accordingly. China has the lion's share of reported PWS in developing countries. Since 2002, China has implemented the largest PWS program in the developing world, called Sloping Land Conversion Program (SLCP) nationwide. The Chinese government allocated a total budget of over US$40 billion to SLCP, under which farmers were subsidized to convert 14.67 million ha of erosion-prone cropland (4.4 million ha of them having slopes greater than 25 degrees) into forests (Bennett, 2008). In 2002, Mexico implemented an innovative nationwide PWS program, the Payments for Hydrological Environmental Services (PSAH), using an earmarked portion of fiscal revenues from water fees to pay for forest owners to protect well-preserved forests in critical watersheds and over-exploited aquifers. As areas with well-preserved forests are inhabited by poor forest owners, the PSAH was also motivated by fairness and thus attempted to avoid pursuing environmental protection at the cost of increasing poverty (Muñoz-Piña et al., 2008).

Recently, China and Latin America (particularly Costa Rica) have increased efforts to provide compensation for watershed services. The Chinese government is experimenting with a wide array of policy and program innovations under the broad heading of 'eco-compensation'.[9] Since 2001, the government has spent more than US$2 billion on the Forest Ecosystem Compensation Fund, which supports payments made to individual households, communities and local governments to protect about 44.53 million ha of key forest areas in 30 provinces (Bennett et al., 2013). In China's most recent Five-Year Plan (2011–2015), eco-compensation is placed in a key position and is garnering increasing government funding support. In 2007, Costa Rica implemented a new project, known as Mainstreaming Market Based Instruments for Environmental Management (MMBIEM), to enhance ES provision and secure long-term sustainability (Pigiola, 2008). Despite the dominance of China and the United States in terms of the number and size of PWS programs, Latin America is regarded as the most innovative in implementing PWS programs not only in terms of payment methods, such as trust funds, but also monitoring, measuring, perfecting and replicating these methods (Stanton et al., 2010).

Payments for biodiversity conservation

Payments for biodiversity conservation (PBC) provide an alternative conservation approach to traditionally used indirect conservation approaches, such as community-based natural resource management or integrated conservation and development projects (ICDPs).[10] Compared to the indirect approach, the PBC approach is more cost-effective, as it is a direct and less complicated approach targeting conservation outcomes (Wendland et al., 2010; Ferraro and Kiss, 2002; Ferraro and Simpson, 2002; Engel et al., 2008). It also can help mobilize additional funding for biodiversity conservation, through the bundling of biodiversity conservation with other ES (such as carbon and watershed services) (Wendland et al., 2010). A typical example for bundling biodiversity conservation with carbon ES is Reducing Emissions from Deforestation and Forest Degradation (REDD) projects. Although the REDD projects are designed mainly for carbon benefits, they also directly provide incentives to local land users for biodiversity conservation (Ferraro and Kiss, 2002).

In 2010, the global annual value of the biodiversity conservation market reached $1.8 billion to $2.9 billion, protecting at least 86,000 ha of land (Madsen, Carroll and Moore Brands, 2010). Madsen et al. (2010) noted at this time that 39 biodiversity offset and compensatory mitigation programs were implemented and 25 programs were in the development stage.[11]

Around the world, current biodiversity markets are in various development stages. The markets are somewhat established in developed countries (e.g. North America and Australia), while they are still at a nascent stage in developing countries (e.g. China and Brazil). The developed countries have used mitigation banking as a key measure to implement compensatory mitigation programs, while the developing countries have resorted to compensation funds.[12] Although the biodiversity markets in developed countries are driven largely by regulatory compliance, government-mediated payments and voluntary provisioning are the major mechanisms used in the developing countries.

The regulatory compliance used in developed countries (particularly North America)[13] is essentially a market-based system. Under this system, the government first sets a (legally binding) limit – or 'cap' – on the impact to a species or habitat with the objective of creating a potential demand for biodiversity from the regulated parties (including the government and the private sector). To mitigate the impacts above the limit, the regulated parties may choose to buy credits from the suppliers who undertake the required mitigation activities. A typical example is the US Endangered Species Act, which limits harm to federally listed endangered species and then requires the following consecutive mitigation obligations: avoiding, minimizing and finally mitigating the impacts that exceed the defined cap.

As for the government-mediated payments adopted in many developing countries, the government (and/or an NGO) acts as a sole 'buyer' of conservation easements or payment programs for biodiversity stewardship activities. For developing countries, the majority of PBC can be classified in this category, which tends toward government compensation rather than a market-based system for offsetting biodiversity compensation. In 2010, there were a total of five existing programs in South America (two in Brazil, one in Colombia and two in Paraguay). In Asia, only China and Laos are implementing offset-like programs. China's offset-like program is called Forest Vegetation Restoration Fee, which requires developers who impact lands zoned for forestry to avoid and/or minimize impacts. To minimize the impact, the developers are required to pay a forest vegetation restoration fee, which is channeled into a dedicated fund used by the government for tree planting and forest restoration activities.

Another mechanism used by developing countries is referred to as the voluntary provision. In Africa and Asia, voluntary and industry initiatives are arising. In Africa particularly, some oil and mining companies are voluntarily compensated for impacts to biodiversity in countries including Ghana, Guinea, Madagascar and South Africa. In Asia, some industry initiatives were created to respond to increasing public criticism of the environmental and social impacts of extractive and agribusiness industries (Madsen et al., 2010). Overall, while the number of PBC projects is rising on the one hand, there is also significant work being done on assuring quality and maintaining transparency on the other, which is equally important (Madsen et al., 2010).

Payments for forest carbon

With the development of forest carbon markets, payments for forest carbon have gradually entered into the central stage of international actions toward climate change mitigation. By 2012, forest carbon markets had evolved into two segments – voluntary and compliance based. The total value of the forestry offsets reached US$237 million, increasing by 33% from US$177 million in 2010. The transaction volume of forest carbon projects totaled 105.9 million tons of CO_2e (MtCO_2e), of which about 77% was executed in the voluntary market, while the remainder was realized in the compliance market (Peters-Stanley, Hamilton and Yin, 2012). Thus far, transactions in the forest carbon marketplace have mainly been realized in the voluntary marketplace, where buyers have voluntarily pursued emission reduction targets or prepared for potential regulations.

Voluntary markets encompassed a wide variety of forest projects, including afforestation/reforestation (A/R) projects, improved forest management (IFM) projects and REDD projects. Currently, only the over-the-counter (OTC) market is active, as the Chicago Climate Exchange (CCX) phased out at the end of 2010. In 2011, REDD projects accounted for the highest market share among all forest project types in the voluntary markets, although this share decreased from 69% in 2010 to 30% in 2011. In 2010, REDD projects supplied 19.5 MtCO_2e out of the total 29.0 MtCO_2e contracted in the primary market (Diaz, Hamilton and Johnson, 2011). However, the transaction volume declined to 7.4 MtCO_2e in 2011. Although projects in the voluntary OTC marketplace added 11 new country locations in 2011, the total number of country locations decreased from 48 in 2010 to 41 in 2011. A number of countries or states, among them Brazil, China, Canada, Netherlands, the UK and California in the United States, have leveraged the voluntary market to allow for and/or fund forest carbon offset projects.

The compliance market has been dominated by Clean Development Mechanism (CDM) forest projects and supplemented with regional markets, such as New South Wales Greenhouse Gas Reduction Scheme (NSW GGAS), New Zealand Emission Trading System (NZ ETS) and British Columbia's Carbon-Neutral Government directive. In 2011, while the total market value of the compliance market for forest carbon projects reached $29 million, the market share of CDM/Joint Implementation (JI) forest projects was 79% ($23 million). Currently, A/R is the only forest project type allowed by the CDM executive board. Since the first CDM A/R project was registered in 2006, the number of registered CDM forest projects has increased to 19 distributed over 16 countries in 2009, 22 in 18 countries in 2010, and 40 in 19 countries in 2011 (Peters-Stanley et al., 2012). In 2011, forest carbon projects under the CDM contracted the largest volume (5.9 MtCO_2e). Regional trends indicate that North America's compliance markets preferred IFM projects.

Factors affecting PES: PES frameworks and ES providers' participation level[14]

A number of frameworks and principles, conceptually and operationally, have been discussed regarding the design of PES schemes. Some focus on the development of criteria used to measure the capability of PES schemes to achieve desired outcomes, i.e. environmental effectiveness, economic efficiency and social equity (Wunder, 2005, 2006), while others highlight the importance of socio-economic and political conditions for achieving the desired outcomes (Kinzig et al., 2011; Jack et al., 2008; Pascual, Muradian, Rodríguez and Duraiappah, 2010).

Jack et al. (2008) presented a more inclusive conceptual framework that emphasizes the interaction between socio-economic, environmental and political contexts and incentive-based mechanisms in determining the outcomes of the schemes. In this conceptual framework, the environmental context essentially involves the nature and types of ES to be generated and the biophysical conditions of ecosystems that generate the services. The institutional and policy context ranges from legal frameworks (such as laws and regulations) to implementing regulations (such as land tenure arrangements and contractual rules). The socio-economic context involves the distribution of resources, price of the ES (i.e. payments) and social system in which a policy occurs (Jack et al., 2008). In essence, the framework in Jack et al. (2008) is focused on the environmental, socio-economic, institutional and political factors that formed necessary conditions for PES schemes to achieve expected outcomes. The conceptual framework used in this section was adapted from Jack et al. (2008) and Pagiola (2008) by adding another perspective, i.e. the ES providers' participation, considering the vital roles played by the ES providers' participation in PES schemes to achieve expected outcomes (Gong et al., 2010).

To identify factors influencing ES providers' decision to participate in PES projects, Pagiola (2007) included factors affecting the ES providers' eligibility, willingness and ability to participate in the PES schemes. Their willingness to participate is largely determined by the profitability from participation: ES providers will only be willing to participate if their participation makes them better off and the payment level is at least equal to their opportunity cost of participation (Pagiola et al., 2005, 2007). Ability to participate is affected by the ES provider's technical and financial capabilities, costs of participation and institutions (both formal and informal) (Pagiola et al., 2005, 2007). Although limited financial and technical capabilities may often constrain ES providers' ability to participate, the constraints can be partly eased by some institutional arrangements (such as land tenure arrangements and contractual rules), increasing accessibility to credit markets and providing technical assistance (Pagiola et al., 2005; Wunder, 2005; Tschakert et al., 2007; Gong et al., 2010). Therefore, in this modified framework, at the ES providers' level, the underlying factors affecting ES providers' participation decision mainly include: (1) those involved in the socio-economic context, such as costs of participation, payments made to the ES providers, among others; and (2) those involved in the institutional and political context, such as land tenure arrangements.

Figure 14.1 presents the modified framework that involves both the PES scheme level and the ES providers' participation level. In Figure 14.1, cost-effectiveness was used to evaluate the efficiency at the PES scheme level, while ES providers' profit from participating was used to evaluate efficiency at the ES providers' level. To evaluate the equity, particular attention was paid to poverty reduction. The following discussions highlight some key factors affecting the expected outcomes on PES scheme level and the ES providers' participation decisions.

Key factors in the PES framework

In the environmental context, the nature and types of ES are essential for the effectiveness of PES schemes (Kinzig et al., 2011; Jack et al., 2008), which are reflected in two aspects. The first is whether ES is a local or global environmental public good. PES schemes that are designed with the actual ES users paying service providers of local public goods (such as the downstream bottled water companies paying for upstream forest land users to adopt land-use practices that

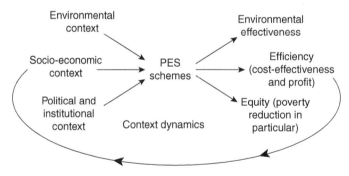

Figure 14.1 A conceptual framework for the design of PES schemes.

Source: Adapted from Jack et al. (2008).

can provide water purification service) might not be effective for generating global public goods (such as carbon sequestration) (Kinzig et al., 2011). The second is related to whether or not multiple types of ES can be bundled in a PES scheme (Kinzig et al., 2011). Some suggest that PES schemes could be designed to generate multiple ES, such as biodiversity conservation and carbon sequestration (Wendland et al., 2010), while others warn that the interdependence of different types of ES could lead to the generation of one type of service at the cost of having negative effects on other types of ES (Kinzig et al., 2011; Foley et al., 2005).

In the institutional and policy context, the legal framework is a critical factor that has particular importance for the effectiveness of PES schemes, as it can create legal certainty, ensure good governance, help scale up the user-financed PES projects implemented at local scales to the national-scale programs and promote PES development (Greiber, 2009). A typical example to demonstrate the importance of legal frameworks in scaling up the locally implemented PES project to a nationwide PES program is Costa Rica's PAS program. Initially, the PAS program was built on relatively small-scale initiatives, including tax rebates on timber plantations, forest credit certificates and forest protection certificates. With the enactment of a national forest law, Forest Law No. 7575, which explicitly specifies the roles of forests for watershed protection, the PAS program was scaled up to a nationwide program (Pagiola, 2008).

In the socio-economic context, the payment method is a key factor that may affect the cost-effectiveness of PES schemes, especially in a situation where the ES providers have heterogeneous opportunity costs. In most cases, determining a socially optimal level of ES and verifying the delivery of that level of ES are often technically infeasible or extremely costly. Thus, a target level of environmental benefits to be delivered, essentially referred to as 'environmental effectiveness', has to be set for PES schemes to be successful financially. The PES schemes that have the lower costs to reach the target are the more cost-effective ones. However, the cost-saving by the program might be attained at the sacrifice of lowering quality and desired levels of ES (Jack et al., 2008).

The previous point can be typically demonstrated by China's SLCP, in which a fixed per-hectare payment method based on of the area of land enrolled was used in order to reduce the implementation costs. Given fixed payments, those ES providers having opportunity costs of participation below the payment level appeared to be willing to participate in the program, but they might not be necessarily those who can provide the higher quality/level of ES. On the contrary, those with higher opportunity costs may decide not to participate, but in fact they may be the ones who can provide the higher quality/level of ES. Therefore, the PES programs designed with a fixed payment method might not necessarily be cost-effective.

Key factors for the ES providers

In the institutional context, land tenure arrangements and contractual rules are the key factors affecting ES providers' participation decisions, as they define the riskiness, transaction costs and potential benefits of the ES providers' participation. Based on evidence from the implementation of the world's first CDM project, Facilitating Reforestation for Guangxi Watershed Management in the Pearl River Basin in China, land tenure security and contractual rules critically affected local land users' participation decisions through their effect on potential risks and transaction costs (Gong et al., 2010).

In the socio-economic context, there are two key factors that may significantly affect ES providers' participation decisions.

Factor 1: The expected net benefits of their participation. The net benefit depends on the costs of providing the services and the payments made to ES providers. The costs include opportunity costs and transaction costs. High transaction costs are the principal obstacle for the land owners and users (especially small-scale owners and users) to be engaged in and benefit from PES projects (Gong et al., 2010; Grieg-Gran, Porras and Wunder, 2005). The opportunity costs associated with the potential reduction in crop yields due to the implementation of an agroforestry-based carbon sequestration project in Mozambique was a major concern expressed by potential participants (Hegde and Bull, 2011).

The costs may be a critical factor affecting ES providers' participation decision, especially in situations where fixed payments are made. Take the world's first CDM forest carbon project as an example. As the project offered a homogeneous and fixed payment to all potential participants, those land users who had higher opportunity costs of participation associated with increasing local land rental prices or higher costs of planting trees on extremely marginal land decided to hold up their participation.

Factor 2: ES providers' ability to participate. The ES providers' socio-economic characteristics also influence their participation decisions mainly through their effect on the ES providers' ability to participate, especially ES providers' financial and technical capabilities to participate. After studying farmers in Michigan, United States, Ma, Swinton, Lupi and Jolejole-Foreman (2012) concluded that farmers' decisions on whether and how much to enroll in a PES program depend more on the payment offer and marginal benefit cost considerations. Although PES programs often require a certain level of investment (labor and capital) and entail a certain level of technical difficulties, the following family and individual characteristics may play important roles in determining their ability to participate: (1) a family's assets and savings, off-farm income and remittance, which determine the ES providers' ability to make the required investment (e.g. tree plantings for forest carbon project); (2) families' land titles and other collateral, which may affect the families' accessibility to the credit market; and (3) the individuals' human capital (e.g. education, skills and experience) and family labor etc., which often determine their technical ability to participate in the PES programs. Indeed, empirical evidence shows that an ES provider's family size, education level, family labor, knowledge and skills have positive effects on their participation capabilities and decisions (Zbinden and Lee, 2005; Hegde, 2010; Pattanayak et al., 2010).

In the institutional and political context, social capital, which helps reduce mutual monitoring costs, increase the credibility of social sanction, facilitate interactions among individuals, organize information sharing and coordinate activities (Sobel, 2002; Besley et al., 1993), can critically affect ES providers' participation decisions through its effect on the local communities' enforcement capacity. For example, Indonesia's weak property rights regime influenced its rural communities' low ability of self-enforcing PES agreements and ultimately decreased the communities' participation in a forest conservation project (Engal and Palmer, 2008). Indonesia is not a unique case. In many developing countries, formal institutions are often weak and rural communities often resort to social capital, which can be broadly defined as the connections among individuals, such as networks, norms, trust, concerns for one's associates and willingness to sanction violators of rules or norms (Bowles and Gintis, 2002; Putnam, 2000), for self-enforcement of their agreements. When social capital is weak, ES providers might be prevented from participating in PES programs that may require collective action and cooperation among communities. For example, in the world's first CDM forest project, because it pooled lands from a number of communities to reduce the average transaction cost, its implementation required the local communities to take collective action. However, with low levels of mutual trust, some communities failed to reach a mutual agreement on cost-sharing and income distribution and thus were prevented from participating in the project that otherwise would have benefited them (Gong et al., 2010).

PES project impacts

A key limitation in the existing PES literature is the lack of quantitative analysis on the impacts of PES projects (Hegde and Bull, 2011). Exceptions include the evaluation of the SLCP in China (Uchida et al., 2007; Bennett, 2008; Uchida et al., 2009) and Nhambita carbon project in Mozambique (Hegde and Bull, 2011). The current lack of quantitative evidence from PES projects may limit further expansion of projects and further development of ES markets, particularly in developing countries (Landell-Mills and Porras, 2002). This section reviews key impacts at the participant household level.

PES projects could have direct and indirect impacts on household income. With the implementation of PES projects, direct cash payments or job opportunities are provided to local communities, having a direct impact on household income of the local communities (Pagiola et al., 2005; Grieg-Gran et al., 2005; Hegde and Bull, 2011). The implementation of PES projects also could lead to changes in the labor allocation strategies of the participating households, which creates an indirect impact on household income. An example of direct impact can be seen in Costa Rica,[15] where cash payments made by the PES program formed over 10% of household income for over a quarter of participants. Likewise, in Bolivia, PES-like schemes generated an annual income between US$77 and US$640 per household (Wunder, 2008). In Mozambique, although the amount of cash payments received by households was not very high (on average about Metical (MTS) 1,498,933 for participating households, equivalent to US$60 per year), the amount of cash payments made by an agroforestry carbon project accounted for about 10% of the total cash income of the participating households (Hegde and Bull, 2011). Regarding potential indirect impacts, it was found that the implementation of the SLCP in China caused the participating households to increasingly shift their labor endowment from on-farm work to the off-farm labor market (Uchida et al., 2009).

The majority of research measured household income impacts exclusively on the cash payments (revenue) received. However, this is an incomplete assessment because it ignores foregone benefits (i.e. opportunity costs) arising from the PES-induced restrictions (Wunder, 2008). When opportunity costs were taken into consideration, the net benefit (i.e. the difference between the cash payments and the opportunity costs) might actually be a net negative. However, there are also some exceptions in the existing literature. For example, Hegde and Bull (2011) attempted to calculate the potential opportunity costs of the households participating in Mozambique's agroforestry carbon project. They found that the agroforestry tree planting activities undertaken in the project could potentially lead to reduced crop yields in the participating households and reduced the subsistence use of resources from the *miombo* woodlands. Bennett (2008) found that the payments made in China's SLCP to the participating farmers did not cover their opportunity costs in some cases.

For the impact of PES programs/projects on family labors' employment, the implementation of SLCP led to an increasing shift of family labor from on-farm work to the off-farm labor market (Uchida et al., 2009). Mozambique's agroforestry carbon project created 100 off-farm jobs for local people (carpentry unit, nursery, village garden). Although the wage rates offered in the project employment were generally about the same as in the local wage market, the project employment served as a regular source of salaried employment to the households that were otherwise dependent on unstable seasonal wage work (Hegde, 2010).

Poverty reduction remains one of the most widely discussed outcomes of PES projects. The PES concept emerged as a mechanism to improve ecosystem conservation, but many proponents argued that the PES model can address both conservation and development objectives (Landell-Mills and Porras, 2002; Pagiola et al., 2005; Grieg-Gran et al., 2005). The anticipated synergy is based on the notion that because there is a high incidence of poverty in developing countries and households are heavily dependent on natural resources, a PES scheme to pay the poor for their environmentally friendly management could generate both benefits (Bulte and Damania, 2008; Muradian et al., 2010; Pagiola et al., 2005; Wunder, 2008). However, the empirical evidence to support this hypothesis is limited. One example exists in the case of Nhambita. After reviewing the Nhambita carbon project and SLCP project, Groom and Palmer (2012) concluded that PES projects can offer income enhancement opportunities, which is consistent with Hegde's (2010) findings in this region.

PES programs and projects also had nonincome impacts at the community or regional scale. They mainly included land-tenure consolidation, increases in human and social capital, increased social profile and flow on impact on local communities as well as effects on local and regional economies (Wunder, 2008; Robertson and Wunder, 2005; Hegde, 2010). Although PES projects had positive nonincome impacts, they could have negative impacts as well. The commonly cited negative impacts included curtailed resource access to low-income populations and increased elite control of lands with insecure land tenure (Kerr, 2002; Pagiola et al., 2005; Wunder, 2005; Landell-Mills and Porras, 2002).

Conclusions

No one should expect that a PES scheme can resolve all of the problems associated with human impacts on the land. The desire to simultaneously obtain a maximum level of environmental benefits, an increase in economic efficiency and a reduction in inequality is a laudable goal, but project developers must realize that there are trade-offs and tough decisions to be made, and that some members of society will 'lose'. The main lessons learned from existing PES programs and projects implemented in developing countries include the following.

The transaction costs of PES programs are often high, and this can affect the cost-effectiveness of the programs. Because many programs often have to involve a large number of small-scale and poor landholders (Wunder, 2005; Gong et al., 2010), innovative institutional arrangements will have to be made to reduce the transaction costs and encourage potential ES providers to participate.

Local communities' social capital and their ability to enforce contracts are critical for PES programs to successfully generate ES, improve social welfare and facilitate equity. Although aggregating individual small-scale landholders' activities can help reduce transaction costs, it also requires the local communities to have sufficient social capital to take successful collection actions. Moreover, with weak governments and property rights in developing countries, if local communities do not have sufficient ability to self-enforce the contracts, the PES programs may fail.

The complexity of PES programs with multiple objectives could easily doom the program to failure. First, trade-offs might have to be made between efficiency and equity (Wunder, 2005). Second, trade-offs between efficiency and environmental effectiveness have to be made. Although different types of ES can be bundled into a single PES scheme to achieve cost-effectiveness or seek additional funding, the increase in one type of ES (e.g. carbon sequestration) might decrease the supply of another type of ES (e.g. biodiversity).

The main challenges for PES in the future will center on the following: scientific measurement of ES, the key characteristics of property rights (duration, transferability, comprehensiveness, benefit conferred and exclusiveness) (Pearse, 1990), innovative management arrangements, the design of the contracts and new institutional arrangements. Measuring ES can require the careful cataloguing of ecosystem structure and function and mapping of the ecosystem goods and services (Heal et al. 2000). The valuation of the property rights conferred is equally complex and requires a careful assessment of resource or ecosystem scarcity and the development of supply and demand curves. Successful implementation also requires the use of innovative partnerships. However, the lack of high social capital could be a major impediment to success in program adoption. The design of contracts is only in its infancy and there remains a great deal of experimentation required. For instance, successful projects will depend on the nature of the ES in question, the length of the contract and the need to manage risk and uncertainty. New institutional designs are required to facilitate partnerships at various geographic scales, to increase the fungibility of projects around the world, to explore the science behind ES and to support the contracts.

With the current economic downturns, PES is facing even bigger challenges. PES pays for the provision of ES (e.g. species preservation, carbon, soil erosion, etc.), which are public goods normally underprovided in a market system. It is not difficult to imagine that the implementation of ES can be even more underprovided in an uncertain economic climate.

Notes

1 Typical examples of ES users are users of clean water in downstream urban areas, whereas land users who take conservation measure in the upstream areas are ES providers.

2 Thus far, two major distinctive perspectives taken to conceptualize PES are the environmental economics approach and the ecological economics approach (for a good reference for these two approaches, readers may refer to one special issue on PES in *Ecological Economics* in 2008 and two special issues on PES in *Ecological Economics* in 2010). Generally speaking, the environmental economics approach focuses on the economic efficiency of PES, whereas the ecological economics emphasizes multiple goals of PES, i.e. ecological sustainability, economic efficiency and equity.

3 Coasian theory describes the economic efficiency of an economic allocation or outcome in the presence of externalities. Coasian theorem posits that regardless of the initial allocation of property rights, individual bargaining will lead to an efficient outcome, if transaction of an externality is possible and transaction costs are zero.

4 It is worth noting that the environmental economics approach clearly sets its scope, i.e. addressing the market failure problem resulting from externalities. It admits the limitations of PES in addressing the market failures arising from other sources (e.g. poorly defined property rights and imperfect capital market imperfection; Engel et al., 2008).

5 In practice, these two types of programs have substantial differences. The user-financed programs are found to outperform the government-financed programs: The user-financed programs are better targeted to landscapes, better able to deliver environmental services, better tailored to local conditions and needs and have better monitoring and many fewer competing side objectives (Pattayanak et al., 2010).

6 In this chapter, we refer to those PES schemes implemented on the national level as 'programs', whereas those implemented on the local level are referred to as 'projects'. Typical examples of national PES programs are the Conservation Reservation Program in the United States and the Sloping Land Conversion Program (SLCP) in China, and regional-scale PES projects are exemplified by those implemented in Brazil, Columbia, Ecuador, El Salvador, Kenya and Vietnam.

7 For example, the world's first Clean Development Mechanism (CDM) forest carbon project implemented in China was designed to provide biodiversity enhancement, carbon sequestration and soil erosion reduction (Gong et al., 2010), and Costa Rica's nationwide PES program was designed to provide multiple ES, including hydrological services, biodiversity and carbon sequestration (Pagiola, 2008).

8 In essence, the Pago por Servicios Ambientale (PSA) program implemented in Costa Rica is a mix of forest conservation, sustainable forest management and reforestation; the Payments for Hydrological Environmental Services (PSAH) implemented in Mexico is a forest conservation program; the SLCP implemented in China is a reforestation program.

9 The funding for eco-compensation is earmarked fiscal transfer. The eco-compensation seeks to compensate opportunity costs borne by land users (or ES providers) due to environmental protection policies. PWS programs have become a key component of eco-compensation.

10 In developed countries, PBC has also been used an alternative conservation approach for national parks and protected areas A typical example is the conservation easements initiated by the Nature Conservancy

in 1961 in the United States. Since land subject to a conservation easement can make the landowner eligible for certain tax benefits, conservation easements provide private landowners with incentives to be involved in protecting the land and water as habitat for native plant and animal species.

11 Compensatory mitigation is loosely defined as the restoration, creation, enhancement and/or, in certain circumstances, preservation of natural resources for the purposes of offsetting adverse impacts which remain after all appropriate and practicable avoidance and minimization have been achieved (Madsen et al., 2010).

12 The mitigation banking requires well-developed market infrastructure and high implementation complexity. In contrast, the compensation funds have low requirements for market structure and implementation complexity.

13 The regulatory compliance has been well developed in North America. It is still a developing concept in Europe.

14 The factors are only limited to environmental, socio-economic, political and institutional features within a certain country, without considering factors within an international context. Thus, factors such as the influence of the Global Financial Crisis, Eurozone crisis, the convoluted institutional requirements for CDM projects versus flexibility with Voluntary Carbon projects and REDD, REDD+ and other conventions that provide for the inclusion of different carbon pools are left out of this discussion.

15 In other research, Miranda et al. (2003) reported that in the Virlilla watershed in Costa Rica, the PES payment was about 16% of cash income, three-quarters of households earned more than $820 per month and thereby households moved out of poverty.

References

Bennett, G., Nathaniel, C. and Hamilton, K. (2013). 'Charting new waters: State of watershed payments 2012'. Washington, D.C. Forest Trends. Retrieved from http://www.ecosystemmarketplace.com/reports/sowp2012

Bennett, M. (2008). 'China's sloping land conversion program: Institutional innovation or business as usual?' *Ecological Economics*, vol. 65, pp. 699–711.

Besley, T., Coate, S. and Glenn, L. (1993). 'The Economics of Rotating Savings and Credit Associations', *American Economic Review*, vol. 83, no. 4, pp. 792–810. Retrieved from http://ideas.repec.org/a/aea/aecrev/v83y1993i4p792-810.html

Bowles, S. and Gintis, H. (2002). 'Social capital and community governance', *The Economic Journal*, vol. 112, pp. F419–F436.

Bulte, E. and Damania, R. (2008). 'Resources for sale: Corruption, democracy and the natural resource curse', *The B.E. Journal of Economic Policy and Analysis*, vol. 8, no. 1, pp. 5.

Coase, R. (1960). 'The problem of social cost', *Journal of Law and Economics*, vol. 3, pp. 1–44.

Diaz, D., Hamilton, K. and Johnson, E. (2011). 'State of the forest carbon markets 2011: From canopy to currency', *Ecosystem Marketplace*. Retrieved from http://forest-trends.org/documents/files/doc_2963.pdf

Engel, S. and Palmer, C. (2008). 'Payments for environmental services as an alternative to logging under weak property rights: The case of Indonesia', *Ecological Economics*, vol. 8, no. 4, pp. 799–809.

Engel, S., Pagiola, S. and Wunder, S. (2008). 'Designing payments for environmental services in theory and practice: An overview of the issues', *Ecological Economics*, vol. 65, pp. 663–674.

Ferraro, P. J. and Kiss, A. (2002). 'Direct payments to conserve biodiversity', *Science*, vol. 298, pp. 1718–1719.

Ferraro, P. J. and Simpson, R. D. (2002). 'The cost-effectiveness of conservation payments', *Land Economics*, vol. 78, no. 3, pp. 339–353.

Foley, J. A., DeFries, R., Asner, G. P., Barford, C., Bonan, G., Carpenter, S. R., . . . Snyder, P. K. (2005). 'Global consequences of land use', *Science*, vol. 309, pp. 570–574.

Gong, Y., Bull, G. and Baylis, K. (2010). 'Participation in the world's first clean development mechanism forest project: The role of property rights, social capital and contractual rules', *Ecological Economics*, vol. 69, pp. 292–1302.

Greiber, T. (Eds.). (2009). *Payments for Ecosystem Services: Legal and Institutional Frameworks*, IUCN, Gland, Switzerland.

Grieg-Gran, M., Porras, I. and Wunder, S. (2005). 'How can market mechanisms for forest environmental services help the poor? Preliminary lessons from Latin America', *World Development*, vol. 13, no. 9, pp. 1511–1527.

Groom, B. and Palmer, C. (2012). 'REDD+ and rural livelihoods', *Biological Conservation*, vol. 154, pp. 42–52.

Heal, G. (2000). 'Valuing ecosystem services.' *Ecosystems*, vol. 3, no. 1, pp. 24–30.

Hegde, R. (2010). 'Performance of an agroforestry-based payments-for-ecosystem-services project in Mozambique: A household level analysis', PhD thesis, University of British Columbia, Vancouver, Canada.

Hegde, R. and Bull, G. Q. (2011). 'Performance of an agroforestry-based payments-for-ecosystem-services project in Mozambique: A household level analysis', *Ecological Economics*, vol. 71, pp. 122–130.

Jack, B. K., Kousky, C. and Sims, K.R.E. (2008). 'Designing payments for ecosystem services: Lessons from previous experience with incentive-based mechanisms', *Proceedings of the National Academy of Sciences of the United States of America*, vol. 105, pp. 9465–9470.

Kerr, J. (2002). 'Watershed development, environmental services, and poverty alleviation in India', *World Development*, vol. 30, no. 8, pp. 1387–1400.

Kinzig, A. P., Perrings, C., Chapin, F. S., III, Polasky, S., Smith, V. K., Tilman, D. and Turner, B. L., II (2011). 'Paying for ecosystem services: Promise and peril', *Science*, vol. 334, pp. 603–604.

Kosoy, N., Martinez-Tuna, M., Muradian, R. and Martinez-Alier, J. (2007). 'Payments for environmental services in watersheds: Insights from a comparative study of three cases in Central America', *Ecological Economics*, vol. 61, pp. 446–455.

Landell-Mills, N. and Porass, I. (2002). *Silver Bullet or Fool's Gold? A Global Review of Markets for Forest Environmental Services: Market-Based Mechanisms for Conservation and Development*, International Institute for Environment and Development, London.

Ma, S., Swinton, S. M., Lupi, F. and Jolejole-Foreman, C. (2012). 'Farmers' willingness to participate in payment-for-environmental-services programmes', *Journal of Agricultural Economics*, vol. 63, no. 3, pp. 604–626.

Madsen, B., Carroll, N. and Moore Brands, K. (2010). 'State of biodiversity markets report: Offset and compensation programs worldwide'. Retrieved from http://ecosystemmarketplace.com/documents/acrobat/sbdmr.pdf

Millennium Ecosystem Assessment (MA). (2003). *Ecosystems and Human Well-being: A Framework for Assessment*, Island Press, Washington, DC.

Millennium Ecosystem Assessment (MA). (2005). *Ecosystems and Human Well-being: Synthesis*, Island Press, Washington, DC.

Muñoz-Piña, C., Guevara, A., Torres, J. M. and Braña, J. (2008). 'Paying for the hydrological services of Mexico's forests: Analysis, negotiations and results', *Ecological Economics*, vol. 65, no. 4, pp. 725–736.

Muradian, R., Corbera, E., Pascual, U., Kosoy, N. and May, P. H. (2010). 'Reconciling theory and practice: An alternative conceptual framework for understanding payments for environmental services', *Ecological Economics*, vol. 69, no. 6, pp. 1202–1208.

Pagiola, S. (2008). 'Payments for environmental services in Costa Rica', *Ecological Economics*, vol. 65, no. 4, pp. 712–724.

Pagiola, S., Arcenas, A. and Platais, G. (2005). 'Can payments for environmental services help reduce poverty? An exploration of the issues and evidence to date from Latin America', *World Development*, vol. 33, no. 2, pp. 237–253.

Pigiola, S., Ramírez, E., Gobbic, J., de Haana, C., Ibrahimc, M., Murgueitiod, E. and Ruíz, J.P. (2007). 'Paying for environmental services of silvopastoral practices in Nicaragua', *Ecological Economics*, vol. 64, no. 2, pp. 374–385.

Pagiola, S., Rios, A. N. and Arcenas, A. (2008). 'Can the poor participate in payments for environmental services? Lessons from the Silvopastoral Project in Nicaragua', *Environment and Development Economics*, vol. 13, pp. 299–325.

Pascual, U., Muradian, R., Rodríguez, L. C. and Duraiappah, A. (2010). 'Exploring the links between equity and efficiency in payments for environmental services: A conceptual approach', *Ecological Economics*, vol. 69, no. 6, pp. 1237–1244.

Pattanayak, S. K., Wunder, S. and Ferraro, P. J. (2010). 'Show me the money: Do payments supply environmental services in developing countries?' *Review of Environmental Economics and Policy*, vol. 4, no. 2, pp. 254–274.

Pearse, P. H. (1990). *'Introduction to forestry economics'*, UBC Press, Vancouver, Canada.

Peters-Stanley, M., Hamilton, K. and Yin, D. (2012). 'Leveraging the landscape: State of the forest carbon markets 2012'. Retrieved from www.forest-trends.org/documents/files/doc_3242.pdf

Pigiola, S., Ramírez, E., Gobbic, J., de Haana, C., Ibrahimc, M., Murgueitiod, E. and Ruíz, J.P. (2007). 'Paying for environmental services of silvopastoral practices in Nicaragua', *Ecological Economics*, vol. 64, no. 2, pp 374–385.

Putnam, R. (2000). *Bowling Alone: The Collapse and Revival of American Community*, Simon and Schuster, New York.

Robertson, N. and Wunder, S. (2005). 'Fresh tracks in the forest assessing incipient payments for environmental services initiatives in Bolivia', Center for International Forestry Research, Bogor, Indonesia.

Sobel, J. (2002). 'Can we trust social capital?' *Journal of Economic Literature*, vol. 40, no. 1, pp. 139–154.

Stanton, T., Echavarria, M., Hamilton, M. and Ott, C. (2010). 'State of watershed payments: An emerging marketplace', *Ecosystem Marketplace*. Retrieved from http://foresttrends.org/documents/files/doc_2438.pdf

Tacconi, L. (2012). 'Redefining payments for environmental services', *Ecological Economics*, vol. 73, no. 1, pp. 29–36.

Tschakert, P., Coomes, O.T. and Potvi, C. (2007). 'Indigenous livelihoods, slash-and-burn agriculture, and carbon stocks in eastern Panama', *Ecological Economics*, vol. 60, no. 4, pp. 807–820.

Uchida, E., Rozelle, S. and Xu, J. (2007). 'Conservation payments, liquidity constraints, and off-farm labor: Impact of the Grain-for-Green Program on rural households in China', in *Proceedings, American Journal of Agricultural Economics Association Annual Meeting*, July 29–August 1, Portland, OR.

Uchida, E., Xu, J., Xu, Z. and Rozelle, S. (2007). 'Are the poor benefiting from China's land conservation program?' *Environment and Development Economics*, vol. 12, no. 4, pp. 593–620.

Uchida, E., Rozelle, S. and Xu, J. (2009). 'Conservation payments, liquidity constraints, and off-farm labor: Impact of the Grain-for-Green Program on rural households in China', *American Journal of Agricultural Economics*, vol. 91, no. 1, pp. 70–86. Retrieved from http://ideas.repec.org/a/oup/ajagec/v91y2009i1p70–86.html

Wendland, K. J., Honzák, M., Portela, R., Vitale, B., Rubinoff, S. and Randrianarisoa, J. (2010). 'Targeting and implementing payments for ecosystem services: Opportunities for bundling biodiversity conservation with carbon and water services in Madagascar', *Ecological Economics*, vol. 69, pp. 2093–2107.

Whittington, D. and Pagiola, S. (2012). 'Using contingent valuation in the design of payments for environmental services mechanisms: A review and assessment', *World Bank Research Observer*, vol. 27, no. 2, pp. 261–287.

Wunder, S. (2005). *Payments for Environmental Services: Some Nuts and Bolts*, Occasional Paper No. 42, CIFOR, Bogor.

Wunder, S. (2006). 'Are direct payments for environmental services spelling doom for sustainable forest management in the tropics', *Ecology and Society*, vol. 11, no. 2, p. 23.

Wunder, S. (2008). 'Payments for environmental services and the poor: Concepts and preliminary evidence', *Environment and Development Economics*, vol. 13, pp. 279–297.

Zbinden, S., and Lee, D. R. (2005). 'Paying for environmental services: An analysis of participation in Costa Rica's PSA program', *World Development*, vol. 33, no. 2, pp. 255–272.

PART 3

Economics of forests, climate change and bioenergy

15

FORESTS AND CLIMATE CHANGE

Economic perspectives

Brent Sohngen

AED ECONOMICS, OHIO STATE UNIVERSITY, 2120 FYFFE RD, COLUMBUS, OH 43210, USA.
SOHNGEN.1@OSU.EDU, 614-688-4640.

Abstract

This chapter examines economic analysis of climate change impacts in the forest sector. It begins with a discussion of the potential effects of climate change on ecosystem and then discusses how those impacts can be introduced into an economic model. One critical issue in economic modeling identified in the paper is that the way in which ecosystem impacts are introduced into the economic model could have important implications for the results. Thus, models that incorporate dieback directly will estimate different impacts than those that incorporate dieback through changes in growth and yield. Given the importance of potential dieback in climate change impacts, this difference in modeling can have implications for measuring climate change adaptation and damages. To illustrate how these modeling choices can affect results, the chapter presents a simple numerical example of climate change impacts. The study then presents a literature review discussing the results of climate change impact studies to date. It concludes with a discussion about potential research topics that could and should be addressed with future research.

Keywords

Climate change, dynamic optimization, dieback, disturbance, ecosystems

Introduction

The world's forest ecosystems are amazingly diverse, ranging from dense tropical rainforests along the equatorial belt to boreal forests covering the northern tier of the world. Without humans, forests would cover over 6 billion hectares (FAO, 2012). Conversion of land to agriculture in the past several centuries, however, has reduced this to less than 4 billion hectares today. In recent decades, the rate of converting forests to agriculture has stabilized in temperate regions, while conversion of forests to agriculture continues in the tropics (Houghton, 1999, 2003; FAO, 2010). Although most expansion of agriculture has occurred in the tropics in recent decades, forest cover loss occurs in virtually all continents for a number of reasons (Hansen, Stehman and Potapov, 2010), including harvesting, forest fires or other disturbances, or urbanization.

There is substantial concern that climate change could have large impacts on forests glob-ally. The impacts projected by many ecosystem models include larger and more intense dis-turbance events, such as forest fires or bug infestations, changes in the distribution of different types of trees and shifts in the rates of growth of species (IPCC, 2007). Such changes would clearly have large-scale ecological and economic implications, from changes in ecosystem ser-vice flows to losses in economic value. As a consequence, there has been substantial research in the past two decades to try to determine how large these ecological and economic impacts may be.

The impacts, of course, will vary depending on location and type of forests. As a rule, timber harvesting has become more sustainable in the past century, shifting from primarily old-growth extraction to a larger share of plantation forestry (Daigneault, Sohngen and Sedjo, 2008). The economic implications of climate change on highly managed plantations will be substantially different than the implications of climate change on old-growth forests. As more and more tim-ber output is derived from plantations, a larger share of forests is left in a natural state with very little management. This trend has important implications for measuring the impacts of climate change in forests and for assessing the scope of adaptation. For example, there are many more opportunities for adaptation in forests that are heavily managed than in regions that are not managed. Regions that are relatively unmanaged may experience large-scale ecosystem changes, with little economic impact.

This chapter examines the implications of climate change and forests. It begins with a discus-sion of the impacts of climate change on ecosystems. The results of ecosystem models dictate what we know about the potential economic impacts of climate change in the forestry sector. The chapter then turns to discuss how these ecological impacts are integrated into economic models. A formal economic model is presented and examples are shown illustrating how eco-system impacts can be linked into the economic model. The results of the economic model change substantially depending on how the ecosystem results are linked into the economic model, illustrating the importance of conducting integrated research. The final section examines the existing literature to discuss the potential impacts of climate change in markets.

Modeling the impacts of climate change in timber markets

Modeling climate change impacts on timberland use and management is substantially more complex than modeling climate change impacts in most other sectors. The impacts in many other sectors often can be modeled econometrically, with reduced form models that link average temperature and precipitation to output (e.g. Mendelsohn, Nordhaus and Shaw, 1994). Forestry, however, is inherently dynamic, and as a result, efforts to model impacts in the forestry sector require a different approach.

The typical approach taken by most analyses thus far involves linking climate models to ecosystem models to economic models (Figure 15.1). The data sets do not currently exist to go straight from climate to economic models, so it is necessary to use biological models to first measure the impacts on ecosystems. The outputs from ecosystem models typically cannot be used directly in most economic models, however. For example, many economic models aggregate inventories across space, and these aggregations occur at different scales than those used by the ecosystem models. Alternatively, many ecosystem models operate at different time scales than economic models. Economic modelers must take additional steps, often in conjunction with ecosystem modelers, to utilize the appropriate results from the ecological models within the economic models.

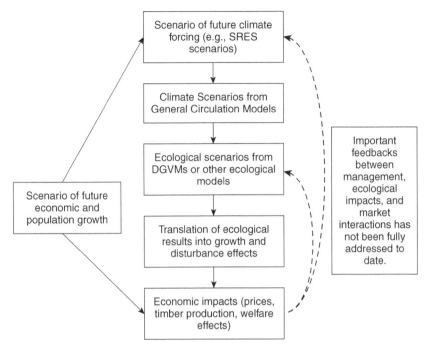

Figure 15.1 Integrated climate-forest modeling systems (from Balgis et al., 2009).

Potential climate change impacts on forested ecosystems

Ecological models suggest a wide range of potential impacts of climate change in forests, including shifts in the rate of forest growth, shifts in the disturbance regimes and changes in the distribution of different species (Balgis et al., 2009). Given that climate change is likely to strengthen over time, these changes will continue to affect forests over long periods of time. This section examines these impacts in more detail.

Growth changes

Tree growth is heavily dependent on both temperature and precipitation, and as climate influences these variables in any given location, trees could start to grow more quickly or more slowly depending on the impact of climate change. Additional precipitation will increase the growth rate of trees in locations where growth is limited by moisture, while higher temperatures or longer growing seasons could enhance growth if accompanied by adequate precipitation. The critical question with climate change will be whether higher temperatures are accompanied by adequate precipitation.

Additional carbon dioxide in the atmosphere should also increase plant growth. Because trees convert carbon dioxide (CO_2) from the atmosphere into woody biomass, higher concentrations of CO_2 are expected to help trees grow. Estimates suggest that a doubling of CO_2 from preindustrial times should increase tree growth by around 20%–25% (Norby et al., 2005). Indeed, some evidence exists now suggesting that forest growth is already accelerating as a result of higher CO_2 concentrations (Boisvenue and Running, 2006). Modeling studies suggest that over the past century, the combined impact of climate change and higher CO_2 concentrations have resulted in increased forest growth of 0.3% to 0.6% per year (Scholze, Knorr, Arnell and Prentice, 2006).

Although climate change and CO_2 fertilization could increase tree growth, that increase may not automatically convert into ecosystems with more live biomass. The effects on the landscape will be complicated by numerous additional factors. For instance, if plant growth is not limited by temperature or precipitation, it is likely to be limited by other nutrients, such as nitrogen. If these other nutrients are not available in adequate amounts in a given location, the effects of climate change will be reduced. Furthermore, one of the most important influences on forests is disturbance. If disturbance regimes change, then even if gross growth is projected to increase, net growth may not increase at all. If disturbances increase enough, then net growth may in fact be negative.

Dieback

As climate changes in a given location, tree species and forested ecosystems will end up in climatic conditions that differ from their optimum. For instance, if temperatures rise, but there is not enough additional precipitation, ecosystems could be susceptible to increases in forest fire activity. As a consequence, dieback from forest fires, windthrow, ice storms or insect infestations represents a bigger concern than climate-induced changes in forest growth (Adams et al., 2009).

Some studies suggest that current observed increases in forest fires may be caused by climate change (Westerling, Hidalgo, Cayan, and Swetnam, 2006). More recently, portions of western Canada have been devastated by large-scale insect infestations, and there is concern that the scales of impacts of recent widespread insect infestations are related to climate change (Kurz et al., 2008). Ecosystem models imply that climate change may cause more and more damage in the future by causing conditions that lead to increases in forest fires and other disturbances (Bachelet, Lenihan, Drapek and Neilson, 2008).

Dieback is one of the most important ecological effects to consider when modeling climate change because it can have substantial impacts if modeled in economic models (Sohngen and Mendelsohn, 1998). It also will have different impacts in different regions depending on the management regime. If climate change increases potential dieback in areas where forests are heavily managed, land owners and managers are more likely to adapt by changing their management strategy, e.g. by salvaging or by changing the date of harvest. If dieback increases in regions where management is sparse or does not occur (e.g. boreal zones), then the dieback may have large ecological consequences but little direct economic impact. Either way, changes in disturbance regimes can have long-term consequences for forest ecosystems by altering the age class distribution of forests for years to come.

Species shift

Individual tree species can live within a wide range of temperature and precipitation levels, but they also have a limited range of temperature and precipitation where they gain competitive advantage over other trees or plants. Thus, the optimal tree types in any given location will be a function of temperature and precipitation. As climate changes, one would expect the optimal ecological mix of tree species to adjust. Most trees will move further north and upslope in mountainous regions. Maps of potential changes in the United States are available from Iverson and Prasad (1998) and Iverson, Prasad, Matthews and Peters (2008). These maps illustrate potentially large changes in tree locations under climate scenarios proposed today.

With climate change, species are generally expected to move northward and upslope. The rate of movement of trees, if left to natural forces alone, such as the spread of seeds by birds or wind, could take long periods of time to occur. If humans assist in the movement of tree species, as they are widely expected to do with our long history of moving trees, the movement of species northward is expected to occur much more rapidly. It is useful to use economic models in addition to ecological models to measure the movement of species, given the important influence humans can have on the process.

Economic modeling

In order to assess the economics of these ecosystem impacts, one must develop economic models that account for several key features. First, the models must be dynamic. Dynamics in economics means not just capturing changes over time, but also modeling economic decision making in a dynamic sense. When humans manage forests, they must do so with one eye on the future. For instance, the harvesting decision is often based not only on the current stock of timber available to harvest, but also on an understanding about the growth of the trees and the likely change in timber prices over the coming year. Landowners will make different decisions depending on whether their trees are currently growing quickly or slowly, and whether they anticipate prices to increase, stay the same or fall over the coming year.

Beyond the harvesting decision, which may require looking forward for only a short time horizon, most planting decisions require very long time horizons. Many species will not mature for 20, 30 or more than 50 years. Any decision to spend resources planting or managing forests that cannot be harvested for such long time horizons require some information or assumptions about what future market conditions will be. When we think of dynamics in economics, models must be developed to account for these long-term considerations of landowners and managers.

Second, models must be clear about whether they assume prices are exogenous or endogenous. On the one hand, climate change is such a widespread phenomenon, which will affect growth and productivity in ecosystems throughout the world, that prices will certainly be affected as climate change occurs. Even regions that do not experience large ecological changes could be affected by climate change if timber prices or land prices change. Although this suggests that it is important to measure the price effects associated with climate change, economic analyses need not focus on global changes to provide insights. For instance, if modelers are interested primarily in understanding how stand management changes when forests are perturbed by climate change, then they may choose to use stand level models with prices fixed. Furthermore, if analysts are interested in conducting stochastic analysis of the effects of changes in forest fires (e.g. Stainback and Alavalapati, 2004; Amacher, Malik and Haight, 2005; Daigneault, Miranda and Sohngen, 2010), they likely will need to assume that prices are exogenous in order to solve the models. The key issue is that modelers should be clear about their assumptions when developing their models.

Third, modelers must be careful when integrating ecosystem impacts into their economic models. The way in which ecological impacts are actually used in economic models can make a large difference to the impacts. For example, suppose climate change causes more disturbance in a given region, and hence a slowdown in the net growth of timber over time. Modelers could directly model the disturbance, or they could simply alter their yield functions to account for the implied changes in net growth. This difference in modeling the same phenomenon would alter the results substantially, as shown subsequently. It is consequently very important for economic modelers to understand the results proposed by ecological models and carefully integrate those results into their models.

Illustrative example of a dynamic forestry model of climate impacts

To illustrate the importance of these features in modeling climate change impacts, this section develops a simple model of climate change and applies it to forestry analysis. The model follows that laid out originally in Sohngen and Sedjo (1998). It is a simplified dynamic timber model that assumes only a single timber type but also assumes that timber prices are endogenous. The economic model is first presented. Then, two of the climate change impacts described previously, changes in growth and dieback, are integrated into the model. Finally, the results of the model are compared assuming different methods of integrating the ecological phenomenon.

The forestry model is assumed to maximize the net present value of consumer's plus producer's surplus in forestry. To develop the model, it is useful to start with the inverse demand function, given as

$$P_t = \alpha_t - \beta(\Sigma_a H_{a,t} V_a) \tag{1}$$

where P_t is timber price, α_t and β are demand function parameters, $H_{a,t}$ is the area of forest harvested in age class a and time period t and V_a is the volume of timber in the age class a. Total annual harvest is $\Sigma_a H_{a,t} V_a$. Given this demand function, annual welfare in timber markets is

$$W_t = \alpha \left(\sum_a H_{a,t} V_a \right) - \left(\frac{1}{2} \right) \beta \left(\sum_a H_{a,t} V_a \right)^2 - c \left(\sum_a H_{a,t} V_a \right) - mG_t \tag{2}$$

where c is the constant marginal cost of harvesting timber, m is the cost of replanting trees and G_t is the area of land replanted each year. The objective of the model is to choose $H_{a,t}$ and G_t so as to maximize the present value of welfare:

$$\sum_{t=1}^{\infty} \rho^t W_t \tag{3}$$

The term ρ^t is discount factor. The function in equation (3) is maximized subject to the following constraints, where $X_{a,t}$ is the area of land in forest in age class a at time period t:

$$X_{a+1,t+1} = X_{a,t} - H_{a,t} \tag{4}$$

$$X_{1,t+1} = G_t \tag{5}$$

$$H_{a,t} \leq X_{a,t} \tag{6}$$

$$H_{a,t}, X_{a,t}, G_t \geq 0 \tag{7}$$

This model is well defined, and the baseline case is one in which there are no climate perturbations. Demand shifts out (i.e. α_t increases over time) as income and population increase, driving demand up. Alternatively, of course, recycling and environmental concerns could slow the rate of growth of timber harvesting. The base case can be solved in a fairly straightforward way, determining the optimal timber price, rotation age and forest stock. One would need to impose terminal conditions on the model, but as long as those terminal conditions are imposed sufficiently far into the future, they will not affect the solution over the period of interest (say the first 50 to 100 years).

The climate change impacts described previously will change various features of this model. The first impact of climate change is a change in timber growth, which alters forest yield, V_a, over time. These changes will not occur all at once, and in fact, they are likely to occur slowly over time. Furthermore, changes in forest growth affect only future growth, not the standing stock of timber. This is a critical distinction to make and requires modelers to take care when introducing the impact of climate change into their model.

For example, suppose tree growth increases by 5% from one period to the next. One cannot simply multiply the yield function by 1.05 to determine the new yield for standing trees. The stock of trees that is already standing is the result of historical growth, which will not be affected by this future change. Modelers must be careful to link changes in tree growth caused by climate change only to annual increments in tree growth, and specifically only to future growth.

The yield function for trees is typically given as the sum of historical annual growth,

$$V_a = \sum_0^a AG_a \tag{8}$$

where AG_a is the annual growth of trees. The volume in any year is the sum of the growth up to time period a. If the impacts of climate change on annual growth are γ_a, then the climate adjusted yield function becomes:

$$V_a^C = \sum_0^a \gamma_a AG_a \tag{9}$$

The effect of climate change on the forest in this model is captured by γ_a. This parameter must be obtained from ecosystem models.

To see how changes in forest growth are incorporated into economic models, consider a southern pine stand that is 20 years old when climate change starts affecting tree growth. If climate change causes the stand to grow 2% more quickly each year, then Figure 15.2 shows the change that should be modeled. The increase in tree growth does not lead instantly to a bigger tree in time period 20. Instead, in time period 20, the annual growth in year 20 is increased by 2%, so the volume in year 21 will be modestly larger than it would have been without climate change. Subsequent annual growth is increased as well, so that the effect of climate change accumulates over time. Faster growth due to climate change does ultimately lead to more tree volume, but only after many years. Of course, if the timber manager harvests the stand and starts over, the new stand will be growing at a significantly faster rate than the original stand.

The dieback effect described previously can be modeled similarly to the yield changes, by modeling net effects, or it can be modeled directly as a dieback effect. The method of actually implementing the ecological change will have important implications for the economic results. One way modelers have accounted for dieback is through net yield effects. Most estimated yield functions used in forestry models are 'net' yield functions, meaning that they model timber volume net of all growth and dieback processes. As a consequence, many economic modelers have simply aggregated the impact of dieback with the growth effects described previously to determine net yield impacts (Joyce et al., 1995; Perez-Garcia, Joyce, McGuire and Xiao, 2002). Modelers using this approach capture the net effects of changes in tree growth and changes in dieback together. If tree growth increases but dieback also increases, the net effect of climate change on biomass on site may actually be negative. That is, dieback may be large enough and strong enough to reduce overall biomass on forested sites.

Alternatively, one can directly incorporate dieback in the previous model through equation (4), as has been done in Sohngen and Mendelsohn (1998) and Sohngen, Mendelsohn and Sedjo

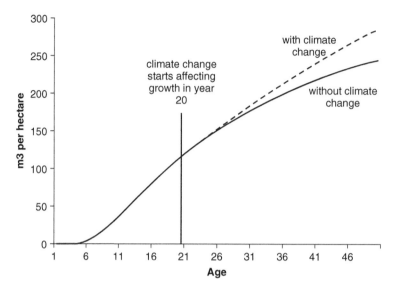

Figure 15.2 Volume of a southern pine stand with and without climate change effects on tree growth. The impact of climate change is an increase in stand growth by 2% per year starting in year 20.

(2001). If the proportion of stock that dies back each year is given as δ_t, then equation (4) can be adjusted to incorporate dieback directly as:

$$X_{a+1,t+1} = X_{a,t} - H_{a,t} - \delta_t X_{a,t} \tag{10}$$

The adjustment in equation (10) is fairly simple in that it assumes that all age classes of trees will be affected similarly by climate change. Ecosystem models may provide data that suggests differing impacts depending on the age of the trees. This could be incorporated into the model in a fairly straightforward way by modeling δ_t also as a function of a.

Modeling dieback via equation (10) rather than as a net effect in the yield adjustment shown in equation (9), even if the net effects from the ecosystem model are the same, will lead to far different estimates of the economic impacts. The perturbation in equation (10) will induce dynamic stock adjustments in a dynamic forestry model, such as incentives to harvest forests before dieback occurs. Modelers who attempt to model the same ecological phenomenon using only net yield changes will not be able to capture these types of adaptations. As a consequence, they likely will show fairly modest impacts of climate change on timber markets, at least initially. Modelers who use equation (10) combined with equation (9) likely will show larger impacts, be they negative or positive, in markets.

Beyond yield changes and dieback, it is important also to account for changes in area, or the effect of climate change on the distribution of tree species. Many species are likely to move northward or upslope with climate change. For commercially important species, the change most likely will be driven by humans who shift the species across space. In natural areas, the changes are likely to occur much more slowly.

A shift in species distribution can be modeled via a change in constraints in the previous model. For example, the total area of the timber type in the model can be constrained to be less than a given amount. The model described previously provides the area planted as a decision variable; therefore, the area of forests can be expanded or reduced over time through replanting decisions. These replanting decisions will be a function of the allowable area for the forest type and the economic efficiency of planting (i.e. whether the present value of replanting at a given time exceeds the marginal costs). The efficient replanting decision can be shown by taking first order conditions on the model with respect to G_t.

Numerical simulation

To illustrate the potential implications of climate change on markets, the model described through previous equations is programmed and simulated for a simple single region forestry sector. The model uses parameters developed originally by Sohngen and Sedjo (1998). The demand function is given as:

$$P_t = 40A_t - 0.084(\Sigma_a H_{a,t} V_a) \tag{11}$$

and the timber yield function is given as

$$V_a = \exp(7.82 - 52.9/a) \tag{12}$$

The term A_t accounts for growth in timber demand over time due to population and income growth in the economy. The forest in this model initially has 500,000 ha in each of 32 age classes. For the purposes of this analysis, $X_{a,t}$ is given in millions of hectares, so each of the timber age classes has 0.5 million ha of trees. The age of 32 years is approximately the Faustmann rotation for the forest if $A_t = 1$, so the forest starts out roughly in steady state if demand is constant. For the purposes of this example, it is assumed that timber demand increases 1% per year, but that the rate of growth in demand slows over time.

For this analysis, the area of timberland is assumed to remain constant. This simplifying assumption means that we impose another constraint on the dynamic model shown previously, namely that the area of timberland replanted each year equals the area of timberland harvested last year:

$$X_{1,t+1} = \Sigma_a H_{a,t} \tag{13}$$

As a result of this simplifying assumption, in the climate analysis it is not possible to consider the effects of changes in forestland area; however, this simplification allows us to focus on the effects of the yield changes and the implications of different methods of modeling forest dieback. The model is programmed and run in GAMS for 200 years. A terminal condition is imposed at that time, but because only the first 100 years of results are shown, this terminal condition has little effect on the results examined.

The base scenario assumes no climate change, and demand grows. Demand increases 1% per year initially, but the rate of increase slows over time so that eventually it is stable. Timber prices rise over time, but the rate of growth slows (Figure 15.3). An interesting dynamic adjustment occurs with the rising demand for timber. The total forest area is fixed, so timberland cannot be expanded to satisfy the rising demand; however, the model can still accommodate additional timber output in the long run by increasing the rotation age. As a result of the increasing demand, the model will shift the rotation age from about 32 years initially to between 33 and 34 years of age during the transition period. In a dynamic model, increasing the rotation age leads initially to lower timber harvests. The only way to increase rotations is to withhold timber from the market initially. Over time, with higher rotation ages, supply will expand and slow the

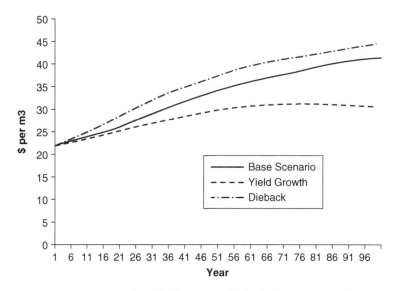

Figure 15.3 Representative price paths in the forestry model for the base scenario, the yield growth scenario, and the dieback scenario.

growth in prices. Thus, in this example, timber supply initially falls to accommodate a rise in rotation age, and this adjustment ultimately expands timber supply and reduces the impact of future demand increases on timber prices.

The first climate scenario assumes a net yield increase of 0.5% per year for 100 years. After 100 years, timber yields are assumed to stabilize at the higher level. Timber prices rise more slowly under this scenario because higher yields offset the demand increases (Figure 15.3). Timber prices do not change very much in the first couple of decades; in fact, the price change is less than half the yield change. For example, in year 10 yields have increased 5%, but prices fall only 1.6% relative to the base. The reason for this is fairly straightforward; although growth has increased substantially, all timber harvested in the first 10 years was already at least 20 years old at the beginning of the simulation period, so the total effect of the yield increase on the volume of timber available for harvest is limited. Ultimately prices begin to fall as the cumulative effects of the yield increases outpace demand growth (recall, demand growth is slowing and ultimately demand is stable).

The second climate scenario assumes that 1% of the forest dies back each year due to forest fires. This proportion of dieback is assumed to remain constant over time. This differs from the way the yield increase discussed previously is modeled (where it is assumed that the yield increase grows over time). It is also assumed that 30% of the forest material that dies back can be salvaged. In the first assessment of this scenario shown in Figure 15.3, it is assumed that there is no yield increase in this scenario, i.e. yields remain at their baseline level. This allows us to examine the implications of the increased disturbance effects in isolation.

With an increase in disturbance, prices fall modestly in the first few years. There are two reasons for this. First, the model incorporates salvage, and salvaged timber enters the market and lowers prices. Second, the economic incentives when disturbance occurs suggest that rotation ages should fall (Reed, 1984). With additional disturbance, landowners would prefer to harvest trees sooner rather than lose a large portion of the stock (70% that is not salvaged) to dieback. In order to reduce rotation ages, more timber has to be harvested initially, so this increases supply in initial periods. By shifting to shorter rotation ages, however, long-term timber supply falls. Thus, the dynamic adjustment to dieback entails at first a reduction in timber prices and then an increase relative to the base.

Thus far, changes in forest yields and dieback are modeled separately. The most likely climate scenario, however, will include some change in yields and some change in the area of land that dies back every year due to an increase in forest fires or other natural disturbances. As discussed previously, economists may choose to model the yield and dieback effects directly, or they may choose to model the net effects of both processes on aggregate timber yields. Whether one models the effects of these two climate impacts as a net effect or as separate impacts will have critical implications for estimating the resulting economic impact.

With the simple model developed previously, it is possible to compare how these alternative methods of modeling climate effects would influence the resulting estimates of the economic impacts. The analysis in Sohngen and Mendelsohn (1998) and Sohngen et al. (2001) used an approach that accounts for the two effects separately, such that they modeled the gross effects of the yield changes (+0.5% per year) and the gross effects of the dieback (1.0% of the stock dies back each year) as separate perturbations in the same model. Alternatively, one could model the net effect of both dieback and yield changes on total carbon stocks. The change in total carbon stocks implied by these two effects would then be used to adjust forestry yield functions used in the economic model. This is the approach used by Joyce et al. (1995) and Perez-Garcia et al. (2002). For our previous analysis, when dieback increases to a 1.0% loss each year, and the yield is increasing at 0.5% per year, total forest carbon will fall at first. Total forest carbon falls because the increase in dieback initially is greater than the yield increases, particularly for existing stocks. Over the long run, total forest carbon will rise at nearly 0.5% per year as the steady annual increases in forest yields ultimately overtake the losses due to dieback.

The results of both of these approaches are presented in Figure 15.4. The first approach is titled 'Yield + Dieback' and the second approach is titled 'Net Effects'. When dieback and yield changes are modeled directly, prices fall initially because additional timber makes its way onto markets through salvage, and harvests increase as foresters reduce the optimal rotation age of their forest. As noted previously, this avoids losing 70% of the stock that dies back near the optimal rotation age. Although dieback reduces the stock modestly, continued increases in yields ultimately overtake the losses due to dieback and there is substantially more stock in forests and greater supply. Hence, prices are lower in the long run when dieback and yield changes are modeled separately.

Under the net effects model, the market takes more time to adapt the forest to climate change (Figure 15.4). As a result of the dieback, net forest yields are projected to fall initially. By year 25, net yields have risen above the baseline and remain greater than in the baseline for the remainder of the scenario. Although dieback is actually occurring in the forest, the market model does not account for it directly; thus, the model has no way to respond to it directly. What the model sees is a reduction in forest growth in the initial periods. It responds to this reduction in forest growth with lower harvests and higher prices, which is exactly the opposite response of the model that incorporates dieback directly. Over the long run, prices fall relative to the baseline, but not as much as in the model that accounts for dieback directly, because it takes longer to adjust the forest to climate change with this approach to economic modeling.

The two different models lead to very different welfare effects as well. To measure welfare, the net present value of consumer's plus producer's surplus is calculated for each of the scenarios (baseline and two climate scenarios). This calculation is given in equation (3). The with and without (baseline) scenarios are then compared to determine the welfare effects of climate change. Under the yield + dieback scenario, welfare declines by $1.4 million, while in the net

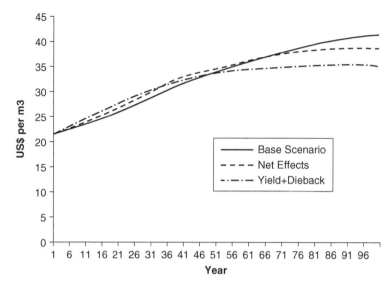

Figure 15.4   Representative price paths in the forestry model for the base scenario, the "yield + dieback" scenario, and the "net effects" scenario.

effects scenario, welfare declines by $2.7 million. Modeling the net effects without directly modeling disturbance potentially leads to a large over-statement of the welfare effects. The reason for this is that the model allows many fewer options for adaptation when only the net effects are modeled.

These results illustrate that the method of introducing climate change into the economic model has as much of an impact on the measurement of the impacts as the scale of the impacts themselves. In both cases, climate change is projected to decrease welfare, but when the effects of climate change are modeled directly (i.e. through the yield + dieback model), more adaptive responses are measured, and the welfare effect is estimated to be smaller. In economic analysis that is reliant on ecological modeling, it is thus critical to attempt to measure the ecosystem effects properly and to incorporate them into the economic model appropriately.

Review of economic estimates in the literature

Compared to agricultural systems, economic impacts and adaptation in forest systems are much more difficult to assess. One reason is that the data sets are not as widely available to assess economic outcomes from climate variation, such as in the Ricardian or hedonic studies (Mendelsohn et al., 1994; Dechenes and Greenstone, 2007). Another reason for this is that forests involve dynamic resources and investments which take many years to provide benefits. One needs dynamic models to assess impacts in forests (Sohngen and Mendelsohn, 1998).

Despite the complexities, there have been a number of economic analyses of climate change impacts in forests to date. The earlier economic analyses focused on the United States and suggested that climate change would increase timber supply and reduce timber prices (Joyce et al., 1995; Sohngen and Mendelsohn, 1998). The largest impacts in the United States occurred in the South and Pacific Northwest, which makes sense given that these regions also have the largest timber sectors. Sohngen and Mendelsohn (1998) directly account for changes in dieback and disturbance in addition to changes in timber yield. They also allow species to shift from one region to another. Joyce et al. (1995) focus on net yield changes and assume that forest types remain in the same location over time.

One problem with the earlier studies is that they were national in scope. Climate change is likely to have global impacts, and the effects of climate change in markets in any given region is a function not only of the underlying ecological impacts, but also of the changes that occur in other regions. For example, if timber supply expands dramatically across the world, adaptations that would otherwise be efficient in the United States when evaluated by a model of only the United States may not be efficient if evaluated with a global model. Sohngen et al. (2001) and Perez-Garcia et al. (2002) both develop dynamic models to address this issue. Sohngen et al. (2001) use similar methods to those in Sohngen and Mendelsohn (1998) to model climate change impacts. They find that climate change in general is likely to increase global timber supply, although subtropical and tropical regions gain more. With shorter rotation periods, subtropical and tropical regions are able to adapt fairly rapidly to climate change.

Temperate and boreal regions, in fact, may experience losses in some climate scenarios because prices fall enough to make many adaptation options inefficient. Furthermore, temperate and boreal regions experience significantly greater dieback with climate change, further adding to economic losses. Over the long run, the global studies suggest that output in northern regions does expand significantly relative to the baseline. Thus, places like Canada, northern Europe and Russia appear to be vulnerable in the short run due to dieback, but they appear to benefit in the long run.

Emerging issues

There are a number of important emerging issues in the adaptation of forests to climate change. Most studies have focused on adaptation in managed forests, and a growing proportion of timber is derived from managed plantation forests. Daigneault et al. (2008) suggest that the amount of timber supplied from managed plantations will continue to increase in the future. Sohngen et al. (2001) illustrate how climate change likely will strengthen this trend by enhancing subtropical forests relative to boreal and temperate regions. Shorter-rotation plantation species can be adapted across space and time more readily than many of the longer-rotation species and unmanaged forests currently used for a large share of the world's timber supply.

Although the results of climate analyses have important implications for adaptation in the timber sector, they have equally important implications for adaptation in unmanaged forested regions. As the share of plantation forests grows, a larger share of forests around the world is being left unmanaged each year due to economic circumstances. Higher productivity in plantation forests is driving down timber prices, and these lower prices are reducing the efficiency of extracting timber in regions that are not managed. Practically, this means that as climate change affects the world's ecosystems, managers may or may not be available to help adaptation along. If there is little economic incentive to manage forests without climate change, the effects of climate change are unlikely to make management more efficient.

Forests are not only being reserved for economic reasons; they are also being reserved for ecological reasons. Many parks and reserves have been established over the past 100 years to protect places with unique features or ecosystem attributes. These locations may have high biodiversity, an abundance of plant and animal life or some other features that provide incentives for government to protect them. With climate change, however, many of these forests may be imperiled. This chapter has not addressed how adaptation may occur in these reserved forests.

It is beyond the scope of this chapter to detail adaptation plans for public forests and preserves, but it is worthwhile noting that it is likely that adaptation in these areas, especially those with high ecosystem value, will be substantially more difficult than adaptation in private forestlands. Many private forestlands are managed with a fairly simple set of objectives, such as maximizing the value of the land in timber, or providing income with enhancing recreational opportunities. Public forests, and particularly those with substantial ecosystem value, are often managed with many objectives. They also have many stakeholders. Developing adaptation plans will be exceedingly difficult for these forests, given the many individuals who will have a say in the plan. Carrying out the plans likely will be even harder.

One potential response to climate change is to sequester carbon in forests by expanding the area of forests, changing forestland management or reducing deforestation. Estimates indicate that forestry could efficiently provide up to 30% of the total reduction in CO_2 this century (Sohngen and Mendelsohn, 2003). None of the studies that have examined carbon sequestration in forests, however, have fully considered the effects of climate change. Climate change will undoubtedly have large implications for carbon sequestration given the potential shifts in dieback, species range and forest growth. Future studies of carbon sequestration should more carefully account for potential climate change impacts.

A final emerging issue to consider relates to the growth in biofuels as an energy source. With higher energy prices in recent years, there has been a re-emergence of forests as a potential source of energy, both for electricity and as a source for biofuels through cellulosic ethanol. To some extent, the use of forests as an input into energy production is promoted by government policy, largely renewable energy laws. These trends, if they continue, could dramatically increase the demand for all forests.

Conclusion

This chapter examines the implications of climate change on forested ecosystems. The chapter begins with a discussion about the potential ecological effects of climate change in forests. These include changes in the rate of growth of trees, changes in disturbance patterns and shifts in the distribution of tree species. The paper then describes how these results can be integrated into economic models. Several different approaches have been discussed in the literature and there is some debate about the best way to approach important issues like the modeling of forest dieback.

To illustrate how climate change impacts can be integrated into an economic model of forestry, a simple model of the forest sector is shown. The differences in the economic effects associated with different methods of perturbing the economic model with the impacts of climate change are then examined. The results show that directly modeling dieback leads to far different estimates of the potential for adaptation and to far different estimates of welfare effects of climate change. These results suggest that modelers need to carefully consider how best to integrate ecology into their economic models. The chapter concludes with a discussion about results in economic analyses to date.

Although a number of studies on climate change impacts in forests have been conducted to date, research in this area is actually fairly limited, and there are a number of areas where additional work could be very useful. First, there is fairly little research examining potential adaptation strategies for individual landowners. The changes described in the chapter suggest that landowners will need to adapt to new disturbance regimes, shifts in the types of species that will grow in their location and changes in timber prices. It would be useful to conduct additional research on the costs and benefits of making different harvesting or planting decisions, given both the ecological and economic uncertainties involved.

Second, beyond adaptation on private lands, a vitally important issue of global concern will be adaptation on common property forestlands or public forestlands. Adaptation in these regions actually will be much more difficult to accomplish given the much more complicated incentive arrangements at play. Common property forests are often managed by groups most effectively when the institutions have had a long period of stable ecological conditions in which to evolve. If ecological conditions are changing and important forest outputs are declining due to exogenous climate-related factors, it may be very difficult for these institutions with long histories to adapt. Understanding adaptation in these regions is another important area for research, given that common property forests do provide a large share of the world's non-timber forest products and fuelwood. Beyond common property lands, many protected zones will be undergoing important climate-related changes, and society will have to decide whether to actively manage the change or let adaptation occur naturally.

Third, policy responses to climate change could have important consequences for forested ecosystems. Carbon sequestration would change the area of forests and the amount of harvesting that occurs in different regions. Understanding how climate change potentially affects forests preserved or planted for carbon sequestration will be important for preserving carbon in the biosphere in the long run. For instance, regions with increasing forest fire potential due to climate change may not be the best places to increase forest area for carbon sequestration. Biofuels also could dramatically affect the landscape by altering timber harvests. Understanding whether biofuels are a net carbon source or sink is actually still an important research question that needs to be addressed with additional work.

References

Adams, H. D., Guardiola-Claramonte, M., Barron-Gafford, G. A., Villegas, J. C., Breshears, D. D., Zou, C. B., . . . Huxman T. E. (2009). 'Temperature sensitivity of drought-induced tree mortality portends increased regional die-off under global change-type drought', *Proceedings of the National Academy of Science*, vol. 106, pp. 7063–7066.

Amacher, G., Malik, A. and Haight, R. (2005). 'Not getting burned: The importance of fire prevention in forest management', *Land Economics*, vol. 81, pp. 284–302.

Bachelet, D., Lenihan, J., Drapek, R. and Neilson, R. (2008). 'VEMAP vs VINCERA: A DGVM sensitivity to differences in climate scenarios', *Global and Planetary Change*, vol. 64, no. 1, pp. 38–48.

Balgis O.-E., Parrotta, J., Adger, N., Brockhaus, M., Pierce Colfer, C. J., Sohngen, B., . . . Robledo, C. (2009). 'Future socio-economic impacts and vulnerabilities', in R. Seppälä, A. Buck and P. Katila (Eds.). *Adaptation of Forests and People to Climate Change – A Global Assessment Report*, International Union of Forestry Research Organizations, World Series Volume 22, Vienna, Austria.

Boisvenue, C. and Running, S. W. (2006). 'Impacts of climate change on natural forest productivity – evidence since the middle of the 20th century', *Global Change Biology*, vol. 12, pp. 862–882.

Daigneault, A., Miranda, M. and Sohngen, B. (2010). 'Optimal forest management with carbon sequestration credits and endogenous fire risk', *Land Economics*, vol. 86, no. 1, pp. 155–172.

Daigneault, A., Sohngen, B. and Sedjo, R. (2008). 'Exchange rates and the competitiveness of the United States timber sector in a global economy', *Forest Policy and Economics*, vol. 10, no. 3, pp. 108–116.

Deschenes, O. and Greenstone, M. (2007). 'The economic impacts of climate change: Evidence from agricultural output and random fluctuations in weather', *American Economic Review*, vol. 97, pp. 354–385.

Food and Agricultural Organization of the United Nations (FAO). (2010). *Global Forest Resources Assessment Main Report*, Forestry Paper 163, Food and Agricultural Organization of the United Nations, Rome.

Food and Agricultural Organization of the United Nations (FAO). (2012). *State of the World's Forests 2012*, Food and Agricultural Organization of the United Nations, Rome.

Hansen, M. C., Stehman, S. V. and Potapov, P. V. (2010). 'Quantification of global gross forest cover loss', *Proceedings of the National Academy of Sciences*, vol. 107, no. 19, pp. 8650–8655.

Houghton, R. A. (1999). 'The annual net flux of carbon to the atmosphere from changes in land use 1850–1990', *Tellus*, vol. 51, pp. 298–313.

Houghton, R. A. (2003). 'Revised estimates of the annual net flux of carbon to the atmosphere from changes in land use and land management 1850–2000', *Tellus*, vol. 55, pp. 378–390.

Intergovernmental Panel on Climate Change (IPCC). (2007). *Mitigation of Climate Change*, Intergovernmental Panel on Climate Change, Cambridge University Press, Cambridge, UK.

Iverson, L. R. and Prasad, A. M. (1998). 'Predicting abundance of 80 tree species following climate change in the eastern United States', *Ecological Monographs*, vol. 68, no. 4, pp. 465–485.

Iverson, L. R., Prasad, A. M., Matthews, S. N. and Peters, M. (2008). 'Estimating potential habitat for 134 eastern US tree species under six climate scenarios', *Forest Ecology and Management*, vol. 254, no. 3, pp. 390–406.

Joyce, L. A., Mills, J. R., Heath, L. S., McGuire, A. D., Haynes, R. W. and Birdsey, R. A. (1995). 'Forest sector impacts from changes in forest productivity under climate change', *Journal of Biogeography*, vol. 22, pp. 703–713.

Kurz, W. A., Dymond, C. C., Stinson, G., Rampley, G. J., Neilson, E. T., Carroll, A. L., . . . Safranyik, L. (2008). 'Mountain pine beetle and forest carbon feedback to climate change', *Nature*, vol. 452, no. 7190, pp. 987–990.

Mendelsohn, R., Nordhaus, W. D. and Shaw, D. (1994). 'The impact of global warming on agriculture: A Ricardian analysis', *The American Economic Review*, vol. 84, pp. 753–771.

Norby, R. J., DeLucia, E. H., Gielen, B., Calfapietra, C., Giardina, C. P., King, J. S. and Ledford, J. (2005). 'Forest response to elevated CO_2 is conserved across a broad range of productivity', *Proceedings of the National Academy of Sciences of the United States of America*, vol. 102, no. 50, pp. 18052–18056.

Perez-Garcia, J., Joyce, L. A., McGuire, A. D. and Xiao, X. (2002). 'Impacts of climate change on the global forest sector', *Climatic Change*, vol. 54, no. 4, pp. 439–461.

Reed, W. J. (1984). 'The effects of the risk of fire on the optimal rotation of a forest', *Journal of Environmental Economics and Management*, vol. 11, pp. 180–190.

Scholze, M., Knorr, W., Arnell, N. W. and Prentice, I. C. (2006). 'A climate-change risk analysis for world ecosystems', *Proceedings of the National Academy of Sciences*, vol. 103, no. 35, pp. 13116–13120.

Sohngen, B. and Mendelsohn, R. (1998). 'Valuing the market impact of large scale ecological change: The effect of climate change on US timber', *American Economic Review*, vol. 88, no. 4, pp. 689–710.

Sohngen, B. and Mendelsohn, R. (2003). 'An optimal control model of forest carbon sequestration', *American Journal of Agricultural Economics*, vol. 85, no. 2, pp. 448–457.

Sohngen, B., Mendelsohn, R. and Sedjo, R. (2001). 'A global model of climate change impacts on timber markets', *Journal of Agricultural and Resource Economics*, vol. 26, no. 2, pp. 326–343.

Sohngen, B. and Sedjo, R. (1998). 'A comparison of timber market models: Static simulation and optimal control approaches', *Forest Science*, vol. 44, no. 1, pp. 24–36.

Stainback, G. A. and Alavalapati, J. R. R. (2004). 'Modeling catastrophic risk in economic analysis of forest carbon sequestration', *Natural Resource Modeling*, vol. 17, no. 3, pp. 299–317.

Westerling, A. L., Hidalgo, H. G., Cayan, D. R. and Swetnam, T. W. (2006). 'Warming and earlier spring increase western US forest wildfire activity', *Science*, vol. 313, no. 5789, pp. 940–943.

16

ECONOMICS OF FOREST CARBON SEQUESTRATION

G. Cornelis van Kooten, Craig Johnston and Zhen Xu

DEPARTMENT OF ECONOMICS, UNIVERSITY OF VICTORIA, CANADA.

Abstract

In this chapter, economic issues related to the creation of forest carbon offset credits are discussed in the context of forest carbon management strategies. Carbon offsets are defined as reductions in CO_2 emissions, or removals of CO_2 from the atmosphere, that are realized outside a compliance market but can nonetheless be used to counterbalance purchases of emission allowances. It is shown that carbon offsets created through forestry activities reduce compliance costs. Such offsets are created by sequestering carbon in living biomass, soil carbon pools and wood products; they also arise when wood biomass is used to produce energy, replacing CO_2 emitted from fossil fuel burning. The greatest potential for carbon offsets from forestry may come when harvested wood products replace steel and/or cement in construction, thereby reducing CO_2 emitted during the production of steel and cement. This potential is demonstrated using an example. Even so, four main problems with forest carbon offsets that militate strongly against their widespread use are discussed: additionality, leakage, duration or impermanence and governance. Other related issues are highlighted, including the use of temporary certified emission reductions and the possibility of including activities that Reduce Emissions from Deforestation and forest Degradation (REDD).

Keywords

Carbon offset credits, forest management, wood product substitution, climate change

Introduction

In order to mitigate projected climate change, leaders of the G8 countries meeting in L'Aquila, Italy, agreed on 8 July 2009 to limit the increase in global average temperature to no more than 2°C above pre-industrial levels. To do this, the leaders set an ambitious target – to reduce global greenhouse gas (GHG) emissions by 50% from 1990 levels by 2050, with rich countries to reduce their aggregate emissions by 80% or more. The European Union's (EU) target is to reduce GHG emissions by 20% from the 1990 level by 2020, while the United Kingdom's Climate Change Act (2008) is even more ambitious, requiring GHG emissions to be cut by 34%

from 1990 levels by 2018–2022, and by 80% by 2050 (see Lea, 2012). Given the draconian and unrealistic nature of the emission reduction targets, countries need to find ways around these targets. This has been done by permitting emission offsets, known simply as carbon offsets. These are defined as reductions in GHG emissions (principally carbon dioxide or equivalent emissions, denoted CO_{2-e}), or an equivalent removal of CO_2 from the atmosphere, that are realized outside a compliance market and can be used in lieu of emissions reductions required under an official target (van Kooten and de Vries, 2012).[1] Thus, reductions in CO_{2-e} emissions in other countries and activities in other sectors that reduce concentrations of CO_{2-e} in the atmosphere can substitute for domestic reductions in CO_{2-e} emissions, thereby providing countries with escape valves that protect their industries and economy.

The motivation for the current chapter is the 1997 Kyoto process that permitted developed countries to meet a portion of their CO_{2-e} emission-reduction targets through the purchase of carbon offsets in developing countries. In essence, rich countries could pay poor countries to reduce their emissions by investing in processes that improve energy efficiency in the developing country (e.g. upgrading power plants, investments in wind turbines or solar panels). Alternatively, rich countries could sponsor activities in developing countries that remove carbon dioxide from the atmosphere and store it in terrestrial ecosystems (e.g. afforestation, conversion of cropland to pasture). Projects that create offsets in developing countries are certified under the United Nations' Clean Development Mechanism (CDM). These are referred to as certified emission reductions (CERs), whether they come from actual emissions reduction or from activities that destroy trifluoromethane (HFC-23) or increase sequestration of carbon in forest ecosystems (Wara, 2007).[2] Developed countries, on the other hand, would be responsible for certifying emission reductions or offset schemes in their own countries, including certifying activities that sequester carbon in forest ecosystems.

The focus of this chapter is on carbon dioxide emissions and, in particular, the potential for forestry activities to contribute to major reductions in atmospheric CO_2. Under Kyoto's rules, activities that affect land use, land-use change and forestry (LULUCF) can generate carbon offset credits, both in developed and developing nations. The only difference relates to certification: LULUCF projects in developing countries are certified under the CDM, while those in developed countries are certified by the relevant national government. Along with limits on the overall use of LULUCF-generated carbon offsets, the certification requirement presumes that the problems associated with such offsets, including additionality, leakages, duration and governance (which are discussed subsequently), are thereby minimized.

The overarching question that we address here is whether it is worthwhile including forestry-generated carbon offset credits in a cap-and-trade scheme that sets a target on CO_2 emissions. Do carbon offsets enable a country to attain its emission reduction targets more efficiently than in the absence of terrestrial sequestration? What are the costs and benefits of forestry-generated carbon offsets? What are the challenges and limits to forestry activities?

We proceed in the next section by demonstrating that carbon offsets reduce the costs to large emitters (countries) of meeting emission reduction targets. Because carbon offsets are meant to substitute for emission reductions and be traded in markets, in the third section, we consider carbon markets in more detail, with particular focus on Europe's Emissions Trading System (ETS) because it is the only such market in existence. Then, we focus specifically on carbon sequestration in forest ecosystems, examining in particular issues related to the additionality of forestry projects, potential for leakages, duration, transaction costs and governance. Along with biological uncertainty, these problems make it extremely difficult to determine the actual carbon flux associated with forestry activities and especially so if avoided deforestation and forest degradation are taken into account. Finally, we illustrate what happens to the overall net carbon flux associated with forestry activities when wood product carbon sinks and the substitution of wood products for steel and/or concrete in construction are included. As indicated in the conclusions, the task of creating valid forest carbon offsets may well exceed our capacity to do so.

Forest carbon sequestration: Theory

Carbon sequestration in forest ecosystems must yield economic benefits or there would be no sense pursuing this option in lieu of CO_2 emissions reduction. In a perfect world with no transaction costs, leakages and governance and duration issues, it is straightforward to demonstrate the benefits of carbon sequestration. Consider Figure 16.1. The emissions reduction and carbon sequestration sectors are shown as back-to-back panels. In the left panel, there is a cap on emissions given by $0E$. In the absence of carbon offsets, the costs of reducing emissions through a combination of emissions trading and abatement of emissions by industrial emitters are given by the area under the marginal cost function, or area $0aE$. At the level of the cap, the marginal cost of abatement is P, which is also the price of purchasing an emission allowance.

Assuming no other means of purchasing offsets, the derived demand for carbon offsets in the forest sector is denoted by DD. Such a carbon offset is referred to as a 'removal unit' (RMU), which is defined under Kyoto rules as an Emission Reduction Unit (ERU) generated by removing a tonne of CO_2 (tCO_2) from the atmosphere by sequestration. If the price a country or large emitter is required to pay for a RMU is P, there is no benefit to purchasing carbon offsets in the forest carbon sequestration sector because firms will abate or buy allowances from firms that exceed their abatement targets. (These allowances are known under Kyoto as Assigned Amount Units, or AAUs.) If, on the other hand, RMUs are costless, emitters will obtain their entire targeted reduction $0E$ in the forest sequestration sector; hence, $0E = 0C$. At other prices for carbon offsets, the derived demand is determined in a similar way, so that the line CP is parallel $0a$.

Now introduce a marginal cost of carbon sequestration as shown in the right panel. The forestry sector would provide $0C^*$ carbon offsets at price P^*. This would then be the marginal cost of abatement, so that $0E^*$ emissions are abated, with $0E^* + 0C^* = 0E$. That is, carbon offsets of amount $0C^*$ would be substituted for E^*E of emissions abatement.

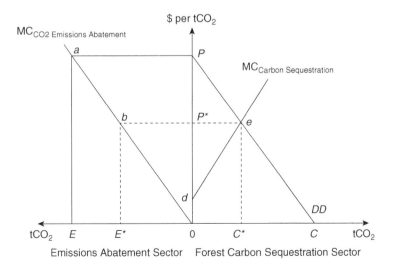

Figure 16.1 The benefits of carbon sequestration.

Given the relationship between the derived demand function *DD* and the marginal cost of abatement, the area under *DD* provides an indication of the net benefit of carbon sequestration in forest ecosystems. Without carbon offsets, the cost of achieving the targeted emission reductions *OE* is given by area $0aE$. When carbon offsets are permitted, the cost of attaining the target is given by area $0bE^* + 0deC^*$. The cost saving is given by $EE^*ba - 0deC^* > 0$; because *CP* is parallel to $0a$, the cost saving is identical to area *deP* in right-side panel.

Clearly, if activities to create carbon offsets in forest ecosystems are too costly (see van Kooten, Eagle, Manley and Smolak, 2004; van Kooten, Laaksonen-Craig and Wang, 2009), then the MC in the right panel might well intersect the vertical axis at or above P $(d \geq P)$, in which case there is no benefit for a country or large emitter to purchase RMUs in the forest carbon sequestration market. What factors affect the marginal costs of creating carbon offsets?

Carbon markets

Economists prefer economic incentives over regulation because they incentivize firms to adopt technical changes that lower the costs of reducing CO_2 emissions. In the case of a cap-and-trade scheme, firms can sell permits or avoid buying them; in the case of carbon taxes, they seek ways to avoid paying the tax. Further, market instruments provide incentives to change products, processes and so on, as marginal costs and benefits change over time. Because firms are always trying to avoid the tax, or avoid paying for emission rights, they tend to respond quickly to technological change.

In the context of climate change, most economists generally favour carbon taxes over cap and trade because the marginal damage (marginal benefit of mitigation) function is likely flatter than the marginal cost of mitigation. In an uncertain world, a tax is a more flexible instrument than an emissions cap. Although an emissions cap guarantees that a target is met (assuming the cap is enforceable), if the cap is set too low, the costs of attaining that emissions level could be unbearably high. With a tax the marginal cost of abatement is known when the tax rate is revealed as firms set the marginal abatement cost equal to the tax. The tax could be increased over time if insufficient abatement occurs and more becomes known about potential damages from climate change. Of course, one could similarly adjust the cap over time in like fashion to avoid unpalatable costs.

There are other drawbacks to emissions trading, of which two are particularly troublesome. First, politicians and extant firms prefer that rights to emit CO_2 are grandfathered. Firms are given permits to emit an amount of CO_2 that is below their current level depending on the domestic or global target. Firms can present permits to enable them to release CO_2 into the atmosphere, or they can reduce their own emissions (e.g. through improvements in energy efficiency, switching to non–fossil fuels or going out of business) and sell permits in carbon markets. Whatever the case, the price that permits fetch in the carbon market is considered a cost of production by all firms that are affected by the trading scheme. To avoid the adverse impacts on the economy (e.g. firms going out of business, permit prices rising too high), carbon offsets are allowed, which effectively negates a true cap-and-trade scheme.

To date few jurisdictions have imposed carbon taxes (one exception is British Columbia) and there have been few carbon markets. The voluntary Chicago Climate Exchange disappeared at the end of 2010, leaving the EU's ETS as the only compliance carbon market in operation. ETS is a mandatory market for large industrial emitters in Europe; these firms have been allocated emission allowances (EUAs) and they must present one EUA for every tCO_2 that they emit. If they emit more CO_2 than their allocated permits allow, they must purchase EUAs on the ETS. However, they could also purchase carbon offsets that are sold on the ETS. Two carbon offsets are available: ERUs that are created in countries of the former Soviet Union through Kyoto's Joint Implementation program and CERs that are created in developing countries through Kyoto's CDM.

CERs are certified strictly under the process developed by United Nations' Framework Convention on Climate Change (UNFCCC), while ERUs are certified by developed countries that invest in the creation of carbon offsets in ex-Soviet states, sharing these offset credits with the host country, which also has an emissions-reduction target under Kyoto. Likewise, EUAs are certified by the EU, although it has delegated this to the individual countries. This, in turn, led to the collapse of the first stage of the ETS as countries permitted their large industrial emitters to overstate their emissions and the number of permits for which they were eligible.

Finally, there has been remarkable growth in voluntary carbon markets, with a number of private companies emerging as certifiers of voluntary emissions reductions (VERs). In this market, forestry activities and especially forest conservation play a large role, accounting for more than 40% of VERs sold globally in 2010 (Peters-Stanley, Hamilton, Marcello and Sjardin, 2011). Certification standards include the 'Gold Standard' (GS), the Climate, Community and Biodiversity Alliance's CCB certification, and the Verified Carbon Standard (VCS). Various (mainly European) sponsors grant the certifying agencies their legitimacy. For example, core sponsors of the Gold Standard include the German Federal Ministry for the Environment, Nature Conservation and Nuclear Safety, WWF International (headquartered in The Netherlands), the European Climate Foundation and Merrill Lynch Commodities (Europe) Limited; the GS standard is endorsed by Renewable Energy and Energy Efficiency Partnership (Austria), MyClimate (Switzerland) and 'astmosfair' (Germany), among others. The market for VERs amounted to US$572 million in 2011, with trades averaging $6.2 per tCO_2 (Peters-Stanley and Hamilton, 2012). The VER market is small compared to global trade in emissions worth $142 billion in 2010 (van Kooten, 2013). However, there is concern that VERs are sold not only in the voluntary market but are also entering the ETS as carbon offsets (see van Kooten, Bogle and deVries, 2012). If that is truly the case, then the existence of a legal carbon offset market facilitates the laundering of

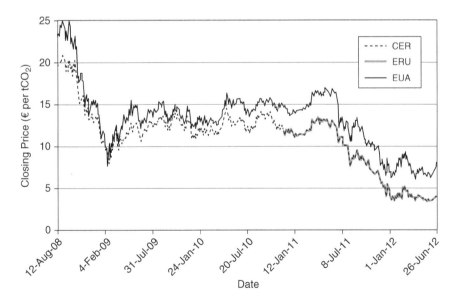

Figure 16.2 Prices of EUAs and carbon offsets (CERs and ERUs), European Trading System, 2008 to mid-2012.

VER credits, aided and abetted by environmental nongovernmental organizations, governments and financial intermediaries.

Increasing reliance on carbon offsets, including illegitimate ones, might help explain the drop in prices on the ETS, as indicated in Figure 16.2. This issue is discussed further in the section 'Governance.' At this stage, we only point out that forestry activities that create carbon offset credits, whether these are legitimate or not, play an important role in the marketplace.

Forest carbon sequestration: Real-world challenges

Society wishes to mitigate climate change at the lowest possible cost, but any activities to achieve emission reduction targets must also be effective in reducing atmospheric concentrations of carbon dioxide and equivalent GHGs. When it comes to carbon sequestration in forest ecosystems, the greatest challenges pertain to the compatibility of carbon offsets (RMUs) and CO_2 emission reductions (AAUs). Issues relate to the additionality of forestry projects, leakages, duration, transaction costs and governance. Although these issues are inter-related, we address each in turn.

Additionality

In principle, carbon offset credits should be earned only for carbon sequestration above and beyond what occurs in the absence of carbon-uptake incentives, a condition known as additionality. Thus, carbon sequestered as a result of incremental forest management activities (e.g. juvenile spacing, commercial thinning, fire control, fertilization) would be eligible for carbon credits only if the activities would not otherwise have been undertaken. Similarly, afforestation projects are additional if they provide environmental benefits (e.g. regulation of water flow and quality, wildlife habitat) not captured by the landowner and would not be undertaken in the absence of economic incentives, such as subsidy payments or the ability to sell carbon credits. Further, if it is demonstrated that a forest would be harvested and converted to another use in the absence of a specific policy (say, subsidies) to prevent this from happening, the additionality condition is met. Demonstrating that the additionality criterion is met is not easy; the problem is that the process is opaque and open to political manipulation and, thus, corruption.

Consider for example the case of zero tillage. Schmitz, Moss, Schmitz, Furtan and Schmitz (2010, pp. 18–19) argue that, as a result of reduced tillage and conversion of cropland to perennial grasses, Saskatchewan farmers sequester annually some 20 million tonnes of carbon, or more than 70 Mt CO_2. However, as Nagy and Gray (2012) point out, 'the development and adoption of zero tillage cropping systems is perhaps the most important agricultural innovation of the past fifty years,' with farmers gaining some $1.7 billion in terms of reduced fuel, labour, machinery and other input costs. Although farmers often argue that they should be compensated for the carbon uptake benefits associated with the adoption of such practices (e.g. Paustian et al., 1997), clearly compensation is unwarranted because zero tillage has been adopted (by over 90% of farmers in Saskatchewan) in the absence of carbon payments. Carbon sequestered as a result of zero tillage clearly fails the additionality test even though policymakers clamour for the acceptance of carbon offsets related to the adoption of zero tillage.

Leakages

Another difficulty is that of assessing leakages – the extent to which carbon sequestration in one place increases harvests and release of stored carbon as CO_2 in another. Estimates indicate that, for forestry activities meant to sequester carbon, leakages range from 5% to 93%, depending on the type of project and its location (Murray, McCarl and Lee, 2004; Wear and Murray, 2004; Sohngen and Brown, 2006). The effect on the marginal cost function in the right-hand panel of Figure 16.1 could be large, with Boyland (2006) finding that a failure to include a 25% leakage factor will underestimate costs by one-third.

Based on the result of a meta-regression analysis by van Kooten et al. (2009), and adding 25% to costs to account for leakage which none of the reported studies took into account, the only forestry activities that might be able to provide carbon offsets at prices below what EUAs trade for on the European ETS (Figure 16.2) are tree-planting projects in the tropics. In essence, if the marginal cost function found by the meta-regression analysis were adjusted upwards to account for leakages, it would likely intersect the vertical axis in Figure 16.1 above point *P*. This would especially be true if duration was also properly taken into account. The 68 studies considered by van Kooten et al. (2009) ignored the problem of duration – the fundamental incompatibility between emissions reduction and terrestrial carbon sequestration credits because of the differing lengths of time that CO_2 is prevented from residing in the atmosphere.

Duration

If carbon offsets can be created via forest carbon sequestration, one must deal with the problem of duration (van Kooten, 2009). Duration refers to the fact that carbon offsets created by sequestering carbon in terrestrial sinks remove CO_2 from the atmosphere over some time period, but eventually release it back to the atmosphere. Because the timing of removal and release are not known with certainty, and varies across projects, it is impossible to determine how many RMUs any project creates. If one assumes that an emissions reduction is permanent – one tCO_2 not released to the atmosphere as a result of taking the bus instead of driving one's car is permanent – but that CO_2 sequestered in a forest ecosystem is temporary, then there needs to be some means to compare the permanent and temporary credits. There needs to be a mechanism for equating an AAU and an RMU – there must be some way to compare a permanent emissions reduction with a temporary carbon offset.

It is no wonder that, while LULUCF activities are eligible as CERs under the CDM, strict conditions apply to have RMUs certified. For one thing, only carbon offsets earned through afforestation or reforestation projects are considered eligible as CERs. Afforestation refers to tree planting on sites that had not previously been forested, while reforestation refers to tree planting on sites that are considered forestland but where no trees are currently growing, perhaps because land has recently been converted to another use.

The certification process dealt with the duration issue by creating a temporary certified emission reduction (tCER) and a long-term certified emission reduction (lCER). The tCER operates like an annual rental of a permanent CER, while the lCER is something between an annual rental and a permanent reduction. Both instruments are a response to the duration problem, but are also designed to reduce transaction costs. For example, a tCER facilitates the sale of carbon offsets from forestry activities, because it allows a firm to purchase tCERs to cover emissions while it makes the necessary investments to reduce emissions permanently.

To understand how tCERs and lCERs have been implemented, consider Figure 16.3, where a landowner plants trees to create carbon offset credits.[3] The landowner chooses the initial time to enrol tCERs for sale, say time T_1. At that time, the number of eligible tCERs for sale is given by $tCER_1$, which is equal to the total carbon sequestered from time 0 to T_1 as a result of tree planting. The owner can sell an amount $tCER_1$ each year for 5 years (the length of the Kyoto commitment period), despite the fact that the site will continue sequestering carbon beyond T_1. After 5 years, the carbon available on the site is re-evaluated, with the

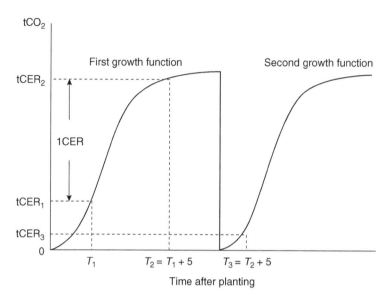

Figure 16.3 Defining tCERs and lCERs from forestry activities.

landowner now eligible to sell whatever carbon is available on the site at time $T_2 = T_1 + 5$; the eligible amount is now $tCER_2 > tCER_1$, which can then be sold for the next 5-year period. Ten years after the initial sale of carbon offset credits, at $T_3 = T_1 + 10$, the tCERs available for sale have fallen dramatically to $tCER_3$ as a result of an intervening harvest. The sequestered carbon subsequently lost to the atmosphere as a result of harvests is completely ignored.

The landowner could also sell more permanent lCERs, which equal the change in carbon over the project life. In the context of Figure 16.3, for example, an lCER might equal $tCER_2 - tCER_1$. The purchaser would be able to claim the CO_2 equivalent of the carbon that is sequestered against any emissions. However, because it is a one-time claim against emissions but the CO_2 is not stored permanently, the firm would then be responsible for buying further carbon offset credits after T_2, or purchase permanent emissions reduction credits (AAUs) to cover the lCERs.

If the landowner wants to participate in LULUCF activities that are eligible for CER offsets, she can choose to sell either tCERs or lCERs to address the impermanence problem. From the point of view of the purchaser, a tCER can be applied against emissions each year for a 5-year period, while an lCER enables the buyer to apply a much larger amount against emissions but only in a given year (or presumably the lCER can be spread across years). The lCER is paid for only once, while a rental payment for a tCER is required each year; however, the price of the former will be greater than that of the latter and an emitter can buy several lCERs. Once an approach is chosen, however, it has to remain fixed for the entire crediting period (UNFCCC, 2006), although it still can be replaced by permanent credits at a future date.

How does one choose between tCERs and lCERs at the beginning of a project? Unlike permanent CERs, there is no universally applicable pricing mechanism for both kinds of expiring CERs (Singh, 2009). Dutschke et al. (2006) and Bird et al. (2005) argue that the value of tCERs and lCERs greatly depends on buyers' expectations about a future market or, more specifically, the prices in the subsequent commitment period. Based on that, Lecocq and Couture (2008) indicate the feasible range of prices of tCERs and lCERs in the current commitment period should be less than or equal to the difference between the price of permanent credits in the current period and its discounted expected value in the next period. Whether expiring CERs are preferable, in that case, depends on the expected change in the future discount rate and the expected price of permanent credits. The choice of tCERs or lCERs then becomes speculative due to risk preferences towards unexpected expiry of a project and financial needs of landowners.

A landowner who sells lCERs should be held responsible for the potential loss of carbon that might occur as a result of a planned harvest or a natural disturbance. Suppose a landowner sells lCERs for the period T_2 to T_3. If the drop in sequestered carbon just prior to T_3 in Figure 16.3 is due to a planned harvest, the landowner is acting dishonestly by selling credits. This is a governance problem that is discussed in more detail subsequently, although it is worth mentioning here that carbon sequestration in forest ecosystems is susceptible to possible bogus carbon uptake claims.

The problem with forest carbon offsets is that, while facilitating trade and enabling large emitters to keep costs down, they are not truly equal to emissions reduction credits. Both are clearly artificial constructs that have little to do with real emissions reduction. In the case of tCERs, harvests are clearly ignored; with lCERs, the time path of carbon uptake is ignored.

Governance

Another major problem with forest carbon sequestration is governance. Measurement, monitoring and enforcement of forest carbon projects are especially problematic, mainly because tree growth is variable and ecosystem carbon fluxes are difficult (and expensive) to measure. Transaction costs are high and there is opportunity to misrepresent the size of the carbon offsets that are generated – projects are particularly vulnerable to corruption (Helm, 2010; van Kooten and de Vries, 2012). The link between an LULUCF project and the creation of carbon offset credits is not always clear. Those who certify CDM forestry projects must rely on computer models and analyses by forest management specialists to forecast future carbon uptake and release, and to identify the counterfactual (business-as-usual alternative). Significant leeway remains for speculation, error and corruption. Further, developing countries may well sell carbon offset credits to developed countries but, because they are not bound by international targets, still credit the activities to their own emissions reduction, resulting in 'double-dipping' (Woodward, 2011).

The principles involved in creating CERs from carbon offsets through LULUCF activities also apply to industrial countries, except certification falls to the government of the rich country. Further, as noted earlier, voluntary markets have circumvented government, enabling forest landowners to earn (voluntary) carbon offsets for potential sale. There already exists a market VERs in developed countries (Peters-Stanley et al., 2011), with some voluntary forest carbon offsets potentially even making their way onto legitimate markets, perhaps because the project certifiers take charge of some of the carbon offsets. This appears to have been the case for a forest conservation project in southeastern British Columbia (van Kooten et al., 2012), even though forest conservation is not a permitted activity for generating carbon offsets under Kyoto (see subsequent discussion). The interaction between the voluntary market and the compliance market, which currently exists only as the EU's ETS (Figure 16.2), is troublesome. The problem is that there is too much room for rent seeking (Helm, 2010).[4]

Governance may become an even bigger problem as a result of other recent initiatives. Although forest conservation activities are currently not eligible as carbon offsets, concerns about tropical deforestation and related CO_2 emissions (which account for perhaps some 20% of total annual GHG emissions) have led many commentators to commend the use of forest conservation in developing countries as a tool for addressing global warming. In international negotiations, activities that Reduce Emissions from Deforestation and Forest Degradation (REDD) are touted as an alternative means for earning CER credits. Indeed, as a result of negotiations at Cancun in December 2010, the narrow role of REDD has been expanded to include sustainable management of forests, forest conservation and the enhancement of forest carbon stocks, collectively known as REDD+. In this way, it is possible to link the UNFCCC and the Convention on Biological Diversity (CBD) – the other agreement signed at the 1992 Earth Summit in Rio de Janeiro. Increasingly, therefore, climate negotiators appear willing to accept REDD+ activities as potential carbon offsets to the extent that these activities also enhance biodiversity. Because deforestation and biodiversity are a greater problem in developing countries and because industrial nations are also interested in providing indirect development aid through the CDM, only REDD+ projects in developing countries merit attention, although these still need to be approved under the CDM.

It is this complexity that fundamentally impacts the carbon price mechanism. That is, by supplying the market with REDD+ carbon offsets, the price mechanism that ensures demand for credits equals supply becomes distorted because sales of credits from other than emissions reduction take place. Instead of dealing only with the sale and purchase of permits to emit CO_2, the market mechanism has to deal with emission reduction credits from sources that have nothing to do with CO_2 emissions from fossil fuel burning. REDD+ credits derive from protection of biodiversity on private forestland and do not contribute explicitly to reductions in CO_2 emissions. By allowing these offsets into the carbon market, the corresponding carbon price does not reflect its true value, i.e. it is distorted, with the price of carbon below what it would otherwise be. This results in inefficiency and reduces the incentive to invest in R&D that conserves energy, results in greater efficiency in the use of fossil fuels or spurs use of alternative energy sources. Thus, credits created by activities that enhance preservation of biodiversity enter the global carbon market without actually contributing to a net carbon reduction.

Although carbon sequestration in terrestrial ecosystems was only meant to be a bridge to provide time for an economy or firm to develop and invest in emission-reducing technologies, the sale of such credits has turned out to be an impediment to the implementation of new technology (as carbon prices are lower than necessary), while creating a larger gap between actual emissions and emission targets in the future (van Kooten, 2009) and doing little if anything to mitigate climate change. One can only conclude that any carbon offset program is a second-best solution that induces rent seeking.

Carbon sequestration: Forest products

Currently, the UNFCCC and the Kyoto Protocol do not include carbon stored in Harvested Wood Products (HWPs) in their carbon accounting guidelines. The current protocol assumption is that additions to the forest product carbon pool are equal to emissions from decomposition.[5]

There are many reasons why we may want to include HWPs in forest carbon accounting. The amount of carbon stored in forest products may remain sequestered for a considerable time. In addition, carbon comprises about one-half the mass of dry wood in structures, furniture and other finished wood products (Sjostrom, 1993). Carbon that is transferred from the living timber into wood products can be considered an addition to the carbon that is stored as a result of forestry activities. The carbon stored in wood products increases with each harvest. Carbon in wood products decreases when the product reaches the end of its life, although it might be recycled (prolonging the carbon storage), used for energy (displacing emissions from fossil fuel energy production)[6] or left to decompose in a landfill (resulting in slow release of CO_2 over time). Thus, accounting for the forest product pool may result in better management of the forest and the various carbon pools, encouraging recycling and energy production through incineration. Accounting for HWPs could assist timber producing countries such as Canada to meet their carbon emission targets, particularly if the carbon sink benefits of forest products can be enhanced by including the abatement of CO_2 emissions resulting from the substitution of lumber for steel and/or concrete in construction (Hennigar, Maclean and Amos-Binks, 2008).

There are reasons for discouraging the consideration of forest products, however, because doing so would increase carbon accounting complexity and introduce greater uncertainty into estimates of carbon flux. There are also concerns that it may incentivize enhanced forest management and harvests, while eroding concerns for forest conservation and protection (e.g. see van Kooten et al., 2012). It might also result in trade disputes related to the responsibility for carbon stored in HWPs. Forest product exporting countries would like to claim the carbon contained in products but consumer countries, such as the United States, China and some EU countries, would not be willing to accept responsibility for CO_2 emissions from imported wood products as these decompose.

The impact of including wood products in forest sector carbon budgets can be analyzed by expanding forest management models to include products. Indeed, forest management models have been used to demonstrate how the inclusion of HWPs and CO_2 emission reductions caused by substituting wood products for steel and/or concrete can affect optimal forest management strategies and how these can significantly increase the forest sector's contributions towards reducing atmospheric carbon dioxide (Hennigar et al., 2008; van Kooten et al., 2012).

In Figure 16.4, we compare levels of carbon abatement in a hypothetical temperature forest when HWPs substitute for more fossil fuel–demanding products in construction. The results provide an indication of the impact on optimal harvest volumes when carbon storage in forest and product pools is considered in conjunction with different levels to which HWPs substitute for steel and concrete. Harvest levels are presented for a carbon price of $5/tCO_2$ and a 200-year

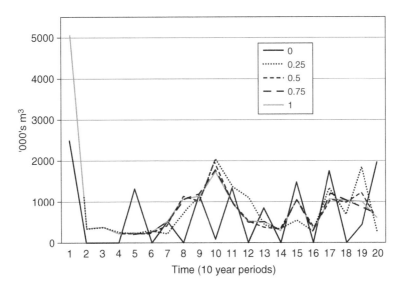

Figure 16.4 Volume harvested over 200 years for five lumber substitution benefit levels ranging from 0 to 1 tC abated from the atmosphere per m³ of lumber used relative to more fossil fuel–demanding products such as steel or concrete.

time horizon. As the substitution value rises, the optimal forest management strategy is to harvest trees as quickly as possible subject to sustainability and growth constraints; an increase in the 'pickling factor' (proportion of carbon in harvested biomass that remains sequestered in wood products for a long period) has a similar impact. Conversely, a scenario with a low substitution benefit encourages a more even pattern of harvests over the time horizon so that more harvesting occurs later in the horizon and less early on. Overall, the substitution benefit of removing one tonne of carbon per cubic meter (1 tC/m^3 or 3.67 tCO$_2$/m^3) from the atmosphere results in an overall increased harvest of approximately 4.8 million m^3 over the 200-year time horizon as compared to the scenario with a substitution benefit of 0 tC.

By recognizing carbon stored in wood products, optimal forest management strategies change considerably for positive carbon prices. Not only do harvest patterns change with the degree of substitution of wood for cement and steel, but also the magnitude of the overall harvest. With positive (nonzero) carbon prices, higher substitution benefits make it more profitable to harvest a greater amount of timber to be used in the production of wood products, particularly lumber. The next logical question is: Does this increase or decrease the amount of carbon sequestered?

In Figure 16.5, we present the amounts of carbon stored in the forest and wood product pools over the 200-year time horizon for our hypothetical forest; the product pool includes CO$_2$ savings from reduced fossil fuel emissions associated with the substitution of wood for steel and/ or concrete. Different degrees of product substitution are assumed in the figure. Similar to the harvest volumes indicated in Figure 16.4, when account is not taken of the reduced fossil fuel emissions by substituting wood for other material in construction, the optimal forest management scheme leads to a small albeit consistent level of carbon storage in the forest and wood products over time (Figure 16.5a). As the degree of substitution increases, forest management strategies change (Figure 16.4) so that greater amounts of carbon are sequestered in both the

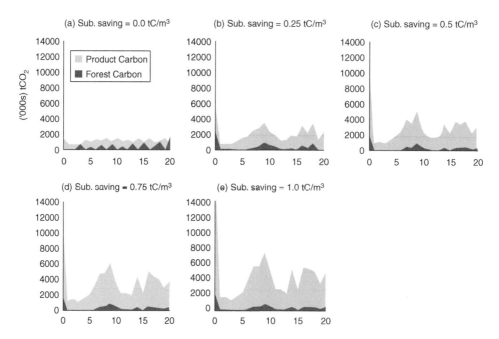

Figure 16.5 Carbon stored in forest and product pools over 200 years, five levels of substitution of lumber for steel and/or concrete, measured as per m^3 of commercial timber harvest.

forest ecosystem and product pools, although the latter simply swamps the former. This is clear from panels (b) through (e) in Figure 16.5.

If the issue of including the carbon stored in HWP were only technical, then it would likely have been included in the global protocols for carbon accounting long ago. International political obstacles to their acceptance remain because different issues need to be resolved. What nation should get credit for carbon stored in internationally traded wood products? Will acknowledging carbon storage in wood lead to more forest harvesting? Does the increase in demand for wood (to displace steel or conventional biofuels) stimulate forest output at the intensive and extensive margins? Will recognizing the longevity of stored carbon in discarded products within landfills encourage waste and discourage durability and recycling? These are a few of the political questions that keep HWPs out of the global protocols.

Conclusions

There is no denying that forestry activities impact the earth's carbon balance, with harvest activities contributing to human emissions of CO_2 and activities such as reforestation, afforestation and silvicultural activities that enhance tree growth removing CO_2 from the atmosphere and storing it in wood biomass. By taking into account the carbon that gets stored in wood product pools and potential substitution of wood for steel and/or concrete in construction, even harvest activities could enhance carbon uptake and storage. Further, forest activities that reduce CO_2 emissions, primarily conservation activities that prevent deforestation, benefit biodiversity and other services provided by forests. None of this can be denied. Nonetheless, none of this is sufficient to make the case that forest activities should be allowed to generate carbon offsets that can be sold in lieu of CO_2 emission reductions. As demonstrated in this chapter, the problems associated with the creation of carbon offsets through forestry activities are simply too complicated.

As one simple example, consider the issue of forest conservation and the creation of REDD or REDD+ credits. The idea is interesting, but, if harvested timber is used to produce wood products that then substitute for concrete in construction, say, the carbon offset benefits would swamp the carbon sink benefits of leaving the forest in its undisturbed state (Figure 16.5). There are other problems. Uncertainty and duration (impermanence) alone prevent the comparison of one forestry activity to sequester carbon with another, and neither can be compared with an emissions reduction. Transaction costs related to the striking of contracts, monitoring, verification and enforcement (how are those who act dishonestly in promoting projects to be punished?), and issues of governance, additionality and leakage simply militate against carbon offsets. Further, although we could demonstrate that carbon offsets lower the costs of meeting emission reduction targets, they also reduce incentives for investing in conservation, R&D and substitution of renewable energy for fossil fuels.

The road that enabled the inclusion of carbon offsets from forestry activities in nations' carbon mitigation arsenal has been a rocky one, and for good reason. Along with the biophysical uncertainty associated with the carbon fluxes accompanying forestry activities (e.g. uncertain growth, natural disturbance), measurement, governance and transaction costs make it difficult to achieve any sort of policy consensus (e.g. agreement on REDD+, agreeing on which country to credit carbon offsets stored in traded wood products). Finally, even if nations managed to achieve consensus about how forest carbon offsets are to be treated, there remains uncertainty regarding whether a forestry project actually reduces the amount of carbon dioxide in the atmosphere. It is simply not possible to take into account all of the economic and biological factors that enable one to make an affirmative statement one way or another.

Notes

1   CO_2 is the most important greenhouse gas, with other GHGs equated to CO_2 using an index of global warming potential (see http://unfccc.int/ghg_data/items/3825.php).
2   Wara (2007) found that 28% of 1,534 CER projects involved destroying HFC-23, primarily in China, because HFC-23 has a global warming potential 9,100 times greater than that of CO_2.
3   This example is adapted from van Kooten (2013, chapter 9).
4   In a review of a forthcoming book by Dieter Helm, *The Economist* quotes him as pointing out that '. . . the entire renewables sector has become an orgy of rent-seeking' (see www.economist.com/news/books-and-arts/21564815-climate-change-needs-better-regulation-not-more-political-will).
5   This is equivalent to assuming that such carbon is discounted at 0% – that the timing or duration of uptake and release of CO_2 does not matter.
6   In many situations, biofuels can make an important contribution to reducing reliance on fossil fuels, but their main drawbacks relate to their low energy intensity on a landscape level and the CO_2 emissions they emit that are subsequently removed from the atmosphere at a future date, if at all. Current evidence relating to the role of biofuels in mitigating climate change is not strongly supportive (see van Kooten, 2013).

References

Bird, D. N., Dutschke, M., Pedroni, L., Schlamadinger, B. and Vallejo, A. (2005). 'Should one trade tCER or lCER?'. Available from http://www.gruporeddperu.net/biblioteca/index.php?option=com_docman&task=doc_download&gid=535&Itemid=94

Boyland, M. (2006). 'The economics of using forests to increase carbon storage', *Canadian Journal of Forest Research*, vol. 36, no. 9, pp. 2223–2234.

Dutschke, M., Kapp, G., Lehmann, A. and Schafer, V. (2006). 'Risks and Chances of Combined Forestry and Biomass Projects under the Clean Development Mechanism'. *CD4CDM Working Paper Series* No. 1. Available from: http://www.cd4cdm.org/Publications/RisksChancesForestryBiomassCDM.pdf

Helm, D. (2010). 'Government failure, rent-seeking, and capture: The design of climate change policy', *Oxford Review of Economic Policy*, vol. 26, no. 2, pp. 182–196.

Hennigar, C., Maclean, D. and Amos-Binks, L. (2008). 'A novel approach to optimize management strategies for carbon stored in both forests and wood products', *Forest Ecology and Management*, vol. 256, pp. 786–797.

Lea, R. (2012). 'Electricity costs: The folly of wind power', Civitas: Institute for the Study of Civil Society, London. Retrieved from www.civitas.org.uk

Lecocq, F. and Couture, S. (2008). 'The Permanence Challenge: An Economic Analysis of Temporary Credits', Chapter 9 in C. Streck, R. O'Sullivan, T. Janson-Smith and R.G. Tarasofsky (Eds.). *Climate Change and Forests: Emerging Policy and market opportunities*, Brookings Institution Press, Washington D.C.

Murray, B. C., McCarl, B. A. and Lee, H.-C. (2004). 'Estimating leakage from carbon sequestration programs', *Land Economics*, vol. 80, no. 1, pp. 109–124.

Nagy, C. and Gray, R. S. (2012). 'Rate of return to the research and development expenditure on zero tillage technology development in Western Canada (1960–2010)', Working Paper, Department of Bioresource Policy, Business & Economics, U of Saskatchewan, Saskatoon, Saskatchewan, Canada. Retrieved from www.kis.usask.ca/publications/index.html

Paustian, K., Andren, O., Janzen, H. H., Lal, R., Smith, P., Tian, G., . . . Woomer, P. L. (1997). 'Agricultural soils as a sink to mitigate CO_2 emissions', *Soil Use and Management*, vol. 13, pp. 203–244.

Peters-Stanley, M. and Hamilton, K. (2012). *Developing Dimension: State of the voluntary carbon markets 2012*, A report by Ecosystem Marketplace & Bloomberg New Energy Finance, Ecosystem Marketplace & Bloomberg New Energy Finance, Washington, DC.

Peters-Stanley, M., Hamilton, K., Marcello, T. and Sjardin, M. (2011). *Back to the future. State of the voluntary carbon markets 2011*, A report by Ecosystem Marketplace & Bloomberg New Energy Finance, Ecosystem Marketplace and Bloomberg New Energy Finance, Washington, DC.

Schmitz, A., Moss, C., Schmitz, T., Furtan, W. H. and Schmitz, C. (2010). *Agricultural Policy, Agribusiness and Rent Seeking Behavior* (2nd ed.), University of Toronto Press, Toronto, Ontario, Canada.

Singh, G. (2009). *Understanding Carbon Credits*. Aditya Books, Pvt. Ltd., New Delhi.

Sjostrom, E. (1993). *Wood Chemistry: Fundamentals and Applications* (2nd ed.), Academic Press, San Diego, CA.

Sohngen, B. and Brown, S. (2006). 'The cost and quantity of carbon sequestration by extending the forest rotation age', Draft, AED Economics, Ohio State University. Retrieved from http://aede.osu.edu/people/sohngen.1/forests/Cost_of_Aging_v2.pdf

UNFCCC, (2006). United Nations Framework Convention on Climate Change Report of the Conference of the Parties serving as the meeting of the Parties to the Kyoto Protocol on its first session, held at Montreal from 28 November to 10 December 2005. Addendum. Part two: Action taken by the Conference of the Parties serving as the meeting of the Parties to the Kyoto Protocol at its first session *FCCC/KP/CMP/2005/8/Add.1. 30 March* (pp. 100).

van Kooten, G. C. (2009). 'Biological carbon sequestration and carbon trading re-visited', *Climatic Change*, vol. 95, no. 3–4, pp. 449–463.

van Kooten, G. C. (2013). *Climate Change, Climate Science and Economics: Prospects for an Alternative Energy Future*, Springer, Dordrecht, The Netherlands.

van Kooten, G. C., Bogle, T. and de Vries, F. P. (2012). 'Rent seeking and the smoke and mirrors game in the creation of forest sector carbon credits: An example from British Columbia', REPA Working Paper #2012–06, Department of Economics, University of Victoria, Victoria, British Columbia, Canada. Retrieved from http://web.uvic.ca/~repa

van Kooten, G. C. and de Vries, F. P. (2012). 'Carbon offsets', chapter 165 in J. Shogren (Ed.), *Encyclopedia of Energy, Natural Resource and Environmental Economics*, Elsevier, Oxford, UK.

van Kooten, G. C., Eagle, A. J., Manley, J. and Smolak, T. (2004). 'How costly are carbon offsets? A meta-analysis of carbon forest sinks', *Environmental Science & Policy*, vol. 7, no. 4, pp. 239–251.

van Kooten, G. C., Laaksonen-Craig, S. and Wang, Y. (2009). 'A meta-regression analysis of forest carbon offset costs', *Canadian Journal of Forest Research*, vol. 39, no. 1, pp. 2153–2167.

Wara, M. (2007). 'Is the global carbon market working?' *Nature*, vol. 445, no. 8, pp. 595–596.

Wear, D. N. and Murray, B. C. (2004). 'Federal timber restrictions, interregional spillovers, and the impact on U.S. softwood markets', *Journal of Environmental Economics and Management*, vol. 47, no. 2, pp. 307–330.

Woodward, R. T. (2011). 'Double-dipping in environmental markets', *Journal of Environmental Economics and Management*, vol. 61, pp. 153–169.

17

ECONOMIC SUPPLY OF CARBON STORAGE THROUGH MANAGEMENT OF UNEVEN-AGED FORESTS

Joseph Buongiorno,[1] *Ole Martin Bollandsås,*[2] *Espen Halvorsen,*[2] *Terje Gobakken*[2] *and Ole Hofstad*[2]

[1]DEPARTMENT OF FOREST AND WILDLIFE ECOLOGY, UNIVERSITY OF WISCONSIN, MADISON, WI 53706, USA. JBUONGIO@WISC.EDU
[2]DEPARTMENT OF ECOLOGY AND NATURAL RESOURCE MANAGEMENT, NORWEGIAN UNIVERSITY OF LIFE SCIENCES, P.O. BOX 5033, N-1432, ÅS, NORWAY.

Abstract

The potential for carbon storage in forests to alleviate greenhouse emissions and attendant climate change, and the option of doing so with uneven-aged multispecies forests, is of current high interest due to their diversity of species and stand structure. This chapter proposes methods of investigating the economics of carbon storage with uneven-aged management at stand level, and especially the trade-off with timber production. Nonlinear programming is used to find sustainable management regimes that maximize timber revenues or CO_2 sequestration only, or timber revenues subject to a CO_2 sequestration floor, or in the presence of markets for CO_2, the combined production of timber with income from carbon storage. The methods allow derivation of a schedule of the supply for carbon storage in terms of the price of CO_2e. Although applicable to many forest ecosystems, the methods are illustrated with an application to forests of Norway spruce, pine, birch and other hardwoods. In this context, there is a clear conflict between carbon storage and diversity of forests as carbon storage is maximized with monospecific stands. Furthermore, although carbon sequestration must ultimately conflict with timber production, such conflict is not significant unless the price of CO_2 increases considerably.

Keywords

Carbon sequestration, uneven-aged forest, continuous cover forestry, economics, optimization, nonlinear programming, biodiversity

Introduction

There is increasing evidence of global climate change, indicated for example by the recession of the global ice cap (NOAA, 2012) and permafrost and ecosystem changes in the arctic (Hinzman et al., 2005). A strong causal effect of carbon dioxide concentration in the atmosphere on climate change is also evident, leading to a rise in global temperatures (IPCC, 2012; Zickfeld et al., 2012).

Several international initiatives are attempting to reduce the amount of CO_2 in the atmosphere by lessening emissions and stimulating carbon storage by diverse means. These proposals give a central role to economic incentives through multiple carbon markets. They also emphasize initiatives at sector level, especially enhancing the role of forestry. For example, the Reduce Emissions from Deforestation and Forest Degradation (REDD) and REDD+ programs, involving several international agencies and countries, have the goal to 'reduce emissions from deforestation and forest degradation and contribute to conservation, sustainable management of forests, and enhancement of forest carbon stocks' (World Bank, 2011).

For field implementation, such initiatives need operational guidelines. To that end, this chapter presents methods for analyzing some of the economic, ecological and carbon storage implications of forest management at stand level. The emphasis is on uneven-aged and mixed species forests. Although the existing literature gives much more attention to carbon sequestration in even-aged forests, as for example in Asante and Armstrong (2012), Daineault, Miranda and Sohngen

Figure 17.1   Single-species, even-aged forest: A stand of Scots pine (*Pinus sylvestris*) in Norway.
Source: Photo courtesy of Erik Næsset.

(2010), Olschewski and Benitez (2002) and Van Kooten, Binkley and Delcourt (1995), for reasons pointed out subsequently and elsewhere (Pukkala and von Gadow, 2012) uneven-aged forests or continuous cover forestry are experiencing a modern revival throughout the world.

The methods of this chapter are general, and they have been adapted to a wide variety of forest ecosystems, including tropical forests (Boscolo, Buongiorno and Panayotou, 1997). However, the specific data and results presented here deal with the specific case of Norwegian forests. Since the 1950s, Norway's forestry, as in most of Scandinavia and many other countries, has been dominated by even-aged, monospecific systems like the pure Scots pine stand shown in Figure 17.1.

Such even-aged forests can be managed simply and efficiently as wood-producing machines, and they can also be used effectively for carbon sequestration (Hoen and Solberg, 1994). Nevertheless, their ecological aridity has led to increasing interest in more diverse uneven-aged forests, such as the rich mixture of all sized trees of spruce, pine, birch and other hardwoods found in some natural Norwegian stands (Figure 17.2).

Uneven-aged forests have many evident nontimber values. Aesthetically, the diverse texture and color of the uneven-aged landscape is attractive to the public. This value is well reflected by the name of 'futaie jardinée' (garden forest) used in France as synonym of 'forêt inéquienne'. In mountainous regions, continuous cover forestry, another name for uneven-aged management reflecting the fact that no stand is ever clearcut, is a natural defense against erosion and avalanches. From the point of view of habitat, the diversity of tree species and size of uneven-aged stands multiplies the feeding and nesting niches for equally diverse birds and animals. Also

Figure 17.2 Multispecies, uneven-aged forest: A mixed stand of aspen (*Populus tremula*) and spruce (*Picea abies*) trees in Norway.
Source: Photo by O. M. Bollandsås.

attractive is the possibility that the high level of stocking per unit of land area maintained in the uneven-aged forest provides great potential as a carbon sink to absorb the ever-increasing amount of carbon dioxide spewed into the atmosphere.

Besides these evident nontimber values, one must keep in mind that the forest remains an important economic agent as a source of wood. This unique renewable raw material, further transformed in sawnwood, wood-based panels, pulp, paper and paperboard and all their derivatives, contributes substantially to the wealth of nations and people's well-being.

The purpose of this chapter is to show how all these different objectives can be combined in a coherent analytical way with concrete data to determine what specific uneven-aged management regimes would be preferable with specific criteria, and what would be the trade-off, if any, among those objectives.

Management criteria

A quantitative analysis requires the formulation of a limited number of well-defined criteria covering the range of management objectives. In this exposition we use three categories of criteria: economic, ecological and climatic.

The simplest economic criterion is the wood production, in m^3/year. A more complete criterion, the net present value (NPV), reflects the price of the wood, the cost of harvesting and the cost of capital reflected by the interest rate. The NPV is the discounted value, over an infinite horizon of the future net revenues obtained by harvesting wood.

The two chosen ecological criteria are the diversity of tree species and the diversity of tree size. Both are based on Shannon's diversity index, a measure of system entropy or disorder (Shannon and Weaver, 1963). For example, the diversity of tree species is:

$$H = -\sum_{i=1}^{n} p_i \ln p_i \tag{1}$$

Where p_i is the fraction of the stand basal area made up of trees of species i, and n is the total number of species. The maximum tree species diversity occurs when basal area is equally distributed across species, i.e. the state of maximum disorder. This maximum value is equal to $\ln(n)$. The index is minimum and equal to 0 when the entire basal area is concentrated in one single species.

A similar index is used to measure tree size diversity, i.e. the diversity of stand structure, with p_i equal to the fraction of basal area in a particular tree size, of which there are m. The lowest theoretical tree size diversity, equal to 0, occurs when all the trees are in a single size class. The maximum tree size diversity, $\ln(m)$, occurs when the stand basal area is equally distributed by size class.

The climatic criterion consists of the amount of carbon or carbon dioxide equivalent stored in trees, expressed in tonne/ha.

Data

The forest data used in the growth model that forms the backbone of the analysis come from 5,524 permanent sample plots of the Norwegian national forest inventory. They were distributed throughout the country (Figure 17.3).

The plots have been remeasured every 5 years, from 1994 to 2005. Among the plots, 416 were selected at random and put aside to serve in a postsample validation of the model. The rest were used for model estimation. Among all the plots, 38% of the trees were Norway spruce (*Picea abies*), 16% were Scots pine (*Pinus sylvestris*), 35% were birch (*Betula pubescens* and

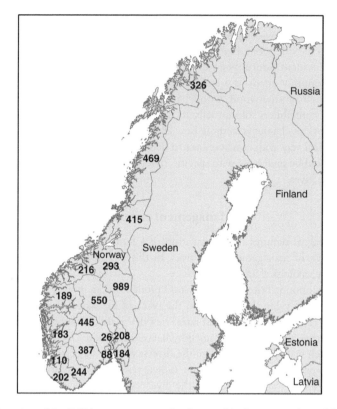

Figure 17.3 Location of the 5,524 permanent sample plots used in forest growth model estimation and testing.

B. pendula) and 11% of the trees were other broadleaves including alder (*Alnus* spp.), ash (*Fraxinus excelsior*) and aspen (*Populus tremula*).

Growth model

The model used to represent stand growth is a recursive dynamic matrix model of the form:

$$\mathbf{y}_{t+5} = \mathbf{G}_t(\mathbf{y}_t - \mathbf{h}_t) + \mathbf{R}_t \tag{2}$$

This model is quite general, and it has been applied to many forest types in temperate zones (e.g. Buongiorno and Michie, 1980; Liang et al., 2005), and in the tropics (e.g. Ingram and Buongiorno, 1996; Boscolo et al., 1997), in linear or nonlinear forms. In the matrix equation (2), $\mathbf{y}_t$ is a vector that defines the stand state at instant t. The elements of this vector are the numbers of trees of a particular species and diameter class. In this application, detailed in the appendix, there are four species groups: spruce, pine, birch and other broadleaves; and 13 diameter classes, ranging from size class 75 mm for trees of 50 to 100 mm diameter to size class 675 mm for trees of diameter 650 mm and above. Thus, $\mathbf{y}_t$ has the dimension $4 \times 13 = 52$. The object of the model is to predict the stand state in 5 years, indicated by $\mathbf{y}_{t+5}$ given the current state $\mathbf{y}_t$, and an eventual harvest $\mathbf{h}_t$. $\mathbf{G}_t$ is a 52×52 matrix of parameters that represent this transfer from the stand state after harvest, $\mathbf{y}_t - \mathbf{h}_t$, to the future stand state $\mathbf{y}_{t+5}$.

The matrix $\mathbf{G}_t$ consists of the probability that trees of a given species stay in the same size class from t to $t + 5$, move up one size class, or die. The subscript t of $\mathbf{G}_t$ indicates that the matrix is a function of the stand state at time t. For example, the probability that a tree moves up a size class decreases with the stand density after harvest, while the probability that a tree dies increases with the stand density.

The vector $\mathbf{R}_t$ describes the recruitment, the number of trees of a particular species that enters the smallest size class from t to $t + 5$. As indicated by the subscript t, $\mathbf{R}_t$ is also state dependent, a function of $\mathbf{y}_t - \mathbf{h}_t$. The state dependence of the growth matrix $\mathbf{G}_t$ and recruitment vector $\mathbf{R}_t$ make the model nonlinear in this application (see 'Appendix' for the specific equations of the model).

After estimating the model parameters with data from 5,108 sample plots, the model was tested on the remaining 416 plots. The projections were done plot by plot over 10 years, i.e. with two iterations of model (2) to predict the plot state at $t + 10$ given the state at t and any eventual harvest. The results in Figure 17.4 suggest that the model predicted well the stand growth over

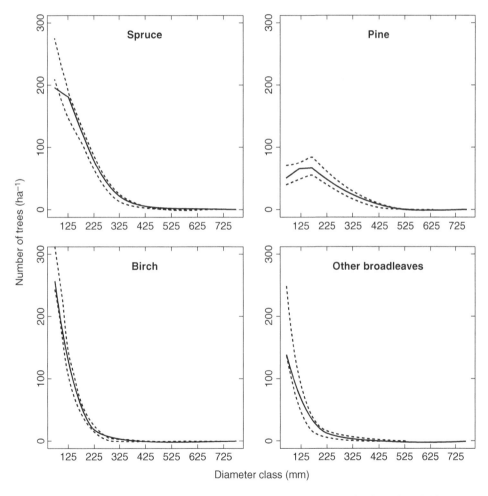

Figure 17.4 Predicted and observed distribution of trees over 416 post-sample plots. The solid line is the average distribution predicted after 10 years with the growth model. The dotted lines indicate the 95% confidence interval of the number of trees observed on the plots.

Source: Bollandsås et al. (2008).

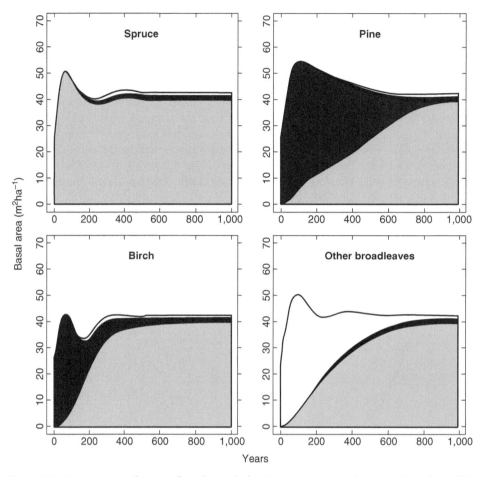

Figure 17.5 Long-term predictions of stand growth showing a common steady state independent of the initial species composition of the stand. Spruce = light gray, pine = dark grey, birch = black, other broadleaves = white.

Source: Bollandsås et al. (2008).

10 years, as the mean predicted number of trees by species and size class tended to fall within the 95% confidence interval of the observed mean number of trees.

However, 10 years is a short time in the life of a forest; thus, another, more theoretical validation test was also executed. The hypothesis was that over a very long time period and without disturbance, a stand on a particular site should converge to a 'climax' state independent of the initial condition. Figure 17.5 shows that for stands initially dominated by spruce, pine, birch or other hardwoods, the model did in fact predict that the long-term stand would be the same, dominated by spruce, with few birches or other broadleaves and practically no pine.

Management policies

As the growth model was deemed acceptable, it was used to predict the consequences of management and to seek the best management according to the criteria laid out previously and for

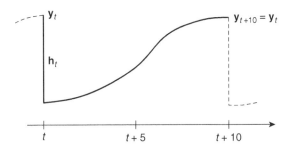

Figure 17.6 In a sustainable management regime, the stand growth over the cutting cycle t to $t + 10$ just replaces the harvest, $\mathbf{h}_t$, and thus restores the stand in a state $\mathbf{y}_{t+10}$ identical to the pre-harvest state $\mathbf{y}_t$.

specific policies. All policies considered exclusively sustainable management regimes. As illustrated in Figure 17.6, this implies that the growth over the cutting cycle, assumed to be 10 years, just replaces the harvest $\mathbf{h}_t$, so that the stand state at the end of the cycle is exactly the same as at the beginning, and the periodic harvest can continue in perpetuity at level $\mathbf{h}_t$.

Within this sustainability constraint, five policies were examined. Two may be viewed as extreme. One was purely timber-oriented and meant to maximize the NPV of the harvest over an infinite horizon. The other ignored timber revenues totally and tried to maximize the amount of CO_2e maintained in the stand.

Two other policies were compromises between timber production and CO_2 storage. One maximized the NPV of timber revenues with a constraint on the amount of CO_2e maintained in the stand. The other symmetrically maximized the CO_2 stored with a constraint on the NPV of timber revenues.

A last policy assumed that there was an active market for CO_2, analog to the timber market, and that the policy was to maximize the NPV from timber and CO_2 revenues combined.

The remainder of the chapter examines the formulation of these policies as optimization problems, and their consequences in terms of financial returns, tree species and size diversity, and carbon sequestration.

Maximizing timber revenues

The harvest, $\mathbf{h}_t$, and the corresponding growing stock, $\mathbf{y}_t$, that maximized the NPV from timber revenues only, in the steady state, without carbon storage limitations, can be found by solving the following optimization problem:

$$\max_{\mathbf{h}_t, \mathbf{y}_t} NPV_h = \mathbf{vh}_t + \frac{\mathbf{vh}_t}{(1+r)^{5n} - 1} - \mathbf{vy}_t \tag{3}$$

Subject to:

$$\mathbf{y}_{t+5} = \mathbf{G}_t(\mathbf{y}_t - \mathbf{h}_t) + \mathbf{R}_t \qquad t = 0, 5, \ldots, n-1 \tag{4}$$

$$\mathbf{y}_{t+5n} = \mathbf{y}_t \tag{5}$$

$$0 \le \mathbf{h}_t \le \mathbf{y}_t \tag{6}$$

where $\mathbf{v}$ is the row vector of tree value for timber alone, by species and size, and NPV_h is the NPV of the timber harvests, in a steady-state regime over an infinite horizon, given the net value of the periodic timber harvest, **vh**$_t$, occurring every $5n$ years, given the annual interest rate, r, and the initial investment in the growing stock valued as timber only, **vy**$_t$. NPV_h is a linear function of **h**$_t$ and **y**$_t$. However, as pointed out previously, the growth model (4) is nonlinear, and the nonlinearity of the constraints increases with the length of the cutting cycle due to the successive iterations (4). The steady-state constraint (5) means that the stand state at the beginning of the cutting cycle must be the same as at the end, before harvest. The constraints (6) insure the feasibility of the solution, as the harvest cannot be negative and must be less than the stock.

Solution of this nonlinear programming problem requires care to avoid a local optimum. That a global optimum has been reached can be ascertained by varying the initial conditions of the optimization. Table 17.1 shows the solution of the problem (3) to (7) with current prices and costs, a real interest rate of 3% per year, and a cutting cycle of 20 years. The steady-state stock contains trees of all species, but mostly spruce and broadleaves, and only a few pines.

The largest trees are spruce reaching 425 mm in diameter. The harvesting rule is very simple, consisting essentially of diameter limit cuts. For example, for spruce, all the trees in diameter class 275 mm and above and about half the trees in diameter class 225 mm are cut every 20 years. All birches in diameter class 125 mm and above are cut, and so are all the other broadleaves in diameter class 175 mm and above and the few pines in diameter class 225 mm and above.

Table 17.2 shows the consequences of this management regime and for a 5-year cutting cycle for the criteria considered in this study. The maximum NPV is 10% lower with the longer

Table 17.1 Sustainable stock and harvest that maximize the net present value of timber revenues with a 20-year cutting cycle and without constraint on carbon sequestration.

		Diameter class (mm)										
		75	12	17	22	27	32	37	42	47	52	57
		Trees/ha										
Spruce	Stock	16	13	12	10	77	40	12	1			
	Harvest				52	77	40	12	1			
Pine	Stock	2	2	2	1	1						
	Harvest				1	1						
Birch	Stock	61	33	10	2							
	Harvest		33	10	2							
Other	Stock	84	54	32	14	4						
	Harvest			32	14	4						

Table 17.2 Sustainable stand characteristics for a management regime that maximizes the net present value of timber without constraint on carbon sequestration. Diversity indices are relative to their maximum theoretical value.

Criteria	Cutting cycle	
	5-year	20-year
NPV ($/ha)	6,800	6,100
CO_2e stored (t/ha)	162	196
Stock value ($/ha)	3,300	4,500
Harvest (m³/ha/yr)	6.2	6.3
Tree species diversity	0.3	0.4
Tree size diversity	0.6	0.7

cutting cycle, due in large part to the larger investment in growing stock that it implies. But the higher growing stock leads to more CO_2e being stored with the 20-year cutting cycle. The average annual timber production is about the same for both cutting cycles, and the species and tree size diversity are also very close.

Maximizing CO_2 sequestration

At the opposite extreme of maximizing NPV from timber production only, we can find the sustainable management regime that maximizes the sequestration of CO_2e per unit area by solving the following optimization problem:

$$\max_{\mathbf{h}_t, \mathbf{y}_t} C = \mathbf{c}(\mathbf{y}_t - \mathbf{h}_t) \tag{7}$$

Subject to:

$$\mathbf{y}_{t+5} = \mathbf{G}_t(\mathbf{y}_t - \mathbf{h}_t) + \mathbf{R}_t \qquad t = 0, 5, \ldots, n-1 \tag{8}$$

$$\mathbf{y}_{t+5n} = \mathbf{y}_t \tag{9}$$

$$0 \le \mathbf{h}_t \le \mathbf{y}_t \tag{10}$$

The objective function (7) is the amount of CO_2e maintained in the stand after harvest, where $\mathbf{c}$ is the vector of CO_2e sequestered in trees of different species and size (Bollandsås, Buongiorno and Gobakken, 2008). The constraints (8) to (10) express the same growth equations, steady state and feasibility conditions as constraints (4) to (6) in the previous problem.

The solution, which is essentially the same for a 5-year or 20-year cutting cycle, is in Table 17.3. It shows that the stand that maximizes CO_2 storage per unit area consists almost purely of spruce trees. Some of them are very large, reaching above 600 mm in diameter. There is practically no pine, and the few birches and other broadleaves are removed as soon as they appear in the stand; this is the only management being applied.

Table 17.4 shows the consequences of this management for the criteria considered in this study. There is a very high financial cost due to the high level of growing stock investment

Table 17.3 Sustainable stock and harvest that maximize carbon sequestration for a 5-year or 20-year cutting cycle.

		Diameter class (mm)											
		75	125	175	225	275	325	375	425	475	525	575	625
		Trees/ha											
Spruce	Stock	264	127	85	67	57	51	47	43	38	30	18	5
	Harvest												
Pine	Stock												
	Harvest												
Birch	Stock	4											
	Harvest	4											
Other	Stock	4											
	Harvest	4											

Table 17.4 Sustainable stand characteristics for a management regime that maximizes carbon sequestration with a 5-year or 20-year cutting cycle. Diversity of tree species and size is a fraction of the theoretical maximum.

Criteria	Value
NPV ($/ha)	−22,500
CO_2e stored (t/ha)	630
Stock value ($/ha)	22,400
Harvest vol (m^3/ha/yr)	0.04
Tree species diversity	0.01
Tree size diversity	0.92

and the lack of any substantial harvest. There is also a very high ecological cost, as the species diversity is practically nil because the stand consists almost entirely of spruce trees. On the other hand, the diversity of tree size is near maximum due to the presence of spruce trees in all the diameter classes.

Maximizing timber production with CO_2 sequestration

The following model can be used to reach a compromise between timber production and CO_2 sequestration:

$$\max_{\mathbf{h}_t, \mathbf{y}_t} NPV = \mathbf{vh}_t + \frac{\mathbf{vh}_t}{(1+r)^{5n} - 1} - \mathbf{vy}_t \tag{11}$$

Subject to:

$$\mathbf{y}_{t+5} = \mathbf{G}_t \left(\mathbf{y}_t - \mathbf{h}_t\right) + \mathbf{R}_t \qquad t = 0, 5, \ldots, n-1 \tag{12}$$

$$\mathbf{y}_{t+5n} = \mathbf{y}_t \tag{13}$$

$$0 \le \mathbf{h}_t \le \mathbf{y}_t \tag{14}$$

$$\mathbf{c}(\mathbf{y}_t - \mathbf{h}_t) \ge C^\star \tag{15}$$

This is the same as model (3) to (6), except for the addition of constraint (15) which maintains the amount of CO_2e stored in the stand at a specified level, $C^\star$. Figure 17.7a shows the maximum NPV obtained from timber revenues at $C^\star$ varying from 0 to 630 t/ha of CO_2e, the maximum possible sequestration level.

According to Figure 17.7a, up to 200 t/ha of CO_2e can be stored in the stand at no cost in terms of foregone timber revenues. Beyond this point, the opportunity cost of carbon sequestration increases at an increasing rate. At about 450 t/ha NPV reaches 0, at which point the present value of timber revenues just balances the investment in growing stock. Viewed another way, the internal rate of return on the capital in growing stock is equal to the interest rate, 3% per year in this case. Further increases in carbon sequestration lead to negative NPV.

The derivative of the NPV curve in Figure 17.7a gives the curve of stored CO_2e as a function of its price, i.e. the supply schedule (Figure 17.7b). Up to 200 t/ha CO_2e is stored at zero price. Within that range, carbon storage is a costless byproduct of timber production. Then, the supply of CO_2e increases linearly with the price, up to 450 t/ha. Beyond that point, the supply

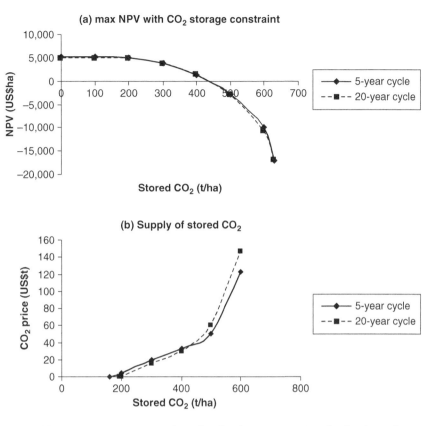

Figure 17.7 (a) Maximum net present value of timber harvests at various levels of stored CO_2e and (b) supply of stored CO_2e as a function of the price of CO_2e.

Source: Adapted from Buongiorno et al. (2012).

curve is much steeper, as higher prices induce less and less additional storage of CO_2e. There is little practical difference in the supply curves derived with cutting cycles of 5 or 20 years.

Maximizing returns from timber and CO_2 sequestration

Assuming the existence of a market for CO_2 as active as the market for timber, the model can be modified to maximize the NPV from combined timber production and the sequestration of CO_2e in uneven-aged stands. The model formulation becomes:

$$\max_{\mathbf{h}_t, \mathbf{y}_t} NPV = \mathbf{v}\mathbf{h}_t + \frac{\mathbf{v}\mathbf{h}_t}{(1+r)^{5n} - 1} - \mathbf{v}\mathbf{y}_t + q\mathbf{c}(\mathbf{y}_t - \mathbf{h}_t) \tag{16}$$

Subject to:

$$\mathbf{y}_{t+5} = \mathbf{G}_t(\mathbf{y}_t - \mathbf{h}_t) + \mathbf{R}_t \qquad t = 0, 5, \ldots, n-1 \tag{17}$$

$$\mathbf{y}_{t+5n} = \mathbf{y}_t \tag{18}$$

$$0 \leq \mathbf{h}_t \leq \mathbf{y}_t \tag{19}$$

The only change with respect to model (3) to (6) is in the objective function (16). To the present value of the perpetual timber income, net of the initial investment in growing stock, is now added the return from the carbon maintained after harvest, $q\mathbf{c}(\mathbf{y}_t - \mathbf{h}_t)$, where q is the price of CO_2e.

Table 17.5 shows the solution with a 20-year cutting cycle and a price of CO_2e of $110/t, a price projected by the International Energy Agency for the year 2030 (IEA, 2009). In contrast with the management that maximized carbon sequestration only, this management maintains a diverse stand of spruce, birch and other hardwoods, but still no pine. As in the other managements obtained previously, the prescription is quite simple. Every 20 years, all the spruce trees in diameter class 475 mm and above are harvested, together with birches in diameter class 175 mm and above, and the other broadleaves in diameter class 325 mm and above.

The consequences for some management criteria of the solutions obtained at varying prices of CO_2e are in Table 17.6. The amount of CO_2e stored in the stand almost triples as the price of CO_2e increases from 0 to $110/t. However, the species diversity becomes three times lower, as expected, because, as observed previously, the stand of maximum carbon storage is a pure spruce stand. Meanwhile, the tree size diversity increases, also as expected, because the highest carbon sequestration occurs in stands with large spruce trees. The maximum NPV of the combined timber production and carbon storage increases more than eight times as the price of CO_2e increases from 0 to $110/t. At a price of $15/t, nearly 65% of the NPV still comes from timber revenues. But at a price of $50/t, most of the NPV comes from carbon storage, and at higher prices the timber aspect of production bears negatively on total NPV as the value of the stock as timber is much higher than the value of the timber output. An unexpected result is the simultaneous increase of the annual average timber harvest and of the amount of carbon sequestered for up to about $50/t of CO_2e.

Table 17.5 Effects of the price of CO_2e on management criteria when maximizing the net present value of timber and carbon sequestered. Diversity of tree species and size is a fraction of the theoretical maximum.

		Diameter class (mm)										
		75	125	175	225	275	325	375	425	475	525	575
Spruce	Stock	232	140	106	88	76	69	62	57	38	14	2
	Harvest	0	0	0	0	0	0	0	0	38	14	2
Pine	Stock	0	0									
	Harvest	0	0									
Birch	Stock	22	12	5	1							
	Harvest	0	0	5	1							
Other broadleaves	Stock	22	9	5	4	3	2	1				
	Harvest	0	0	0	0	0	2	1				

Table 17.6 Effects of the price of CO_2e on management criteria when maximizing the net present value of timber and carbon sequestered. Diversity of tree species and size is a fraction of the theoretical maximum.

CO_2e price ($/t)	CO_2e stored (t/ha)	Species diversity	Size diversity	NPV from timber and CO_2e ($/ha)	NPV from timber alone ($/ha)	Harvest (m³/ha/yr)
0	177	0.30	0.72	6,300	6,300	6.1
15	237	0.29	0.73	9,600	6,100	7.0
50	421	0.16	0.80	21,800	700	7.6
110	556	0.10	0.85	52,300	−8,800	5.6

Conclusion

The results obtained in this study suggest that even in contexts where it has been scarcely practiced, as in Norway, uneven-aged management of mixed species stands can be sustainable and financially rewarding. Short cutting cycles lead to higher NPVs, while longer cutting cycles induce more carbon sequestration and higher diversity of stand species and size.

Regardless of objectives, the management regimes suggested by the optimization models turn out to be simple diameter limit cuts. Naturally, this would have to be applied prudently and supplemented by silvicultural prescriptions such as the removal of cull trees to avoid high grading and the inducement of poor-quality stands.

The pursuit of maximum carbon sequestration leads to high financial losses because it implies practically no harvest. It is also ecologically unattractive, as it induces almost pure spruce stands of very low species diversity. Thus, regardless of financial consideration, there is a definite conflict between the goal of carbon storage and that of stand diversity, with negative implications for total ecosystem diversity, wildlife habitat and aesthetics of forests.

From a purely economic point of view, the results suggest that the first 200 t/ha of stored CO_2e have little cost; within that range there is no conflict between carbon sequestration and timber revenues. In fact, both carbon sequestration and annual timber output increase for prices of CO_2e of up to \$50/t.

Consequently, as the recent prices of CO_2e have been less than \$15/t, it appears that there is currently little conflict between timber supply and CO_2 sequestration. Indeed, the participation of forest landowners in carbon sequestration programs is still limited (Dickinson, Sevens, Lindsay and Kittredge, 2012; Pajot, 2011). However, some projections, such as IEA (2009), predict that the price of CO_2e may increase to \$50/t by 2020 and to \$110/t in 2030. Synapse Energy Economics Inc. (2011) forecasts are more modest, but still increase to \$80/t by 2030. If prices of CO_2e and attendant carbon offset payments were to reach those levels, then a conflict between timber production and carbon sequestration would indeed appear. At that point the methods presented in this chapter would be useful in designing optimum forest management policies in this new environment.

One limitation of this study is that it is deterministic. As pointed out previously, the future price of CO_2 is highly uncertain, and so is the price of wood, although to a lesser extent, as wood markets are more established. Furthermore, the growth of forests is itself subject to risk due to climate variations, insects and disease attacks and catastrophic events such as fires and hurricanes. A useful field of future research would then be to determine optimal decisions in such a risky environment. In this respect, Markov decision process models, which have already been applied widely in forestry to deal with risk in markets and stand growth, with discounted economic objectives or undiscounted ecological objectives (Zhou and Buongiorno, 2006; Zhou, Buongiorno and Liang, 2008), should be a promising approach to developing adaptive decisions to store carbon in forests while producing wood efficiently and sustaining forests of high ecological diversity.

Acknowledgments

This chapter is a synthesis and summary of material originally published in two articles in the *Scandinavian Journal of Forest Research*, Bollandsås et al. (2008) and Buongiorno, Halvorsen, Bollandsås, Gobakken and Hofstad (2012). This research and the preparation of this chapter were supported in part by the US Department of Agriculture Forest Service Southern Research Station through a cooperative research agreement with Joseph Buongiorno.

References

Asante, P. and Armstrong, G. W. (2012). 'Optimal forest harvest age considering carbon sequestration in multiple carbon pools: A comparative statics analysis', *Journal of Forest Economics*, vol. 18, pp. 145–156.

Bollandsås, O., Buongiorno, J. and Gobakken, T. (2008). 'Predicting the growth of stands of trees of mixed species and size: A matrix model for Norway', *Scandinavian Journal of Forest Research*, vol. 23, pp. 167–178.

Boscolo, M., Buongiorno, J. and Panayotou, T. (1997). 'Simulating options for carbon sequestration through improved management of a lowland tropical rainforest', *Environmental and Development Economics*, vol. 2, pp. 241–263.

Buongiorno, J., Halvorsen, E. A., Bollandsås, O. M., Gobakken, T. and Hofstad, O. (2012). 'Optimizing management regimes for carbon storage and other benefits in uneven-aged stands dominated by Norway spruce, with a derivation of the economic supply of carbon storage', *Scandinavian Journal of Forest Research*, vol. 27, pp. 460–473.

Buongiorno, J. and Michie, B. (1980). 'A matrix model of uneven-aged forest management', *Forest Science*, vol. 26, no. 4, pp. 609–625.

Daineault, A. J., Miranda, M. J. and Sohngen, B. (2010). 'Optimal forest management with carbon sequestration credits and endogenous fire risk', *Land Economics*, vol. 86, no. 1, pp. 155–172.

Dickinson, B. J., Sevens, T. H., Lindsay, M. M. and Kittredge, D. B. (2012). 'Estimated participation in U.S. sequestration programs: A study of NIPF landowners in Massachussetts', *Journal of Forest Economics*, vol. 18, pp. 36–46.

Hinzman, L. D., Bettez, N. D., Bolton, R. W., Chapin, F. S., Dyurgerov, M. B., Fastie, C. L., . . . Yoshikawa, K. (2005). 'Evidence and implications of recent climate change in northern Alaska and other arctic regions', *Climatic Change*, vol. 72, no. 3, pp. 251–298.

Hoen, H. F. and Solberg, B. (1994). 'Potential and economic efficiency of carbon sequestration in forest biomass through silvicultural management', *Forest Science*, vol. 40, no. 3, pp. 429–451.

Ingram, D. and Buongiorno, J. (1996). 'Income and diversity tradeoffs from management of mixed lowland dipterocarps in Malaysia', *Journal of Tropical Forest Science*, vol. 9, no. 2, pp. 242–270.

Intergovernmental Panel on Climate Change (IPCC). (2012). *Meeting Report of the Intergovernmental Panel on Climate Change Expert Meeting on Economic Analysis, Costing Methods, and Ethics*, C. B. Field, V. Barros, O. Edenhofer, R. Pichs-Madruga, Y. Sokona, M. D. Mastrandrea, . . . C. von Stechow (Eds.), IPCC Working Group II Technical Support Unit, Carnegie Institution, Stanford, CA.

International Energy Agency (IEA). (2009). *How the Energy Sector Can Deliver on a Climate Agreement in Copenhagen*, Special early excerpt of the World Energy Outlook 2009 for the Bangkok UNFCCC meeting, IEA.

Liang, J., Buongiorno, J., and Monserud, R. A. (2005). 'Growth and yield of all-aged Douglas-fir–western hemlock forest stands: a matrix model with stand diversity effects. *Forest Science*, vol. 35, pp. 2368–2381.

National Oceanic and Atmospheric Administration (NOAA). (2012). 'Arctic report card: Update for 2012, tracking recent environmental changes'. Retrieved from www.arctic.noaa.gov/reportcard/index.html

Olschewski, R. and Benitez, P. C. (2002). 'Optimizing joint production of timber and carbon sequestration of afforestation projects', *Journal of Forest Economics*, vol. 16, pp. 1–10.

Pajot, G. (2011). 'Rewarding carbon sequestration in South-Western French forests: A costly operation?' *Journal of Forest Economics*, vol. 17, pp. 363–377.

Pukkala, T. and von Gadow, K. (Eds.). (2012). *Continuous Cover Forestry* (2nd ed.), Springer, Dordrecht, The Netherlands.

Shannon, C. E. and Weaver, W. (1963). *The Mathematical Theory of Communication*, University of Illinois Press, Urbana, IL.

Synapse Energy Economics Inc. (2011). *2011 Carbon Dioxide Price Forecast*, Synapse Energy Economics Inc., Cambridge, MA.

Van Kooten, G. C., Binkley, C. S. and Delcourt, G. (1995). 'Effect of carbon taxes and subsidies on optimal forest rotation age and supply of carbon services', *American Journal of Agricultural Economics*, vol. 77, pp. 365–374.

World Bank. (2011). *State and trends of the carbon market 2011*. Environment Department. World Bank, Washington DC.

Zhou, M. and Buongiorno, J. (2006). 'Forest landscape management in a stochastic environment, with an application to mixed loblolly pine-hardwood forests', *Forest Ecology and Management*, vol. 223, pp. 170–182.

Zhou, M., Buongiorno, J. and Liang, J. (2008). 'Economic and ecological effects of diameter caps: A Markov decision model of Douglas-fir/western hemlock forests', *Forest Science*, vol. 54, no. 4, pp. 397–407.

Zickfeld, K., Arora, V. K. and Gillett, N. P. (2012). 'Is the climate response to CO2 emissions path dependent?' *Geophysical Research Letters*, vol. 39, L05703.

Appendix

The matrix **G** and the vector **R** in the growth model (2) have the following form (Bollandsås et al., 2008):

$$\mathbf{G} = \begin{bmatrix} \mathbf{G}_1 & & & \\ & \mathbf{G}_2 & & \\ & & \mathbf{G}_3 & \\ & & & \mathbf{G}_4 \end{bmatrix}, \mathbf{G}_i = \begin{bmatrix} a_{i1} & & & & \\ b_{i1} & a_{i2} & & & \\ & \ddots & \ddots & & \\ & & b_{i,17} & a_{i,18} & \\ & & & b_{i,18} & a_{i,19} \end{bmatrix}$$

$$\mathbf{R} = \begin{bmatrix} \mathbf{R}_1 \\ \mathbf{R}_2 \\ \mathbf{R}_3 \\ \mathbf{R}_4 \end{bmatrix}, \qquad \mathbf{R}_i = \begin{bmatrix} R_i \\ 0 \\ \vdots \\ 0 \end{bmatrix}$$

where a_{ij} is the probability that a tree of species i and diameter class j stays alive and in the same diameter class between t and $t + 5$ years; $i = 1$ for Norway spruce, 2 for Scots pine, 3 for birch and 4 for other broadleaves; b_{ij} is the probability that a tree of species i and diameter class j stays alive and grows into diameter class $j + 1$ and R_i is the number of trees per ha of species group i recruited in the smallest diameter class between t and $t + 5$.

The b_{ij} probability is equal to the 5-year tree diameter growth, g_{ij}, in mm, divided by the width of the diameter class, 50 mm. Diameter growth is a function of tree diameter D_j (mm), stand basal area of tree larger than class j, BAL_j (m²ha⁻¹), site index SI (m), total stand basal area BA and latitude LAT (degrees and minutes north, minutes scaled by 100/60 so that e.g. 50°30′ north = 60.5°):

$$g_{1j} = 17.839 + 0.0476D_j - 11.585\mathrm{E} - 5D_j^2 - 0.3412BAL + 0.906SI$$
$$+ -0.024BA - 0.268LAT$$

$$g_{2j} = 25.543 + 0.0251D_j - 5.660\mathrm{E} - 5D_j^2 - 0.216\ BAL + 0.698SI$$
$$- 0.123BA - 0.336LAT$$

$$g_{3j} = 11.808 + 9.616\mathrm{E} - 5D_j^2 - 9.585\mathrm{E} - 8D_j^3 + 0.519SI - 0.152BA$$
$$- 0.161LAT$$

$$g_{4j} = 2.204 + 0.0631D_j - 8.320D_j^2 + 0.359SI - 0.177BA$$

The expected recruitment per ha of species i, R_i, is the product of the probability of recruitment in 5 years, π_i, and of the expected recruitment if there is any, CR_i. The probability of recruitment depends on the stand basal area, the site index and the percent of basal area in species i, PBA_i, according to the logistic equations:

$$\pi_1 = (1 + e^{2.291+0.018BA-0.066SI-0.019PBA_1})^{-1}$$

$$\pi_2 = (1 + e^{3.552+0.062BA-0.031PBA_2})^{-1}$$

$$\pi_3 = (1 + e^{0.904+0.037BA-0.016PBA_3})^{-1}$$

$$\pi_4 = (1 + e^{-3.438+0.029BA-0.123SI-0.048PBA_3})^{-1}$$

And the expected recruitment per ha in 5 years, conditional on positive recruitment, is a function of the same variables, according to:

$$CR_1 = 43.142BA^{-0.157}SI^{0.368}PBA^{0.051}$$

$$CR_2 = 67.152BA^{-0.076}$$

$$CR_3 = 64.943BA^{-0.161}SI^{0.143}PBA^{0.104}$$

$$CR_4 = 31.438BA^{-0.1695}SI^{0.442}PBA^{0.193}$$

The probability of tree mortality per year, m_{ij}, is a species-dependent function of tree size and stand basal area:

$$m_{1j} = (1 + e^{2.492+0.020D_j-3.2E-5D_j^2-0.031BA})^{-1}$$

$$m_{2j} = (1 + e^{1.808+0.027D_j-3.3E-5D_j^2-0.055BA})^{-1}$$

$$m_{3j} = (1 + e^{2.188+0.016D_j-2.7E-5D_j^2-0.030BA})^{-1}$$

$$m_{4j} = (1 + e^{1.551+0.011D_j-1.4E-5D_j^2-0.016BA})^{-1}$$

The probability that a tree stays alive and in the same size class from t to $t + 1$ is, then: $a_{ij} = 1 - b_{ij} - m_{ij}$.

18

ECONOMICS OF FOREST BIOMASS-BASED ENERGY

Pankaj Lal[*][1] *and Janaki R. R. Alavalapati*[2]

[*]CORRESPONDING AUTHOR

[1]ASSISTANT PROFESSOR, EARTH AND ENVIRONMENTAL STUDIES, MONTCLAIR STATE UNIVERSITY, MONTCLAIR, NJ, USA. 973-655-3137 (PHONE), 973-655-4072 (FAX), LALP@MAIL.MONTCLAIR.EDU

[2]PROFESSOR AND HEAD, DEPARTMENT OF FOREST RESOURCES AND ENVIRONMENTAL CONSERVATION, COLLEGE OF NATURAL RESOURCES AND ENVIRONMENT, VIRGINIA TECH, BLACKSBURG, VA, USA.

Abstract

Woody bioenergy is an option to reduce dependency on fossil fuels, increase the share of renewable energy and improve the sustainability of forests. Forest economists have analyzed various issues, including economic and social consequences of dramatic increases in growing, harvesting and utilizing forest feedstocks for energy purposes and the ensuing impacts on family forest landowners and other key stakeholders along the supply chains as well as the society at large. We review the problems, applications of economic techniques and ensuing results and outcomes from the literature, focusing more on the ways in which bioenergy issues have been approached by economists, the tools and techniques that have been applied, assumptions therein and the ensuing results. Some of the key research areas, such as biomass supply, public preferences for woody bioenergy, competition with traditional forest industries, land-use change and greenhouse gas emissions, are highlighted. The use of multiple approaches and tools with varied focus, planning periods, feedstock considerations, models and parameter assumptions results in outputs and outcomes that might not be amenable for comparison; some examples are biomass supply estimates, price premiums for woody bioenergy and extension of rotation age to account for 'carbon saving'. The woody bioenergy markets offer a plethora of research opportunities.

Keywords

Biomass supply, land-use change, willingness to pay, emissions, public preferences

Introduction

Forest economists have applied economic principles and techniques to study bioenergy and have made forays into research areas like rotation age, life cycle analyses and greenhouse gas (GHG) emissions, willingness to pay (WTP) for bioenergy, carbon sequestration, socioeconomic availability and supply, trade-offs with existing forest industries, land-use change, fire and pest risks, trade, biodiversity, soils, hydrology and so forth. The most prolific of these efforts has been to examine bioenergy production as new 'forest biomass for energy' markets. This chapter addresses economics of forest bioenergy assessments. We review the problems, applications of economic techniques, and ensuing results and outcomes from the literature, focusing more on the ways in which bioenergy issues have been approached by economists, the tools and techniques that have been applied, assumptions therein and the ensuing results. We focus less on the methodological aspects and rely more on comparative results and outcomes.

Sustainable production and use of bioenergy can increase energy security, reduce dependence on foreign oil, decrease GHG emissions (Hill, Nelson, Tilman, Polasky and Tiffany, 2006) and enhance socioeconomic benefits to rural communities. It could be produced from multiple forest feedstocks, including energy crops, logging residue, mill residue, wood boles and urban wood waste. Products from forest biomass may include biopower, liquid fuels and bioproducts. Burning biomass produces electricity through a steam turbine in power plants. This could be done through dedicated cellulosic power plants, co-firing biomass with coal in existing plants, and developing combined heat and power plants (Alavalapati et al., 2009). Biobutanol and ethanol can be used in engines as substitutes for gasoline and can also be combined with gasoline, with E10 and E85 being popular combination ratios. A well-marketed bioproduct is wood pellets. As such, pellet sales have risen considerably over the past years, largely in response to European demand (Junginger et al., 2008).

The rest of the chapter is organized as follows. First, we discuss forest biomass supply assessments, focusing on some of their differences and similarities. We then turn to public preferences and forest biomass-based energy in the second section. In the third section we focus on the competitive context of forest bioenergy vis-à-vis traditional forest industries such as pulpwood and sawtimber. The fourth section is focused toward forest bioenergy and economics of land-use change and GHG emissions, and finally, in the last section we briefly provide our perspective for future research.

Forest biomass supply

Forest economists have applied economic approaches to improve upon the biophysical estimates of forest biomass availability and supply. Designing policies to aid woody bioenergy development and investment decisions by enterprising agents relies considerably on an estimate of biomass availability. The literature on biomass availability estimates reveals a host of approaches with differences in assumptions, including those on yield and land use, feedstock considerations, spatial and temporal scale, technological considerations and policy settings. The biophysical approach takes a static view of the inventory of available biomass resources and forest growth and has been used by researchers like Perlack et al. (2005), Milbrandt (2005) and Alavalapati et al. (2009). These studies have used Forest Inventory and Analysis (FIA) data on current and historic estimates. The historical growth per unit area (i.e. acre) is estimated generally in term of forest types and ownership and is estimated based on the basic growth relationship represented by:

$$RG_{ijk} = \sum_{i=1}^{N} \frac{G_{ijk}}{AC_{ijk}} AC_{ijk}$$

where i represents one of the geographic regions, j represents one of the forest types, k is two ownership categories (e.g. corporate or noncorporate), N represents number of geographic regions, RG represents regional growth and AC represents area (number of acres). The information so gathered is fed into regional bioeconomic models discussed later in this section.

The results from these estimates can be scaled up from local to state and regional levels and altered for scenarios arrived at by changing rates of residual biomass retention, rotation age, harvest levels, planning period, thinning regimes and other stand management practices. The researchers develop biomass estimates of the amount of biomass available based on biomass attributes such as respective share of tops and limbs to total living biomass, along with biomass available as bole wood, stumps and roots. Existing biomass demand is accounted for and the excess biomass that is over and above the existing demand is estimated. The excess biomass so calculated can be diverted for energy production. Such considerations allow this approach to be consistent with existing industrial demand while developing biomass availability estimates.

The historic data of biomass availability, and information regarding terrain features and distance from roads, assist biorefineries in estimating biomass that can be made available from nearby regions. The spatial aggregation-based biomass county level estimates represented through geographic information system (GIS) maps have been especially useful. A case in hand is the county-level biomass potential map developed by Milbrandt (2005) for the contiguous United States. The assumptions of biophysical estimates by and large ignore significant disturbances (insect, disease, fire) and land-use change and assume constant harvest and management inputs. Though some of the assumptions have been contested by some economists, the biophysical approach provides valuable information in cases where information on production costs, social acceptability, legal and political considerations are not available.

Biophysical estimates, however, ignore the effect of economic and policy factors on supply. Economists have typically used information on stumpage, harvest and transportation to estimate the supply of biomass for bioenergy production. Historic data for a given planning period and results are sensitive to species type and equipment used. The fixed operating and labor costs are aggregated to develop a static supply curve that describes the cumulative amount of biomass that can be made available at differing prices. This approach thus enables a more reliable result compared to mere biophysical availability. Such estimates show a wide range, starting from simplistic short-term static supply curves and moving to large process-based models that estimate woody biomass supply across multiple time scales through linear programming-based optimizations (e.g. the Policy Analysis System [POLSYS] by De La Torre Ugarte, Ray and Tiller, 1998; and the Forest and Agriculture Sector Optimization Model [FASOM] by Adams, Alig, Callaway, McCarl and Winnett, 1996). Walsh (2008) and Kumarappan, Joshi and MacLean (2009) used short-term static supply curves to estimate woody biomass quantities that can be supplied at exogenously determined market prices. These static economic estimates are much more conservative than the biophysical ones, as they estimate only economically available forest biomass that can be diverted for bioenergy production.

The forest biomass supply depends on a multitude of factors such as future demand for woody bioenergy markets, advancements in conversion technologies, different product prices and elasticities, as well as productivity increases. The POLSYS and FASOM-based analyses project woody biomass supply under different policy and productivity scenarios. In the southern United States, the Subregional Timber Supply (SRTS) Model (Abt, Cubbage and Pacheco, 2000) has been applied in the forest bioenergy context to assess policies such as contribution of forest biomass toward meeting renewable fuel targets (e.g. Abt, Abt, Cubbage and Henderson, 2010; Galik, Abt and Yun, 2009), trade-offs with existing sawtimber and pulpwood industries, land-use change occurring due to development of forest biofuel markets (Alavalapati, Lal, Susaeta, Abt and Wear, 2013) and supply response in terms of planting forest bioenergy feedstocks in nonforested areas, as well as increased inputs to increase productivity on existing forestlands (Abt, Abt and Galik, 2012; Alavalapati, Lal, Susaeta, Abt and Wear, 2013).

Increased consumption of biomass by a new bioenergy industry can be expected to result in supply side adjustments such as expansion of short-rotation woody crops (e.g. eucalyptus and poplar) and increased productivity due to intensive management and genetic improvements. The increase in plantation area in the United States in the last few decades points toward the feasibility of biomass supply for energy production. Genetic modifications have also resulted in notable change in forest species type and abundance over the past few decades. In addition, recent years have seen increased investment in silviculture treatments such as thinning and fertilization (BRDI, 2008). However, these practices have raised questions about future forested landscapes and the ecological footprint of woody bioenergy markets. Lal (2011) analyzed supply response under productivity increase scenarios for the southern United States and found that increases in productivity result in lower levels of private forest acreage and prices.

Economists have factored in the environmental impact of feedstock harvests for bioenergy by moderating the supply estimates based on county-level soil erosion data (Walsh, 2008). Generally, the two-thirds harvest rule is suggested, with the objective of conserving biodiversity and maintaining ecological functions of forests (Perlack et al., 2005; Lal et al., 2011). However, the impact of the collection of residual biomass on soil organic matter content and moisture and erosion, and economic considerations such as the cost of replacing most solid nutrients through use of fertilizers in place of biomass residues and associated externalities, have not been appropriately addressed. The economic supply estimates mimic the market process that allocates resources; however, economic supply models face challenges of incorporating supplier return and risk into feedstock prices. The factors also raise effective minimum viable prices and change the supply estimates altogether. The removal of biomass depends not only on the biomass products demand, but also the actual amount the consumers are willing to pay for it. The ways by which federal and state policies and programs affect landowners' behavior adds to the complexity. One of the problems observed in forest biomass availability estimates is that most of the approaches model just one side of the market, the demand side or the supply side. Furthermore, the economic approaches do not necessarily say which forest landowners will supply biomass, at what price, how much they will supply and what socioeconomic characteristics can be used as predictors of their willingness to supply forest biomass for energy production.

Landowners' and public preferences for forest bioenergy

An understanding of landowners' preferences and their differences based on their characteristics is quite critical in understanding supply of forest biomass for energy. Generally, forest landowners are motivated by monetary as well as nonmonetary factors. Koontz (2001) argued that financial motivations led to choice of management practices geared toward maximizing potential timber harvest and regeneration potentials, whereas nonfinancial motivations tend to maximize the nonconsumptive use of the forest, focusing on aesthetic values and environmental stewardship of the forests. Butler et al. (2007) distinguished among four different types of landowners to identify their management preferences: those who own the land for retreat, those who work the land, those who use the land for supplemental income and those who are ready to sell their land. Specific to woody bioenergy, Majumdar, Teeter and Butler (2008) argued that the likelihood of supplying forest biomass for energy depends on attributes like forest management objectives, and differentiated between the landowners who managed their land for timber and nontimber production. Joshi and Arano (2009), based on a survey of landowners in West Virginia, found that landowner, ownership and management characteristics influence their forest management decisions.

Feedstock supply for energy use will be moderated by forest landowners' participation in the woody bioenergy markets; therefore, supplier preferences in terms of 'probability' of participation may be a useful economic measure. Paula, Bailey, Barlow and Morse (2011), based on a survey of Alabama forest owners, found that size of the woodland, active management and price offers acted as important indicators of willingness to harvest and that 61% of the respondents were willing to supply stumpage while 73% were willing to supply timber harvest residues for biofuels. Mehmood and Shivan (2010) found that for forest landowners in Florida, Arkansas and Virginia, woodland size, species composition, management objectives, education level and age of landowners acted as significant predictors. Results of a similar survey in Massachusetts suggested that a considerably lower percentage (about 17%) of foresters were willing to participate in forest bioenergy markets (Markowski-Lindsay et al., 2012).

Buchholz and Canham (2011) suggested that social factors are the least understood, yet potent, determinants of feedstock supply. Conrad, Bolding, Smith and Aust (2011) noted that landowners do not have different outlooks to committing their produce for traditional timber-based industries as compared to bioenergy markets. Markowski-Lindsay et al. (2012) found that determinants of landowners' willingness to supply to the same industry differ across states, as evidenced by their comparative analyses of Massachusetts and Minnesota landowners. Beach, Pattanayak, Yang, Murray and Abt (2005) suggested that size of forest and previous experience of landowners may be linked to their decision to use their land for bioenergy resource plantations. These results underline the presence of inter-regional variations in landowners' preferences and the problem of generalizing results.

Economists have also computed the 'extra' values attached to woody bioenergy factors like forest risk reduction, climate change modulation, lower susceptibility to pest and pathogens, ensuring reliable energy flow and security and decreased competition with food-based energy sources. For example, Li, Jenkins-Smith, Silva, Berrens and Herron (2009) found that people in the United States are willing to pay price premiums for renewable sources of energy. Roe, Teisl, Levy and Russell (2001) found that median WTP ranged between $0.38 and $5.66 per year for green electricity, leading to a 1% decrease in US GHG emissions. In Scotland, Bergmann, Hanley and Wright (2006) estimated that per-household WTP for renewable energy projects that do not increase air pollution was £14.03 per year. Though the general pattern shows similarities in term of positive WTP values, these studies have adopted a broader definition of renewable energy that includes bioenergy.

Soliño, Prada and Vázquez (2010) found that woody bioenergy, as compared to conventional energy, possesses desirable environmental outcomes and that people are willing to pay price premium. Soliño, Farizo, Vázquez and Prada (2012) used the choice experiment method to study how woody bioenergy contributes to reduced fire risk and found that people attach a marginal positive value to this attribute. They estimated an average WTP value of €38 on an annual basis to reach a 10% woody electricity generation goal. Neuwahl, Löschel, Mongelli and Delgado (2008) noted that among a portfolio of renewable alternatives, bioenergy feedstocks are valued higher by consumers, as they do not have dual demands in the food markets. Susaeta, Alavalapati, Lal, Matta and Mercer (2010) and Susaeta, Lal, Alavalapati and Mercer (2011) noted that the contribution of bioenergy to reduction of GHG emissions is the attribute that has been most studied in the last few years and found that individuals in the southern United States are willing to pay price premiums for both E10 and E85 blends.

Solomon and Johnson (2009), in order to assess the variations in WTP measured by different methods, used contingent valuation (CV) and fair share survey approaches. They focused on mitigation of global climate change attributes in the Midwestern United States and found that mean total WTP for cellulosic ethanol ranged between $252 and $556 per capita per year using the CV method as compared to the range of $192–$472 using the fair share approach. However, when they expanded the portfolio of alternative feedstocks to include energy crops, they found that the difference in terms of WTP was less significant.

Although most studies find an association between higher WTP and respondent income and educational level, the effect of location on WTP appears to be less conclusive. For example, Solomon and Johnson (2009) found that the geographic location of respondents in three states of the United States does not make a difference. Skipper et al. (2009) found significant differences in people's perceptions about the potential impacts of biofuels in the United States and Belgium. Hence, this may merit further investigation to identify if a uniform or region-specific approach would be a more effective policy to consider.

The commonly used stated preference approach, if implemented within the prescribed guidelines (NOAA Panel prescription), can produce reliable results (Arrow et al., 1993). The cost-based approach, on the other hand, determines the value of benefit streams from forests and estimates the amount of money it would take to maintain or replace forest goods and services. Within this context, replacement costs and defensive expenditure approaches find their relevance in terms of how people will pay to secure the continued benefit or pay to reinstate lost benefits that are bestowed on them by the forest resource.

Criticisms of such approaches in the context of bioenergy are not different than the ones that have been leveled broadly for forest resources. These include the argument that some functions, such as preservation, are priceless and that not all of the value people attach to forest resources can be monetized (Diamond and Hausman, 1994). Certain services are private (timber value) and others are public (flood protection), and aggregating such values on common currency terms may mean treating them as substitutes and additives (Kant, 1997). The WTP estimates are also susceptible to potential bias and moral judgment (Diamond and Hausman, 1994). For all these challenges, application of these approaches eases the trade-off involved in making choices. In addition, various aspects related to preferences have not been explored. These include people's beliefs regarding changes and/or variations, long-term or short-term expectations, determinants of consumer opinion about woody bioenergy policy and public opinion on woody bioenergy.

Competition from other industries, including other forest-based industries

Different forest-based industries compete for the same raw material, and the current forest growth and harvest rate may determine the synergistic and competitive process among these industries. The synergy between the bioenergy and traditional wood-based industry can be in terms of urban wood waste and mill and logging residues for energy production. However, woody bioenergy industry, in its nascent stage, has to compete for feedstock with an established traditional forest industry, and staying competitive may prove tasking. Furthermore, the supporting infrastructure for the woody bioenergy industry in North America has yet to be established. Ensuring efficient supply chain components may take some time. Although some strides have been made in innovation, affordable technologies have yet to surface (Dwivedi, Alavalapati and Lal, 2009). This partly takes away from enabling the bioenergy sector to be competitive. The incentives for biopower and biofuel are playing important roles in modulating the competitive outcome. These incentives render the bioenergy sector less elastic in relation to feedstock price changes compared to the sectors that do not benefit from these incentives.

There is a debate in the economic literature on the impact of the woody bioenergy industry and associated trade-offs. There are conflicting results concerning the socioeconomic impact of this industry. The argument in support of the woody bioenergy industry has been made by studies like that of Gan and Smith (2007), who estimated economic impacts of bioenergy in East Texas and found that benefits of utilizing logging residues for electricity production included the environmental benefits in term of CO_2 emissions displaced by substituting logging residues for coal in power generation, reduced site preparation costs during forest regeneration and increases in employment and income in local communities. In a computable general equilibrium (CGE) study simulating a policy scenario of increased woody biomass-based power generation in the state of Florida, Hodges, Stevens and Rahmani (2010) found the outcome to be a mixed bag. In a scenario where capital was assumed to be mobile and with a biomass supply level of 40 million tons, they found that forestry sector sales would increase by 69% while output of forestry sector manufacturing would decrease by 6.7%. The employees' income would increase by $1.61 billion, while state government tax revenues would increase by $108 million.

The expansion of woody bioenergy might push feedstock prices up and might work against bioenergy and competing industries alike (Galik et al., 2009). It is hard to determine the impact of the woody bioenergy industry on traditional forest industries with certainty; however, based on the simulation results using the SRTS model in the southern United States, it can be concluded that the impact is more pronounced for pulp-based industries than for sawtimber industries and that landowners get higher returns for their forest biomass (Alavalapati, Lal, Susaeta, Abt and Wear, 2013). White (2010) suggested that as feedstock prices increase (e.g. $25 to $40/dry ton), it is likely that milling residues would be drawn away from existing production along with more timber harvest residues, while biomass from energy crops such as poplar would likely be diverted for biofuels at moderate prices of less than $50 per ton. At high enough prices of biofuels, even merchantable timber can be diverted for biofuel production (Alavalapati, Lal, Susaeta, Abt and Wear, 2013; Alavalapati and Lal, 2009; White, 2010).

In terms of modeling the competitive context of forest-based industries, the profit function maximization approach has been used. In the Scandinavian context, which has established bioenergy markets, Ankarhem, Brännlund and Sjöström (1999) and Geijer, Bostedt and Brännlund (2011) modeled the effect of woody biomass for bioenergy markets on traditional forest industries. Susaeta, Lal, Carter and Alavalapati (2012) assessed the impact of bioenergy markets on forest landowners and found that an increased price of woody bioenergy would increase pulpwood production and demand for forestry labor. Susaeta, Lal, Alavalapati and Carter (2013) and Lal (2011) used a four-sector model composed of landowners, sawmills, pulpmills and the bioenergy industries and kept capital quasi-fixed and variable, respectively. Their results suggest that stumpage for the three industries act as substitutes, supporting the notion that higher diameter sawtimber or pulpwood can be diverted for bioenergy production at high enough prices.

Linear programming models have also been used to delineate the impact of bioenergy markets on traditional industries. Abt, Abt and Galik (2012) allowed for land competition (against growing other crops) vis-à-vis forest biomass that is supplied for traditional forest industries under exogenously determined price conditions. Alavalapati, Lal, Susaeta, Abt and Wear (2013) estimated diversion of biomass from existing sawtimber and pulpwood industries under different demand and productivity improvement scenarios in the southern United States for the next four decades. They found that productivity increases tend to lead to higher removals and lower displacement of biomass from forest industries for energy use.

Competition with sectors that sell energy to the same consumers also shapes the future of wood bioenergy. The price of substitutes such as fossil fuel might prove significant, given that its current and short-term projected production is limited and woody biofuels do not have the capacity to influence market price. The woody bioenergy market is likely to remain a price taker both in its feedstock and product markets. As such, this may affect the timing and extent of integration of advanced biofuels in the energy system. In addition, the forest bioenergy industry may also compete with other industries for other resources such as land. Huang, Alavalapati and Banerjee (2012), using a General Equilibrium model, suggested that policy instruments create new market opportunities for woody biomass and tend to increase the demand for woody bioenergy, resulting in land shifting from agricultural production to forestry production. Another set of models include the integration of GIS with optimization models. Examples of such studies include the least-cost biomass quantity estimates for Northern Spain (Panichelli and Gnansounou, 2008) and optimal allocation of a plant in Italian mountainous areas (Freppaz et al., 2004). Some researchers, such as De La Torre Ugarte, Walsh, Shapouri and Slinsky (2003), have analyzed bioenergy impacts by restraining the diversion of lands allocated for other purposes such as food crops, industrial demand and exports, and they found that at a price of $42.32/dry ton for willow and $43.87 for poplar, an estimated 28.95 million acres of agricultural, idle and pasture land in the United States could be diverted for bioenergy crop production.

The current state of the art in woody bioenergy forecasting is such that the research community focuses on multiple scenarios, and there is no incentive for the private sector to support research. Understanding the feasibility of the woody bioenergy industry and its impacts on other forest industries, local economies and vulnerable groups is important to understand the implications of new technologies.

Land-use change and GHG emissions

Growing woody feedstocks for bioenergy production can result in land-use changes that are direct (e.g. diverting existing forestland for growing feedstocks that can be used for energy production) and indirect (e.g. biomass production in one part of the world might result in land-use changes elsewhere). This change in land use and cover may have notable short-term and long-term effects that are ecological, biophysical, economic and social. Scholars have projected future trends of such impacts and have also analyzed the underlying processes to explain past impacts. Drummond and Loveland (2010) reported the declining agricultural land cover in the eastern United States as a result of conversion to forest, urbanization, mining and reservoir construction during the years 1973–2000.

Using life cycle analysis (LCA), Blottnitz and Curran (2007), Eriksson et al. (2007) and Gustavsson et al. (2007) have reported that the use of biofuels, including woody biomass, results in overall GHG reductions. However, Searchinger et al. (2008) argued that diversion of croplands or forestlands for biofuels in United States could result in adverse land-use effects elsewhere due to indirect land-use change effects, complicating the GHG emission issue. They used GHG emissions and energy use to estimate emission levels associated with land-use change. However, these results have been challenged by researchers like Wang and Haq (2008), who critiqued the unclear baseline definition of global food demand and supply and overestimation of production capacity. Indirect land-use changes are better captured by a general equilibrium type of model as opposed to partial equilibrium studies and ones that are undertaken at a global scale (Birur, Hertel and Tyner, 2007). What complicates the matter further is the fact that indirect land-use change effects are very difficult to assess, and today there is no generally accepted methodology for determining such effects.

Land-use change models vary in terms of their ability to capture underlying processes, precision and ability to be replicated under different conditions (Grimm et al., 2005). Differences in the results could emerge from how models vary in terms of model type, unit of analyses, intended user, temporal and spatial considerations, feedback frequencies (if any), accounting for uncertainty, data requirement and complexity of models, ability to simulate policy changes, ability to model emergent behavior and room for modelers' judgment (Agarwal, Green, Grove, Evans and Schweik, 2002).

Spatial land-use models, which account for the spatial heterogeneity of the land, are the most popular models. In these models, probability of a change given expected payoff is modeled by using information on the selling price of alternative land uses and the cost of converting from one discrete land-use class to another (e.g. Bockstael, 1996). Additionally, by fitting economic variables and statistical models such as hazard and survival models, the time period when conversion takes place is modeled along with the probability of its change (see Irwin and Geoghegan, 2001 for review of spatial land-use models). This approach allows for scenario analyses for different changes in policy such as tax and energy policy. The FASOM developed by Adams, Alig, Callaway, McCarl and Winnett (1996) is one such model. Though it does not predict exactly where the changes will happen, it can predict land-use changes occurring at a regional level. The challenges associated with these spatially explicit econometric approaches relate to the intensive data requirement and lack of parcel contagion criterion. As such, possible changes acceptable for a given land parcel are not accounted for.

Contagion property-based land-use changes are better captured by the spatial allocation models, used by geographers like Clark, Gaydos and Hoppen (1997). This approach predicts the amount of land-use change based on a given parcel's neighborhood conditions and what they are typically correlated with. Parcel characteristics such as existence of other developed sites, proximity to roads, soil type and terrain feed into predicting the probability of potential conversion of land use from one class to another. This information is used to develop the transition matrix that determines potential candidates for land-use conversion. Observed patterns and historic norms guide the transition rule used in this approach. Probability estimates can be used to indicate the type and probability of conversion. These options are iteratively fit into the model until results consistent with the historic norm and contagion specification are produced. The spatial allocation approach benefits from a merger of quantitative and qualitative approaches. However, this approach is less adept for economic policy change scenario simulations and provides results that are a continuation of a historic pattern and cannot account for nonhistoric or new development. A study that used a similar approach was Rokityanskiy et al. (2007), who assessed the effects of policies geared toward inducing landowners to change land-use and management patterns with a view to sequester carbon or reduce deforestation at a global scale. Land-use decisions are made at a specific location, based on attributes like land prices, cost of forest production and harvesting, site productivity, population density and estimates of economic growth.

Dynamic interactions among multiple factors impacting decisions are better captured by models that fall under agent-based land-use change approaches. Heterogeneous landscape attributes and historic patterns are used to project future changes. The transition rules, a probability matrix or a table describing the likelihood of one land use being converted to another, are used to select possible alternatives that can be either probabilistic (like the type produced by the spatially explicit econometric model or Bayesian) or deterministic (multicriteria and analytical hierarchy processes to select land-use classes). These models have feedback between decisions and environmental variables and can be used to simulate effects of policy changes. The downside is that these models are complex, based on use of information from multiple sources whose magnitude of errors upon interaction is difficult to quantify and extrapolate. An example of such a study is the one by Scheffran, BenDor, Wang and Hannon (2007), who used an agent-based model to explore the process by which individual farmers optimize profits through bioenergy crop selection and cost minimization in the state of Illinois. They assessed optimal spatial arrangement of crops throughout Illinois, considering factors like subsidies, costs and the introduction of new ethanol production plants.

The other focus of these studies is ecosystem benefits gained by reducing GHG emissions vis-à-vis fossil fuels. In this regard, carbon, methane and nitrous oxide balances are considered. Studies have approached this issue by evaluating the amount of CO_2 equivalent emissions avoided. Besides assumptions regarding feedstock sourcing to life cycle consideration, the displaced fossil fuel type (coal, oil, gas) impacts determination of how much CO_2 is avoided. The interaction of bioenergy and carbon sequestration and the ensuing effect on rotation age has been primarily modeled using a Faustmann-Hartman framework (Hartman, 1976) whereby the impacts of 'woody bioenergy benefits' are treated as amenity values. The optimal rotation age that maximized the land expectation value (LEV) is estimated in the presence of woody bioenergy amenities. The woody bioenergy amenities are typically fed into the model as 'carbon savings' arrived at either through back-of-the-envelope estimations or through a rigorous LCA (e.g. Dwivedi, Alavalapati, Susaeta, and Stainback 2009). The LEV in a particular time period is estimated by a standardized framework.

Researchers have imputed carbon value drawn from voluntary carbon markets (e.g. Chicago Climate Exchange values) and estimated the impact in term of rotation age. The implications of carbon subsidies and taxes on the optimal harvest decision (Plantinga and Birdsey, 1994; Creedy and Wurzbacher, 2001) holds true in the case of woody bioenergy scenarios as well. The parameters of interest have largely been amenity value functions, forest stand value, volume and harvesting decisions. The amenity values fed into the model have largely been based on the societal value, which tends to be much higher than the individual value, where carbon markets by and large do not exist. The literature also suffers, as there are hardly any studies that quantify impacts of harvesting wood for bioenergy on biodiversity, hunting leases, soil and water quality, or impacts on traditional forest industries. Bioenergy research and public assistance to woody bioenergy stakeholders can be viewed as future-oriented public goods. The LEV literature has focused on time preferences, treating them as private goods. Public policy decisions involve public goods, and generally the market discount rates calibrated in real terms are often used to proxy social discount rates. The Hartman framework is no exception, as it tends to follow the assumption that markets tend to aggregate private good discount rates, resulting in a social discount rate that equilibrates across goods and services choices. Furthermore, the probability of compulsory carbon markets and their timing could impact forest stand value and landowner decisions. It is not clear how landowners will react to woody bioenergy markets in the future and how they will prioritize different types of amenities while making management decisions. Existing studies tend to suggest that the optimal harvest age is extended when the economic value of sequestered carbon is considered. The bioenergy option, increased revenues, other management and amenity options and carbon sequestration impacts muddle the optimal rotation age estimation. The parameter of private and social discount rates calibrated for expectations and risk can provide important insights for public policy development.

The way ahead

The ways through which economists have approached woody bioenergy issues through a multiplicity of approaches and tools suggest wide-ranging focus, planning periods, feedstock considerations, models, parameter assumptions and results that might not necessarily be amenable to comparison. However, the literature also suggests that land-use change, GHG emissions, public preferences, biomass supply estimation and competition with existing forest industries are dimensions of woody bioenergy when assessed for sustainability merit.

This chapter is not exhaustive in terms of capturing the whole breadth of literature that applies economic principles for forest bioenergy assessments; rather, it highlights some of the key research areas. In fact, we limit this chapter to key themes such as biomass supply, public preferences for woody bioenergy, competition with traditional forest industries, land-use change and carbon sequestration. This said, we acknowledge the literature diversity and how forest economists have innovatively applied market versus government failures concepts in the domain of forest bioenergy and have undertaken quantitative and qualitative analyses, arguing both for and against expansion of woody bioenergy sources. However, we strongly argue that even when many papers are being published on the economics of woody bioenergy and the decision makers and social planners do not know much about landowner perception, acceptability and responses, the trade-offs with other forest-based industries, or the conversion plants that are almost on the cusp of gaining market traction for several years, then market features might unravel that we have not yet studied.

The forest conditions will also depend on factors like forest growth, afforestation of agricultural or pasture lands, intensive management of forest land and increased plantations of fast-growing species. Existing literature attempts to account for some these factors, but future conditions are clouded by great uncertainties about market conditions, giving rise to significant avenues for further research. As such, it may be important to identify, continually update and clearly communicate the social, economic and ecological factors that affect biomass supply with all the relevant stakeholders. Time series data could be used to forecast future production of woody bioenergy and its trade-offs with other forest products. The inclusion of spatial distribution of forests and its implications for conversion plants would also provide further insights to the overall wood-related industries. Another plausible option would be to assume that capital stocks adjust over time to see the long-run responses regarding change in input and output prices. The potential effects of capital substitutability and their impact on forest industries can contribute significantly in the future.

Future research could necessarily focus on the economic and ecosystem trade-offs arising from woody biomass diversion for energy use, and the level at which woody bioenergy might become ecologically, economically and socially undesirable. Policy implications based on changing the price ceiling or bioenergy production targets can be undertaken in the future as well. Indirect land-use change impact is an interesting avenue for further research. Engaging and educating consumers about the positive externalities associated with the use of woody bioenergy and the role of incentives focused on sustainable forest practices for woody bioenergy production can be researched further. Research is needed to disaggregate consumers' preferences toward woody bioenergy. Periodic revisions of existing studies are required to ensure that policies that incentivize use of such energy sources reflect changing public attitudes and preferences. Landowners' decisions regarding supplying residual woody biomass may also depend on unobserved reservation price, which represents the minimum acceptable bid offer. This price is specific to each forestland owner, it is linked to their preferences and expectations, and it may be elastic to changing bid offers. This is a matter that warrants further investigation. Ecosystem service analyses based on feedback gathered through field evidence and stakeholder responses on issues such as GHG balance, energy balance, sustainable forestry, biodiversity impact, job creation, equity issues, social impacts, transaction costs, transparency and fair trade will also contribute to the literature. Results can be used to inform government policy and act as guiding principles for woody bioenergy stakeholders.

Acknowledgment

The authors gratefully acknowledge support for this chapter from US Department of Agriculture National Institute of Food and Agriculture Grant 2012-67009-19742.

References

Abt, K. L., Abt, R. and Galik, C. (2012). 'Effect of bioenergy demands and supply response on markets, carbon, and land use', *Forest Science*, vol. 58, no. 5, pp. 523–539.

Abt, R., Abt, K., Cubbage, F. and Henderson, J. (2010). 'Effect of policy-based bioenergy demand on southern timber markets: A case study of North Carolina', *Biomass and Bioenergy*, vol. 34, no. 12, pp. 1679–1686.

Abt, R. C., Cubbage, F. W. and Pacheco, G. (2000). 'Southern forest resource assessment using the subregional timber supply (SRTS) model', *Forest Products Journal*, vol. 50, no. 4, pp. 25–33.

Adams, D. M., Alig, R. J., Callaway, J. M., McCarl, B. A. and Winnett, S. M. (1996). *The Forest and Agricultural Sector Optimization Model (FASOM): Model Structure and Policy Applications*, Research Paper PNW-RP-495, USDA Forest Service, Pacific Northwest Research Station, Portland, OR.

Agarwal, C., Green, G. M., Grove, J. M., Evans, T. P. and Schweik, C. M. (2002). *A Review and Assessment of Land-Use Change Models: Dynamics of Space, Time and Human Choice*, General Technical Report NE-297, USDA Forest Service, Northeastern Research Station, Newton Square, PA.

Alavalapati, J. and Lal, P. (2009). 'Woody biomass for energy: An overview of key emerging issues', *Virginia Forests*, Fall, pp. 4–8.

Alavalapati, J., Lal, P., Susaeta, A., Abt, R. C. and Wear, D. (2013). 'Forest biomass-based energy', in D. Wear and J. Greis (Eds.), *Southern Forest Futures Project: technical report*, General Technical Report SRS-178, USDA Forest Service, Southern Research Station, Asheville, NC.

Alavalapati, J. R. R., Hodges, A. W., Lal, P., Dwivedi, P., Rahmani, M., Kaufer, I., . . . Stevens, T. J. (2009). *Bioenergy Roadmap for Southern United States*, Southeast Agriculture and Forestry Energy Resources Alliance (SAFER), Research Triangle Park, NC.

Ankarhem, M., Brännlund, R. and Sjöström, M. (1999). 'Biofuels and the forest sector', in A. Yoshimoto and K. Yukutake (Eds.), *Global Concerns for Forest Resource Utilization; Sustainable Use and Management*, Kluwer Academic Publishers, Dordrecht, The Netherlands.

Arrow, K. R., Solow, P. R., Portney, E. E., Leamer, R., Radner, R. and Schuman, H. (1993). *Report of the NOAA Panel on Contingent Valuation*. Retrieved from www.cbe.csueastbay.edu/~alima/courses/4306/articles/NOAA%20on%20contingent%20valuation%201993.pdf

Beach, R. H., Pattanayak, S. K., Yang, J.-C., Murray, B. C. and Abt, R. C. (2005). 'Econometric studies of non-industrial private forest management: A review and synthesis', *Forest Policy and Economics*, vol. 7, no. 3, pp. 261–281.

Bergmann, A., Hanley, N. and Wright, R. (2006). 'Valuing the attributes of renewable energy investments', *Energy Policy*, vol. 34, pp. 1004–1014.

Biomass Research and Development Initiative (BRDI). (2008). *Increasing Feedstock Productions for Biofuel: Economic Drivers, Environmental Implications, and the Role of Research*. Retrieved from www.usbiomass board.gov/pdfs/increasing_feedstock_revised.pdf

Birur, D., Hertel, T. W. and Tyner, W. E. (2007). *The Biofuels Boom: Implications for World Food Markets*, Center for Global Trade Analysis, Department of Agricultural Economics, Purdue University, West Lafayette, IN.

Blottnitz, H. and Curran, M. A. (2007). 'A review of assessments conducted on bio-ethanol as a transportation fuel from a net energy, greenhouse gas, and environmental life cycle perspective', *Journal of Cleaner Production*, vol. 15, pp. 607–619.

Bockstael, N. (1996). 'Modeling economics and ecology: The importance of a spatial perspective', *American Journal of Agricultural Economics*, vol. 78, no. 5, pp. 1168–1180.

Buchholz, T. and Canham, C. D. (2011). *Forest Biomass and Bioenergy: Opportunities and Constraints in the Northeastern United States*, Cary Institute of Ecosystem Studies, Millbrook, NY.

Butler, B. J., Tyrrell, M., Feinberg, G., VanManen, S., Wiseman, L. and Wallinger, S. (2007). 'Understanding and reaching family forest owners: Lessons from social marketing research', *Journal of Forestry*, vol. 105, pp. 348–357.

Clark, K. C., Gaydos, L. and Hoppen, S. (1997). 'A self-modifying cellular automaton model of historical urbanization in the San Francisco Bay area', *Environment and Planning B: Planning and Design*, vol. 24, pp. 247–261.

Conrad, J. L., IV, Bolding, M. C., Smith, R. L. and Aust, W. M. (2011). 'Wood-energy market impact on competition, procurement practices, and profitability of landowners and forest products industry in the U.S. south', *Biomass and Bioenergy*, vol. 35, no. 1, pp. 280–287.

Creedy, J. and Wurzbacher, A. D. (2001). 'The economic value of a forested catchment with timber, water and carbon sequestration benefits', *Ecological Economics*, vol. 38, pp. 71–83.

De La Torre Ugarte, D. G., Ray, D. E. and Tiller, K. H. (1998). 'Using the POLYSYS modeling framework to evaluate environmental impacts in agriculture', in T. Robertson, B. C. English and R. R. Alexander (Eds.), *Evaluating Natural Resource Use in Agriculture,* Iowa State University Press, Ames, IA.

De La Torre Ugarte, D. G., Walsh, M. E., Shapouri, H. and Slinsky, S. P. (2003). *The Economic Impacts of Bio-energy Crop Production on U.S. Agriculture,* Oak Ridge National Laboratory. Retrieved from www.usda.gov/oce/reports/energy/AER816Bi.pdf

Diamond, P. A. and Hausman, J. A. (1994). 'Contingent valuation: Is some number better than no number?' *Journal of Economic Perspectives,* vol. 8, no. 4, pp. 45–64.

Drummond, M. A. and Loveland, T. R. (2010). 'Land-use pressure and a transition to forest-cover loss in the eastern United States', *Bioscience,* vol. 60, no. 4, pp. 286–298.

Dwivedi, P., Alavalapati, J. R., Susaeta, A. and Stainback, A. (2009). 'Impact of carbon value on the profitability of slash pine plantations in the southern United States: An integrated life cycle and Faustmann analysis', *Canadian Journal of Forest Research,* vol. 39, no. 5, pp. 990–1000.

Dwivedi, P., Alavalapati, J. R. R. and Lal, P. (2009). 'Cellulosic ethanol production in the United States: Conversion technologies, current production status, economics, and emerging developments', *Energy for Sustainable Development,* vol. 13, pp. 174–182.

Eriksson, E., Gillespie, A., Gustavsson, L., Langvall, O., Olsson, M., Sathre, R. and Stendahl, J. (2007). 'Integrated carbon analysis of forest management practices and wood substitution', *Canadian Journal of Forest Research,* vol. 37, no. 3, pp. 671–681.

Freppaz, D., Minciardi, R., Robba, M., Rovatti, M., Sacile, R. and Taramasso, A. (2004). 'Optimizing forest biomass exploitation for energy supply at a regional level', *Biomass and Bioenergy,* vol. 26, no. 1, pp. 15–25.

Galik, C. S., Abt, R. and Yun, W. (2009). 'Forest biomass supply in the southeastern United States – Implications for industrial roundwood and bioenergy production', *Journal of Forestry,* vol. 107, no. 2, pp. 69–77.

Gan, J. and Smith, C. T. (2007). 'Co-benefits of utilizing logging residues for bioenergy production: The case for East Texas, USA', *Biomass and Bioenergy,* vol. 31, no. 9, pp. 623–630.

Geijer, E., Bostedt, G. and Brännlund, R. (2011). 'Damned if you do, damned if you do not – Reduced climate impact vs. sustainable forests in Sweden', *Resource Energy Economics,* vol. 33, no. 1, pp. 94–106.

Grimm, V., Revilla, E., Berger, U., Jeltsch, F., Mooij, W. M., Railsback, S. F., . . . DeAngelis, D. L. (2005). 'Pattern-oriented modeling of agent-based complex systems: Lessons from ecology', *Science,* vol. 30, pp. 987–991.

Gustavsson, L., Holmberg, J., Dornburg, V., Sathre, R., Eggers, T., Mahapatra, K. and Marland, G. (2007). 'Using biomass for climate change mitigation and oil reduction', *Energy Policy,* vol. 35, no. 11, pp. 5671–5691.

Hartman, R. (1976). 'The harvesting decision when standing forest has a value', *Economic Inquiry,* vol. 14, pp. 52–58.

Hill, J., Nelson, E., Tilman, D., Polasky, S. and Tiffany, D. (2006). 'Environmental, economic, and energetic costs and benefits of biodiesel and ethanol biofuels', *Proceedings of the National Academy of Sciences,* vol. 103, no. 30, pp. 11206–11210.

Hodges, A. W., Stevens, T. J. and Rahmani, M. (2010). *Economic Impacts of Expanded Woody Biomass Utilization on the Bioenergy and Forest Product Industries in Florida,* Report submitted to Florida Department of Agriculture and Consumer Services Division of Forestry. Retrieved from https://test.gru.com/Pdf/futurePower/TestimonialsExhibits/Schroeder/RMS-7%20Div%20of%20Forestry%202-23-10.pdf

Huang, M. Y., Alavalapati, J. R. R. and Banerjee, O. (2012). 'Economy-wide impacts of forest bioenergy in Florida: A computable general equilibrium analysis', *Taiwan Journal of Forest Science,* vol. 27, no. 1, pp. 81–93.

Irwin, E. G. and Geoghegan, J. (2001). 'Theory, data, methods: Developing spatially explicit economic models of land use change', *Agriculture, Ecosystems and Environment,* vol. 85, no. 1, pp. 7–24.

Joshi, S. and Arano, K. G. (2009). 'Determinants of private forest management decisions: A study on West Virginia NIPF landowners', *Forest Policy and Economics,* vol. 11, no. 2, pp. 118–125.

Junginger, M., Bolkesjø, T., Bradley, D., Dolzan, P., Faaij, A., Heinimö, J., . . . Walter, M. W. (2008). 'Developments in international bioenergy trade', *Biomass and Bioenergy,* vol. 32, no. 8, pp. 717–729.

Kant, S. (1997). 'Integration of biodiversity conservation in tropical forest and economic development of local communities', *Journal of Sustainable Forestry,* vol. 4, pp. 33–61.

Koontz, T. M. (2001). 'Money talks – but to whom? Financial v. non-monetary motivations in land use decisions', *Society and Natural Resources,* vol. 14, pp. 51–65.

Kumarappan, S., Joshi, S. and MacLean, H. (2009). 'Biomass supply for biofuel production: Estimates for the United States and Canada', *BioResources,* vol. 4, no. 3, pp. 1070–1087.

Lal, P. (2011). *Economic Modeling of Forest Biomass Supply for Bioenergy Production in Southern United States*, PhD dissertation, University of Florida, Gainesville, FL.

Lal, P., Alavalapati, J., Marinescu, M., Matta, J. R., Dwivedi, P. and Susaeta, A. (2011). 'Developing sustainability indicators for woody biomass harvesting in the United States', *Journal of Sustainable Forestry*, vol. 30, no. 8, pp. 736–755.

Li, H., Jenkins-Smith, H., Silva, C. L., Berrens, R. P. and Herron, K. G. (2009). 'Public support for reducing US reliance on fossil fuels: Investigating household willingness-to-pay for energy research and development', *Ecological Economics*, vol. 68, pp. 731–742.

Majumdar, I., Teeter, L. and Butler, B. (2008). 'Characterizing family forest owners: A cluster analysis approach', *Forest Science*, vol. 54, no. 2, pp. 176–184.

Markowski-Lindsay, M., Stevens, B., Kittredge, B., Butler, J., Catanzaro, P. and Damery, D. (2012). 'Family forest owner preferences for biomass harvesting in Massachusetts', *Forest Policy and Economics*, vol. 14, no. 1, pp. 127–135.

Mehmood, S. R. and Shivan, G. C. (2010). 'Factors influencing nonindustrial private forest landowners' policy preference for promoting bioenergy', *Forest Policy and Economics*, vol. 12, no. 8, pp. 581–588.

Milbrandt, A. (2005). 'A geographic perspective on the current biomass resource availability in the United States', Technical Report NREL/TP-60-39181. Retrieved from www.nrel.gov/docs/fy06osti/39181.pdf

Neuwahl, F., Löschel, A., Mongelli, I. and Delgado, L. (2008). 'Employment impacts of EU biofuels policy: Combining bottom-up technology information and sectoral market simulations in an input-output framework', Center for European Economic Research, Mannheim, Germany. Retrieved from ftp://ftp.zew.de/pub/zew-docs/dp/dp08049.pdf

Panichelli, L. and Gnansounou, E. (2008). 'GIS-based approach for defining bioenergy facilities location: A case study in Northern Spain based on marginal delivery costs and resources competition between facilities', *Biomass and Bioenergy*, vol. 32, no. 4, pp. 289–300.

Paula, A. L., Bailey, C., Barlow, R. J. and Morse, W. (2011). 'Landowner willingness to supply timber for biofuel: Results of an Alabama survey of family forest landowners. *Southern Journal of Applied Forestry*, vol. 35, pp. 93–97.

Perlack, R., Wright, L., Turhollow, A., Graham, R., Stokes, B. and Erbach, D. (2005). 'Biomass as feedstock for a bioenergy and bioproducts industry: The technical feasibility of a billion-ton annual supply Oak Ridge National Laboratory'. Retrieved from http://feedstockreview.ornl.gov/pdf/billion_ton_vision.pdf

Plantinga, A. J. and Birdsey, R. A. (1994). 'Optimal forest stand management when benefits are derived from carbon', *Natural Resource Modeling*, vol. 8, no. 4, pp. 373–387.

Roe, B., Teisl, M. F., Levy, A. and Russell, M. (2001). 'US consumers' willingness to pay for green electricity', *Energy Policy*, vol. 29, pp. 917–925.

Rokityanskiy, D., Benítez, P. C., Kraxner, F., McCallum, I., Obersteiner, M., Rametsteiner, E. and Yamagata, Y. (2007). 'Geographically explicit global modeling of land-use change, carbon sequestration, and biomass supply', *Technological Forecasting and Social Change*, vol. 74, no. 7, pp. 1057–1082.

Scheffran, J., BenDor, T., Wang, Y. and Hannon, B. (2007). 'A spatial-dynamic model of bioenergy crop introduction in Illinois', *International Conference of the System Dynamics Society*, vol. 25. Retrieved from http://gisthal.ornl.gov/sites/default/files/nagendra/KC_090916160525.pdf

Searchinger, T., Heimlich, R., Houghton, A., Dong, F., Elobeid, A., Fabiosa, J., . . . Yu, T. H. (2008). 'Use of U.S. Croplands for ethanol increases greenhouse gases through emissions from land use change', *Science*, vol. 319, pp. 1238–1240.

Skipper, D., Van de Velde, L., Popp, M., Vickery, G., Van Huylenbroeck, G. and Verbeke, W. (2009). 'Consumers' perceptions regarding tradeoffs between food and fuel expenditures: A case study of U.S. and Belgian fuel users', *Biomass and Bioenergy*, vol. 33, pp. 973–987.

Soliño, M., Farizo, B. A., Vázquez, M. X. and Prada, A. (2012). 'Generating electricity with forest biomass: Consistency and payment timeframe effects in choice experiments', *Energy Policy*, vol. 41, pp. 798–806.

Soliño, M., Prada, A. and Vázquez, M. X. (2010). 'Designing a forest-energy policy to reduce forest fires in Galicia (Spain): A contingent valuation application', *Journal of Forest Economics*, vol. 16, pp. 217–233.

Solomon, B. D. and Johnson, N. H. (2009). 'Valuing climate protection through willingness to pay for biomass ethanol', *Ecological Economics*, vol. 68, no. 7, pp. 2137–2144.

Susaeta, A., Alavalapati, J., Lal, P., Matta, J. R. and Mercer, E. (2010). 'Assessing public preferences for forest biomass based energy in the Southern United States', *Environmental Management*, vol. 45, no. 4, pp. 697–710.

Susaeta, A., Lal, P., Alavalapati, J. and Carter, D. R. (2013). 'Modelling the impacts of bioenergy markets on the forest industry in the Southern United States', *International Journal of Sustainable Energy*, doi: 10.1080/14786451.2013.774003

Susaeta, A., Lal, P., Alavalapati, J. and Mercer, E. (2011). 'Public preferences towards environmental externalities: A choice experiment study of woody biomass based electricity in the Southern United States', *Energy Economics*, vol. 33, no. 6, pp. 1111–1118.

Susaeta, A., Lal, P., Carter, D. R. and Alavalapati, J. (2012). 'Modeling nonindustrial private forest landowner behavior in face of woody bioenergy markets', *Biomass and Bioenergy*, vol. 46, pp. 419–428.

Walsh, M. E. (2008). 'U.S. cellulosic biomass feedstock supplies and distribution'. Retrieved from www.ageconsearch.umn.edu/bitstream/7625/2/U.S.%20Biomass%20Supplies.pdf

Wang, M. Q. and Haq, Z. (2008). 'Letter to Science, Science Express'. Retrieved from www.science express.org

White, E. M. (2010). *Woody Biomass for Bioenergy and Biofuels in the United States – A Briefing Paper*, General Technical Report PNW-GTR-825, USDA Forest Service, Pacific Northwest Research Station, Portland, OR.

19

THE ECONOMICS OF REDD+

Arild Angelsen

PROFESSOR, SCHOOL OF ECONOMICS AND BUSINESS, NORWEGIAN UNIVERSITY OF LIFE
SCIENCES, ÅS, NORWAY. ARILD.ANGELSEN@NMBU.NO

Abstract

REDD+ (Reducing Emissions from Deforestation and Forest Degradation) is a broad set of
policies, programs and projects at all scales aiming to reduce greenhouse gas emissions from
forests in developing countries. The launching of REDD+ in 2007 initiated studies on designing and implementing effective, efficient and equitable REDD+ measures. The initial idea of
REDD+ was to establish a market for the public services provided by forests by incentivizing
and compensating forest owners and users through a system of Payments for Environmental
Services (PES). The recent policy debate has increasingly focused on broader Policies and Measures (PAM), e.g. to reduce the relative profitability of agricultural expansion into forests. The
implementation of both PES and PAM is fraught with challenges: defining performance criteria,
establishing reference levels, designing contracts, balancing trade-offs, building necessary institutions and so forth. This chapter takes stock of four broad REDD+ themes: REDD+ credits in
international carbon markets, REDD+ as performance-based aid, national and local PES systems
and other national policy approaches to curb deforestation (PAM).

Keywords

Carbon market, climate, conservation policies, deforestation, development aid, economics of
land use, forests, PES, REDD+

Taking the textbook to the forest

In 2007, the annual conference of the parties (COP) of the UN Framework Convention on
Climate Change (UNFCCC) decided to fully integrate forests in developing countries into
negotiations on a new climate agreement. REDD (Reducing Emissions from Deforestation and Forest Degradation) or REDD+ (including carbon stock enhancements) is seen as
critical to limit global warming to 2°C. Donors have pledged billions of dollars, countries
have developed and started implementing national REDD+ strategies and nongovernmental
organizations (NGOs) and other proponents are engaged in hundreds of REDD+ projects
locally.

Table 19.1 A classification of key issues in the economics of REDD+.

Level	Policy types	
	Market-based policies (PES)	*Other policies (PAM)*
International	*REDD+ in a carbon market:* Balancing demand and supply Reference levels Uncertainty	*Performance-based aid:* Spending pressure Performance criteria
National	*PES at the national-local level:* Defining emission reductions Who are the sellers? Who are the buyers? Contract design	*National policies:* Uncertain policy impacts Trade-offs

Forest economists should be pleased to see that in many respects, REDD+ follows text-book recommendations. Forests produce climate services in the form of carbon sequestration and storage. These are public goods that currently have no large-scale markets or market-like mechanisms to incentivize the forest owners and users to factor the value of these services into their land-use decisions. The initial REDD+ idea was to create a multilevel (global → national → local) system of Payments for Environmental Services (PES) (Angelsen and Wertz-Kanounnikoff, 2008), potentially making REDD+ the world's largest experiment in PES (Corbera, 2012).

Designing and implementing a market-based multilevel PES system has proved much more difficult than anticipated. No global carbon market has emerged, and creating national and local PES systems presents a number of challenges. National REDD+ strategies therefore rely heavily on other policy instruments, referred to as 'Policies and Measures' (PAMs). Today, REDD+ should be understood as a broad set of policies, programs and projects at all scales that aims to reduce emissions and increase removals (sequester carbon) from forests in developing countries.

This chapter is organized along two dimensions: scale (international and national/local) and types of policies (PES and PAM) (Table 19.1). There is no sub-field called 'REDD+ econom-ics', but rather a wide range of economic models and approaches – mainly outside the realm of conventional forest economics – that is relevant to analyze REDD+. Yet, a body of literature on the economics of REDD+ is emerging. One starting point for an overview of REDD+ and the issues involved is the three edited volumes from the Center of International Forestry Research (CIFOR) (Angelsen, 2008b, 2009; Angelsen, Brockhaus, Sunderlin and Verchot, 2012b). Two special issues of economic journals give state-of-the-art overviews: *Environment and Develop-ment Economics* (introduction by Bosetti and Rose, 2011) and *Review of Environmental Economics and Policy* (Angelsen and Rudel, 2013; Kerr, 2013; Lubowski and Rose, 2013; Pfaff, Amacher and Sills, 2013). Finally, Murray, Lubowski and Sohngen (2009) provide a readable introduction to key economic issues of REDD+.

REDD+ credits in carbon markets

Introduction

The 2007 Bali Action Plan (UNFCCC, 2007) was, in the view of key actors, a plan to make REDD+ part of a global climate agreement in which REDD+ credits could be used as offsets in a global cap-and-trade (CAT) system. COP 15 in Copenhagen (2009) failed to deliver that agreement. The Durban Platform (COP 17, 2011) says an agreement should be ready by 2015 and take effect from 2020. Parallel to the UNFCCC process, national and regional carbon markets are evolving. Yet it remains highly uncertain whether carbon markets will ever become a major source of international (and possibly national) funding for REDD+.

The idea of introducing REDD+ as an offset option in an international carbon market is simple. Countries commit to an emission cap and can, in part, fulfill their obligation by buying REDD+ credits. This will provide large-scale funding for REDD+. And, with relatively cheap REDD+ credits integrated in the carbon markets, higher global emissions reductions (ERs) can be achieved for any given level of total mitigation costs.

Issue 1: Market flooding and additionality

A major concern of including REDD+ credits as an offset option in a carbon market is market flooding. REDD+ inclusion, as argued by some environmental NGOs in particular, should be additional to existing efforts or, alternatively, to what would have occurred in a (hypothetical) situation without REDD+ inclusion.

Figure 19.1 gives the basic demand-supply framework to help understand the impact of REDD+ credits in a carbon market. The vertical demand curve represents the global emissions cap, while the supply curve represents the aggregate marginal costs of emissions reductions. Including REDD+ credits in the carbon market shifts the supply curve to the right. The impact depends on the assumptions made for the cap (demand): No change in the cap leads to a lower price (P^1), perfect (100%) crowding out mitigation in other sectors and no additionality; changing the global cap, keeping the carbon price constant, causes the REDD+ efforts to become 100% additional (no crowding out); and an intermediate case increases the cap, but less than in the second case. For example, the cap can be modified such that global mitigation cost is kept constant.

A few simulation studies have explored the effects of including REDD+ credits in a global carbon market (Anger and Sathaye, 2008; Bosetti, Lubowski, Golub and Markandya, 2011; den Elzen, Beltran, Piris-Cabezas and Vuuren, 2009; Dixon, Anger, Holden and Livengood, 2008; Eliasch, 2008; Murray et al., 2009). These studies suggest overall cost reduction of 7%–40% and carbon price reduction of 22%–60%. The variation of the studies and impact depends on the scope and rules for REDD+ credit inclusion and the assumed changes in the global cap.

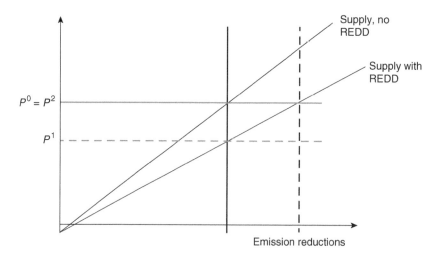

Figure 19.1   The introduction of REDD+ credits in a carbon market.

The study by Angelsen, Gierløff, Beltrán and Elzen (2013) explores ten scenarios based on different assumptions regarding the global emission target and the form and extent of REDD+ credit inclusion. Fully including REDD+ credits without changing the global cap leads to substantial (probably politically unacceptable) reductions in carbon price. For example, in the high pledge scenario, 2020 carbon price is reduced from US$19 to US$7 per tCO_2. On the other hand, bringing emissions down to a path compatible with a 2°C target seems unrealistic without harnessing the REDD+ potential, e.g. by including REDD+ in the carbon markets. With a 2°C target, including REDD+ implies that carbon prices are reduced from US$108 to US$63 and overall mitigation costs reduced by 36%.

How could a balanced introduction of REDD+ into a carbon market be achieved? First, the agreement can include restrictions on the supply and/or demand (offsetting rules) of REDD+ credits. One option is to set a limit on the extent to which REDD+ credits can offset reductions in developed countries (Annex I, Kyoto Protocol), as is the case for CDM credits. Another option is to set the trade ratio such that more than one REDD+ credit (1 tCO_2) is needed to offset 1 tCO_2 of emissions in Annex I countries. This option of discounted REDD+ credits is explored theoretically in Murray, Jenkins, Busch and Woodward (2012) and simulated in Angelsen et al. (2013). Such a system has two effects on overall ERs, compared to no discounting: REDD+ suppliers are paid less, and REDD+ trading increases overall greenhouse gas (GHG) reductions. Angelsen et al. (2013) find the latter effect to dominate.

Second, banking of REDD+ credits can limit the risk of market flooding. A country (or other entity) can buy carbon credits beyond the cap in the current period and put the surplus credits in 'the bank' to comply with the cap in later periods (Murray et al., 2009; Piris-Cabezas, 2010). Banking should increase demand for REDD+ credits if prices are low, and carbon prices should increase at a rate equal to the interest rate (Hotelling price path). Piris-Cabezas and Keohane (2008) find that, assuming a fixed cap, the 2020 carbon price without banking is US$11/$tCO_2$, compared to US$30/tCO_2 when banking is permitted. A banking option should also reduce price volatility and help achieve earlier reductions, increasing the climate benefit. It does, however, hinge on a credible system of long-term caps for countries or other entities.

Third, flexible caps might be introduced as a hybrid regime as proposed by McKibbin and Wilcoxen (2002). A price floor and ceiling are introduced, and will reduce risks for both the buyer (ceiling) and seller (floor). There are, however, organizational challenges. For example, what mechanisms should kick in when the price hits the floor, e.g. in the form of restrictions in supply and a central buyer with sufficient resources to keep the minimum price?

Fourth, the risk of flooding increases if the reference level is set 'generously'.

Issue 2: Reference levels

The term reference level (RL), or baseline, in the REDD+ debate has two fundamentally different meanings (Angelsen, 2008a). First, it can refer to the projected *business as usual* (BAU) baseline, or the counterfactual scenario used in impact assessments. Second, RL can refer to the *crediting baseline* (CB), which is comparable to an emission quota. Two questions arise: how to predict emissions from deforestation and forest degradation, and how to set the CB.

There is much literature on the causes of deforestation, including economic and statistical models trying to identify and quantify factors causing higher forest clearing (reviews by Angelsen and Kaimowitz, 1999; Geist and Lambin, 2002; Rudel, 2007). These models try to find causal patterns, while a predictive model is simply concerned with making a good prediction. Almost all proposals and UNFCCC submissions in the REDD+ RL debate suggest using historical deforestation as the starting point for setting RLs (Guizol and Atmadja, 2008). The reference period is typically set to the average deforestation rate of the last 10 years, and updated every 3 or 5 years (Santilli et al., 2005). This formula was used by the Amazon Fund and adopted in the Norway-Brazil agreement on REDD+ in 2008.

Deforestation can be highly variable from year to year, but can also display systematic trends over longer periods (5–10 years) which depart from historical deforestation. The forest area change may follow a pattern suggested by the forest transition (FT) theory (Mather, 1992; Rudel et al., 2005; Angelsen, 2007). A simple extrapolation of historical rates tends to underestimate (overestimate) future BAU deforestation for countries at early (late) stages in the transition.

UNFCCC debates have also focused on drivers and 'national circumstances', country-specific factors that can predict deforestation. A key question is what 'national circumstances' are sufficiently robust across time and space to be included in a RL formula. Meridian Institute (2011) concludes that at this stage, the empirical evidence is too weak to make such a generalized recommendation. Consistent historical deforestation estimates are not available for most countries. As better time series data of land-use change at lower scales become available, identifying robust predictors of deforestation should be a priority research area, with high policy relevance.

The second main question is how to set the CB, which forms the basis for rewarding countries, projects or other entities for successful REDD+ efforts. A basic principle of REDD+ as introduced in UNFCCC negotiations was voluntary participation and 'positive incentives'. This might be interpreted as a 'no-lose' principle, i.e. REDD+ countries should have a non-negative net benefit (total international REDD+ transfers less the real costs of REDD+) from any REDD+ agreement they enter. A country will reduce emissions up to the point where the marginal costs equal the price (realized REDD+) (Figure 19.2). This might be the credit price in a carbon market or the agreed price in a bilateral agreement (US$5/tCO$_2$e in the agreements between Norway and Brazil/Guyana). The total cost of these reductions is equal to the area $A + B$. With the CB given, the country receives revenue from selling REDD+ credits for reductions beyond CB, equal to the area $B + C$. Thus, the country's net gain equals $C - A$. If CB is set equal to BAU, the country will gain the area $C + D$, i.e. the REDD+ rent.

A key question is how large the CB must be for the country to have a positive REDD+ rent. If the marginal cost curve is linear, the CB must be more than one-half realized REDD+ rent. But empirical studies show that the marginal cost curve for REDD+ is convex, thus the CB

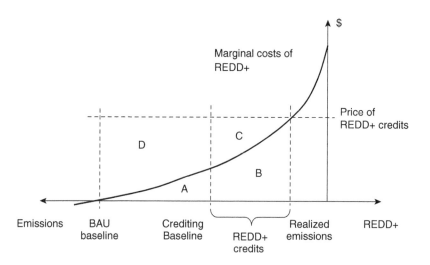

Figure 19.2 REDD+ marginal costs and setting CBs.

can be set further to the right; it can be less than half the realized REDD+ and the country still benefits. In short, 'no lose' can imply significant uncredited ERs by developing countries. The participation constraint or no-lose principle is a requirement that CB is set such that $A \leq C$.

The participation constraint is linked to the question of international leakage (displaced emissions), i.e. successful REDD+ in one country leading to higher emissions in other countries. This can be avoided if all potential REDD+ countries benefit from participation (the concern for international leakage is taken care of by the participation constraint).

Given the participation constraint (no lose), how should concerns about effectiveness and efficiency influence the way CB are set? In a system with a fixed amount of money to buy REDD+ credits, the total transfers equal the area $(C + B)$. A tighter RL implies that the carbon price can be set higher and therefore higher ERs achieved. Thus, the participation constraint and the effectiveness are maximized when the CB is set such that $A = C$.

Issue 3: Uncertainty

Designing a system for including REDD+ has uncertainties. First, a BAU baseline has several inherent uncertainties: The future values of drivers of deforestation and degradation are not known (e.g. prices of palm oil and soybean), and the relationship between such drivers and agricultural land expansion into forests is uncertain. Second, REDD+ costs are uncertain, e.g. foregone agricultural income. Third, effectiveness of REDD+ policies implemented is uncertain, e.g. farmers' responses to particular incentives aimed to constrain forest clearing. Setting RLs too low compared to a true RL gives over-compensation and reduced effectiveness, and potentially no additionality and the risk of 'hot air' in a market-based system. Setting RLs too high gives under-compensation and potential dropout if the participation constraint is no longer met, with risk of international leakage.

Several options exist to deal with this. One solution is an ex post adjustment of the RL, initially proposed as 'Compensated Successful Efforts' (Combes Motel, Pirard and Combes, 2009). Deforestation pressures in, for example, the Brazilian Amazon, are closely linked to profitability of cattle and soybean production, and adjusting RLs based on prices of these commodities during the relevant period should better reflect the true BAU scenario and better measure the domestically determined emissions reductions.

Another way to deal with uncertainty is the corridor approach (Schlamadinger et al., 2005). It recognizes that any point estimate of RL is uncertain. A discount factor is therefore introduced, where deeper ERs get an increasingly lower discount factor (higher compensation). The approach defines an interval (corridor) around the point estimate of RL, with the discount factor increasing from 0 to 1 (no to full compensation) within this interval. Thus, a REDD+ country will get some compensation even if it is unlucky and faces strong deforestation drivers, making policies less successful in reducing deforestation. A donor country will, on the other hand, not pay full compensation in the opposite case, i.e. deforestation is reduced for other reasons than successful REDD+ policies. Another attractive feature of the corridor approach (from a donor's perspective) is that the payment scheme mimics the marginal cost curve.

REDD+ as performance-based aid (PBA)

Introduction

So far, about two-thirds of international support for REDD+ has been from bilateral and multilateral development aid budgets (Streck and Parker, 2012). This 'aid-ification' of REDD+

(Seymour and Angelsen, 2012) can be explained by several factors. First, UNFCCC has failed to establish a global carbon market integrating REDD+ credits. Second, many donors were involved in REDD-relevant sectors (forest, conservation, rural development, institutional building), and ongoing activities could with light modifications be relabeled as REDD+. Third, aid already provided a mechanism and modality to transfer fresh money to REDD+ countries. Fourth, REDD+ money was labeled as aid to reach international targets for aid as share of GDP of donor countries.

PBA or conditional aid is not new and was part of the Structural Adjustment Programs (SAP) from the mid-1980s. Disbursement of aid money from the IMF and World Bank was supposed to be conditioned on deep policy reforms. 'This is indeed the core of what conditionality is supposedly about – aid buys reform. Unfortunately, it does no such thing' (Collier, 1997, p. 56). Since the SAP era, PBA has appeared under different names ('output-based aid', 'result-based aid', 'performance-based payments', 'aid on delivery', 'cash on delivery'), but with the same underlying idea: pay (only) for results.

Issue 1: Donors willing to spend and recipients unwilling to reform

The theoretical basis for PBA is the principal-agent framework. The principal (donor) has an interest in getting an outcome (policy reform), which is costly to undertake for the recipient. A contract of conditional payment is made. Conditionality often fails *in practice* due to the 'budget-pressure system' of donors (Svensson, 2003). His analysis of conditional lending by the World Bank found that degree of compliance with conditions had *no* impact on actual disbursement. Spending budget has become a key goal, and underspending is viewed within bureaucracies and by the public as poor planning and performance. Underspending carries a high risk of cuts in future budgets.

The following three-stage game, inspired by Mosley, Harrigan and Toye (1991), can explain why conditionality does not work (Mosley et al., 1991; Svensson, 2003). The donor (*D*) and recipient (*R*) have the following utility functions:

$$D = d(M, P); d_M > 0, d_p > 0 \tag{1}$$

$$R = r(M, P); r_M > 0, r_p < 0 \tag{2}$$

The donor wants policy reforms (*P*), but the recipient does not (otherwise, reforms would have been implemented). The recipient also gets positive utility from receiving money (*M*), but – and this is critical – the donor is also interested in spending. Thus, they have a common interest in spending, and conflicting interests in undertaking reforms. The game is played in three stages.

Stage 1 – Negotiation of contract: Donor and recipient agree on a set of reforms and a sum of money to be paid after implementation (M^1, P^1).
Stage 2 – Implementation: Recipient chooses the level of policy reforms to implement (P^2).
Stage 3 – Disbursement: Donor decides how much money to disburse (M^3).

Using backward induction, the sub-game perfect Nash equilibrium is straightforward but perhaps surprising: The parties agree on the contract, recipient undertakes *no* policy reforms ($P = 0$) and donor pays the agreed amount ($M^3 = M^1$). Conditionality does not work.

There are two principal ways out of this dilemma. Donors could undertake a range of reforms to make conditionality more credible (change from $d_M > 0$ to $d_M < 0$). Within the organization, performance criteria of staff could change from disbursement to documented results, perhaps even with penalties for not being tough on performance-based disbursements.

Budget pressure could also be reduced if disbursement is untied from the annual budget process. Establishing multiyear funds would reduce spending pressure and create an opportunity cost of spending money on country x in year y (reduced spending in future). In most bureaucracies, that opportunity cost is negative; *not* spending leads to future reduced budget allocations. Another way of creating a positive opportunity cost for disbursements is creating competition among recipients for scarce REDD+ funds.

A second route to soften the dilemma is to weaken domestic resistance to policy reforms needed to implement REDD+ (reduce the absolute value of r_p, and even get $r_p > 0$, where the dilemma ceases to exist). In donor circles, this is known as recipient country governments assuming 'ownership' of policy reforms (Gibson, Andersson, Ostrom and Shivakumar, 2005). In interactions with policymakers from REDD+ donor countries, this point is often stressed. They also hope that REDD+ aid will provide financial arguments to proponents of policy reforms in domestic political struggles.

Issue 2: Performance criteria and measurement

A second major issue is identification of performance criteria and measurements. This can be framed within a modified logical framework approach (LFA) (Table 19.2). This is also linked to the phased approach of REDD+ implementation (Meridian Institute, 2009), now adapted by UNFCCC: (i) readiness, (ii) policy implementation and (iii) payment for results. Within the LFA,

Table 19.2 REDD+ performance criteria.

Level	Input	Activity or process	Output	Outcome	Impact
Focus	Quantities of various inputs, in values or time	Activities undertaken to produce specific outputs	Immediate/ technical results of intervention	Intermediate and midterm effects (observable behavioral, institutional and societal changes)	Broader and long-term effects, often captured in sectoral statistics
Terms	Input indicators	Process indicators; milestones	Output indicators	Results indicators; outcome indicators	Impact indicators; goal indicators
REDD+ examples	Resources spent (US$); technical assistance (person-days)	National REDD+ plan completed; Free Prior Informed Consent (FPIC) consultations conducted	Policies adopted and enforced; no. of loggers adopting reduced impact logging practices	Reductions in deforestation; reductions in unsustainable timber harvest	Certified/ verified changes in GHG emissions

Source: Based on Wertz-Kanounnikoff and McNeill (2012); see also Klingebiel (2012).

moving through these phases implies that donors move from supporting inputs and activities to outputs and finally outcomes and impacts.

There are strong theoretical arguments for selecting performance criteria as far to the right in the table as possible. The primary goal of REDD+ is reduced emissions, and performance should be measured as directly as possible. Input- or process-based measures are generally poor indicators of final impact (Mumssen, Johannes and Kumar, 2010).

Moving toward impact criteria is nevertheless demanding. The actual measurement is more challenging further to the right. For example, whether a forest has been legally designated as protected area (an output) is easy to verify. Measuring area of deforestation (output) over a specific time period is more demanding, but doable with time series of satellite images. To measure emissions (impact) one also needs emissions factors (emissions/ha), requiring field measurements from plots.

Yet, we are interested in ERs, actual emissions minus a benchmark (BAU baseline). Establishing that benchmark is the critical issue in all impact analysis, also for REDD+. This problem can be viewed as an *attribution* problem, i.e. knowing whether the output/outcome/impact results from intervention or external factors. In general, the REDD+ country (agent or service provider) has less control the further right the performance criteria is located. This implies that output/outcome/impact-based aid shifts risk to the service providers (Mumssen et al., 2010).

National and local PES systems

Introduction

This section complements the discussion by Gong et al. (2014) elsewhere in this volume, by focusing on specific aspects of PES-related REDD+. The theoretical rationale for a PES system is linked to the core idea of REDD+. Sequestration and storage of carbon in forests has positive externalities, creating a market failure. To achieve an optimal solution the forest owners/users must be incentivized and compensated. A state-of-the-art review of critical issues is provided by Engel, Pagiola and Wunder (2008).

PES is commonly defined as a voluntary transaction between a service producer (seller) and a buyer, with a contract specifying environmental service, payment and other terms (Wunder, 2005). This is illustrated in Figure 19.3, and some challenges are discussed subsequently.

Issue 1: Defining and measuring the service

The term REDD+ suggests a focus on ERs. This is the service being traded in voluntary carbon markets (Verified Emissions Reductions, VERs) and in compliance markets under the Kyoto

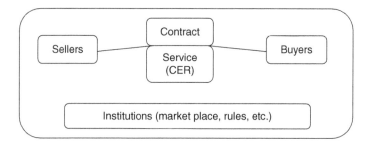

Figure 19.3   Key elements of a PES system for REDD+.

Protocol (Certified Emissions Reductions, CERs). Emissions are calculated as product of an activity (change in land use/cover) and an emission factor (change in carbon per area unit). There are uncertainties related to both variables, and up to 60% of uncertainty in total emissions may be due to the emission factor (Baccini et al., 2012).

There are trade-offs between costs and accuracy and precision of measurement, and between costs and spatial resolution. This raises at least two interesting, largely unexplored, economic questions. First, if we are to design a PES system within a project or a larger (up to national) scale, how much should we invest in producing more accurate estimates of emissions? Or, what are the costs of being wrong? Second, the key idea of REDD+ (and PES) is to create incentives for individual forest owners/users, implying measuring emission at very small scales. What is the trade-off facing REDD+ designers and policymakers between the effectiveness of more precise incentives and the costs of measurement?

The second part of the equation defining ER is the RL. Most issues that apply to RL at the international level are relevant at the national, sub-national and project levels. However, informational problems become greater the lower the scale.

A key concern in the REDD+ (and climate) debate is 'additionality', ensuring that payments are for *real* ERs only (beyond what would happen without the PES scheme). For example, according to Kaimowitz (2008, p. 492), PES programs in Costa Rica and Mexico are heavily oversubscribed, and 'payments went largely to landowners with little inclination to clear or exploit their forests'. The additionality of PES programs is also questioned by Pattanayak, Wunder and Ferraro (2010).

The problem of additionality is one of setting realistic RLs or targeting those who plan to deforest. Estimating RLs for individual service providers is one of the greatest challenges for PES programs. With asymmetric information and poor data, PES projects face a dilemma between overspending and insufficient participation (Busch et al., 2009).

Issue 2: Who are the sellers?

In the REDD+ debate, the question of who the sellers are has included legal issues of forest and carbon ownership and moral issues of 'carbon rights'. Most countries make no legal distinction between land rights and carbon rights (Streck, 2009). There are, however, a number of actors involved in REDD implementation, and a major debate concerns REDD+ 'benefit sharing'. This debate has both effectiveness and equity perspectives.

To generate ERs, it might be necessary to provide incentives to all actors involved, not just forest owners and (potential) users, but also, for example, various levels of government and project proponents. The equity perspective comes in the form of several different rationales of what represents a fair distribution of REDD+ benefits (Luttrell et al., 2013).

Interestingly, if REDD+ is designed and implemented efficiently from the international donor or national government, then just actual (opportunity, transaction and implementation) costs should be compensated, the REDD+ rent is minimized and there are few net benefits to distribute. Yet, a perception of large REDD+ (net) benefits to be allocated appears to dominate the REDD+ debate.

Issue 3: Who are the buyers?

The buyers in PES programs are either 'user-financed' or 'government-financed' (Engel et al., 2008). User-financed programs are considered more efficient, as users have better information and stronger incentives to ensure delivery. Given the global public goods nature of REDD+, this argument is weaker for REDD+ compared to environmental services. The ultimate beneficiary of climate stabilization is the global community. The actual buyers of REDD+ credits can be domestic or foreign governments, private sector companies, or NGOs, with three principal sources of funding:

1 Public funding: Payments can be made as part of a nationally funded PES program or as part of development aid directly funding VERs.
2 Compliance markets: REDD+ credits can be bought as part of a national or international CAT regime, accepting REDD+ credits as offsets. Buyers are either foreign governments or domestic/foreign companies.
3 Voluntary markets: Demand is from individuals or companies supporting REDD+ for non-commercial reasons.

Current REDD+ funding is dominated by domestic public sources and development aid (Streck and Parker, 2012), and is likely to remain so until a strong international climate agreement is reached. The voluntary forest carbon market is small, with a value of US$185 million in 2011 (Peters-Stanley, Hamilton and Yin, 2012), although it has played an important role in developing standards and safeguards.

Issue 4: Contract design

Design and implementation of a PES system for forest carbon raises most of the issues found in general economic theory of contracts (Bolton and Dewatripont, 2005). The problem of asymmetric information in PES contracts is reviewed in Ferraro (2008). The two classical asymmetric information problems concern *moral hazard* (hidden action) and *adverse selection* (hidden information).

The hidden action problem in REDD+ contracts can arise because there is noise in links between actions taken by forest owners/users and emissions reductions. This might be due to natural factors, e.g. forest fires and pests. Property rights might be insecure, and illegal encroaching by outsiders might happen. At higher scales, pressure on forest conversion depends on future agricultural commodity prices.

The hidden information problem in REDD+ PES contracts relates particularly to opportunity costs of forest conservation, ranging from negative (no plans for forest conversion, deriving private benefits from standing forests) to positive (realistic plans for removing valuable timber and converting land to highly profitable crops). Forest owners would obviously like to send signals of high opportunity costs in order to get high compensation payments. Ferraro (2008) discusses three mechanisms to reduce this informational rent: gathering more information, using screening contracts and auctioning. For example, a rough categorization based on geographical differences in opportunity costs between actors and Brazilian Amazon states can reduce implementation costs by over half (Börner and Wunder, 2008).

REDD+ as PAM: A broad approach to changing incentives

Introduction

Given the many challenges of implementing PES-based REDD+ policies, a broader set of policies is warranted to implement REDD+ at the national level. Angelsen (2009) gives a broad survey of various policy options.

The economics of land use takes as its starting point that land is allocated to the use with highest land rent: farmers, companies or other land users deforest because nonforest land uses have higher rent than forest uses. The process of deforestation can be considered a race between agricultural and forest rents.

Building on Angelsen (2007) and Angelsen (2010), agricultural rent (r^a) is defined as:

$$r^a = p^a y^a - wl^a - qk^a - v^a d \tag{3}$$

y^a = yield (agricultural production per ha); p^a = price of homogenous agricultural product at a central market; l^a and k^a = fixed labor and capital costs/ha; w = wage rate; q = annual costs of capital. The fixed wage assumption implies that labor can move freely in and out of agriculture, and is an essential assumption (Angelsen, 1999; Pfaff et al., 2013). v^a = transport costs/km, and d = distance from market.

Forest rent can be divided into three components: (1) private forest products (e.g. timber and a large number of nontimber forest products [NTFPs]), (2) local public goods (e.g. water catchment and pollination services) and (3) global public goods (e.g. carbon sequestration and storage and biodiversity maintenance). The first type is *extractive* forest rent, and the latter two are *protective* forest rents. Total forest rent is given by the sum:

$$r^f = (p^t y^t - wl^t - qk^t - v^t d) + p^l y^l + p^g y^g \tag{4}$$

with notation similar to equation 3. Agricultural and forest rents, as a function of distance from the center (or aggregate deforestation), are illustrated in Figure 19.4. Three direct forest conservation policies can be distinguished: reduce agricultural rent (at the frontier), increase forest rent and – equally or more important – its capture by land users and directly regulate.

Agricultural rent is boosted by higher output prices and technologies that increase yield and/or reduce cost, making expansion more attractive (Angelsen, 2010; Pfaff et al., 2013). Lower costs of capital in the form of better access to credit and lower interest rates pull in the same direction. Reduced travel costs in the form of new or better roads also stimulate deforestation. Higher wages (or opportunity costs of family labor) reduce agricultural rent and deforestation.

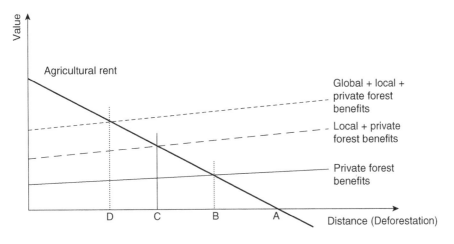

Figure 19.4 Agricultural and forest rents.

Forest rent can be increased in a similar manner, by higher prices and/or higher yield, but with the opposite effect on deforestation. The critical issue is, however, the capture of forest rent. Large tracts of tropical forests are characterized by weak, unclear and contested property rights, making them de facto open access. Land users have no incentives to include any forest rent in their decisions, and we still end up in point *A*, or even higher if we get land races (Angelsen, 1999). In our simple model, introducing private property rights to forests moves the equilibrium solution from point *A* to *B*.

With more secure property rights, factors influencing productive forest rent are relevant to consider. The impact of higher timber is complex. First, higher timber prices increase the value of standing forests for timber production, giving incentives to protect forest against agricultural encroachment. But higher timber prices will – particularly in a situation with insecure property rights – increase short-term profit of unsustainable timber harvesting, possibly by clearcutting forests ('cut-and-run' practices). Also considering forest degradation, effects on carbon storage can be negative. According to the standard Faustmann model, higher timber prices will shorten rotation period and reduce average forest carbon stock (e.g. Lofgren and Johansson, 1985).

Within this framework, Community Forest Management (CFM) attempts to move decisions from individual land users to the community in order to incorporate community-level negative externalities from deforestation (*C*). Communities may apply a different set of sanctions, as resource management is embedded in larger social systems (Aoki, 2001). However, achieving collective REDD+ outcomes is difficult, particularly when the user group is large, heterogeneous and poor, and the forest benefit flow and economic environment unstable (Agrawal and Angelsen, 2009).

The key idea of REDD+ is to provide incentives and compensation for carbon services provided by forests, and neither secure property rights nor CFM will, alone, make land users take that service into account. The problems of establishing a PES system have been discussed previously.

Finally, governments can limit deforestation through direct regulation of land use. Roughly one-fifth of all rainforest is protected areas (PAs) (Schmitt et al., 2009). PAs have less deforestation than nonprotected forests, but this can in part be explained by 'passive protection' (remote areas with lower deforestation pressure). Recent research efforts have tried to control for this using marching techniques (Andam, Ferraro, Pfaff, Sanchez-Azofeifa and Robalino, 2008; Gaveau et al., 2009). In a global study, Joppa and Pfaff (2011) found that PAs reduce forest conversion in about 75% of countries, but controlling for land characteristics (passive protection) reduces impact by 50% or more in 80% of these countries.

Although PAs have been criticized for not being effectively implemented ('paper parks'), and simple comparison of means exaggerates the impact, PAs are increasingly seen as an effective instrument. A World Bank forest policy review (Chomitz, Buys, De Luca, Thomas and Wertz-Kanounnikoff, 2007, p. 162) concluded that 'protected areas may be more effective than is commonly thought'.

Issue 1: Uncertain policy impacts

Research on causes of deforestation has stressed that broader, macro-level economic trends and policies, particularly in agriculture, often trump forest-specific policies in explaining changes in deforestation rates (Angelsen and Kaimowitz, 1999). But although such broad policies are critical for deforestation, impact pathways are complex with contradictory influences, with net effects often highly context specific.

For example, new agricultural technologies and intensified production are seen as ways to contain agricultural land expansion (Fisher et al., 2011). Within the partial equilibrium and perfect market von Thünen model of Figure 19.4, a technological change increasing yield (γ^a) or reducing costs (l^a, k^a) will spur deforestation as agricultural profitability goes up. But a number of factors can modify this result (Angelsen and Kaimowitz, 2001). Technological change can be more labor (capital) intensive, and therefore constrain land expansion if farmers are labor (capital) constrained and in a situation with limited geographical mobility of labor (capital). Further, supply increase can dampen land expansion by lowering output prices.

Issue 2: Trade-offs and political economy

Some forest conservation policies, particularly those related to depressing agricultural rent, may be considered anti-development. Ignoring rural road building, depressing agricultural prices (e.g. export taxes) and not introducing new technologies are probably effective ways of containing deforestation, but they will not promote local income and reduce rural poverty. Wunder (2003, p. 368) refers to such policies as 'the "improved Gabonese recipe" for forest conservation'.

The degree of compatibility between forest conservation and development objectives varies greatly among policies (Angelsen, 2010). Policies that depress agricultural rent present the strongest trade-off between conservation and agricultural production (food security, farm income, poverty), although the trade-off can be minimized if favorable agricultural policies are targeted to nonfrontier areas, nondeforesting (often intensive) agricultural systems and nondeforesting crops. Policies that aim to increase forest rent and its capture are more likely to be win-win but can also have heavy budgetary costs.

Concluding remarks

REDD+ is an idea from the environmental economics textbook that has met reality. The core idea of REDD+ was a multilevel PES system, where a global willingness to pay for carbon sequestration and storage services provided by forests was to be channeled to service providers (forest owners, users and other stakeholders). The reality check came in many forms: how to extract that global willingness to pay (create a carbon market), set up institutions to channel funding and define and measure ERs. The REDD+ idea has also faced political reality, where the concept has become multiobjective and more diluted, and strong political economy interests can block necessary policy reforms.

To economists, the REDD+ (ad)venture represents a great research opportunity. Angelsen, Brockhaus, Sunderlin and Verchot (2012a) outline three generations of REDD+ research. The first generation concerns REDD+ mechanism *design*, mainly drawing on past relevant experiences rather than actual REDD+ examples. The second generation deals with *implementation* issues, including the political economy. The third generation will look into the *impact* of REDD+, and ask, '*Does REDD+ work?*' Most REDD+ research, including this review, is generation 1, while some generation 2 research is emerging.

REDD+ is a large-scale experiment, not just on PES, but also on creating new international markets, conditional or performance-based aid and new national policies. A broad and exciting research agenda lies ahead, and at this stage there are more good questions than precise answers.

Acknowledgments

I thank the book editors, Caroline Wang Gierløff, Ruben Lubowski, Desmond McNeill and Sven Wunder, for constructive comments and criticism on the first draft of this chapter.

References

Agrawal, A. and Angelsen, A. (2009). 'Using community forestry management to achieve REDD+ goals', in A. Angelsen (Ed.), *Realizing REDD: National Strategy and Policy Options*, Center for International Forestry Research, Bogor, Indonesia.

Andam, K. S., Ferraro, P. J., Pfaff, A., Sanchez-Azofeifa, G. A. and Robalino, J. A. (2008). 'Measuring the effectiveness of protected area networks in reducing deforestation', *Proceedings of the National Academy of Sciences of the United States of America*, vol. 105, no. 42, pp. 16089–16094.

Angelsen, A. (1999). 'Agricultural expansion and deforestation: Modelling the impact of population, market forces and property rights', *Journal of Development Economics*, vol. 58, pp. 185–218.

Angelsen, A. (2007). *Forest Cover Change in Space and Time: Combining von Thünen and the Forest Transition*, World Bank, Washington, DC.

Angelsen, A. (2008a). 'How do we set the reference levels for REDD payments?' in A. Angelsen (Ed.), *Moving Ahead with REDD: Issues, Options and Implications*, Center for International Forestry Research, Bogor, Indonesia.

Angelsen, A. (Ed.). (2008b). *Moving Ahead with REDD: Issues, Options and Implications*, Center for International Forestry Research, Bogor, Indonesia.

Angelsen, A. (Ed.). (2009). *Realising REDD+: National Strategy and Policy Options*, CIFOR, Bogor, Indonesia.

Angelsen, A. (2010). 'Policies for reduced deforestation and their impact on agricultural production', *Proceedings of the National Academy of Sciences of the United States of America*, vol. 107, no. 46, pp. 19639–19644.

Angelsen, A., Brockhaus, M., Sunderlin, W. D. and Verchot, L. (2012a). 'Introduction: Analysing REDD+: Challenges and choices', *Analysing REDD+: Challenges and Choices*, Center for International Forestry Research, Bogor, Indonesia.

Angelsen, A., Brockhaus, M., Sunderlin, W. D. and Verchot, L. V. (Eds.) (2012b). *Analysing REDD+: Challenges and Choices*, Center for International Forestry Research, Bogor, Indonesia.

Angelsen, A., Gierløff, C. W., Beltrán, A. M. and Elzen, M. D. (2013). *REDD Credits in a Global Carbon Market: Options and Impacts*, NOAK, Nordic Council, Helsinki.

Angelsen, A. and Kaimowitz, D. (1999). 'Rethinking the causes of deforestation: Lessons from economic models', *World Bank Research Observer*, vol. 14, no. 1, pp. 73–98.

Angelsen, A. and Kaimowitz, D. (Eds.). (2001). *Agricultural Technologies and Tropical Deforestation*, CAB International, Wallingford, UK.

Angelsen, A. and Rudel, T. K. (2013). 'Designing and implementing effective REDD+ policies: A forest transition app. roach', *Review of Environmental Economics and Policy*, vol. 7, no. 1, pp. 91–113.

Angelsen, A. and Wertz-Kanounnikoff, S. (2008). 'What are the key design issues for REDD and the criteria for assessing options?' in A. Angelsen (Ed.), *Moving Ahead with REDD: Issues, Options and Implications*, Center for International Forestry Research, Bogor, Indonesia.

Anger, N. and Sathaye, J. (2008). *Reducing Deforestation and Trading Emissions: Economic Implications for the post-Kyoto Carbon Market*, Discussion Paper No. 08-016, Centre for European Economic Research (ZEW), Mannheim, Germany.

Aoki, M. (2001). 'Community norms and embeddedness: A game theoretic approach', in M. Aoki and Y. Hayami (Eds.), *Communities and Markets in Economic Development*, Oxford University Press, Oxford.

Baccini, A., Goetz, S. J., Walker, W. S., Laporte, N. T., Sun, M., Sulla-Menashe, D., . . . Houghton, R. A. (2012). 'Estimated carbon dioxide emissions from tropical deforestation improved by carbon-density maps', *Nature Climate Change*, vol. 2, no, pp. 182–185.

Bolton, P. and Dewatripont, M. (2005). *Contract Theory*, MIT Press, Cambridge, MA.

Börner, J. and Wunder, S. (2008). 'Paying for avoided deforestation in the Brazilian Amazon: from cost assessment to scheme design', *International Forestry Review*, vol. 10, no. 3, pp. 496–511.

Bosetti, V., Lubowski, R., Golub, A. and Markandya, A. (2011). 'Linking reduced deforestation and a global carbon market: Implications for clean energy technology and policy flexibility', *Environment and Development Economics*, vol. 16, no. 4, pp. 479–505.

Bosetti, V. and Rose, S. K. (2011). 'Reducing carbon emissions from deforestation and forest degradation: issues for policy design and implementation', *Environment and Development Economics*, vol. 16, no. 4, pp. 357–360.

Busch, J., Strassburg, B., Cattaneo, A., Lubowski, R., Bruner, A., Rice, R., . . . Boltz, F. (2009). 'Comparing climate and cost impacts of reference levels for reducing emissions from deforestation', *Environmental Research Letters*, vol. 4, no. 4, pp. 044006.

Chomitz, K. M., Buys, P., De Luca, G., Thomas, T. and Wertz-Kanounnikoff, S. (2007). *At Loggerheads? Agricultural Expansion, Poverty Reduction, and Environment in the Tropical Forests*, World Bank, Washington, DC.

Collier, P. (1997). 'The failure of conditionality', in C. Gwin, J. M. Nelson, E. Berg, M. Bruno, P. Collier, M. Ravallion and L. Squire (Eds.), *Perspectives on Aid and Development*, Overseas Development Council, Washington, DC.

Combes Motel, P., Pirard, R. and Combes, J. L. (2009). 'A methodology to estimate impacts of domestic policies on deforestation: Compensated successful efforts for "avoided deforestation" (REDD)', *Ecological Economics*, vol. 68, no. 3, pp. 680–691.

Corbera, E. (2012). 'Problematizing REDD+ as an experiment in payments for ecosystem services', *Current Opinion in Environmental Sustainability*, vol. 4, no. 6, pp. 612–619.

Den Elzen, M.G.J., Beltran, A. M., Piris-Cabezas, P. and Vuuren, D.P.V. (2009). *Analysing the International Carbon Market and Abatement Costs by 2020 for Low Concentration Targets: Policy Choices and Uncertainties*, Netherlands Environmental Assessment Agency (PBL), Bilthoven, The Netherlands.

Dixon, A., Anger, N., Holden, R. and Livengood, E. (2008). *Integration of REDD into the International Carbon Market: Implications for Future Commitments and Market Regulation*, New Zealand Ministry of Agriculture and Forestry, Wellington.

Eliasch, J. (2008). *Climate Change: Financing Global Forests. The Eliasch Review*, Office of Climate Change, London.

Engel, S., Pagiola, S. and Wunder, S. (2008). 'Designing payments for environmental services in theory and practice: An overview of the issues', *Ecological Economics*, vol. 65, no. 4, pp. 663–674.

Ferraro, P. J. (2008). 'Asymmetric information and contract design for payments for environmental services', *Ecological Economics*, vol. 65, no. 4, pp. 810–821.

Fisher, B., Lewis, S. L., Burgess, N. D., Malimbwi, R. E., Munishi, P. K., Swetnam, R. D., . . . Balmford, A. (2011). 'Implementation and opportunity costs of reducing deforestation and forest degradation in Tanzania', *Nature Climate Change*, vol. 1, pp. 161–166.

Gaveau, D.L.A., Epting, J., Lyne, O., Linkie, M., Kumara, I., Kanninen, M. and Leader-Williams, N. (2009). 'Evaluating whether protected areas reduce tropical deforestation in Sumatra', *Journal of Biogeography*, vol. 36, no. 11, pp. 2165–2175.

Geist, H. J. and Lambin, E. F. (2002). 'Proximate causes and underlying driving forces of tropical deforestation', *Bioscience*, vol. 52, no. 2, pp. 143–150.

Gibson, C. C., Andersson, K., Ostrom, E. and Shivakumar, S. (2005). *The Samaritan's Dilemma: The Political Economy of Development Aid*, Oxford University Press, Oxford, UK.

Guizol, P. and Atmadja, S. (2008). 'Overview of REDD proposal submitted to UNFCCC', in A. Angelsen (Ed.), *Moving Ahead with REDD*, Center for International Forestry Research, Bogor, Indonesia.

Joppa, L. N. and Pfaff, A. (2011). 'Global protected area impacts', *Proceedings of the Royal Society B Biological Sciences*, vol. 278, no. 1712, pp. 1633–1638.

Kaimowitz, D. (2008). 'The prospects for Reduced Emissions from Deforestation and Degradation (REDD) in Mesoamerica', *International Forestry Review*, vol. 10, no. 3, pp. 485–495.

Kerr, S. C. (2013). 'The economics of international policy agreements to reduce emissions from deforestation and degradation', *Review of Environmental Economics and Policy*, vol. 7, no. 1, pp. 47–66.

Klingebiel, S. (2012). *Results-Based Aid (RBA): New Aid Approaches, Limitations and the Application to Promote Good Governance*, German Development Institute/Deutsches Institut für Entwicklungspolitik (DIE), Bonn, Germany.

Lofgren, K. G. and Johansson, P. O. (1985). *Forest Economics and the Economics of Natural Resources*, Basil Blackwell, Oxford, UK.

Lubowski, R. N. and Rose, S. K. (2013). 'The potential for REDD+: Key economic modeling insights and issues', *Review of Environmental Economics and Policy*, vol. 7, no. 1, pp. 67–90.

Luttrell, C., Loft, L., Gebara, F., Kweka, D., Brockhaus, M., Angelsen, A. and Sunderlin, W. (2013). 'Who should benefit from REDD+? Rationales and realities', *Ecology and Society* 18(4): 52. URL: http://www.ecologyandsociety.org/vol18/iss4/art52/.

Mather, A. (1992). 'The forest transition', *Area*, vol. 24, pp. 367–379.

McKibbin, W. J. and Wilcoxen, P. J. (2002). 'The role of economics in climate change policy', *Journal of Economic Perspectives*, vol. 16, no. 2, pp. 107–129.

Meridian Institute. (2009). *Reducing Emissions from Deforestation and Forest Degradation: An Options Assessment Report*, Prepared for the Government of Norway by A. Angelsen, S. Brown, C. Loisel, L. Peskett, C. Streck and D. Zarin, Washington, DC. Retrieved from www.REDD-OAR.org

Meridian Institute. (2011). *Modalities for REDD+ Reference Levels: Technical and Procedural Issues*, Prepared for the Government of Norway by A. Angelsen, D. Boucher, S. Brown, V. Merckx, C. Streck and D. Zarin, Washington DC. Retrieved from www.REDD-OAR.org

Mosley, P., Harrigan, J. and Toye, J. (1991). 'Conditionality as a bargaining process', *Aid and Power: The World Bank and Policy-based Lending: Volume 1. Analysis and Policy Proposals*, Routledge, London.

Mumssen, Y., Johannes, L. and Kumar, G. (2010). *Output-Based Aid: Lessons Learned and Best Practices*, The World Bank, Washington, DC.

Murray, B. C., Jenkins, W. A., Busch, J. M. and Woodward, R. T. (2012). *Designing Cap and Trade to Correct for 'Imperfect' Offsets*, Duke University, Durham, NC.

Murray, B. C., Lubowski, R. and Sohngen, B. (2009). *Including International Forest Carbon Incentives in Climate Policy: Understanding the Economics*, Nicholas Institute for Environmental Policy Solutions, Duke University, Durham, NC.

Pattanayak, S. K., Wunder, S. and Ferraro, P. J. (2010). 'Show me the money: Do payments supply environmental services in developing countries?' *Review of Environmental Economics and Policy*, vol. 4, no. 2, pp. 254–274.

Peters-Stanley, M., Hamilton, K. and Yin, D. (2012). *Leveraging the Landscape: State of the Forest Carbon Markets 2012*, Forest Trends, Washington, DC.

Pfaff, A., Amacher, G. S. and Sills, E. O. (2013). 'Realistic REDD: Improving the forest impacts of domestic policies in different settings', *Review of Environmental Economics and Policy*, vol. 7, no. 1, pp. 114–135.

Piris-Cabezas, P. (2010). 'REDD and the global carbon market: The role of banking', in V. Bosetti and R. Lubowski (Eds.), *Deforestation and Climate Change: Reducing Carbon Emissions from Deforestation and Forest Degradation*, Edward Elgar, Cheltenham, UK.

Piris-Cabezas, P. and Keohane, N. (2008). *Reducing Emissions from Deforestation and Degradation in Developing Countries (REDD): Implications for the Carbon Market*, Environmental Defense Fund, Washington, DC.

Rudel, T. K. (2007). 'Changing agents of deforestation: From state-initiated to enterprise driven processes, 1970–2000', *Land Use Policy*, vol. 24, no. 1, pp. 35–41.

Rudel, T. K., Coomes, O. T., Moran, E., Achard, F., Angelsen, A., Xu, J. C. and Lambin, E. (2005). 'Forest transitions: Towards a global understanding of land use change', *Global Environmental Change-Human and Policy Dimensions*, vol. 15, no. 1, pp. 23–31.

Santilli, M., Moutinho, P., Schwartzman, S., Nepstad, D., Curran, L. and Nobre, C. (2005). 'Tropical deforestation and the Kyoto Protocol', *Climatic Change*, vol. 71, no. 3, pp. 267–276.

Schlamadinger, B., Ciccarese, L., Dutschke, M., Fearnside, P. M., Brown, S. and Murdiyarso, D. (2005). 'Should we include avoidance of deforestation in the international response to climate change?' in D. Murdiyarso and H. Herawati (Eds.), *Carbon Forestry. Who Will Benefit?* Center for International Forestry Research, Bogor, Indonesia.

Schmitt, C. B., Burgess, N. D., Coad, L., Belokurov, A., Besançon, C., Boisrobert, L., . . . Winkel, G. (2009). 'Global analysis of the protection status of the world's forests', *Biological Conservation*, vol. 142, no. 10, pp. 2122–2130.

Seymour, F. and Angelsen, A. (2012). 'Summary and conclusions: REDD+ without regrets', in A. Angelsen, M. Brockhaus, W. D. Sunderlin and L. V. Verchot (Eds.), *Analyzing REDD+: Challenges and Choices*, Center for International Forestry Research, Bogor, Indonesia.

Streck, C. (2009). 'Rights and REDD+: Legal and regulatory considerations', in A. Angelsen, M. Brockhaus, M. Kanninen, E. Sills, W. D. Sunderlin and S. Wertz-Kanounnikoff (Eds.), *Realising REDD+: National Strategy and Policy Options*, Center for International Forestry Research, Bogor, Indonesia.

Streck, C. and Parker, C. (2012). 'Financing REDD+', in A. Angelsen, M. Brockhaus, W. D. Sunderlin and L. V. Verchot (Eds.), *Analyzing REDD+: Challenges and Choices*, Center for International Forestry Research, Bogor, Indonesia.

Svensson, J. (2003). 'Why conditional aid does not work and what can be done about it?' *Journal of Development Economics*, vol. 70, no. 2, pp. 381–402.

United Nations Framework Convention on Climate Change (UNFCCC). (2007). *Decision 2/CP.13: Reducing Emissions from Deforestation in Developing Countries: Approaches to Stimulate Action*. Retrieved from http://unfccc.int/resource/docs/2007/cop13/eng/06a01.pdf#page=8

Wertz-Kanounnikoff, S. and McNeill, D. (2012). 'Performance indicators and REDD+ implementation', in A. Angelsen, M. Brockhaus, W. D. Sunderlin and L. V. Verchot (Eds.), *Analyzing REDD+: Challenges and Choices*, Center for International Forestry Research, Bogor, Indonesia.

Wunder, S. (2003). *Oil Wealth and the Fate of the Forest*, Routledge, London.

Wunder, S. (2005). *Payments for Environmental Services: Some Nuts and Bolts*, Center for International Forestry Research, Bogor, Indonesia.

PART 4

Economics of risk, uncertainty and natural disturbances

20

RISK AND UNCERTAINTY IN FOREST RESOURCES DECISION MAKING

Gregory S. Amacher[*][1] *and Richard J. Brazee*[2]

[*]CORRESPONDING AUTHOR
[1]JULIAN N. CHEATHAM PROFESSOR, DEPARTMENT OF FOREST RESOURCES AND ENVIRONMENTAL CONSERVATION, VIRGINIA TECH, BLACKSBURG VA, USA. 540-231-5943 (PHONE), 540-231-3698 (FAX), GAMACHER@VT.EDU
[2]ASSOCIATE PROFESSOR, DEPARTMENT OF NATURAL RESOURCES AND ENVIRONMENTAL SCIENCES UNIVERSITY OF ILLINOIS, URBANA IL, USA.

Abstract

Risk has been an important topic in forest economics since the early 1980s. This chapter will address problems of risk, although problems of pure uncertainty will be discussed as a future research area. We will survey the problems, models, and results from the literature, focusing more on how problems have been approached, the assumptions under which these problems have been studied, and the types of results that have been found. The discussion is separated into three parts. First, there are cases in which forest stands, once established, are subject to the arrival of natural or catastrophic natural events such as fire or disease that may occur before trees are harvested. Second are cases in which future market parameters are unknown and landowners must make land-use or forest management decisions in the present. And finally, we discuss key opportunities for future work regarding realistic situations and risks faced by forest landowners.

Keywords

Risk and uncertainty, forest rotation, landowner behavior

Introduction

Risk has been an important topic in forest economics since the early 1980s, and in fact Paul Samuelson, in his defining statement of forest economics and one of the earliest statements on the condition of the field, proposed understanding better the web of uncertainties and risks facing forest landowners, arguing these are perhaps the greatest complicating factors in starting a forest (Samuelson, 1976, Amacher, 2012). Since his lecture, there has been a relative explosion of articles modeling the uncertainties facing forest landowners. Two basic results have emerged. First we know that risk of arrival of natural disasters that destroy trees before harvesting often decreases optimal rotation ages. Second, we know for risks involving market parameters that high variability *in these parameters* can create incentives for landowners to harvest later. These two basic results were discovered only recently, by Reed (1984) in the first case and Brazee and Mendelsohn (1988) *in the second*. These results can also hold when prices evolve according to general stochastic processes (e.g. Willassen, 1998; Insley, 2002; Chang, 2005).

In forest economics, the words 'risk' and 'uncertainty' are often used interchangeably. The term 'risk' is meant to describe cases in which the realization of future variables is uncertain but the distribution governing these realizations is known – an example is a case in which future prices are not known, but the price observed in each future period comes from a normal distribution with a known mean and variance. In this case it is easy to define an expected value of a forest investment or operation and then examine how landowner risk preferences affect their behavior. Pure, or Knightian, 'uncertainty' refers to cases such that the distribution of a random variable is unknown. This is especially likely in forests for future amenities that may have spatial dependencies that change over time as forests age, or for a government that must choose policies with incomplete knowledge of the preferences governing forest landowner behavior. In both cases, a distribution defining possible states of nature cannot be defined.

This chapter will survey some of the more important articles on risk and forest landowner decisions published during the past few decades of research, focusing on how problems have been approached by economists, the assumptions under which these problems have been studied, and the types of results that have been found. Our main point is to uncover where results are similar and where they are different. Although our survey will be relatively complete, for lack of space we will focus less on the methods required to solve these problems. For a detailed discussion of these methods and deeper analysis of the concepts, the reader is referred to Amacher, Ollikainen, and Koskela (2009, chapters 10–12). We will also not focus specifically on two-period or life-cycle approaches to risk, more complex forest age class modeling beyond the single even-aged stand assumption that the vast majority of work has used, or financial portfolio analyses of landowner forest investment decisions. However, we will discuss how some of this work is relevant in the context of the general literature.

There are three themes of research on risk and forest landowner decisions. First, there are cases in which future market parameters are unknown and landowners must make land-use or forest management decisions in the present. The second theme is one in which forest stands, once established, are subject to the arrival of natural or catastrophic natural events such as fire or disease that occur before trees are harvested. A third theme has been to focus on uncertain future stand growth or growth in amenities that may be dependent on stand condition and age or that have unknown paths throughout a rotation. In a few cases there have been studies that have combined more than one theme, although these are relatively few in number.

Catastrophic events are those that decrease rents during a forest rotation when they arrive. These processes are often studied by assuming that the arrival of the event in any one year is independent of arrival in another year. This is not universally true for market risks such as unknown future demand, costs, or interest rates. Because these can evolve over time as markets shift, and because market equilibrium places some bounds on the range in observed parameters, it is sometimes true that market risk makes landowners better off. Unknown market parameters realized from distributions with high variances may actually serve to increase rent in future periods if a landowner waits to harvest in a period with an exceptionally high price.

There are several hundred articles published in the area of forest landowner decisions under risk, but there are some basic results that have repeatedly been shown, which is not surprising because many of the same assumptions continue to be used. First, given that natural risk often makes landowners worse off, the presence of such risk induces shorter rotation ages when landowners are risk neutral, even when landowners receive both timber and amenity benefits from their forests (Englin, Boxall, and Hauer, 2000). Longer rotations are possible only if costly stand protection or another management tool can be applied that reduces future damage should an event arrive. Classic examples are thinning that reduces ice damage from future storms or fuel removal that increases salvageable timber once a fire starts (Amacher, Malik, and Haight, 2005; Busby, Amacher, and Haight, 2013). There are different results for rotation age choice when market price risk is considered. In these cases, risk-neutral landowners could become better off harvesting later than a Faustmann model would predict. This result has held in discrete time models with normal distributions governing prices and also more complex stochastic processes in which prices evolve over time subject to drift or volatility (Brazee and Mendelsohn, 1988; Willassen, 1998; Insley, 2002; Chang, 2005). There are few examples in which natural and market risks have been combined, but generally these approaches generate results that are similar to those found in models where the risks are considered separately (Reed, 1993, Meilby, Strange, and Thorsen, 2001). Forest economists have also examined market and forest yield risks beyond the single even-aged stand, including an examination of the optimal rotation age when forests consist of uneven-aged forests or multiple age classes, and carbon price risk (Haight, 1987; Tahvonen, 2004; Chladna, 2007). There are also examples of risk in bioenergy markets and carbon sequestration (Stainback and Alavalapati, 2004; Hallman and Amacher, 2012).

Whatever the parameters assumed unknown in future periods, a basic Markov assumption that probabilities are uncorrelated through time is almost universal. This is true even though there are conflicting econometric results concerning the persistence of timber price processes. Almost all of the articles that visit natural or catastrophic events have assumed a Poisson jump process in modeling the arrival of these events during a rotation. In most cases these articles have assumed an open loop Faustmann, Hartman, or single rotation structure to the landowner's decision problem (although there are a few exceptions, such as Yin and Newman, 1995a, 1995b; Thorsen and Helles, 1998). The literature concerned with market price risk has assumed other types of stochastic processes, and in some cases different parameters are assumed to follow separate stochastic processes, or more than one parameter is assumed unknown but there is some degree of contemporaneous correlation. The earliest articles on price risk assumed a dynamic programming type structure in which a future period is entered and a price is observed drawn from a simple distribution such as the normal one.

The most recent articles on market risk involve modeling market parameters as stochastic processes in which the parameter of interest evolves over time subject to drift and volatility. Parameters of interest in this sub-literature are price, interest rates, amenity value functions, and forest stand value and volume. The types of stochastic processes chosen have been simple and geometric Brownian motion, similar to a random walk in econometrics, and mean reversion that has elements of autoregressive persistence and structure. Forest stand value has been modeled using Brownian motion or geometric Brownian motion. Prices have been assumed to follow geometric or simple Brownian motion or mean reversion, the latter defined as a case where perturbations in prices eventually revert to a long-term mean that is more consistent with timber market equilibrium. Interest rates, although rarely studied, have been assumed to follow mean reversion. One common observation of the literature, which to some extent is not justified, has been to lump stand risk and price risk into a single parameter defined as unknown 'stand value'. When this has been done, stand value has been assumed to evolve according to either shifts in biological timber outputs or price changes during a rotation. Stand value is typically modeled as evolving according to simple or geometric Brownian motion.

Stochastic amenities in this literature have been referred to differently, in some cases associated with a market value (i.e. through hunting leases) and in others associated with a private value important to the landowner and/or a public value important to society in general. Very little is known empirically about how amenities evolve or are produced in a forest stand, but most economics studies to date assume that they evolve according to either a simple felicity function as in Hartman (1976) or as simple Brownian motion over time.

Finally, we should point out the choice of decision maker most often studied and assumptions about their preferences. Broadly speaking, forest landowner decisions such as rotation age, planting intensity, and land use have been considered under the various risks for a risk-neutral landowner. Only relatively few articles have considered risk aversion for the landowner making rotation age choices, likely because this assumption complicates tractability (Clarke and Reed, 1989; Alvarez and Koskela, 2007; Lien et al., 2007).

Catastrophic and natural events

Catastrophic events that can arrive during the life of a stand were perhaps the first area of risk and uncertainty studied seriously by forest economists – early work includes Martell (1980), Routledge (1980), Lohmander and Helles (1987) and Anderson, Guidin, and Vaseivich (1987). These events can arrive in any time period and damage a stand of trees. Examples include wind, snow, or ice; fire; drought; and shifts in land productivity. Without doubt, though, the most cited article, and one of the highest cited articles in forest economics to this day, is Reed (1984), who incorporated risk of fire using a Poisson jump process and developed an open loop problem similar to Faustmann. The basic result in Reed's paper was interesting and important and has been applied more than any other risk model in forest economics. He showed under certain conditions that risk is capitalized into the interest rate and therefore affects many components of the cost of continuing a rotation, including land rent.

To illustrate a basic Poisson arrival problem, we first assume all Faustmann constancy conditions hold for all relevant parameters such as prices, interest rates, forest growth, and costs of planting ad infinitum. Suppose X is a Poisson random variable representing the time between successive arrivals of an event. The probability that the event arrives before rotation age T is $\Pr(X < T) = 1 - e^{-\gamma T}$, while the probability that no event occurs up to harvest time is $\Pr(X = T) = e^{-\gamma T}$, where γ is the Poisson parameter and represents the rate at which the event arrives. Notice here that the arrival of the event is constant and not dependent on the age of the stand.

All studies that introduce such a probability and event must then describe how forest rents are affected by the event's arrival. For this example, assume that destruction of the stand is total, so that the current value of the stand in rotation n, R_n depends on whether the event arrives before the rotation age or not, so that for the nth rotation,

$$R_n = \begin{cases} -c \text{ if } X_n < T \\ pF(T) - c \text{ if } X_n = T \end{cases} \tag{1}$$

where we use c to define the cost of starting a stand at time 0 and of restarting a stand at any time during a future rotation after an event arrives (these could be different without loss but we assume they are the same for simplicity), and $F(T)$ is harvest yield at rotation age T. We then assume a risk-neutral landowner and write the net present value of successive rotations, where (1) holds in each rotation, as:

$$V = E\left[e^{-rX_1}R_1 + e^{-r(X_1+X_2)}R_2 + \cdots\right] = E\left[\sum_{n=1}^{\infty}e^{-r(X_1+\cdots+X_n)}R_n\right]$$

After manipulation, and using (1), this becomes

$$V = \frac{E\left(e^{-rX}Y\right)}{1 - E(e^{-rX})}$$

Evaluating the expectations in the numerator or denominator is not trivial, and the reader is referred to Reed (1984) and Amacher et al. (2009, chapter 10) for explanations. After this evaluation, we arrive at the basic objective function that has appeared in many articles that derive from Reed's model,

$$V = \frac{(r+\gamma)\left(pF(T)-c\right)e^{-(r+\gamma)T}}{r\left(1-e^{-(r+\gamma)T}\right)} - \frac{\gamma}{r}c \tag{2}$$

The significance of this equation cannot be overemphasized. This is stochastic analog of Faustmann under assumptions required of that model, and it says simply that the Poisson arrival risk parameter is capitalized into the real interest rate, and into the cost of restarting a stand after it is destroyed (last RHS [right hand side] term). Because risk of arrival increases the effective discount rate, the rotation age is shortened under this type of risk, as the marginal cost of delaying harvest is increased more than the marginal benefit of waiting to harvest one more period.

Reed's model was simple but powerful, and it is no surprise that it has been extended in a great number of ways. We now discuss some of these, and in doing so review the theoretical literature on catastrophic risk in forest economics. The first important extension was to introduce amenities and derive an analog to (2) in a Hartman type of model. This was accomplished by adding a felicity function describing amenities that develop over time and depend only on time, $B(t)$. This is a relatively recent modification and due to Reed (1993) and to Englin et al. (2000). In these articles, as in others of this type, arrival of the natural event reduces amenities produced in the forest stand in various ways. For example, Englin et al. modify the rent equations as,

$$R_n = \begin{cases} -c + e^{rX_n}\int_0^{X_n} B(s)e^{-rs}ds & \text{if } X_n < T \\ pF(T) - c + e^{rT}\int_0^{T} B(s)e^{-rs}ds & \text{if } X_n = T \end{cases}$$

where the amenity function has similar properties to that in Hartman (1976). Englin et al. find once again that an increase in the effective interest rate shortens rotation age compared to the risk-free case.

Another extension that is relevant at least for some tree species is the case in which arrival rates are not constant over time, but rather could change over time due to forest structure changes in a stand. In this case, the Poisson parameter now depends on the timing of the event during the rotation – this is the case of a Poisson process that is nonhomogeneous the way the statistical literature defines these probability distributions. If the probability of arrival increases as the stand ages, then the rotation age would be even shorter than in the case in which the arrival rate is uniform over time. The case in which arrival rates decrease as stands age and forest stocking decreases is more complex, though, because the marginal costs and benefits of delaying harvesting depend on where in the rotation the event arrives. This all said, there is some debate in the literature concerning how relevant nonhomogeneous Poisson distributions really are. For example, many have argued that the arrival rate is not dependent on stand age, but rather that the damage to the stand once a natural damaging event arrives is what depends on stand structure or age. The classic example would be a fire that arrives through a lightning strike whose resulting damage depends on the fuel loading at the time of the strike once the fire ignites, or fires that arrive in fronts that have started elsewhere so that the amount salvageable or lost depends on the age and condition of the stand. Examples of this type of assumption are Amacher et al. (2005) and Yoder (2004).

Other relatively recent extensions assume that the loss to the stand is not total as we wrote in (1). Rather, there could be random damage that is less than 100%, such as in Reed (1984), or there could be damage that is dependent on the level of costly management effort a landowner chooses to apply before the arrival of an event, such as a fire. Examples are Reed (1987), Reed and Apaloo (1991), Meilby et al. (2001) and Amacher et al. (2005) for single stands, and Crowley, Malik, Amacher, and Haight (2009) for landowners with adjacent stands. Collectively, what these studies suggest is that risk need not shorten rotations if protection effort is employed, but there is an important cost-risk comparison that determines the ultimate effect of damaging events on rotation age.

Market risks

The introduction of unknown future market parameters in modeling landowner behavior is fundamentally different than modeling those of natural risks. Natural risks, when they arrive, imply discrete jumps in forest returns or amenities during a rotation, nearly always making forest landowners worse off through partial or total destruction of the standing forest. Such an assumption makes little sense for market parameters such as prices or costs because they are not as likely to jump rapidly in either positive or negative directions through time. In most cases, price and cost changes are damped or centered around a drift defined through equilibrium in stumpage markets. The risk to the landowner also differs with market parameters, and landowners are not always worse off because of these risks. A low price or high replanting cost could exist at the time of harvesting, making landowner returns lower than expected. However, high prices or low costs could also exist, especially if the standard deviations governing price distributions are high, meaning there might be rents from postponing harvesting and waiting for these periods. Indeed, a forest landowner could be better off following a flexible harvesting schedule rather than following a Faustmann decision framework.

For the most part, the forest landowner decision problem under market risk has been handled by proposing that rotation age choice is the result of solving an optimal stopping problem, in which each period a landowner faces a decision to harvest or wait and continue a rotation, with information about unknown parameters revealed at the start of each period. There are both single rotation and ongoing rotation interpretations of this problem in the literature. But there are two distinct differences in the types of frameworks used for solving them. In the first and earliest versions, the stopping problem is written as a dynamic stochastic discrete time programming problem. In this case, unknown parameters are resolved when the landowner arrives at a given period in the life of a forest, so that the decision becomes a Bellman related problem in which the landowner decides whether to continue or to harvest by comparing the returns from harvesting in the current period and possibly a land sale at bare land value with the expected returns in all future periods that arise from continuing to grow the forest and assuming that the landowner will behave optimally from the current period onward. In these problems, parameters are resolved in each period in a discrete way as a draw from a known distribution, and this defines an endogenous reservation price such that the landowner harvests at the point in time that an observed harvest price exceeds the reservation price.

The second version of stopping problems in forestry is to model the decision within a continuous time variant that results in a similar continuation and harvesting decision at each instant in time but also introduces a continuous state stochastic process that governs the evolution of market parameters. This allows specification of drift and volatility as separate features defining how unknown parameters evolve over time. We will address both types of problems in this section, beginning first with the stochastic process-based stopping problem and then discussing the reservation price approach.

The common way to write a continuous time stochastic rotation problem is:

$$max_{x,T} \, Ee^{-rT}H(x) \tag{3}$$

where E is the expectations operator, r is the real interest rate, T is the stopping time (harvest age), $H(x)$ is the harvest payoff value, and x is defined as a stochastic stand value that evolves in general terms according to the following Ito (or diffusion) process:

$$dx = a(x, t)dt + b(x, t)dz \tag{4}$$

where $a(x, t)$ and $b(x, t)$ are nonrandom functions and dz is a continuous time increment of a Weiner process such that the following properties hold: $E(dz) = 0$, and $E(dx) = a(x, t)dt$ represents drift, while $var(dx) = b^2(x, t)dt$ represents drift neutral volatility in the process. The problem in (3) is often subject to some initial condition $x(0)$, defined at time zero when the stand is planted at zero cost (in this example). The function $H(x)$ is defined according to the problem and could in principle include amenities, harvest value, or adjacent site effects. The modeling of x could also be assumed to equal forest stock at a given time, and this compares somewhat with catastrophic risk studies discussed earlier that examine shocks for forest growth during a rotation. The single rotation variant in (2) has also been extended to allow for ongoing rotations (e.g. see Chang, 2005), and multiple sources of stochasticity have been introduced, notably amenities and stand value (Reed, 1993).

The decision to harvest, or more precisely, exercise the option to harvest in (3) could be assumed to be undertaken by a fixed time specified according to another condition, or an infinite time horizon could be assumed. The stopping time is usually solved for by defining boundaries such that when the critical level of stand value $x^\star$ is reached, the landowner stops growing the trees and harvests. In this type of case, the rotation age is not specifically solved for, but the stopping boundary is then used to define a corresponding 'expected' rotation age, and it is this expected rotation that is used to evaluate the impact of drift and volatility in the stochastic process. Also, because the single rotation analog of (3) is most common, we stick with that here.

Perhaps the most important assumption in (3) is what is assumed to govern the underlying stochastic process, and it is here that most articles deviate depending on what interpretation of stand value is used. The most common form of stochastic stand value is due to price, and there are the most differences in what has been assumed for its evolution. Early work assumed prices changed according to simple Brownian motion (equivalent to a random walk in econometrics),

$$dx = adt + bdz$$

where a and b are simple known drift and volatility terms, respectively. Because Brownian motion suffers by its underlying implication that prices could increase without bound, two different processes have been used, including geometric Brownian motion, in which changes in price are proportional to the price value,

$$x = axdt + bxdz$$

so that drift and volatility depend on the value of x, and more recently, mean reversion, in which prices are assumed to return to some long-term mean $\bar{x}$ once perturbed,

$$dx = a(\bar{x}-x)dt + bdz$$

With mean reversion, if x is above its mean, then we expect it to decrease, $E(dx) < 0$ and for x below its mean, we have $E(dx) > 0$. Given the notion of persistence in the evolvement of prices here, mean reversion is consistent with an autoregressive process assumption in econometrics.

What is assumed for price has been subject to much debate, and empirical studies attempting to validate one or another stochastic process have been inconclusive. Early theoretical work assuming Brownian and geometric Brownian motion includes, among others, Yin and Newman (1995b), Thomson (1992) and Clarke and Reed (1989). Early studies assuming mean reversion include Insley (2002) and Plantinga (1998). What is interesting here is what we know about stumpage price data, and unfortunately the choice of stochastic process may in fact be very specific to a species or market type. For example, there is some evidence of autoregression in prices (Hultkrantz, 1995; Brazee, Amacher, and Conway, 1999; Saphores, Khalaf, and Pelletier, 2002), but also other evidence of mean reversion (Gjolberg and Guttormsen, 2002). Some softwood prices have even inexplicably been found consistent with random walks supporting both versions of Brownian motion (Prestemon, 2003). Still other work suggests that what is usually assumed for price processes is far too simplistic, and in fact a more complex process may better explain price trends and volatility over time, and this may differ from what has been assumed in the literature (Yoshimoto and Shoji, 2002).

Regardless of what is assumed for the stochastic evolution of x, and regardless of whether stand value risk is due to stand volume, amenity values, or price, there is a standard procedure that is followed for all stochastic rotation models that begin as continuous time stopping problems. Although we will not go through all derivations required of these solutions (because Ito calculus is needed and this is beyond the scope of the chapter), the reader is referred to Amacher et al. (2009) for the methods involved in solving forest harvesting problems like (1). First we set up the problem in a continuous time Bellman equation analog, and using a stochastic stand value interpretation for the process defining x by assuming x follows an Ito process in (2):

$$V(x) = max\{H(x), B(x) + \beta E[V(x'|x)]\}$$

Where x' is an increment of x in the next instant of time, expectations E are taken over x, and $B(x)$ is the value to holding the stand and not harvesting, such as known amenities the landowner receives similar to Hartman (1976) and Englin et al. (2000). The first element of the

RHS is the harvest payoff at a stopping value for x, while the second argument is the value of continuation and not harvesting in the stopping problem. Following procedures in Amacher et al. (2009) and Dixit and Pindyck (1994), the Bellman equation can be rewritten by defining it in a small increment of time, dt:

$$rV(x) = max\left\{B(x) + E\frac{dV(x,t)}{dt}\right\}$$

which implies that the return from harvesting in the next period depends on additional growth that yields higher amenities $B(x)$ plus the value (or cost) to the landowner from another instant step forward in the stochastic process. The second RHS term $dV(x)$ is a function of a stochastic process and is itself one. In order to evaluate this term and write it as a function of the drift and volatility of the process x, which is the unknown parameter we are interested in, we must make use of Ito's lemma. Applying Ito's lemma and taking expectations, we can write the form of the harvesting equation that many articles focusing on stochastic forest rotation problems have derived:

$$rV(x,t) = max\left\{B(x) + V_t(x,t) + a(x,t)V_x(x,t) + \frac{1}{2}b(x,t)^2 V_{xx}(x,t)\right\} \qquad (5)$$

A value of setting up the problem this way is that we have decomposed changes to the Bellman equation in terms of time (second RHS term), drift (third term), and volatility (fourth term). A closed form solution for this equation requires solving a partial differential equation. Thus, researchers have relied on simulations to understand how volatility and drift in the process for x affect expected rotation age. To solve these simulations, additional boundary conditions must be imposed that hold at the optimal stopping point, called smooth pasting and value matching. These conditions define the rotation age continuation region and thus expected rotation age. The value matching implies at the optimal stopping point that the value function of the Bellman equation in (5) simply equals the harvesting returns at that instant or from (3), $V(x^\star, t^\star) = H(x^\star)$. The smooth pasting condition requires, at the instant of stopping, that the marginal change in the value function equals the change in harvesting returns evaluated at the optimal stopping time, $V_x(x^\star, t^\star) = H_x(x^\star)$.

Many different versions of (5) exist in the literature depending on what is assumed for $B(x)$ and x, with many authors omitting $B(x)$ and considering only the effect of stochastic prices on the option value of harvesting (Insley and Rollins, 2005), or modeling price series for harvesting and carbon storage (Chladna, 2007). The basic stochastic rotation problem in (5) has also been extended in other directions. In most of this literature, higher price volatility, regardless of the basic stochastic process assumption, leads to longer waiting times to harvest unless there are catastrophic risks of continuing a rotation in addition to market risks. Interest rates have been considered under mean reversion (Alvarez and Koskela, 2002). This work finds that higher volatility tends to lengthen optimal harvesting thresholds and thus expected rotation ages.

Risk aversion for landowners has been rarely studied or modeled in (1) because it greatly complicates characterization of the optimal stopping region. One exception is Motoh (2004), who finds that such aversion can have profound effects on stopping regions and expected rotation ages when forest stocks evolve according to Brownian motion and when catastrophic losses due to a Poisson arrival process are also jointly present. His work and that of others such as Thorsen and Helles (1998) and Yin and Newman (1995a, 1995b) effectively combine stochastic evolution in market parameters with the types of discrete jump processes that natural events involve. In this work it proves critical whether stand damage allows harvesting after a natural event or not. One area that is still open is how to model amenities in stochastic rotation models. At present, nearly all of the work either lumps the notion of amenities into a general stochastic stand value, modeled simple or geometric Brownian motion, or assumes amenities themselves evolve according to Brownian motion. Because this is essentially the same as assuming amenities develop over time as random walks, it is not a very satisfying link to the ecology literature, in which the development of amenities is known to depend on stand structure or adjacent forests.

We would also argue that more is needed concerning the interaction of risks and importance of natural risk versus market risk in developing landowner decision models. At present, there has not been enough study to know what types of risks are more important, or how their interaction affects forest management. Such knowledge is critical to policy choice problems in which landowner responses must be modeled correctly in order to achieve various goals with policy instruments. Another issue with the literature in stochastic rotation models is the fact that stand value is typically modeled using Brownian type processes, yet we know stand value includes components of price and forest stock changes, and thus assuming that the same process governs each of these variables is problematic. It may be that we have seen most of the easy advancements in modeling these types of risks, and now future evolution of research will be much slower even though we still have critical unresolved questions.

There are many other variations of the basic stopping problem, and one in particular that has garnered much attention is the reservation price approach. Here, the landowner is assumed to solve a dynamic programming problem in which a new price is revealed in each period, and this problem is solved for a reservation price that induces harvesting and thus defines the optimal stopping point. This type of approach is similar to (3), but it has advantages in that it can be formulated as an open loop problem in which rotation age is solved at time zero according to expectations of future market parameters, or a closed loop problem in which the landowner updates information over time as the stand ages, re-optimizing as market parameters continue to be drawn from independently and identically distributed random variables in each period.

The first instance of the reservation price stopping problem in forestry is due to Brazee and Mendelsohn (1988) and Lohmander (1987). The landowner is assumed to begin a stopping problem in (3) written in discrete time. Usually, a terminal time is given when harvesting must occur, and in periods prior to terminal time the landowner will harvest when he arrives at a time point when the price observed is greater than his reservation price. This price is solved endogenously by considering the decision faced at each time period, either to harvest at the observed price and sell the land at bare land value, or to continue the rotation and wait to observe a new price in the next period. Because this is a dynamic programming problem, the principle of optimality holds and it is assumed that all uncertainty is resolved at the beginning of each period.

The reservation price is solved endogenously using a Bellman equation defined by the dynamic programming stopping problem. Using a discrete version of (1) and additionally assuming that a land sale occurs at harvest time defined by its expected value $E(L)$, where expectations are taken over the next period price random variable, the Bellman equation is given by:

$$p'(t)V(T-1)+E(L)=\frac{1}{(1+r)}\left[Ep'(T)V(T)+E(L)\right] \qquad (6)$$

where $V(T)$ is the value function of the Bellman equation defined at time T, and the reservation price $p'(T)$ is a function of time and equates the marginal cost of delaying harvesting as foregone rents in the current period (LHS [left hand side]) with the marginal benefit of waiting to harvest, factoring in the value of land (RHS). The optimal reservation price path through time is solved here simultaneously with land value. The expected harvesting time is solved for by considering when a price is observed above the path, and numerous articles have shown that such behavior outperforms the Faustmann age when prices are not known.

Several extensions of the basic model in (6) have been used in the literature to understand landowner behavior in a variety of settings. Gong (1998) uses adaptive dynamic programming to solve for a reservation price beyond the simple independent arrival assumption of prices over time used in the early literature. He shows that the reservation price approach can outperform other rotation age solutions when prices are unknown but auto-correlated through time. Gong, Boman, and Mattsson (2005) introduces amenities as a function of time in the basic problem of (3) and (6), finding that reservation price paths are sensitive to these landowner preferences. Gong and Lofgren (2003, 2007) consider several variations, including landowners who switch from Faustmann to reservation price approaches under risk of unknown prices and forest stocks, and that reservation price behavior increases the short-term timber supply elasticity with respect to price as landowners attempt to dampen the impact of price risk.

What is needed?

As forest economists we are far from a perfect understanding of risk and uncertainty. The majority of the literature still retains the basic assumptions of Reed (1984) that parameters are constant, and thus adjustment to the risk-free interest rate is valid for expected land value. Even in more dynamic modeling of risk, we still find Reed's same basic result concerning shorter rotations with higher risk of loss. Stochastic rotation problems also suffer from simplistic construction of stochastic processes for tractability.

It is likely that risks facing forest landowners are correlated to some degree, yet we tend to isolate one type of risk when examining decisions, and few studies have considered multiple risks, let alone correlated ones. Consider that fire can make forest stands susceptible to insect infestations and vice versa, and certainly wind and storms can increase fuel loadings and thus the damage to a stand should fire arrive in the periods following a storm. There are also certainly cases where amenity production in a stand is correlated with other events or even the amenity production of adjacent landowners with unknown preferences.

The modeling of amenities in risk problems is also too basic. We know much more about amenities than is revealed by the way we tend to introduce them into our models. In catastrophic event problems they are usually simple functions of time or forest stock size, while in stochastic process problems they are even simpler, as the Brownian motion assumption often used simply means that amenities are random walks. This makes the risk literature completely disjoint from other literature in forest economics, in which adjacent stands are modeled in describing amenities that depend on forest structure (e.g. see Amacher, Koskela, and Ollikainen, 2004, for a review).

A side effect of inadequacies in modeling landowner behavior means that our policy design problems are overly simplistic, and it may be important to revisit these problems and consider the impact of policies or their design when the government is not certain about how forest landowners respond. To assume we know reaction functions of landowners in these problems is to ignore what we have found through our review of risk here. If governments and landowners all face the same risks when making policy and forest management decisions, then the impact of risk will be less important. However, if risk is not aggregate and the government knows less about landowner preferences or decisions than the landowners themselves, then serious social costs may arrive that we have not yet studied in forest policy problems.

Finally, more needs to be evaluated under the condition that distributions for unknown parameters may not be known. We imagine that, should this be studied in management decisions, a precautionary type of behavior could be discovered in the face of uncertainty that might lead to different conclusions than the literature we discussed previously, which models risk in the traditional way assuming the distribution defining states of nature is known.

References

Alvarez, L., and Koskela, E. (2002.) 'On the forest rotation under interest rate variability', *International Tax and Public Finance*, vol. 10, pp. 489–503.

Alvarez, L., and Koskela, E. (2007). 'The forest rotation problem with stochastic harvest and amenity value', *Natural Resource Modeling*, vol. 20, pp. 477–509.

Amacher, G. (2012). 'Samuelson's economics of forestry in an evolving society: Still an important and relevant article thirty-six years later', *Journal of Natural Resources Policy Research*, vol. 4, pp. 197–201.

Amacher, G., Koskela, E., and Ollikainen, M Gong. (2004). 'Forest rotations and stand interdependency: Ownership structure and timing of decisions', *Natural Resource Modeling*, vol. 17, pp. 1–43.

Amacher, G., Malik, A., and Haight, R. (2005). 'Not getting burned: The importance of fire protection in forest management', *Land Economics*, vol. 81, pp. 284–302.

Amacher, G., Ollikainen, M., and Koskela, E. (2009). *Economics of Forest Resources*, MIT Press, Cambridge, MA.

Anderson, W., Guidin, R., and Vaseivich, J. (1987). 'Assessing the risk of insect attack in plantation investment', *Journal of Forestry*, vol. 85, pp. 46–47.

Brazee, R., Amacher, G., and Conway, C. (1999). 'Optimal harvesting with autocorrelated stumpage prices', *Journal of Forest Economics*, vol. 5, pp. 201–216.

Brazee, R., and Mendelsohn, R. (1988). 'Timber harvesting with fluctuating timber prices', *Forest Science*, vol. 34, pp. 359–372.

Busby, G., Amacher, G., and Haight, R. (2013). 'The social costs of homeowner decisions in fire-prone communities: Information, insurance, and amenities', *Ecological Economics*, vol 92, pp. 104–113.

Chang, F. (2005). 'On the elasticities of harvesting rules', *Journal of Economic Dynamics and Control*, vol. 29, pp. 469–485.

Chladna, Z. (2007). 'Determination of the optimal rotation period under stochastic wood and carbon prices', *Forest Policy and Economics*, vol. 9, pp. 1031–1045.

Clarke, H., and Reed, W. (1989). 'The tree cutting problem in a stochastic environment: The case of age dependent growth', *Journal of Economic Dynamics and Control*, vol. 13, pp. 569–595.

Crowley, C., Malik, A., Amacher, G., and Haight, R. (2009). 'Adjacency externalities and forest fire prevention', *Land Economics*, vol. 85, pp. 162–185.

Dixit, A. and Pindyck, R. (1994). *Investment under uncertainty*, Princeton University Press, Princeton, NJ.

Englin, J., Boxall, P., and Hauer, G. (2000). 'An empirical investigation of optimal rotations in a multiple use forest in the presence of fire risk', *Journal of Agriculture and Resource Economics*, vol. 25, pp. 14–27.

Gjolberg, O., and Guttormsen, A. (2002). 'Real options in the forest: What if prices are mean reverting?' *Forest Policy and Economics*, vol. 4, pp. 13–20.

Gong, P. (1998). 'Risk preferences and adaptive harvest policies for even-aged stand management', *Forest Science*, vol. 44, 496–506.

Gong, P., Boman, M., and Mattsson, L. (2005). 'Nontimber benefits, price uncertainty, and optimal harvest of an even aged stand', *Forest Policy and Economics*, vol. 7, pp. 283–295.

Gong, P., and Lofgren, K. (2007). 'Market and welfare implications of the reservation price strategy for forest harvest decisions', *Journal of Forest Economics*, vol. 13, pp. 217–243.

Gong, P. and Löfgren, K-G. (2003). 'Risk aversion and the short-run supply of timber', *Forest Science*, vol. 49, pp. 647–656.

Haight, R. (1987). 'Evaluating the efficiency of even- and uneven-aged management', *Forest Science*, vol. 31, pp. 957–974.

Hallman, F., and Amacher, G. (2012). 'Forest bioenergy adoption for a risk averse landowner under uncertain emergence of markets', *Natural Resource Modeling*, vol. 25, no. 3, pp. 482–510.

Hartman, R. (1976). 'The harvesting decision when a forest stand has value', *Economic Inquiry*, vol. 14, pp. 52–55.

Hultkrantz, L. (1995). 'The behavior of timber rents in Sweden 1909-1990', *Journal of Forest Economics*, vol. 1, pp. 165–180.

Insley, M. (2002). 'A real options approach to the valuation of forestry investment', *Journal of Environmental Economics and Management*, vol. 44, pp. 471–492.

Insley, M., and Rollins, K. (2005). 'On solving the multirotational timber harvesting problem with stochastic prices: A linear complementarity formulation', *American Journal of Agricultural Economics*, vol. 87, pp. 735–755.

Lien, G., Stordal S., Hardaker, J., and Asheim, L. (2007). 'Risk aversion and optimal forest replanting: A stochastic efficiency study', *European Journal of Operational Research*, vol. 181, pp. 1584–1592.

Lohmander, P. (1987). *The Economics of Forest Management under Risk*, Report 79 SE-901, Swedish University of Agricultural Sciences, Umea, Sweden.

Lohmander, P. and Helles, F. (1987). 'Windthrow probability as a function of stand characteristics and shelter', *Scandinavian Journal of Forest Research*, vol. 2, pp. 227–238.

Martell, D. (1980). 'The optimal rotation of a flammable forest stand', *Canadian Journal of Forest Research*, vol. 10, pp. 30–34.

Meilby, H., Strange, N., and Thorsen, B. (2001). 'Optimal spatial harvest planning under risk of windthrow', *Forest Ecology and Management*, vol. 149, pp. 15–31.

Motoh, T. (2004). 'Optimal natural resources management under uncertainty with catastrophic risk', *Energy Economics*, vol. 26, pp. 487–499.

Plantinga, A. (1998). 'The optimal timber rotation: An option value approach', *Forest Science*, vol. 44, pp. 192–202.

Prestemon, J. (2003). 'Evaluation of US southern pine stumpage price market informational efficiency', *Canadian Journal of Forest Research*, vol. 33, pp. 561–572.

Reed, W. (1984). 'The effects of risk of fire on the optimal rotation of a forest', *Journal of Environmental Economics and Management*, vol. 11, pp. 180–190.

Reed, W. (1987). 'Protecting a forest against fire: Optimal protection patterns and harvest policies', *Natural Resource Modeling*, vol. 2, pp. 23–53.

Reed, W. (1993). 'The decision to conserve or harvest old growth forest', *Ecological Economics*, vol. 8, pp. 45–69.

Reed, W., and Apaloo, J. (1991). 'Evaluating the effects of risk on the economics of juvenile spacing and commercial thinning', *Canadian Journal of Forest Research*, vol. 21, pp. 1390–1400.

Routledge, R. (1980). 'The effect of potential catastrophic mortality and other unpredictable events on optimal forest rotation policy', *Forest Science*, vol. 26, pp. 389–399.

Samuelson, P. (1976). 'The economics of forestry in an evolving society', *Economic Inquiry*, vol. 14, pp. 466–492.

Saphores, J., Khalaf, L., and Pelletier, D. (2002). 'On jumps and ARCH effects in natural resource prices: An application to Pacific Northwest stumpage prices', *American Journal of Agricultural Economics*, vol. 84, pp. 387–400.

Stainback, G. A., and Alavalapati, J. (2004). 'Modeling catastrophic risk in economic analysis of forest carbon sequestration', *Natural Resource Modeling*, vol. 17, no. 3, pp. 299–317.

Tahvonen, O. (2004). 'Optimal harvesting of forest age classes: A survey of some recent results', *Mathematical Population Studies*, vol. 11, pp. 205–232.

Thomson, T. (1992). 'Optimal forest rotation when stumpage prices follow a diffusion process', *Land Economics*, vol. 68, pp. 329–342.

Thorsen, B., and Helles, F. (1998). 'Optimal stand management with endogenous risk of sudden destruction', *Forest Ecology and Management*, vol. 108, pp. 287–299.

Willassen, Y. (1998). 'The stochastic rotation problem: A generalization of Faustmann's formula to stochastic forest growth', *Journal of Economic Dynamics and Control*, vol. 22, pp. 573–596.

Yin, R., and Newman, D. (1995a). 'A note on the tree cutting problem in a stochastic environment', *Journal of Forest Economics*, vol. 1, pp. 181–190.

Yin, R., and Newman, D. (1995b). 'The effect of catastrophic risk on forest investment decisions', *Journal of Environmental Economics and Management*, vol. 31, pp. 186–197.

Yoder, J. (2004). 'Playing with fire: Endogenous risk in resource management', *American Journal of Agricultural Economics*, vol. 86, pp. 933–948.

Yoshimoto, A., and Shoji, I. (2002). 'Comparative analysis of stochastic models for financial uncertainty in forest management', *Forest Science*, vol. 48, pp. 755–766.

21

MODELING NATURAL RISKS IN FOREST DECISION MODELS BY MEANS OF SURVIVAL FUNCTIONS

Thomas Burkhardt,[1] *Bernhard Möhring*[2] *and Johannes Gerst*[2]

[1]ABTEILUNG FINANZIERUNG, FINANZDIENSTLEISTUNGEN UND EFINANCE, UNIVERSITÄT KOBLENZ-LANDAU, CAMPUS KOBLENZ, UNIVERSITÄTSSTR. 1, D-56070 KOBLENZ, GERMANY.
[2]ABTEILUNG FORSTÖKONOMIE UND FORSTEINRICHTUNG, GEORG-AUGUST-UNIVERSITÄT GÖTTINGEN, BÜSGENWEG 3, D-37077 GÖTTINGEN, GERMANY.

Abstract

Forests are subject to survival risks, usually caused by calamities like insects, fire or storm. If a calamity hits a forest stand, it will reduce its value. Survival functions can be used to provide a probabilistic description of such an event. Economic models for optimal forest management should incorporate this information. Available modeling approaches are usually based on quite specific stochastic assumptions, and most require simulation studies. Here, we provide a more general approach. It is only assumed that the survival function is known, but no assumptions on specific properties are made. It is then shown how survival risks will change optimal management decisions in a stochastic, Faustmann-type model in discrete time. Furthermore, the costs associated with survival risks are exemplified based on real-world data and a Weibull-type survival function. The costs are found to be substantial, dependent on age and particularly on tree species. Consequently, ignoring survival risks in forest management is likely to result in substantial losses.

Keywords

Age-dependent risk, hazard rate, juvenile risk, land expectation value, land rent, old-age risk, optimal rotation, stochastic Faustmann model, survival function, Weibull function

Survival as a fundamental problem in forest management

On a very fundamental level, forestry deals with the growing of trees. Growth may come to a planned end at the intended harvesting time, or to an unplanned end caused by the realization of survival risks. The purpose of this chapter is to consider the description and economic consequences of the latter. Essentially, survival risks are described by a probability assessment for future harmful events, which formally cause a state transition from 'alive' to 'dead'. In forestry, such risks are caused by calamities, e.g. fire, insects, snow or storm.

A survival function is a probability function that describes the probability of survival as a function of time. Therefore, it is monotonically decreasing, but not necessarily strictly so. In many cases, one would like to link the survival function to more risk factors than just time. This is possible, and may improve respective models, if suitable data are available.

From an economic point of view, we are not only interested in mere probabilistic descriptions of survival risks, but rather in the economic consequences. Hence it is important to incorporate probabilistic knowledge into economic models (see Brazee and Newman, 1999). Hanewinkel, Hummel and Albrecht (2010) surveys more than 200 scientific publications regarding these issues, organized according to four steps of analysis:

1. *Analysis of framework*: Risk factors have to be identified for the application at hand, like tree, stand and site characteristics.
2. *Probabilities for hazards* have to be assessed. Here, a wide array of statistical modeling techniques is applied, such as traditional regression models (see König, 1996) or more recent models that integrate nonparametric smoothers such as generalized additive models 'GAM' (see Schmidt, Hanewinkel, Kändler, Kublin and Kohnle, 2009). Output-oriented empirical models affiliate with this step, like survival probabilities (see Kouba, 2002; Staupendahl and Möhring, 2011) or transition probabilities (Suzuki, 1971, 1983), which are based on Markov chains and were also described by Kurth, Gerold and Dittrich (1987), Deegen (1994) and others.
3. *Estimation of costs* of potential damage, combined with the information of the respective probability, provide the basis for further economic analysis. Different approaches were used in the context to evaluate the costs of risks in forest production; for example, Routledge (1980) and Reed (1984) further developed the classical Faustmann approach by integrating the probability of destruction by fire or other catastrophic events. Dieter (1997), Bräunig and Dieter (1999), Kuboyama and Oka (2000), Beinhofer (2007) and others used Monte-Carlo simulation techniques to evaluate and quantify the economic impact of natural risks on forest management. Knoke and Wurm (2006) added risks like price fluctuation into the simulation model.
4. *Choice of action*: Optimized management strategies are derived which take individual risk preferences into account. Possible actions to cope with risks include changing the rotation length (Price, 2011; Möhring, Burkhardt, Gutsche and Gerst, 2011, Möhring, Staupendahl and Leefken, 2010), adapting the thinning strategy (Thorsen and Helles, 1997), changing the tree species distribution (Knoke, Ammer, Stimm and Mosandl, 2008) or contracting an insurance policy (Holecy and Hankewinkel, 2006).

The available literature is usually based on rather specific assumptions regarding the stochastic processes which govern survival and is mostly based on simulation studies. Although this research provided substantial insight, both methodologically and for the practical applications considered, there is a lack of understanding of how survival risks can be modeled and incorporated into forestry decisions in a more general way. It is the objective of this chapter to provide a more general approach to include survival risks in economic models and to derive some fundamental relations.

The next section covers the modeling of natural risks (calamities) by survival functions. Section 3: Integrating survival risks into economic decision models is devoted to the integration of survival risks into economic decision models. We use a discrete time formulation that is well suited to practical application, and follows the well-known Faustmann approach. This provides a general understanding of how survival risks will change optimal forest decision making, and allows development of the underlying intuition. Section 4: Numerical example based on real-world data covers the implementation of this model, with some extensions of practical importance, and provides real world examples. Section 5: Discussion and conclusion concludes.

Modeling natural risks using survival functions

The use of survival functions in modeling natural risks is much like the use of these functions in life insurance mathematics. In both cases, survival functions provide descriptive means to characterize the probability of survival of a certain entity. In life insurance, this entity is usually thought to be a single creature. Although it might be possible to model the survival of a single tree, if we are interested in economic consequences of risks, we are more interested in what happens in a forested area. We will follow this approach here.

From an empirical point of view, this means that we estimate the survival function based on observations of the survival of the trees in a predefined unit area that can reasonably be treated as the entity of survival, which implies that we approximate reality by assuming that either all or none of the trees in the area survive. This assumption allows for a useful interpretation of survival probabilities: They then correspond to the fraction of forest land that will survive up to some point in time.

Survival functions are purely descriptive tools. They describe survival over time, but do not explain why an entity survives or dies. The possible causes or risk factors which govern survival are only implicitly represented by the survival function, and may or may not be based on structural models.

These issues must be distinguished from the possible choice of a certain parametric form of the survival function. In most applications we have to face limited data and structural information on the underlying risk factors. In this case, it may well be that the available information can reasonably be described by several alternative parametric specifications of the survival function, which calls for an additional modeling effort to come up with a good choice for the respective application.

Fundamentals of the survival function

Survival is a question of time. We assume that ultimate death is unavoidable, and denote the time of death by τ. We usually choose the time scale such that time zero is the time of birth; in forestry that is the time when the stand is created. In this case, the time of death corresponds to the age of death and can be interpreted as the random variable of lifetime.

The survival function $S(t)$ is then defined to be the probability of the event $\{\tau > t\}$, short $S(t) = \mathrm{Prob}[\tau > t]$. Clearly, $t \geq 0$, $S(0) = 1$ and $S(\infty) = 0$. The survival function is a cumulative probability distribution function, and complementary to the probability distribution of lifetime, $W(t) = \mathrm{Prob}[\tau \leq t] = 1 - S(t)$.

Both cumulative distributions may be the result of modeling considerations in either continuous or discrete time. In continuous time, lifetimes are nonnegative real variables; in discrete time, they are restricted to a countable set $\mathfrak{T}$, where the admissible values are usually chosen with fixed time increments like a year or a decade. For analytical purposes, continuous time models tend to be more convenient, whereas in real-world planning scenarios we almost exclusively encounter discrete time horizons.

Although both approaches may convey essentially the same information, the mathematical formulation is quite different. In continuous time models, we can make use of continuous probability density functions $w(t) = W'(t) = -S'(t)$, whereas in discrete time we have to work with discrete probability mass functions, $w(t) = w_k$ if $t \in \mathfrak{T}$ and zero otherwise, as their counterpart.

It is often important to have a clear understanding of the relationship between the continuous time and discrete time formulations. In continuous time, we have $W(t) = \int_0^t w(\tau) d\tau$ and $S(t) = \int_t^\infty w(\tau) d\tau$, compared to $W_k = \sum_{j=1}^k w_j$ and $S_k = \sum_{j=k+1}^\infty w_j$ in discrete time, assuming $k = 0, 1, 2, \ldots$ and $\tau_0 = 0$. In most applications, the set $\mathfrak{T}$ of admissible lifetimes will be modeled with constant time increments, i.e. $\tau_k = \Delta t \cdot k$ with $k = 0, 1, 2, \ldots\ldots$.

Most important in practice, and an important starting point for modeling survival, is the 'local' survival behavior, expressed by the conditional probability of death in the next period, given survival up to the beginning of the respective period. This gives rise to the notion of the hazard rate h_k, sometimes also called the rate of mortality. More formally, we have to consider $h_k = \text{Prob}[\tau \le \tau_k | \tau > \tau_{k-1}] = \text{Prob}[\tau \le \tau_k \text{ and } \tau > \tau_{k-1}] / \text{Prob}[\tau > \tau_{k-1}] = (W_k - W_{k-1}) / \text{Prob}[\tau \ge \tau_{k-1}] = w_k / S_{k-1} = 1 - S_k / S_{k-1}$. Clearly, the probability to also survive period k, given survival up to the beginning of this period, is the complement $1 - h_k$.

To summarize, we have the following useful relations:

$$W_k = \sum_{j=1}^k w_j = 1 - S_k \tag{1}$$

$$S_k = \sum_{j=k+1}^\infty w_j = S_{k-1} - w_k = (1 - h_k) \cdot S_{k-1} = \prod_{j=1}^k (1 - h_j) \tag{2}$$

$$h_k = w_k / S_{k-1} = 1 - S_k / S_{k-1} \tag{3}$$

$$w_k = h_k \cdot S_{k-1} \tag{4}$$

From a continuous time perspective, the discrete time formulation may be interpreted as a model with limited information with respect to the exact time of death, because survival is observed only at discrete points in time. From this point of view, the discrete unconditional probability w_k of dying at the time corresponding to k may be interpreted as the probability of dying in the respective time interval, i.e. in $\tau \in (\tau_{k-1}, \tau_k)$, so we have $w_k = \int_{\tau_{k-1}}^{\tau_k} w(\tau) d\tau$ linking these two points of view, or approximately $w_k = w(\tau_{k-1})\Delta t$. The discrete time hazard rate relates to its continuous time counterpart accordingly: $h_k \approx h(\tau_{k-1})\Delta t$, with $h(t) = w(t)/S(t) = -\dfrac{d \log S(t)}{dt}$.

The Weibull function to model survival of forest stands

When actually modeling natural risks for a specific application, one needs to specify an appropriate survival function. This may be done in a nonparametric or parametric form, and may be linked to an explanatory model based on appropriate risk factors. Usually, it is advantageous to work with a parametric form whenever possible. Several distribution types are available to do so. We focus our exposition on the Weibull distribution function (Weibull, 1951). It has the advantages of requiring only two parameters and being quite flexible, and has a proven history of useful application in modeling natural risks in forestry (Pienaar and Shiver, 1981;

Kouba, 2002; Holecy and Hanewinkel, 2006; Staupendahl and Möhring, 2011; Möhring et al., 2011). If τ is Weibull-distributed with scale parameter β and shape parameter α, the density, distribution, survival and hazard rate functions are given as follows, with $t \geq 0$ (Klein and Moeschberger, 1997, p. 37):

$$w(t) = \frac{\alpha}{\beta} \cdot \left(\frac{t}{\beta} \right)^{\alpha-1} \cdot exp \left[-\left(\frac{t}{\beta} \right)^{\alpha} \right] \tag{5}$$

$$W(t) = 1 - exp \left[-\left(\frac{t}{\beta} \right)^{\alpha} \right] \tag{6}$$

$$S(t) = exp \left[-\left(\frac{t}{\beta} \right)^{\alpha} \right] \tag{7}$$

$$h(t) = \frac{\alpha}{\beta} \cdot \left(\frac{t}{\beta} \right)^{\alpha-1} \tag{8}$$

Figure 21.1 illustrates the relations of these functions, which are directly deducible from each other.

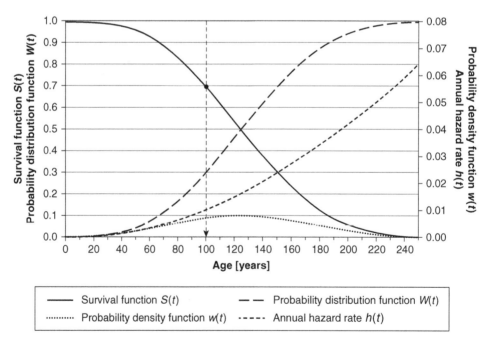

Figure 21.1 The survival function, density function, distribution function and annual hazard rate of the Weibull-distributed time of death τ for $\alpha = 3.0$ and $\beta = 141$. The probability that a stand survives at least the age of 100 years is here 70%, so the risk level S_{100}, is 0.7.

If the survival probability of the forest stand is modeled by a parametric function, it seems advantageous to use a structure where the parameters can be interpreted intuitively. Similar to the approach of measuring the productivity of forest stands by the mean or dominant height at a certain age (site index), Staupendahl and Möhring (2011) proposed that the risk level should be described by the probability that a forest stand reaches the age of 100 years. This parameter was named S_{100} because it is equivalent to the value of the survival function at age 100. Solving equation (7) for β, inserting 100 for t and S_{100} for $S(100)$ and substituting the resulting expression for β back into equation (7), leads to the following, newly parameterized survival function of the Weibull model:

$$S(t) = S_{100}^{\left(\frac{t}{100}\right)^{\alpha}} \tag{9}$$

with the corresponding hazard rate function

$$h(t) = \frac{\alpha}{100} \left(-\log S_{100}\right) \cdot \left(\frac{t}{100}\right)^{\alpha-1} \tag{10}$$

The parameter S_{100} can be interpreted as a quantitative measure of the level of the survival risk. The lower the value of S_{100}, the higher the risk level (the lower the proportion of the stand reaching the age of 100 years). In the limit $S_{100} = 1$, there is no survival risk.

Additionally, the Weibull function comprises the parameter α which characterizes the shape of the survival function. If $\alpha > 0$, the Weibull distribution shows growing ($\alpha > 1$), decreasing ($\alpha < 1$) or constant ($\alpha = 1$) hazard rates over time. Therefore, the parameter α can be interpreted as an indicator of the risk type of the survival function, and we distinguish old-age risk (e.g. windblow) with growing hazard rate, juvenile risk (e.g. deer damage) with decreasing hazard rate, and age-independent risk (e.g. damage by meteorites) with constant hazard rate over time. Figure 21.2a exemplarily illustrates survival functions for all three risk types, but a common risk level ($S_{100} = 0.7$). Figure 21.2b illustrates the corresponding hazard rates.

Empirical estimates for the survival function

Once a specific survival function has been chosen to model survival risk in a given application, it remains to estimate the corresponding parameters. The availability of appropriate data will usually be an issue, and limitations in this regard may influence the overall modeling process.

Subsequently, we report on tree species-specific parameter estimates of the Weibull model for a German site. Based on rather extensive time series data of a forest damage survey in Rhineland-Palatine, Germany, Staupendahl and Zucchini (2011) provided the estimates of S_{100} and α as shown in Table 21.1. Figures 21.3a and 21.3b show the corresponding survival and hazard rate functions. It can clearly be seen that spruce stands have the highest risk of destructive events. This is reflected by the value of S_{100}, which indicates that on average, only 73% of the spruce stands survive the age of 100 years without lethal destruction by natural risks. Figures 21.3a and 21.3b show that the other tree species differ considerably less with respect to S_{100} at this age. The value of α is found to be greater than 1 for all tree species, which shows that survival risk generally increases with age. However, the degree of this increase shows major differences among tree species.

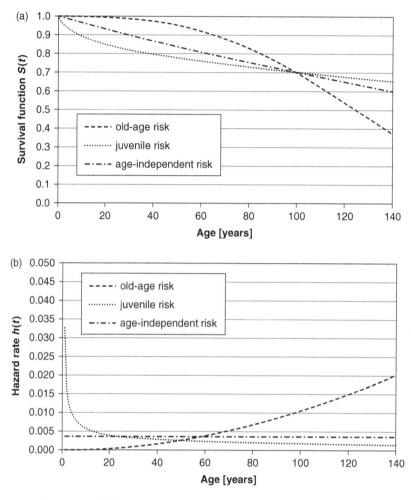

Figure 21.2a and 21.2b Weibull survival function and hazard rate for all three risk types, with a common risk level (S_{100} = 0.7). The α-parameter used for old-age risk is 3.0, for age-independent risk 1.0 and for juvenile risk 0.5.

Table 21.1 Estimated Weibull parameters for different tree species, based on data from a forest damage survey in Rhineland-Palatine (Germany) (from Staupendahl and Zucchini, 2011, table 5, p. 139).

	S_{100}	α
Oak	*0.971*	*2.75*
Beech	*0.967*	*1.76*
Norway spruce	*0.726*	*2.78*
Douglas fir	*0.916*	*3.11*
Scots pine	*0.923*	*2.45*
Larch	*0.940*	*2.3338*

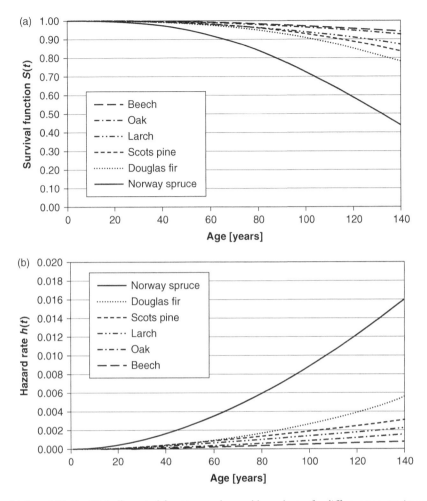

Figure 21.3a and 21.3b Weibull survival functions and annual hazard rates for different tree species according to the parameter estimates of Table 21.1.

Integrating survival risks into economic decision models

Starting points

To integrate survival risks into economic decision models, we start with the standard Faustmann approach, which explains land value by the present value of the annuity that results from infinite identical repetition of forest rotations. This very fundamental approach can be generalized to cover a wide range of applications (Chang and Gadow, 2010). The resulting land value depends on the length of the rotation, and the economically optimal rotation can be found by maximizing the present value.

The standard Faustmann approach is a deterministic model. Our goal is to augment it by integrating survival risks, which obviously requires the development of a stochastic model. In doing so, the fundamental ideas underlying the Faustmann approach remain valid. Although it remains reasonable to assume an infinite repetition of identically planned rotations, the essential difference is that the actual rotations are no longer constant and known, but of random length. This random length is the lifetime of each rotation, whose stochastic behavior is assumed to be known by the assumed survival function. But although actual rotations will depend on the realization of a priori unknown risk factors, it will prove sensible to define a planned rotation length T, after which we will harvest in case our stand survives until then.

The result of this approach will necessarily be a stochastic present value, analogous to the Faustmann deterministic one. But although it is straightforward to interpret the deterministic value as a land value, it is not obvious how to interpret its stochastic counterpart. The task at hand is to demonstrate within a very well-known model context how the information represented by the survival function can fruitfully be incorporated into economic decision models. The main purpose of the Faustmann model is to provide a conceptual framework for the valuation of forest land and its relation to the choice of optimal forest management (e.g. optimal rotation). From this point of view it seems appropriate to consider the expected present value as the land value in our model. It is crucial to note that by this we mean the mathematical expectation of this random present value of our model. This note is important to avoid any confusion with the traditional notion of land expectation value, which is commonly used for the deterministic present value attributed to the land value in the standard Faustmann approach – which is unfortunate, as no expectation in a mathematical sense enters the standard model. In our model, we attribute the mathematical expectation of the present value to the land value, so we consider a land expectation value in a mathematical sense.

Using an expectation in a valuation model under risk may seem problematic at first sight. Indeed, that would certainly be so, as the use of the expectation implies risk neutrality on behalf of the valuator, which is a quite crude behavioral assumption. In the case at hand, this assumption is much less problematic. This is so because we consider a single stand. Given that a forest owner owns many similar stands subject to independent risks, the standard deviation of the value sum is proportional to the inverse square root of the number of stands, so the remaining variation will be negligible if the number of stands is sufficiently high. Subsequently, we assume that the expectation provides a reasonable proxy for the overall value.

More specific assumptions

As mentioned, we do not make any specific assumption on the survival function, except that it is defined for the complete stand as the underlying entity, so the stand will either survive or die in whole if hit by some sort of natural risk. By construction, the survival considered is the survival within a given rotation, and we add the quite plausible assumption that the lifetimes of the subsequent rotations are independent and identically distributed.

It might be worthwhile to reconsider the latter assumption if in the future more advanced survival models become available, e.g. to reflect possible time dependencies of survival probabilities which will be caused by environmental changes like climate change, but we are not aware of such models yet. The inclusion of rotation-dependent survival functions in the model is straightforward and will not change the structure in an essential way, whereas the assumption of independence is more essential, but we do not see any evidence to challenge it.

In accordance with most planning systems in practice, we formulate the model in discrete time, where lifetimes are considered as multiples of a given planning time period Δt. The resulting probability mass function for a given planned rotation time $T = n \times \Delta t$ is $m(\tau_k) = w_k$ for $k = 1, 2, \ldots n$.

To keep the focus on the essentials, we will consider only two payments per rotation: At the beginning, a payout of L is required for afforestation or reforestation. At the end of the rotation, we get an incoming payment $f(T)$, if the stand survives up to the planned rotation time, or $g(\tau)$ if the stand is subject to a calamity before that time. $f(T)$ may be considered as the net revenue (stumpage value) function. In either case, we assume that the harvest is followed by an immediate reforestation. Quite naturally, we assume $g(\tau) < f(\tau)$. All payments are normalized to be per unit of area. For the ease of notation, we indicate the time dependence of these functions by the time index where appropriate.

We capture the time value of money by a continuously compounded interest rate i, which corresponds to the discount factor e^{-it} over a time span t. It will prove convenient to alternatively use a per-period interest rate r defined by $(1 + r) = e^{i \cdot \Delta t}$ for the period Δt.

For primarily technical reasons we assume the usual time dependence of $f(T)$ according to the law of diminishing returns to guarantee the existence of a single optimum rotation, and, for the same reason, we assume that at least one rotation T exists for which forestry is profitable, that is, for which $f(T)e^{-iT} > L$ holds, and we restrict our discussion to the set of profitable rotations so defined.

Analytic model formulation for land value

We consider first the single period case, and then build on these results to develop the infinite series result.

One rotation

For just one rotation, the equation for the stochastic present land value is

$$\widetilde{LV}_1\left(k,n\right) = -L + e^{-i \cdot k \cdot \Delta t} \cdot \tilde{z}_k \tag{11}$$

with $\tilde{z}_k = g_k$ if the calamity hits at $k \leq n$ and $\tilde{z}_n = f_n$ if the stand survives up to the planned rotation time. We use the tilde to denote random variables.

These settings are due to the discrete time formulation. The planned rotation time corresponds to the time index n. In continuous time, we would consider risks up to but not including time n. In discrete time, we may choose to consider such risks only until time $n - 1$, which will neglect risks in the last period, or to include risks up to time n. We decided to do the latter, and assume therefore that the planned use will take place immediately after time point n, given survival until then.

We use the mathematical expectation $LV_1 = E\left[\widetilde{LV}_1\right]$ of this equation to value the land. This land expectation value obviously depends on the planned rotation period T corresponding to time index n and is given by

$$LV_1\left(n\right) = -L + \sum_{k=1}^{n}(1+r)^{-k}g_k w_k + (1+r)^{-n}f_n S_n \tag{12}$$

Analogous to the deterministic case, we ask for the optimal rotation, which now is the optimal planned rotation, that is, we need to find the time index n which maximizes $LV_1(n)$. In the time discrete model, we cannot easily take a derivative to find the first order condition. Although it is easy to find the optimal value of n numerically, it is preferable to derive at least an approximate optimality criterion to illuminate the underlying economic intuition. We do just that. Obviously, in analogy to the usual first order condition on $f'(T)$ the first difference must vanish at least approximately in the optimum, that is, $LV_1(n) - LV_1(n - 1) \approx 0$. This yields the optimality condition for the first difference $\Delta f_n = f_n - f_{n-1}$:

$$\Delta f_n = r \cdot f_{n-1} + (f_n - g_n) \cdot h_n \tag{13}$$

Proof: The first difference is $LV_1(n) - LV_1(n - 1) = (1 + r)^{-n}g_n w_n + (1 + r)^{-n}f_n S_n - (1 + r)^{-(n-1)} f_{n-1}S_{n-1}$. Setting to zero and canceling $(1 + r)^{-n}$ yields: $0 = g_n w_n + f_n S_n - (1 + r)f_{n-1}S_{n-1}$. Then substitute $S_n = (1 - h_n)S_{n-1}$ to get $0 = g_n w_n - f_n S_{n-1}h_n + (f_n - f_{n-1})S_{n-1} - rf_{n-1}S_{n-1}$. Then divide by S_{n-1}, use $w_n/S_{n-1} = h_n$ and rearrange terms to get the assertion †.

This result allows for an intuitive economic interpretation: The increase in value Δf_n in the nth period must compensate the forester for the foregone interest on the stumpage value at the beginning of that period, $r \cdot f_{n-1}$, and the expected loss due to a possible calamity during that period, $(f_n - g_n) \cdot h_n$. If we compare this to the solution for the corresponding deterministic model, exactly this latter term is added to the optimality criterion to reflect the survival risk. Please note that the single rotation model does not contain, as usual, a term related to the scarcity of land.

Obviously, the land value in the stochastic model is strictly lower than in the corresponding deterministic one, because of a strictly positive probability of a calamity before the planned rotation time.

We can also draw a quite general conclusion with respect to the optimal rotation time compared to the deterministic case: Ceteris paribus, the stochastic case has a shorter optimal rotation time. The reason is the added positive term $(f_n - g_n) \cdot h_n$ on the right-hand side of the optimality criterion (13), which requires a higher growth rate corresponding to a shorter rotation compared to the deterministic case.

Infinitely many rotations

For infinitely many rotations, the resulting equation for the stochastic present land value generalizes the well-known Faustmann result to incorporate survival risk. It is important to see that the present values for all subsequent rotations are identical in distribution if considered at the beginning of each rotation, and that the discount factors for subsequent periods are independent of each other by assumption.

In the infinite sequence of rotations, the rotation k starts at time t_k with $t_0 = \tau_0 = 0$ und $t_k = \sum_{i=1}^{k}\tau_i$ with $k = 0,1,2, \ldots$, so that the durations of all rotations are unknown a priori, as are the beginnings of all rotations but the first.

By definition we have $\tau_j = j\Delta t$ for the lifetime within a given rotation, so the corresponding random lifetime can be represented by the random time index j. For each rotation k, we have to consider the corresponding lifetime random variable, so we index the time index by the rotation index, giving j_k. The present land value corresponding to rotation k at the beginning of this rotation is given by the random present value $\widetilde{LV}_1(j_k,n)$, which we abbreviate as $\widetilde{LV}_1(j_k)$. Obviously, the distribution of τ_{jk} is given by the probability mass function as noted before, which also describes the distribution of the random time indices j_k for rotations $k = 1,2, \ldots$.

The present value of the infinite series is therefore

$$\widetilde{LV}(n) = \widetilde{LV}_1(j_1) + (1+r)^{-j_1} \cdot \widetilde{LV}_1(j_2) + (1+r)^{-(j_1+j_2)} \cdot \widetilde{LV}_1(j_3) + \cdots \tag{14}$$

$$= \sum_{k=0}^{\infty} (1+r)^{-\sum_{\ell=0}^{k} j_\ell} \cdot \widetilde{LV}_1(j_{k+1}) \tag{15}$$

with j_0 set to zero. By assumption, all j_k are independent and identically distributed according to m. Therefore, the expectation is as simple as

$$LV(n) = LV_1(n) + E\left[(1+r)^{-j_1}\right] \cdot LV_1(n)$$
$$+ E\left[(1+r)^{-(j_1+j_2)}\right] \cdot LV_1(n) + \cdots \tag{16}$$

$$= \sum_{k=0}^{\infty} E\left[(1+r)^{-\sum_{\ell=0}^{k} j_\ell}\right] \cdot LV_1(n) \tag{17}$$

$$= \sum_{k=0}^{\infty} v(n)^k \cdot LV_1(n) \tag{18}$$

$$= LV_1(n) \cdot \frac{1}{1-v(n)} \tag{19}$$

where

$$v(n) = E[(1+r)^{-j}] \tag{20}$$

$$= \sum_{j=0}^{n} (1+r)^{-j} w_j + (1+r)^{-n} S_n \tag{21}$$

Please note the structural equivalence to the corresponding equation for the deterministic case. In the stochastic case, all that happens structurally is the replacement of the deterministic one-rotation present value by its stochastic counterpart LV_1 and the replacement of the deterministic discount factor $(1+r)^{-n\Delta t}$ by its stochastic counterpart $v(n)$ to reflect the positive probability of premature calamity use.

As in the one-rotation case, we are interested in the optimal (planned) rotation n, and approach this by again using our approximate argument for the optimality condition based on the first difference of land value. From $LV(n) - LV(n-1) \approx 0$ we get the first order optimality criterion

$$\Delta f_n = r \cdot f_{n-1} + (f_n - g_n) \cdot h_n + r \cdot LV(n-1) \tag{22}$$

Proof: The first difference is $LV(n) - LV(n-1) = LV_1(n)/(1-v(n)) - LV_1(n-1)/(1-v(n-1))$. Using the same line of argument as in the one-period case, we want this difference to vanish. Setting to zero and eliminating the common denominator yields $LV_1(n) \cdot (1-v(n-1)) - LV_1(n-1) \cdot (1-v(n)) = 0$. Now we introduce the first difference $\Delta LV_1(n) = LV_1(n) - LV_1(n-1)$ known from the one-period model and $\Delta v(n) = v(n) - v(n-1)$ to get the equivalent formulations $\Delta LV_1(n) - LV_1(n) \cdot v(n-1) + LV_1(n-1) \cdot v(n) = \Delta LV_1(n) - (LV_1(n-1) + \Delta LV_1(n)) \cdot v(n-1) + LV_1(n-1) \cdot v(n) = \Delta LV_1(n) + LV_1(n-1) \cdot \Delta v(n-1) - \Delta LV_1(n-1) \cdot v(n-1) = \Delta LV_1(n) \cdot (1-v(n-1)) + LV_1(n-1) \cdot \Delta v(n) = 0$. Dividing the last equation by $1 - v(n-1)$ gives

$$\Delta LV_1(n) + LV(n-1) \cdot \Delta v(n) = 0$$

and recovers the land value for infinitely many rotations LV from the corresponding one-period value LV_1.

Now observe that $\Delta v(n) = (1+r)^{-n} w_n + (1+r)^{-n} S_n - (1+r)^{-(n-1)} S_{n-1}$. Using $w_n/S_{n-1} = h_n$ and $S_n/S_{n-1} = 1 - h_n$ and rearranging yields $\Delta v(n) = -r(1+r)^{-n} S_{n-1}$.

Substituting this expression in the previous equation and then rearranging terms as in the proof of the corresponding one-period condition yields the assertion †.

Again, the economic interpretation is quite intuitive. The first two terms on the right-hand side of condition (22) are exactly the same as in the corresponding condition (13) for the single-period case, and have the same interpretation. What is added is the last term, which reflects the foregone interest on the land value due to its continued use to let the currently planted trees grow one more time increment. Therefore, the stochastic model captures the scarcity of land in exactly the same way as the standard Faustmann model does. As a result, we see a remarkable structural equivalence between the classic deterministic approach and the stochastic model just presented. Nevertheless, despite the structural equivalence, there are important differences with regard to actual land values, and those differences vary with the type of risk faced by a specific stand. This will be demonstrated further in the next section, using a numerical example based on actual data. But before doing so, we consider how survival risk influences optimal rotation and add some remarks on the land rent in our stochastic setting.

The optimality criterion for the stochastic one-period model (13) was different than the corresponding one-period deterministic model by one added positive term $(f_n - g_n) \cdot h_n$ on the right-hand side, which caused ceteris paribus a shorter optimal rotation time. The optimality criterion (22) has one additional positive term, the land value, causing a further reduction of the optimal rotation time compared to the one-period stochastic model. But when compared to the standard deterministic Faustmann model, the optimal rotation time may be shorter or longer. The reason is as follows: The new term $(f_n - g_n) \cdot h_n$ is positive, so ceteris paribus a shorter rotation time would be optimal. But both the deterministic Faustmann and our stochastic model contain an additional land value term on the right-hand side of the optimality criterion. In the stochastic model, this land value term is lower than in the deterministic Faustmann model. So, the optimal rotation in the stochastic model will be longer than in the deterministic case if the decrease in land value overcompensates the term $(f_n - g_n) \cdot h_n$. This depends on the shape of the survival function, that is, on the data of the application.

Analytic model formulation for land rent

In the deterministic Faustmann model, it is conceptually easy to define a land rent. In the stochastic model, the concept of rent is not as clear as in the deterministic case. In contrast to that, there is no well-defined annuity attributable to any one of the rotations, because the actual rotation length is stochastic.

We assume that the land rent to be defined should provide a useful description of the potential of consumption per year attributable to the land value. Given this perspective, and assuming that the model just presented provides a reasonable description of land value, we can define the yearly land rent simply as the product of land value times yearly interest rate. Defined this way, it is not necessary to relate the notion of rent explicitly to the stochastic properties of the payment stream.

We use this notion of land rent in the following examples.

Numerical example based on real-world data

This section will illustrate how the shape of the survival function relates to the land rent as defined previously, which we have chosen as the economic target. The land rent is more intuitive than the land value, despite the simple proportionality between the two values. The analysis is based on actual empirical data. That is, all the results indicate the actual order of magnitude of the described effects, and their economic relevance – at least for German conditions.

The numerical example also demonstrates that it is straightforward and easy to include additional payments of practical importance in the model outlined so far. A standard spreadsheet is sufficient for the implementation. We used Microsoft Excel.

Basic production, revenue and cost models

The example is based on a Norway spruce stand (Wiedemann, 1936/1942, I. yield class, heavy thinning, diameters adjusted according to Wollborn and Böckmann, 1998) mapped in 5-year periods. Thinning volumes (extending our model as presented here) and volumes from final harvesting were valuated with costs and revenues according to Offer and Staupendahl (2009) and Hessen-Forst (2011). We included pre-commercial thinnings at the ages of 10 and 20 years, with costs of 250 EUR/ha for each of them. To get a steady development of the stumpage value and the yields from thinnings over time, the discrete, and a bit erratic, data were smoothed by polynomial functions. The function of the stumpage value corresponds to the net revenue function f_n in our analytical model. The annual value increment corresponding to Δf_n in the optimality criterion (22), however, cannot simply be derived from the stumpage value function in the extended model, but has to be derived from the total financial yield, because yields from thinnings are taken into account (Table 21.4).

Next, we have to specify the loss of net revenue if a calamity hits the stand. At the age of n, this loss corresponds to the difference $(f_n - g_n)$ in our analytical model. Based on Dieter (2001) we assume a proportional reduction of 40% from the stumpage values. In addition, a general increase in costs of 1500 EUR/ha was assumed for calamity-caused final harvest, due to additional logging residues and difficult replanting (based on Hessen-Forst, 2011). Altogether, we assume $g_n = 0.6 \cdot f_n - 1500$ EUR/ha.

Furthermore, we assume $L = 2250$ EUR/ha for stand establishment and an interest rate of $r = 1.5\%$ *p.a.* This is interpreted as a real interest rate, as all prices and costs were also assumed to be in real terms.

Otherwise, all model assumptions have been kept. Particularly, we stick to the assumption that the economic environment remains constant over time, that after calamity and planned harvest reforestation takes place immediately, and that all costs are also payments.

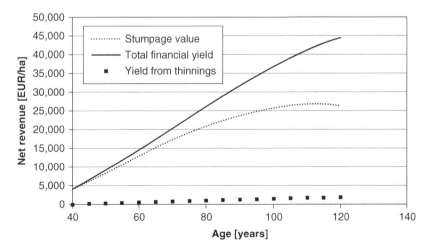

Figure 21.4 Development of stumpage value, yields from periodic thinnings and total financial yield over time for the exemplary Norway spruce stand.

Influence of survival risk on forest management

Next, we analyze the optimal forest management for various realistic risk scenarios. Therefore, we consider the land rent as a function of planned rotation time. Figures 21.5 and 21.6 show how this relation changes with varying risk levels and risk type parameters S_{100} and α. The maxima correspond to the optimal planned rotation time, which should be realized by forest management. Both figures also contain the graph for the classic deterministic Faustmann model, that is, the risk-free case. For that, we find an optimal rotation time of about 95 years with corresponding (maximum) land rent of about 119 EUR/ha/a.

Figure 21.5 additionally shows the results for two risk levels (high risk $S_{100} = 0.5$, moderate risk $S_{100} = 0.7$), keeping the risk type constant (old-age risk with $\alpha = 3.0$). We observe that the land rent decreases with increasing risk levels. Simultaneously, the optimum rotation period also decreases. Under these circumstances it obviously is profitable to reduce the planned

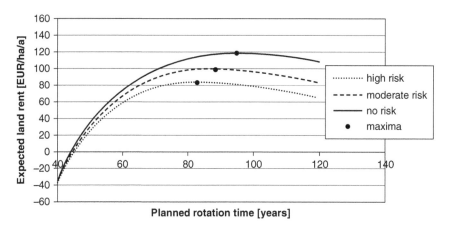

Figure 21.5 The land rent for "no risk" and different risk levels ($S_{100} = 0.7$ "moderate risk" or 0.5 "high risk") with old–age risk type ($\alpha = 3.0$).

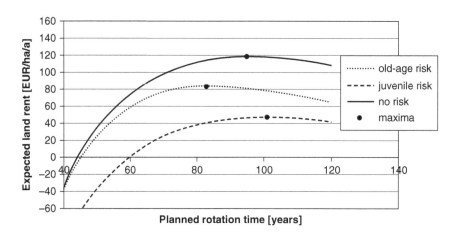

Figure 21.6 The land rent for "no risk" and different risk types ($\alpha = 0.3$ "juvenile risk", $\alpha = 3.0$ "old-age risk") with high-risk level ($S_{100} = 0.5$).

rotation period when survival risks occur. The difference between the land rent that could be gained without survival risk and the corresponding value taking risk into account can be interpreted as average annual costs of the survival risk in EUR/ha/a. When comparing those costs for different risk levels, one should compare the corresponding maxima, because forest management should choose the optimal rotation time for each risk level. In our example, Figure 21.5 shows a difference between no risk and high risk of about 35, − EUR/ha/a, that is, about one-quarter of the rent attainable in the absence of risk.

Figure 21.6 shows the results for different risk types (old-age risks $\alpha = 3.0$ versus juvenile risks $\alpha = 0.3$) for a given risk level (high risk $S_{100} = 0.5$) in comparison to the risk-free case. We observe that the risk type has a major influence on the land rent and the optimal rotation time.

If we compare the optimal rotation time for the two risk types with the risk-free case, we find that old-age risks reduce the optimal rotation time, whereas juvenile risks increase optimal rotation time. In our example, the difference is a decrease to about 85 years or an increase to about 100 years, respectively.

Besides that, the expected land rent is lower for juvenile risk than for old-age risk, which means that juvenile risk seems to be more harmful from an economic point of view. The reason is that in the case of calamities in young stands, no salable timber comes up, whereas old stands usually deliver salable timber, even if timber prices are lower and costs are higher in the case of calamities.

In forest management practice, however, it is often more important to answer the question if a given stand is mature, than to define an optimal planned rotation time. Our model also provides the economic intuition and a guideline for this decision. Note at first that the optimality criterion (22) is a purely local one in time, that is, only the relations of the next time period enter (the future enters only indirectly via the land value term). At the beginning of each period, there are two options: Either harvesting or keeping the current stand for the additional period.

Harvesting the stand immediately gains the net revenue and enables replanting for the next production cycle, so the interest on the stumpage value plus the land rent (the interest on the land value) will be earned (clearly, the land value has to then be calculated with respect to the given risk level). Keeping the stand will result in an expected increase in value, which depends on the two possible events within the considered time period: The calamity may hit (with probability h_n) or not (with probability $[1 - h_n]$). If it hits, one has to face a loss of $(f_n - g_n)$; if not, one gets a gain of Δf_n. If one expects a value increase that is higher than the sum of the expected loss and the interest attainable after harvesting, one should keep the stand; otherwise, not. These thoughts may be considered as an alternative way to derive the optimality criterion (22) for a risk-neutral decision maker.

Please note that survival risks have a negative effect not only on the development of the current stand, but also on all subsequent stands. That is, not only is the expected profit from keeping the current stand lower compared to the risk-free case, but also the land value, which reflects the effect of survival risks on all future stands. Therefore, the survival risk also reduces the possible interest profit in the case of harvesting, which makes the replacement of the current stand less attractive.

These considerations provide another, maybe more intuitive, explanation of the already-stated fact that the optimal rotation time may be either shorter or longer compared to the risk-free case:

Depending on the relationship between the actual risk of the current stand and the average risk of the following stands, an earlier or a later time to harvest is economically optimal. Old-age risks, by which the hazard rate grows with age, lead to an earlier optimal harvest time. Typical juvenile risks, however, which an elder stand already has left behind, lead to later harvest times, because the following stands will again be exposed to these juvenile risks, which makes the harvest of the current stand and its replacement by a following stand economically less attractive. Between these two cases there are risks which lead only to a reduction of the land rent, but do not influence the optimal harvest age compared with a risk-free situation.

Discussion and conclusion

The presented results clearly show that risks of survival have great influence on economic decision criteria and that they should be considered explicitly in forest valuation and decision making.

In line with the classic deterministic Faustmann approach, we developed a stochastic model that incorporated survival risk in a completely general way and allowed the calculation of land value, land rent and optimal rotation time, if these terms are defined as mathematical expectations, suitable for a risk-neutral decision maker, or the management of a forest enterprise that is diversified over sufficiently many stands.

This approach has at least two advantages. First, it supports the development of the economic intuition for practical forest management and illustrates much structural equivalence to known deterministic models. Second, it allows the calculation of the quantities of interest analytically. It is not necessary to use Monte-Carlo simulations with many thousands of variations, followed by tedious data analysis.

For practical applications, a survival function must be specified. By means of the two-parameter Weibull function, the risk level and the risk type can be modeled in a quite flexible way. Using the parameters S_{100} and α to characterize risk level and risk type allows for an intuitive interpretation of the resulting model, which we consider as an advantage in forest practice.

The model shows how the economic target figure, the expected land value or equivalently the land rent, changes with different risk levels and risk types. The differences between the risk-adapted and the risk-free land rent can be interpreted as annual average risk costs, which have to be expected in the long run average. Thus, the presented valuation scheme provides a foundation for cost-benefit considerations, if decisions on risk mitigation like forest protection, tending of stands, changing of tree species, etc., have to be assessed.

The model also indicates that under risk, different forest management strategies are required depending on the risk type (old-age risks versus juvenile risks). Typical juvenile risks (such as damage caused by game or snow breakage, etc.) lead to longer optimal rotation times, whereas typical old-age risks (such as windblow) lead to shorter optimal rotation times compared to the risk-free case. These relations are intuitive as survival risks affect both the expected gain from keeping the current stand for one more time period and the potential gain from harvesting immediately.

The availability of appropriate data will likely remain an issue in applications. But even a qualitative incorporation of survival risks in decision making is likely to improve the status quo. Furthermore, some fundamental relationships like the susceptibility of Norway spruce especially to storm damage are known quite well. And although it is true that the problem of risk management will get even more involved if we consider the effect that forest management itself has on survival risks (e.g. by the frequency and intensity of thinnings, as stressed by Thorsen and Helles, 1997), the conceptual approach used here is likely to remain useful, as forest management decisions will basically be reflected in survival functions as explanatory variables.

Future research should address two main issues, the improvement of descriptive models of survival with the explicit inclusion of risk factors and scale of damage, and the development of more advanced forest management models to adapt to risks. Flexibility in management and capital market restrictions are of fundamental importance, but in forest economics there is a gap in knowledge of how they relate to survival risk.

References

Beinhofer, B. (2007). 'Zum Einfluss von Risiko auf den optimalen Zieldurchmesser der Fichte', *Forstarchiv*, vol. 78, pp. 117–124.

Bräuning, R. and Dieter, M. (1999). 'Waldumbau, Kalamitätsrisiken und finanzielle Erfolgskennzahlen. Eine Anwendung von Simulationsmodellen auf Daten eines Forstbetriebes', *Schriften zur Forstökonomie*, vol. 18, J. D. Sauerländer's, Frankfurt a.M.

Brazee, R. J. and Newman, D. H. (1999). 'Observations on recent forest economics research on risk and uncertainty', *Journal of Forest Economics*, vol. 5, pp. 193–200.

Chang, S. J. and Gadow K. v. (2010). 'Application of the generalized Faustmann model to uneven-aged management', *Journal of Forest Economics*, vol. 16, pp. 313–325.

Deegen, P. (1994). 'Beitrag zur Analyse und Berechnung von Risiko am Einzelbestand', *Forstarchiv*, vol. 65, pp. 280–285.

Dieter, M. (1997). 'Berücksichtigung von Risiko bei forstbetrieblichen Entscheidungen', *Schriften zur Forstökonomie*, vol. 16, J. D. Sauerländer's, Frankfurt a.M.

Dieter, M. (2001). 'Land expectation values for spruce and beech calculated with Monte Carlo modelling techniques', *Forest Policy and Economics*, vol. 2, pp. 157–166.

Hanewinkel, M., Hummel, S. and Albrecht, A. (2010). 'Assessing natural hazards in forestry for risk management: A review', *European Journal of Forest Research*, vol. 130, no. 3, pp. 329–351.

Hessen-Forst, H. F. (2011). *Einschlags- und Verkaufsstatistik 2007–2009 (für Waldbewertungszwecke aufbereitet)*, Gießen.

Holecy, J. and Hanewinkel, M. (2006). 'A forest management risk insurance model and its application to coniferous stands in southwest Germany', *Forest Policy and Economics*, vol. 8, no. 2, pp. 161–174.

Klein, J. P. and Moeschberger, M. (1997). *Survival Analysis: Techniques for Censored and Truncated Data*, Springer, New York.

Knoke, T., Ammer, C., Stimm, B. and Mosandl, R. (2008). 'Admixing broadleaved to coniferous tree species: A review on yield, ecological stability and economics', *European Journal of Forest Research*, vol. 127, pp. 102–116.

Knoke, T. and Wurm, J. (2006). 'Mixed forests and a flexible harvest policy: A problem for conventional risk analysis?' *European Journal of Forest Research*, vol. 125, no. 3, pp. 303–315.

König, A. (1996). *Sturmgefährdung von Beständen im Altersklassenwald. Ein Erklärungs- und Prognosemodell*, J. D. Sauerländer's, Frankfurt a.M.

Kouba, J. (2002). 'Das Leben des Waldes und seine Lebensunsicherheit', *Forstwissenschaftliches Centralblatt*, vol. 121, pp. 211–228.

Kuboyama, H. and Oka, H. (2000). 'Climate risks and age-related damage probabilities – effects on the economically optimal rotation length for forest stand management in Japan', *Silva Fennica*, vol. 34, no. 2, pp. 155–166.

Kurth, H., Gerold, D. and Dittrich, K. (1987). 'Reale Waldentwicklung und Zielwald–Grundlagen nachhaltiger Systemregelung des Waldes', *Wissenschaftliche Zeitschrift der Technischen Hochschule Dresden*, vol. 36, pp. 121–137.

Möhring, B., Burkhardt, T., Gutsche, C. and Gerst, J. (2011). 'Berücksichtigung von Überlebensrisiken in den Modellen der Waldbewertung und der forstlichen Entscheidungsfindung', *Allgemeine Forst- und Jagd-Zeitung*, vol. 182, pp. 160–171.

Möhring, B., Staupendahl, K. and Leefken, G. (2010). 'Modellierung und Bewertung naturaler forstlicher Risiken mit Hilfe von Überlebensfunktionen', *Forst und Holz*, vol. 65, no. 4, pp. 26–30.

Offer, A. and Staupendahl, K. (2009). 'Neue Bestandessortentafeln für die Waldbewertung und ihr Einsatz in der Bewertungspraxis', *Forst und Holz*, vol. 64, no. 5, pp. 16–25.

Pienaar, L. V. and Shiver, B. D. (1981). 'Survival functions for site prepared slash pine plantations in the flatwoods of Georgia and northern Florida', *Southern Journal of Applied Forestry*, vol. 5, pp. 59–62.

Price, C. (2011). 'When and to what extent do risk premia work? Cases of threat and optimal rotation', *Journal of Forest Economics*, vol. 17, pp. 53–66.

Reed, W. J. (1984). 'The effects of the risk of fire on the optimal rotation of a forest', *Journal of Environmental Economics and Management*, vol. 11, pp. 180–190.

Routledge, R. D. (1980). 'The effect of potential catastrophic mortality and other unpredictable events on optimal forest rotation policy', *Forest Science*, vol. 26, no. 3, pp. 389–399.

Schmidt, M., Hanewinkel, M., Kändler, G., Kublin, E. and Kohnle, U. (2009). 'An inventory-based approach for modeling single tree storm damage – experiences with the winter storm 1999 in southwestern Germany', *Canadian Journal of Forest Research*, vol. 40, no. 8, pp. 1636–1652.

Staupendahl, K. and Möhring, B. (2011). 'Integrating natural risks into silvicultural decisions models: A survival function approach', *Forest Policy and Economics*, vol. 13, pp. 496–502.

Staupendahl, K. and Zucchini, W. (2011). 'Schätzung von Überlebensfunktionen der Hauptbaumarten auf der Basis von Zeitreihendaten der Rheinland-Pfälzischen Waldzustandserhebung', *Allgemeine Forst- und Jagd-Zeitung*, vol. 182, pp. 129–145.

Suzuki, T. (1971). 'Forest transition as a stochastic process', *Mitt. Forstlichen Bundesversuchsanstalt (FBVA)*, vol. 91, pp. 137–150.

Suzuki, T. (1983). 'Gentan-Wahrscheinlichkeit, Vorhersagemodelle für die Entwicklung des Normalwaldes und für die Planung des Holzaufkommens', *Schriften a. d. Forstl. Fak. d. Univ. Göttingen*, vol. 76, J. D. Sauerländer's, Frankfurt a.M.

Thorsen, B. J. and Helles, F. (1997). 'Optimal stand management with endogenous risk of sudden destruction', *Forest Ecology and Management*, vol. 108, pp. 287–299.

Weibull, W. (1951). 'A statistical distribution of wide applicability', *Journal of Applied Mechanics*, vol. 18, pp. 293–297.

Wiedemann, E. (1936/1942): Fichtenertragstafel pp. 62–75. in R. Schober (1995) *Ertragstafeln wichtiger Baumarten bei verschiedener Durchforstung* (2nd ed.), J. D. Sauerländer's, Frankfurt a.M.

Wollborn, P. and Böckmann, T. (1998). 'Ein praktikables Modell zur Strukturierung des Vorrates aus Ertragstafelschätzung', *Forst und Holz*, vol. 53, no. 18, pp. 547–550.

22

REAL OPTION ANALYSIS IN FOREST MANAGEMENT DECISION MAKING

Rajendra Prasad Khajuria

IFS, CHIEF CONSERVATOR OF FORESTS, ANDHRA PRADESH, INDIA, 500004. RPKHAJURIA@GMAIL.COM

Abstract

An overview of the importance of real options analysis (ROA), its relevance and applications to forest management decision making is provided. The significance of ROA in valuing the managerial flexibility under stochastic situation is highlighted. The application of ROA is divided into two categories – forest harvesting and forest conservation decisions. In the case of forest harvesting decisions, the main features of modeling of timber prices as the geometric Brownian motion (GBM), mean reversion (MR), mean reversion with jumps (MRJ) and mean reversion with varying long-run marginal cost process are discussed, and the main findings of the studies using these processes are analyzed. The key findings of the studies focused on forest conservation decisions are highlighted. Some suggestions for future research on the relevant issues are proposed.

Keywords

Forestry, forest conservation, geometric Brownian motion, mean reversion, real options analysis, stochastic price, timber price, uncertainty

Introduction

Forest management involves investments which require long gestation periods to yield returns. This makes forest investment decisions prone to various risks like fires, storms, insect attacks and uncertainties associated with prices of timber and other forest products, interest rates and future rates of growth of timber. In literature, sometimes risk and uncertainty are used interchangeably. However, it is important to understand that risk occurs due to uncertainty about the future. If future events were known in advance, then there would be no risk at all. Uncertainty, therefore, needs to be accounted for while making investment or other forest management decisions. Further, most forest management decisions are irreversible in nature. For instance, once a forest stand is harvested, it cannot be restored. Ignoring uncertainty could seriously affect the quality of management decisions, resulting in losses to the investments made, and this could be detrimental to the cause of forest management. Hence, efforts should be made to use uncertainty to one's advantage rather than being at its mercy. Traditional approaches in dealing with uncertainty do not capture the flexibility of the administration to adapt appropriately and to review decisions in response to unexpected changes in the market (Trigeorgis, 1996) and other factors. It is inherently assumed that once begun, management cannot deviate from its initial plan, even if market and other factors turn out to be significantly different than originally anticipated. Decision making under uncertainty should be adaptive so that as time passes and uncertainty gets resolved with availability of new information, decisions can be altered depending upon the new situation. The ability to adapt to uncertainty will be better if the time span considered is shorter. This implies, in essence, that decision makers should keep their options open so that the response to changing scenarios is better. This assumes more significance when decisions are irreversible in nature. Forestry is a perfect example of such situations. Flexibility in decision making is key in the presence of uncertainty, and specifically in the environment of irreversible decisions (Dixit and Pindyck, 1994).

Real options analysis (ROA) or real options valuation (ROV) offers the required flexibility in decision making. ROA is the application of option value techniques from finance to capital budgeting decisions influencing real investments (Trigeorgis, 1996). The theoretical foundations of real options are rooted in uncertainty and irreversible decisions, and that is why it is also referred to as the theory of irreversible investment under uncertainty. The risk and uncertainty associated with management decisions are included in the formulation of real options problems (Dixit and Pindyck, 1994).

Real options are investments in real assets which confer the investor with the right, but not the obligation, to undertake certain actions in the future (Schwartz and Trigeorgis, 2004). A real option is defined as 'the value of being able to choose some characteristic (e.g. the timing) of a decision with irreversible consequences, which affects a real asset (as opposed to a financial asset)' (Saphores and Carr, 2000, p. 255). Risk due to uncertainty (e.g. stochasticity of an asset's value) is exploited to the advantage by adaptive decision making for an optimal decision. This is akin to accounting for the value of the future uncertainty.

Arrow and Fischer (1974) measured the impact of uncertainty on some economic activities by using the quasi-option value concept. Conrad (1980) determined the option value of preserving or developing a wilderness area. In forestry literature of the early 1970s, the major thrust was on studying the effect of some stochastic variable on the rotation age. Norstorm (1975) determined the optimal rotation age in the presence of stochastic timber prices. That study was subsequently extended under various scenarios by other researchers such as Brazee and Mendelsohn (1988), Lohmander (1988), Clarke and Reed (1989) and Reed and Clarke (1990) to study the impact of stochastic timber and stumpage prices on determination of harvesting age. These studies, however, do not value the flexibility that is the prime feature of ROA. Dixit's (1989) study on the entry and exit decisions of a generic firm has wider applications in forest economics because many decisions related to forest management, such as harvesting age of a stand, protection of biodiversity and development of a wilderness area, can be conceptualized as entry or exit decisions. Subsequently, ROA received attention in forestry as well, owing to its emergence in finance and other natural resources studies. Forest economists have used ROA profoundly to analyze forest harvesting decisions by incorporating the stochastic nature of standing timber or lumber prices and tree growth and volumes. Some studies have used ROA to analyze forestry conservation decisions such as wilderness preservation, option value of an old-growth forest, biodiversity conservation and land conversion decisions (e.g. Conrad, 1997, 2000; Bakshi and Saphores, 2004; Marwah and Zhao, 2002).

In this chapter, an overview of the application of ROA to forest management decisions regarding harvesting as well as conservation is provided. In the case of the application of ROA to forest harvesting decisions, the key issue is the characterization of the stochastic timber price process; therefore, different aspects associated with characterization of the price process are discussed, followed by key findings of these studies. There follows a discussion of the application of ROA to forest conservation decisions. Finally, the chapter concludes with some thoughts on the future road map for research in this area.

ROA and forest harvesting decisions

Before the advent of the application of ROA to estimate option values and capture flexibility as explained in the introduction, forest harvesting decisions with stochastic variables were studied by Miller and Voltaire (1983), Brazee and Mendelsohn (1988), Lohmander (1988) and Clarke and Reed (1989). However, the actual application of the ROV concept in forestry took some time due to the mathematical complexity of the techniques (as compared to traditional approaches of net present value and land expectation value) and higher data requirements. Since the 1990s, researchers such as Morck, Schwartz and Stangeland (1989), Zinkhan (1991), Thomson (1992), Plantinga (1998), Yin (2001), Insley and Rollins (2005), Khajuria, Kant and Laaksonen-Craig (2009) and others have used these techniques in forestry while studying risk and uncertainty.

The real option models are based on several assumptions, including the assumption regarding the stochastic price process of the underlying asset. In order to assess the various options, assumptions regarding evolution of future prices need to be made very carefully. The results obtained are sensitive to the type of price process assumed in such models. Brazee and Mendelsohn (1988) solved a multirotation model wherein the stochastic price process was assumed to be a random draw from a normal distribution. Haight (1993), in his study on effects of stochastic stumpage price trends of pulpwood and sawtimber of loblolly pine, assumed the price trends to be independent random variables with triangular probability distributions. Lohmander (1988) modeled the price as a random draw from a uniform distribution in his study of the effect of stochastic prices on optimal harvesting in a single rotation problem. In these studies, it was assumed that the timber price in any given period is statistically independent of its level in any other period. Many researchers argue that the assumption of serially uncorrelated prices is not realistic; for instance, Washburn and Binkley (1990) found serial correlation in monthly southern pine stumpage market prices. However, these articles provide some useful intuition of the impact of stochastic stationary prices.

Some of the researchers conducted detailed statistical tests to gain insights into the behavior of the prices. For instance, Haight (1990), Hultkrantz (1995), Yin and Newmann (1997), Insley and Rollins (2005) and Khajuria et al. (2009) have observed stationarity in the data series used by them. Prestemon (2003), using the data series of some earlier studies and some new data sets, and applying powerful tests, found evidence of nonstationarity in the quarterly price series. Hultkrantz (1995) and Khajuria et al. (2009) rejected the unit-root hypothesis for their data series and observed structural breaks in the prices. However, these tests are not conclusive, mainly due to the short span of time-series data available for stumpage and timber prices and also due to the shortcomings of the statistical tests used in the studies.

Broadly speaking, these studies can be segregated into two different categories depending upon the price process adopted while modeling the stochastic process. The first one models the price as some random walk process, usually a geometric Brownian motion (GBM) process, and the other one uses a stationary process like the mean reverting (MR) Ornstein-Uhlenbeck process.

The MR process can be modeled by assuming the long-run equilibrium price to be fixed or variable, and also by incorporating the sudden shocks in the long-run price formation process. Next, we discuss the main features of these different price formation processes and studies that used these processes.

Evolution of timber prices as GBM process

The GBM model is convenient and simpler to use and is based on the assumption of efficient markets. For the GBM process, percent change in timber price (P) is given by

$$\frac{dP}{P} = \mu dt + \sigma dz \tag{1}$$

where μ is the constant drift rate, σ is the constant variance rate, also referred to as the volatility of the price process, and dz is the increment of a standard Wiener process $= e\sqrt{dt}$ (where e is the standard normal distribution).

Under stochastic prices, the optimal decision is to hold the asset as long as it is expected to earn the required return. The option value $F(P,t)$ arises from the flexibility to respond to the future values of the stochastic variable. If revenue is V and growing stock of timber is Q, then $V = P \times Q$, where P follows a known GBM stochastic process and Q may also be stochastic or may be assumed to be deterministic. If S is the strike price per cubic meter, T is the terminal period, ρ is the discount rate and decisions are binary, i.e. harvest or wait, then the solution procedure can be viewed as a maximization problem under uncertainty. Using the Bellman equation (Dixit and Pindyck, 1994, chapter 4):

$$F(P,t) = Max[V(P) - S, E[F (P + dp,t + dt)e^{-\rho dt}]]; \forall t \leq T \tag{2}$$

Using Ito's lemma (Dixit and Pindyck, 1994, p. 117), equation (2) can be converted into a partial differential equation (PDE), which can be solved by a finite difference method for numerical solution of the maximization problem.

Morck et al. (1989) adopted a GBM model for the price process and used a PDE approach to determine the optimal harvesting rate. They concluded that the flexible price harvest policy significantly increases the present value of expected returns over the rigid Faustmann model. Thomson (1992) determined land rent endogenously treating stumpage prices as a GBM process and compared the optimal rotation ages of the Faustmann model with fixed timber prices with a binomial option-pricing model. Clarke and Reed (1989) and Reed and Clarke (1990) derived an optimal harvesting rule treating the price as a GBM and growth rate as a function of stand age plus random Brownian motion. In their study, ignoring harvesting and management costs, the age at which a stand is harvested becomes a random variable that is independent of the absolute level of timber prices. Yin and Newman (1997) compared the optimal harvest time at which both the growth and price are treated as stochastic. Yoshimoto and Shoji (1998) used the binomial tree approach to model a GBM process for timber prices in Japan and solved for the optimal rotation ages. Manley and Niquidet (2010), assuming that the timber price follows a GBM, compared the Faustmann method with three ROV methods, namely the binomial option pricing model, the stochastic dynamic programming model and the abandonment adjusted price model. Their findings indicate that when prices follow a random walk, there is little to no option value.

The GBM model assumption suggests that prices may go on increasing without bounds, which is not possible in a competitive market. For example, in the case of a GBM process, when the mean and variance grow without bounds, it implies the scope for infinite profits. The economic logic will tell us that in a competitive market, if the prices significantly decrease, the profit in the industry may dwindle, leading to the exit of high-cost producers, which will reduce supply, and prices will subsequently rise again. On the other hand, if prices significantly rise, new suppliers will enter, supply will increase and prices will decrease. Further, equation (1) implies that if the price of a commodity under GBM ever becomes zero, then it will remain zero forever, which seems implausible. Hence, many authors have argued for the use of the mean reversion (MR) process to model timber prices.

Evolution of timber prices as an MR process

It is generally agreed that in the long run, commodity prices should be related to marginal production costs. The adjustment to long-run marginal production costs often takes time. Consequently, prices can temporarily rise above or fall below the long-run marginal cost, but subsequently should drift back toward the long-run equilibrium. Hence, in recent times many economists, including Plantinga (1998), Gjolberg and Guttormsen (2002), Insley (2002), Insley and Rollins (2005) and Khajuria et al. (2009, 2011) modeled timber prices as an MR process in their studies of the application of ROA to forest harvesting.

MR is typically captured by assuming price follows an Ornstein-Uhlenbeck process (Dixit and Pindyck, 1994). An Ornstein-Uhlenbeck process has independent increments, meaning that the change in price between two consecutive periods is independent of the change between any two other consecutive periods.

For a standard Ornstein-Uhlenbeck MR process (Dixit and Pindyck, 1994, p. 161) proportional change in timber price (P) is given by

$$\frac{dP}{P} = \eta\left(\bar{P} - P\right)dt + \sigma dz \tag{3}$$

where $\bar{P}$ is the long-run mean price which the prices tend to revert, η is the rate of reversion to mean; dz and σ are the same as in equation (1). The instantaneous variance σ^2 grows with P, so that the variance is zero if P is zero. The term on the right-hand side, $\eta(\bar{P} - P)$, as a whole is referred to as the instantaneous drift of the asset.

The general finding of most researchers is that an increase in mean drift rate leads to an increase in the asset's value as well as the threshold limit for harvesting. So there is an incentive in delaying the harvest if prices are likely to face an upward trend, and vice versa. Plantinga (1998), using the mean reverting process for prices, found that option values increase, which demonstrates the role of option values in choosing the optimal timing of harvests. Therefore, the option value becomes a premium over the expected value of a timber stand, reflecting the opportunity cost of harvesting now and foregoing the option to delay harvest until information on future stand values is revealed. In his study, it is observed that expected timber values are higher with a reservation price policy when timber prices are stationary compared to the Faustmann model with expected prices. When timber prices are nonstationary, the expected timber values are identical to Faustmann values, thereby implying that when prices follow a random walk, there is little to no option value. Gjolberg and Guttormsen (2002) applied real options to a tree cutting problem, assuming that the stumpage prices follow a mean reverting process, and observed consequences for mean reverting prices compared to the traditional random-walk assumption. They argued that what traditionally has been seen as irrational pricing with discount rates that are too low may represent rational pricing of relatively low-risk, long-term investments. Insley

(2002) investigated the role of the timber price process on the rotation length in a single-rotation model. A dynamic programming approach and a general numerical solution technique were used to determine the value of the option to harvest a stand of trees and the optimal cutting time when timber prices follow a known stochastic process. It was observed that for prices below the mean, a process of MR implies option values higher than that of the GBM. Further, during the early years of growth, the MR process has lower critical prices as compared to the GBM, implying that under MR, it is more likely that harvesting a forest will be optimal even if it is not fully mature. Timber prices should eventually revert to some mean, reflecting long-run marginal costs. Duku-Kaakyire and Nanang (2004) compared a forestry investment using the Faustmann Net Present Value (NPV) model and the real options approach. They investigated four options: an option to delay deforestation, an option to expand the size of the wood processing plant, an option to abandon the processing plant if timber prices fall below a certain level and an option that included all three of these individual options. This analysis was conducted using the binomial tree method. The results show that while the Faustmann analysis rejected investments as unprofitable, the ROA showed that all four options were highly valuable. In Insley and Rollins (2005) the authors extended the single-rotation work of Insley (2002) to multiple rotations, and analyzed forest stand value with stochastic timber prices. In their work, it was argued that the prices of timber should revert to some long-run mean which represents the marginal product costs. Guthrie and Kumareswaran (2009) assumed a mean reverting price process for studying the optimal harvest decision of forests in Oregon (USA) using the binomial tree method. They compared results for real options and NPV/Land Expectation Value (LEV) for both single and infinite rotations. The increased valuation of real options compared to fixed rotations is also a highlight of the study.

Evolution of timber prices as an MR process with structural breaks

Both the MR and the GBM consider only continuous diffusion processes. Most market prices are outcomes of two types of information: normal information, which results in smooth variation in prices, and abnormal information, which results in discrete jumps in prices (Merton, 1976). For instance, the entry by a new big firm into a market or the introduction of a new substitute product may result in sudden price changes. Sometimes natural hazards, uncertainties related to outcomes of elections in democratic countries and similar other uncertainties may also result in sudden, but temporary, price changes. These abrupt changes in prices, known as jumps, can be incorporated into mathematical models as a Poisson discrete time process. Yin and Newman (1996) studied the effect of catastrophic risk through a Poisson jump process on forest investment decisions by employing a forest-level model where the output price is specified to follow a stochastic process and found that the presence of catastrophic risk always results in a reduced production value but an increased investment threshold for a forestry project. Saphores, Khalaf and Pelletier (2002) considered jump effect for a GBM process, with respect to an old-growth cutting problem in the Pacific Northwest. They found evidence of jumps and ARCH effects in the data and also studied the consequences of ignoring these jumps. Their study demonstrates the importance of modeling jumps explicitly. Khajuria et al. (2009) incorporated jumps with an MR process in their study of option values for a pine forest in Ontario, Canada. In their specification, if P follows a known mean reversion with jumps (MRJ) stochastic process (Dixit and Pindyck, 1994, p. 85), then equation (3) can be modified as:

$$\frac{dP}{P} = \left[\eta(\bar{P} - P) - \lambda k \right] dt + \sigma dz + dq \tag{4}$$

where $\bar{P}$ is the long-run equilibrium level (or the long-run mean price to which the prices tend to revert); η is the speed of reversion to mean; dz is a Wiener increment $= e\sqrt{dt}$ (where e is the standard normal distribution) and σ is the volatility of P. The instantaneous variance σ^2 grows with P, so that the variance is zero if P is zero. The term on the right-hand side, $\eta(\bar{P} - P)$, as a whole is referred to as the instantaneous drift of the asset; dq is the Poisson (jump) term; λ denotes the mean arrival rate of an event during an infinitesimally small time interval dt. The probability that an event will occur is given by λdt and the probability that an event will not occur is given by $1 - \lambda dt$. Here, the mean reverting drift has been compensated by λk, for the Poisson jump expected value. The jump is of size Φ, which can be a random variable, defined as follows:

$$dq = \begin{cases} 0 & \text{if} \quad probability\ 1 - \lambda dt; \\ \Phi & \text{if} \quad probability\ \lambda dt; \end{cases}$$

Generally, it is assumed that the Wiener (dz) and Poisson (dq) processes are not correlated. The jump size and direction can be random. In the previous equation, Φ is the jump size probability distribution. The numerical solution can be obtained using the explicit finite difference method. Khajuria et al. (2009) found that options values are higher for the jump model as compared to the simple mean reverting model, which means that under the jump scenario, waiting is beneficial, whereas the regular mean reverting model would suggest harvesting the forest. The thresholds determined for the jump model are also higher than those for the MR model. The results indicate that jumps have the potential to enhance option values, and ignoring jumps may lead to suboptimal decisions. This has clear implications for management decision making.

Evolution of timber prices as a MR process with stochastic trend

Generally, the studies using the MR process in timber prices have assumed that timber prices tend to revert to a fixed long-run equilibrium price (Insley, 2000; Insley and Rollins, 2005). However, this assumption may not hold ground under a variety of scenarios. In reality, the long-run marginal costs will be subject to change over time due to future technological progress, market variations and government policies. Perron (1989) developed a stochastic switching model and showed that there are discrete shifts in the slope or level of the long-run trend line. Pindyck (1999), while forecasting long-run energy prices, incorporated MR with trend lines that have both the slope and the level shifting over time. As observed by Pindyck (1999), the shifts themselves may be mean-reverting but ignoring them or assuming instead that there are a few discrete shifts in the trend line could be misleading and can lead to suboptimal forecasts. Khajuria et al. (2011), deriving from the work of Pindyck (1999), developed a model of long-run timber price evolution incorporating two key characteristics: (1) reversion to an unobservable long-run mean (equivalent to total marginal cost) which follows a trend and (2) continuous random fluctuations in both the level and slope of that trend. Assuming that the log price follows a simple trending Ornstein-Uhlenbeck process, the quadratic trend was incorporated in continuous time – similar to Pindyck (1999):

$$d\bar{P} = -\eta\bar{P}dt + \sigma dz \qquad (5)$$

where $\bar{P} = P - \alpha_0 - \alpha_1 t - \alpha_2 t^2$; is the detrended price.

In terms of the price level, this is equivalent to:

$$dp = [-\eta(P - \alpha_0 - \alpha_1 t - \alpha_2 t^2) + \alpha_1 + 2\alpha_2 t] + \sigma dz \tag{6}$$

The multivariate Ornstein-Uhlenbeck process is a simple generalization of equation (5). In the bivariate case, when the level and the slope fluctuations are included, we may write the process as:

$$d\bar{P} = [-\eta\bar{P} + \lambda_1 x + \lambda_2 yt]dt + \sigma dz_p \tag{7}$$

where x and y are themselves Ornstein-Uhlenbeck processes given by:

$$dx = -\delta_1 x dt + \sigma_x dz_x \tag{8}$$

$$dy = -\delta_2 y dt + \sigma_y dz_y \tag{9}$$

where dz_p, dz_x and dz_y may be correlated. Equation (7) simply states that $\bar{P}$ reverts to $(\lambda_1 x + \lambda_2 yt)/\eta$ rather than 0 and x is mean reverting around 0 if either $\delta_1 > 0$ or $\delta_2 > 0$ and a random walk if both δ_1 and δ_2 are zero. If one of them is zero, then the previous process becomes a univariate stochastic process. In general, the variables x and y could be observable economic variables or could be unobservable. Their results indicate that the stochastic trend model outperforms the fixed trend model. They used time series data of timber prices up to 2007 and through their forecasting model generated a 20-year price forecast. The actual prices revealed from 2008 to 2010 were compared with the model forecasts and found to be within a 4%–5% prediction margin. The option values and the thresholds for triggering harvest were observed to have lower values compared to the fixed trend model, implying that the fixed trend model may be overvaluing the asset value.

Some studies have attempted to consider stochastic timber prices and some other stochastic variable in a real options setting. Morck et al. (1989) used a contingent claims approach to determine the optimal cutting schedule, assuming stochastic output prices and stochastic natural growth rate as well as stochastic timber inventories. Insley and Rollins (2005) used the real options approach, assuming stochastic timber prices and stochastic timber growth. Chladná (2007) and Petrasek and Perez-Garcia (2010), in similar studies, considered stochastic timber prices and stochastic carbon prices in real options models to study the impact on harvesting schedules. Reed (1993) considered future timber volume as well as future amenity values as stochastic processes in a wildlife conservation study.

ROA and forest conservation decisions

Originally, the concept of option value for a park was developed by Weisbrod (1964). Arrow and Fischer (1974) used a similar option value concept for preserving a wilderness area when development benefits are uncertain. However, it took some time to extend the applications of ROA to forest conservation decisions such as biodiversity conservation and conservation of old-growth forests. Reed (1993), assuming that the future timber volume as well as future amenity values are stochastic, addressed the issue of whether to conserve or harvest an old-growth forest. His comparison of the results through cost-benefit analysis implied that under the latter, it was suggested to harvest the old-growth forest, basically ignoring the option of preserving the forest for future amenity value. In a similar type of study, Conrad (1997) determined the option value of an old-growth forest, assuming amenity values evolve stochastically as a GBM, whereas the future volume is deterministic. He derived the critical amenity value necessary to justify continued preservation of the forest. In another study on wilderness, Conrad (2000) used option pricing to evaluate the sequence and timing of wilderness preservation, resource extraction and development. He observed that resource extraction or development results in the permanent destruction of wilderness and the loss of an amenity dividend. Resource extraction does not preclude subsequent development. If the wilderness is directly developed without prior extraction of resources, wilderness and resource extraction options are both killed. Starting from a state of wilderness, there are two stochastically evolving barriers: one for the price of the resource, and the other for the return on development. Wilderness is preserved provided the price of the resource never catches the price barrier and the return on development never catches the return barrier. Building on the study of Conrad (1997), a study was done by Forsyth (2000) wherein the option value of a wilderness area was calculated assuming that the stochastic process for the amenity value is a logistic process rather than a GBM. The critical amenity value is also calculated which is necessary to justify the continued preservation of the wilderness area. Bakshi and Saphores (2004), using a simple real options framework, analyzed wildlife management policies that account for ecological uncertainty and the risk of extinction. The application to wolves provides an economic justification for their reintroduction and highlights the importance of existence value. They used a GBM process for the wilderness area. Their results demonstrated that the optimal management policy depends on the growth rate, the volatility and the minimum viable density of the wolf population, but little on damages, existence value and the discount rate for the parameters considered. Kassar and Lasserre (2004) assessed biodiversity conservation issues using a real options model. In their model, they assumed that the processes governing species values are symmetrical GBMs, when the species are in existence, and used numerical solution techniques. The most usable species is exploited and the remaining unproductive species are preserved at some cost. The options are whether to pay for the continued preservation or abandon one or all of them and harvest them. They observed that the option to replace the currently productive species with a similar species increases the need for biodiversity conservation. Leroux, Martin and Goeschl (2007), using real options, studied the trade-offs in land conversion and conservation. They explored the implications on the rules of allowing for feedback between conversion decisions and the stochasticity of conservation benefits, using the well-known ecological mechanism of extinction debt as an illustration. This yields a model with a controlled-diffusion process at its core that is solved using a real options approach and that leads to the conventional conversion rule as a special case. They assumed that conservation benefits are dependent on species richness and marginal economic value of biodiversity. The paper demonstrated the presence of an augmented quasi-option value depending on the strength of the feedback. This results in quantifiable changes in land values and the amount of conservation. The option for biodiversity conversion leads to reduced rate of land conversion.

Conclusion

The application of ROA in forestry has been mainly in forest harvesting decisions and, to a limited extent, in forest conservation decisions. In the case of harvesting decisions, the main research progress has been in terms of modeling the evolution of timber and lumber prices, starting with the assumption of a random draw from the assumed distribution of prices to GBM and MR process. In the case of conservation decisions, the developments have been in the areas of preservation of wilderness areas, old-growth preservation for wildlife species habitat, biodiversity conservation, land conversion decisions and studying some amenity value of a wilderness area.

The ROA applications have remained focused on timber as a product from the forests. There are other marketable products from the forest such as pulpwood, saw-wood, wood chips and so forth. Their joint production from standing timber needs to be incorporated to make analysis more realistic. This will require price time series for each of these products and generating separate price processes for them. Similarly, incorporation of growth curves for different species and their respective prices with spatial distribution of the species in ROA will enhance the applicability and reliability of the outcomes of ROA. Inclusion of thinning operations, in addition to final harvests, will also enhance the applicability of ROA in forest management decisions. The incorporation of these features will increase the complexity of models that will require model-specific solutions and estimation techniques.

On the forest conservation side, wildlife and biodiversity conservation issues need to be studied in depth using the theory of real options to get any meaningful and pragmatic insights into the economics of conservation forestry, such as the introduction of new floral or faunal species in a specific area, relocation of species to new areas, preservation of flagship species, creating new conservation zones or biosphere reserves and so forth.

Another area for current active research is the incorporation of ecosystem services into the analysis. Ecosystem services such as carbon sequestration, which has become a market entity, have been studied recently. Some scholars, such as Chladná (2007) and Petrasek and Perez-Garcia (2010), have included uncertain revenues from carbon trading in a real options model and determined optimal harvest schedules. However, there is need for more work on this as well as studying the plethora of ecosystem services that are still nonmarketable. The application of ROA will be very useful in understanding tradeoffs between timber and these services ultimately enhancing the returns from forestry investments. It will help in giving forestry its due in terms of profitable investment options as well as strengthening the hands of policy framers for providing incentives to investors and levying taxes on the sectors that exploit these nonmarketable ecosystem services in their production processes.

On the econometric side, there is further need to extend the MR process models. Random variables may be modeled as some mixed process such as the MRJ process incorporating Poisson jumps. Further, the long-run trend also needs to be treated as stochastic. This will require numerical solution techniques to approximate value functions for determination of the option values and the thresholds for triggering the decisions. Researchers in the future must explore these aspects. This applies not only to studying stochastic timber prices in these models, but also to incorporating site-specific, species-wise stochastic yields from timber, and in a variety of management scenarios as well, for incorporating a multitude of timber products and ecosystem services. This will be a daunting task, but it would be a big step forward in recognizing the role of forests and giving them their due share in the economic development of nations.

Finally, considering many things simultaneously can very easily lead to a lot of complexity, requiring interdisciplinary approach. The results obtained can be even harder to explain as well. Therefore, it is necessary that any enhancement in the analyses is done carefully with proper objectives in mind.

References

Arrow, K. J. and Fisher, A. C. (1974). 'Environmental preservation, uncertainty, and irreversibility', *The Quarterly Journal of Economics*, vol. 8, no. 2, pp. 312–319.

Bakshi, B. and Saphores, J.-D. M. (2004). 'Grandma or the wolf? A real options framework for managing human-wildlife conflicts'. Retrieved from www.realoptions.org/papers2004/SaphoresHA.pdf

Brazee, R. and Mendelsohn, R. (1988). 'Timber harvesting with fluctuating prices,' *Forest Science*, vol. 34, no. 2, pp. 359–372.

Chladná, Z. (2007). 'Determination of optimal rotation period under stochastic wood and carbon prices', *Forest Policy and Economics*, vol. 9, no. 8, pp. 1031–1045.

Clarke, H. and Reed, W. (1989). 'The tree-cutting problem in a stochastic environment', *Journal of Economic Dynamics Control*, vol. 13, no. 4, pp. 569–595.

Conrad, J. M. (1980). 'Quasi-option value and the expected value of information', *Quarterly Journal of Economics*, vol. 94, no. 4, pp. 813–820.

Conrad, J. M. (1997). 'On the option value of old-growth forest', *Ecological Economics*, vol. 22, no. 2, pp. 97–102.

Conrad, J. M. (2000). 'Wilderness: Options to preserve, extract or develop', *Resource and Energy Economics*, vol. 22, no. 3, pp. 205–219.

Dixit, A. K. (1989). 'Entry and exit decisions under uncertainty', *Journal of Political Economy*, vol. 97, no. 3, pp. 620–638.

Dixit, A. K. and Pindyck, R. S. (1994). *Investment under Uncertainty*, Princeton University Press, Princeton, NJ.

Duku-Kaakyire, A. and Nanang, D. (2004). 'Application of real options theory to forestry investment analysis', *Forest Policy and Economics*, vol. 6, no. 6, pp. 539–552.

Forsyth, M. (2000). 'On estimating the option value of preserving a wilderness area', *Canadian Journal of Economics*, vol. 33, no. 2, pp. 413–434.

Gjolberg, O. and Guttormsen, A. (2002). 'Real options in the forest: What if prices are mean-reverting?' *Forest Policy and Economics*, vol. 4, no. 1, pp. 13–20.

Guthrie, G. A. and Kumareswaran, D. (2009). 'Carbon subsidies, taxes and optimal forest management', *Environmental and Resource Economics*, vol. 43, no. 2, pp. 275–293.

Haight, R. G. (1990). 'Feedback thinning policies for uneven-aged stand management with stochastic prices', *Forest Science*, vol. 36, no. 4, pp. 1015–1031.

Haight, R. G. (1993). 'Optimal management of loblolly pine plantations with stochastic prices', *Canadian Journal of Forest Research*, vol. 23, no. 1, pp. 41–48.

Hultkrantz, L. (1995). 'The behaviour of timber rents in Sweden 1909–1990', *Journal of Forest Economics*, vol. 1, no. 3, pp. 165–180.

Insley, M. (2002). 'A real options approach to the valuation of a forestry investment', *Journal of Environmental Economics and Management*, vol. 44, no. 3, pp. 471–492.

Insley, M. and Rollins, K. (2005). 'On solving the multi-rotational timber harvesting problem with stochastic prices: A linear complementarity formulation', *American Journal of Agricultural Economics*, vol. 87, no. 3, pp. 335–755.

Kassar, I. and Lasserre, P. (2004). 'Species preservation and biodiversity value: A real options approach', *Journal of Environmental Economics and Management*, vol. 48, no. 2, pp. 857–879.

Khajuria, R. P., Kant, S. and Laaksonen-Craig, S. (2009). 'Valuation of timber harvesting options using a contingent claims approach', *Land Economics*, vol. 85, no. 4, pp. 655–674.

Khajuria, R. P., Kant, S. and Laaksonen-Craig, S. (2011). 'Modelling of timber harvesting options using timber prices as a mean reverting process with stochastic trend', *Canadian Journal of Forest Research*, vol. 42, no. 1, pp. 179–189.

Leroux, A. D., Martin, L. and Goeschl, T. (2007). 'Real options in biodiversity conservation', *Proceedings in Applied Mathematics and Mechanics*, vol. 7, no. 1, pp. 1080801–1080802.

Lohmander, P. (1988). 'Pulse extraction under risk and a numerical forestry application', *Systems Analysis Modelling Simulation*, vol. 5, no. 4, pp. 339–354.

Manley, B. and Niquidet, K. (2010). 'What is the relevance of option pricing for forest valuation in New Zealand?' *Forest Policy and Economics*, vol. 12, no. 4, pp. 299–307.

Marwah, S. and Zhao, J. (2002). 'Irreversibility, uncertainty and learning in land preservation decisions'. Retrieved from www.sls.wau.nl/enr/conference/papers/long/Marwah_long.pdf

Merton, R. C. (1976). 'Option pricing when underlying stock returns are discontinuous', *Journal of Financial Economics*, vol. 3, no. 1–2, pp. 125–144.

Miller, R. A. and Voltaire, K. (1983). 'A stochastic analysis of the tree paradigm', *Journal of Economic Dynamics and Control*, vol. 6, no. 1, pp. 371–386.

Morck, R. Schwartz, E. and Stangeland, D. (1989). 'The valuation of forestry resources under stochastic prices and inventories', *The Journal of Financial and Quantitative Analysis*, vol. 24, no. 4, pp. 473–487.

Norstorm, C. J. (1975). 'A stochastic model for the growth period decision in forestry', *The Swedish Journal of Economics*, vol. 77, no. 3, pp. 329–347.

Perron, P. (1989). 'The great crash, the oil price shock, and the unit root hypothesis', *Econometrica*, vol. 57, no. 6, pp. 1361–1401.

Petrasek, S. and Perez-Garcia, J. (2010). 'A Monte Carlo methodology for solving the optimal timber harvest problem with stochastic timber and carbon prices', *Mathematical and Computational Forestry and Natural-Resource Sciences*, vol. 2, no. 2, pp. 67–77.

Pindyck, R. S. (1999). 'The long-run evolution of energy prices', *The Energy Journal*, vol. 20, no. 2, pp. 1–27.

Plantinga, A. J. (1998). 'The optimal timber rotation: An option value approach', *Forest Science*, vol. 44, no. 2, pp. 192–202.

Prestemon, J. P. (2003). 'Evaluation of U.S. southern pine stumpage market informational efficiency', *Canadian Journal of Forest Research*, vol. 33, no. 4, pp. 561–572.

Reed, W. J. (1993). 'The decision to conserve or harvest old growth forest', *Ecological Economics*, vol. 8, no. 1, pp. 45–69.

Reed, W. J. and Clarke, H. J. (1990). 'Harvest decisions and asset valuation for biological resources exhibiting size-dependent stochastic growth', *International Economic Review*, vol. 31, no. 1, pp. 147–169.

Saphores, J., Khalaf, L. and Pelletier, D. (2002). 'On jumps and arch effects in natural resource prices: An application to Pacific Northwest stumpage prices', *American Journal of Agricultural Economics*, vol. 84, no. 2, pp. 387–400.

Saphores, J.-D. M. and Carr, P. (2000). 'Real options and the timing of implementation of emission limits under ecological uncertainty', in M. J. Brennan and L. Trigeorgis (Eds.), *Project Flexibility, Agency and Competition: New Developments in the Theory and Applications in Real Options*, Oxford University Press, New York.

Schwartz, E. and Trigeorgis, L. (2004) *Real Options and Investment under Uncertainty: Classical Readings and Recent Contributions*, Massachussetts Institute of Technology Press, Cambridge, MA.

Thomson, T. A. (1992). 'Optimal forest rotation when stumpage prices follow a diffusion process', *Land Economics*, vol. 68, no. 3, pp. 329–342.

Trigeorgis, L. (1996). *Real Options: Managerial Flexibility and Strategy in Resource Allocation*, Massachusetts Institute of Technology Press, Cambridge, MA.

Washburn, C. L. and Binkley, C. S. (1990). 'Informational efficiency of markets for stumpage', *American Journal of Agricultural Economics*, vol. 72, no. 2, pp. 394–405.

Weisbrod, B. A. (1964). 'Collective-consumption services of individual-consumption goods', *Quarterly Journal of Economics*, vol. 78, no. 3, pp. 471–477.

Yin, R. (2001). 'Combining forest-level analysis with options valuation approach – A new framework for assessing forestry investment', *Forest Science*, vol. 47, no. 4, pp. 475–483.

Yin, R. and Newman, D. H. (1996). 'The effect of catastrophic risk on forest investment decisions', *Journal of Environmental Economics and Management*, vol. 31, no. 2, pp. 186–197.

Yin, R. and Newman, D. H. (1997). 'When to cut a stand of trees?' *Natural Resource Modeling*, vol. 10, no. 3, pp. 251–61.

Yoshimoto, A. and Shoji, I. (1998). 'Searching for an optimal rotation age for forest stand management under stochastic log prices', *European Journal of Operations Research*, vol. 105, no. 1, pp. 100–112.

Zinkhan, C. F. (1991). 'Option pricing and timberland's land-use conversion option', *Land Economics*, vol. 67, no. 3, pp. 317–325.

23

ECONOMICS AND PLANNING OF BIODIVERSITY CONSERVATION UNDER UNCERTAINTY OF CLIMATE CHANGE

Niels Strange,[*1] *Jette Bredahl Jacobsen,*[1] *David Noguès-Bravo,*[2] *Thomas Hedemark Lundhede,*[1] *Carsten Rahbek*[2] *and Bo Jellesmark Thorsen*[1]

[*]CORRESPONDING AUTHOR, NST@LIFE.KU.DK
[1]DEPARTMENT OF FOOD AND RESOURCE ECONOMICS, CENTER OF MACROECOLOGY, EVOLUTION AND CLIMATE, FACULTY OF SCIENCE, UNIVERSITY OF COPENHAGEN, ROLIGHEDSVEJ 23, DK-1958 FREDERIKSBERG C, DENMARK.
[2]THE NATURAL HISTORY MUSEUM OF DENMARK, CENTER OF MACROECOLOGY, EVOLUTION AND CLIMATE, FACULTY OF SCIENCE, UNIVERSITY OF COPENHAGEN, UNIVERSITETSPARKEN 15, DK-2100 COPENHAGEN Ø, DENMARK.

Abstract

This chapter discusses the need for biologically and economically well-reasoned conservation decisions in a dynamic world, and stresses the potential insights gained using sound and theoretically consistent approaches combining biodiversity, valuation and decision modelling under uncertainty. Recent research on the potential long-term influence of climate change on biodiversity distributions and persistence is revisited. Environmental valuation may provide insight into how this may affect human well-being. The methodological challenges of measuring how society values biodiversity effects under climate change and the inherent scarcity and uncertainty are discussed. This indeed imposes changes in conservation priorities and the chapter illustrates and discusses adaptive management and the value of flexibility. The chapter concludes with an address to the need for geopolitical coordination of conservation efforts to halt the decline of biodiversity under climate change.

Keywords

Climate change, dynamic, policy, environmental valuation, reserve selection, uncertainty, welfare

Introduction

At the global level, there is an increasing awareness about the value of biodiversity and other eco-system services and their importance for sustaining human welfare (TEEB, 2010). Worldwide, societies are giving greater importance to the protection of overall diversity of the landscape in terms of its species, scenery and cultural history. However, the degradation and conversion of natural landscapes is resulting in an unprecedented loss of biological diversity (Thomas et al., 2004). Conservation authorities are thus facing the problem of how to target their biodiversity protection actions so that they accomplish the most with limited budgets, while acknowledging the uncertainty of the future states of areas and of the environment globally.

One of the most important drivers for landscape degradation is intensive agriculture in terms of nitrification and land use. Although less productive areas in the European Union (EU) in partic-ular are set aside or made more extensive, fertile agricultural areas are being intensively exploited. Furthermore, climate change may cause in shifts in the distribution and abundance of species and their habitats. As species and habitats cannot necessarily adjust spontaneously, it may accelerate the risk of extinction (Doak and Morris, 2010). This may also impose large socio-economic costs in terms of lost or damaged biodiversity and ecosystem services. The valuation of biodiversity and ecosystem services needs to include these dynamics and uncertainties in making appropriate preference studies. Consequently this should be addressed by policies and decision makers so that the prioritization of future conservation investments explicitly accounts for these dynamic and uncertain impacts of degradation and climate change (Carroll et al., 2010; Strange, Thorsen, Bladt, Wilson and Rahbek, 2011). It is a prevailing question whether the current policy supports a long-term provision of services and the current designation of conservation areas holds an efficient portfolio in terms of species protected and management options. This topic is at the very forefront of forest and nature economics and planning.

This chapter discusses the need for biologically and economically well-reasoned conser-vation decisions in a dynamic world, and stresses the potential insights gained using sound and theoretically consistent approaches combining biodiversity and valuation modelling under uncertainty. Following the introduction, the chapter reviews recent research on the potential long-term influence of climate change on biodiversity distributions and persistence. Next it is discussed how the value of biodiversity is affected not only by the abundance of biodiversity but also by how society values the inherent scarcity and uncertainty and what the implications of benefits, costs and climate change are on conservation priorities. Finally, the chapter con-tinues the discussion of adaptive management and flexibility by using a small two-period, two conservation-site model to illustrate the potential gains of adaptive management and how this may be affected by changing geographical distributions of biodiversity.

Climate change and changing biodiversity distributions

Planet Earth is currently experiencing a period of global climatic changes that are impact-ing biological diversity and the ecosystem services that sustain our societies. Although climatic changes and their influence on the environment are not confined to recent times, the magni-tude and speed of current climate change have not been preceded for the last 2 million years. Projections estimate that global surface temperature is likely to rise a further 1.1°C to 2.9°C for their lowest emissions scenario (2.4°C to 6.4°C for their highest; IPCC, 2007). Regional trends show large variation in spatial patterns, with high-latitudinal regions like arctic and circumpolar ecosystems as the ones expected to warm more than tropical ones. Species can cope with these climatic changes by following two main strategies: adapting, for example, changing phenology

(Charmantier et al., 2008), or tracking climatic changes by adjustment of their geographical ranges.

Changes in the distribution of species responding to climate change have been already reported for different taxa (Parmesan and Yohe, 2003). Two general patterns arise from the scientific literature: high-latitude migrations at the global scale and species moving up along elevational gradients in mountain ranges. A recent study (Chen, Hill, Ohlemüller, Roy and Thomas, 2011) shows that species belonging to 32 different taxonomic groups have moved away from the equator at a median rate of 16.9 km per decade, exceeding previous attempts to summarize latitudinal shifts of geographical ranges of species by more than 10 km per decade (Parmesan and Yohe, 2003). Similarly, species on average migrated to higher elevations at a median rate of 11.0 m per decade (Chen et al., 2011). Comparing the altitudinal distribution of 171 forest plant species between 1905 and 1985 and 1986 and 2005 along the entire elevation range in western Europe, Lenoir, Gegout, Marquet, de Ruffray and Brisse (2008) showed that climate warming has resulted in a significant upward shift in species optimum elevation averaging 29 m per decade. However, responses of geographical range to climate change are sometimes more complex than upward shifts. By comparing the altitudinal distributions of 64 plant species between the 1930s and the present day within California, Crimmins, Dobrowski, Greenberg, Abatzoglou and Mynsberge (2011) showed that climate change has resulted in a significant downward shift in the species' optimal elevation by 88.2 m since 1930. These counterintuitive patterns may be the result of complex interaction of water-energy dynamics, empty climatic niches and species interactions, which point out the need for enhancing our knowledge of the responses of full ecological communities to climate change (Nogués-Bravo and Rahbek, 2011).

Even those observed trends may already be pushing species to their limit. Climate change projections for the twenty-first century increase significantly those magnitudes, with potential extinctions forecast across taxa and regions. However, forecasting exercises are still facing major challenges (Bellard, Bertelsmeier, Leadley, Thuiller and Courchamp, 2012), and projections significantly differ among studies. Although some studies project that less than 0.3% of the world's 8,750 species of land birds would be committed to extinction (e.g. Jetz, Wilcove and Dobson, 2007), other studies project that the potential extinctions could reach 30% of all land bird species in the Western Hemisphere (Sekercioglu, Schneider, Fay and Loarie, 2008). Acknowledging the discrepancies among studies and uncertainty of the potential risk of future extinctions and shifts in the geographical ranges of species, it is likely that the ability of current conservation networks to harbor and protect species under climate change will be highly compromised. In fact, there is a risk that the current protected areas may retain climate suitability for species no better and sometimes less effectively than unprotected areas. This is indicated in a recent evaluation of the European Natura 2000 network (Araújo, 2011).

Valuation of environmental change

Passive and use values of biodiversity

Climate changes the future of habitats and biodiversity and potentially also the value society assigns to them. Valuation of environmental changes has often been made in terms of improvement when dealing with biodiversity, and it has the aim of helping decision makers estimate the benefits of management decisions (Carson, 2012). With the introduction of choice experiments to environmental economics (Adamowicz, Boxall, Williams and Louviere, 1998), focus has changed to extracting the values of attributes/characteristics of the good being valued, including different utility components of use and passive use values (e.g. Juutinen et al., 2011).

This has been of great importance for the understanding of what constitutes environmental values, not the least for multipurpose forest management where environmental values often are jointly produced. One of the findings is that passive use values seem to make up the larger share of the values from biodiversity conservation, at least in developed countries (Jacobsen, Lundhede, Martinsen, Hasler and Thorsen, 2011). Another is that rare species are valued higher than unthreatened species (Jacobsen, Lundhede and Thorsen, 2012), as are more familiar species (Christie et al., 2006). Biodiversity conservation tends to focus on the rare species, but not necessarily on the familiar ones. Jacobsen, Boiesen, Thorsen and Strange (2008) find that the general public have a quite high willingness to pay (WTP) for preserving even small and unknown species when these are rare and part of the preservation of a habitat. This indicates that not only charismatic and known species have a high preservation value, as demonstrated clearly by Jakobsson and Dragun (2001), but that the general public may also associate large values with more unknown species. To what degree this is an issue of their being informed of the existence of the species or an issue of association of a more holistic approach is less clear and is a research direction that has to be followed in the coming years. This is crucial because it may shed light on priorities for conservation of biodiversity.

Regardless of which value components of biodiversity are being assessed, they change over time and depend on the current status of biodiversity. From an economic point of view, we would expect diminishing marginal utility, and thus the lower the biodiversity level, the higher is the value of preserving it. In accordance with this, in a survey conducted in Denmark in 2011 (in the project www.newforex.org), 65% of the respondents reported that they found it most important to improve conditions for biodiversity in forests with a relatively low level of biodiversity, regardless of the distance to respondents' homes. This contrasts with the large emphasis on protection of species-rich areas (hot spots) (Myers, Mittermeier, Mittermeier, da Fonseca and Kent, 2000).

Biodiversity values and loss aversion

Another factor that makes values of biodiversity state-dependent is loss aversion. Theoretically it originates from prospect theory (Kahneman and Tversky, 1979), which questions the neoclassical economic approach and argues that the value of a loss is higher than the value of a gain. Biodiversity conservation can relate both to obtaining gains (e.g. improvements in nature quality on locations where it has been abandoned) and to preventing losses (e.g. preserving species from extinction). For the general public, the latter may be of larger importance, even if the number of protected species would be the same. In a climate change setting this becomes important because it may cause conservation strategies preferred most by a society to be strategies that tend to preserve species that will go extinct if nothing is done, even if something else of 'equal' biological value could be preserved more cheaply. Despite speculations based on theory that suggest these findings, no empirical evidence exists as to whether this is the case. Thus, future research is needed to increase the knowledge of loss aversion of biodiversity. What complicates this issue is that it might also depend on citizens' opinions on reasons for climate change and causes of species and habitat extinction as well as on the cultural values of the species and habitats. A few studies find diminishing value of nature values with increasing human intervention (Czajkowski, Buszk-Briggs and Hanley, 2009). If these results carry over to the case of climate change, people believing that climate change is occurring naturally may find it more acceptable that species disappear from given areas and others take over. Finally, the cultural values of species and habitats are important for the assessment of values (Hanley et al., 2009) – again a factor that tends to favor the protection of current species distributions.

Provision uncertainty

When formulating biodiversity conservation initiatives to be valued by survey participants, the potential management change outcome is often – implicitly – assumed or communicated as certain. In a valuation study, respondents may in many cases believe the often-implicit assumption of certainty in provision. However, they may also be equally likely to perceive some environmental changes as more likely to come true than others. One such example related to varying degrees of belief in the realism of a large-scale change scenario relative to a smaller change is provided by Powe and Bateman (2004), who showed that taking into account scale-correlated prior beliefs about the realism of proposed environmental changes improved scope sensitivity of underlying valuation measures. Furthermore, again building on Prospect Theory (Kahneman and Tversky, 1979), several studies have produced evidence that people factor in their own perceptions of risk when evaluating choices involving specific risks and uncertainty, e.g. within research on consumer choices involving gradients in product safety or anglers' health risks of consuming the fish they catch (Jakus and Shaw, 2003).

This confirms a more general result on risk aversion and its sensitivity to the scope of the outcomes (Andersen, Harrison, Lau and Rutstrom, 2008), which is also reflected in several contingent valuation studies of environmental change associated with stated or perceived uncertainty (Isik, 2006). And more recently, in choice experiments, Wielgus, Gerber, Sala and Bennett (2009) found that explicitly stating a high provision probability improved goodness of fit of choice models and concluded that omitting information on scenario risk may contribute to hypothetical bias.

The relation between an environmental policy or a proposed management change and the postulated outcomes may in many cases be better described by a probability distribution or perhaps even by broader terms of uncertainty. Thus, it seems relevant to evaluate what happens to people's valuation of policy alternatives if they are in fact informed that the policy outcome is to some degree uncertain. This motivated Roberts, Boyer and Lusk (2008) to investigate how respondents in a split-design CE reacted to stated probabilities of experiencing (un)pleasant water qualities and levels at a recreational lake visit. They document that respondents did not interpret stated probabilities in a standard linear weighted utility manner, but rather underweight low-probability events as compared to high-probability events. This differs from findings by, e.g. Tversky and Fox (1995) who find overweight of low-probability events. Accounting for uncertainty, Roberts et al. (2008) found little difference in WTP among attributes. In a choice experiment-based valuation, exercise of agri-environmental measures to increase soil carbon sequestration (with two additional co-benefits provided along with it: enhanced biodiversity and job creation), they evaluated the effect of introducing half way through the choice sets a new attribute describing the provision uncertainty – in quantitative likelihood measures – for the change in one attribute: soil carbon sequestration. They found a negative WTP for increasing uncertainty of provision, but otherwise found the WTP for other attributes insignificantly affected, including the soil carbon sequestration attribute. The conclusion from these studies and the general literature on uncertainty is that stated provision probabilities will affect the valuation of the attributes concerned, though the impact will often not be according to a linear weighted utility, and may be influenced by people's priors.

This inherent uncertainty may be even more pronounced when dealing with climate change. Therefore, future research needs to include this aspect in order to analyze not only the direct value, but also people's risk perception and preferences across types of goods which are subject to change.

Use of the valuation studies

An approach that has been followed to make use of valuation studies more directly is benefit transfer (Johnston and Rosenberger, 2010), where the reasoning is that performing a valuation study is often too expensive, or impossible to do ex post, so if we could transfer estimated benefits from one location to another, it would be easier to base policies on it. However, an aspect that often limits this approach is to find study and policy cases sufficiently similar across both the environmental good being valued and the people valuing it. Therefore, we need to get a better knowledge, through primary studies as well as transfer studies, of what components of a good are being valued by people, who it is valued by, and what their relationship is toward the good and its surroundings. Getting a better knowledge of these perspectives is crucial to increasing the understanding of which attributes of biodiversity represent a value for respondents. This may improve the usefulness of the valuation studies in the real-world decision process and supply advice on management and how to improve welfare, and thereby how we can best manage the nature around us. Additional components in this setting arrive from the biodiversity being dynamic rather than static – both in terms of the characteristics of the good and in terms of the society around it.

Prioritizing conservation investments in a changing world

A large number of economic valuation studies on biodiversity are found in the literature (Carson, 2012). So far, however, only a few results from this literature have been successfully applied to designing conservation strategies (Naidoo and Iwamura, 2007). An important future research question is the potential of environmental valuation research in guiding forest and nature conservation strategies. Because only limited resources are available for investment in nature protection, allocating the resources efficiently would be helped by an increased understanding of not only the economic costs of conservation strategies, but also of the associated benefits. Such cost-benefit analysis is indeed common in many project appraisals, but rather limited within conservation biology. Decision makers are often left with quite reliable estimates of conservation costs (effort) but less reliable estimates on the value of such effort. In many cases, the marginal cost of supplying more protected biodiversity will be very different than the marginal benefits of such provision. Figure 23.1 illustrates a simple example of potential implications for identifying priorities for approximately 40 conservation areas. Each area has a potential conservation value and cost. The most cost-effective areas (highest B/C ratio) are located in quadrant 1 and are the ones which should be prioritized first. Areas in quadrants 2a and 2b are inferior to areas in quadrant 1. Most likely 2a, if budgets are available, is preferable to 2b because of its higher benefits. Finally, the least prioritized areas with the lowest conservation benefits and still-high conservation costs are in quadrant 3.

Figure 23.1 also illustrates how climate change may require larger investments in ensuring suitable habitat, but could in principle make the same point for other environmental changes. Some species may move from left to right (e.g. from quadrant 1 to 2a or even further out of 2a) and others may move from right to left because of improved habitat quality. It is even more interesting that the priorities may change if the preference for a species decreases. So far, most research has focused on stochastic costs and species distributions (Strange et al., 2011). Much less research has been invested in understanding the dynamics of biodiversity preferences. One example is the high political interest in protecting national breeding species, and much less interest in reserving areas for species, which may become national breeding species in the future

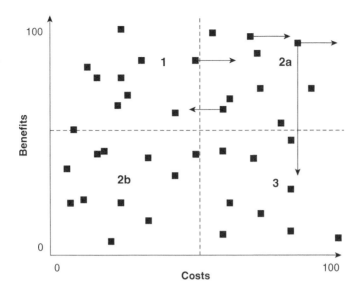

Figure 23.1 Changes in priorities when costs and benefits are dynamic.
Source: Modified from Margules and Pressey (2000).

due to, e.g. climate change. The European Natura 2000 network as well as many other global protection areas has been selected for its historical high biodiversity values.

Typically, the management objective for these areas is to maintain and preserve the habitats and species populations in their existing condition, and to improve their biodiversity value over time. This makes sense, but also holds the risk of focusing on preservation with a risk of allocating higher shares of the budget to species which may be under high pressure and risk of disappearing despite the conservation effort. Other attributes (e.g. changing status as an icon species, changing protection status at an EU level) may of course change the preferences or benefits of protecting particular species or areas.

Species 'moving out' may then at some point move not only to the right but also downwards from quadrant 2a to quadrant 3. This leaves us with an interesting question: Should we maintain investing in conservation of sites that may lose their value in the future or shift to conservation of other areas that may represent higher future potential? To answer this prevailing question, there is a need to combine disciplines within macro-ecology and environmental economics.

Conservation policies in a changing world

Static versus dynamic conservation policies

The field of systematic conservation planning has developed increasingly robust and sophisticated decision-support tools that can account for the dynamic nature of ecosystems (e.g. Strange, Thorsen and Bladt, 2006; Visconti, Pressey, Segan and Wintle, 2010). Climate change introduces additional dynamism, such as climate-induced migration of species (Parmesan, 2006), that is likely to change optimal priorities in both space and time. As was described in Prioritizing conservation investments in a changing world, climate change may also cause significant uncertainty about future species distributions and the effectiveness of conservation actions. Ignoring uncertainty and treating conservation investments as static will lead to poor management decisions (Regan et al., 2005). With a few exceptions (Araújo et al., 2004; Carroll et al., 2010) little attention has been directed to the prioritization of conservation investments under climate change, and there has been limited exploration of the consequences of current conservation policies in the context of predicted climate change impacts. Furthermore, these studies have typically compared the value of protected site networks using current species distributions and snapshot projections of future species distributions. This may bear the risk of leading to inadequate decisions because budgets are limited and most conservation needs to be made sequentially rather than upfront, based on future predictions. Strange et al. (2011) confirmed that conservation strategies employing updated and system-wide information about the presence of species performed better than systems relying on the more limited information in reserve networks. Thus, the more fieldwork information is generated, the higher the effectiveness of conservation actions – but of course at a potentially significant inventory cost with implications for cost-effectiveness (Gardner et al., 2008). If information collection is expensive and delays species protection, it could even come with an opportunity cost (Grantham, Wilson et al., 2009).

It may be possible to obtain a higher performance of the conservation networks by supplementing species distribution data with estimates of future species distributions and habitat suitability. Improved information on these factors can be crucial for the mitigation of adverse climate change effects on biodiversity. However, there may be numerous sources of species model uncertainties which may influence species coverage within the conservation network (Araújo and Guisan, 2006). Modelling the climatic impact on species survival and migration is just one of these sources. To date, most species distribution modelling is considering a temporal scale of at least 50 years. Evidently, this is much longer than what is needed for making real-world management decisions based on the models. Even if it was possible, the expected coverage heuristic additionally depends on precision in the estimations of climatic suitability, immigration, conversion and habitat degradation probabilities. Alternatively, resources could be spent on collecting biodiversity data and updating presence/absence records. Obviously there is a cost in collecting this additional data and performing these estimations. Simulations (Strange et al., 2011) show that if the estimations can be performed with a fairly low level of uncertainty (up to 10%–20%), the estimations may still be valuable in designing the future networks of protected sites. However, if uncertainty is too high, estimations are worthless or even misleading, leading to protected site networks that perform no better than randomly selected sites. Although the estimations may be valuable even with a moderate level of uncertainty, it will be difficult to determine in a real-world scenario whether the uncertainty is too high. One solution could be to supplement presence/absence data for most species with predictive models for the most well-studied and easily modelled species. This could significantly reduce uncertainty, although it is impossible to know the precise level.

Modelling the value of more conservation options under climate change

As more flexibility is called for to mitigate adverse climate change effects on biodiversity, options for replacing existing protected sites or reallocating conservation investments with other sites more suitable for immigrating species become necessary. In the following, we present a small two-period two-site model to illustrate the potential gains from adaptive conservation management under budget constraints. It raises the issue of protection benefits being secured by allowing decision makers to consider swapping protected sites that have for some reason lost their protection value for other nonprotected sites that are more attractive.

Specifically, environmental change included not only the risk of degradation of protected sites, but also the probability of climate change-induced stochastic immigration into protected as well as nonprotected sites.

The problem is formulated as a stochastic dynamic integer programming problem, and in this section, we present a formal model of the optimization problem to be addressed. Let the number of potential protected sites j be J, i.e. $j = 1, 2, \ldots, J$. Assume that each site may host a set of currently native species i^n and potentially all I^n species in the network at $t = 0$, i.e. $i^n = 1, 2, \ldots, I^n$. Define an additional set of immigration species I^f, and a subset of the $i^f = 1, ., I^f$, species may immigrate into any nondegraded, nonconverted site in J. Letting $I = I^n + I^f$, we define a $J \times I$ matrix, A_t, where an element of the matrix $a_{ij} = 1$ if species i is present in area j at time t, and equals 0 otherwise. For $t = 0$ this matrix represents the initial state of the sites' distribution of species and hence for the last I^f rows $a_{ij} = 0$ initially. As we allow for stochastic degradation, there is a site-specific probability $prob(d_j)$ that in any time step t a reserve or nonreserve site j is for some reason degraded. In that event, the initial set of species present at the jth site, I_j is reduced to a predefined smaller set $I_j^d \subset I_j$. Furthermore, for nondegraded sites there is a site- and species-specific probability, $prob(m_{ij})$, of immigration to site j of a subset of species $i^f 5 I^f$. Thus, the matrix A_t is stochastic and evolves over the number of time steps T. Nonreserve status implies that the area has neither been converted nor selected as a reserve. In any time period t, nonreserve areas may be converted at the end of period t with the site-specific probability $prob(con_j)$. Once the area is converted, any suitable habitat within the area is destroyed and we assume that all species are removed from area j. Conversion as well as degradation are assumed to be irreversible, but immigration of new species does not exclude future risk of degradation or conversion. At the start of each time period t, every site is in one of several possible states: converted, reserved and potentially degraded or containing new immigrated species, nonreserved and potentially degraded or containing new immigrated species.

The cost of selecting area j at time t as a reserve could involve time- and site-specific variations in costs, but in our simplified case we take the costs as constant across time and site, i.e. the same cost, c, for any site, at any time. We let R_t be a $J \times 1$ vector where R_{jt} equals 1 if site j is part of the reserve network at the beginning of period t, and 0 otherwise. Let X_t be a $J \times 1$ vector where X_{jt} equals 1 if site j is selected as a reserve in period t, and 0 otherwise. Let L_t be a $J \times 1$ vector where L_{jt} equals 1 if the protected area j is sold (swapped) in period t, and 0 otherwise. Therefore, the total set of reserve areas R_{t+1} in period $t + 1$ equals the set of reserve areas at the beginning of period t plus the selected reserve areas in period t minus the sold reserve areas in period t. That is, $R_{t+1} = R_t + X_t - L_t$. We also define N_t as a $J \times 1$ vector where N_{jt} equals 1 if area j is nonreserve at the beginning of period t, and 0 otherwise. S_t is defined as a $J \times 1$ random vector where element S_{jt} equals 1 if stand j is converted in period t (following the allocation decision in that period), and 0 otherwise. Hence, the equation of motion of nonreserve areas between period t and $t + 1$ is $N_{t+1} = N_t - X_t - S_t + L_t$.

In each period, the authority faces a budget constraint, which consists of the funds supplied, b_t, plus the amount achieved by selling – if optimal – and swapping protected areas from the existing reserve network. To keep the budget dynamics simple, we assume that protected areas are sold at identical market prices, m, equaling the cost c. We also let $m = c = b_t$, i.e. the new budget available in each period allows for buying exactly one site. This is a simple structure and it would be more realistic to assume variations in selling prices and acquisition costs for sites; hence, a dynamic budget constraint (in the absence of borrowing) could be suggested. This, however, would increase the computational complexity of the problem considerably. For convenience we define two vectors C and M of size $J \times 1$, with each element equaling c and m, respectively.

Thus, altogether, there are two J period t state variables in this model, N_t and R_t, and two J period t control variables, X_t and L_t. At the beginning of period t, the planning authority observes N_t and R_t. The planner receives a budget payment b_t sufficient for buying one site and then chooses $X_t \leq 1 + L_t \leq N_t$ and $L_t \leq R_t$. Elements of N_t that have not been selected as reserve sites are then subject to possible conversion, and all elements not yet converted are subject to possible degradation according to $prob(d_j)$. The dynamic optimization problem, which maximizes the number of species covered within the selected reserves at the end of the planning horizon (i.e. the beginning of period T), is as follows:

$$V(N_{T-1}, R_{T-1}) = \max_{\substack{X_{T-1} \leq N \\ L_{T-1} \leq R}} E(V(N_T, R_T)) \tag{1}$$

s.t.

$$X'_{T-1} c \leq b_{T-1} + L'_{T-1} m \tag{2}$$

$$N_T = N_{T-1} - X_{T-1} - S_{T-1} + L_{T-1} \tag{3}$$

$$R_T = R_{T-1} + X_{T-1} - L_{T-1} \tag{4}$$

$$E(V(N_T, R_T)) = \min\left[\in_{1 \times I}, R_T' E(A_{t+1}) \right] \in_{I \times 1} \tag{5}$$

E is the expectation operator with respect to the stochastic evolution of A_{t+1}. The first constraint represents the budget (equation 2), whereas the second (equation 3) and third (equation 4) include the equations of motion for the state variables between periods. The matrix of ones sorts out the relevant sets of species from the matrix between them; the minimum operator ensures against double-counting of species (equation 5). The problem is solved recursively, so as to maximize the diversity at $T + 1$. For all possible states of the world at T, we estimate the value of the optimal program at the end of the planning period. As a stochastic dynamic optimization problem, in principle, it is possible to solve it using backward recursion to determine the best decision strategy at any state and time, applying Bellman's principle (1957). In practice, the combinatorials of an integer problem like this can be daunting and calculation times challenging. It is common to search for heuristics that can do a reasonably good job at targeting an optimal policy, while providing it at much lower calculation costs.

Sensitivity analysis of the value of adaptive management in the small conceptual model

The value of adaptive management under uncertainty, i.e. the option to swap and replace existing protected areas that have lost protective value for other nonprotected areas of higher protective value, will depend on the empirical context: how risk is across sites, how species are distributed, what likely immigration patterns are and so on.

Core parameters that have been introduced in this illustration are, however, the degradation and immigration probabilities. The following performs a sensitivity analysis of how these parameters affect the value of the option to swap protected sites for unprotected sites. A small two-period two-site model is applied, and in all cases start with site A as the protected site. The parameters shown in Table 23.1 are used in the example. All parameters are constant but the probability of degradation and immigration for the initially nonprotected site B vary

Table 23.1 Configuration as well as probabilities of conversion, environmental degradation and immigration.

	Area A	Area B
Initial species present without degradation	1,2,3,4	5,6,7
Species present with degradation	4	7
Probability of conversion	0.4	0.5
Probability of degradation	0.5	Varies
Probability of immigration of species no. 8	0.5	Varies

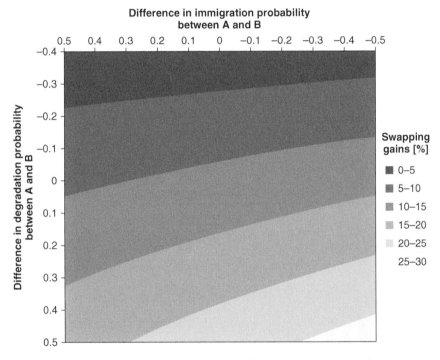

Figure 23.2 The gain in expected species coverage from the option to swap and replace depends on the differences in degradation and migration probabilities between the protected site A and the nonprotected site B. The degradation and immigration probabilities in site A are 0.5 and 0.5, respectively.

systematically between 0 and 1, creating cases where degradation and immigration are much more or less likely on protected than nonprotected sites.

Figure 23.2 shows how the gain in overall expected species coverage arising from the option to swap and replace site A for site B depends on the differences in degradation and migration probabilities between the protected site A and the nonprotected site B. Along the axis of degradation, an increasing difference implies that site A has a higher likelihood of degrading than does site B. This of course increases the gain from having the option to swap, as it increases the likelihood for the states where site A will have only one species left, but site B will have at least three species. Conversely, along the axis of immigration probability difference, an increasing difference implies a decreasing probability of immigration to the nonprotected site B, and hence a slowly decreasing value of the option to swap, as it is less likely to be relevant. The gain for this two-period, two-site problem is in the range of 0%–26% and increases with difference in degradation probabilities between A and B, but decreases with difference in immigration probabilities between A and B. The degradation and immigration probabilities in site A are 0.5 and 0.5, respectively; hence, differences are caused only by varying the probabilities for site B. In our example here, only one species can immigrate, whereas several species can be lost from degradation; hence, the value function is steeper along the degradation axis of Figure 23.2.

How much the degradation probability matters when several species can be lost relative to new species immigrating is illustrated in Figure 23.3. Here, the gain from the option to swap and replace protected site A is shown for cases where the degradation probability of the protected site A is close to one (90%) and immigration probability for site A remains 0.5, and the probabilities for site B are varied systematically across the {0,1} grid. The value is in the range of 0%–100% and still increases with difference in degradation probabilities between A and B, but decreases with difference in immigration probabilities between A and B. It illustrates that not only the gradient in degradation probability between protected and unprotected areas is important to consider, but also the levels of degradation probabilities in the protected areas.

From a theoretical viewpoint, there is much resemblance to the questions analyzed in the real options literature, which treat management planning decisions with long-ranging and irreversible consequences. These properties are well-known in other economic disciplines where

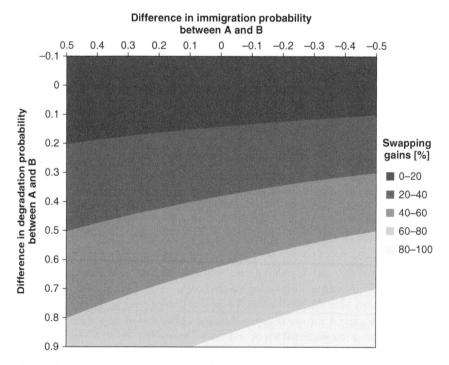

Figure 23.3 The value of swapping when degradation and migration probabilities vary. The degradation and immigration probabilities in site A are 0.9 and 0.5, respectively.

economists have for more than three decades worked on how uncertainty about possible future states in a dynamic world can be explicitly modelled and analyzed. In particular, studies on uncertainty, irreversibility and flexibility have attracted much attention (Insley, 2002; Alvarez and Koskela, 2006).

We recognize the existence of high transactions costs to trading in land, demanding negotiations and significant processing time which may at the end imply that the costs outweigh the potential gains from such adaptive strategies. We argue that if transaction costs can be estimated a priori, it is perfectly possible to include them in the decision evaluation as an element in the budgeting constraint (equation 1). Alternatively, consider them simply as threshold levels of swapping gains above which replacement is relevant.

Geopolitical coordination

A number of studies have been made on the importance of geopolitical units for the identification of priority areas for nature conservation (e.g. Bladt, Strange, Abildtrup, Svenning and Skov, 2009). In general, when the number of geopolitical planning units increases, the efficiency of priority area networks decreases, as a larger number of protected areas are required to meet conservation targets. For example, within a certain country a high priority may be given to a species that is atypical and rare in that country, even if the species is widespread and common in the neighboring countries. Such species may therefore play a disproportionally strong role in choosing which areas to protect, whereas the species would be less influential for area priority setting when operating on a broader geographical scale (Erasmus, Freitag, Gaston, Erasmus and van Jaarsveld, 1999). On the other hand, a global analysis across all regions is likely to focus resources on regions with highest diversity or endemism, while other regions may be left entirely without resources, which may not be politically acceptable (Moilanen and Arponen, 2011). This suggests that increased efficiency, i.e. the difference between the benefits of conservation and the costs of conservation, may be achieved by geopolitical coordination between units. However, from a welfare economic point of view it may still be socio-economically justifiable to maintain protection of locally rare species despite their high abundances in neighboring countries, if the local value of such species is very high.

Conclusions

Human activities are rapidly increasing the number of species threatened with extinction, thereby increasing the conservation needs globally. Yet, resources available for conservation are very limited. This stimulates economic thinking for identifying which areas of habitat are the most important to protect in order to preserve biodiversity and its value for the present and future. However, as discussed in this chapter, the condition of land areas may change in the future, thereby influencing their conservation value. If the main aim is to maximize biological effectiveness of nature, protection policies should probably be dynamic. However, loss aversion may in fact argue for a more static conservation policy. Public preferences are also contingent on the current status of the species. Conservation authorities are thus facing the problem of how to target their actions so that they accomplish the most with limited budgets, while acknowledging the uncertainty of the future states and values of areas locally and globally. Still, there are no studies which have successfully modelled climate change impacts and their temporal and spatial distributions at sufficiently high resolution to support real-world conservation decisions. This issue is at the very forefront of environmental valuation and conservation planning under uncertainty. Climate change adds to the complexity of assessing people's risk perception and preferences across types of goods which are subject to change. Therefore, future research should also find ways to cope with the complexity which forms public preferences for environmental goods. A mix of qualitative and quantitative methods, improving the design of questionnaires and description of attributes may contribute to more appropriate assessments of relevant components of a good which are being valued by people. An increased understanding of this may also improve the usefulness of the valuation studies in making real-world decisions on how to manage nature to enhance human welfare.

Acknowledgments

The authors would like to thank the Danish National Research Foundation for supporting the research at the Center for Macroecology, Evolution and Climate.

References

Adamowicz, W., Boxall, P., Williams, M. and Louviere, J. (1998). 'Stated preference approaches for measuring passive use values: Choice experiments and contingent valuation', *American Journal of Agricultural Economics*, vol. 80, pp. 64–75.

Alvarez, L. and Koskela, E. (2006). 'Irreversible investment under interest rate variability: Some generalizations', *Journal of Business*, vol. 79, pp. 623–644.

Andersen, S., Harrison, G. W., Lau, M. I. and Rutstrom, E. E. (2008). 'Eliciting risk and time preferences', *Econometrica*, vol. 76, pp. 583–618. doi: 10.1111/j.1468-0262.2008.00848.x

Araújo, M. B., Alagador, D., Cabeza, M., Nogues-Bravo, D., Thuiller, W. (2011). 'Climate change threatens European conservation areas', *Ecology Letters*, no 14, pp. 484–492.

Araújo, M. B., Cabeza, M., Thuiller, W., Hannah, L. and Williams, P. H. (2004). 'Would climate change drive species out of reserves? An assessment of existing reserve-selection methods', *Global Change Biology*, vol. 10, pp. 1618–1626.

Araújo, M. B. and Guisan, A. (2006). 'Five (or so) challenges for species distribution modelling', *Journal of Biogeography*, vol. 33, pp. 1677–1688.

Bellman, R. (1957). *Dynamic Programming*, Princeton University Press.

Bellard, C., Bertelsmeier, C., Leadley, P., Thuiller, W. and Courchamp, F. (2012). 'Impacts of climate change on the future of biodiversity', *Ecology Letters*, vol. 15, no. 4, pp. 365–377.

Bladt, J., Strange, N., Abildtrup, J., Svenning, J. C. and Skov, F. (2009). 'Conservation efficiency of geopolitical coordination in the EU', *Journal for Nature Conservation*, vol. 17, pp. 72–86.

Carroll, C., Dunk, J. R. and Moilanen, A. (2010). 'Optimizing resiliency of reserve networks to climate change: Multispecies conservation planning in the Pacific Northwest, USA', *Global Change Biology*, vol. 16, pp. 891–904.

Carson, R. T. (2012). *Contingent Valuation: A Comprehensive Bibliography and History*, Edward Elgar, Cheltenham, UK.

Charmantier, A., McCleery, R. H., Cole, L. R., Perrins, C., Kruuk, L.E.B. and Sheldon, B. C. (2008). 'Adaptive phenotypic plasticity in response to climate change in a wild bird population', *Science*, vol. 320, pp. 800–803.

Chen, I.-C., Hill, J. K., Ohlemüller, R., Roy, D. B. and Thomas, C. D. (2011). 'Rapid range shifts of species associated with high levels of climate warming', *Science*, vol. 333, no. 6045, pp. 1024–1026.

Christie, M., Hanley, N., Warren, J., Murphy, K., Wright, R. and Hyde, T. (2006). 'Valuing the diversity of biodiversity', *Ecological Economics*, vol. 58, pp. 304–317.

Crimmins, S. M., Dobrowski, S. Z., Greenberg, J. A., Abatzoglou, J. T. and Mynsberge, A. R. (2011). 'Changes in climatic water balance drive downhill shifts in plant species' optimum elevations', *Science*, vol. 331, pp. 324–327.

Czajkowski, M., Buszk-Briggs, M. and Hanley, N. (2009). 'Valuing changes in forest biodiversity', *Ecological Economics*, vol. 67, pp. 2910–2917.

Doak, D. F. and Morris, W. F. (2010). 'Demographic compensation and tipping points in climate-induced range shifts', *Nature*, vol. 467, pp. 959–962.

Erasmus, B., Freitag, S., Gaston, K., Erasmus, B. and van Jaarsveld, A. (1999). 'Scale and conservation planning in the real world', *Proceedings of the Royal Society of London Series B Biological Sciences*, vol. 266, pp. 315–319.

Gardner, T. A., Barlow, J., Araujo, I. S., Avila-Pires, T. C., Bonaldo, A. B., Costa, J. E., . . . Peres, C. A. (2008). 'The cost-effectiveness of biodiversity surveys in tropical forests', *Ecological Letters*, vol. 11, pp. 139–150.

Grantham, H. S., Wilson, K. A., Moilanen, A., Rebelo, T. and Possingham, H. P. (2009). 'Delaying conservation actions for improved knowledge: How long should we wait?' *Ecological Letters*, vol. 12, pp. 293–301.

Hanley, N., Ready, R., Colombo, S., Watson, F., Stewart, M. and Bergmann, E. A. (2009). 'The impacts of knowledge of the past on preferences for future landscape change', *Journal of Environmental Management*, vol. 90, pp. 1404–1412. doi: 10.1016/j.jenvman.2008.08.008

Insley, M. (2002). 'A real options approach to the valuation of a forestry investment', *Journal of Environmental Economics and Management*, vol. 44, pp. 471–492.

Intergovernmental Panel on Climate Change (IPCC). (2007). *Contribution of Working Group I to the Fourth Assessment Report of the Intergovernmental Panel on Climate Change, 2007*, S. Solomon, D. Qin, M. Manning, Z. Chen, M. Marquis, K. B. Averyt, . . . H. L. Miller (Eds.), Cambridge University Press, Cambridge, UK.

Isik, M. (2006). 'An experimental analysis of impacts of uncertainty and irreversibility on willingness-to-pay', *Applied Economic Letters*, vol. 13, pp. 67–72. doi: 10.1080/13505485050019203

Jacobsen, J. B., Boiesen, J. H., Thorsen, B. J. and Strange, N. (2008). 'What's in a name? The use of quantitative measures versus "iconised" species when valuing biodiversity', *Environmental and Resource Economics*, vol. 39, pp. 247–263.

Jacobsen, J. B., Lundhede, T. H., Martinsen, L., Hasler, B. and Thorsen, B. J. (2011). 'Embedding effects in choice experiment valuations of environmental preservation projects', *Ecological Economics*, vol. 70, pp. 1170–1177.

Jacobsen, J. B., Lundhede, T. H. and Thorsen, B. J. (2012). 'Valuation of wildlife populations above survival', *Biodiversity and Conservation*, vol. 21, pp. 543–563.

Jakobsson, K. and Dragun, A. K. (2001). 'The worth of a possum: Valuing species with the contingent valuation method', *Environmental and Resource Economics*, vol. 19, pp. 211–227.

Jakus, P. and Shaw, W. (2003) 'Perceived hazard and product choice: An application to recreational site choice', *Journal of Risk and Uncertainty*, vol. 26, pp. 77–92. doi: 10.1023/A:1022202424036

Jetz, W., Wilcove, D. S. and Dobson, A. P. (2007). 'Projected impacts of climate and land-use change on the global diversity of birds', *PLoS Biology*, vol. 5, pp. 1211–1219.

Johnston, R. J. and Rosenberger, R. S. (2010). 'Methods, trends and controversies in contemporary benefit transfer', *Journal of Economic Surveys*, vol. 24, pp. 479–510. doi: 10.1111/j.1467-6419.2009.00592.x

Juutinen, A., Mitani, Y., Mantymaa, E., Shoji, Y., Siikamaki, P. and Svento, R. (2011). 'Combining ecological and recreational aspects in national park management: A choice experiment application', *Ecological Economics*, vol. 70, pp. 1231–1239. doi: 10.1016/j.ecolecon.2011.02.006

Kahneman, D. and Tversky, A. (1979). 'Prospect theory–analysis of decision under risk', *Econometrica*, vol. 47, pp. 263–291. doi: 10.2307/1914185

Lenoir, J., Gegout, J.-C., Marquet, P. A., de Ruffray, P. and Brisse, H. (2008). 'A significant upward shift in plant species optimum elevation during the 20th century', *Science*, vol. 320, pp. 1768–1771.

Margules, C. R. and Pressey, R. L. (2000). 'Systematic conservation planning', *Nature*, vol. 405, no. 6783, pp. 243–253.

Moilanen, A. and Arponen, A. (2011). 'Administrative regions in conservation: Balancing local priorities with regional to global preferences in spatial planning', *Biological Conservation*, vol. 144, pp. 1719–1725.

Myers, N., Mittermeier, R. A., Mittermeier, C. G., da Fonseca, G.A.B. and Kent, J. (2000). Biodiversity hotspots for conservation priorities', *Nature*, vol. 403, no. 6772, pp. 853–858.

Naidoo, R. and Iwamura, T. (2007). 'Global-scale mapping of economic benefits from agricultural lands: Implications for conservation priorities', *Biological Conservation*, vol. 140, pp. 40–49.

Nogués-Bravo, D. and Rahbek, C. (2011). 'Communities under climate change', *Science*, vol. 334, no. 6059, pp. 1070–1071.

Parmesan, C. (2006). 'Ecological and evolutionary responses to recent climate change', *Annual Review of Ecology Evolution and Systematics*, vol. 37, pp. 637–669.

Parmesan, C. and Yohe, G. (2003). 'A globally coherent fingerprint of climate change impacts across natural systems', *Nature*, vol. 421, pp. 37–42.

Powe, N. and Bateman, I. (2004). 'Investigating insensitivity to scope: A split-sample test of perceived scheme realism', *Land Economics*, vol. 80, pp. 258–271.

Regan, H. M., Ben-Haim, Y., Langford, B., Wilson, W. G., Lundberg, P., Andelman, S. J. and Burgman, M. A. (2005). 'Robust decision-making under severe uncertainty for conservation management', *Ecological Applications*, vol. 15, pp. 1471–1477.

Roberts, D. C., Boyer, T. A. and Lusk, J. L. (2008). 'Preferences for environmental quality under uncertainty', *Ecological Economics*, vol. 66, pp. 584–593.

Sekercioglu, C. H., Schneider, S. H., Fay, J. P. and Loarie, S. R. (2008). 'Climate change, elevational range shifts, and bird extinctions', *Conservation Biology*, vol. 22, pp. 140–150.

Strange, N., Thorsen, B. J. and Bladt, J. (2006). 'Optimal reserve selection in a dynamic world', *Biological Conservation*, vol. 131, pp. 33–41.

Strange, N., Thorsen, B. J., Bladt, J., Wilson, K. A. and Rahbek, C. (2011). 'Conservation policies and planning under climate change', *Biological Conservation*, vol. 144, pp. 2968–2977.

TEEB. (2010). *The Economics of Ecosystems and Biodiversity: Mainstreaming the Economics of Nature: A Synthesis of the Approach, Conclusions and Recommendations of TEEB*. Retrieved from www.teebweb.org/TEEBSynthesisReport/tabid/29410/Default.aspx

Thomas, C. D., Cameron, A., Green, R. E., Bakkenes, M., Beaumont, L. J., Collingham, Y. C., . . . Williams, S. E. (2004). 'Extinction risk from climate change', *Nature*, vol. 427, pp. 145–148.

Tversky, A. and Fox, C. (1995). 'Weighing risk and uncertainty', *Psychological Review*, vol. 102, pp. 269–283. doi: 10.1037//0033-295X.102.2.269

Visconti, P., Pressey, R. L., Segan, D. B. and Wintle, B. A. (2010). 'Conservation planning with dynamic threats: The role of spatial design and priority setting for species' persistence', *Biological Conservation*, vol. 143, pp. 756–767.

Wielgus, J., Gerber, L. R., Sala, E. and Bennett, J. (2009). 'Including risk in stated-preference economic valuations: Experiments on choices for marine recreation', *Journal of Environmental Management*, vol. 90, pp. 3401–3409. doi: 10.1016/j.jenvman.2009.05.010

24

ECONOMIC ANALYSIS OF BIOLOGICAL INVASIONS IN FORESTS

Thomas P. Holmes,[1] *Juliann Aukema,*[2] *Jeffrey Englin,*[3] *Robert G. Haight,*[1] *Kent Kovacs*[4] *and Brian Leung*[5]

[1]USDA FOREST SERVICE.
[2]NATIONAL CENTER FOR ECOLOGICAL ANALYSIS AND SYNTHESIS.
[3]ARIZONA STATE UNIVERSITY.
[4]UNIVERSITY OF ARKANSAS.
[5]MCGILL UNIVERSITY.

Abstract

Biological invasions of native forests by nonnative pests result from complex stochastic processes that are difficult to predict. Although economic optimization models describe efficient controls across the stages of an invasion, the ability to calibrate such models is constrained by lack of information on pest population dynamics and consequent economic damages. Here we describe economic approaches for analyzing pre-invasion and post-invasion management of biological invasions under conditions of risk and uncertainty and emphasize the need for new microeconomic and aggregate studies of economic damages across gradients of forest types and ownerships.

Keywords

Risk, uncertainty, optimization, timber, nonmarket values, biosecurity

Introduction

Biological invasions by forest insects, pathogens and plants are a type of externality (or biological pollution: Horan, Perrings, Lupi and Bulte, 2002) from trade that has altered the productivity, structure and species composition of many forest ecosystems around the world. Economists have argued that an efficient solution to reducing the threat of biological invasions would be to require parties benefiting from trade in products that pose risks to terrestrial or aquatic ecosystems to pay the costs associated with environmental degradation (Perrings, Dehnen-Schmutz, Touza and Williamson, 2005). However, until policies such as tariffs and taxes or improved trading standards are implemented, current practices will persist and nonnative pests will continue to threaten the integrity of native forests (Holmes, Aukema, Von Holle, Liebhold and Sills, 2009).

Biological invasions are complex stochastic processes that are difficult to predict. History has shown that many accidental introductions of nonnative species fail to become established in new habitats and, of those that become established, few species ultimately cause widespread economic damage (Williamson, 1996). In the United States, more than 450 nonnative insect species and 16 pathogens have colonized US forest and urban trees since European settlement. Of this number, roughly 60 insect species, and all pathogens, have caused notable damage to trees via increased mortality or reductions in reproduction and growth, and only a few species have caused major ecological disruptions (Aukema et al., 2010).

In this chapter, we provide an overview of a suite of economic approaches that can be used to analyze pre-invasion and post-invasion management of biological invasions in forests. Because forest managers and policymakers are typically confronted by a pervasive lack of knowledge about the factors that ultimately shape the establishment and spread of nonnative organisms in native ecosystems, we begin with a review of decision-making frameworks under conditions of risk and uncertainty. This is followed by a review of theoretical developments in the natural resource economics literature describing optimal prevention and control activities across the stages of a biological invasion. Next, we discuss empirical approaches for quantifying market and nonmarket impacts of nonnative forest pests. Although most empirical economic studies have focused on post-invasion analysis of individual pest species, we next present an innovative method for aggregating economic damages across multiple species of forest pests that can be used to forecast future damages from new invasions. Finally, we present our conclusions and suggestions for future research.

Managing risk and uncertainty in biological invasions

When faced with threats to the integrity of native forest ecosystems, public and private forest owners generally take actions to mitigate ecological and economic impacts. Forest protection is a public good (nonrival and nonexcludable), and if forest owners fail to account for the costs and benefits generated by their individual protection efforts on other forest owners, the socially optimal level of forest protection will not be provided (Alavalapati, Jose, Stainback, Matta and Carter, 2007; Holmes et al., 2009). Therefore, governmental programs and cooperation among landowners play an essential role in the provision of forest protection.

Economic analysis supporting decisions about forest protection can be targeted at pre-invasion or post-invasion controls. Pre-invasion control focuses attention on preventing the introduction of new forest pests in native ecosystems, and economic analysis is needed to help balance the costs of improved biosecurity with the benefits of avoiding damage to forest ecosystems. Economic analysis of pre-invasion control is challenged by a lack of knowledge about factors contributing to successful invasions:

1. To the degree that each biological invasion in forests is a novel event, the past may provide limited scope for predicting future outcomes, and probabilistic damage functions based on historical invasions may be misleading (Williamson, 1996).
2. Characteristics that make newly arrived organisms invasive are poorly understood for most taxonomic groups (Kolar and Lodge, 2001).
3. As nonindigenous forest species accumulate, interactions among introduced species may facilitate the establishment of new species or magnify impacts on native species (Von Holle, 2011), causing super-additive damages.

4. Significant lags typically occur between the time at which an organism arrives in a new environment and when it becomes widely established and causes economic damage (Essl et al., 2011).
5. Forest protection from biological invasions is only as good as the weakest link in the chain of defensive actions (Perrings et al., 2002).

Economic analysis of post-invasion strategies, which focus attention on the costs and benefits (reduced damages) of controlling individual pest species, faces many of the sources of uncertainty listed previously for pre-invasion control, as well as the following:

1. Efficacy of management interventions used to eradicate or control pest outbreaks are difficult to predict and linkages between control costs and specific levels of damage reduction are highly uncertain.
2. Changes in global climate are causing some native forest pests to invade forest ecosystems beyond their historical range, and may alter future forest conditions in ways that increase forest vulnerability from nonnative forest species (Weed, Ayres and Hicke, 2013).

Most economic analyses recognize biological invasions as stochastic processes, due to either environmental (e.g. Olson and Roy, 2002) or demographic (e.g. Jaquette, 1970) stochasticity. The decision maker is characterized as either facing conditions of risk (relying upon probability distributions) or uncertainty (lack of knowledge regarding probability distributions). If a decision maker believes that biological invasions are replicable – factors causing varying degrees of invasion severity (e.g. infested area) could be known by designed or natural experiments – then the control problem can be viewed from the perspective of risk management. In this case, objective or subjective assessments of the probability of various degrees of invasion severity, at some future time, would yield a latent severity probability density function (PDF) such as shown in Quadrant I (Figure 24.1).[1] Such a PDF might be based on historical observations, simulation models or expert opinion. As more is learned about a pest, the PDF could be modified using methods such as Bayesian updating (Kelsey and Quiggan, 1992; Prato, 2005).

Many biological invasions cause little or no economic damage, so the threshold of economic damage is depicted to the right of the origin (the dashed vertical line). Quadrant II traces out the relationship between latent invasion severity and the level of market and/or nonmarket economic damages, which is the sum of costs and economic losses. This function could be estimated using new economic analysis or prior studies (via benefit-transfer analysis). Quadrant III translates economic damages on the vertical axis to the horizontal axis. Quadrant IV plots economic damage as a function of the associated latent invasion severity PDF. Integration of the area beneath an economic damage PDF yields the expected value of economic damages.

Risk management interventions (controls) are conducted with the intention of shifting the latent severity PDF (the state variable) toward smaller values (leftward pointing arrow in Figure 24.1).[2] Alternative controls incur different management costs (Quadrant II) and the cost associated with any feasible control should be less than the expected loss with no control. Given information on the relationship between the cost of a feasible control and the anticipated shift in the latent invasion severity PDF, the associated shift in the economic damage function can be computed (rightward pointing arrow in Figure 24.1). Under the expected value framework, the preferred alternative is selected from a set of feasible controls as the alternative that minimizes expected economic damages. If the decision space is continuous over alternative controls, the optimal solution is found where marginal cost equals expected marginal benefits (expected reduction in economic loss) (Herrick, 1981; Horan et al., 2002). In a dynamic model, marginal benefits include the reduction of

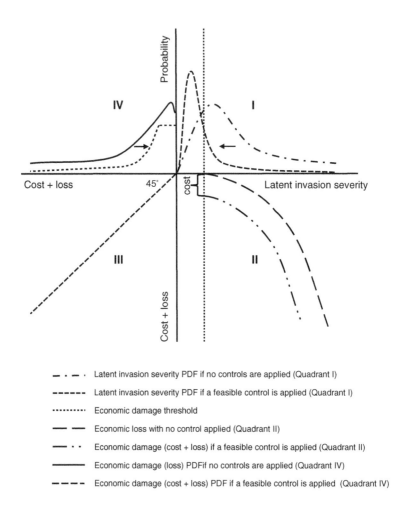

Figure 24.1 A stylized risk management model shows how invasion controls shift the latent invasion severity PDF and the economic damage PDF towards smaller values.

damage in current and (discounted) future periods (e.g. Olson and Roy, 2002). This learn-then-act strategy is most appropriate for post-invasion controls where established nonnative forest organisms are slow-spreading, the efficacy of control options is well understood, and reasonable estimates of economic losses are available. In contrast, this strategy is less appropriate for severe, fast-moving invasions or for pre-invasion control policies when little is known about the latent invasion severity, the efficacy of control strategies, or potential economic damages for a suite of potential invaders.

The risk management framework for controlling biological invasions in forests is limited in many applications by its stringent information requirements. Further, it is relatively insensitive to potentially catastrophic outcomes which may be difficult to characterize in a probabilistic framework. When the parameters of either the latent invasion severity or economic damage PDFs are unknown or highly uncertain, or when invasions are severe and fast-moving, other decision-making frameworks can be used.

A framework for making decisions under ignorance (when the decision maker has no knowledge of relevant probabilities) is the *maximin* rule. Under this rule, control *x* is preferred to another control *y* if and only if the worst possible outcome from *x* is better than the worst possible outcome from *y* (Kelsey and Quiggan, 1992).

An alternative framework for making decisions under ignorance is based on the idea that decision makers characterize possible outcomes in terms of how surprised they would be if each outcome came true (Shackle, 1966; Katzner, 1990). Potential surprise functions are similar to inverse subjective probability functions, although they do not need to sum to one over the range of possible outcomes (Horan et al., 2002). Several possible outcomes can be associated with each control, and decision makers choose the control that minimizes the degree of surprise for attention-gaining (ascendant) outcomes (the *focus loss*).[3] Similar to *maximin*, the *focus-loss* framework shifts the decision-makers' attention from expected outcomes toward the catastrophic tail of the outcome distribution.

Stages of a biological invasion

A key point not illustrated in Figure 24.1 is that biological invasions proceed by stages where each stage is associated with one or more management actions and a vector of economic costs and losses (Figure 24.2). Economic analysis proceeds by seeking efficient strategies either within a stage or across stages.

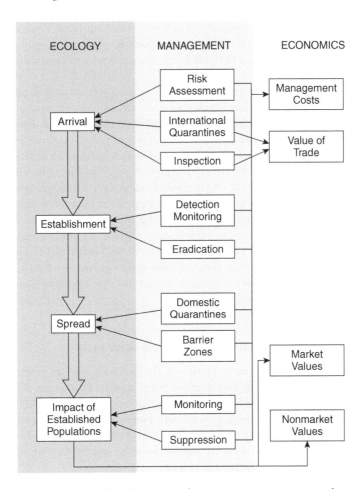

Figure 24.2   Each stage of a biological invasion induces management responses and economic impacts.

Stage 1: Arrival and introduction of nonnative forest pests

International trade is the major pathway for the introduction of nonnative forest pests. The importation of live plants is the most probable pathway of introduction for most damaging forest insects and pathogens established in the United States, and nearly three-quarters of plant shipments infested with exotic organisms pass through US ports undetected (Liebhold, Brockerhoff, Garrett, Parke and Britton, 2012). Wood packing materials are the most common pathway of introduction for wood-boring forest insects, and the use of these shipping materials is an increasing concern (Aukema et al., 2010; Strutt, Turner, Haack and Olson, 2013).

Economic models: Preventing arrival and introduction

A general economic strategy for preventing the introduction of invasive species is to internalize the costs of biological invasions using tariffs combined with improved port inspections (Perrings et al., 2005). Economic optimization reveals that the importing country should set the tariff at the Pigouvian level, equal to the sum of expected damages from contaminated units not detected during inspections plus the costs of inspections (McAusland and Costello, 2004). When it is possible to estimate the probability of a successful invasion, each biosecurity facility should optimally set the marginal cost of undertaking preventive measures equal to marginal expected benefits, taking into account the probability that a species will invade through a different facility (Horan et al., 2002).

Economic models focusing on a single stage of the invasion process cannot provide globally optimal solutions because they ignore potential trade-offs among defensive actions across the stages of an invasion. Optimal allocation among prevention and control depends on the nature of prevention and control cost curves and the decision-maker's preferences over risky events. Research has shown that under some conditions, invasive species can be managed most cost-effectively using greater investments in prevention relative to control because damages can be catastrophic (Leung et al., 2002). Other research has shown that if decision makers are risk-averse and if control options are thought to be more certain than prevention, then control may be preferred to prevention (Finnoff, Shogren, Leung and Lodge, 2007). Recent innovations in the analysis of trade-offs among invasion stages include the development of spatial models of prevention, detection and control (Sanchirico, Albers, Fischer and Coleman, 2010). The primary lesson is that focusing on a subset of transmission pathways, on only one or two controls, or on a single region, misses important interactions that are critical in identifying cost-effective policy recommendations.

Stage 2: Establishment of nonnative pests

The probability of successful establishment depends on the frequency and size of arrivals (propagule pressure), spatial habitat suitability and temporal environmental fluctuations (Leung, Drake and Lodge, 2004; Von Holle and Simberloff, 2005) which are highly stochastic. Most preventative strategies are based on reducing propagule pressure. However, if new species are repeatedly introduced through similar or novel invasion pathways, Allee effects and stochastic population dynamics are much less likely to cause initial populations to go extinct, thereby increasing the likelihood that isolated populations become established.

Economic models: Surveillance and eradication
of newly established populations

Surveillance systems to detect newly established species that evade port inspections are critical to reducing the potential for ecological and economic damage. Cost-effective surveillance systems for newly established populations balance the intensity and cost of surveillance (which increase with the level of effort) with the costs of damage and eradication of newly detected populations (which may be less if detected early) (Epanchin-Niell and Hastings, 2010). Economic models that account for this trade-off have assumed the pest location is unknown (Mehta, Haight, Homans, Polasky and Venette, 2007), small invasive populations establish ahead of an advancing front (Homans and Horie, 2011) or that the likelihood of detection increases with the size of an infestation (Bogich, Liebhold and Shea, 2008).

Research effort has also focused on the properties of optimal one-time surveillance across multiple sites when species' presence is uncertain prior to detection, accounting for heterogeneity in species presence and detectability across sites (Hauser and McCarthy, 2009). Other models of one-time surveillance have investigated the impact of uncertainty regarding the extent (rather than simply the presence) of an infestation (Horie, Haight, Homans and Venette, 2013).

Economic models of long-term surveillance programs have been developed using dynamic optimization algorithms and indicate that greater surveillance effort is warranted in locations that have higher establishment rates, higher damage and eradication costs or lower sampling costs (Epanchin-Niell, Haight, Berec, Kean and Liebhold, 2012). Active research is underway in which invasion dynamics are uncertain. This line of research recognizes that surveillance may not provide correct information, and researchers have used partially observable Markov decision process to address optimal invasive species surveillance (Regan, McCarthy, Baxter, Panetta and Possingham, 2007) and monitoring and control strategies (Haight and Polasky, 2010). More generally, partially observable decision models have been used to allocate management resources for networks of cryptic diseases, pests and threatened species (Chadès et al., 2011).

Stage 3: Spread of nonnative pests

The spread of a biological invasion results from the combination of three factors: (1) pest population growth, (2) dispersal of organisms and (3) spatial characteristics of the environment. The classic reaction-diffusion model of a biological invasion predicts circular traveling waves that spread outwards from the point of invasion origin and a linear relationship between the square root of the invaded area and time – that is, the range expands at a constant rate (Holmes, Lewis, Banks and Veit, 1994; Shigesada, Kawasaki and Takeda, 1995). Despite the apparent simplicity of this model, it has been successfully used to explain the spread of nonnative forest species such as the gypsy moth (Liebhold, Halvorsen and Elmes, 1992).

Several modeling approaches have been developed that take into account local and long-distance dispersal due to factors such as transportation networks. The stratified diffusion model describes range expansion in terms of the effective range radius, which is the square root of the invaded area (integrated across all colonies) divided by the square root of pi (Shigesada et al., 1995). Building on this idea, the temporal and spatial dynamics of economic damages resulting from a biological invasion were analyzed using the area of economic damage (AED) occurring in 'economic colonies' (Figure 24.3) (Holmes, Liebhold, Kovacs, and Von Holle, 2010). Although applications of the stratified diffusion model are largely backward looking, gravity models and random utility models have been used to make predictions of invasions when

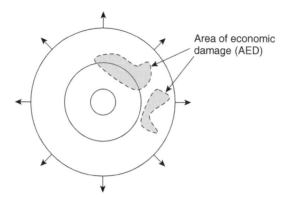

Figure 24.3 Economic damage disperses by local and long-distance pathways over successive time periods. The integrated AED in 'economic colonies' is used to compute the effective range radius and the rate of spread of economic damage associated with a biological invasion.

human-mediated dispersal is important (Chivers and Leung, 2012). Each takes into account distance as well as the 'attractiveness' of alternative locations, and therefore can incorporate differential traffic to each site and their consequences on patterns of spread.

Economic models: Slowing the spread of established populations

When a nonnative species becomes established, various strategies can be used to affect the expansion of its range, including reduction of the chances of accidental movement of organisms to uninfested areas via domestic quarantine, detection and eradication of isolated colonies or control activities to slow or stop the spread of the core population. Research has focused on developing optimal control strategies for slowing or eradicating population growth, addressing questions of when, where and how much control should be applied (see Epanchin-Niell and Hastings, 2010, for a review).

Invasive species control models generally include pest population dynamics with an objective of minimizing the sum of discounted control costs and invasion damages over time. The most basic models of invasive species dynamics focus on the numbers of individuals or area of infestation and ignore spatial description (Sharov and Liebhold, 1998; Eiswerth and Johnson, 2002; Saphores and Shogren, 2005). A general principle emerging from this research is that, if the invasive species stock is initially greater than its optimal equilibrium level, then the highest level of management effort should be initially applied and then decline over time until the steady state is reached (Eiswerth and Johnson, 2002). When controlling a population front, the optimal strategy changes from eradication to slowing the spread to doing nothing as the initial area occupied by the species increases, the negative impact of the pest per unit area decreases or the discount rate increases. Stopping population spread is not an optimal strategy unless natural barriers to population spread exist (Sharov and Liebhold, 1998).

These basic population models have been extended to account for uncertainty in invasion growth. The optimal control strategy is obtained using discrete-time stochastic dynamic programming (Eiswerth and van Kooten, 2002; Olson and Roy, 2002) or a real options framework in continuous time (Saphores and Shogren, 2005; Marten and Moore, 2011).

Recently, spatially explicit models of invasive species dynamics have gained prominence. These models define the landscape as a set of discrete patches, define control activities for each patch and predict the growth and dispersal of the invasive species among patches as a function of the selected controls (e.g. Hof, 1998; Albers, Fischer and Sanchirico, 2010; Blackwood, Hastings and Costello, 2010; Epanchin-Niell and Wilen, 2012; Kovacs, Haight, Mercader and McCullough, 2013). Although these spatial dynamic models are complicated to solve, they can provide pragmatic guidance to forest managers. For example, managers should (1) use landscape features that alter the shape of the initial invasion in order to reduce the length of exposed invasion front and look forward over space to slow the spread and (2) steer the invasion front away from the direction of greatest potential damages or in the direction where the costs of achieving control are low (Epanchin-Niell and Wilen, 2012).

Economic impacts from nonnative forest pests

One of the primary challenges of applying optimization models to post-invasion management of biological invasions in forests is that the analyst needs first to specify pest population dynamics and second to describe how pest dynamics are coupled with a proper measure of economic impacts. Welfare economic theory should guide the choice of empirical methods used to measure economic impacts, which may be transitory, cyclical or persistent.[4] Determining the temporal relationship between a nonnative forest pest, its host and the flow of market or nonmarket goods and services is necessary to establish scope for economic analysis.

Timber market losses

If reduction in the volume of timber harvest is a small percentage of total harvest volume and can be fully offset by timber harvest of nonimpacted timber species, and if compensatory growth on nonimpacted healthy trees will eliminate all lost timber harvest volumes from a biological invasion within a given period, timber prices will remain fixed and only forest landowners with impacted stands will experience economic losses (Aukema et al., 2011). In contrast, if a biological invasion is severe enough to shift timber supply (e.g. via pre-emptive or salvage harvest), timber demand (e.g. via substitution of alternative species) or both, prices will be variable and three types of models may be used for analysis: (1) market trends, (2) partial equilibrium models and (3) computable general equilibrium (CGE) models.

Timber market trends

The simplest approach to evaluating timber market dynamics is to qualitatively describe market forces using time-series data on timber prices and quantities (Zivnuska, 1955). If in successive years timber prices increase and quantity decreases, economic theory stipulates that market supply has shifted back (and it is not possible to state what has happened to demand). Other shifts in supply and demand are implicit with other combinations of price and quantity changes.

Although the market trend model is simple to implement, its application can provide useful insights. For example, consider data on chestnut lumber prices and quantities (Figure 24.4) for years spanning the onset and spread of the chestnut blight (1904–1943).[5] The chestnut blight was first identified in New York City during 1904, and at that time, chestnut accounted for about one-quarter of the lumber produced in New England and about 15 percent of lumber production in the Appalachian region. As can be seen in Figure 24.4, the volume of chestnut lumber produced in the United States increased rapidly from 1904 to 1909 and then gradually decreased as chestnut timber inventories were exhausted. Although real chestnut lumber prices varied during this period, they remained remarkably stable. These trends are

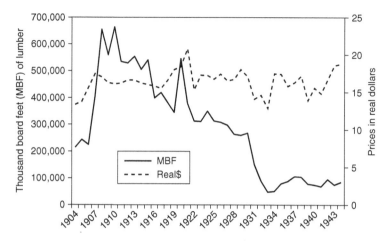

Figure 24.4 Time-series of chestnut lumber prices and quantities in the United States, 1904–1943.
Source: Steer (1948).

consistent with a reduction in lumber supply due to a declining timber inventory and a gradual reduction in chestnut lumber demand as hardwood products firms substituted other timber species.

PARTIAL EQUILIBRIUM MARKET MODELS

A more informative analysis of the impacts of forest pests on timber markets can be conducted in cases where shifts in timber supply and demand can be estimated or simulated using empirical functions provided by prior studies. These models estimate quasi-rents for timber producers and consumers in an intermediate market and do not include impacts on other groups, such as final consumers.

A conceptual model describing timber market impacts of catastrophic mortality from forest insects and other forest disturbances has been developed and used to provide empirical estimates of supply shocks on producer and consumer surplus (Holmes, 1991; Prestemon and Holmes, 2000). The market model describes a short-run, outward shift in timber supply as damaged timber stocks are salvaged, and then a backwards shift in supply as damaged stocks are exhausted and supply is provided by a diminished timber inventory. As shown in Figure 24.5, the market equilibrium immediately preceding a disaster corresponds with price P_0 and quantity Q_0 (point *a*) based on supply curve $S_0(I_0)$, a function of initial inventory I_0 and an initial demand curve, $D_0(P)$. The volume of timber salvage, V, induces an inelastic salvage supply curve as the opportunity cost of holding damaged stocks is very low. During salvage operations, market supply shifts outwards to P_T and Q_T (point *b*). The volume supplied from undamaged stands (Q_u) is found where P_T intersects the undamaged supply curve $S_1(I_1)$, point *d*. The salvage volume, V, gradually shifts back as salvage volumes are exhausted, leaving a new equilibrium of supply and demand at P_1 and Q_1 (point *c*). In the short run, economic surplus is transferred from forest owners with damaged stands to owners with undamaged stands and to wood-consuming firms. However, in the longer run, if standing inventories are significantly reduced, wood-using firms may lose economic surplus to the point that they go out of business.

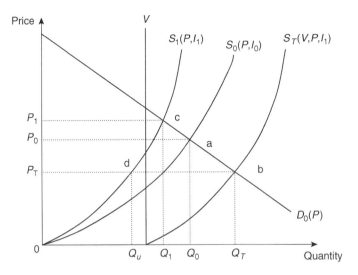

Figure 24.5 Timber market impact of a biological invasion where timber mortality is salvaged (V) after being killed, shifting supply from S_0 to S_T. Timber supply then shifts backwards (S_1) as damaged timber stocks are exhausted and inventories of undamaged timber are diminished.

CGE MODELS

CGE models provide a mathematical representation of monetary flows through an economy as firms and households interact through the markets for inputs and products. It is generally assumed that consumers maximize utility, producers maximize profit and consumers and producers can substitute alternative inputs and final consumption goods as relative prices change and equilibrium is attained in each period. These models are critically dependent on the many assumptions that are made to operationalize them. CGE models have been used to estimate the regional economic impacts of indigenous (Patriquin, Wellstead and White, 2007) and nonnative forest pest outbreaks (McDermott, Finnoff and Shogren, 2013).

Nonmarket losses from nonnative forest pests

Forest ecosystems generate a constellation of benefits that are valued by people but are not bought and sold in markets. The proper economic measure of the value of a change in nonmarket forest benefits is what people are willing to pay to access, protect, enhance or restore a flow of benefits. The total value of nonmarket benefits associated with a change in forest conditions is the sum of use, option and passive use values. Although quantification of the economic benefits of controlling biological invasions in forests has historically emphasized protection of commodity production, recent research demonstrated that the economic benefits of invasion control efforts in US forests would largely result from the protection of nonmarket goods and services (Aukema et al., 2011).

Several economic tools are available for measuring the nonmarket value of natural resources (Champ, Boyle and Brown, 2003). Approaches for estimates are categorized by methods that use observed behavior, known as revealed preference methods, and methods based on responses to hypothetical questions, known as stated preference methods. Because revealed preference methods are based on observations of how people behave in situations that are linked with markets, these methods are not able to provide estimates of passive use value. In contrast, stated preference methods are able to uncover estimates of total value.

Revealed preferences for forest protection

Two revealed preference methods have been used to study the economic impacts of forest pests – the travel cost method and the hedonic price method. If an insect or disease outbreak changes the condition of a forest recreational site and alters visitation rates, then the travel cost method can be used to estimate changes in economic welfare. Because several recreational sites may have related demand functions, the demand linkages among sites must be accounted for (Englin, Holmes and Sills, 2003). Several studies have used the travel cost method to estimate economic impacts of forest insects via the demand for forest recreation (reviewed in Rosenberger, Bell, Champ and Smith, 2012), although we are unaware of forest pest impact studies that consider substitution across recreational sites.

The hedonic price method is based on the idea that the price of a good represents the sum of values associated with the qualities or attributes that comprise the good (Champ et al., 2003). Where localized changes in forest health affect a relatively small proportion of properties in a housing market, the hedonic price function for the market remains unaffected, and the first-stage hedonic price function can be used to compute marginal changes in economic welfare.[6] This method has become increasingly popular for estimating the nonmarket impacts of nonnative forest pests in the United States, largely as a result of the increasing availability of spatially referenced remote sensing data on forest health and the availability of electronic records on housing prices and attributes. One of the key discoveries has been that the loss of forest health on one property causes economic spillovers onto neighboring properties (Holmes, Murphy, Bell and Royle, 2010). Estimates of percentage losses in property value from declines in tree health vary widely (Holmes, Murphy, et al., 2010; Kovacs, Holmes, Englin and Alexander, 2011; Price, McCollum and Berrens, 2010), and it is not yet understood what causes this variation.

Stated preferences for forest protection

Several studies have used stated preferences (contingent valuation or choice experiments) to estimate willingness to pay for forest health protection programs. The contingent valuation method was used to evaluate the benefits of programs protecting forests from nonnative forest insects in residential forests (Miller and Lindsay, 1993; Jakus, 1994) and in public forests (Holmes and Kramer, 1996; Moore, Holmes and Bell, 2011). These studies concluded that (1) the benefits of forest protection programs are generally several times larger than their costs, and (2) passive use values constitute a substantial proportion of the total value of forest protection in public forests.

The choice experiment method differs from contingent valuation in that it focuses attention on trade-offs between various levels of environmental attributes and money (Champ et al., 2003). This method has been used to estimate willingness to pay for multiple attributes of state parks in Florida (United States), including preventing invasive plants from becoming abundant (Adams, Bwenge, Lee, Larkin and Alavalapati, 2011). Similar to the general conclusion from contingent valuations studies of forest health protection, statewide benefits of protecting parks from invasive plants were found to be several times larger than current expenditures on these programs.

Aggregation of multiple economic impacts

Relatively few studies have been conducted that estimate aggregate economic impacts of invasive species at a national scale. Despite the success of these studies in drawing attention to the economic significance of biological invasions (Pimental, Lach, Zuniga and Morrison, 2000; Colautti, Bailey, van Overdijk, Amundsen and MacIsaac, 2006), their policy relevance has been limited by lack of a theoretically consistent economic framework (Born, Rauschmayer and Brauer, 2005; Holmes et al., 2009).

Recent research (Aukema et al., 2011) has reported an improved method for estimating aggregate economic impacts from multiple biological invasions based on the idea that economic damages are random variables that can be depicted using economic damage PDFs (Figure 24.6A). The level of economic damage associated with each historically occurring pest is categorized as low, medium or high, and it is assumed that the probability that a new pest will fall into any of the damage categories is equal to the integral of the corresponding area under the damage PDF. Because the shape and scale of the damage PDFs are unknown, alternative functions are generated using alternative parameter values (Figure 24.6B). The relative probability of

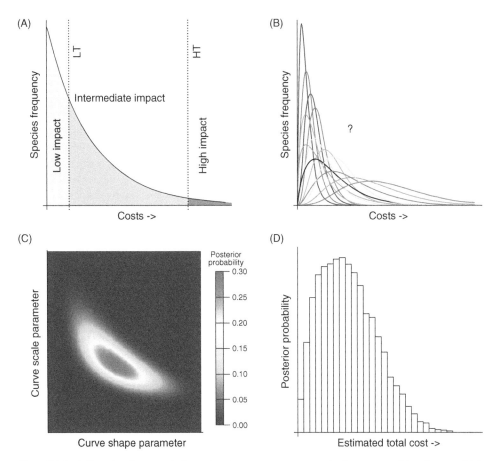

Figure 24.6 A Bayesian framework for estimating aggregate economic damages resulting from multiple biological invasions.

Source: Aukema et al. (2011).

each set of parameters being true is determined using Bayesian statistical methods (comparing model predictions to the observed data on pest frequencies and economic damage estimates). The resulting collection of posterior PDFs then yields estimates of parameter uncertainty (Figure 24.6C). Additionally, in order to include model or structural uncertainty, different families of curves can be considered (e.g. gamma, log-normal, power or Weibull distributions) and integrated into the analysis using Bayesian model averaging. Finally, the output from these analyses is used to generate the posterior predictive distribution of damage (Figure 24.6D).

Using this aggregation method, the authors found that the greatest impacts of recent biological invasions in US forests are largely borne by local governments and residential landowners. Wood-boring insects were found to cause the largest annual economic impacts, inducing nearly $1.7 billion in local government expenditures and approximately $830 million in lost residential property values. Timber impacts were typically an order of magnitude smaller than impacts on local governments and residential households.

Conclusions and discussion

Biological invasions have historically caused major disruptions in the flow of valued goods and services provided by forest ecosystems. Unless trans-boundary biosecurity programs are improved, new invasions are anticipated to impact ecological and economic systems well into the future. Although optimization models have recently been developed that focus attention on economic efficiency across the stages of a biological invasion, calibration of these models is severely limited by the dearth of theoretically consistent measures of economic damages. Better information on economic costs, losses and the efficacy of control programs across multiple spatial scales is needed to provide improved information to decision makers.

Several research themes are suggested. First, the importance of protecting nonmarket values provided by healthy forests has recently been recognized in both microeconomic and aggregate economic studies and substantial research effort is warranted to address many unanswered questions. Topics include (1) understanding sources of variation in first-stage hedonic functions and estimation of second-stage hedonic demand functions for healthy residential forests, (2) understanding the importance of passive use forest protection values in relation to use values across a broad spectrum of public forest ecosystems and (3) understanding whether preferences for forest health protection are stable or variable across generations.

Innovative methods are also needed for estimating timber market impacts of biological invasions. One approach that (to our knowledge) has yet to be utilized by forest economists is estimation of econometric general equilibrium supply and demand functions, as described by Just and Hueth (1979).

A greater number of microeconomic studies of market and nonmarket values associated with protecting forest health are needed for improving models of aggregate economic impacts of biological invasions in forests. Although progress has been made in developing aggregate economic damage functions, it is not known if aggregate damage functions are stable or, if not, what factors shift damage functions over time. The development of systems for updating aggregate damage estimates as new information becomes available is recommended.

Linkages between the costs of pre-invasion and post-invasion controls and forest ecosystem damages avoided are poorly understood and deserve greater research attention. Until more knowledge is gained on this topic, balancing costs and benefits as prescribed by economic analysis will remain an elusive goal.

Finally, almost nothing is known about how forest health protection decisions are actually made. Research is needed to shed light on the degree to which, and under what circumstances, decision makers think in probabilistic terms or rely on alternative modes for making decisions. We suggest that stated preference methods (such as choice experiments) or the use of economic experiments might be fruitfully employed to understand how decision makers assign weights to alternative pest management programs and outcomes, and how they update their thinking as knowledge is gained about the nature of individual or multiple biological invasions.

Notes

1   In practice, nonsmooth PDFs could be used, such as a triangular distribution which only requires estimates of the minimum, maximum, and mode.
2   Control effort may also reduce the economic loss associated with each level of invasion severity, causing a change in the shape of the loss function.
3   The focus loss is determined by the tangency between the potential surprise function and the corresponding iso-ascendency contour (Katzner 1990).
4   Transitory economic impacts can occur when a pest functionally eradicates its host, forest ecosystems provide compensatory growth of other tree species, and people substitute alternative goods or services as conditions of relative scarcity change. Cyclical economic impacts result from oscillatory pest population dynamics, and forests recover between outbreaks. Persistent impacts occur when a pest and its host coexist over long periods of time, due to factors such as slow rates of spread or genetic improvement in host trees.
5   Lumber prices were deflated to 1913 prices using the consumer price index, which was initiated in that year. Prices prior to that year shown in Figure 24.4 are nominal prices.
6   If data on forest conditions are available for several housing markets, then it may be possible to estimate the second-stage demand function for forest conditions.

References

Adams, D. C., Bwenge, A. N., Lee, D. J., Larkin, S. L. and Alavalapati, J.R.R. (2011). 'Public preferences for controlling upland invasive plants in state parks: Application of a choice model', *Forest Policy and Economics*, vol. 13, pp. 465–472.

Alavalapati, J.R.R., Jose, S., Stainback, G. A., Matta, J. R. and Carter, D. R. (2007). 'Economics of cogongrass control in slash pine forests', *Journal of Agricultural and Applied Economics*, vol. 39, pp. 61–68.

Albers, H. J., Fischer, C. and Sanchirico, J. N. (2010). 'Invasive species management in a spatially heterogeneous world: Effects of uniform policies', *Resource and Energy Economics*, vol. 32, pp. 483–499.

Aukema, J. E., Leung, B., Kovacs, K., Chivers, C., Britton, K., Englin, J., . . . Von Holle, B. (2011). 'Economic impacts of non-native forest insects in the continental United States', *PLoS ONE*, vol. 6, no. 9, p. e24587. doi:10.1371/journal.pone.0024587

Aukema, J. E., McCullough, D. G., Von Holle, B., Liebhold, A. M., Britton, K. and Frankel, S. J. (2010). 'Historical accumulation of nonnative forest pests in the continental United States', *Bioscience*, vol. 60, no. 11, pp. 886–897.

Blackwood, J., Hastings, A. and Costello, C. (2010). 'Cost-effective management of invasive species using linear-quadratic control', *Ecological Economics*, vol. 69, pp. 519–527.

Bogich, T. L., Liebhold, A. M. and Shea, K. (2008). 'To sample or eradicate? A cost minimization model for monitoring and managing an invasive species', *Journal of Applied Ecology*, vol. 45, pp. 1134–1142.

Born, W., Rauschmayer, F. and Brauer, I. (2005). 'Economic evaluation of biological invasions – A survey', *Ecological Economics*, vol. 55, pp. 321–336.

Chadès, I., Martin, T. G., Nicol, S., Burgman, M. A., Possingham, H. P. and Buckley, Y. M. (2011). 'General rules for managing and surveying networks of pests, diseases, and endangered species', *Proceedings of the National Academy of the United States of America*, vol. 108, no. 20, pp. 8323–8328.

Champ, P. A., Boyle, K. J. and Brown, T. C. (2003). *A Primer on Nonmarket Valuation*, Kluwer Academic Publishers, Dordrecht, The Netherlands.

Chivers, C. and Leung, B. (2012). 'Predicting invasions: Alternative models of human-mediated dispersal and interactions between dispersal network structure and Allee effects', *Journal of Applied Ecology*, vol. 49, pp. 1113–1123.

Colautti, R. I., Bailey, S. A., van Overdijk, C.D.A., Amundsen, K. and MacIsaac, H. J. (2006). 'Characterised and projected costs of nonnative species in Canada', *Biological Invasions*, vol. 8, pp. 45–59.

Eiswerth, M. E. and Johnson, W. S. (2002). 'Managing nonnative invasive species: Insights from dynamic analysis', *Environmental and Resource Economics*, vol. 23, no. 3, pp. 319–342.

Eiswerth, M. E. and van Kooten, G. C. (2002). 'Uncertainty, economics, and the spread of an invasive plant species', *American Journal of Agricultural Economics*, vol. 84, no. 5, pp. 1317–1322.

Englin, J. E., Holmes, T. P. and Sills, E. O. (2003). 'Estimating forest recreation demand using count data models', in E. O. Sills and K. L. Abt (Eds.), *Forests in a Market Economy*, Kluwer Academic Publishers, Dordrecht, The Netherlands.

Epanchin-Niell, R. S., Haight, R. G., Berec, L., Kean, J. M. and Liebhold, A. M. (2012). 'Optimal surveillance and eradication of invasive species in heterogeneous landscapes', *Ecology Letters*, vol. 15, pp. 803–812.

Epanchin-Niell, R. S. and Hastings, A. (2010). 'Controlling established invaders: Integrating economics and spread dynamics to determine optimal management', *Ecology Letters*, vol. 13, pp. 528–541.

Epanchin-Niell, R. S. and Wilen, J. E. (2012). 'Optimal spatial control of biological invasions', *Journal of Environmental Economics and Management*, vol. 63, pp. 260–270.

Essl, F., Dullinger, S., Rabitsch, W., Hülme, P. E., Jarošik V., Kleinbauer, I., . . . Pyšek, P. (2011). 'Socio-economic legacy yields an invasion debt', *Proceedings of the National Academy of Sciences*, vol. 108, no. 1, pp. 203–207.

Finnoff, D., Shogren, J. F., Leung, B. and Lodge, D. (2007). 'Take a risk: Preferring prevention over control of biological invaders', *Ecological Economics*, vol. 62, no. 2, pp. 216–222.

Haight, R. G. and Polasky, S. (2010). 'Optimal control of an invasive species with imperfect information about the level of infestation', *Resource and Energy Economics*, vol. 32, pp. 519–533.

Hauser, C. and McCarthy, M. (2009). 'Streamlining "search and destroy": Cost-effective surveillance for invasive species management', *Ecology Letters*, vol. 12, pp. 683–692.

Herrick, O. W. (1981). 'Forest pest management economics – Application to the gypsy moth', *Forest Science*, vol. 27, no. 1, pp. 128–138.

Hof, J. (1998). 'Optimizing spatial and dynamic population-based control strategies for invading forest pests', *Natural Resource Modeling*, vol. 11, pp. 197–216.

Holmes, E. E., Lewis, M. A., Banks, J. E. and Veit, R. R. (1994). 'Partial differential equations in ecology: Spatial interactions and population dynamics', *Ecology*, vol. 75, pp. 17–29.

Holmes, T. P. (1991). 'Price and welfare effects of catastrophic forest damage from southern pine beetle epidemics', *Forest Science*, vol. 37, pp. 500–516.

Holmes, T. P., Aukema, J. E., Von Holle, B., Liebhold, A. and Sills, E. (2009). 'Economic impacts of invasive species in forests: Past, present, and future', *Annals of the New York Academy of Sciences*, vol. 1162, pp. 18–38.

Holmes, T. P. and Kramer, R. A. (1996). 'Contingent valuation of ecosystem health', *Ecosystem Health*, vol. 2, no. 1, pp. 1–5.

Holmes, T. P., Liebhold, A. M., Kovacs, K. F. and Von Holle, B. (2010). 'A spatial-dynamic value transfer model of economic losses from a biological invasion', *Ecological Economics*, vol. 70, pp. 86–95.

Holmes, T. P., Murphy, E. A., Bell, K. P. and Royle, D. D. (2010). 'Property value impacts of hemlock woolly adelgid in residential forests', *Forest Science*, vol. 56, no. 6, pp. 529–540.

Homans, F. and Horie, T. (2011). 'Optimal detection strategies for an established invasive pest', *Ecological Economics*, vol. 70, pp. 1129–1138.

Horan, R. D., Perrings, C., Lupi, F. and Bulte, E. (2002). 'Biological pollution prevention strategies under ignorance: The case of invasive species', *American Journal of Agricultural Economics*, vol. 84, no. 5, pp. 1303–1310.

Horie, T., Haight, R. G., Homans, F. R. and Venette, R. (2013). 'Optimal strategies for the surveillance and control of forest pathogens', *Ecological Economics*, vol. 86, pp. 78–85.

Jakus, P. M. (1994). 'Averting behavior in the presence of public spillovers: Household control of nuisance pests', *Land Economics*, vol. 70, no. 3, pp. 273–285.

Jaquette, D. L. (1970). 'A stochastic model for the optimal control of epidemics and pest populations', *Mathematical Biosciences*, vol. 8, pp. 343–354.

Just, R. E. and Hueth, D. L. (1979). 'Welfare measures in a multimarket framework', *American Economic Review*, vol. 69, no. 5, pp. 947–954.

Katzner, D. W. (1990). 'The Shackle-Vickers approach to decision-making in ignorance', *Journal of Post-Keynesian Economics*, vol. 12, no. 2, pp. 237–259.

Kelsey, D. and Quiggan, J. (1992). 'Theories of choice under ignorance and uncertainty', *Journal of Economic Surveys*, vol. 6, no. 2, pp. 133–152.

Kolar, C. S. and Lodge, D. M. (2001). 'Progress in invasion biology: Predicting invaders', *Trends in Ecology and Evolution*, vol. 16, no. 4, pp. 199–204.

Kovacs, K., Holmes, T. P., Englin, J. E. and Alexander, J. (2011). 'The dynamic response of housing values to a forest invasive disease: Evidence from a sudden oak death infestation', *Environmental and Resource Economics*, vol. 49, pp. 445–471.

Kovacs, K. F., Haight, R. G., Mercader, R. J. and McCullough, D. G. (2013). 'A bioeconomic analysis of an emerald ash borer invasion of an urban forest with multiple jurisdictions', *Resource and Energy Economics*, vol. 36, no. 1, pp. 270–289.

Leung, B., Drake, J. M. and Lodge, D. M. (2004). 'Predicting invasions: Propagule pressure and the gravity of Allee effects', *Ecology*, vol. 85, pp. 1651–1660.

Leung, B., Lodge, D. M., Finnoff, D., Shogren, J. F., Lewis, M. A. and Lamberti, G. (2002). 'An ounce of prevention or a pound of cure: Bioeconomic risk analysis of invasive species', *Proceedings of the Royal Society of London, Biological Sciences*, vol. 269, no. 1508, pp. 2407–2413.

Liebhold, A. M., Brockerhoff, E. G., Garrett, L. J., Parke, J. L. and Britton, K. O. (2012). 'Live plant imports: The major pathway for forest insect and pathogen invasions of the US', *Frontiers in Ecology and the Environment*, vol. 10, no. 3, pp. 135–143.

Liebhold, A. M., Halvorsen, J. A. and Elmes, G. A. (1992). 'Gypsy moth invasion in North America: A quantitative analysis', *Journal of Biogeography*, vol. 19, pp. 513–520.

Marten, A. L. and Moore, C. C. (2011). 'An options based bioeconomic model for biological and chemical control of invasive species', *Ecological Economics*, vol. 70, pp. 2050–2061.

McAusland, C. and Costello, C. (2004). 'Avoiding invasives: Trade-related policies for controlling unintentional exotic species introductions', *Journal of Environmental Economics and Management*, vol. 48, no. 2, pp. 954–977.

McDermott, S. M., Finnoff, D. C. and Shogren, J. F. (2013). 'The welfare impacts of an invasive species: Endogenous vs. exogenous price models', *Ecological Economics*, vol. 85, pp. 43–49.

Mehta, S. V., Haight, R. G., Homans, F. R., Polasky, S. and Venette, R. C. (2007). 'Optimal detection and control strategies for invasive species management', *Ecological Economics*, vol. 61, pp. 237–245.

Miller, J. D. and Lindsay, B. E. (1993). 'Willingness to pay for a state gypsy moth control program in New Hampshire', *Journal of Economic Entomology*, vol. 86, no. 3, pp. 828–837.

Moore, C. C., Holmes, T. P. and Bell, K. P. (2011). 'An attribute-based approach to contingent valuation of forest protection programs', *Journal of Forest Economics*, vol. 17, pp. 35–52.

Olson, L. J. and Roy, S. (2002). 'The economics of controlling a stochastic biological invasion', *American Journal of Agricultural Economics*, vol. 84, pp. 1311–1316.

Patriquin, M. N., Wellstead, A. M. and White, W. A. (2007). 'Beetles, trees, and people: Regional economic impact sensitivity and policy considerations related to the mountain pine beetle infestation in British Columbia, Canada', *Forest Policy and Economics*, vol. 9, no. 8, pp. 938–946.

Perrings, C., Dehnen-Schmutz, K., Touza, J. and Williamson, M. (2005). 'How to manage biological invasions under globalization', *Trends in Ecology and Evolution*, vol. 20, pp. 212–215.

Perrings, C., Williamson, M., Barbier, E. B., Delfino, D., Dalmazzonne, S., Shogren, J. . . . Watkinson, A. (2002). 'Biological invasion risks and the public good: An economic perspective', *Conservation Biology*, vol. 6, no. 1, [http://www.consecol.org/vol6/iss1/art1].

Pimental, D., Lach, I., Zuniga, R. and Morrison, D. (2000). 'Environmental and economic costs of nonnative species in the United States', *Bioscience*, vol. 50, pp. 53–65.

Prato, T. (2005). 'Bayesian adaptive management of ecosystems', *Ecological Modelling*, vol. 183, pp. 147–156.

Prestemon, J. P. and Holmes, T. P. (2000). 'Timber price dynamics following a natural catastrophe', *American Journal of Agricultural Economics*, vol. 82, pp. 145–160.

Price, J. I., McCollum, D. W. and Berrens, R. P. (2010). 'Insect infestation and residential property values: A hedonic analysis of the mountain pine beetle epidemic', *Forest Policy and Economics*, vol. 12, pp. 415–422.

Regan, T. J., McCarthy, M. A., Baxter, P. W. J., Panetta, F. D. and Possingham, H. P. (2007). 'Optimal eradication: When to stop looking for an invasive plant', *Ecology Letters*, vol. 9, no. 7, pp. 759–766.

Rosenberger, R. S., Bell, L. A., Champ, P. A. and Smith, E. L. (2012). *Nonmarket Economic Values of Forest Insect Pests: An Updated Literature Review*, General Technical Report RMRS-GTR-275WWW, USDA Forest Service, Albany, California.

Sanchirico, J. N., Albers, H. J., Fischer, C. and Coleman, C. (2010). 'Spatial management of invasive species: Pathways and policy options', *Environmental and Resource Economics*, vol. 45, pp. 517–535.

Saphores, J. and Shogren, J. (2005). 'Managing exotic pests under uncertainty: Optimal control actions and bioeconomic investigations', *Ecological Economics*, vol. 52, pp. 327–339.

Shackle, G.L.S. (1966). 'Policy, poetry, and success', *The Economic Journal*, vol. 76, no. 304, pp. 755–767.

Sharov, A. A. and Liebhold, A. M. (1998). 'Bioeconomics of managing the spread of exotic pest species with barrier zones', *Ecological Applications*, vol. 8, no. 3, pp. 833–845.

Shigesada, N., Kawasaki, K. and Takeda, Y. (1995). 'Modeling stratified diffusion in biological invasions', *The American Naturalist*, vol. 146, no. 2, pp. 229–251.

Steer, H. B. (1948). *Lumber Production in the United States 1799–1946*. Government Printing Office, Washington, DC.

Strutt, A., Turner, J. A., Haack, R. A. and Olson, L. (2013). 'Evaluating the impacts of an international phytosanitary standard for wood packaging material: Global and United States trade implications', *Forest Policy and Economics*, vol. 27, pp. 54–64.

Von Holle, B. (2011). 'Invasional meltdown', in D. Simberloff and M. Rejmánek (Eds.), *Encyclopedia of Biological Invasions*, pp. 360–364. University of California Press, Berkeley, CA.

Von Holle, B. and Simberloff, D. (2005). 'Ecological resistance to biological invasion overwhelmed by propagule pressure', *Ecology*, vol. 86, pp. 3212–3218.

Weed, A. S., Ayres, M. P. and Hicke, J. A. (2013). 'Consequences of climate change for biotic disturbances in North American forests', *Ecological Monographs*, vol. 83, no. 4, pp. 441–470.

Williamson, M. (1996). *Biological Invasions*, Chapman and Hall, London.

Zivnuska, J. (1955). 'Supply, demand, and the lumber market', *Journal of Forestry*, August, pp. 547–553.

PART 5

Economics of forest property rights and certification

25

ECONOMIC IMPLICATIONS OF FOREST TENURES

M. K. (Marty) Luckert[1]

[1]PROFESSOR, DEPARTMENT OF RESOURCE ECONOMICS AND ENVIRONMENTAL SOCIOLOGY, UNIVERSITY OF ALBERTA, EDMONTON, T6G 2H1, CANADA. MARTY.LUCKERT@UALBERTA.CA

Abstract

Governments worldwide, from nations to local villages, regulate the use of forest resources with property rights known as forest tenures. These tenures are made up of combinations of rights and rules that influence the economic behavior of property holders. Forest tenures may be defined and conceptualized in ways that allow them to be studied using economic tools. Such an approach involves considering forest tenures within the context of market versus government failures, economic rents, economic behavior and Pareto efficiency. But the study of forest tenures is fraught with challenges. Forest tenures often grant rights to multiple forest resources, each of which may have different, but interrelated, property rights and values. Forest tenures are also dynamic, influencing the security of forest tenures and the subsequent economic behavior of tenure holders. Identifying impacts of forest tenures is also complicated by the presence of market and regulatory forces that combine to influence economic behavior. Such complexity contributes toward the problem of conservative reinforcement that can inhibit policy change. Finally, challenges in identifying economic impacts of tenures in developing countries is made more difficult by complications that arise from thin or missing markets, and the presence of multiple layers of formal and informal rules.

Keywords

Forest tenures, property rights, economic analysis, market failures, government failures

Introduction

Much of the theory of forest economics assumes that firms operate under conditions of exclusive, perpetual, transferable and otherwise unfettered property rights, which are frequently referred to as 'private property'. Indeed, a cornerstone of forest economics theory, the Faustmann optimal economic rotation, assumes that forest land embodies all of these characteristics. But in practice, we see numerous different types of property rights which govern the use and management of forest resources. That is, governments worldwide have decided that various aspects of forest resources should not be held as private property. These different types of property rights that govern the use of forest resources are frequently referred to as variations in forest tenure.[1]

The pervasive role of governments in regulating the use and management of forests arises from concerns of market failures. In a general sense, market failures arise when economies, made up of firms pursuing their private interests, fail to further broader social objectives. Forests are thought to be particularly susceptible to market failures, because they are frequently large contributors to livelihoods of local people, and supply many types of goods and services, many of which are not traded in markets. Therefore, many governments play active roles in trying to regulate the actions of private firms operating in markets. In developed countries, there are various types of forest tenure agreements that regulate what small and large private firms can and cannot do in forests, whether on private or public land (e.g. Boyd and Hyde, 1989). Forest tenures are also thought to be important in pursuing objectives in developing countries that focus on enhancing livelihoods of rural households (e.g. Deininger, 2003).

But regulating the use of forest resources with forest tenures is fraught with complexity. Therefore, governments seeking to design forest tenures may run into unintended consequences of their policies, which undermine their policy objectives. In cases where government policies do more harm than good, government failures can be said to have occurred.

The consideration of market versus government failures is central to the analysis of policy and the understanding of economic implications of forest tenures. Forest tenures represent variants of government policies which attempt to correct market failures by regulating firms, while attempting to avoid unintended consequences that can lead to government failures.

At the core of considering market versus government failures is the concept of economic behavior, which assumes that firms behave in predictable ways in the presence of various incentives. Economists are frequently interested in property rights because these frameworks have the potential to influence the economic behavior of firms, which is a crucial tool for understanding impacts of policies. Given the vast variety of forest tenures that we see across jurisdictions, it sometimes becomes difficult to understand economic behavior, because such actions occur within the context of various types of forest tenures. For example, governments in the United States and Canada have vastly different systems of forest tenures for pursuing similar objectives embodied in Sustainable Forest Management. Therefore, an understanding of forest tenures is a prerequisite for most policy analyses in forest economics.

The objective of this chapter is to provide an overview of concepts and challenges associated with understanding economic implications of forest tenures. In pursuit of this objective, the next section begins by discussing ways of conceptualizing forest tenures as property rights. Next, a number of economic concepts are reviewed that are necessary for understanding the economic significance of forest tenures. I then turn to a number of challenges currently being faced in analyzing impacts of forest tenures, and conclude by summarizing the chapter.

Definitions of property rights and forest tenures

Luckert, Haley and Hoberg (2011) review a number of definitions of property rights and note that common to most concepts of property rights are two key aspects: (1) an asset or service and (2) prohibitions and conditions on the use of the asset or service. Taken together, these two concepts create a stream of benefits which the holder of the property may receive, within the context of social rules that society imposes and enforces (Bromley, 1991). Along these lines, Arnot, Luckert and Boxall (2011) define a property right as

$$U^R = \sum_{t=0}^{T} \delta u^R_t (R_t, C_t) \qquad (1)$$

where:

U^R = the present value of utility derived from resource R, or the total value of a given resource property right to the holder over some time horizon T;

δ = a discount factor;

R_t = resource R at time t which is a representation of the physical state of the resource including its mass, age structure, location and other physical features;

C_t = a vector of social rules at time t that may constrain, or accommodate securing the benefit stream that the property right provides;

$u^R_t (.)$ = utility derived from resource R at time t which is a function of R_t and C_t.

Equation 1 describes a property right to a given forest resource in the form of utility received as a benefit stream, conditioned by social rules. Note that the benefit stream is defined for a specific forest resource because social rules may be different for different forest resources. Accordingly, it can be important to avoid thinking about forest tenures in terms of property rights to land. It is frequently important to think separately about property rights to trees, recreational resources, water and so forth.[2]

In equation 1, note that C_t includes conditions such as commands and controls, taxes and subsidies; regulatory tools commonly studied as part of environmental economics. But C_t also includes conditions such as limited duration or transferability of rights, realms that are more often considered in economic studies of property rights and are discussed further subsequently. Moreover, C_t can include formal legislated conditions, or informal social, cultural and equity-related norms that may influence the behavior of households, communities and the general public. As such, property rights may be looked upon as being a framework which can include the consideration of numerous regulatory possibilities.

From the previous discussion, it is evident that forest tenures are a subset of property rights that allow individuals or organizations (i.e. various types of firms) access to benefit streams from forest resources. Forest tenures serve as a framework within which governments can allow members of society to benefit from forest resources, while maintaining control over what property right holders may or may not do. Along these lines, Luckert et al. (2011, p. 51) define forest tenures as 'property rights to forest resources granted to individuals or organizations by governments'.

Economic concepts of forest tenures

Economic implications of forest tenures arise from a number of concepts that have relevance to economic analysis – namely, rules as attenuations and subsidies, economic behavior, economic rents, Pareto efficiency and market and government failures.

Rules as attenuations and subsidies

The specification of C_t (equation 1) is important from an economic point of view because it conditions the benefits that tenure holders receive. In most types of property right structures, C_t typically represents a vector of costs to the property right holder. That is, these rules are thought to constrain tenure holders in one, or both, of two ways: (1) they require the property holder to do something that he or she would not voluntarily do (e.g. planting trees), or (2) they keep the property holder from doing something that he or she would like to do (e.g. harvest close to streams). In either case, there are costs to tenure holders (frequently referred to as shadow prices) associated with these types of constraints. For example, Boyd and Hyde (1989) estimate the costs of regulations that require forestland owners to maintain seed trees. In forest tenure and/or property rights jargon, C_ts that constrain the rights of tenure holders are frequently referred to as *attenuations*.

In addition to representing costs to a tenure holder, C_t can also represent benefits. That is, C_t may condition benefit streams in such a way that property rights are augmented, rather than attenuated. For example, rules of forest tenures may allow seedlings to be provided to the holder at a subsidized cost. In sum, C_t may either increase or reduce the benefit stream, and the corresponding value of the property right, that tenure holders receive.

Recognizing the costs and benefits associated with C_t is important for policymakers, because rules influence who gets benefits from, and who bears costs of, forest policies. For example, in the case where C_t represents attenuations, adding a regulatory constraint will reduce the value of the forest tenure to the holder. But society overall may gain if the constraint is well designed, such that the correction of a market failure yields benefits greater than the cost of the attenuation to the property right holder. Whether or not benefits of regulatory actions exceed costs are largely determined by the behavior which forest tenures incent for their holders.

Forest tenures and economic behavior

By influencing the benefits and costs that tenure holders face, forest tenures influence economic behavior. Though the underlying behavior of firms incented by forest tenures can become somewhat complex, there are some simple relationships between social rules and economic behavior. For example, if some type of economic production behavior is taxed, or constrained, then there are incentives to produce less. Conversely, if some type of economic production behavior is subsidized, then there are incentives to produce more. But complications emerge when we attempt to understand the incentives associated with the large numbers of social rules that define forest tenures – a situation that will be discussed further subsequently.

In short, forest tenures may be thought of as incentive frameworks which influence the behavior of firms. Those incentives may further the pursuit of policy objectives, such as Sustainable Forest Management, or they may stand in the way of social progress. The economic analysis of such frameworks is meant to provide insights regarding which of these outcomes is more likely to occur.

Property rights, forest tenures and economic rents

From the previous concepts of property rights and forest tenures, it is evident that there are significant similarities between concepts of property rights and concepts of economic rents. Economic rents are returns to a factor of production above and beyond the cost of bringing that factor into production.[3] We use the concept of economic rent frequently in forest economics to refer to values of standing trees (i.e. stumpage values as economic rents of a stock resource) and to refer to values of land (i.e. Faustmann values as economic rents of flows). In both cases, the concept is that a resource is valued in terms of the benefits that are left over after all costs of production have been paid. From equation 1, it is evident that property rights are similar to economic rents, in that they arise if benefits exceed costs (including costs of social constraints) over time (Luckert, 2007).

But there are two key differences with respect to rents and rights. First, estimates of values of rents frequently do not include all of the costs associated with C_i. For example, calculating a Faustmann land value would not generally include the loss of options associated with restrictions on the transferability of property rights. Second, the estimation of values of rents frequently does not include all of the benefits associated with U^R. That is, U^R includes aspects of utility derived from benefit streams of forests that can be difficult to quantify. For example, a Faustmann land value calculation would not generally include values associated with passing on forest resources to future generations. In sum, values of property rights may be interpreted as being expanded notions of economic rents that include costs of social rules and wider concepts of benefits that include utility.

Given the close similarities between rents and rights, it is not surprising that forest rents take on a great importance in designing property rights. Take, for example, forest tenures that are focused around rights from harvesting stock rents associated with standing trees. A frequent problem that public owners of resources face is how much to charge firms for the right to cut down trees. These stumpage fees are a key aspect of forest policy in many jurisdictions. In theory, the government puts in place a stumpage collection system that estimates the economic rental value of standing trees, and then designs a system to collect that value. But the complications posed by forest tenures in such a system are vast. As mentioned previously, every attenuation of property rights has a shadow price. Therefore, although governments may wish to add new attenuations to forest tenures over time, doing so has a cost; the more attenuated a firm's property right to harvest trees, the less value will be available to collect in stumpage fees. The importance of forest tenure conditions in influencing the economic rent that remains to be collected in stumpage fees is a key part of the controversy regarding the softwood countervailing duty case between Canada and the United States (Luckert, 2007). The United States, which has alleged that Canada does not charge enough for its standing timber, has attempted to support its case by comparing stumpage prices in Canada and the United States. But such comparisons frequently ignore the differences in forest tenure that influence the value of property rights to harvesting trees that tenure holders possess.

Instead of focusing on forest rents associated with rights to harvest standing trees, forest tenures may also be structured around land productivity rents associated with rights to *grow* trees. In order to provide property right holders with incentives to grow trees, such tenures must typically resemble concepts of private property, as even minimal attenuations can eliminate the benefits associated with growing trees, which require immediate investments with frequent long periods before returns are realized at harvest. Within such forest tenure systems, attenuations are often focused on taxation issues associated with income derived from growing and harvesting trees (e.g. Gamponia and Mendelsohn, 1987).

Forest tenures, market failures and Pareto efficiency

One of the cornerstones of welfare economics is a proof which shows that, under a given set of circumstances, economic transactions will result in a situation where resources are allocated, such that no reallocation could make a person better off without making somebody else worse off. This result is referred to as a Pareto efficient solution. But the value of this theory is *not* that markets always lead us toward social objectives. Rather, the value of this theory lies in identifying those conditions which cause markets to deviate from Pareto efficiency. These conditions are referred to as market failures.

Boyd and Hyde (1989) refer to three categories of market failures. The first is stability. Markets are not inherently stable, so governments attempt to correct for this problem with fiscal and monetary policies. With regard to forestry, sustained yield policies have often been used as part of forest tenure systems in attempts to stabilize economies. But these policies have been criticized by economists in that they potentially increase instability at a substantial cost.[4] The second category of market failure is allocations. When economies receive the wrong price signals, such as the case when pollution externalities are present, resources that are used to produce goods and services are misallocated. Accordingly, governments frequently regulate the pollution of firms with taxes and commands and controls. In forest tenure systems, the numerous interrelated forest resources pose serious challenges for pricing systems. Many resources are not marketed, and so have no price, while others are interrelated with one another. Interrelated resources are frequently assigned to multiple tenure holders, causing external effects of forest management among holders. Though cooperation among tenure holders can theoretically fix misallocations (Coase, 1960), externalities frequently persist in forestry settings, such as cases when multiple tenure holders manage different species within a given forest (Luckert, 1993). The final market failure category is distributions.[5] Markets are mechanisms that determine who in society gets what. But there is no guarantee that such distributions, created by markets, will be considered to be equitable or ethical by society. As such, we see government policies such as progressive income taxes that attempt to tax higher income-earning people at higher rates. In forest tenures, distributional concerns are frequently evident in policies that attempt to preserve jobs in the forestry sector, such as log export restrictions designed to promote value-added manufacturing in local economies.

From the previous discussion, it is evident that forest tenures may be viewed as government attempts to regulate market failures. From an economics perspective, government policies should be motivated by the need to correct for market failures, because the absence of market failures implies that social objectives are being furthered by market processes. The logic is that if something in a market is not broken, there is no need to fix it. But if there are one or more market failures, then the government may choose to regulate with numerous types of policy tools.[6] These sets of tools represent the social rules, C_t, that we identified in equation 1.

But how can the efficacy of such forest tenure policies be evaluated? To begin with, we need some criteria to judge one policy against another. One frequently used criterion can be derived from the concept of Pareto efficiency discussed previously. That is, we can consider the Pareto improvement criterion which a policy will satisfy if, as a result of the policy: (1) nobody is made worse off, and (2) at least one person is made better off.

Figure 25.1 shows a means of considering changes to forest tenure policies, and whether a given change results in a Pareto improvement. The figure depicts a utility possibility frontier for a hypothetical two-person economy.[7] Utility possibility frontier 1 indicates the potential combinations of utility (a measure of well-being) that the economy could create for each of the two people for a given organization of productive resources. Point A on the figure indicates that distributions in this economy are such that person A is receiving a utility level of U_A, while person B is receiving a utility level of U_B. For these economic conditions (i.e. for utility possibility frontier 1) any movement would not be a Pareto improvement, because at least one person would be made worse off. But forest policies could create conditions that would allow for a Pareto improvement. Assume that a tenure policy was changed. Further to the previous discussion, such a change could influence the economic behavior of firms, thereby causing the economy to produce a new utility possibility frontier. If the government responds with an effective policy to correct a market failure, then it is possible that a Pareto improvement could result. In Figure 25.1, the new curve is depicted as an outward shift in the curve (utility possibility

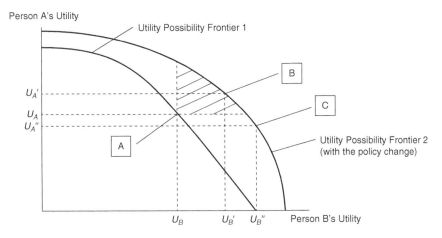

Person A's Utility

Utility Possibility Frontier 1

U_A'

U_A
U_A''

B

C

A

Utility Possibility Frontier 2
(with the policy change)

U_B U_B' U_B'' Person B's Utility

Figure 25.1   Impacts of forest tenure policy changes on benefits to society in a two-person economy.

frontier 2), indicating numerous opportunities to meet the Pareto improvement criterion. Given the starting point A, the hashed area indicates the potential changes that could yield Pareto improvements. In the figure, point B indicates one place where the conditions of the criterion are met, where the people in the economy get U_A' and U_B' levels of utility.

But what happens if we try to apply this conceptual framework in the real world? To begin with, expanding this model to consider large parts of society creates situations where it is difficult, if not impossible, to find a policy change that would not make at least one person worse off. Moreover, it would be difficult to find a means to measure and aggregate the utility of all members of society that are influenced by the policy change. For these and other reasons, the Pareto improvement criterion frequently gets watered down to a less stringent criterion referred to as a potential Pareto improvement. This criterion is met if the gains in utility of those peoples who are positively affected by a policy change are greater than the losses in utility from those negatively affected by a policy change. The logic behind the potential Pareto criterion is that if overall gains are greater than overall losses, then those who gain could compensate those who lose, thereby creating a Pareto improvement. Note, however, that compensation does not have to occur in order to satisfy the conditions of a potential Pareto improvement, thus the word *potential*.

In Figure 25.1, a potential Pareto improvement is depicted by a change from point A to point C. With this change, person A loses $U_A'' - U_A$, and person B gains $U_B'' - U_B$. Because the loss to person A, $(U_A'' - U_A)$, is less than the gain to person B, $(U_B'' - U_B)$, a potential Pareto improvement has occurred. Though switching to a potential Pareto improvement allows us to address the problem of not being able to identify policies where no one is harmed, we still have the problem of measuring, and aggregating, all project impacts in terms of utility. In practice, utility is frequently substituted with dollar estimates. When dollar values are used with the potential Pareto improvement criterion, we are effectively conducting a cost-benefit analysis on a policy change. In these cases, nonmarket goods and services are frequently ignored.

From Figure 25.1, we not only see how tenure policies can influence the benefits to society, but we also see the mechanism for such changes. That is, a change to forest tenures causes changes in the economic behavior of tenure holders, which in turn changes the goods and services that are produced that provide members of society with utility. Therefore, changes to forest tenure, or alternatively, variations in forest tenure policies across jurisdictions, imply differential economic outcomes. This concept is referred to as the 'non-uniqueness of Pareto efficiency' (Randall, 1987). That is, conceptualizing impacts of forest tenure policy in such a framework makes us realize that each different forest tenure system is likely to have varying impacts on benefits to society. The challenge becomes identifying those changes to forest tenures that are likely to do more good than harm.

Government failures

The previous discussion is based on the concept of gains to be had from governments regulating market failures with forest tenures. Such an approach assumes that governments are willing and able to identify problems and address them effectively with new tenure policies. But what if governments fail in these efforts? Wolf (1988) argues that in order for policy analysis to be objective, it is necessary to consider both market and government failures. Wolf points out that governments may fail because of a number of problems with political processes that distort incentives of government employees and thereby may fail to facilitate regulatory practices that are best for society. Governments may also fail because of the sheer complexity of the problem associated with designing forest tenure policies that avoid unintended consequences. We now turn our attention to some of these complexities.

Challenges in analyzing economic implications of forest tenures

Tenures and multiple forest resource values

Forest tenures are generally meant to provide incentive frameworks that facilitate the use and sustainability of values derived from forest resources. But each type of forest resource could have different market failures and different potential regulatory responses. As discussed previously, this is one important reason for considering property rights to individual forest resources (i.e. U_j), rather than analyzing property rights to forest land.

Further to our previous discussion, multiple forest resources are frequently associated with market failures because they are not independent of one another and are frequently nonmarketed. The first problem of nonindependence can cause problems when multiple forest resources are held by independent firms. Under such conditions, one firm's operations can have external effects on another, thereby preventing integrated management. For example, if one forestry firm holds rights to grow trees, and another firm holds rights to harvest moose, and if moose are eating young trees, then the two firms may be at cross purposes. The firm with the rights to trees sees the moose as a pest, while the firm with the hunting rights sees the moose as a valuable resource. In some cases, one firm may choose to purchase rights to all forest resources on a given piece of land, thereby internalizing all decisions into one firm such that trade-offs associated with integrated management are considered.[8,9] Continuing our previous example, if both hunting and tree growing rights were held by the same firm, then the firm could consider trade-offs associated with increasing moose habitat in order to increase moose numbers for hunting, versus reducing moose habitat and numbers. But it may not work well for one firm to hold multiple rights to forest resources. Some firms may be specialized in the management and production of one resource, but have little expertise in managing another. For example, in Alberta, Canada, rights to oil and gas are granted to energy companies, while rights to harvest trees on top of these deposits are granted to forest products companies. Considering how many forest resources to grant to a single firm is referred to as specifying the *comprehensiveness* of forest tenures and is explored further in Luckert et al. (2011).

The second problem of nonmarketed forest resources can create difficulties because firms do not receive price signals from these resources, so they may be omitted in management decisions. For example, forestry firms do not generally receive monetary benefits for managing forests for biodiversity, so such aspects may be neglected. Internalizing such values can be approached with a number of different types of forest tenure structures. A common approach is to regulate forestry firms with commands and controls designed to facilitate the production and sustainability of nonmarketed forest resources. But such approaches can be difficult to employ because:[10] there are a great variety of forest conditions (i.e. what should the command be?); there are disincentives associated with such an approach (i.e. how can I minimize the cost of the requirement?); figuring out how to enforce requirements can be complex (i.e. how often should I monitor, and what should the penalty be?).

Another approach to internalizing nonmarketed values is to attempt to create markets. For example, markets for forest carbon are currently a large concern in forestry. But for such policies to be successful, it is important to consider why markets for carbon were not present in the first place. The general problem is that benefits of carbon sequestration are received by society in general, while costs of reducing carbon are borne by individual firms. So how can markets be created in such a situation? One approach is to rely on voluntary carbon markets in which consumers, out of the goodness of their hearts, buy carbon offsets for some of their carbon-producing activities. Such approaches are currently common, but demand through such voluntary actions is likely to result in too little carbon reduction without regulations and penalties for noncompliance. In terms of equation 1, the voluntary approach does not impose a new C_t on a property right unless there are social norms regarding the importance of voluntary controls that influence behavior. Instead, purchasing carbon credits provides the buyer with a new resource right, U_r: a benefit stream derived from knowing that their carbon is being offset.

Another approach for establishing forest carbon markets is to impose regulatory limits on carbon emissions, and allow emissions to be offset by increased carbon sequestration from forests. In such cases, by putting a new social rule, C_t, on the rights of all firms emitting carbon in an economy, a new property right to carbon, U_r, is created. But such approaches also face a number of challenges, including difficulties in creating offsets with forest resources that both sequester and emit carbon. Establishing protocols for tracking carbon has proven challenging.

Forest tenures and multiple behavioral impacts

In the previous discussions, we spoke of tenures as creating incentive frameworks that influence the economic behavior of firms. But a key question is, exactly how do tenures influence behavior? Though we discussed some general concepts previously (i.e. about subsidies increasing production and taxes and regulations decreasing production), there are potentially many nuanced effects of different types of social rules on tenure behavior. For example, if forest tenures are attenuated such that the property right may not be sold, how does that influence a forestry firm's investment? Similarly, if the duration of a property right is limited to 10 years with uncertain renewability, what incentives does a forestry firm face? Although studies have investigated these types of questions (e.g. Nautiyal and Rawat, 1986), most such studies have been empirical, with little formal theoretical guidance (e.g. Zhang and Pears, 1996).

Each of these questions is difficult enough to address by itself. But the economic analysis of forest tenure frequently requires that we consider packages of property rights (i.e. combinations of U_r's) with a plethora of social rules (i.e. many C_t's). Such analyses are necessary because forest tenures, in practice, are packages of rights and rules. This type of complication poses theoretical and empirical difficulties. Ideally, our goal is to identify how changes to an individual aspect of a forest tenure (i.e. a change in one C_i) influences economic behavior and social welfare outcomes, while holding all other aspects of tenures constant. This information is frequently desired by policymakers who are interested in making marginal adjustments to forest tenures rather than introducing whole-scale change. On the theoretical side, such an approach requires a model inclusive of the many social rules associated with forest tenures, and then an understanding of how outcomes change when just one of these rules is changed. Similarly, on the empirical side, such an approach requires data that allow us to identify effects of individual attributes of forest tenures. But case studies that allow such identification are difficult to find. In many situations, all firms in a given case study may share a common type of forest tenure, thereby preventing the identification of cause and effect. For example, it is not possible to discern the impacts of increased duration of tenure on forestry investment if everybody has the same tenure duration. If the case study is broadened to encompass other types of forest tenures, then we frequently find that many social rules are different between two different types of tenures. For example, one tenure may have a duration of 25 years, be freely transferable and have reforestation requirements, while another may have a duration of 10 years, have restricted transferability, and have no reforestation requirements. Such differences make it difficult to identify effects of individual rules. Because of these difficulties, many studies do not disaggregate individual attributes of forest tenures. Rather, differences in economic behavior are identified for differences among types of forest tenures.[11] As such, we frequently do not know what aspect of a given tenure is responsible for the behavior that we witness.

The identification of economic behavior effects of forest tenures is further complicated by issues of reverse causality. In order to identify impacts of forest tenures, the statistical methods employed frequently assume unidirectional causality. That is, as discussed previously, forest tenures are assumed to impact economic behavior, and subsequently social welfare outcomes. But economic behavior may also impact the structure of forest tenures. For example, forestry firms may make investments in lobbying governments for changes to social rules. Such situations can cause analysis to be representative of correlations, rather than causations, thereby limiting their usefulness in informing policy.

Changes to forest tenures and tenure security

In our discussion of forest tenures thus far, we have treated the rules which define property rights as being static. That is, we have assumed that there is a given set of C_i's that do not change. But governments frequently change forest tenures in response to changes in social values and conditions of forests. For example, increasing social concerns over clearcuts have caused governments to reduce the size of harvest blocks and impose adjacency requirements. Likewise, increased risks of forest fires have caused governments to regulate harvesting practices that can increase risks of ignition.

Though the need for governments to change forest tenures is evident, there are consequences for the firms who hold these property rights. Changing the rules that condition values from forests can cause uncertainty in future benefit streams to tenure holders who are considering various investments in forests. In other words, changes in government policy affect tenure security and subsequent economic behavior. Therefore, understanding the impact of tenure security on economic behavior involves trying to ascertain the expectations of tenure holders, and connecting such expectations to management practices. But success in linking tenure security to economic behavior has been mixed, partially because there has been confusion over concepts and impacts of tenure security (Arnot et al., 2011).

Identifying market versus government failures

Previous discussions indicate that the behavior of tenure holders is conditioned by a number of government and market forces. Therefore, if we witness economic behavior among forest tenure holders that is not in line with public interests, is such behavior the result of market or government forces? Addressing this question is important for policy analysis because if we seek to fix a problem, it is important to know its cause.

An example of mixed market and government failures may occur when investigating the behavior of forest tenure holders regarding the planting of trees. Assume that we observe a failure to sufficiently reforest cutover lands for a given forest tenure. Is this failure due to market or government failures? Resources for reforestation could be under-allocated because of a market failure, where positive externalities associated with planting trees, such as the sequestration of carbon, may not be part of the property rights of tenure holders. But limited reforestation could also be due to government failures. When governments structure forest tenures, they may fail to grant rights to grow trees (Luckert and Haley, 1990). That is, instead of allowing tenure holders benefit streams that could incent investments in new forests, governments frequently rely on commands and controls. In such cases, the potential for economic investment behavior has been eliminated by government restrictions of property rights. As discussed previously, reliance on such policies can result in undesirable management practices.

In sum, forest practices that we observe under forest tenures are frequently the result of incentives that are created by the combined forces of markets and rules established by governments. Therefore, it becomes important to try and tease out how each set of forces is contributing to the resultant economic behavior.

The problem of conservative reinforcement

In the previous discussion, we noted how variations in types of forest tenures could create incentive frameworks that result in differing patterns of economic behavior and social welfare outcomes. That is, we considered how outcomes in an economy are endogenous to a given set of property rights. This basic observation creates a potential problem when it comes to considering changes to forest tenures. Forestry firms are frequently able to identify problems with current forest tenure policies, but have difficulties imagining how to make policies better. In terms of Figure 25.1, utility possibility frontier 1 may be known, but what does utility possibility frontier 2 look like? With their current benefit streams and economic decisions being defined by current forest tenures, it can be difficult to imagine the impacts of changes in social rules. Randall (1987) has referred to this phenomenon as the problem of conservative reinforcement, because our inability to understand, a priori, impacts of changes in policy can create inertia for policy change.

The problem of conservative reinforcement highlights the importance of theoretical and empirical work in helping to improve our understanding of the consequences of policy changes before such changes are made. To the extent that we have a better understanding of how variants in forest tenures influence social objectives, the inertia associated with the problem of conservative reinforcement may be reduced.

Forest tenures in developing countries

The importance of forest tenures in influencing economic behavior has been widely investigated in developed and developing economies. But applying the concepts discussed previously in developing country settings brings about additional challenges. Whereas studies in developed countries are frequently based on the economic behavior of large forest companies, studies in developing countries frequently focus on the behavior of low-income households struggling within impoverished environments. Likewise, whereas developed country forest tenures are frequently based on large areas of forests allocated to one or a few firms,[12] forest tenures used by households in developing countries' forests are relatively small areas with complicated arrangements among groups of households.

Developing countries are frequently characterized by thin or missing markets (e.g. de Janvry, Fafchamps and Sadoulet, 1991). Consequently, reliance on resources within the household (as opposed to resources being available for purchase) is common. For example, a household may have to rely on their own family members to undertake most household activities if labor markets are thin or absent. Such situations imply that specific characteristics of households and local markets may play important roles in understanding impacts of forest tenures on economic behavior of households.

The social conditions of forest tenures may also be difficult to understand in developing country situations. As mentioned previously, C_t can refer to formal or informal rules. In many developing jurisdictions, forest tenures are defined by complex combinations of both types. For example, formal rules regarding restrictions on the selling of forest tenures may be enacted in national legislation, or lie under the jurisdiction of a chief in a village. Informal rules may include showing due courtesy in the collecting of fuelwood by following local norms and customs.[13]

These features of developing country economies and forest tenures create extra challenges for researchers. In many cases, the complexity of forest tenures in these settings makes it difficult just to describe the property rights – a crucial prerequisite for attempting to understand economic behavior. As a result, we seem to know the least about forest tenures in those areas where the stakes for local people, whose lives are frequently heavily dependent on forest resources, are highest.

Summary

Forest tenures are a pervasive feature of forest management landscapes globally. Their prevalence makes understanding the impact of forest tenures on economic behavior an important area of study in forest economics. Forest tenures may be considered to be various types of property rights to forest resources. Conceptually, forest tenures supply benefit streams to their holders, conditioned by social rules that attenuate or augment benefit streams. Concepts of economic rents are central to concepts of forest tenures. Both concepts involve considering benefits remaining after all costs associated with economic behavior are subtracted, whether the forest resource in question is standing trees, land capable of growing trees or some other type of nontimber resource. But studying forest tenures frequently involves the consideration of a broader set of costs, associated with shadow prices of social rule constraints, and may also include broader concepts of benefit streams as depicted by utility. The specific impacts of these social rules on benefit streams can influence the incentives, and subsequently the economic behavior, of forestry firms.

The justification for regulating markets with social rules embodied in forest tenures arises from a recognition that unfettered markets may fail to further social objectives. The specific types of market failures that apply to forest resources indicate potential means of correcting for failures with social rules. Investigating the impact of forest tenures on social welfare outcomes involves tracing impacts of different types of social rules on economic behavior which determines how the utility derived from goods and services is distributed in society. Ideally, forest tenures would correct market failures such that nobody is harmed, and more utility is derived by one or more people. But such situations are difficult to find, so analyses tend to consider whether benefits to those who gain are greater than costs to those that are harmed by policy changes. Moreover, there is also the potential for governments to fail in their attempts to correct market failures.

The study of forest tenures is fraught with challenges. Forest tenures can grant rights to multiple forest resources, each of which can have numerous social rules associated with its use and management. It is important for each set of rules to be tailored to the market failures associated with the specific resource in question. These resources and multiple rules create complex packages of incentives, from both a theoretical and an empirical perspective, when their impacts are considered in concert. Forest tenures are also dynamic, influencing the security of forest tenures, which can be an important factor in influencing economic behavior. Identifying impacts of forest tenures is also complicated by the presence of both market and regulatory forces that combine to influence economic behavior. As such, it can be difficult to identify whether undesirable economic behavior is the result of market and/or government failures. The complexity of analyzing forest tenures also contributes toward the problem of conservative reinforcement, where a lack of knowledge about alternative states of forest tenures, other than those that currently govern economic transactions, creates inertia that can inhibit change. Finally, challenges in identifying economic impacts of tenures in developing countries is made more difficult by complications that arise from thin or missing markets, and the presence of multiple layers of formal and informal rules.

In sum, economic implications of forest tenures are of significant importance to any practicing forester, and their analysis is a challenging endeavor for forest economists. This combination, of policy importance, yet challenges in understanding their outcomes, will ensure that the study of forest tenures will play a pivotal role in forest economics for the foreseeable future.

Notes

1 Concepts of property rights and forest tenures will be further defined in the next section.
2 For the case of forest land, there are frequently cases where multiple benefit streams are granted in concert. One way to express these types of property rights would be to consider the aggregations of utility across multiple property rights to forest resources. But the utility associated with these resources may not be strictly additive. Some combinations of forest resources may be compliments or substitutes, implying the presence of interaction terms when benefit streams are combined.
3 For more on concepts of economic rents and forests, see Pearse (1990).
4 For a literature review on sustained yield, see Luckert and Williamson (2005).
5 Note that, when taken together, the first letters of the three categories reveal that market failures are SAD.
6 Economists have long contemplated situations where regulatory actions are being undertaken in response to multiple market failures. This line of thinking is referred to as the 'theory of the second best' (Lipsey and Lancaster, 1948). The general problem involves difficulties associated with addressing any one market failure without addressing all failures simultaneously. The practical difficulties associated with fixing all market failures raises doubts about whether piecemeal corrections (i.e. fixing some but not all failures) increase the welfare of society.
7 Note that this framework could be increased to reflect more than two people, but increasing the number of people in the economy makes it more difficult to find solutions, as will be discussed subsequently.
8 For a review of techniques for considering multiple values, see Bowes and Krutilla (1989).
9 Yet another option is for firms to negotiate with one another such that values of each are internalized into decision making. But these so-called Coasian solutions (Coase, 1960) may not emerge or optimally internalize values for a variety of reasons (see, e.g. Randall, 1987).
10 For a more in depth discussion of such issues, see Luckert et al. (2011).
11 See for example Zhang and Pearse (1996) who compare the performance of tenure types in British Columbia.
12 There are some cases in developing countries where rights to forest resources have been granted to large forestry firms, similar to situations in developed countries. These arrangements are frequently referred to as forest concessions. But there are remarkably few studies of these forest tenures.

13 Hegan et al. (2003) investigate the role of local norms in the collection of fuelwood in Zimbabwe. A more general discussion on social norms and economic behavior can be found in Sethi and Somanathan (1996).

References

Arnot, C., Luckert, M. K. and Boxall, P. (2011). 'What is tenure security? Conceptual implications for empirical analysis', *Land Economics*, vol. 87, no. 2, pp. 297–311.

Bowes, M. D. and Krutilla, J. V. (1989). *Multiple Use Management: The Economics of Public Forest Land*, Resources for the Future, Washington DC.

Boyd, R. G. and Hyde, W. F. (1989). *Forest Sector Intervention: The Impacts of Regulation on Social Welfare*, Iowa State University Press, Ames.

Bromley, D. (1991). *Environment and Economy: Property Rights and Public Policy*. Basil Blackwell, Cambridge, MA.

Coase, R. H. (1960). 'The Problem of Social Cost', *Journal of Law and Economics*, 3 October, pp. 1–46.

Deininger, K. (2003). *Land Policies for Growth and Poverty Reduction*, World Bank and Oxford University Press, Washington DC.

de Janvry, A., Fafchamps, M. and Sadoulet, E. (1991). 'Peasant household behaviour with missing markets: Some paradoxes explained', *The Economic Journal*, vol. 101, no. 409, pp. 1400–1417.

Gamponia, V. and Mendelsohn, R. (1987). 'The economic efficiency of forest taxes', *Forest Science*, vol. 33, no. 2, pp. 367–378.

Hegan, L., Hauer, G. and Luckert, M. K. (2003). 'Is the tragedy of the commons likely? Investigating factors preventing the dissipation of fuelwood rents in Zimbabwe', *Land Economics*, vol. 79, no. 2, pp. 181–197.

Lipsey, R. G. and Lancaster, K. (1948). 'The general theory of the second best', *Review of Economic Studies*, vol. 24, pp. 11–32.

Luckert, M. K. (1993). 'Property rights for changing forest values: A study of mixed wood management in Canada', *Canadian Journal of Forest Research*, vol. 23, no. 4, pp. 688–699.

Luckert, M. K. (2007). 'Property rights, forest rents, and trade: The case of US countervailing duties on Canadian softwood lumber', *Forest Policy and Economics*, vol. 9, no. 6, pp. 581–590.

Luckert, M. K. and Haley, D. (1990). 'The implications of various silvicultural funding arrangements for privately managed public forest land in Canada', *New Forests*, vol. 4, no. 1, pp. 1–12.

Luckert, M. K., Haley, D. and Hoberg, G. (2011). *Policies for the Sustainably Managing Canada's Forests: Tenure, Stumpage Fees and Forest Practices*, University of British Columbia Press, Vancouver.

Luckert, M. K. and Williamson, T. B. (2005). 'Should sustained yield be part of sustainable forest management?' *Canadian Journal of Forest Research*, vol. 35, no. 2, pp. 356–364.

Nautiyal, J. C. and Rawat, J. K. (1986). 'Role of forest tenure in the investment behaviour of integrated Canadian forestry firms', *Canadian Journal of Forest Research*, vol. 16, no. 3, pp. 456–463.

Pearse, P. H. (1990). *Introduction to Forestry Economics*, University of British Columbia Press, Vancouver.

Randall, A. (1987). *Resource Economics: An Economic Approach to Natural Resource and Environmental Policy* (2nd ed.), John Wiley and Son, Toronto.

Sethi, R. and Somanathan, E. (1996). 'The evolution of social norms in common property resource use', *The American Economic Review*, vol. 86, pp. 766–788.

Wolf, C. (1988). *Markets or Governments: Choosing between Imperfect Alternatives*, Massachusetts Institute of Technology, Cambridge, MA.

Zhang, D. and Pearse, P. (1996). 'Differences in silvicultural investment under various types of forest tenures in British Columbia', *Forest Science*, vol. 42, no. 4, pp. 442–449.

26

ECONOMICS OF THE EVOLUTION OF THE AMAZON FRONTIER

Erin Sills

PROFESSOR OF FOREST RESOURCE ECONOMICS, NORTH CAROLINA STATE UNIVERSITY, USA.
ERIN_SILLS@NCSU.EDU

Abstract

The deforestation frontier in the Brazilian Amazon is driven by the demand for agricultural land, which in turn reflects the distribution of land rents and the characteristics of farmers arriving in the region. Historically, the Brazilian government increased agricultural rents by building roads and subsidizing farmers through the tax and credit systems. Although the government has scaled back policies that directly encourage deforestation, these have been effectively replaced by new drivers, including international commodity markets, improvements in agricultural production technologies and the socio-cultural as well as economic value of cattle ranching. In contrast, forest management has not been a competitive land use from a private perspective, and old-growth timber has essentially been 'mined' by the logging sector. This reduces the net cost of deforestation, through both improved access and significant revenues from sale of old-growth timber. Current policy initiatives seek to change the incentives facing landowners by increasing tenure security for forest land, imposing penalties for illegal deforestation and creating new opportunities to earn revenue from standing forest.

Keywords

Brazilian Amazon, tropical deforestation, land rents, land tenure

Introduction

The Amazon frontier attracts international concern because its advance comes at the expense of global public goods such as biodiversity and carbon storage (Torras, 2000). Although the Amazon has long been occupied by indigenous peoples and subject to periodic waves of settlement and resource exploitation such as the 'rubber boom' of the last century (Bunker, 1985), the recent frontier development process has imposed more significant and visible (by satellite) changes on the landscape. This is largely due to the demand for agricultural land, which drives deforestation, defined as the permanent conversion of forest to another land use. Continued rapid deforestation is predicted to result in undesirable and irreversible changes in climatic and hydrologic systems, further compounding biodiversity loss and carbon emissions (Nepstad, Stickler, Soares-Filho and Merry, 2008). From an economic perspective, these global environmental costs are the externalities of land-use decisions by agents who act to maximize their welfare given local conditions.

Studies of tropical deforestation generally distinguish between (1) the 'sources' of deforestation, including various types of 'agents' who clear forest for different land uses, and (2) the 'drivers' or 'causes' of deforestation. Kaimowitz and Angelsen (1998) divide this second category into (2a) agents' characteristics and decision parameters that are 'direct drivers' of deforestation, and (2b) 'underlying causes' or indirect drivers of deforestation that determine agents' characteristics and decision parameters. This framework suggests that neither the blame nor the solution for excessive deforestation lies with the agents; rather, the key policy questions are about the drivers of deforestation. This chapter focuses on those drivers, reviewing the conceptual frameworks, empirical models, and findings of economic studies, after first defining the study region and briefly sketching the key underlying causes, direct drivers and agents of deforestation.

The Amazon

Although precise definitions of 'the Amazon' vary, it is clearly the world's largest remaining intact tropical forest landscape (Hansen, Stehman and Potapov, 2010), harboring a large portion of the world's terrestrial biodiversity and carbon stocks (Nepstad et al., 2008). The region can be defined by biophysical characteristics as either the river basin or the forest biome, both estimated to cover 6–7 million km^2 and more than half of the territories of Bolivia, Brazil and Peru. Recent modeling exercises (e.g. Soares-Filho et al., 2006) have considered a broader 'pan-Amazon' region covering 8 million km^2 in eight countries and the territory of French Guiana. However, much of the economic literature on the Amazon analyzes the administratively defined Brazilian 'Legal Amazon'. The current definition of the Legal Amazon was established by the Brazilian government in 1966 (as amended in 1977) for planning and administrative purposes. Spanning over 5 million km^2 (nearly 60% of the Brazilian territory) and home to over 24 million inhabitants, the Legal Amazon is made up of 775 municipalities in ten Brazilian states (IBGE, 2011). Economists study the Legal Amazon because of its size (containing more than half of 'the Amazon', whether defined as the river basin, forest biome, or 'pan-Amazon'), its relatively high deforestation rates (both for the past half-century (Hansen et al., 2010) and predicted into the future (Soares-Filho et al., 2006), and its publicly available remote sensing and census data.

In many ways, the theory, methods and findings from studies of the Brazilian Legal Amazon also apply to the rest of the Amazon. Deforestation throughout the region reflects the relative returns to agricultural versus forest land. These returns are shaped by markets, by public investments in transportation infrastructure and agricultural development and by both the letter of the law and enforcement efforts. Returns to deforestation are also affected by interactions with other resource frontiers such as mining and hydroelectricity, which have attracted capital and labor to the region, and by conflicts associated with coca production and trafficking as well as military insurgencies (Perz, Aramburú and Bremer, 2005). In the Brazilian Amazon, perhaps the most important interaction is between the agricultural and logging frontiers, with logging both expanding the road network and generating substantial revenues from sale of valuable old growth timber (Veríssimo et al., 1992).

Causes, drivers and agents of deforestation

Underlying causes of deforestation

Brazilian government policy and macroeconomic factors shape local conditions in the Amazon, including both the types of agents present in the region and the parameters that drive their decisions. Over the past half century, the Brazilian government has sought to promote the development and integration of the Amazon into the nation. The initial impetus for this came from the military government's desire to defuse growing social conflict over unequal land distribution in other parts of the country and to clearly establish national sovereignty over the Amazon. These concerns led to road construction, establishment of colonization projects and tax incentives for agricultural development, which have been characterized as 'perverse incentives' for deforestation.

Although federal government policy was the key underlying cause of deforestation in the early decades of frontier development, it is increasingly overshadowed by macroeconomic conditions, including inflation, exchange rates and commodity demand driven by both urbanization and international trade (Defries, Rudel, Uriarte and Hansen, 2010; Rudel, 2007). In particular, recent studies point to international demand for soy, beef and bioenergy as key causes of deforestation (Morton et al., 2006; Arima, Richards, Walker and Caldas, 2011; Lapola et al., 2010). The demand for beef is directly reflected in the sources of deforestation: Cattle pasture is by far the most common land use in recently deforested areas in the Amazon (EMBRAPA/INPE, 2011). This is partly due to its displacement from other regions where it has become more lucrative to cultivate sugar cane and soybeans for bioenergy and animal feed, thus increasing demand for new cattle pasture, which directly drives deforestation.

Since 2006, the deforestation rate in the Brazilian Amazon has remained below 15,000 km²/year, substantially less than the nearly 20,000 km²/year average forest loss in the previous decade (INPE, 2012). One underlying cause of the decline may be the global economic recession and reduced demand for commodities from deforested land. However, at the same time, the Brazilian government has professed a new commitment to forest conservation, encouraged by global concerns about climate change. Economists are seeking to understand how much of the decline in deforestation can be attributed to changes in government policy and law enforcement efforts versus changes in economic conditions (Hargrave and Kis-Katos, 2013).

This is challenging to estimate empirically, because of limited spatial and temporal variation in the underlying policy and macroeconomic causes of deforestation. Underlying causes have been empirically tested in models of deforestation rates across countries, and incorporated into simulation models that predict the total area deforested annually in the Brazilian Amazon (e.g. Soares-Filho et al., 2006). However, most of the economic literature on the Amazon focuses on understanding how agents respond to the direct drivers of deforestation created by these underlying policy and economic conditions.

Direct drivers of deforestation

The direct drivers of deforestation are the parameters that define returns to deforestation relative to alternative uses of land and labor. For example, government road construction changes the effective costs of inputs and prices of outputs at 'farm gates' accessed via those roads. Agents may respond by migrating to the newly accessible land or by adjusting how much and how quickly they deforest land along the roads. In addition to economic conditions largely shaped by access, deforestation drivers include biophysical conditions and characteristics of the agents themselves, all discussed further subsequently.

Agents of deforestation

In the Amazon, the most significant agents of deforestation are farmers who have recently migrated to the region, including both smallholders (families claiming 100 ha or less) and largeholders (with properties ranging into the tens of thousands of hectares). Over half of the privately owned land in the Amazon is in properties of 1,000 ha or larger (Pacheco and Poccard-Chapuis, 2012). However, the relative importance of smallholders versus largeholders as agents of deforestation remains contentious and is subject to ongoing debate, partly because of variation over time and space (Börner et al., 2010; de Souza, Miziara and De Marco, 2013; Pacheco and Poccard-Chapuis, 2012), and partly because largeholders often acquire land cleared by smallholders, thus making it difficult to identify the initial agent of deforestation (Binswanger, 1991; Margulis, 2003).

These migrant farmers are distinctly different from the Amazon's indigenous and traditional populations, who typically practice shifting cultivation. Rather than contributing to the deforestation frontier, traditional shifting cultivation results in a patchwork of small agricultural fields and forest at different stages of regeneration. Studies have found significantly lower deforestation rates where the traditional tenure rights of these populations are officially recognized, e.g. in indigenous territories or sustainable use protected areas such as extractive reserves (Nepstad et al., 2006).

Loggers are generally not considered agents of deforestation in the Brazilian Amazon. Because of the high biodiversity of the forest, including the large number of tree species, most logging is very selective, removing only the most valuable trees in operations that have been called 'high-grading' or 'creaming' the forest. Conventional logging operations increase access to and damage the residual forest, leaving it more vulnerable to fire and clearing by farmers and less valuable for future timber harvests. However, loggers typically do leave behind forest, even if degraded; they rarely if ever fully deforest a site.

Economic frameworks and methods

The empirical literature on deforestation in the Amazon models the probability or rate of deforestation in spatial units ranging from entire states to single pixels of remote sensing images, as a function of local conditions in those units. Drawing on theories of land rents, economists have focused on two dimensions of local conditions: (1) market access, which determines farm gate prices, as suggested by von Thünan; and (2) biophysical characteristics, which determine productivity in the Ricardian framework. Government policy, production technology and macroeconomic factors shape these conditions and moderate their influence on agents' land-use decisions. Where markets are incomplete, the characteristics of the agents also influence their decisions, as posited by household production theory. Collectively, these conditions are considered direct drivers of deforestation, because they determine agents' returns to alternative land uses in particular places.

Economists have taken different empirical approaches to analyzing these direct drivers. Some have calculated returns to alternative production technologies and land uses under different conditions by applying financial, capital budgeting, and cost-benefit analysis (e.g. Mattos and Uhl, 1994, on cattle ranching; Holmes et al., 2002, on reduced impact logging [RIL]). Others have estimated how conditions influence deforestation using large secondary spatial data sets. To model agent decisions in the context of incomplete markets, economists apply the household production framework to household survey data. When they can obtain credit, buy and sell inputs (such as labor) and sell outputs (such as crops) in the market, these households' land-use and other production decisions can be modeled as if they were separable from their consumption decisions. However, many smallholders are not fully integrated into markets; thus, their preferences and assets influence their land-use decisions, in addition to and interacting with market access and biophysical conditions (Pfaff, Amacher and Sills, 2013). This is often the starting point for studies of deforestation in government colonization (or 'land reform') settlements.

Regardless of the modeling framework, economists typically assume that agents choose the land use that results in the highest net present value of expected future returns. This suggests that the deforestation frontier reflects agents' choices between (a) maintaining the native forest to produce timber and nontimber goods or (b) clearing the forest for agricultural production. Where markets are complete, this choice depends on the net present values of profits from the two options, which in turn depend on prices of agricultural and forest outputs and inputs, the quality of the land, the timing of agricultural and forest production, and the interest rate. Where markets are incomplete, the choice is also influenced by agents' productive capital, discount rates and preferences over consumption of goods and leisure (vs. labor).

In either case, another important factor is the relative security of tenure over agricultural versus forest land (Alston, Libecap and Mueller, 2000). In addition to ensuring that land remains available to its owner for production, secure tenure improves access to credit and increases the potential sale price of the land, both of which can be particularly important in the context of high inflation rates. In the Brazilian Amazon, tenure is typically less secure for forested land than for agricultural land (Araujo, Bonjean, Combes, Combes Motel and Reis, 2009). Together with high interest rates and the slow growth of timber, this has suppressed expected returns to long-term forest management. Therefore, economists sometimes make the simplifying assumption that maintaining land in forest is not competitive, and agents deforest when and where the net present value of expected future returns to cleared land is greater than the initial cost of clearing the forest.

Starting from this insight that deforestation is fundamentally about the demand for agricultural land, the next section first reviews empirical findings on the deforestation drivers suggested by theories of economic rent and household production, then considers forestry (logging and forest management) and forest conservation policy.

Empirical findings on deforestation drivers

Market access

The pattern of first clearing areas with the best market access appears at multiple scales in the Amazon. Historically, deforestation has been concentrated in an 'arc' extending across the southern part of the Amazon that is closest to markets in southern Brazil, where there is strong demand for commodities produced on deforested land. In addition, most deforestation has taken place along major roads (75% within 50 km of paved roads according to IPAM, 2000). And family farmers typically start clearing from the road frontage of their lots.

This pattern is predicted by the von Thünen model of land rent, which posits that land use is determined by distance to market. Although both agricultural and forest production are most profitable closest to markets, the value of agricultural production starts higher and declines faster with increasing distance from market. Thus, forest is cleared for agricultural production closer to markets, i.e. closer to cities, where farm gate prices for production are higher. In these more central areas, there is both greater incentive and greater possibility of obtaining secure title to land. Title reduces the private costs of defending land, allows long-term investments and creates the possibility of selling the land. The cost of obtaining and defending title is higher and the incremental benefits are lower in more remote regions. In fact, Alston, Libecap and Schneider (1995) define the economic frontier as 'the point where the net present value of claiming the land just covers the opportunity cost of the claimant'. Loggers may operate beyond this frontier, but solely to extract valuable timber, without attempting to lay claim to the forest or the land.

Although the correlation of deforestation and roads is obvious in both satellite imagery and cross-sectional statistical analysis of the Brazilian Amazon, there is continuing debate over the causality and heterogeneity of that relationship. For example, Weinhold and Reis (2008) conclude that roads follow clearing, rather than the reverse. They argue that roads only cause deforestation in municipalities with low initial levels of deforestation, e.g. new federal roads into areas of relatively undisturbed forest. In contrast, using census tract data, Pfaff et al. (2007) find that roads promote deforestation in both the census tract that they pass through and neighboring census tracts within a 100-km radius. Aguiar, Câmara and Escada (2007) and Mann et al. (2010) argue that the profitability of agriculture is affected by distance to markets over the relevant transportation network, rather than simply distance to the nearest road. Specifically, Aguiar et al. (2007) find that road distance to major markets in southern Brazil is a key determinant of land use in the arc of deforestation, but distance to urban centers and ports is more important in the central and western Amazon. Mann et al. (2010) find that land use is better predicted by a measure of rent that includes transportation costs, rather than distance to roads itself. Pacheco and Poccard-Chapuis (2012) note that the cattle processing industry is increasingly moving into ranching areas in the Amazon, effectively bringing the market closer to ranchers.

Biophysical characteristics

Although distance to market drives variation in economic conditions across the landscape, biophysical characteristics such as climate, soil quality and slope – sometimes called Ricardian factors – also influence returns to alternative land uses. In one of the first economic analyses of deforestation in the Amazon that combined satellite data on land cover with secondary data on biophysical and economic conditions, Pfaff (1999) found that the proportion of deforestation in a municipality (the smallest jurisdiction of Brazilian government) was significantly and positively related to (1) the density of paved roads in that and neighboring municipalities, (2) soil quality (specifically availability of nitrogen) and (3) the number of development projects supported by the federal government in that municipality. In a similar study at the finer scale of census units, Chomitz and Thomas (2003) confirmed that the extent of deforestation is related to market access and soil quality, as well as to precipitation, which has a strong negative effect. The effects of biophysical factors can change over time as a result of research and development of new agricultural technologies, as discussed next.

Technology

Cattle ranching in the Amazon is frequently criticized as low-productivity and land-extensive, raising two concerns. First is that low productivity implies low benefits from deforestation, making it even less likely to pass a social benefit-cost test; second is that low levels of agrarian technology are associated with high levels of deforestation (de Souza et al., 2013). These concerns underlie many efforts to promote intensification of agricultural production through improved farming systems, crop varieties and other technologies.

Economists have shown that the effects of intensification vary across spatial and temporal scales. Globally, better technology that decreases the cost of agricultural production is likely to increase supply and suppress prices (depending on the demand elasticity), thus lowering the net present value of agricultural land and the incentive to deforest (Angelsen 2010). At the local scale, the short-run response to a new technology may be to concentrate resources on implementation of that technology in a smaller area, but over the longer run, agents are likely to marshal more resources (e.g. labor, machinery) for deforestation if they can earn higher profits from cleared land with more intensive production. This dynamic was in evidence in Mato Grosso in the early 2000s, when better adapted varieties and techniques made soybean cultivation in the Amazon more profitable (CGIAR, 2011). This induced deforestation both for soybean cultivation (Morton et al., 2006) and for cattle pasture displaced by soybean cultivation (Arima et al., 2011). In recent years, farmers have adopted more intensive production systems, which, together with market and policy initiatives to slow deforestation, appear to have de-coupled demand for newly cleared land from demand for soybeans (Macedo et al., 2012).

Agent characteristics

Although the standard von Thünian and Ricardian models of land use assume profit-maximizing agents, many 'agents' in the Amazon are better described as maximizing utility subject to household resource constraints. This is because markets are incomplete on the forest frontier, requiring households to supply their own production inputs (e.g. rely on household labor) and to produce for their own consumption (e.g. produce crops for household subsistence) (Pfaff et al., 2013). Households on the forest frontier are rarely if ever completely autarkic (operating independent of the market), but they also rarely have perfect access to markets for outputs or for inputs such as labor, land, machinery and credit. As a result, although their decisions about *where* to deforest follow the logic of the von Thünan and Ricardian models, their decisions about *how much* to deforest and *what* farming systems to adopt are also influenced by their consumption preferences, production capabilities and access to credit, labor and product markets.

In the economics literature, this behavior is usually conceptualized in the agricultural household production framework (Singh, Squire and Strauss, 1986), which recognizes that households' production decisions are shaped by both their productive capital (e.g. the household labor force) and their consumption needs and preferences (e.g. dependency ratio and preference for household activities like childcare or 'leisure' vs. labor in fields or forests). In this context, deforestation is the manifestation of derived demand for agricultural land, or an investment in future agricultural production and potentially land sales, both influenced by household capabilities and preferences.

Building on the household production model, Walker, Stephen, Caldas and Silva (2002) proposed a family 'life cycle hypothesis', predicting that the choice of farming system evolves over time with household demographics, largely because of their influence on risk aversion and discount rates. In their study region, Walker et al. (2002) find that choice of farming system is affected by family work force, dependency ratio, education and external income. These factors collectively encourage families to focus on production of staple annual crops when they first arrive in the Amazon. In parts of the region, these crops have not done well, contributing to high rates of lot turnover and further deforestation as families move on to the new frontier, while their former lots are incorporated into large ranches. In other cases, small farmers have quickly shifted into commercial crops and cattle, exemplified by profitable small-scale dairy farming in the state of Rondônia (Caviglia-Harris, 2005). Walker (2003) presents data from surveys of colonist farmers across the Amazon basin, finding that they have consistently diversified production but, at the same time, converted most of their cleared land into pasture. In fact, cattle ranching is pervasive throughout the Amazon, even in sustainable use protected areas occupied by traditional rural populations (Salisbury and Schmink, 2007). This has been attributed to both the profitability of cattle and their desirability as stores of savings and cultural status symbols (Hoelle, 2011).

Public incentives

Early studies (e.g. Hecht, 1985; Binswanger, 1991; Fearnside, 1993) of deforestation in the Brazilian Amazon found that it was largely driven by public incentives for agriculture, including reinvestment tax credits, very low tax rates on agricultural income and subsidized agricultural loans. The livestock sector alone received US$5.1 billion (in 1990 US$) in fiscal incentives and subsidized credit from 1971 to 1987 (Pacheco and Poccard-Chapuis, 2012). Further, land tenure policies required deforestation (as evidence of beneficial use) to gain or defend land title. In combination with rapid inflation, these policies made it profitable to deforest in order to gain access to financial incentives and land titles, even when agricultural production on the deforested land was not in itself very profitable (Margulis, 2003).

In the 1990s, in response to international criticism of deforestation and economic constraints, the Brazilian government reduced subsidies for the agricultural sector in the Amazon. At the same time, researchers were developing new varieties and producers were testing new farming systems better adapted and more profitable in the Amazon (Mattos and Uhl, 1994; Margulis, 2003). Recent literature (e.g. Pacheco and Poccard-Chapuis, 2012) argues that the profitability of agriculture – along with the low opportunity cost of clearing forest – is now the most important driver of deforestation in the Amazon, reinforced but not driven by public incentives for agriculture and insecure tenure for forest land.

The value of land and profitability of cattle ranching still depend on public investments in transportation, research and development, and services such as the phytosanitary program that has eliminated hoof-and-mouth disease from most of the Amazon. These public investments help producers in the Amazon access both urban markets in the region and international commodity markets (Faminow, 1998). Exchange rate policy also has a significant impact, with a large devaluation in 1999 encouraging an increase in beef exports and reducing competition from dairy imports.

The government has also directly promoted deforestation by establishing and inviting farm families to settle in agricultural colonization projects. In the 1980s and 1990s, landless farmers forced this process to move faster by invading and occupying both public and private lands. In the Amazon, these 'invasions' typically focused on forested land, which is generally considered 'unimproved' and therefore easier for the government to expropriate and re-distribute – even when it is part of the legally mandated forest reserve on a private property. Although this was initially a highly conflicted process, by the late 1990s, a new pattern of settlement emerged: Groups of organized landless families identify and invade forest areas, followed by government expropriation, allocation of the land to the occupying families and support for farm establishment such as subsidized credit (Simmons et al., 2010). With this new model, colonization proceeded at a more rapid pace, with 90,000 families per year being settled from 1995 to 2002 (Pacheco and Poccard-Chapuis, 2012).

Although there has been some experimentation with alternative settlement designs – including different sized and shaped lots, and collective forest reserves rather than required reserves on each property – most settlements have a standard layout of 100-ha plots on regular road grids, resulting in the so-called fishbone pattern of deforestation as settlers clear forest starting at the front of their lots and work their way back, replicating the von Thünen pattern at the micro scale (Tucker, Holben and Goff, 1984; Caviglia-Harris and Harris, 2011).

Forestry

The forestry sector, including harvest and management of timber, can play two distinctly different roles in tropical deforestation. Under the right conditions, forest management competes with agriculture as a land use. But more typically, logging of old-growth forest reduces the cost of clearing that forest by providing access along logging roads, removing some of the trees and generating stumpage fees. Thus, although profitable long-term forest management (or payments for forest conservation) would compete with agriculture and thereby discourage deforestation, logging the mature forest can facilitate deforestation.

There is clear evidence that logging has facilitated deforestation in the Brazilian Amazon. Based on remote sensing data, Asner et al. (2006) find that forest 5 to 25 km from main roads is two to four times more likely to be deforested if it has been logged. They suggest that this is the range in which logging roads effectively increase access to market, by providing a link to the main road that would otherwise not be financially viable. There is a vast network of logging and other unofficial roads in the Amazon, sometimes called 'endogenous roads' because private actors construct them in response to economic opportunities such as the harvest of high-value timber like mahogany (Veríssimo et al., 1995).

These roads could also provide access for future timber harvests, except that the forest is rarely managed for additional harvests. One reason is that the initial logging operation usually substantially damages the residual stand, both increasing its vulnerability to fire and decreasing the profitability of any future harvests (Veríssimo et al., 1992). This raises the question of why forests are mined for valuable old-growth timber without any concern for future harvests – that is, why is forest management not a competitive land use? One explanation is that much timber is harvested illegally from public lands that are effectively open-access. In recent policy reforms, the government has sought to address this by offering concessions in national forests for timber harvest following best practices, and at the same time increasing enforcement efforts against illegal logging.

Even private forest owners with land title rarely concern themselves with logging damages to the residual stand. Explanations include (a) the vulnerability of forest land to invasion by illegal loggers or by squatters, partly because of the greater difficulty of monitoring activities in the forest and partly because government policy does not recognize forest management as productive use of land (Araujo et al., 2009), and (b) lack of information on how to profitably manage natural tropical forests for sustained timber yield. As a first step toward remedying this lack of information, several institutions have developed and promoted RIL methods that do less damage to the soils and residual stand. Comparisons of RIL with conventional logging have found that its additional costs are largely if not entirely offset by greater productivity and less waste, which in turn are due to advance planning of harvest and training of logging crews (Holmes et al., 2002). However, firms may be reluctant to invest in planning and training before the harvest, in the same way that they are reluctant to invest in long-term management such as enrichment planting, because of high discount rates and tenure insecurity (Putz, Dykstra and Heinrich, 2000; Keefe, Alavalapati and Pinheiro, 2012).

Forest certification could provide an additional incentive for sustainable forest management. Although forest certification has helped some producers gain access to niche international markets, its impacts are constrained by the nature of timber production in the Amazon. More detailed discussion of forest certification is provided in Chapter 29 of this book.

Forest conservation policy

Over the past decade, the Brazilian government has increased efforts to control deforestation in the Amazon (May, Millikan and Gebara, 2011). The Brazilian Forest Code of 1965 requires private landowners to maintain forest on a portion of their properties, previously 50% and currently 80% in the Amazon, in addition to protecting riparian forests (in a 5- to 30-m buffer depending on property and river size) and steep slopes. The Forest Code, however, had been widely ignored. To address this, in 2004 the federal government launched an action plan to control deforestation with better coordination across agencies, including more strategic use of remote sensing. Access to benefits such as agricultural credit became conditional on compliance with the Forest Code.

In 2007, the government 'blacklisted' or 'embargoed' the municipalities with the highest deforestation rates and withheld credit from all producers in those municipalities until deforestation rates fall and at least 80% of the properties register in a 'Rural Environmental Cadastre' or CAR, which in turn requires reforestation plans to meet the minimum 80% forest cover requirement. De Souza et al. (2013) confirm that the blacklist targets municipalities with high deforestation rates. This targeting of the intervention creates a challenge for evaluating its impact, making it difficult to find comparable municipalities to construct a counterfactual. Hargrave and Kis-Katos (2013) face a similar problem with evaluating the impact of enforcement efforts. Using instrumental variables, they conclude that more environmental fines lead to less deforestation.

The Brazilian government also piloted a land titling program, called *Terra Legal*, in the blacklisted municipalities. The program grants titles to smallholders who have not benefited from land reform but who have occupied and cultivated public land (*terra devoluta*) since 2004. The titles issued under this program are conditional on compliance with the Forest Code, including registration in the CAR. These efforts have been supported by nongovernmental organizations (NGOs) with funding for REDD+ (reduced emissions from deforestation and forest degradation) because clear land tenure is critical for assigning credit and liability for greenhouse gas emissions from land-use change (Börner et al., 2010; Duchelle et al., 2013). Assunção, Gandour and Rocha (2012) argue that the impact of these policies depends on both output prices and the availability of forest land that can be legally cleared; where the law imposes an effective constraint on deforestation, they find a significant impact. Details of REDD+ are available in Chapter 19 of this book.

Another way to conserve forests is to establish protected areas. As of December 2010, federal and state conservation units covered 22.2% of the Legal Amazon and indigenous territories covered 21.7%. Of the 307 conservation units in the Amazon (132 federal and 175 state), 196 are for sustainable use (e.g. national forests where timber can be harvested, and extractive reserves where traditional rural people live and collect forest products), while 111 are for preservation (e.g. national and state parks). As with the blacklist, a key challenge for evaluating protected areas is to identify an appropriate comparison group. Unlike the blacklist, which targeted municipalities with high levels of deforestation, protected areas are generally established in less accessible and less desirable areas for clearing (Joppa and Pfaff, 2009).

Even after controlling for site selection, analysts have found protected areas to be effective barriers to both fire and deforestation (Soares-Filho et al., 2010). Some studies have suggested that indigenous reserves, with a local population vested in protecting the area, are most effective (Nepstad et al., 2006). Others argue that sustainable use areas with a sustained yield of timber are critical to reducing displacement of demand to other regions, either directly or through market responses (Soares-Filho et al., 2010). Although there is mounting evidence that protected areas can effectively deter deforestation within their boundaries, there remains debate over whether and how they will be credited for reduced deforestation under REDD+ (Ricketts et al., 2010).

Finally, government policy has long been complemented by conservation initiatives funded by bilateral agencies and NGOs, e.g. the Pilot Program for the Protection of Tropical Forests in Brazil (PPG7) launched in 1993. In recent years, government policy has been reinforced by market initiatives such as forest certification and bans on beef and soy from recently deforested areas. These bans were implemented in coordination with the government, essentially requiring producers to obtain environmental certification before they could access the market

(Pacheco and Poccard-Chapuis, 2012). The soybean moratorium has been given partial credit for reduced deforestation in Mato Grosso since 2006 (Brannstrom, Rausch, Brown, de Andrade and Miccolis, 2012).

Conclusion

The deforestation frontier in the Brazilian Amazon has been driven primarily by agricultural markets and policies, facilitated by logging but with no effective competition from forest production as an alternative land use. Deforestation decisions are based on comparisons of the present cost of clearing forests to the net present value of future returns to deforested land, ignoring the possibility of long-term forest management because of high discount rates, weak forest tenure and lack of markets for many forest benefits. Deforestation agents are concerned only with their private costs and benefits, so for example, they consider the present cost of clearing forest to be offset by stumpage fees, but they do not consider externalities that occur at multiple scales from regional (e.g. public health costs of smoke from fires set to clear the slash; de Mendonça et al., 2004) to global (e.g. biodiversity and carbon). On the other hand, they consider the returns to deforested land to include both agricultural production and access to other benefits, such as more secure land tenure and financial incentives. This is reflected in land values: Economists have consistently found that only the proportion of land cleared for agriculture (and not the forested land) contributes positively to the value of a property in the Amazon (Alston, Libecap and Mueller, 1999; Merry, Amacher and Lima, 2008; Sills and Caviglia-Harris, 2008).

However, the conservation value of forests is increasingly becoming a decision parameter for deforestation agents, as environmental NGOs influence global consumers and supply chains, the Brazilian government more effectively enforces environmental regulations and both tap into new funding streams associated with REDD+. Although those funding streams are limited by the lack of an international agreement to cap carbon emissions, Brazil has still managed to capture significant funds from sources such as the Government of Norway's International Climate and Forest Initiative and voluntary carbon offset markets (Peters-Stanley and Yin, 2012). A key issue facing Brazil (and the other Amazonian countries) is how best to deploy those funds to reduce deforestation without undercutting development. Options being discussed include decentralization of forest governance and intensification of agricultural production. In both cases, prior research has shown that effects depend on the exact design, timing and local conditions. Further research is needed to map out variation across the entire Amazonian landscape in how agents are likely to respond to changes in these drivers and evaluate their impacts on deforestation.

Acknowledgments

The author thanks Jill Caviglia-Harris for her insightful review of the manuscript and Simon Hall for research assistance.

References

Aguiar, A.P.D., Câmara, G. and Escada, M.I.S. (2007). 'Spatial statistical analysis of land-use determinants in the Brazilian Amazonia: Exploring intra-regional heterogeneity', *Ecological Modelling*, vol. 209, pp. 169–188.

Alston, L., Libecap, G. and Mueller, B. (1999). *Titles, Conflict, and Land Use: The Development of Property Rights and Land Reform on the Brazilian Amazon Frontier*, University of Michigan Press, Ann Arbor, MI.

Alston, L. J., Libecap, G. D. and Mueller, B. (2000). 'Land reform policies, the sources of violent conflict, and implications for deforestation in the Brazilian Amazon', *Journal of Environmental Economics and Management*, vol. 39, no. 2, pp. 162–188.

Alston, L. J., Libecap, G. D. and Schneider, R. (1995). 'Property rights and the preconditions for markets: The case of the Amazon frontier', *Journal of Institutional and Theoretical Economics*, vol. 151, no. 1, pp. 89–107.

Angelsen, A. (2010). 'Policies for reduced deforestation and their impact on agricultural production', *Proceedings of the National Academy of Sciences*, vol. 107, no. 46, pp. 19639–19644.

Araujo, C., Bonjean, C., Combes, J., Combes Motel, P. and Reis, E. J. (2009). 'Property rights and deforestation in the Brazilian Amazon', *Ecological Economics*, vol. 68, no. 8–9, pp. 2461–2468.

Arima, E. Y., Richards, P., Walker, R. and Caldas, M. M. (2011). 'Statistical confirmation of indirect land use change in the Brazilian Amazon', *Environmental Research Letters*, vol. 6, p. 024010. doi:10.1088/1748-9326/6/2/024010

Asner, G., Broadbent, E., Oliveira, P., Keller, M., Knapp, D. and Silva, J. (2006). 'Condition and fate of logged forests in the Brazilian Amazon', *Proceedings of the National Academy of Sciences*, vol. 103, no. 34, pp. 12947–12950.

Assunção, J., Gandour, C. and Rocha, R. (2012). *Deforestation Slowdown in the Legal Amazon: Prices or Policies?* CPI Working Paper, Climate Policy Initiative, Rio de Janeiro. Retrieved from http://climatepolicyini tiative.org/wp-content/uploads/2012/03/Deforestation-Prices-or-Policies-Working-Paper.pdf

Binswanger, H. P. (1991). 'Brazilian policies that encourage deforestation in the Amazon', *World Development*, vol. 19, no. 7, pp. 821–829.

Börner, J., Wunder, S., Wertz-Kanounnikoff, S., Tito, M. R., Pereira, L. and Nascimento, N. (2010). 'Direct conservation payments in the Brazilian Amazon: Scope and equity implications', *Ecological Economics*, vol. 69, pp. 1272–1282.

Brannstrom, C., Rausch, L., Brown, J. C., de Andrade, R.M.T. and Miccolis, A. (2012). 'Compliance and market exclusion in Brazilian agriculture: Analysis and implications for "soft" governance', *Land Use Policy*, vol. 29, pp. 357–368.

Bunker, S. G. (1985). *Underdeveloping the Amazon: Extraction, Unequal Exchange, and the Failure of the Modern State*, University of Chicago Press, Chicago.

Caviglia-Harris, J. (2005). 'Cattle accumulation and land use intensification by households in the Brazilian Amazon', *Agricultural and Resource Economics Review*, vol. 34, no. 2, pp. 145–162.

Caviglia-Harris, J. and Harris, D. (2011). 'The impact of settlement design on tropical deforestation rates and resulting land cover patterns', *Agricultural and Resource Economics Review*, vol. 40, no. 3, pp. 451–470.

CGIAR Independent Science and Partnership Council (CGIAR). (2011). *Measuring the Environmental Impacts of Agricultural Research: Theory and Applications to CGIAR Research*, Independent Science and Partnership Council Secretariat, Rome.

Chomitz, K. M. and Thomas, T. S. (2003). 'Determinants of land use in Amazônia: A fine-scale spatial analysis', *American Journal of Agricultural Economics*, vol. 85, no. 4, pp. 1016–1028.

Defries, R. S., Rudel, T., Uriarte, M. and Hansen, M. (2010). 'Deforestation driven by urban population growth and agricultural trade in the twenty-first century', *Nature Geoscience*, vol. 3, pp. 178–181.

de Mendonça, M., Vera Diaz, M., Nepstad, D., da Motta, R., Alencar, A., Gomes, J. and Ortiz, R. (2004). 'The economic cost of the use of fire in the Amazon', *Ecological Economics*, vol. 49, no. 1, pp. 89–105.

de Souza, R. A., Miziara, F. and De Marco, P., Jr. (2013). 'Spatial variation of deforestation rates in the Brazilian Amazon: A complex theater for agrarian technology, agrarian structure and governance by surveillance', *Land Use Policy*, vol. 30, no. 1, pp. 915–924.

Duchelle, A., Cromberg, M., Gebara, M. F., Melo, T., Larson, A., Cronkelton, P., . . . Sunderlin, W. (2014). 'Linking forest tenure reform, environmental compliance, and incentives: Lessons from REDD+ initiatives in the Brazilian Amazon', *World Development*, vol. 55, pp. 53–67.

EMBRAPA/INPE. (2011). 'TerraClass: Levantamento de informações de uso e cobertura da terra na Amazônia'. Retrieved from www.inpe.br/cra/projetos_pesquisas/terraclass.php

Faminow, M. D. (1998). *Cattle, Deforestation and Development in the Amazon: An Economic, Agronomic and Environmental Perspective*, CAB International, New York.

Fearnside, P. M. (1993). 'Deforestation in Brazilian Amazonia: The effect of population and land tenure', *Ambio*, vol. 22, no. 8, pp. 537–545.

Hansen, M. C., Stehman, S. V. and Potapov, P. V. (2010). 'Quantification of global gross forest cover loss', *Proceedings of the National Academy of Sciences*, vol. 107, no. 19, pp. 8650–8655.

Hargrave, J. and Kis-Katos, K. (2013). 'Economic causes of deforestation in the Brazilian Amazon: A panel data analysis for the 2000s', *Environmental and Resource Economics*, vol. 54, no. 4, pp. 471–494.

Hecht, S. (1985). 'Environment, development and politics: Capital accumulation in the livestock sector in eastern Amazonia', *World Development*, vol. 13, no. 6, pp. 663–684.

Hoelle, J. (2011). 'Convergence on cattle: Political ecology, social group perceptions, and socioeconomic relationships in Acre, Brazil', *Culture, Agriculture, Food and Environment*, vol. 33, no. 2, pp. 95–106.

Holmes, T. P., Blate, G. M., Zweede, J. C., Pereira, R., Barreto, P., Boltz, F. and Bauch, R. (2002). 'Financial and ecological indicators of reduced impact logging performance in the eastern Amazon', *Forest Ecology and Management*, vol. 163, no. 1–3, pp. 93–110.

IBGE. (2011). *Geostatistics of Natural Resources of the Legal Amazon*, Estudos e Pesquisas Informação Geográfica (8), IBGE, Rio de Janeiro.

INPE. (2012). Taxas anuais do desmatamento Retrieved from www.obt.inpe.br/prodes/prodes_1988_2012.htm

Instituto de Pesquisa Ambiental da Amazônia (Amazon Institute for Environmental Research, IPAM). (2000). *Forward Brazil: Environmental Costs for Amazonia*, Belém, Brazil. Retrieved from www.ipam.org.br

Joppa, L. and Pfaff, A. (2009). 'High & far: Biases in the location of protected areas', *PLoS ONE*, vol. 4, no. 12, p. e8273.

Kaimowitz, D. and Angelsen, A. (1998). Economic models of tropical deforestation: a review. Bogor, Indonesia: CIFOR. Available from: www.cifor.org/publications/pdf_files/Books/model.pdf

Keefe, K., Alavalapati, J.A.A. and Pinheiro, C. (2012). 'Is enrichment planting worth its costs? A financial cost-benefit analysis', *Forest Policy and Economics*, vol. 23, pp. 10–16.

Lapola, D. M., Schaldach, R., Alcamo, J., Bondeau, A., Koch, J., Koelking, C. and Priess, J. A. (2010). 'Indirect land-use changes can overcome carbon savings from biofuels in Brazil', *Proceedings of the National Academy of Sciences*, vol. 107, pp. 3388–3393.

Macedo, M. N., DeFries, R. S., Morton, D. C., Stickler, C. M., Galford, G. L. and Shimabukuro, Y. E. (2012). 'Decoupling of deforestation and soy production in the southern Amazon during the late 2000s', *Proceedings of the National Academy of Sciences*, vol. 109, pp. 1341–1346.

Mann, M., Kaufmann, R., Bauer, D., Gopal, S., Vera-Diaz, M.D.C., Nepstad, D., . . . Amacher, G. S. (2010). 'The economics of cropland conversion in Amazonia: The importance of agricultural rent', *Ecological Economics*, vol. 69, pp. 1503–1509.

Margulis, S. (2003). *Causes of Deforestation of the Brazilian Rainforest*, World Bank Working Paper no. 22, The World Bank, Washington, DC.

Mattos, M. M. and Uhl, C. (1994). 'Economic and ecological perspectives on ranching in the eastern Amazon', *World Development*, vol. 22, no. 2, pp. 145–158.

May, P. H., Millikan, B. and Gebara, M. F. (2011). *The Context of REDD+ in Brazil: Drivers, Agents, and Institutions* (2nd ed.), Occasional Paper 55, CIFOR, Bogor, Indonesia.

Merry, F., Amacher, G. and Lima, E. (2008). 'Land values in frontier settlements of the Brazilian Amazon', *World Development*, vol. 36, no. 11, pp. 2390–2401.

Morton, D. C., DeFries, R. S., Shimabukuro, Y. E., Anderson, L. O., Arai, E., del Bon Espirito-Santo, F., . . . Morisette, J. (2006). 'Cropland expansion changes deforestation dynamics in the southern Brazilian Amazon', *Proceedings of the National Academy of Sciences*, vol. 103, pp. 14637–14641.

Nepstad, D., Schwartzman, S., Bamberger, B., Santilli, M., Ray, D., Schlesinger, P., . . . & Rolla, A. (2006). 'Inhibition of Amazon deforestation and fire by parks and indigenous lands', *Conservation Biology*, vol. 20, pp. 65–73.

Nepstad, D. C., Stickler, C. M. Soares-Filho, B. and Merry, F. (2008). 'Interactions among Amazon land use, forests and climate: Prospects for a near-term forest tipping point'. *Philosophical Transactions of the Royal Society B*, vol. 363, no. 1498, pp. 1737–1746.

Pacheco, P. and Poccard-Chapuis, R. (2012). 'The complex evolution of cattle ranching development amid market integration and policy shifts in the Brazilian Amazon', *Annals of the Association of American Geographers* vol. 102, no. 6, pp, 1366–1390.

Perz, S. G., Aramburú, C. and Bremer, J. (2005). 'Population, land use and deforestation in the Pan-Amazon Basin', *Environment, Development and Sustainability*, vol. 7, pp. 23–49.

Peters-Stanley, M. and Yin, D. (2013). Maneuvering the mosaic. State of the voluntary carbon markets 2013. A Report by Forest Trends' Ecosystem Marketplace & Bloomberg New Energy Finance, Washington, DC, pp. 126.

Pfaff, A. (1999). 'What drives deforestation in the Brazilian Amazon?' *Journal of Environmental Economics and Management*, vol. 37, no. 1, pp. 26–43.

Pfaff, A., Amacher, G. and Sills, E. (2013). 'Realistic REDD: Improving the forest impacts of domestic policies in different settings', *Review of Environmental Economics and Policy*, vol. 7, no. 1, pp. 114-135.

Pfaff, A., Robalino, J. A., Walker, R., Reis, E., Perz, S., Bohrer, C., . . . Kirby, K. (2007). 'Road investments, spatial intensification and deforestation in the Brazilian Amazon', *Journal of Regional Science*, vol. 47, pp. 109–123.

Putz, F. E., Dykstra, D. P. and Heinrich, R. (2000). 'Why poor logging practices persist in the tropics', *Conservation Biology*, vol. 14, no. 4, pp. 951–956.

415

Ricketts, T. H., Soares-Filho, B., da Fonseca, G.A.B., Nepstad, D., Pfaff, A., Petsonk, A., ... & Victurine, R. (2010). 'Indigenous lands, protected areas, and slowing climate change', *PLoS Biology*, vol. 8, no. 3, p. e1000331.

Rudel, T. K. (2007). 'Changing agents of deforestation: From state initiated to enterprise driven processes, 1970–2000', *Land Use Policy*, vol. 24, no. 1, pp. 35–41.

Salisbury, D. and Schmink, M. (2007). 'Cows versus rubber: Changing livelihoods among Amazonian extractivists', *Geoforum*, vol. 38, pp. 1233–1249.

Sills, E. and Caviglia-Harris, J. (2008). 'Evolution of the Amazonian frontier: Land values in Rondônia, Brazil', *Land Use Policy*, vol. 26, pp. 55–67.

Simmons, C., Perz, S., Aldrich, S., Caldas, M., Pereira, R., Leite, F. and Fernandes, L. C. (2010). 'Doing it for themselves: Direct action land reform in the Brazilian Amazon', *World Development*, vol. 38, no. 3, pp. 429–444.

Singh, I., Squire, L. and Strauss, J. (Eds.). (1986). *Agricultural Household Models*, Johns Hopkins University Press, Baltimore.

Soares-Filho, B. S., Moutinho, P., Nepstad, D., Anderson, A., Rodrigues, H., Garcia, R., . . . Maretti, C. (2010). 'Role of Brazilian Amazon protected areas in climate change mitigation', *Proceedings of the National Academy of Sciences*, vol. 107, no. 24, pp. 10821–10826.

Soares-Filho, B. S., Nepstad, D., Curran, L., Cerqueira, G., Garcia, R., Ramos, C., ... McGrath, D. (2006). 'Modeling conservation in the Amazon Basin', *Nature*, vol. 440, no. 7083, pp. 520–523.

Torras, M. (2000). 'The total economic value of Amazonian deforestation, 1978–1993', *Ecological Economics*, vol. 33, pp. 283–297.

Tucker, C. J., Holben, B. N. and Goff, T. E. (1984). 'Intensive forest clearing in Rondonia, Brazil, as detected by satellite remote sensing', *Remote Sensing of Environment*, vol. 15, pp. 255–261.

Veríssimo, A., Barreto, P., Mattos, M., Tarifa, R., and Uhl, C. (1992). 'Logging impacts and prospects for sustainable forest management in an old Amazonian frontier – The case of Paragominas', *Forest Ecology and Management*, vol. 55, nos. 1–4, pp. 169–199.

Veríssimo, A., Barreto, P., Tarifa, R., and Uhl, C. (1995). 'Extraction of a high value natural resource in Amazonia: The case of mahogany', *Forest Ecology and Management*, vol. 72, no. 1, pp. 39–60.

Walker, R., Stephen, P., Caldas, M. and Silva, L.G.T. (2002). 'Land use and land cover change in forest frontiers: The role of household life cycles', *International Regional Science Review*, vol. 25, no. 2, pp. 169–199.

Walker, R. (2003). 'Mapping process to pattern in the landscape change of the Amazonian frontier', Annals of the Association of American Geographers, vol. 93, no. 2, pp. 376–398.

Weinhold, D. and Reis, E. J. (2008). 'Transportation costs and the spatial distribution of land use in the Brazilian Amazon', *Global Environmental Change*, vol. 18, pp. 54–68.

27

CHINA'S FORESTLAND TENURE REFORMS

Redefining and recontracting the bundle of rights

Yaoqi Zhang,[1] *Yueqin Shen,*[2] *Yali Wen,*[3] *Yi Xie*[3] *and Sen Wang*[4]

[1]SCHOOL OF FORESTRY AND WILDLIFE SCIENCES, AUBURN UNIVERSITY, USA.
[2]ZHEJIANG AGRICULTURE AND FORESTRY UNIVERSITY, CHINA.
[3]BEIJING FORESTRY UNIVERSITY, CHINA.
[4]CANADIAN FOREST SERVICE, CANADA.

Abstract

Focused on how changes in the bundle of rights to forests and forestland and particularly the use rights are separated from ownership through various contractual arrangements, this chapter reviews and examines China's forestland tenure reforms in the past three decades. The new challenges of the reform are discussed. China's tenure reforms not only offer lessons in practice and insights regarding land tenures in China and other countries where similar reforms are carried out, but also herald a new economic theory of land tenure and property rights.

Keywords

Commons, collective forestland, land transaction, economic reform, privatization, use rights

Introduction

Land tenure is defined as the bundle of rights which a person or community holds to land, trees or other land-based resources. The rights are governed by the rules established by the state or by custom. Land tenures are usually categorized as private, communal, open access and state, based on to whom the rights are given. In general, the rules tend to change over time from informal and customary rules to formal and statutory systems. Land tenure is widely regarded as critical to rural economic development as it guides the resource allocation (Feder and Feeny, 1991). It is often argued that well-defined and secure property rights would increase investment and economic growth. The problem is that defining and securing property rights might involve significant transaction costs. More importantly, land tenure matters not just to economic efficiency, but also to income distribution. Therefore, land tenure as well as its evolution should not be assessed by economic efficiency alone. The political powers often play an important role in the land tenure arrangement (Schmid, 1987).

Study of property rights and land tenure has been a very important part of economic investigation (North and Thomas, 1973; Posner, 1991; Barzel, 1997). Land tenure and property rights regarding grassland and forestland have received particular attention (Bromley, 1989) because many of them have often been recognized for their nature as public goods and common resources. It was widely believed that 'commons' led to tragedy, ever since Hardin (1968) revealed the tragic consequences of overgrazing and rent dissipation from the 'commons'. Later work, particularly by Elinor Ostrom, has shown that commons, which are different from open access, can be successfully managed with various management arrangements (Ostrom, 1990; Ostrom, Gardner and Walker, 1994). For example, there was no evidence of any tragedy in medieval English common fields or in community-based natural resource management in many parts of the world (Pagdee, Kim and Daugherty, 2006; Shahi and Kant, 2007).

Compared to other sectors, the share of public forestland is larger in both developing and developed countries (White and Martin, 2002). Due to the generally more successful management on private forestland and grassland in western society than on communal and public forestland in developing countries, policies to promote privatization of communal and state-owned forestland are often suggested to the developing countries by various international organizations. In fact, sustainable forest management can be carried out differently (Wang and Wilson, 2007). However, some evidence also indicates that privatization does not work and communal land tenure might work better. For example, co-management of the forest resource and fisheries has been quite popular in many places (Klooster, 2000; Pretty, 2003). Each type of ownership or land tenure has its own merits in terms of forest management, and no clear evidence has indicated that one is necessarily better than another (Ostrom, 1990).

Property rights are becoming the center of economic analysis, and economics offers important insights into property rights (Posner, 1991; Barzel, 1997). This is particularly true of forest management. In the general context of the new institutional economics and the economics of property rights, many earlier studies that were concerned with questions about land tenure in forestry were interested in how land tenure or property rights would have an impact on tropical deforestation (e.g. Deacon, 1999), timber supply (Newman and Wear, 1993; Liao and Zhang, 2008), silviculture and forest investment (Besley, 1995; Wang and van Kooten, 2001; Zhang 2002; Zhang, Uusivuori and Kuuluvainen, 2000).

Altogether, the area of privately owned forests increased by an average of 2.7 million ha per year from 1990 to 2000 (FAO, 2006). In the midst of the current trend of privatization of forestland in many developing countries (Enters, Durst and Victor, 2000), there is an urgent need to assess the impact of forestland tenure reforms. China's land reforms, including forestland reforms, have received significant attention (Yin and Xu, 2002; Ho, 2005; Hyde, Belcher and Xu, 2003). However, unlike the apparent success in agriculture, the reforms in forestry seem more controversial and contentious. Considering the unprecedented scale and comprehensiveness of the forestland tenure reform taking place in recent decades (Wang, van Kooten and Wilson, 2004; Table 27.1), China seems to serve as a good case for examination of the problems and can provide insights and lessons to the rest of the world.

In this chapter, two conceptual approaches (bundle of rights and contracting) to land tenure reform are applied. The bundle of rights is used to explain how a property can simultaneously be 'owned' by multiple parties. Although the bundle of rights concept grew out of a long-standing and serious philosophical debate about legal rights and legal liberties, the bundle of rights as

Table 27.1 China's forest tenure changes since the late 1940s.

Period	Institutional change	Tenure system
Pre-1949		Warlords, landlords, bureaucrats, merchants, self-sufficiency farmers, common and open access in practice.
1950–52	Land reform and economic recovery	Government confiscated all forestlands owned by landlords and bureaucrats and redistributed to the local peasants. The rest, mostly in remote and sparsely populated areas, were claimed by the state.
1953–55	Initiation of collectivization	Private trees, forest and forestland, while allowed, were encouraged to be merged into the mutual aid teams and elementary peasants' cooperatives.
1956–58	Advancing collectivization to socialism	Upgraded to higher agricultural producers' cooperatives and transitioning to collective owned and People's Commune.
1958–81	People's Commune system	Forests and land, regardless of ownership by households or collectives, were mostly owned and managed by village-level community, Brigade and the Commune.
1982–92	Privatization and decentralization of use rights	Use rights were allocated and contracted to individual households; a shareholding system largely replaced the collective forestland regime; the state-owned forestland reform was initiated.
1993–2002	Market liberalization	Free market for timber and wood products, longer-term contract for forest and forestland, auction of nonforestland for long-term holding.
2003–present	A new phase of tenure reforms	Taxes on agricultural and forest products removed, promoting the forestland transaction, mature forests as equity for loan.

a theory of property does present an analytical and descriptive one (Johnson, 2007). Following Cheung (1969, 1983), Libercap (1989) and Allen and Lueck (1992, 1993), this chapter also uses contracting theory to understand the property rights arrangements in forestland tenure reform.

In organizing this chapter, we first provide three tree tenures and two major land tenures from the perspectives of historical evolutions and relative scarcity. Next, we analyze forestland tenures through examination of the bundle of rights before investigating the core reform of separating use rights and ownership and contracting use rights. In the concluding section, we discuss the effects of tenure reforms in terms of their lessons for policy making and the new challenges for forest management practices.

Heterogeneity of forest, forestland value and diverse tenures

China has 195.45 million ha of forest area with a timber stock of 14.91 billion m^3. Forest area accounts for 20.36% of China's total land (SFA, 2009). China's forests are mainly distributed in the northeast, south central, southwest and southeast. The structure of forestland tenures varies across regions. The northeast and southwest are dominated by state ownership and are primarily

Table 27.2 Ownerships of forests, economic trees and bamboo.

	State	Collective	Private
By volume (million m³)	8,788	2,904	1,670
By land area (million ha)	71	52	58
Timber forests	70	45	37
Economic trees	1	2.5	16.9
Bamboo	0.24	1.04	4.1

Source: SFA (2009).

managed by state-owned forestry enterprises. The central, south and southeast are dominated by collective-owned forestland.

The diversity of the tenures of land and trees reflects the heterogeneity of land and tree values in China (Table 27.2). Trees and land possess different values depending on the quality of the forests, the soil and climate, the topography and often, very importantly, the distance from market. Trees command positive economic value when harvesting and transportation costs are less than the wood market value, but the value for land to grow trees is different and needs to justify all costs, including growing, protecting and finally harvesting.

The different values of trees and land lead to different tree tenure and land tenure. It must be noted that private ownership of land has never been accepted (only state and collective ownerships), but private ownership of trees is acceptable. Table 27.2 suggests that forests on about one-third of the forestland by area in China were privately owned by late 2000s, although the land supporting the private forests was collectively owned. The majority of plantation forests are privately or collectively owned, while natural forests are mostly owned by the state. It is interesting to note that the private own about 80% of the economic trees, which are the tree species mainly planted and managed to generate nontimber products like tree fruits, nuts and oils, and more than 70% of bamboo. Because economic trees and bamboo carry more economic values, this suggests that forests of higher economic value have a greater propensity to be owned by individuals. The economic explanation could be that higher-valued trees can justify the higher costs of establishing and enforcing property rights.

Such a situation has also resulted and evolved from historical change. The forestland in remote and sparsely populated areas was mainly owned by the royal families and warlords. However, these lands were often without clear ownership and in reality were in an open or semi-open access state. Due to poor accessibility and low population density in these remote areas, the forest resources were not economically accessible. Historically, no individuals except some powerful families were able to control and get benefits from owning the forests. Consequently, the state became the default owner of the forests and forestland. The first Forest Law and Hunting Law in China were enacted by the Republic of China in 1915 (Chen, 1982, pp. 71–95). The law stipulated: 'All mountains and forests within the country, unless they are owned by private persons under the rule of local government, belong to the state and are administered by the Ministry and the local government' (Article 1).

After the People's Republic of China (PRC) came into being in 1949, the central government claimed the land and forests under state ownership. The large areas of forests and forestland are mostly located in the northeast and southwest, characterized by low population density and poor road access. At the beginning, only the trees had received attention, and over 130 state logging enterprises were established to harvest the timber resources, which became important for economic development in the early stages of the PRC. Because the land value was hardly justified, the firms were in a mining mode and basically 'cut and run'. It was not until the 1970s that reforestation received attention and became the business of the state-owned firms after the scarcity of trees led to scarcity of the land that can grow trees.

In addition to the forest enterprises, the state is also the owner of more than 4,200 forest farms, managing 46 million ha of land across the collective forest regions in south and south-central China. But this story is very different. The establishment of state-owned forest farms in the 1950s and the 1960s was mostly motivated by the desire to plant trees on land that was low in value and often abandoned without any interest by individual and the collective. Considering the importance of environmental values, particularly in terms of soil and water conservation as well as job creation opportunities, the state was motivated to establish the forest farms primarily through tree planting. After decades of effort, those areas currently have pretty good forest cover, which not only plays an important role in ecosystem functions, but also has tremendous timber and nontimber values.

Both the forests and forestland were much more valuable in central, south and southeast China, thanks to the high population density and close proximity to the villages, and consequently private ownership by landlords and peasants was widespread. However, informal institutions, such as customary norms, conventions, beliefs, religions and ethics of the local communities, largely governed the way in which forests were managed (Zhang, 2001). Quasi-governmental organizations, such as 'mountain councils' (or 'village councils'), were quite commonly involved in governing forestland in southern China. The mountain councils designated and regulated the boundaries of land, hunting rights, harvest and even output sharing (Wu, 1962).

In the long history of China's agrarian economy, almost all favorable and flat land has been cultivated for agriculture, leaving forests mostly to mountains or sloping land and remote sites. Therefore, forestlands are usually referred to as mountain areas, and the owners of mountains and forests used to be called mountain lords. In the late 1940s and the beginning of the 1950s, land was allocated to peasants in recognition of their support for the revolution. During the land reform, all land owned by landlords was confiscated and mostly redistributed to local peasants. However, the period of private land ownership was very brief. In the name of the revolution, the government expropriated the land shortly afterwards.

The subsequent Socialist Transformation, or the first phase of collectivization from 1952 to 1955, vested the ownership of private forestland into the collectives, but individual peasants were still allowed to hold some forestland. After 1956, however, private forests and forestland (except for a small number of dispersed trees around the homes) were gradually folded under collective ownership – initially, into the Agricultural Producers' Cooperatives, and later taken over by the People's Communes (usually several villages comprising one commune). Since then, private forestland has disappeared. Apart from a large number (about 8,000) of collective forest farms having been established up to 1960 (Ministry of Forestry, 1987), the majority of collective forests and forestland, both the ownership and use rights, were controlled by three levels of collectives: People's Communes, Brigades (a group of villages) and natural villages. In many cases, one village was broken down into several production teams. This structure did not change much in the 1960s and the 1970s, even though specific configurations might have slightly differed from time to time and from one region to another. Table 27.1 provides a chronology of China's major forest tenure changes since the late 1940s.

China's land tenures evolved to accommodate the country's economic realities and new imperatives of change in resource scarcity. Since the late 1970s, dramatic changes in the political and economic systems, particularly the shift from a centrally planned economy to a market economy, have created the conditions for significant changes in land tenures. As a part of the land tenure reform, China's forest tenure reforms largely entail how the bundles of rights change, and particularly how the use rights are contracted to households.

Redefining the bundle of rights

Like any other property, forests and forestland are composed of a bundle of rights and can be used, bought and sold, leased, mortgaged, subdivided and so forth. The completeness of the bundle is measured by four characteristics, namely, comprehensiveness, exclusiveness, transferability and enforceability (Randall, 1975; Bromley, 1989; Feder and Feeny, 1991). Strictly speaking, there is no complete public or private ownership because the public can access, to a certain degree, private land and trees (e.g. governments, acting on behalf of the public, can tax private property and take property for conservation purposes with or without compensation), while private individuals can access the public land and trees. This is particularly true of the collective. To better understand the evolution of forest tenures, it is necessary to assess how the sticks of the bundle are changed, added or taken away. China's forests and forestland provide a good example in understanding the bundle of rights.

Rights to access benefits

All these rights are associated with and fundamentally reflect the way in which benefits flow from the property and how the benefits are accessed. In China's case, the government exercised the right to collect what was known as Forestation Fund, following the Russian system in collecting forest rent. The fundamental difference from the stumpage prices in other countries was the calculation method: Forest Fund is based on accumulated costs of growing timber, while stumpage is based on residual value from the end product in the United States and Canada. For a detailed comparison, see Xu and McKetta (1986). Through collecting the forestation fund, the state obtained the rents of forest resources from the revenue of forest harvesting and circled it back to tree planting on logging sites and on bare mountains owned by the state.

Taxation has been used as an alternative way to access the benefits. China's forest taxation approach was introduced at the very beginning of forest economic reforms and has changed significantly with time. At the beginning and for a substantial amount of time, heavy taxation and various fees were widely used (Chen, 1995; Zhang and Kant, 2005). Not only had the provincial government set the taxes and fees; in fact, counties, townships and even villages levied a multitude of local taxes and fees. The heavy tax burden and the complexity of the tax system were major impediments to the operation of a free market system in the forest sector. The complicated tax system indicates the ambiguity of the property rights and land tenure, and it seemed that all levels of government could access some of the benefits associated with the forests. The taxes associated with forest products were removed during the early to mid-2000s for all provinces.

Rights to sell products

Timber is the major output of forestland. For a long period of time, timber could not be sold on the free market, and government controlled the right through price controls and harvesting quotas. Prior to the economic reforms initiated in the late 1970s, the government controlled timber production, price and allocation. Government control on timber sales was lifted in southern China in 1985. Unfortunately, one of the biggest episodes of deforestation resulted from market instrument liberalization together with forest property reforms. As a result, in 1987, the government reinstated the traditional government timber procurement agencies and restored its monopoly on the timber market in major timber production counties. But as the market mechanism and the socio-economic environment improved, all price controls were finally lifted and free competition for the wood from collective forests has been practiced since 1993. The timber market was opened much later for wood produced by the state-owned forests. Initially, the approach was to gradually increase the timber price closer to the market price. The state-owned forest enterprises were gradually allowed to sell their timber into the free market upon delivering a fixed quota of their products to the state. The proportion of timber sold in the free market has been gradually increasing.

Rights to harvest

The governments, on behalf of the public, exercise the rights through controls on harvesting quotas. Although the logging quota system that was initiated in 1987 has been adjusted many times, the requirement to apply for a harvesting permit from the local government remains significantly restricted, especially for villagers who still rely on timber as their major management objective. The system of distributing logging quotas is also regarded as nontransparent and unequal for forest farmers. In this sense, farmers have no idea about how many quotas are achieved by their neighbors and why they cannot get quotas as required. Survey results in Shen, Zhang, Xu, Zhu and Jiang (2009) indicated that 44.3% of farmers regarded the logging quota system as an obstacle to forestry development. The farmers in Fujian and Jiangxi regarded it as the biggest obstacle, and this was not surprising, because Fujian and Jiangxi have a much larger proportion of forestland used for timber production than other provinces. Only in those villages where timber is not the major output is the quota system not viewed as a big issue in the management of forestland.

Rights to use forest as equity for loan

Only recently, private forests became eligible to be considered as equity for loans from financial institutions, and various aspects of equity loans using forest and forestland are discussed by Zhang (2010). Loans using forest and forestland as equity currently amount to 53 billion yuan involving 38 million mu of forestland (i.e. 1,400 yuan per mu, with 15 mu = 1 ha). According to a survey of 3,500 households in 70 counties administered by the State Forestry Administration, forestland as equity accounted for 4.77% of total forestland as of the end of 2011, the average loan duration was 3.63 years and the average loan amount was 300,000 yuan per household. It was reported that 70% of the loan was used in forestry activities. The survey indicated that more than 40% of the households wish to use the land as equity for loans, but only 3.28% of the households were approved (SFA, 2013).

Contracting out the use rights

China's forestland reform essentially has been carried out through contractual arrangement for the property rights, starting from separating use rights from ownership. The contracting has been first made between the collective and the households (called 'collectively owned, privately managed') and between the state and the enterprises (called 'state-owned, collectively managed'), and in the next stage various arrangements for contracting between households, enterprise managers and the employees were developed.

The 1982 Constitution and the Land Administration Law both stipulate that the administrative villages that inherit the ownership of the land were allowed to allocate the land to smaller units or households for use and management. Although the ownership is still kept in the public domain due to political and legal circumstances as well as other considerations, one of the most important characteristics of the collective forestland tenure reform is the separation of ownership and use rights, which we can interpret as contracting out the use rights. Although land leasing and sharecropping are very common land tenure types, the idea of separating land ownership and use rights (possession) is not new either (e.g. Osborne, 1893; George, 1871, 1894; Pullen, 2001; Barzel, 1997). Still, the large scale of separating use rights from ownership on China's collective forestland is unprecedented.

In 1981, it was officially declared that: (1) for all the forest stands and forestland with clear ownership, the government, at or above the county level, should issue a property certificate to the owner to recognize the property right and guarantee its stability; (2) in accordance with the needs of the local forest farmers, plots of forestland, including waste hills, river beds or beaches, should be allocated to the local farmers for the purpose of long-term planting of trees and grass; and (3) the trees planted by the farmers around their houses, on allocated land and on the land designated by the village committee, shall be the everlasting property of the farmers and can be inherited. The main purpose of this policy was not only to allow privately held forest management rights, but to further clarify what collective (e.g. the villages or communities) owns the forestland.

The policy was implemented for 4 years following 1981. With the exception of Shanghai and Tibet, 1,781 counties, or 77.5% of all the counties that participated in the implementation of the so-called 'Three Stabilizations', completed their requirements. Private management certificates were issued covering about 100 million ha of forestland for confirmation of their rights, while approximately 1.3 million disputes on forest and forestland were settled (90% of all disputes). The main reason for the disputes was the lack of clearly defined property due to historical reasons. The disputes often led to removal of trees or even fighting between villages or communities. A total area of 31 million ha of collective land was re-allocated to about 56 million households under the household responsibility system, with an average of 0.56 ha per household (Ministry of Forestry, 1987).

Decentralization and privatization of forestland-use rights have been instituted through two kinds of arrangements or contracting (i.e. *Ziliu-shan* and *Zeren-shan*) during the early 1980s. *Ziliu-shan* (household-held mountain land) was distributed from the collective forestland primarily based on household size. The land is similar to lease freehold household land but is held only for a specific duration of time. Households can usually enjoy a full stream of benefits from the land during this duration. Village authorities assigned *Zeren-shan* (household's responsible mountain land) to be managed. The amount of land assigned was primarily based on the labor force of the household and their willingness to partake in the program. Households were either paid for providing the service, mostly in contracting to monitor and prevent illegal access, or for sharing the benefit during final harvest, depending on the quality and the stages of the trees. The remaining land still under collective management is called *Tongguan-shan* (collective-managed mountain land).

At the beginning of the economic reforms, the separation of ownership and use rights was in all likelihood a political and ideological compromise to avoid full-fledged privatization. As Ho (2001) pointed out that the law maintains a deliberately vague definition of collective ownership to avoid an escalation of land disputes between the various levels of collective ownership. However, over time, it appears that the privatization of land property rights has become more acceptable. In this sense, the amended Forestry Law, effective since 1998, reflects the rising legal status of private property rights compared to the Forest Law of 1984. The 1998 Forest Law clearly recognizes the rights of the owners and users of the forests, trees and forestland, and includes provisions for the protection against infringements on these rights (Article 3, Forestry Law of China, 1998).

It must be noted that there was another tenure type of forestland, called 'share-holding tenure'. In the beginning, it was particularly popular in Fujian Province but later gained popularity in many other provinces due to widespread deforestation in connection with *Ziliu-shan* and *Zeren-shan*. The purpose of the share-holding tenure was to keep forestland intact through collective management and to let villagers share returns based on the shares held by each household. Usually a special management team was established and leading managers were elected by the villagers. The land tenure arrangements were meant to create incentives for better management but not changing the land tenure (Zhang, 2001).

After a few years, *Ziliu-shan* and *Zeren-shan* were becoming harder to distinguish as time went on and neither a certificate nor written contract was made in the first place. According to my survey in Jiangxi province, Ziliu-shan and Zeren-shan can ben identified throught notes of use right times on the certificate. In terms to Ziliu-shan, the item of time is noted as :long term. While in terms to Zeren-shan, the item of time will be 30 or 50 years. In many cases, the owners of *Zeren-shan* also took full responsibility for management and enjoyed the full benefits of the forests and land. Therefore, at the later stages of tenure reforms, the two tenures merged into what is known as the household managed land, to be distinguished from the collectively managed land. The actual procedure of separating the land-use rights and collective ownership varied from place to place and from time to time (Shen et al. 2009), and the impact on forest management also differed according to various studies (e.g. Yin and Newman, 1997; Kant and Chiu, 2000; Zhang et al., 2000).

In the course of evolving forest tenure reforms, the authorities intended to expand household use rights to larger tracks of forestland and extend the use right holding duration. For instance, in 1983 it was stipulated that (1) the plots of land allocated to the farmers should be enlarged, (2) the acreage of wasteland under a contract should not be limited and the term of contract could prolong to 30 or 50 years and (3) the contracts can be transferred to other persons. At that time, the consequences of, and even the political ramifications on such 'privatization', were not clear at all. No one in the central government was in a position to provide clear direction for the forestland reforms. The actual implementation was very much dependent on the local interpretation and perception of the grassroots officials, which resulted in huge regional variations.

Transaction of the use rights

After some years of various experimental reforms associated with separating use rights from ownership, and tremendous increases in productivity and food production, several problems with the land reforms emerged, in particular: (1) the risk of efficiency loss, fragmentation of land and diminution of plot size for each household; (2) excessively short duration of contract terms to attract long-term investment in land quality improvement and management, including soil quality, irritation and tree plantation; (3) mismatch of holders of the land-use rights with other changing factors such as labor force, capital, and management capability; and (4) long distance between residential homes and plots where their lands are located.

To address these problems stemming from land-use rights reform, both administrative reallocation and market exchange have been in practice. Although administrative reallocation might promote adjustment to the change in household population, it also creates some uncertainty and may facilitate smaller holdings. The potential corruption and abuse of power by the village leaders are big concerns as well. In contrast, market exchange seems to be getting wider support as market mechanisms and transactions are believed to promote better combinations of the input factors and an efficient scale of land management. Therefore, land reallocation has attracted the attention of various scholars for almost two decades, and the scholars have addressed the issue related to equity and the scale of the holding sizes, productivity, the driving force of the market and administrative reallocation.

Through investigation of the type and frequency of farm land-use reallocation among various provinces in the late 1990s, Krusekopf (2002) found a significant amount of heterogeneity in land reallocation. Yao (2000) claimed that scant land rental arrangement was due to the imperfections in other related factor markets (e.g. labor) in the late 1990s. By testing the timing between off-farm labor activities and the emergence of land rental markets, Kung (2002) argued that local off-farm employment is exogenous to land rental decisions. The increasingly active land rental market since the late 1990s was caused by the emerging off-farm labor market. The gap between off-farm activity and farm income was so large that off-farm opportunities would be taken before dealing with the contracted land. Regarding the two kinds of reallocation, Brandit, Huang, Li and Rozelle (2002) found that administrative reallocations were smaller whereas rental markets were more active. Regarding land fragmentation on agricultural productivity, Tan, Heerink, Kruseman and Qu (2008) found that changes in the number of plots and plot size distribution in Jiangxi Province did not affect total production costs per unit output but caused a shift, instead, in cost composition between labor and modern technologies.

The new wave of tenure reforms that began in the early 2000s was often called a new stage of reform (Shen et al., 2009). Like earlier reforms in the early 1980s, a few counties in four provinces (Fujian, Jiangxi, Zhejiang and Liaoning) were selected to conduct pilot experiments in 2003. The reforms were soon extended to Yunnan, Anhui and Hebei, and to more than 80 counties in other provinces. The experiment was widely believed to be successful (Jia, 2007). It was estimated that more than 33 million ha, or one-fifth of the collective forestland, had been assigned or contracted to individual households (Jia, 2007). The core element of the new stage of reform was to foster a market for land-use rights. The proposal, called Promoting Collective Forest Tenure Right System Reform, was made by the State Council in June 2008 and officially encouraged transaction of the use rights in accordance with legal, voluntary and market rules.

Several kinds of transaction have been used for trading of use rights including re-sale of the use rights, leasing, sub-contracting, exchanging and share in investment. It was reported about 40% was sub-contracting out the use rights, and another 40% was sale of the use rights. The price for the use rights is about 250 yuan/ha per year varying by different trading (SFA, 2013).

Final remarks

Between returning the land to households and using land market versus only contracting out the use rights and establishing the use right market, China have chosen the latter. Significant progress has been made in the past three decades and the economic impacts have been widely reported (e.g. Yin and Xu, 2002). Through changing the bundle of the rights, contracting out the use rights, active transactions of the rights have all expanded the rights of the households into the collective and state-owned forestland.

This cautious forestland reform derived from the constraints of the overall socio-economic frame, but also from numerous lessons learned throughout Chinese history in which the collapse of power and society usually resulted from the failure of food production and land tenure (Wen, 2009). To overcome the uncertainty and unforeseen consequences in connection with potential political controversy of private land ownership, recent reforms only attempt to decentralize and privatize the land-use rights rather than ownership per se. There are still various unanswered questions related to China's reforms.

One consequence of contracting out the collective forestland is an increase in small-scale forestry that may raise many questions. Will the small-scale forestry have a negative impact on the productivity or efficiency? Can the small-scale forestry evolve to large and more efficient scale? Throughout history, farm land has primarily been owned by numerous peasants or managed by individual households through crop-sharing. The longevity of small-scale land ownership has its own economic causes (Cheung, 1969) and political causes (Wen, 2009). Small-scale ownership is associated with markets for capital, labor, and other factors (Yao, 2000). It is argued that the current slow aggregation of the small land holdings might have resulted from less developed labor and capital markets in rural areas as claimed by Yao (2000) and Kung (2002). Most recent policy is suggested to promote cooperative arrangement of the household forest management, and serious research efforts are needed to examine its economic efficiency and social implications.

There are many unanswered questions related to the trading of use rights. Does the limitation of the use rights slow down the land use to higher and better uses? To promote the land to a higher-value use, or enable economies of scale, we might need to explore how to attract investors who has better information regarding land-use value. More studies are needed to know the land-use rights market including the buyers and sellers, transaction price and procedures, and the land-use management changes after transaction, particularly the regional variation and trend of the changes.

Finally, devolution and decentralization of forestland tenure has been the trend in most developing countries (Enters et al., 2000; FAO, 2006), and small-scale family forestry is also growing in many developed countries (Zhang, Liao, Butler, and Schelhas, 2009). Hence, comparative studies on economic efficiency and social implications of forestland reforms across countries, particularly the eastern European countries where the land ownership has been privatized and China will be very useful. It would be also interesting to compare the economic efficiency of the family forestry in China with other countries specifically developed countries. Such comparison will help us to have better ideas of policy to direct forestland reforms in China and other countries.

References

Allen, D.W., and Lueck, D. 1992. Contract choice in modern agriculture: Cropshare versus cash rent. *Journal of Law and Economics* 35 (2): 397–426.

Allen, D.W., and Lueck, D. 1993. Transaction costs and the design of cropshare contracts. *Rand Journal of Economics* 24 (1): 78–100.

Barzel, Y. 1997. *Economic Analysis of Property Rights* (2nd ed.). Cambridge University Press, New York.

Besley, T. 1995. Property rights and investment incentives: Theory and evidence from Ghana. *Journal of Political Economy* 103 (5): 903–937.

Brandit, L., Huang, J., Li, G., and Rozelle, S. 2002. Land rights in rural China: Facts, fictions and issues. *The China Journal* 47: 67–97.

Bromley, D.W. 1989. *Economic Interests and Institutions*. Basil Blackwell, New York.

Chen, R. 1982. *Historical Materials of China's Forests*. China's Forestry Publishing House, Beijing. [In Chinese]

Chen, Y. 1995. Levy taxes on forest area instead of on timber. *Forestry Economics* (No. 6 of 1995): 44–47.

Cheung, S.N.S. 1969. *A Theory of Share Tenancy*. University of Chicago Press, Chicago.

Cheung, S.N.S. 1983. The contractual nature of the firm. *Journal of Law and Economics* 2 (1): 1–22.

Collective Forest Tenure Reform Group. 2012. *Monitoring Report of the Collective Forest Tenure Reform 2011*. China's Forestry Publishing House, Beijing.

Deacon, R. T. 1999. Deforestation and ownership: Evidence from historical accounts and contemporary data. *Land Economics* 75 (3): 341–359.

Enters, T., Durst, P. B., and Victor, M. (Eds.). 2000. *Decentralization and Devolution of Forest Management in Asia and the Pacific*. RECOFTC Report No. 18 and RAP Publication 2000/1. RECOFTC, Bangkok, Thailand.

Feder, G., and Feeny, D. 1991. Land tenure and property rights: Theory and implications for development policy. *World Bank Economic Review* 5 (1): 135–153.

Food and Agriculture Organization of the United Nations (FAO). 2006. *Global Forest Resources Assessment 2005.* FAO Forestry Paper 147. FAO, Rome.

George, H. 1871. *Our Land and Land Policy: Nation and State.* White & Bauer, San Francisco.

George, H. 1894. *Progress and Poverty: An Inquiry into the Cause of Industrial Depressions and of Income of What with Increase of Wealth – The Remedy.* William Reeves, London, UK.

Hardin, G. 1968. The Tragedy of the Commons. Science 162: 1243–1248. Ho, P. 2001. Who owns China's land? Policies, property rights and deliberate institutional ambiguity. *The China Quarterly* 166: 394–421.

Ho, P. (Ed.). 2005. *Developmental Dilemmas: Land Reform and Institutional Change in China.* Routledge, New York.

Hyde, W. F., Belcher, B., and Xu, J. (Ed.). 2003. *China's Forests: Global Lessons from Market Reform.* Johns Hopkins University Press, Baltimore.

Jia, Z. 2007. New important reform in China's rural economy and administration. *Truth Seeking* 2007 (17): 27–29.

Johnson, D. R. 2007. Reflections on the bundle of rights. *Vermont Law Review* 32: 246–272.

Kant, S., and Chiu, M. 2000. Forestry reforms and the local Economy of Linan County, Zhejiang, China. *Forest Policy and Economics* 1: 283–299.

Klooster, D. 2000. Institutional choice, community, and struggle: A case study of forest co-management in Mexico. *World Development* 28 (1): 1–20

Krusekopf, C. C. 2002. Diversity in land-tenure arrangements under the household responsibility system in China. *China Economic Review* 13 (2002): 297–312.

Kung, J. K. 2002. Off-farm labor markets and the emergence of land rental markets. *Journal of Comparative Economics* 30: 395–414.

Liao, X., and Zhang, Y. 2008. An econometric analysis of softwood production in the U.S. South: A comparison of industrial and nonindustrial forest ownerships. *Forest Products Journal* 58 (11): 68–74.

Libercap, G. D. 1989. *Contracting for Property Rights.* Cambridge University Press, New York.

Ministry of Forestry (MoF). 1987. *Forestry Yearbook.* China's Forestry Publishing House. Beijing. [In Chinese]

Newman, D. H., and Wear, D. N. 1993. Production economics of private forestry: A comparison of industrial and nonindustrial forest owners. *American Journal of Agricultural Economics* 75: 674–684.

North, D., and Thomas, R. 1973. *The Rise of the Western World: A New Economic History.* Cambridge University Press, New York.

Osborne, G. P. 1893. *Principles of Economics: The Satisfaction of Human Wants, in so Far as Their Satisfaction Depends on Material Resources.* Robert Clarke and Company, Cincinnati, OH.

Ostrom, E. 1990. *Governing the Commons: The Evolution of Institutions for Collective Action.* Cambridge University Press. Cambridge, UK.

Ostrom, E., Gardner, R., and Walker, J. (Eds.). 1994. *Rules, Games, and Common Pool Resources.* University of Michigan Press, Ann Arbor.

Pagdee, A., Kim, Y., and Daugherty, P. J. 2006. What makes community forest management successful: A meta-study from community forests throughout the world. *Society & Natural Resources* 19 (1): 33–35.

Posner, R. A. 1991. *Economic Analysis of Law* (4th ed.). Little, Brown & Co., Boston.

Pretty, J. 2003. Social capital and the collective management of resources. *Science* 302: 1912–1914.

Pullen, J. 2001. Henry George's land reform: The distinction between private ownership and private possession. *The American Journal of Economics and Sociology* 60 (2): 547–556

Randall, A. 1975. Property rights and social microeconomics. *Natural Resources Journal* 15: 729–747.

Schmid, A. A. 1987. *Property, Power and Public Choice.* Praeger Publishers. New York.

Shahi, C., and Kant, S. 2007. An evolutionary game-theoretic approach to the strategies of community members under joint forest management regime. *Forest Policy and Economics* 9 (7): 763–775.

Shen, Y., Zhang, Y., Xu, X., Zhu, Z., and Jiang, X. 2009. Towards decentralization and privatization of China's collective forestland: A study of 9 villages in 3 provinces. *International Forestry Review* 11 (4): 28–35.

State Forestry Administration (SFA). 2009. *Forest Resources in China – the 7th National Forest Inventory.* State Forestry Administration, Beijing.

State Forestry Administration (SFA). 2013. *Monitoring Report of the Collective Forest Tenure Reform 2012.* China's Forestry Publishing House, Beijing.

Tan, S., Heerink, N., Kruseman, G., and Qu, F. 2008. Do fragmented landholdings have higher production costs? Evidence from rice farmers in Northeastern Jiangxi province, P.R. China. *China Economic Review* 19 (2008): 347–358.

Wang, S., and van Kooten, G. C. 2001. *Forestry and the New Institutional Economics.* Ashgate, Aldershot, England.

Wang, S., van Kooten, G. C., and Wilson, B. 2004. Mosaic of reform: Forest policy in post-1978 China. *Forest Policy and Economics* 6 (1): 71–83.

Wang, S., and Wilson, B. 2007. Pluralism in the economics of sustainable forest management. *Forest Policy and Economics* 9 (7): 743–750.

Wen, T. 2009. *The Problems of 'Agriculture, Farmers and Rural Area' and Institutional Evolution.* China's Economic Publishing House, Beijing. [In Chinese]

White, A., and Martin, A. 2002. *Who Owns the World's Forests? Forest Tenure and Public Forests in Transition.* Forest Trends, Washington, DC.

Wu, X. 1962. *A Compendium of Chinese Popular Customs.* Wenxing Publishing House, Taipei. [In Chinese]

Xu, Z., and McKetta, C. W. 1986. Understanding log and stumpage price in China: A primer for capitalist forest economists. *Canadian Journal of Forest Research* 16: 1123–1127.

Yao, Y. 2000. The development of the land lease market in rural China. *Land Economics* 76 (2): 252–266.

Yin, R., and Newman, D. 1997. Impacts of rural reforms: The case of the Chinese forest sector. *Environment and Development Economics* 2: 289–303.

Yin, R., and Xu, J. 2002. Welfare assessment of China's rural forestry reform in the 1980s. *World Development* 30 (10): 1755–1767.

Zhang, X. 2010. Mortgage loan of forest resource asset reviews. *Issues of Forestry Economics* 30 (4): 318–325.

Zhang, Y. 2001. Institutions in forest management: Special reference to China. In M. Palo, J. Uusivuori and G. Mery (Eds.), *World Forests, Market and Policy.* Kluwer Academic Publishers, Dordrecht, The Netherlands. pp. 353–364.

Zhang, Y. 2002. The impacts of economic reforms on the efficiency of silviculture: A non-parametric approach. *Environment and Development Economics* 7 (1): 107–122.

Zhang, Y., and Kant, S. 2005. Collective forests and forestland: Physical asset rights versus economic rights. In P. Ho (Ed.), *Developmental Dilemmas Land Reform and Institutional Change in China.* Routledge, New York. pp. 283–307.

Zhang, Y., Liao, X., Butler, B. J., and Schelhas, J. 2009. The increasing importance of small-scale forestry: Evidence from family forest ownership patterns in US. *Small-Scale Forestry* 8: 1–14.

Zhang, Y., Uusivuori, J., and Kuuluvainen, J. 2000. Impacts of economic reforms on rural forestry in China. *Forest Policy and Economics* 1 (1): 27–40.

28

ECONOMICS OF CONSERVATION EASEMENTS

Anna Ebers[1] and David Newman

SUNY COLLEGE OF ENVIRONMENTAL SCIENCE AND FORESTRY.
[1]CORRESPONDENCE SHOULD BE ADDRESSED TO ANNA EBERS, DEPARTMENT OF FORESTRY AND
NATURAL RESOURCE RESOURCES MANAGEMENT, 320 BRAY HALL, SUNY COLLEGE OF ENVIRONMENTAL
SCIENCE AND FORESTRY, 1 FORESTRY DRIVE, SYRACUSE NY 13210

Abstract

This chapter examines characteristics of conservation easements (CEs) in general and working forest conservation easements (WFCEs) in the United States in particular. It looks at landowner incentives for instituting easements, outlines the methods for easement appraisal and evaluates ways to measure easement performance. The chapter discusses the challenges for CEs that arise due to market imperfections, such as information asymmetry, limited transparency, governmental involvement, principal–agent problems, intergenerational equity, transaction costs and private amenity rents. The chapter concludes with a review of easement effects on land prices and opportunities for future research.

Keywords

Conservation easements, working forests, perpetuity, landowner incentives, easement appraisal, market inefficiency, easement performance, land prices

Introduction: Conservation easements in the United States and beyond

An estimated 86% of the world's 4 billion ha of forestland is held by the public (Siry, Cubbage, Newman and Izlar, 2010); therefore, governments have the bulk of the responsibility to protect forests for enhanced public benefits. These range from maintaining biodiversity, to providing ecosystem services, to securing the availability of timber products. Privately owned forestland also yields significant public and private benefits, and the adoption of conservation easements (CEs) encourages its conservation. The broader goal of CEs is to address concerns over the extensive use of natural resources, urban sprawl and development pressures that lead to a decline in the availability of open space, species habitat and other environmental amenities (Byers and Ponte, 2005).

In the legal literature, private property rights have been likened to a bundle of sticks, in which each stick is a strictly defined and enforceable right (Barlowe, 1978). CE refers to the transfer of one or several sticks from the property owner to the easement holder. A CE is a binding legal agreement between a property owner and an easement holder that lays out the allowed and forbidden uses of a property and which serves a certain conservation objective (for example, to preserve silvicultural, agricultural, recreational, ecological, open-space, scenic, historical, architectural, archaeological or cultural values as defined by the Uniform Conservation Easement Act of 1981). The easement holders also have the obligation to enforce and monitor the easements they hold in perpetuity.

Working forest conservation easements (WFCEs) are a type of easement applied toward actively managed forests, in order to keep the property in its highest and best use while ruling out the option of development. WFCEs have the potential to be implemented in countries where the ownership of forestland is largely private or communal (where both private and public ownership characteristics are combined). Even though conservation covenants, easements or servitudes are used in Canada, Latin America, Australia, New Zealand and the Pacific (Fishburn, Kareiva, Gaston and Armsworth, 2009), their application toward working forests is still in the early stages. For instance, most Canadian provinces, including Quebec, have adopted legislation that enables the use of CEs for the conservation of private lands (Atkins, Hillyer and Kwasniak, 2004). Yet, the first forest CE in Quebec was only placed in 2012 (The Nature Conservancy of Canada, 2012). In Australia, conservation provisions for private lands vary by state or territory, from nonbinding to binding and permanent (Stephens, 2001), which might be difficult to reconcile with long-term forest management objectives.

The widest implementation of WFCEs is in the United States, where all states have passed statutes enabling WFCEs and 57% of the forestland (172 million ha) is held in private hands (USDA Forest Service, 2012). CEs are the most widespread conservation tools currently used for private lands in the United States, protecting 17 million ha (LTA, 2011). Given the wide use of WFCEs, this chapter will focus on CEs placed on working forests in the United States.

In the following sections, we examine characteristics and trends of easements and land trusts and review governmental programs that support CEs, paying special attention to WFCEs. Next, we scrutinize the economic benefits of CEs and review the incentives for owners to encumber their land with an easement as identified by existing research studies. We proceed to discuss cost-benefit analysis as an approach to measure the performance of CEs and look at the changes of the total economic value of forestland in conjunction with easement restrictions. Then, we outline the appraisal procedure that determines the value of an easement and discuss the limitations of this appraisal mechanism in connection to market inefficiencies. We conclude this chapter by examining the impact of CEs on the valuation of the encumbered and adjacent properties.

Characteristics of CEs

Easements possess a number of characteristics that have made them a popular means for land conservation. As opposed to regulations, CEs are flexible, voluntary, perpetual agreements that utilize scientific knowledge to protect the environmental characteristics of a land parcel while creating incentives for landowners to participate through market-based mechanisms of compensation (Byers and Ponte, 2005). Easement agreements result from a private decision of a landowner to enter a conservation deal with a private or public easement holder that meets qualifying standards (Bick and Haney, 2003). Many easements have a local appeal and are often negotiated at 'the kitchen table' between a landowner and an easement holder. They are also flexible in balancing the private landowner's interests with the conservation objectives of the easement holder and the public.

CEs are designed to protect land in perpetuity and have the symbolic (and often direct) meaning of passing along a current landowner's legacy to his or her heirs (Mahoney, 2002). Only perpetual easements that meet certain conservation purposes qualify for federal tax deductions. Property tax benefits accrue not only to current landowners, but also to their heirs. By reducing the land value, CEs alleviate tax burdens and keep forest and agricultural land in active use, which benefits the land-rich, cash-poor foresters, farmers and ranchers who might otherwise be forced to sell their land to pay taxes (Byers and Ponte, 2005). CEs are generally viewed as cost-effective because purchasing the development rights for the purpose of conservation is less costly than purchasing fee simple rights for a property. Nevertheless, transaction and monitoring costs can considerably reduce the public benefits from CEs (Parker, 2004).

In any case, CEs provide an alternative conservation policy to zoning, eminent domain, direct regulation, leases or term agreements (Bick and Haney, 2003). Depending on the state, most CEs do not require the landowner to provide public access unless the easement has a public recreational or educational use, which is sometimes true for scenic easements (Byers and Ponte, 2005). In sum, CEs respect private property rights and promote land stewardship based on scientific knowledge.

Land trusts as easement holders

Easements can be held by private land trusts (e.g. The Nature Conservancy), Native American tribes (e.g. the Navajo) and governmental agencies (e.g. US Department of Agriculture) (NCED, 2012). Land trusts are local, state, regional or national charitable organizations that serve the public by conserving land either through buying the parcels outright or receiving a donation of CEs (Byers and Ponte, 2005). Figure 28.1 shows that the United States has witnessed a continued growth in the total hectares protected, which includes both an increase in fee simple acquisition and CEs (LTA, 2011). CEs emerged as the preferred method for land conservation,

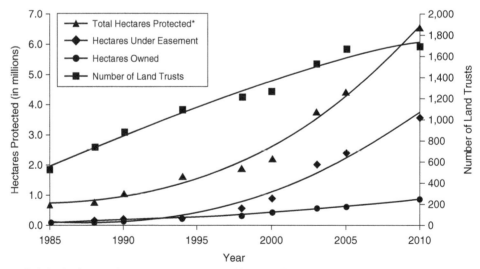

*Includes land conveyed to government and other entities, as well as land conserved by other means e.g. deed restrictions.

Figure 28.1 Trends in private land conservation by state and local land trusts.
Source: LTA (2006, 2011).

overtaking fee simple acquisition in 1994 as measured by total hectares protected, a change explained by Parker (2004) and Yandle (2004).

Another trend is stabilization in the number of land trusts at around 1,700 in 2010, which followed a dramatic increase from 535 in 1985 (see Figure 28.1) (LTA, 2006). Because land trusts are often small, volunteer-run organizations, there have been concerns about their ability to monitor easements for illegal logging, hunting, trapping or trespassing and to enforce the CEs in perpetuity. Recognizing the need to secure long-term endowments, land trusts have sought to increase funding for monitoring, stewardship and legal enforcement, which has resulted in a doubling of such funding from 2005 to 2010 (LTA, 2011). Although numerous land trusts operate on tight budgets, a handful of trusts have managed to consolidate significant funding. A skewed distribution of funds points out the dominance of a small number of larger land trusts, which is manifested in a sevenfold difference between the mean ($460,800) and the median ($62,000) operating budget of state and local land trusts in 2010 (LTA, 2011).

In order to improve conservation efforts among such a great number of organizations, the land trusts formed a coalition called the Land Trust Alliance (LTA) in 1981. The LTA establishes standards for the accreditation of land trusts, offers the trust administrators training and compiles and publishes information and reports on CEs (Gustanski and Squires, 2000; Korngold, 2007).

Governmental support for CEs

Governments can support the CEs by allocating funding for CE programs and by offering tax incentives. In recent years, public funding for CE programs has been rather volatile and tax policy quite uncertain, mainly due to cuts in governmental spending and congressional lags with reauthorization of tax benefits. As of March 2013, Congress reauthorized an enhanced easement incentive for 2013, and for 2012 retroactively. The enhanced easement incentive allows CE donors to deduct up to 50% (or up to 100% for qualified farmers and ranchers) of their adjusted gross income from their federal income tax with a 15-year carryover period (Halperin, 2011). If this enhanced incentive is not reauthorized in the future, the deduction returns to 30% of the donor's adjusted gross income (or up to 50% for qualified farmers and ranchers) with a 5-year carryover period (Durant, 2011).

Some states also offer state income tax incentives and property tax deductions for CE donors. If the easement meets qualification requirements, estate tax incentives allow a deduction of 40% of the value of the encumbered land, not to exceed $500,000 (Byers and Ponte, 2005). Moreover, land trusts enjoy a tax-exempt status as charities if they adhere to their charter and maintain best practices.

In addition to tax incentives, significant public funding has been pledged for governmental agencies and federal or state programs to acquire easements. Since its inception in 1991, the Forest Service's Forest Legacy Program (FLP) has spent at least $585 million to protect forestland with CEs (Trust for Public Land, 2009; LTA, 2012). The FLP matches 75% of funds for a working forest easement, with another 25% of funds raised from private, state or local sources (Trust for Public Land, 2010). By 2012, the FLP program had enrolled 908,592 ha of forestlands, and the program is active in all US states and territories (USDA Forest Service, 2012). The US Farm Bill of 2008 allocated an additional $4 billion for conservation funding, $39 million of which was allocated to the Healthy Forest Reserve Program (HFRP) for 2009–2012 (Food, Conservation and Energy Act, 2008).

In order to ensure that public spending serves its dedicated purpose, the success of the WFCE programs should be evaluated with respect to both financial and biological criteria. Unfortunately, evaluations are often limited to quantitative outcomes of a certain program, such as the number of hectares protected per dollar spent (Lerner, Mackey and Casey, 2007). Biological evaluations of forest ecosystems, such as assessments of forest health (Lindenmayer and Franklin, 2002), are more costly, complicated and less common (Kiesecker et al., 2007). While acknowledging the importance of biological evaluation, this chapter focuses on economic aspects of CEs due to their prominence in conservation decision making.

Working Forest CEs

Protection of working forests is one of the top ten conservation priorities for land trusts in the United States (LTA, 2011). Landowners currently face strong economic incentives to develop their forestland because residential housing often yields much higher returns than timber production. Yet, forests provide public benefits that are currently not incorporated into private decision-making processes. Forests provide biodiversity, and 95% of all federally threatened and endangered species are found on private lands, 19% of them solely on private parcels (Merenlender, Huntsinger, Guthey and Fairfax, 2004). Forests also provide timber and nontimber products and other ecosystem services of value to humans. Many communities economically rely on working forests for jobs and income from recreation and tourism. Forests offer amenity values for on-site uses that forest owners either enjoy themselves or monetize through leases.

Recognizing the importance of forest protection and developmental pressures, WFCEs were designed to conserve forestland by restricting certain development practices on the property and promoting sustainable forest practices and long-term land stewardship. In addition to tax advantages, WFCEs secure a landowner's income stream by allowing active management for timber, recreation and water supply protection (Lind, 2001).

Nonindustrial private forest (NIPF) landowners are one of the most important types of noncorporate forestland owners who take advantage of WFCE tax breaks. NIPF landowners can engage both in nonindustrial timber processing activities (they supply half of the wood in the United States) and provide nontimber goods and services (Newman and Wear, 1993; 16 U.S.C. 2101, 2009). Another prominent group of forestland owners are timberland investment management organizations (TIMOs) and real estate investment trusts (REITs) that primarily invest in timberland on behalf of institutional investors seeking to diversify their portfolios, hedge against inflation and capitalize on the forest asset's low historic volatility (Healey, Corriero and Rozenov, 2005).

If private landowners choose to encumber their forestland with a WFCE, they are often required to add a forest management plan (FMP) to the easement. The FMP is usually drafted by a forestry professional and provides a certain degree of flexibility for the landowner with respect to forest management practices, while also including provisions on forbidden activities (Byers and Ponte, 2005). The FMP reduces the need for detailed prescriptive language in the WFCE and provides clear guidance for monitoring. Unlike an easement, an FMP can be easily updated to accommodate new information, advances in forest management science and changed forest conditions (Lind, 2001).

In addition to an FMP, the landowners are generally encouraged to adhere to best management practices (BMPs) in forestry designed to limit environmental loss and protect the landscapes that could be threatened by intensive management operations, such as timber harvesting, road construction or pesticide application (Lind, 2001; Siry et al., 2010). BMPs vary by state and are periodically updated. Adherence to FMPs and BMPs can alleviate the risks of fragmentation and parcelization of the forestland, which could result in habitat loss, as well as the increased risk of flood and erosion (Mortimer, Richardson, Huff and Haney, 2007).

WFCEs have a number of common elements with forest certification programs like the Forest Stewardship Council (FSC) and the Sustainable Forestry Initiative (SFI). Both forest certification programs and CEs establish compliance standards and require monitoring, input from professional foresters, baseline documentation and an FMP, while focusing on long-term land stewardship and conservation (Newsom, 2004). Integrating WFCEs and forest certification into a single package could improve consistency of management and biodiversity practices, ease monitoring, decrease cost for planning and implementation and boost public credibility (Newsom, 2004; Hagan, Irland and Whitman, 2005).

Along with the advantages of WFCEs, the parties in a conservation deal often face a number of challenges. Some widely recognized challenges are deficiencies in baseline documentation, inadequate record keeping and the need for professionally developed management plans (Mortimer et al., 2007). Because WFCEs tend to increase in sophistication and complexity as they provide for multiple conservation purposes, landowners may put themselves at risk of misunderstanding the easement agreement and creating potential future liability. Many need to obtain professional forestry advice, which is available, for example, through such cost-share programs as the Forest Land Enhancement Program or the Wildlife Habitat Incentives Program (Oehler, 2003).

Placing WFCEs on large tracts of forestland increases the cost of negotiation and monitoring, creates coordination problems between easement parties and often leads to controversy about the best conservation and forest management strategies (Lind, 2001; Newsom, 2004). To put matters in perspective, the Plum Creek CE in Maine protects 146,900 ha and links timberland parcels into 809,370 ha of protected forest (Turkel, 2012).

Other difficulties with WFCEs could stem from reconciling the economic factors of forest productivity with such things as conservation goals, WFCE monitoring and enforcement, sharing of easement appraisal information and managing for public access (Lind, 2001). Finally, when WFCEs considerably reduce land values, communities may witness an erosion of their property tax base (Newman, Brooks and Dangerfield, 2000).

Incentives for CEs

Given the advantages and challenges of CEs, it is vital to understand the motivations of the landowners to engage in conservation (Merenlender et al., 2004). The studies have found that the perceptions, attitudes and subjective norms of the landowners were better predictors for a conservation agreement than the attributes of the land (Brain, 2008; Jacobson, 2002). Landowners were found to be motivated by the following factors:

- emotional attachment to the land and community or a 'sense of place' (Elconin and Luzadis, 1998; Farmer, Knapp, Meretsky, Chancellor and Fischer, 2011; Keske, Hoag, Bastian and Cross, 2011),
- legacy and preservation of traditional lifestyle (Miller, Bastian, McLeod, Keske and Hoag, 2011; Oliver, 2011),
- strong environmental ethic (LeVert, Stevens and Kittredge, 2009),
- desire to conserve species habitat and open space (Brain, 2008; Ernst and Wallace, 2008; Miller et al., 2011),
- preservation of economic resources such as protection of traditional output of land and cultural heritage of the community (Ernst and Wallace, 2008; Keske et al., 2011; Miller et al., 2011),
- aversion to fragmentation and development (Jacobson, 2002; Kabii and Horwitz, 2006).

The aspect of perpetuity raises concerns among some landowners (Miller et al., 2011; Oliver, 2011), while others fear term easements would not provide appropriate protection to their land

(Duke and Lynch, 2007). Increased enrollment is more likely when landowners trust future easement holders (Brain, 2008; Keske et al., 2011; Oliver, 2011) and receive information about CEs from them (Kaetzel, Hodges, Houston and Fly, 2009). Property rights protection under easements remains a great concern, and as a result, landowners tend to be skeptical of providing public access to their property and fear the loss of control over it (Miller et al., 2011; Oliver, 2011).

Demographically, a typical landowner who engages in conservation is an older, educated, affluent male (Farmer et al., 2011; Keske et al., 2011). Female landowners are more likely to participate in easement programs, but they own less property (Kaetzel et al., 2009). Increased parcel size, reliance on timber production, proximity to other eased lands, as well as concern about heirs' willingness to engage in traditional land use tend to be positively related to participation in easement mechanisms (Lynch and Lovell, 2001; Kaetzel et al., 2009; Oliver, 2011). Property that is further from metropolitan areas and faces less development pressure appears to have a greater likelihood to be placed in an easement (Oliver, 2011). Studies offer mixed evidence whether the duration of property ownership affects easement participation (Joshi and Arano, 2009; Kaetzel et al., 2009).

Surprisingly financial considerations do not rank as a top reason for conservation (Elconin and Luzadis, 1998; Ernst and Wallace, 2008; Farmer et al., 2011). Yet, financial incentives cannot be disregarded as they are often perceived as a means to important ends, and not the end goal in themselves; they are crucial for the placement of easements that otherwise would not be attainable (Ernst and Wallace, 2008; Farmer et al., 2011). For example, 30% of landowners surveyed by Farmer (2009) would not place an easement on property without monetary compensation, while about one-half of landowners in Massachusetts and one-third in Vermont would conserve land for $283 per ha (LeVert et al., 2009).

Methods to measure the performance of CEs

The performance of WFCEs is most often measured through the use of cost-benefit analysis (CBA). Despite the popularity of CBA, the estimation of both costs and benefits can be challenging due to aggregation problems, transparency issues and double-counting of the public investment. The benefit-cost ratio can vary, even on a relatively small landscape, depending on the type of land protected (Naidoo and Ricketts, 2006). Forestland was estimated to yield $400 of mean annual benefits per ha in ecosystem services, whereas the benefits from wetland protection were valued 40 times higher (Sauer, 2002).

Cost categories that are difficult to measure are, for example, the opportunity costs of conservation and alternative management strategies, potential liabilities, foregone tax revenues and the administrative costs incurred by all parties in perpetuity. The projection of public and private benefits from CEs requires the monetization of a variety of ecosystem services and nontraded benefits like the preservation of natural heritage or scenic views. On the benefit side, forests provide species habitat while improving air and water quality and reducing ultraviolet levels, air temperatures and noise (Nowak, Wang and Endreny, 2007). There are economic benefits of working forests that are linked to job creation and tourism, as well as indirect benefits of conservation, such as decreased cost of drinking water treatment (Ernst, Gullick and Nixon, 2007).

Although the valuation of ecosystem services has received substantial attention in the literature, the estimation of the effectiveness of WFCEs is limited to a few studies. Sauer (2002) and Sargent-Michaud (2009) estimated that the benefits from CEs outweigh the cost by factors of 6 and 7, respectively. These findings establish a reference point for conservation projects, but also raise questions. Economic theory suggests that high benefit-cost ratios would lead to a major expansion of conservation activity, until the benefits equal the costs at the margin. In addition, the policymakers would be expected to be proactive and generous with CEs to secure considerable public benefits. If these predictions do not materialize, it is likely that the benefits measured by CBA have little connection to current market realities.

Another CBA technique to estimate the effectiveness of WFCEs is comparison of the total economic value (TEV) of the forestland with and without a WFCE. TEV is an analytical method that incorporates direct-use value, indirect use value, option value and existence value of the forestland (Pearce, 2001). Prices for timber and nontimber (e.g. carbon storage, recreation, edibles) products determine the direct-use value of forestland, which may be reduced by easement restrictions (Nunamaker, Rodrigues and Nakamura, 2007). At the same time, WFCEs could increase the indirect use value linked to the ecosystem services of the property. Concurrently, the option value of restricted land may increase if investors are willing to pay for potential future uses of preserved forests (e.g. for the development of forest-derived vaccines). WFCEs secure the existence value of environmental assets on protected forestland for future generations.

TEV estimation has been applied to forestland in Alaska, Mexico and the Mediterranean, but not in conjunction with WFCEs (Adger, 1994; Merlo and Croitoru, 2005; Phillips, Silverman and Gore, 2008). Notwithstanding the challenges of estimation, TEV analysis could provide a comprehensive estimation of the effectiveness of WFCEs.

Appraisal methods and market inefficiencies

Development restrictions often reduce property value, which in turn determines easement value. The appraiser must calculate the difference between two fair market values (FMVs) of a property: the FMV before and after the easement has been placed (Boykin, 2000). The FMV calculation must consider current uses of the property, limitations due to zoning, historic preservation or conservation laws, other restrictions and the likelihood of imminent development (Byrne and Minck, 2000). The courts have consistently applied objective, market-based, highest and best use principles to property valuation (Byrne and Minck, 2000). Other easement valuation methods include identifying comparable easement sales or finding sales transactions where zoning restrictions on the land parcel are similar to the restrictions of the easement (Boykin, 2000).

After an appraisal is completed, a landowner can usually choose among three forms of payment for the easement. The first option is to donate a qualifying easement to the holder and enjoy a tax deduction for the charitable contribution. Alternatively, the landowner can sell the easement and get monetary compensation for the full easement value. Another popular payment option is called a 'bargain sale' in which a landowner receives cash for part of the value of the easement and donates the other part to receive tax benefits (Byers and Ponte, 2005).

Appraisals of CEs determine the market value of the relinquished rights, yet they are carried out in a market plagued with inefficiencies. The first challenge is that private markets tend to undersupply public and quasi-public goods like conservation. Then, agricultural and timber markets protect the property against the pressures of the real estate and international agricultural markets, obstructing in many cases the highest and best use of the property for the sake of conservation. There is also enormous potential for information asymmetry, as the landowners have more information about the characteristics of their property than future easement holders. The public that is supposedly receiving conservation benefits is often oblivious to the value of what is being conserved.

Transparency issues with CEs arise due to limited availability of high-quality data on easement transactions (Merenlender et al., 2004; Korngold, 2007; Lerner et al., 2007; Anderson and Weinhold, 2008). Nondisclosure of conservation details creates so-called (invisible forests), which are created when neither researchers nor public planners have information about conserved parcels (Olmsted, 2011). Lack of information about rejected easement offers prevents conservation agencies from learning from their mistakes and addressing common concerns of landowners (Shultz, 2005). At the moment, the appraisal process obscures how characteristics of a property such as location, future carbon credits or forest management practices may influence its easement value (Lind, 2001).

Several databases like the Conservation Almanac and the National Conservation Easement Database (NCED) have been launched recently to consolidate easement data. They provide information on conservation financing, public conservation policies, acreage conserved and the primary conservation value of properties. The NCED, which was initiated in 2011, holds records on over 800 WFCEs covering 690,706 ha. On the one hand, such databases do not include all transactions and can raise fears of privacy intrusion with more accurate data. On the other hand, disclosure could improve land-use planning and record-keeping, decrease the possibility of 'orphan easements' when land trusts dissolve and provide researchers and scientists with accurate data (Olmsted, 2011).

Even though CEs are recognized as a private conservation tool, public funding plays an important role in these markets. Tax deductions provide an indirect subsidy for landowners, while a number of federal and state programs also subsidize easement acquisition (Boyd, Caballero and Simpson, 1999). Some states, like Massachusetts, require public consultation (Bray, 2010), which is intended to ensure that the public benefits from an easement justify the cost (Korngold, 2007).

The next CE challenge is the principal-agent problem that arises when the public interest is represented by private land trusts. Concerns have been raised that land trusts might have an incentive to act as governmental agents to conduct a 'prearranged flip' on the conserved land (Yandle, 2004; Gattuso, 2008). Being able to act quickly and without excessive public scrutiny, land trusts can acquire a desired property below market value and subsequently resell it to the government for a profit. Another principal-agent problem might arise when easement holders act in accordance with the preferences of their board members and employees, which are not necessarily identical to the preferences of the public. For example, a land trust (or a governmental agency) could be motivated to conserve larger parcels of low-cost, low-quality land if their performance is judged on the basis of hectares protected per dollar of funding.

A number of challenges are connected to the perpetual nature of an easement (Mahoney, 2002; McLaughlin, 2005). A primary rationale for perpetual conservation is to provide a 'buffer' of undeveloped land for the benefit of future generations. It is questionable whether tying up valuable land rights in a perpetual conservation agreement is in line with posterity's best interests. Both land characteristics and land-use preferences are likely to change with time, and a perpetual easement might clash with future economic, social and/or environmental goals. The doctrine of changed conditions and other arguments can be put forward to legally challenge conservation restrictions, but considerable transaction costs will be borne by all parties (Mahoney, 2002).

Another market inefficiency that is connected to easement appraisal is the private amenity rent (PAR). The PAR is the satisfaction that landowners obtain from their current lifestyle and the amenities of their conserved property, while acting in the public interest (Hoag, Bastian, Keske-Handley, McLeod and Marshall, 2005). If PARs are high, compensating a landowner for an easement with its full market price would result in overpaying for conservation. An overpayment also occurs when the landowner is compensated for giving up development rights that he or she did not value in the first place. Theoretically, conservation-minded landowners could be offered lower compensation than landowners motivated by other values (Gustanski and Wright, 2011).

Impact of CEs on land prices

Although there are many potential influences of easements on timber, housing, agricultural and future carbon markets, this section discusses the impact of CEs on land prices. Generally, the price of an unrestricted parcel is expected to be higher than the price of a restricted parcel. The potential reduction in land value can be better understood by decomposing the value of the land into two components: a current use component (e.g. timber harvesting) and a future development component (Plantinga, Lubowski and Stavins, 2002). In the case of rural forestland, the drop in land price would be predictably low as parcels have limited potential for development, and the WFCEs would allow current land uses, like management for timber. In regions with high development pressure, the prices for restricted parcels might drop more dramatically.

Current studies, however, have yielded mixed results. The effect of CEs on land prices varied among studies with respect to sign, magnitude and statistical significance depending on the estimation technique (Nickerson and Lynch, 2001; Taff, 2004; Anderson and Weinhold, 2008). Anderson and Weinhold (2008) did not find a significant effect analyzing the whole sample; yet, by analyzing the parcels with high development potential, the study found a strong, statistically significant, negative effect on land prices. Thus, it is reasonable to assume that many properties placed under an easement have little or no current development value.

Additional studies suggest that a spillover effect clarifies the discrepancy between the expected reduction in land prices and the failure of studies to empirically determine it. Easements can considerably increase the value of adjacent properties, because they conserve environmental amenities of the area in perpetuity (Geoghegan, 2002; Chamblee, Colwell, Dehring and Depken, 2011). That area would likely witness an increase in property values, which in turn might put upward pressure on the price of the encumbered property. As a result, conservation does not necessarily reduce the value of encumbered land by the appraised value of the easement (Irwin and Bockstael, 2001). If this is the case, then private or public easement holders are overpaying for CEs.

Conclusion

CEs have received a large amount of attention in the current academic research, but it is mostly limited to the United States. The most frequently addressed topics included perpetuity concerns (Mahoney, 2002; Thompson, 2004), motivations of landowners to enter into an easement (e.g. Farmer et al., 2011; Miller et al., 2011) and tax implications (McLaughlin, 2004; Parker, 2004). Judging by the sheer number of research papers, open space CEs have been more rigorously researched than WFCEs. In addition, a large number of papers have focused on the legal analysis of CEs, rather than on economic or ecological analyses. This can be explained by the lack of primary data about transactions necessary for economic or ecological research on CEs. Online databases with information on CEs (like the NCED) might provide the input for a new wave of publications probing under-researched aspects.

There are a number of issues that have attracted much less debate than they probably merit. Out of more than 200 research articles located for this chapter, only two have looked into economic performance of CEs by means of CBA (Sauer, 2002; Sargent-Michaud, 2009). It would be helpful to investigate whether the public dollars spent on CE yield sufficient public benefit, which parties appropriate the most benefits from conservation spending and what the reasons were behind failed conservation transactions (Farmer et al., 2011).

Many research papers have proposed improvements to the current structure of CEs. These include proposals to make easements nonperpetual (Mahoney, 2002; Owley, 2012), to introduce tradable easements (Weeks, 2011), to ban prearranged acquisitions of CEs (Gattuso, 2008), to change public funding mechanisms for land trusts (Parker, 2004) and to create dynamic easements in line with 'smart conservation' (Rissman, 2011). As CEs become more entrenched, there will be increasing concerns about their limitations. Yet, if future research is able to clearly demonstrate their public benefits, CEs will be in a secure position as a preferred conservation policy.

Acknowledgments

We would like to thank Bruce Yandle, Steven Bick, Richard Smardon and Daniel Claussen for their thoughtful comments on the paper. We are also grateful to Travis Fisher, Mark Kanchukov and Patrick O'Reilley for their insights and to Katie Chang of the LTA for information on land trusts used in this chapter.

References

Adger, W. N. (1994). *Towards Estimating Total Economic Value of Forests in Mexico*, Centre for Social and Economic Research on the Global Environment (CSERGE) Working Paper, CSERGE. Retrieved from http://millenniumindicators.un.org/unsd/envaccounting/ceea/archive/Forest/TEV_Mexican_Forest.PDF

Anderson, K. and Weinhold, D. (2008). 'Valuing future development rights: The costs of conservation easements', *Ecological Economics*, vol. 68, no. 1–2, pp. 437–446.

Atkins, J., Hillyer, A. and Kwasniak, A. (2004). *Conservation Easements, Covenants and Servitudes in Canada: A Legal Review*, Report No. 04-1, North American Wetlands Conservation Council, Ottawa, Ontario.

Barlowe, R. (1978). *Land Resource Economics: The Economics of Real Estate*, Prentice-Hall, Englewood Cliffs, NJ.

Bick, S. and Haney, H. (2003). *Landowner's Guide to Conservation Easements*, Kendall/Hunt Publishing Company, Dubuque, IA.

Boyd, J., Caballero, K. and Simpson, R. (1999). *The Law and Economics of Habitat Conservation: Lessons from an Analysis of Easement Acquisitions*, Discussion Paper 99–32, Washington, DC. Retrieved from http://papers.ssrn.com/soL3/papers.cfm?abstract_id=166968

Boykin, J. H. (2000). 'Valuing scenic land conservation easements', *Appraisal Journal*, vol. 68, no. 4, pp. 420–426.

Brain, R. (2008). 'Predicting engagement in a conservation easement agreement', PhD thesis, University of Florida, Gainesville, FL.

Bray, Z. (2010). 'Reconciling development and natural beauty: The promise and dilemma of conservation easements', *Harvard Environmental Law Review*, vol. 34, pp. 119–593.

Byers, E., and Ponte, K. M. (2005). *The Conservation Easement Handbook*, Land Trust Alliance, Washington, DC.

Byrne, C. L. and Minck, M. (2000). 'Understanding the evolution of conservation easement appraisal through case law', *Appraisal Journal*, vol. 68, no. 4, pp. 411–419.

Chamblee, J., Colwell, P. F., Dehring, C. A. and Depken, C. A. (2011). 'The effect of conservation activity on surrounding land prices', *Land Economics*, vol. 87, no. 3, pp. 453–472.

Duke, J. and Lynch, L. (2007). 'Gauging support for innovative farmland preservation techniques', *Policy Sciences*, vol. 40, no. 2, pp. 123–155.

Durant, M. (2011). 'The changing landscape of conservation easements', *The Tax Adviser*, vol. 42, no. 3, pp. 166–172, 174–177.

Elconin, P. and Luzadis, V. (1998). 'Landowner satisfaction with conservation easements', *Wild Earth*, vol. 8, pp. 49–51.

Ernst, C., Gullick, R. and Nixon, K. (2007). 'Protecting the source: Conserving forests to protect water', in C. de Brun (Ed.), *The Economic Benefits of Land Conservation*, Trust for Public Land, Denver, CO, pp. 24–28. Retrieved from http://cloud.tpl.org/pubs/benefits_econbenefits_landconserve.pdf

Ernst, T. and Wallace, G. N. (2008). 'Characteristics, motivations, and management actions of landowners engaged in private land conservation in Larimer County Colorado', *Natural Areas Journal*, vol. 28, no. 2, pp. 109–120.

Farmer, J. R. (2009). 'Motivations for the adoption of a conservation easement: A midwestern perspective', PhD thesis, Indiana University, Bloomington, IN.

Farmer, J. R., Knapp, D., Meretsky, V. J., Chancellor, C. and Fischer, B. C. (2011). 'Motivations influencing the adoption of conservation easements', *Conservation Biology*, vol. 25, no. 4, pp. 827–834.

Fishburn, I. S., Kareiva, P., Gaston, K. J. and Armsworth, P. R. (2009). 'The growth of easements as a conservation tool', *PLoS ONE*, vol. 4, no. 3. doi:10.1371/journal.pone.0004996

Food, Conservation and Energy Act. (2008). *U.S. Statutes at Large, Public Law 110–234*. Retrieved from www. usda.gov/documents/Bill_6124.pdf

Gattuso, D. J. (2008). 'Conservation easements: The good, the bad, and the ugly', *National Policy Analysis*, The National Center for Public Policy Research, Washington, DC. Retrieved from www.nationalcenter.org/NPA569.html

Geoghegan, J. (2002). 'The value of open spaces in residential land use', *Land Use Policy*, vol. 19, no. 1, pp. 91–98.

Gustanski, J. A. and Squires, R. H. (2000). *Protecting the Land: Conservation Easements Past, Present, and Future*, Island Press, Washington, DC.

Gustanski, J. A. and Wright, J. B. (2011). 'Exploring net benefit maximization: Conservation easements and the public-private interface', *Law and Contemporary Problems*, vol. 74, no. 4, pp. 109–143.

Hagan, J., Irland, L. and Whitman, A. (2005). *Changing Timberland Ownership in the Northern Forest and Implications for Biodiversity*, MCCS-FCP-2005-1, Manomet Center for Conservation Sciences, Brunswick, MN. Retrieved from http://standards.nsf.org/apps/group_public/download.php/10102/Cassie%20Phillips%202005-06%20Manomet%20ForestOwnerChangeReport-011006[1].pdf

Halperin, D. (2011). 'Incentives for conservation easements: The charitable deduction or a better way', *Law and Contemporary Problems*, vol. 74, no. 4, pp. 29–50.

Healey, T., Corriero, T. and Rozenov, R. (2005). 'Timber as an institutional investment', *The Journal of Alternative Investments*, vol. 8, no. 3, pp. 60–74.

Hoag, D. L., Bastian, C., Keske-Handley, C., McLeod, D. and Marshall, A. (2005). 'Evolving conservation easement markets in the west', *Western Economics Forum*, vol. 4, no. 1, pp. 7–14. Retrieved from www.waeaonline.uwagec.org/WEForum/WEF-Vol.4-No.1-Spring2005.pdf#page=9

Irwin, E. G. and Bockstael, N. E. (2001). 'The problem of identifying land use spillovers: Measuring the effects of open space on residential property values', *American Journal of Agricultural Economics*, vol. 83, no. 3, pp. 698–704.

Jacobson, M. G. (2002). 'Factors affecting private forest landowner interest in ecosystem management: Linking spatial and survey data', *Environmental Management*, vol. 30, no. 4, pp. 577–583.

Joshi, S. and Arano, K. G. (2009). 'Determinants of private forest management decisions: A study on West Virginia NIPF landowners', *Forest Policy and Economics*, vol. 11, no. 2, pp. 118–125.

Kabii, T. and Horwitz, P. (2006). 'A review of landholder motivations and determinants for participation in conservation covenanting programmes', *Environmental Conservation*, vol. 33, no. 1, pp. 11–20.

Kaetzel, B. R., Hodges, D. G., Houston, D. and Fly, J. M. (2009). 'Predicting the probability of landowner participation in conservation assistance programs: A case study of the Northern Cumberland Plateau of Tennessee', *Southern Journal of Applied Forestry*, vol. 33, no. 1, pp. 5–8.

Keske, C. M., Hoag, D. L., Bastian, C. and Cross, J. E. (2011). 'Adoption of conservation easements among agricultural landowners in Colorado and Wyoming: The role of economic dependence and sense of place', *Landscape and Urban Planning*, vol. 101, no. 1, pp. 75–83.

Kiesecker, J. M., Comendant, T., Grandmason, T., Gray, E., Hall, C., Hilsenbeck, R., . . . Rissman, A. (2007). 'Conservation easements in context: A quantitative analysis of their use by the Nature Conservancy', *Frontiers in Ecology and the Environment*, vol. 5, no. 3, pp. 125–130.

Korngold, G. (2007). 'Solving the contentious issues of private conservation easements: Promoting flexibility for the future and engaging the public land use process', Case Legal Studies Research Paper, No. 07-24; NYLS Legal Studies Research Paper, No. 08/09-3. Retrieved from http://papers.ssrn.com/sol3/papers.cfm?abstract_id=1004363

Land Trust Alliance (LTA). (2006). *2005 National Land Trust Census Report*, LTA, Washington, DC. Retrieved from www.landtrustalliance.org/land-trusts/land-trust-census/2005-national-land-trust-census/2005-report.pdf

Land Trust Alliance (LTA). (2011). *2010 National Land Trust Census Report: A Look at Voluntary Land Conservation in America*, LTA, Washington, DC. Retrieved from www.landtrustalliance.org/land-trusts/land-trust-census/national-land-trust-census-2010/2010-final-report

Land Trust Alliance (LTA). (2012). *Previous Funding Updates and Historical Chart*, LTA, Washington, DC. Retrieved from www.landtrustalliance.org/policy/public-funding/update-archive#chart

Lerner, J., Mackey, J. and Casey, F. (2007). 'What's in Noah's wallet? Land conservation spending in the United States', *BioScience*, vol. 57, no. 5, pp. 419–423.

LeVert, M., Stevens, T. and Kittredge, D. (2009). 'Willingness-to-sell conservation easements: A case study', *Journal of Forest Economics*, vol. 15, no. 4, pp. 261–275.

Lind, B. (2001). *Working Forest Conservation Easements: A Process Guide for Land Trusts, Landowners and Public Agencies*, Land Trust Alliance, Washington, DC.

Lindenmayer, D. B. and Franklin, J. F. (2002). 'Conserving forest biodiversity: A comprehensive multi-scaled approach', *Australian Geographical Studies*, vol. 41, no. 2, pp. 210–220.

Lynch, L. and Lovell, S. (2001). 'Factors influencing participation in agricultural land preservation programs', Working Paper 01-05, University of Maryland, College Park, MD.

Mahoney, J. D. (2002). 'Perpetual restrictions on land and the problem of the future', *Virginia Law Review*, vol. 88, no. 4, pp. 739–787.

McLaughlin, N. A. (2004). 'Increasing the tax incentives for conservation easement donations – A responsible approach', *Ecology Law Quarterly*, vol. 31, no. 1, pp. 1–115.

McLaughlin, N. A. (2005). 'Rethinking the perpetual nature of conservation easements', *Harvard Environmental Law Review*, vol. 29, pp. 421–521.

Merenlender, A. M., Huntsinger, L., Guthey, G. and Fairfax, S. K. (2004). 'Land trusts and conservation easements: Who is conserving what for whom?' *Conservation Biology*, vol. 18, no. 1, pp. 65–76.

Merlo, M. and Croitoru, L. (Eds.). (2005). *Valuing Mediterranean Forests: Towards Total Economic Value*, CABI Publishing, Wallingford, UK.

Miller, A. D., Bastian, C. T., McLeod, D. M., Keske, C. M. and Hoag, D. L. (2011). 'Factors impacting agricultural landowners' willingness to enter into conservation easements: A case study', *Society & Natural Resources: An International Journal*, vol. 24, no. 1, pp. 65–74.

Mortimer, M., Richardson, J., Huff, J. and Haney, H. (2007). 'A survey of forestland conservation easements in the United States: Implications for forestland owners and managers', *Small-Scale Forestry*, vol. 6, no. 1, pp. 35–47.

Naidoo, R. and Ricketts, T. H. (2006). 'Mapping the economic costs and benefits of conservation', *PLOS Biol*, vol. 4, no. 11, pp. 1–12. Retrieved from www.plosbiology.org/article/info:doi/10.1371/journal.pbio.0040360

National Conference of Commissioners on Uniform State Laws. (1981). *Uniform Conservation Easement Act*, US Code, Title 26, Section 170 (h). Retrieved from www.cals.ncsu.edu/wq/lpn/PDFDocuments/uniform.pdf

National Conservation Easement Database. (2012). *Open Space–Forest as Easement Purpose Query*, US Endowment for Forestry and Communities. Retrieved from http://nced.conservationregistry.org

The Nature Conservancy of Canada. (2012.) 'Nature Conservancy of Canada establishes the first forest conservation easement in Quebec', The Nature Conservancy. Retrieved from www.natureconservancy.ca/en/where-we-work/quebec/news/nature-conservancy-of-canada.html

Newman, D. H., Brooks, T. A. and Dangerfield, C. W. (2000). 'Conservation use valuation and land protection in Georgia', *Forest Policy and Economics*, vol. 1, no. 3, pp. 257–266.

Newman, D. H. and Wear, D. N. (1993). 'Production economics of private forestry: A comparison of industrial and nonindustrial forest owners', *American Journal of Agricultural Economics*, vol. 75, no. 3, pp. 674–684.

Newsom, D. (2004). *Forest Certification and Working Forest Conservation Easements: Common Elements and First Thoughts on a Combined System*, Rainforest Alliance, Sustainable Forestry Division. Retrieved from www.rainforest-alliance.org/forestry/documents/easementpaper-nov04.pdf

Nickerson, C. J. and Lynch, L. (2001). 'The effect of farmland preservation programs on farmland prices', *American Journal of Agricultural Economics*, vol. 83, no. 2, pp. 341–351

Nowak, D., Wang, J. and Endreny, T. (2007). 'Environmental and economic benefits of preserving forests within urban areas: Air and water quality', in C. de Brun (Ed.), *The Economic Benefits of Land Conservation*, Trust for Public Land, Denver, CO. Retrieved from http://cloud.tpl.org/pubs/benefits_econbenefits_landconserve.pdf

Nunamaker, C., Rodrigues, K. and Nakamura, G. (2007). *Economic Considerations in Forest Stewardship*, Report no. 8251, Forest Stewardship Series 21, University of California, Davis, CA. Retrieved from http://anrcatalog.ucdavis.edu/pdf/8251.pdf

Oehler, J. D. (2003). 'State efforts to promote early-successional habitats on public and private lands in the northeastern United States', *Forest Ecology and Management*, vol. 185, no. 1–2, pp. 169–177.

Oliver, M. D. (2011). 'An evaluation of West Virginia's non-industrial private forest landowner participation in conservation easements', MS thesis, West Virginia University, Morgantown, WV.

Olmsted, J. L. (2011). 'The invisible forest: Conservation easement databases and the end of the clandestine conservation of natural lands', *Law and Contemporary Problems*, vol. 74, no. 4, pp. 51–82.

Owley, J. (2012). 'Neoliberal land conservation and social justice', *International Union for Conservation of Nature Academy of Environmental Law e-Journal*, no. 1, pp. 6–17. Retrieved from http://papers.ssrn.com/sol3/papers.cfm?abstract_id=2040827

Parker, D. P. (2004). 'Land trusts and the choice to conserve land with full ownership or conservation easements', *Natural Resources Journal*, vol. 44, no. 2, pp. 483–518.

Pearce, D. W. (2001). 'The economic value of forest ecosystems', *Ecosystem Health*, vol. 7, no. 4, pp. 284–296.

Phillips, S., Silverman, R. and Gore, A. (2008). *Greater Than Zero: Toward the Total Economic Value of Alaska's National Forest Wildlands*, Wilderness Society, Washington, DC.

Plantinga, A. J., Lubowski, R. N. and Stavins, R. N. (2002). 'The effects of potential land development on agricultural land prices', *Journal of Urban Economics*, vol. 52, no. 3, pp. 561–581.

Rissman, A. R. (2011). 'Conservation easements: New perspectives in an evolving world: Evaluating conservation effectiveness and adaptation in dynamic landscapes', *Law & Contemporary Problems*, vol. 74, pp. 145–279.

Sargent-Michaud, J. (2009). *A Return on Investment: The Economic Value of Colorado's Conservation Easements*, Trust for Public Land, Denver, CO.

Sauer, A. (2002). *The Value of Conservation Easements: The Importance of Protecting Nature and Open Space*, West Hill Foundation for Nature, World Resources Institute, Wilson, WY.

Shultz, S. D. (2005). 'Evaluating the acceptance of wetland easement conservation offers', *Applied Economic Perspectives and Policy*, vol. 27, no. 2, pp. 259–272.

Siry, J., Cubbage, F., Newman, D. H. and Izlar, R. (2010). 'Forest ownership and management outcomes in the U.S., in global context', *International Forestry Review*, vol. 12, no. 1, pp. 38–48.

Stephens, S. (2001). 'Visions and viability: How achievable is landscape conservation in Australia?' *Ecological Management & Restoration*, vol. 2, no. 3, pp. 189–195.

Taff, S. J. (2004). *Evidence of a Market Effect from Conservation Easements*, Staff Paper 04-9, University of Minnesota, Minneapolis, MN. Retrieved from http://purl.umn.edu/13611

Thompson, B. H. (2004). 'The trouble with time: Influencing the conservation choices of future generations', *Natural Resources Journal*, vol. 44, no. 2, pp. 601–620.

Trust for Public Land. (2009). 'TPL praises proposed forest legacy funding increase', Trust for Public Land, Denver, CO. Retrieved from www.tpl.org/news/press-releases/tpl-praises-proposed-forest-legacy.html

Trust for Public Land. (2010). 'Conservation almanac: Federal, state, local & private lands', Trust for Public Land, Denver, CO. Retrieved from www.conservationalmanac.org/secure/federal.shtml

Turkel, T. (2012). 'Massive Plum Creek easement touches off celebration', *Morning Sentinel*, May 16. Retrieved from www.onlinesentinel.com/news/conservationists-developerscelebrate-plum-creek-easement_2012-05-15.html

US Code. (2009). *Cooperative Forestry Assistance, Title 16, Section 2101*. Retrieved from www.gpo.gov/fdsys/pkg/USCODE-2009-title16/html/USCODE-2009-title16-chap41-sec2101.htm

USDA Forest Service. (2012). 'USDA Forest Service–Forest Legacy Program'. Retrieved from www.fs.fed.us/spf/coop/programs/loa/flp.shtml

Weeks, W. W. (2011). 'A tradable conservation easement for vulnerable conservation objectives', Indiana Legal Studies Research Paper No. 198, pp. 239–248. Retrieved from http://papers.ssrn.com/sol3/papers.cfm?abstract_id=1975769

Yandle, B. (2004). 'Comments on land trusts and the choice to conserve land with full ownership or conservation easements', *Natural Resources Journal*, vol. 44, no. 2, pp. 519–527.

29

THE ECONOMICS OF FOREST CERTIFICATION AND CORPORATE SOCIAL RESPONSIBILITY

Anne Toppinen,[1] *Frederick W. Cubbage*[*2] *and Susan E. Moore*[2]

*CORRESPONDING AUTHOR: FREDCUBBAGE@YAHOO.COM
[1]DEPARTMENT OF FOREST SCIENCES, UNIVERSITY OF HELSINKI, HELSINKI, FINLAND.
[2]DEPARTMENT OF FORESTRY AND ENVIRONMENTAL RESOURCES. NORTH CAROLINA
STATE UNIVERSITY, RALEIGH, NORTH CAROLINA, USA.

Abstract

Forest certification is probably the best-known effort under the broad umbrella of corporate social responsibility (CSR) in global forestry and the forest industry. The economics of forest certification and CSR are fairly similar, but not entirely amenable to quantification. The benefits of CSR and forest certification are manifold, including development of strategic organizational capabilities, enhancing 'green organizational reputation' and creating market benefits of better prices, customer loyalty or increased market shares. Costs from adopting certification arise mainly from organizational and standard fulfillment and maintenance expenses, auditing costs and establishment of community education and social support programs. Forest certification has prompted changes in environmental, economic and social forest management practices. Although CSR and forest certification have become accepted means to demonstrate sustainability, the challenge now is to extend these approaches to smaller organizations and developing countries at costs that are covered by the often less tangible and longer-term benefits.

Keywords

Corporate social responsibility, forest certification, sustainability, forest economics, sustainable development

Introduction

Forest certification is probably the best-known effort under the broad umbrella of corporate social responsibility (CSR) in global forestry and the forest industry. Indeed, it was one of the first environmental efforts developed so that corporations and government organizations could demonstrate CSR. CSR covers the full suite of sustainable development components of economic, environmental and social responsibility, as does forest certification, which is the primary means that private and public forestry organizations now use to demonstrate their sustainability. Beyond certification, CSR programs now cover the range of social, community, environmental, worker and public interactions. Certainly, CSR has become one of the most rapidly expanding sectors in modern business practices, with almost all major firms now having CSR policies, departments, web pages and specific programs. In addition, sustainability programs and rankings have become a core metric of corporate nonfinancial performance, with indices such as the Dow Jones Sustainability index, FTSE4GOOD and others being published continually.

Forest certification provides a means to ensure that forests are managed to achieve the tripartite economic, environmental and social goals that are the foundation of sustainable development. This approach provides a set of standards that certified forests must meet, and which are independently audited. Forest certification has been analyzed considerably by (forest) political scientists in the last decade, but the economics of it has received much less attention. Similarly, policy researchers have examined forestry firm CSR to a lesser extent, but even fewer publications have addressed the economics of CSR. The subjects are closely intertwined and becoming more so; thus, we examine them in this chapter in order to provide an integrated synthesis.

Forest certification

Forest certification has been termed nonstate market-driven (NSMD) governance mechanism (Cashore, Auld and Newsom, 2004), indicating its market-based orientation, rather than government intervention. Lister (2011) extended this view, terming forest certification as co-governance, noting that while the private sector served as the program administration body, there were many levels of government involvement including educating, promoting and funding certification, and even being certified.

Forest certification developed as a response to the lack of binding international forestry accords at the United Nations Commission on Environment and Sustainable Development (UNCED) in 1992 (Humphreys, 2006). Environmental nongovernment organizations (ENGOs), social and community organizations and some private sector firms responded with the development of the global Forest Stewardship Council (FSC) in 1993, which was followed slightly later by many individual country certification programs, such as the Sustainable Forestry Initiative (SFI) in the United States and Canada in 1995 and the Finnish Forest Certification System in 1999.

The European forest certification programs were integrated into an organization termed the Pan-European Forest Certification program (PEFC) in 1999, which serves as an umbrella program linking individual European country programs. PEFC now has been expanded to endorse individual country programs throughout much of the world as well, while changing its name to the Programme for Endorsement of Forest Certification, and keeping the same acronym. During the same period, FSC has expanded throughout most of the world as well, retaining a core set of ten principles for all individual country standards.

Voluntary Environmental Programs (VEPs) such as forest certification remain somewhat unique in that this set of programs has very specific sets of standards and third-party auditors who ensure that companies and public organizations comply with those standards. In comparison, the entity of general CSR components remains complex and with the exception of ISO14000 family of environmental management standards, SA8000 labor standards or standardization of corporate reporting under the Global Reporting Initiative, many of them also appear less quantifiable or independently auditable. At a minimum, forest certification includes a large amount of public forests and nongovernment organization (NGO) forests, which are of course not corporations. However, governments also like to demonstrate environmental and social responsibility, and do much to promote that, such as through green procurement practices and funding for sustainable development. Forest certification is really just an extension of these approaches of public and private CSR to forest lands, as suggested by Lister (2011).

Extent of forest certification

Forest certification developed steadily since 1993, and more than 475 million ha, or about 11% of the world's 3.9 billion ha of forests, were certified as of January 2013 (Table 29.1). Approximately 24% of the 614 million ha of forests in Canada and the United States are certified.

In addition to forest land certification, FSC, PEFC and some member PEFC organizations have developed their own chain of custody (CoC) certification to track and ensure sustainable wood procurement and forest products production systems. As of January 2013, FSC had approved 24,619 certificates in 109 countries. PEFC had approved 9,069 certificates, presumably in its 27 member countries.

Issues in forest certification

The costs and benefits of forest certification systems were one of the principal issues affecting their adoption. For example, Laband (2005) criticized the social burden of forest certification, stating it foists substantial costs and regulatory interference on small landowners without any benefits. Chen et al. (2011) found that less than 10% of Canadian consumers requested certified wood products.

On the other hand, many studies have found that consumers and businesses favor certification, and many express a willingness to pay some premium (e.g. Jensen, 2003; Aguilar and Vlosky, 2007; Chen et al., 2011). Cashore et al. (2004) found that forest certification has provided a means for social and environmental groups to promote sustainable forestry through market-based 'soft law' and governance, rather than through direct government regulation. Furthermore, contemporary practitioners have become convinced that forest certification has become a permanent addition to our set of forestry principles and tools, and will continue to expand slowly (e.g. Lowe et al., 2011; Berg, 2012).

Table 29.1 Global forest area certified by major forest certification systems, January 2013.

System	Area (million ha)
Forest Stewardship Council (FSC)	170.5
Programme for Endorsement of Forest Certification (PEFC)[1]	247.0
Sustainable Forestry Initiative (SFI)	78.7
Canadian Standards Association (CSA)	55.4
PEFC Finland	21.1
PEFC Sweden	11.0
American Tree Farm System (ATFS)	10.9
Australian Forestry Standard	10.1
Total, all systems	417.5

Sources: FSC (2013): https://ic.fsc.org/facts-figures.19.htm

PEFC (2013): www.pefc.org/images/stories/documents/Global_Stats/2012-05_PEFC_Global_Certificates.pdf

[1]SFI includes 54.0 million ha in Canada and 24.7 million ha in the United States.

CSR

Forest certification and CSR developed somewhat concurrently, with certification being most prominent initially in the forestry sector, and the more general concept corporate responsibility[1] providing the principal foundations in most other businesses. There are numerous definitions of what constitutes CSR (e.g. Dahlsrud, 2008). According to a commonly used definition by the European Commission (EC, 2001, p.6), CSR is a concept whereby *'companies integrate social and environmental concerns in their business operations and in their interaction with their stakeholders on a voluntary basis'*. This implies that corporate social and environmental behavior which goes beyond legal (regulatory) requirements is present in the relevant markets. More recently, the definition by the EC has been streamlined to consist of the simple statement: *'impacts that corporations have on society'*.

Emerging in the 1970s, the term corporate social performance (CSP) has become an attempt to offer a managerial framework to deal with the challenging measurement of corporate responsibility. According to Wood (1991, p. 693), CSP is 'a business organization's configuration of principles of social responsibility, processes of social responsiveness, and policies, programs, and observable outcomes as they relate to the firm's societal relationship'. CSP thus focuses on the impacts and outcomes for society, stakeholders and the firm, and the types of relevant outcomes are determined by the firm's linkages, both general and specific, as defined by the structural principles of CSR (Wood, 2010).

Economic and management theories on CSR and forest certification

Various concepts and general economic principles could describe forest certification and CSR. The theories are similar, but many studies examine only forest certification or corporate responsibility in isolation, so we review them in that manner as well, although some overlap exists.

The *trade-off hypothesis* on the purely economic impact of CSR on companies reflects Friedman's (1970) neoclassical argument that the social responsibility of business is to increase profits, and the increasing cost of CSR investment inevitably reduces corporate profitability. As such, CSR to meet the demands of various stakeholder groups places additional constraints on the corporate pursuit of success by incurring greater costs in terms of management time, capital investment and operating costs.

According to Crifo and Forget (2014), forces driving CSR are based on three types of market imperfections: the existence of externalities and public goods, consumer heterogeneity and imperfect market competition, and existence of imperfect contracts with key stakeholders, such as investors, employees or managers (agency problem). According to Kitzmueller (2008), CSR has been traditionally connected with the provision of public goods by or the origin of externalities within a firm. Reinhardt et al. (2008) identified conditions that facilitate the production of public goods and services benefiting individuals beyond customers as imposition of regulatory constraints requiring a firm and its competitors to carry out responsible actions, and at times these actions may be cost neutral or even cost saving (such as adoption of energy or other resource-saving technologies). According to Benabout and Tirole (2010, p. 2), an organization's CSR may also be an outcome of individual behavior of a manager 'sacrificing profits in the social interest', driven by intrinsic altruism, material incentives or social or self-esteem concerns.

These broader anticipated benefits of CSR actions are seen as an intended consequence of *implementing company strategy* – to gain market share, receive better prices, seek better reputations, enhance employee relations or even improve production processes – and are not merely a by-product of company-level altruism (Drucker, 1982). Today, as it has become increasingly difficult to compete by the traditional means of product differentiation, intangible resources are more often promoted as a source of sustainable competitive advantage. Thus, in line with the prevailing theory of the firm, the resource-based view (RBV) (Barney, 1991) claims that such intangible resources include, among other things, reputation, brand value, skilled employees and creation of innovation and knowledge (Branco and Rodriguez, 2006).

The third theoretical perspective, the *CSR social impact hypothesis*, advocates even more strongly that meeting the needs and expectations of various stakeholders affects firms positively through, for instance, better employee retention, decreased business risk or providing access to ethical investment funds. Stakeholder theory by Freeman (1984) has been used extensively as a background model for analyzing the interactions between companies and society. Donaldson and Preston (1995) proposed a justification of stakeholder theory in management based on its descriptive accuracy, instrumental power and normative validity. They argue that CSR can add to the bottom line of the firm by improving stakeholder relationships. Intangible resources such as corporate reputation, brand value, retention of skilled employees and the creation of innovation and knowledge have risen in importance as a source of competitive advantage, because they are more likely to prove valuable, rare, inimitable and nonsubstitutable.

Such social impact capabilities can easily be associated with sustainability-related issues high on the agendas of twenty-first-century corporate life, including higher-order learning and continuous innovation in product design and development, habitat preservation, resource management, waste reduction and energy conservation (Hart, 1995; Hart, Arnold and Day, 2000), stakeholder engagement, improved stakeholder consideration, ethical awareness and management issues (Panwar and Hansen, 2007).

The three previous theories behind CSR also underlie the benefits and costs of forest certification. Certification may just represent a cost that reduces firm profits and adds an unnecessary social burden. Or, it may help a firm gain a better brand reputation, attract skilled employees and build consumer confidence. Modern consumers and employees are motivated by a social conscience and environmental ethic and surely prefer benevolent organizations rather than ruthless ones. Forest certification may help firms differentiate their product in the marketplace, thus gaining market share, and in particular, be an important imprimatur for large firms to demonstrate that they are indeed socially responsible and garner more sales. In addition, as noted by Lister (2011), governments and trade associations also support forest certification in order to demonstrate sustainable forestry and procurement practices.

Sustainable development and resource use has become the accepted paradigm by this century, and CSR and forest certification offer the clearest method for firms, government and NGOs to demonstrate their commitment to the principle. Forest products firms and large forest landowners and investors also seek capital, and forest certification is seen as particularly important to demonstrate sustainability, especially in poor countries that perhaps have less institutional and government capacity to protect forest resources. There also may be broad economic impacts of CSR and forest certification on timber and forest product markets, prices and society, which economists term as welfare analysis – who gains, who pays and how much, as measured by changes in consumer and producer surplus.

CSR: Economics research

Conceptual costs and benefits of CSR

Monetary accounting and distinguishing between CSR- and non-CSR-related costs at the corporate level are difficult, which inhibits the evaluation of total revenues or calculating cost-benefit ratios. Typically, benefits arising from CSR activities are more long-term in nature by comparison to CSR costs. According to a review by Wood (2010), findings in literature using event studies on stock market reactions or impacts of product recalls are quite consistent, indicating direct and indirect negative effects of harmful events on stakeholder trust, for example unwillingness to continue buying from or investing in the irresponsible company.

Sprinkle and Maines (2010) point out that the costs associated with CSR arise from the CSR activities performed – but not all are easily measured. Activities requiring cash outlays, product donations, contributions of employees' time and the like can be measured as such. Other costs, such as reduced productivity and opportunity costs, may be more difficult to quantify. They also posit that the relationship between CSR costs and benefits is concave, so that returns perhaps diminish as the level of CSR increases.

Costs from adopting CSR-related standards or certificates (e.g. ISO14001, FSC/PEFC forest certification, GRI reporting or SA8000 employee standard) can be one-time or continuous nature, depending on the scheme being used. The recently launched ISO26000 standard on CSR includes voluntary guidelines for CSR. Although ISO26000 is not yet certifiable, its aim is to determine the common social and environmental expectations toward an organization, thereby lowering transaction costs.

The rising awareness of the link between social consciousness and action to a firm's success has also inspired much research into the linkage between CSR and economic/financial performance, especially in the area of socially responsible investment (SRI), (for reviews of studies, see e.g. Wood, 1991; Orlitzky, Schmidt and Rynes, 2003; Salzmann, Ionescu-Somers and Steger, 2005). The majority of empirical research findings to date has supported the hypothesis that the corporate responsibility-economic (or financial) performance relationship is likely to be nonnegative. Others such as Vogel (2005) or Lee (2008) have, however, been more critical toward the overemphasized focus on 'business case of CR?' in academic research and have pointed out the importance of understanding better conceptual tools and theoretical mechanisms that explain changing organizational behavior from a broader societal perspective.

CSR outcomes: Empirical evidence

In general, forest certification is one type of CSR, but for a clearer outcome of this analysis, it is meaningful to separate the two, and in the following review, CSR is used to cover all activities other than forest certification. A pioneering effort by Näsi, Näsi, Phillips and Zyglidopoulos (1997) explored corporate responsibility in the case of two boreal forestry-rich countries (Canada, Finland). Li and Toppinen (2011) identified 23 peer-reviewed studies between 1997 and 2009 in the forest industry focusing on corporate responsibility. According to Li and Toppinen (2011), the global forestry sector has moved toward a more holistic and encompassing approach to CSR and sustainability initiatives (e.g. Panwar, Rinne, Hansen and Juslin, 2006; Vidal and Kozak, 2008a, 2008b), and it is evident that the largest forest companies shape their social performance strategies to fit their geographical profiles (e.g. Mikkilä and Toppinen, 2008).

Based on the review in Li and Toppinen (2011), forest industry companies covered in the previous studies appear to have adopted CSR activities mainly from a profit-maximizing motive (i.e. an instrumental approach to CSR) with a more limited emphasis on their overall social impacts. However, Bouslah, M'Zali, Turcotte and Kooli's (2010) results certainly indicate that firms are choosing to certify for reasons beyond shareholder value. They perceive CSR initiatives as a worthy investment that serves to demonstrate their commitment to sustainability and secure their legitimacy, whether for ethical reasons or for the achievement of strategic and economic objectives.

Table 29.2 Drivers and key processes of CSR and their importance in the forest industry.

Driver	Key company or industry process involved	Importance in the forest industry context
Cost-benefit ratio	Standards, certification (ISO 14001, GRI, SA8000, ISO26000, etc.)	Medium
External control	Tool for risk management	Medium
Sensitivity to local stakeholders	Tool for reputation management, achieving license to operate, prevention of conflicts	Medium
Geographic spread	Industry internationalization	Medium, increasing
Internal control	Tool for risk management, resource and capability development	Low to medium
Following industry forerunners	Industry isomorphism, conformity with competitors	Low to medium?
Sensitivity to public perceptions	Tool for reputation management, active reshaping of market conditions, prevention of negative media visibility	High (for especially for multinational corporations)
Anticipating future regulation	Tool for reputation management, overcoming less active competitors	Low to medium

Table 29.2 summarizes the concept of CSR in the context of the forest industry, based on the key drivers and barriers approach used in Laudal (2011). We assessed the respective key processes, as well as the level of importance of each driver (using a three-point scale from low to medium to high). Based on our analysis, drivers for adopting activities under the domain of corporate responsibility were affected more by external factors than by internal ones. The drivers also appear to be more market than regulatory driven. In spite of this, the strategic role of CSR as a means of overcoming less proactive competitors is not strongly present in the forestry and forest industry.

However, the scope, depth and level of implementation of what constitutes corporate responsibility are still developing; thus, the ability to conduct any rigorous cost-benefit analysis is lacking. Little can be quantified about the actual impacts of CSR in the forest industry. Nevertheless, there is clearly progress in the area of sustainability reporting, adoption of international standards and certificates on CSR, and in managing sustainability and measuring its business performance, as well as recognizing the potentially strong impacts on natural ecosystems (see, e.g. Hanson, van der Lugt and Ozmet, 2011).

The differential capacity of different-sized organizations does also offer challenges for implementation of CSR. Jaffee and Howard (2010) found that large corporations have often co-opted, captured and weakened these standards. This includes strategies such as removing rules not in concordance with large scale industrial practices, making some rules so challenging only large-scale producers could meet them, simplifying standards or eroding price premiums.

Forest certification: Economics research

Conceptual costs and benefits of certification

Various authors have estimated direct and indirect costs and benefits of forest certification, although they have classified them somewhat differently (Nussbaum and Simula, 2005; Cubbage et al. 2003, 2010b). In reality, the classification is perhaps not as important as the correct enumeration of costs and benefits. Cubbage, Moore, Henderson and Araujo (2010a) reviewed the benefits and costs of forest certification in the Americas, which we draw on here for some principles and results regarding the overall economics of forest certification and CSR. Broad social impacts also are relevant for forest certification, and some studies have examined this.

Bass, Thornber, Markopoulos, Roberts and Grieg-Gran (2001) noted that the largest direct costs of forest management certification vary according to the type of certification (forest management or CoC), the size of the enterprise and the distance that certifiers have to travel. Indirect costs of changes in practices depend primarily on the existing quality of management. In tropical natural forests, the costs of certifying natural forest management were comparatively large unless the area is substantial.

Forest certification outcomes: Empirical evidence

Bass et al. (2001) reported that one study in Latin America recorded costs of $0.26 to $1.1 per m^3 and up to $4 for small areas. In contrast, Polish state forests had costs of about $0.02 to $0.03 per m^3. A South African company reported costs of $0.19 per m^3 or 0.03% of the logging cost. US costs of $0.20 to $0.75 per ha were reported for the initial assessment and about $0.03 to $0.15 per ha for the annual surveillance – the same order as the per m^3 cost in Poland.

A study by Gan (2005) stated that forest certification would include direct costs of assessing forest certification and monitoring the CoC. He classed indirect costs as those for implementing higher forest management standards. Based on various proceedings and organizational reports, Gan assumed that implementing a higher forest management standard could increase forest management costs by 5% to 40%, with an average of 5% to 25%.

The experience throughout the Americas reported by Cubbage et al. (2003, 2010a, 2010b), however, suggests that Gan's cost estimates are much greater that those commonly experienced – many firms have not needed to make costly forest management changes, and the estimates of around $1.00 per ha per year are possible for large industrial firms. Smaller firms, however, are apt to have much higher costs per area.

Cubbage et al. (2003, 2010b) estimated the total direct costs for receiving certification of university and state forest lands for three institutions in North Carolina ranged from $1.24 per ha per year for the largest state forest (10,900 ha) to $19.08 per ha per year for the smallest university forest (1,800 ha). The direct costs to maintain SFI certification ranged from $0.96 per ha per year (the state forest) to $9.56 per ha per year (Duke University). For FSC, the costs ranged from $1.04 (the state) to $7.19 (NC State University) per ha per year.

Goetzl (2006) surveyed SFI firms regarding their forest certification costs. He found that much greater costs occurred when the SFI standard was first implemented by an organization. In most organizations that were certified, one or two additional full-time equivalent employees could be attributed to adopting the standard, not including SFI standard-related activities that were integrated into other functions. When spread over a firm's total certified area, the external audit costs were relatively small. For firms that were dual-certified with FSC, the opportunity costs for forest reserves and set-asides seemed to be greater, but FSC program participation costs appeared to be lower than for SFI.

Average total forest certification costs in the Americas

Cubbage et al. (2010a) estimated forest certification costs in the largest cross-sectional study performed, with data from Argentina, Brazil, the United States and Canada, and Chile, through a series of personal interviews and email surveys. The relevant survey questions asked respondents to list any costs they could related to (1) total certification expenses, (2) internal audit preparation fees or consultants, (3) external audit fees, (4) ongoing certification preparation costs, (5) community education and support programs, (6) management changes required to get/maintain certification or (7) participation in implementation committees or FSC promotion activities.

In the Americas, the median and mean reported average total costs for small forest tracts of less than 4,000 ha were expensive, at $6.45 per ha per year for FSC-US to $39.31 per ha per year for SFI, which surely would deter adoption for small ownerships. Costs decreased significantly with increasing tract size, becoming as little as $0.27 per ha per year for SFI in the 40,001-ha to 400,000-ha size class, and $2.40 for FSC-US, and less than $0.50 per ha per year for ownerships greater than 400,000 ha. Reported costs for the FSC system were less for small ownerships than SFI, and vice versa. Costs in the Southern Cone countries in South America were generally greater than in North America. However, these median costs by ownership size class proved not to be statistically significant with regression analysis. Thus, average total costs among systems and countries were a statistically significant function only of ownership size, not type of certification system.

Table 29.3 provides summary statistics for median and mean (average) total costs reported. Total area is simply the total area of the certified forest ownership for all responding organizations in the surveys and interviews. Average total costs is the economic term for the total costs divided by the total unit of production, which is hectares in this case. Note the mean of the average total costs is generally much greater than the median of the average total costs. This is due to the fact that there are several organizations that reported large costs, while a greater number of organizations reported moderate costs. The median may be a more representative gauge of typical average total costs of forest certification, given this result.

In addition to the estimates of financial costs, various authors have performed economic welfare analyses of the impacts of forest certification. Previous studies have consistently found that intervening in theoretical free markets with forest certification generates a welfare cost in consumer and producer surplus (e.g. Brown and Zhang, 2005; Gan, 2005) but none have estimated the commensurate benefits from sustainable practices.

Forest certification benefits

Nussbaum and Simula (2005) provide a useful list of potential benefits from forest certification as summarized in Table 29.4. These include economic, social and environmental benefits.

Bouslah et al. (2010) used forest certification as a proxy for environmental performance to measure the impact on firms' financial performance in the United States and Canada and found overall a negative relationship, indicating that the market does not yet recognize the benefits of forest certification. Canadian forest products retailers surveyed by Chen et al. (2011) felt that forest certification had minimal impacts on social aspects of forest management, but perceived the economic and environmental aspects as changed more by certification. These retailers also reported up to a 20% price premium on certain certified products.

The scant literature regarding financial benefits of forest certification may be due to strategic behavior by forest products firms, which do not want to reveal price advantages they receive. In addition to better prices, many organizations have stated that having forest certification has provided them with better access to 'green' markets in Europe and Japan, where either consumers or government agencies request certified products (Araujo, 2007). Chen, Innes and Tinkina (2010) predict that government procurement policies will play an increasingly important role as a driver of certification. Some companies in the United States have recently stated that they will

Table 29.3 Summary of forest ownership size class and average total costs per hectare by region and certification system in the Americas, 2007.

System/size class	Area (ha)			Cost ($/ha)		
	N	Median	Mean	N	Median	Mean
All responses						
FSC – United States	45	22,258	251,392	14	0.91	3.24
SFI – North America	36	133,727	992,675	17	0.37	4.92
South America	58	23,469	95,832	25	1.56	6.89
<4,000 ha						
FSC – United States	9	2,543	2,112	3	6.45	8.43
SFI – North America	3	1,974	1,967	2	39.31	39.31
South America	9	1,811	1,883	5	23.75	17.52
4,001–40,000 ha						
FSC – United States	16	13,557	14,663	7	0.54	2.05
SFI – North America	5	30,479	25,797	3	0.91	1.06
South America	27	13,409	16,011	10	2.00	2.97
40,001–400,000 ha						
FSC – United States	14	107,244	152,098	2	2.40	2.40
SFI – North America	15	79,725	144,465	6	0.27	0.89
South America	18	90,531	137,583	9	1.21	6.08
>400,000 ha						
FSC – United States	6	1,416,431	1,488,275	2	0.42	0.42
SFI – North America	12	1,395,682	2,785,867	7	0.07	0.22
South America	4	603,183	658,131	1	0.49	0.49

Source: Cubbage et al. (2010a).

Notes: FSC responses include only the United States; SFI includes the United States and Canada; South America includes FSC in Argentina, Brazil and Chile, combined FSC, Cerflor and Certfor. Total number of replies and statistics are indicated in the area rows; total number of cost responses and statistics are indicated in the cost rows.

at least give procurement preference to certified wood. But these market benefits have not been scientifically quantified. There also may be a plethora of environmental and social benefits, which are more difficult to quantify and less studied.

Rickenbach and Overdevest (2006) surveyed expectations and satisfaction of FSC certificate holders in the United States and categorized benefits into three classes based on possible social and business motivations of certified organizations. They found that 'signaling' benefits of getting better recognition for one's forest practices and public relations were ranked as having the highest satisfaction, exceeding expectations. The numerical differences among these categories developed by Rickenbach and Overdevest (2006) were moderate. Most large landowners were satisfied with forest certification, and small landowners had neutral opinions.

Cubbage et al. (2010a) asked respondents in the Americas about their perceived benefits of forest certification, using a similar schema to Rickenbach and Overdevest and a five-point Likert scale ranking system. Improved market prices were the least important realized perceived benefit of forest certification, but increased market share was important. Organizational strategic reasons were considered the most important, ranging from 3.6 to 4.3 by system and region.

Table 29.4 Potential benefits from forest certification.

Economic benefits

Improved performance standards

Enhanced control of resources

Improved management systems, including internal mechanisms of planning, monitoring, evaluation and
   reporting

Reduced regulatory control

Permanent economic viability and opening of new markets

Improved market access and occasional higher prices

Improved enterprise image and business practice

Social benefits

Addressing the public's environmental and social concerns about forest management

Balancing the objectives of forest owners, other stakeholders and society

Empowering the poor and less favored

Poverty alleviation

Community participation

Environmental benefits

Environmental conservation

Maintenance and enhancement of biodiversity

Maintenance and enhancement of high conservation value forests

Improved workers' rights and living conditions

Source: Nussbaum and Simula (2005).

Signaling to external interest groups was moderately important, as was improved organizational learning. The mean Likert scores seemed to indicate more of a difference among the rankings by certification system than by category of benefits. The benefits perceived for FSC-US ranked slightly below SFI, which ranked slightly below South America, with the means for all groups ranging from 2.8 to 4.3.

Several authors have examined whether forest certification actually prompts changes in forest management practice on certified forests, for both FSC and for PEFC organizations, both public and private (e.g. Federation of Nordic Forest Owners' Organisations, 2005; Newsom, Volker and Cashore, 2006; Moore, Cubbage and Eicheldinger, 2012). This research has consistently found that forest certification has changed environmental, social, economic and forest management practices, leading to improved forest practices. This supports the premise that certification has substance, as well as process and propaganda benefits.

Conclusions

Forest certification and CSR have become integral components of business for major domestic and international forestry companies and indeed have extended their impact to many government organizations that own forest land. These approaches stem from two decades of adoption of sustainable development as a widely accepted paradigm for doing public and private business and land management. The economic rationale stemming from a neoclassical free market private enterprise view would posit that these nonprice CSR and forest certification efforts are merely clumsy interventions in those free markets, which reduce social welfare as measured by reduced consumer and producer surplus. This view, of course, does not account for any negative externalities of exploitive production, or any positive externalities of sustainable development, or imperfections in product markets. Instead, the corporate strategy and social impact hypotheses suggest that CSR and forest certification are appropriate responses to meet the public demand for sustainable development – social, environmental and economic. Corporations and governments have recognized public tastes and desires in this respect and pursue a strategy that will position their products and services as being sustainable, or simply 'green' as commonly used. Thus, the degree of greenness – environmentally and socially – can be used to differentiate products and gain market advantages. The work on quantification and valuation of these largely indirect benefits, however, is modest.

Also, there still may be significant risk that corporations are only using forest certification and CSR as 'greenwash' or for public relations. Large corporations may try to co-opt the sustainability standards so that they favor their production process or make it too burdensome for small organizations to participate. These risks require scrutiny by social and consumer advocates to ensure a level playing field.

Based on our review, the benefits of CSR and forest certification include (1) strategic organizational capabilities through better relationships among management, employees, and communities; (2) signaling benefits of better recognition for one's environmental practices and public relations; (3) market benefits of better prices, customer loyalty or increased market share; and (4) learning benefits of teaching professionals about new management practices.

CSR and forest certification costs could include (1) organizational expenses to develop and maintain the programs, (2) internal standards development or forest certification audit preparation fees or consultants, (3) external audit fees, (4) ongoing CSR/certification maintenance costs, (5) community education and social support programs including stakeholder consultations and capacity building, (6) management changes required to obtain and implement CSR and forest certification or (7) participation in CSR/certification program activities.

Modest empirical evidence existing to date suggests that forest product firms are pursuing strategies to have better CSR programs and greener images, with nascent success. But there is a clearly rising bar, so implementation of CSR policies, programs and rankings are all progressing in major firms. The same race for greenness is occurring in forest certification systems, with both FSC and PEFC improving their standards, auditability, credibility and consumer lobbying to gain more enrolled lands and market share. One challenge in these rapidly developing CSR, forest certification and CoC programs is extending them to small organizations, especially in developing countries, or even specifically not favoring large organizations with more capacity.

Economic analysis of CSR and forest certification is in early stages, but does indicate economic incentives for certification. Currently, FSC-certified firms seem to show some financial market reward for their efforts. The merits of forest certification at least, and perhaps CSR, are further reinforced by the steady increase in forest certification in developed and developing countries.

Global adoption of environmental management systems and sustainability standards also has been on a steady rise despite recent economic and financial turbulence. The company strategy and social impact theories also might be reinforced by the increasing demand pull for FSC certification in the United States due to support from the US Green Building Council's Leadership in Energy and Environmental Design (LEED) green building standard and purchasing preferences from some paper producers, office supply retailers and other major buyers.

Future research

There are myriad opportunities for further research regarding the value of CSR and forest certification to firms, and to public and NGO landowners. Future studies should include use of a wider set of quantitative sustainability indicators of corporate practices in order to assess how internal and external stakeholder perceptions are linked to actual practices and economic performance over time. The geographic scope of analysis should be expanded to the little-studied emerging producer regions (with the exception of some well-known case studies such as the Botnia project in Uruguay, e.g. Aaltonen and Kujala, 2010).

Research that focuses on small forest owners or small and medium enterprises (SMEs) and on the impacts of group certification is scant. Similarly, there is little research focused on the benefits and costs of forest certification and (government) social responsibility for government and community-based organizations. There also is considerable controversy regarding the merits of FSC versus PEFC certification, and very few empirical tests of those claims.

CoC certification for forest products manufacturing has become quite important in just a few years, and research on its impacts and effectiveness is extremely limited. Little research on linkages between CoC certification, CSR, and forest certification has been published (see, however, Vidal, Kozak and Cohen, 2005); the links between direct action campaigns and these linkages also is unknown. Similarly, they are a major driving force of green building codes and government green standards on forest certification implementation.

As this review suggests, forest certification and CSR have had considerable success and show further promise in promoting, measuring and enhancing sustainable development of forests, companies and government organizations. There is a rich literature on these subjects, and there are considerable opportunities for new research lines as the practice of CSR, forest certification and CoC expands. We hope that the review in this chapter provides an economic context for subsequent research and analyses, as well as benchmarks on the path for achieving progress in terms of CSR and forest certification.

Note

1 Corporate social responsibility is often used in conjunction with other terms such as 'corporate responsibility', 'corporate sustainability', 'corporate citizenship', and 'sustainable development', 'corporate social initiative', 'corporate social responsiveness', or as a synonym of other concepts such as triple bottom line (economic, environmental and social) and the three Ps (profits, planet and people).

References

Aaltonen, K. and Kujala, J. (2010). 'A project lifecycle perspective on stakeholder influence strategies in global projects', *Scandinavian Journal of Management*, vol. 26, pp. 381–397.

Aguilar, F. X. and Vlosky, R. P. (2007). 'Consumer willingness to pay price premiums for environmentally certified wood products in the U.S.', *Forest Policy and Economics*, vol. 9, no. 8, pp. 1100–1112.

Araujo, M. (2007). 'Forest certification in Brazil: Choices and impacts', MS thesis, University of Toronto, Toronto, Ontario, Canada.

Barney, J. (1991). 'Firm resources and sustained competitive advantage', *Journal of Management*, vol. 17, no. 1, pp. 99–120.

Bass, S., Thornber, K., Markopoulos, M., Roberts, S. and Grieg-Gran, M. (2001). *Certification's Impacts on Forests, Stakeholders, and Supply Chains*, Earthprint Limited, Hertfordshire, England.

Benabout, R. and Tirole, J. (2010). 'Individual and corporate social responsibility', *Economics*, vol. 77, no. 1, pp. 1–19.

Berg, S. (2012). 'Group certification of family forests: What are the myths? What are the realities?' *Forest Operations Review*, American Forest Resource Association, vol. 14, no. 1, pp. 9–13.

Bouslah, K., M'Zali, B., Turcotte, M.-F. and Kooli, M. (2010). 'The impact of forest certification on firm financial performance in Canada and the U.S.', *Journal of Business Ethics*, vol. 96, no. 4, pp. 551–572.

Branco, M. and Rodriguez, L. (2006). 'Corporate social responsibility and resource-based perspectives', *Journal of Business Ethics*, vol. 69, no. 2, pp. 111–132.

Brown, R. and Zhang, D. (2005). 'The Sustainable Forestry Initiative's impact on stumpage markets in the US South', *Canadian Journal of Forest Research*, vol. 35, no. 8, pp. 2056–2064.

Cashore, B., Auld, G. and Newsom, D. (2004). *Governing through Markets: Forest Certification and the Emergence of Non-State Authority*, Yale University Press, New Haven, CT.

Chen, J., Innes, J. L. and Tinkina, A. (2010). 'Private cost-benefits of voluntary forest product certification', *International Forestry Review*, vol. 12, no. 1, pp. 1–12.

Chen, J., Tinkina, A., Kozak, R., Innes, J., Duinker, J. P. and Larson, B. (2011). 'The efficacy of forest certification: Perceptions of Canadian forest products retailers', *Forestry Chronicle*, vol. 87, no. 5, pp. 636–643.

Crifo, P. and Forget, V. (2012). *The Economics of Corporate Social Responsibility: a Firm-Level Perspective Survey*, Journal of Economic Surveys, in press, pp. 1–19. DOI: 10.1111/joes.12055.

Cubbage, F., Cox, J., Moore, S., Henderson, T., Edeburn, J., Richter, D., . . . Rohr, H. (2010b). 'Management impacts and costs of forest certification of state and university lands in North Carolina, USA', *Forest Certification*, vol. 12, pp. 3–17. [English and Russian]

Cubbage, F., Moore, S., Cox, J., Jervis, L., Edeburn, J., Richter, D., . . . Chesnutt, M. (2003). 'Forest certification of state and university lands in North Carolina', *Journal of Forestry*, vol. 101, no. 8, pp. 26–31.

Cubbage, F., Moore, S., Henderson, T. and Araujo, M. (2010a). 'Costs and benefits of forest certification in the Americas', in J. Pauling (Ed.), *Natural Resources: Management, Economic Development and Protection*, Nova Publishers New York, Chapter 5, pp. 155–185.

Dahlsrud, A. (2008). 'How corporate social responsibility is defined: An analysis of 37 definitions', *Corporate Social Responsibility and Environment Management*, vol. 15, no. 1, pp. 1–13.

Donaldson, T. and Preston, L. (1995). 'The stakeholder theory of the corporation: Evidence and implications', *Academy of Management Review*, vol. 20, no. 1, pp. 65–91.

Drucker, P. (1982). 'The new meaning of corporate social responsibility', *California Management Review*, vol. 26, pp. 53–63.

European Commission. (2001). *Green Paper: Promoting a European Framework for Corporate Social Responsibility*, COM, Brussels.

Federation of Nordic Forest Owners' Organisations. (2005). *Effectiveness and Efficiency of FSC and PEFC Forest Certification on Pilot Areas in Nordic Countries*, Final report, Savcor Indufor Oy, Helsinki, Mimeo/PDF.

Forest Stewardship Council (FSC). (2013). 'Program statistics'. Retrieved from https://ic.fsc.org/facts-figures.19.htm

Freeman, R. (1984). *Strategic Management: A Stakeholder Approach*, Pitman, Marshfield, MA.

Friedman, M. (1970). 'The responsibility of business is to increase its profits', *The New York Times*, 13 September, p. 1.

Gan, J. B. (2005). 'Forest certification costs and global forest products markets and trade: A general equilibrium analysis', *Canadian Journal of Forest Research*, vol. 35, no. 7, pp. 1731–1743.

Goetzl, A. (2006). 'Sustainable Forestry Initiative Standard (SFI) ® cost assessment', speech presented at the SFI Annual Meeting, Toronto, Ontario, Canada, 3–5 October.

Hanson, C., van der Lugt, C. and Ozmet, S. (2011). *Nature in Performance: Initial Recommendations for Integrating Ecosystem Services into Business Performance Systems*, World Resources Institute, Washington, DC.

Hart. S. (1995). 'A natural-resource-based view of the firm', *Academy of Management Review*, vol. 20, no. 4, pp. 986–1014.

Hart, S., Arnold, M. and Day, R. (2000). 'The business of sustainable forestry: Meshing operations with strategic purpose', *Interfaces*, vol. 30, pp. 234–250.

Humphreys, D. (2006). *LogJam: Deforestation and the Crisis of Global Governance*, Earthscan, London/Sterling, VA, Chapter 6, pp. 116–141.

Jaffee, D. and Howard, P. (2010). 'Corporate cooptation of organic and fair trade standards', *Agriculture and Human Values*, vol. 27, pp. 387–399.

Jensen. K. (2003). 'Market participation and willingness to pay for environmentally certified products', *Forest Science*, vol. 49, no. 4, pp. 632–641.

Kitzmueller, M. (2008). 'Economics and corporate social responsibility', EUI Working Papers ECO 2008/37. European University Institute, Firenze.

Laband, D. (2005). 'Why I oppose forest certification', *The Consultant*, pp. 21–22. Annual Journal of the Association of Consulting Foresters of America.

Laudal, T. (2011). 'Drivers and barriers of CSR and size and internationalization of the firm', *Social Responsibility Journal*, vol. 7, no. 2, pp. 234–256.

Lee, M. (2008). 'A review of the theories of corporate social responsibility: Its evolutionary path and the road ahead', *International Journal of Management Reviews*, vol. 10, no. 1, pp. 53–73.

Li, N. and Toppinen, A. (2011). 'Corporate responsibility and sustainable competitive advantage in the forest-based industry: Complementary or conflicting goals?' *Forest Policy and Economics*, vol. 13, pp. 113–123.

Lister, J. (2011). *Corporate Social Responsibility and the State: International Approaches to Forest Co-Regulation*, UBC Press, Vancouver, British Columbia, Canada.

Lowe, L., Brogan, S., Nowak, J., Oates, B., Preston, D. and Tucker, W. (2011). *Forest Certification Programs: Status and Recommendations in the South*, A Report of the Southern Group of State Foresters. Available from Mike Zupko, sgsfexec@zup-co-inc.com

Mikkilä, M. and Toppinen, A. (2008). 'Corporate responsibility reporting by large pulp and paper companies', *Forest Policy and Economics*, vol. 10, pp. 500–506.

Moore, S., Cubbage, F. and Eicheldinger, C. (2012). 'Impacts of FSC and SFI Forest Certification in North America', *Journal of Forestry*, vol. 114, no. 3, pp. 79–88.

Näsi, J., Näsi, S., Phillips, N. and Zyglidopoulos, S. (1997). 'The evolution of corporate social responsiveness: An exploratory study of Finnish and Canadian forestry companies', *Business and Society*, vol. 36, no. 3, pp. 296–321.

Newsom, D., Volker, B. and Cashore, B. (2006). 'Does forest certification matter? An analysis of operation-level changes required during the SmartWood certification process in the United States', *Forest Policy and Economics*, vol. 9, pp. 197–208.

Nussbaum, R. and Simula, M. (2005). *The Forest Certification Handbook* (2nd ed.), Earthscan, London.

Orlitzky, M., Schmidt, F. L. and Rynes, S. L. (2003). 'Corporate social and financial performance: A meta-analysis', *Organization Studies*, vol. 24, no. 3, pp. 403–441.

Panwar, R. and Hansen, E. (2007). 'The standardization puzzle: An issue management approach to understand corporate social responsibility standards for the forest products industry', *Forest Products Journal*, vol. 57, no. 12, pp. 86–90.

Panwar, R., Rinne, T., Hansen, E. and Juslin, H. (2006). 'Corporate responsibility: Balancing economic, environmental, and social issues in the forest products industry', *Forest Products Journal*, vol. 56, no. 2, pp. 4–12.

Programme for the Endorsement of Forest Certification (PEFC). (2013). 'Program statistics'. Retrieved from www.pefc.org/images/stories/documents/Global_Stats/2012-05_PEFC_Global_Certificates.pdf

Reinhardt, F., Stavinsand, R. and Vietor, R. (2011). 'Corporate social responsibility through an economic lens', *The Review of Environmental Economics and Policy*, vol. 2, no. 2, pp. 219–239.

Rickenbach, M. and Overdevest, C. (2006). 'More than markets: Assessing Forest Stewardship Council (FSC) certification as a policy tool', *Journal of Forestry*, vol. 104, no. 3, pp. 143–147.

Salzmann, O., Ionescu-Somers, A. and Steger, U. (2005). 'The business case for corporate sustainability: Literature review and research options', *European Management Journal*, vol. 23, no. 1, pp. 27–36.

Sprinkle, G. B. and Maines, L. A. (2010). 'The benefits and costs of corporate social responsibility', *Business Horizons*, vol. 53, pp. 445–453.

Vidal, N. and Kozak, R. (2008a). 'The recent evolution of corporate responsibility practices in the forestry sector', *International Forestry Review*, vol. 10, no. 1, pp. 1–13.

Vidal, N. and Kozak, R. (2008b). 'Corporate responsibility practices in the forestry sector: Definitions and the role of context', *The Journal of Corporate Citizenship*, vol. 31, pp. 59–75.

Vidal, N. G., Kozak, R. A. and Cohen, D. (2005). 'Chain of custody certification: An assessment of the North American solid wood sector', *Forest Policy and Economics*, vol. 7, pp. 345–355.

Vogel, D. (2005). 'Is there a market for virtue?' *California Management Review*, vol. 47, no. 4, pp. 19–45.

Wood, D. (1991). 'Corporate social performance revisited', *Academy of Management Review*, vol. 16, no. 4, pp. 691–718.

Wood, D. (2010). 'Measuring corporate social performance: A review', *International Journal of Management Reviews*, vol. 12, no. 1, pp. 50 84.

PART 6

Emerging issues and developments

30

FORESTRY AND THE NEW INSTITUTIONAL ECONOMICS

Sen Wang,[*][1] *Tim Bogle*[2] *and G. Cornelis van Kooten*[3]

[*]CORRESPONDING AUTHOR: SENWANG@NRCAN.GC.CA
[1]CANADIAN FOREST SERVICE, NATURAL RESOURCES CANADA.
[2]BRITISH COLUMBIA MINISTRY OF FORESTS, LANDS AND NATURAL RESOURCE OPERATIONS,
VICTORIA, CANADA.
[3]DEPARTMENT OF ECONOMICS, UNIVERSITY OF VICTORIA, CANADA.

Abstract

Forestry activities are characterized by an array of complex institutional constraints and contractual arrangements. The new institutional economics (NIE) offers an appropriate framework for examining the intricacies of the institutional environment and contractual structures in the forest sector, and for understanding the implications of alternative institutional arrangements for economic outcomes and the behaviour of the contractual parties involved. The chapter provides an overview of the genesis, scope and main developments of NIE, with a particular emphasis on several important strands that include property rights and contracting, transaction cost economics, moral hazard and information and principal-agent relationships. A British Columbia case study is provided to illustrate how principal-agent (PA) theory can be applied to a problem in forestry. The NIE-PA analysis yields insights into the patterns of behaviour on the part of forest companies and other players, and the influence of institutional arrangements and policy choices on silvicultural outcomes.

Keywords

Contractual arrangements, institutions, moral hazard, principal-agent theory, property rights, stumpage, transaction cost economics

Introduction

A forest is much more than a bunch of trees. A myriad of biota within a complex biophysical environment are found below the surface of the forested landscape. When forests are viewed from the perspectives of society and the economy, a complex network of relationships and institutions comes onto the scene. It is this dynamic network of institutional arrangements that provides necessary governance structures for an array of activities related to the management and use of forest resources.

A variety of economic approaches have been utilized to analyze forestry problems (van Kooten, 1993; van Kooten and Bulte, 2000; Hyde, 2012). This chapter focuses on a particular approach, known as the new institutional economics (NIE), and how it can be applied to forestry. NIE builds upon and extends neoclassical economic theory by incorporating into mainstream economics a body of theory pertaining to institutions (Coase, 1984, 1998; Eggertsson, 1990; Furubotn and Richter, 1997). Since the 1970s, NIE has impacted the social sciences, especially economics and political science. Uniting theoretical and empirical research to examine the role of institutions in economic activities, NIE comprises work regarding property rights, transaction costs, hierarchy and organization, public choice and so on. In recent years, NIE has increasingly provided fresh insights into various sectors of the economy, including forestry (Wang and van Kooten, 2001). In this chapter, we provide an overview of what NIE contributes to economic practice and illustrate the potential contribution of NIE in terms of forest policy using a Canadian case study.

In the next section, we examine NIE from the perspective of its origin, scope and development, synthesizing the major components that constitute NIE. We identify three main strands in the economics literature: property rights and contracting, transaction cost economics (TCE) and information exchange and the principal-agent (PA) problem. We then show how NIE has evolved to incorporate these overlapping strands, thereby shedding light on issues related to economic activities and, in the context of forestry and natural resources, common property resources. This is followed by a case study from British Columbia (BC), Canada, where public ownership of forestlands dominates; the case study illustrates how NIE can help the public landowner improve the way in which private companies manage forests to better achieve the government's objectives. The chapter ends with a few concluding remarks.

NIE: Genesis, scope and developments

As a distinct field of economics, the NIE traces its origins to the earlier or 'old' institutional economics found in the writings of Thorstein Veblen, John Commons, Wesley Mitchell, Clarence Ayres and others. These writers were discouraged by what they perceived to be the lack of explanatory power in neoclassical economics, and its failure to take account of institutions. However, the old institutional economics sought to jettison much of neoclassical theory, while offering little to take its place except descriptive analyses that took each situation as somehow unique.

The term 'new institutional economics' was coined by Oliver Williamson to distinguish it from the 'old' institutional economics (Coase, 1998). There is now a consensus among scholars that Ronald Coase's 1937 paper, 'The Nature of the Firm', provided the original inspiration for the development of NIE. With a focus centring around the theory of the firm and transaction costs, NIE benefited from the ideas of the Austrian school of economics (Hayek, 1937, 1945) and built upon developments pertaining to the economics of information (Stigler, 1961; Stiglitz, 1975), human behaviour and cognitive science (Simon, 1957, 1962), organizations and markets (Williamson, 1975, 1985; Simon, 1991), the theory of property rights (Demsetz, 1967; Alchian and Demsetz, 1972; Barzel, 1989; Pejovich, 1995), institutions (North, 1990), TCE (Williamson, 1979, 1998) and social norms and collective action (Ostrom, 1990, 2000, 2005).

Institutions, property rights and contractual arrangements

Institutions are the humanly devised constraints that structure political, economic and social interactions. They consist of both informal constraints (sanctions, taboos, customs, traditions and codes of conduct), and formal rules (constitutions, laws, property rights) (North, 1991). Specifically, institutions are defined as the legal, administrative and customary arrangements for repeated human interactions. Their major function is to enhance the predictability of human behaviour (Pejovich, 1995). The growing literature around institutions points to a distinction between the institutional environment and institutional arrangements. Collectively, the institutional environment and institutional arrangements constitute governance structures. Broadly speaking, institutions provide a system of rules plus the instruments that serve to enforce the rules. In daily life, we observe explicit and implicit contractual frameworks, which include markets, firms and mixed modes within which transactions occur.

The notion of property rights is central to institutions. Property rights refer to the socially sanctioned and enforceable claims that an individual or a group has to the benefits associated with certain physical assets, subject to the conditions that society places on the use of these assets. Property rights have a number of dimensions, including comprehensiveness, duration, transferability, benefits, exclusiveness and security. In economic activities, the property rights over an asset indicate the individual's (group's) ability to consume the good or receive the services of the asset directly, or to consume it indirectly through exchange. Property rights include: (1) the right to use an asset, (2) the right to earn income from an asset and contract over terms of use with other individuals and (3) the right to transfer ownership to another party (Demsetz, 1967; Barzel, 1989). It is important to note that property rights are claims that are recognized and enforced by authorities, most notably the government (Furubotn and Pejovich, 1972).

Production and exchange involve contractual arrangements. As a legally enforceable agreement between two parties, a contract is a legal commitment to which each party gives express approval (generally in written form) and to which a particular body of law applies. Contractual activities take place not only for the purpose of accomplishing the exchange of goods and services, but also to permit the exchange of bundles of property rights (Furubotn and Pejovich, 1972). From the standpoint of markets and organizations, contract terms are influenced by a number of factors, including the access that contractual parties have to information, the costs of negotiating and the opportunities for cheating (Simon, 1991). Although it is important to examine the institutional environment and the property rights structure surrounding an economic activity, it would be a mistake to neglect contractual arrangements. In fact, analysis at the level of contractual terms will often yield deep insights regarding economic incentives and transaction costs.

TCE

In spite of the fundamental role of markets in coordinating economic activities, the firm has been recognized as a primary coordination mechanism. According to Coase (1937), the nature of the firm is to reduce the number of transactions for the purpose of producing a more efficient outcome. As long as the firm can coordinate a transaction at a lower cost than the market, it pays to internalize the function. Otherwise, a firm will allocate the function to the market.

The theory of the firm is viewed as the core of NIE, and TCE is at the heart of that theory and at the centre of the economics of organization. The term 'transaction costs' has many definitions. Generally speaking, transaction costs refer to costs incurred for the creation, maintenance, use and change of institutions and organizations. They include the costs of defining rights, the costs of utilizing and enforcing the rights specified and the costs of information and negotiation.

Humans are assumed to be rational economic agents to a certain degree and people are opportunistic. Bounded rationality and opportunism are two key assumptions underlying the TCE theory. Although the opportunistic aspect of human nature is easy to imagine, the concept of bounded rationality needs elaboration (Simon, 1957). Simply, individuals cannot contemplate and contract for every contingency that might arise over the course of the transaction. According to Williamson (1975), the central concern for economic organizations is to devise contracts and governance structures that have the purpose and effect of economizing on bounded rationality, while simultaneously safeguarding transactions against the hazard of opportunism.

From a TCE standpoint, firms and markets are alternative means of economic organization. Whether transactions are organized within a firm (hierarchically) or across a market between autonomous firms is a decision variable. Which mode is adopted depends on the transaction costs that attend each. The basic premise of TCE is that transactions tend to be organized in ways that maximize the net benefits they provide. Transactions differ in their attributes and are thus aligned with governance structures that differ in their costs and competence in a transaction-cost economizing way. Differential transaction costs give rise to discriminating institutional alignment. Each mode of governance is defined by a series of attributes, whereupon each displays discrete structural differences with respect to both cost and competence.

Transaction costs are difficult to quantify, but this difficulty is mitigated by the fact that transaction costs are always assessed in a comparative way, in which one mode of contracting is compared with another. Accordingly, it is the difference between rather than the absolute magnitude of transaction costs that matters (Williamson, 1985). Empirical research on transaction costs focuses on the question of whether contracting practices and governance structures line up with the attributes of transactions as predicted (Williamson, 1985).

The key technical, human and behavioural dimensions of transactions correspond to asset specificity, bounded rationality and opportunism, respectively (Table 30.1). Generally speaking, the more specialized an asset, the higher is the possibility of appropriating the benefits arising from its designated and intended uses, but the less adaptable is the asset (or skill) to employment in an alternative use. Asset specificity is positively correlated with firm size, with small companies more likely to have general purpose plant and equipment. The market is the main governance structure for nonspecific transactions in the case of both occasional and recurrent contracting.

Table 30.1 Important parameters of transaction cost economics (TCE).

Dimension	Attributes
Specificity	*Physical asset specificity* is a measure of asset redeployability of investments. Special-purpose investments may be risky because specialized assets cannot be redeployed if contracts are interrupted or prematurely terminated. General-purpose investments do not pose the same difficulties. Hence, the more specific a physical asset or skill, the lower is its opportunity cost in its best alternative use, and the cost of transacting to redeploy the asset is higher.
	Human asset specificity is a measure of relation-specific investments, such as bilateral dependency.
Uncertainty	The capacity of the governance structure to adapting to disturbances is measured by the probability of continuation (durability of firm-specific assets). Contract length is important because long-term contracts mitigate inefficiencies associated with ex ante underinvestment and ex post opportunism.
Frequency	The degree to which transactions recur is positively related with the specialization of governance structures. For idiosyncratic investment, when frequency reaches a certain point, unified governance comes into being.

Moral hazard and information: The PA problem

Contracts are important because of their indispensable role in coordinating the exchange of products and services (Hart, 1995). With contracts, the issues of moral hazard and agency should be dealt with via appropriate specification of the contractual terms. The incidence and effect of moral hazard have been analyzed extensively in the context of businesses and other organizations (Jensen and Meckling, 1976; Holmström, 1979). The notion of moral hazard may best be understood from the perspective of human behaviour, using a PA setting.

Economic activities are associated with various types of available information that may be limited, incomplete or even irrelevant. Strictly speaking, there are two aspects to the information problem, namely, information deficiency and information asymmetry (Ross, 1973; Stiglitz, 1974, 1975). The recognition that the transfer of information between principal and agent is a key element to their effective relationship has led to significant theoretical research into the economics of information (Stigler, 1961; Campbell, 2006).

The agent has an incentive to hide crucial information from the principal in order to exploit opportunities for gain – to choose actions that are in the agent's interest, but not necessarily in the interests of the principal. It is entirely possible that the agent possesses hidden information or goals that are detrimental to the principal (Holmström, 1979) – there is a 'propensity [for] human agents to behave opportunistically' (Williamson, 1985, p. 51). Ross (1973) was one of the first to formally model PA relationships, while Fama (1980) elaborated the agency problem using the theory of the firm.

A PA relationship usually develops because it is mutually beneficial. If the parties share a common objective and understanding, then the agent's actions will bring about outcomes desired by the principal. Stiglitz (1974) produced a unified description of the PA problem in natural resources; in the case of the landlord–tenant relation in agriculture, he concluded that, when direct supervision is either costly or ineffective, the use of sharecropping has an incentive and risk-sharing effect that leads to greater efficiency because it internalizes the activity. Sterner (2003) notes that information asymmetry in natural resources between the principal and agent can be so severe that simple rental agreements may be the only appropriate instrument.

As an example of the PA relationship in forestry, consider a forest landowner who hires a logging firm to salvage timber damaged by wind, pests or fire. If the logging firm has the ability to select which timber to harvest, it will harvest the least damaged timber, leaving more questionable timber standing. By removing more valuable trees and not addressing proximate areas, the agent reduces the value of the principal's resource base. A proper design of contractual terms with careful consideration of potential agency problems will help diminish transaction costs and achieve efficient outcomes from an economic standpoint.

Institutions, social norms and economic development

Problems of economic development and common property cannot be addressed using the incomplete tools of neoclassical economics (e.g. Ostrom, 2000).[1] Neoclassical economics lacks an appropriate theory for addressing economic institutions, the role of the state and social capital (Fukuyama, 2002). Yet, these factors are important for economic development and the resolution of what Ostrom (1990) refers to as social dilemmas and what economists frequently call externalities.

A region must have a set of institutions within which policy change can occur. Institutions consist of formal rules (constitutions, laws and property rights) and include such things as commercial and criminal courts. Research in economic development now stresses the need for good institutions, as some institutions retard rather than promote growth (La Porta et al., 1999), or become an obstacle to resolving social dilemmas in resource management (Ostrom, 2000). To remain effective, institutions need to evolve over time in response to changing circumstances.

The government's role and performance are essential to economic development and the resolution of social dilemmas (La Porta et al., 1999). Good governments protect property rights and individual freedom, keep regulations to a minimum, provide an adequate level of public goods (e.g. infrastructure, schools, health care) and are run by bureaucrats who are generally competent and ethical.

The third factor needed to resolve social dilemmas is social capital (Putnam, 2000), or 'the proper cultural predispositions on the part of economic and political actors' (Fukuyama, 2002, p. 24). The 'cultural factor' constitutes informal constraints (sanctions, taboos, customs, traditions and norms or codes of conduct) that structure political, economic and social interactions. Social capital has both individual and aggregate components (Gelauff, 2003). Individual social capital consists of intrinsic aspects (charisma, values) and aspects in which one can invest (trustworthiness, personal networks), although these two aspects are difficult to separate. Aggregate social capital, on the other hand, constitutes the total of the social capital of the individuals in society, varying by form (trust in people, trust in government, level of participation in society), place (firm, region, neighbourhood) and group (ethnic and religious groups, service or sport organizations, gangs). It is difficult for society to invest in aggregate social capital because the manner in which the social capital of individuals is aggregated is not clear. A society can only invest in culture by somehow affecting individuals who do the investing. For example, society can encourage couples to stay together longer by making divorce more difficult, or encourage church attendance by providing tax incentives for charitable giving, but both actions fail to address culture directly.

Trust is perhaps the most important component of social capital: 'Virtually every commercial transaction has within itself an element of trust, certainly any transaction conducted over a period of time' (Arrow, 1972, p. 357). Trust is not social capital, but a manifestation of it; trust is related to institutions and affects the costs of transacting. If confidence in an enforcement agency falters, one may not trust others to fulfil their agreements and thus enter into fewer agreements. There is an element of trust in any transaction in which one has to decide before being able to observe the action of the other party to the transaction. One must assume that the other person is not acting with guile and is transparent in information sharing. Like other components of social capital, trust makes an economy function more efficiently (Fukuyama, 1999).

In addition to trust, other elements of social capital include social norms, or behavioural strategies (e.g. always do p if q occurs) subscribed to by all in society, and networks of civic engagement (membership in clubs, church organizations, etc.) that enhance cooperation. Ostrom (2000) shows how social norms of reciprocity and trust, combined with local enforcement and graduated sanctions, result in effective resource management regimes.

Ostrom (1990, 2005) demonstrated that common property could be successfully managed by the users of a resource, given a set of conditions that includes trust and carefully designed cooperation-enhancing incentives. Her work in the area of collective action and political economy provides a link between the fields of organizational theory and political science. Her writings have made significant contributions to advancing our understanding of the evolution of natural resource institutions. In the context of forest ecosystems, for example, Ostrom (1998) rejected simple, large-scale, centralized governance units, arguing that forest biodiversity needs to be matched by institutional diversity. Ostrom also showed that, when communities are constructively connected in decision making regarding forest resource rules, there is not only a higher likelihood of the rules being followed but also a greater community commitment to the forest as evidenced by self-monitoring within the community than when an authority determines a forest policy without community involvement (Ostrom and Nagendra, 2006).

Applications of the NIE framework to forestry

Economists with an interest in forestry are keenly aware that the institutional framework plays a major role in how forests are managed. Forest ecosystems provide timber and nontimber products, as well as a variety of environmental goods and services. The latter include such things as carbon uptake, waste receptor, water quality, water run-off control and wildlife habitat services, and a place to recreate, view wildlife and so on. Regardless of who owns the forestland, the government plays a role to ensure that environmental goods and services are properly taken into account in forest management, although this function has sometimes defaulted to forest certifiers (see van Kooten and Folmer, 2004, pp. 397–409). If forests are publicly owned, the tenure arrangements significantly influence both the policy choices available to the public landowner and the responses of the forestry firms (Nelson, 2007).

Forest policy and tenure in BC

Consider, for example, BC, where some 95% of forestland is publicly owned. Wang and van Kooten (2001) applied an NIE framework to investigate forest companies' behaviour concerning the requirement to restore logged forests to a free-to-grow state. One component of their research focused on transaction costs and companies' decisions on whether to outsource silvicultural activities. A model was developed to test the relationship between a firm's choice of contractual forms and (a) the attributes of the activity (e.g. specificity of technical skills, frequency of operations and uncertainty in controlling performance quality), and (b) the characteristics of the firm (e.g. company size). The empirical results confirmed the transaction cost logic that silvicultural activities performed in-house are likely those that are complex to manage, have a low degree of seasonality, require extraordinary human skills and involve highly specialized physical assets. As asset specificity or specialized skill increased and the duration of the activity decreased, contracting an activity became more attractive to the firm.

Laffont and Martimort (2002) point out that the principal may sometimes *screen* agents as a tactic to get desired results. Until recently, for example, provincial governments in Canada regularly required forest companies to operate a mill as a condition for obtaining access to large timber quota – known as the appurtenancy. Appurtenancy introduced a commitment from a company and served many purposes, including investment in infrastructure, the employment of local people and the increased likelihood that the company would take a longer rather than shorter term view of the forest resource. However, this screening mechanism was more recently seen as an impediment to a competitive forest industry and was eliminated under the BC government's 2003 Forestry Revitalization Plan (Niquidet, 2008).

Clearly, the most common PA relationship likely pertains to the timber disposition on public forestland. The principal must decide on the tenure arrangement and the bundle of property rights and responsibilities to allocate to a forestry firm, which, in turn, influences the complexity of the PA relationship. BC has one of the highest proportions of public forestland ownership in the world, but the government lacks the internal capacity, capital and industry experience to operate logging operations and chooses instead to contract this out to specialized forestry firms. To manage this PA relationship, the BC government uses standing timber sale licenses administered through BC Timber Sales (TSL), two types of volume-based tenures – nonreplaceable (NR) and replaceable (R) – and an area-based tenure. The latter tenure consists primarily of tree farm licenses (TFL) operated by a forest company, although there are smaller area-based tenures such as woodlots managed by individuals and community forests managed by community groups. In Figure 30.1, we provide an indication of the control over management that the principal grants the agent. Three management characteristics are displayed in the figure: (1) the exclusivity of property rights enjoyed by the license holder (vertical axis), (2) the term of the license agreement (horizontal axis) and (3) the size of the extent to which an agent can impact the forest footprint (with a larger font indicative of a greater footprint).

A timber sales license provides complete exclusive rights within the physical boundaries of the timber sale area developed for harvest by the principal, but rights are short-lived. This tenure maximizes control that the principal can exert over the management of harvests. The agent is provided exclusive rights to the area defined by the TSL, but the principal must still be aware of incentive constraints to ensure the best possible outcome from its perspective. Given the repeatability of these transactions and their short duration, the principal gains knowledge of the various agents so as to easily modify future contracts and even refuse certain bidders (Leffler and Rucker, 1991). While providing the highest level of control, in managing TSLs for multiple values, the public landowner may be criticized for failing to achieve the best financial benefit; indeed, administrative cost may even be incurred when the agent decides not to harvest a site (Rucker and Leffler, 1988). To address concerns raised by US softwood lumber producers concerning lack of transparency (market forces) in setting stumpage fees, the provincial government shifted 20% of the volume allocated under long-term R and TFL tenure forms to TSLs sold at auction, thereby also creating a government timber development agency, BC Timber Sales (Niquidet, 2008).

At the other extreme, forest management activities on a particular forest may be completely delegated to the holder of a TFL. A TFL grants exclusive timber rights over a much larger area and for a longer duration than a TSL. Mathey and Nelson (2010) considered optimal decision making within an area-based tenure when mountain pine beetle (MPB) struck, concluding that the tenure-holder's most profitable strategy would actually achieve the government's risk reduction strategy on public land. It is the exclusivity of operations that is assumed to protect the value

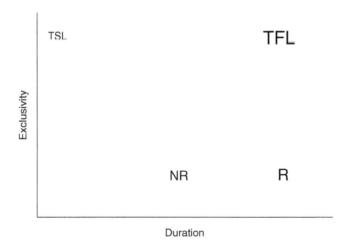

Figure 30.1 Tenures compared by exclusivity, duration and influence on the forest estate: timber supply license (TSL), non-replaceable license (NR), replaceable license (R) and tree farm license (TFL).

of the forest and, indeed, evidence indicates that TFL holders spend more on silviculture than volume-based holders, but not as much as private landowners (Zhang and Pearse, 1996).

A replaceable (R), volume-based tenure holder may experience the same duration of access to timber as with a TFL, simply because the holder operates a wood processing facility that creates local employment and the government is committed to maintaining employment and community stability. However, the R holder may share the area with other tenure holders and thus lacks exclusive rights. The principal needs to create the most effective performance measures in delegating work to agents to ensure that the outcome meets the principal's desires as closely as possible.

Similarly, the NR tenure holder may have harvesting rights in the same general area as the replaceable holder, but such rights have a fixed duration. The NR tenure holder generally does not own a manufacturing facility. It turns out that these two tenure types actually encourage different types of agent behaviour because the objectives of the agents differ. BC has often used the NR license for salvage harvesting, as its fixed duration implies that the license is not meant to be sustainable in perpetuity and conditions can be tailored to describe the timber types eligible for harvest, allowing the principal more discretion in influencing the harvest choices of the licensee. However, the discretion does not allow the principal the ability to change the contractual relationship at a later date.

Catastrophic natural disturbance and forest policy in BC

If we consider how well-designed this continuum of timber tenures is to changing forest conditions and catastrophic natural disturbance, several recent studies provide insights into the PA dilemma faced in BC and other jurisdictions. BC's recent MPB epidemic is unprecedented in the province's history. In response to the catastrophe, the provincial government created a geographic salvage area, and obtained valuable funding from the federal government. Existing tenure holders quickly realized that pine salvage was a key provincial priority, with the government interested in salvaging damaged pine while maintaining an adequate timber supply for the future.

In response to concerns that agents had begun to include more nonpine in the harvest mix in regions affected by the MPB epidemic, the BC government implemented 'partitioning', distributing the harvest level between stand types in an attempt to regulate the harvest activities of the agents. This increased the principal's monitoring requirements. However, as the difficulty of monitoring increases, the principal may find it more beneficial to incentivize truth telling. The historic (pre-2006) use in BC of a dead timber grade (Grade 3), which was charged a nominal fee of $0.25, is an example.[2] Although the principal lost some resource rent if a dead tree was nonetheless valuable, the principal used Grade 3 to encourage the salvage of deteriorating logs while obtaining a clear signal of the agent's activity.

Bogle and van Kooten (2012) use the PA theory to derive a simple monitoring rule for the principal to use in determining the efficacy of the agent's actions towards achieving the principal's post-salvage forest objectives. They highlight the very real risk to the principal's salvage and sustainability objectives when the agent can privately survey and select the most profitable stands to harvest. In an area-based tenure, the agent can confidently manage the entire forest to the company's strategic advantage, an incentive not enjoyed by the agent with quota-based tenure. Lack of coordination between quota-based tenure holders can not only be expected to affect the principal's objective, but the outcome will also be adversely impacted by the multiplicity of agents operating in the region, especially when natural disturbance is considered (Cumming and Armstrong, 2001).

Stumpage allowances are the predominant means used to fund silviculture in BC – that is, silvicultural costs are claimed against stumpage fees. Bogle and van Kooten (2013) use the PA theory to describe the influence of agent responses to stumpage allowances on silvicultural outcomes. They find that allowing agents to manage silvicultural budgets to attain the future timberland productivity outcomes, rather than simply enforcing regeneration standards of harvested stands, could lead to improvements in the future productivity of the forest as it incentivizes agents to include silviculture decisions as part of their harvesting decision.

In PA analysis, it is important to separate the policy variables controlled by the principal from the behavioural variables controlled by the agents, and both from impact variables such as beetle damage and lumber markets that are uncontrollable. Bogle and van Kooten (2014) developed a bilevel mathematical programming approach in the context of the catastrophic disaster caused by the MPB. Using stumpage fees and harvest rates as policy variables, they examined outcomes under two agent types – vertically integrated R license holders and NR license holders, who are primarily market loggers. The authors surmise that R types will generally have built a mill as a result of appurtenancy and thus minimize the costs of harvesting their quota, while NR license holders are considered to be market loggers that maximize net income, or difference between log value and log cost, for the short duration of their license. They find that the cost minimization behaviour of the R tenure holder, rather than the objective of the NR holder, can lead to an outcome that is better aligned with the principal's objectives.

The tenure relations used in the management of BC's public forestlands lead to an interesting array of PA concerns. Under a TSL, the principal is clearly in charge when it comes to forest management, deciding on which areas to harvest and managing the actions of the agent with straightforward timber auctions and short-term contracts. However, this requires the principal to be continually active in preparing forestlands for potential timber sales. At the other extreme, by shifting forest management to the exclusive purview of a single tenure holder, as in the case of a TFL, many of the transactions that occur are eliminated. But the principal has less knowledge about the forest and logging operations than the agent, potentially causing the private company to benefit from informational asymmetries. Other tenure arrangements also have their benefits and drawbacks. The government must act wisely in how it chooses institutions and tenure arrangements in managing public forestlands if it is to efficiently and effectively extract the greatest benefits for citizens.

Conclusion

The NIE complements standard neoclassical economics by drawing attention to the importance of institutions. In this regard, we saw that good governance and rule of law protect property rights and thereby facilitate contracts, while social capital (especially trust) is a key to reducing transaction costs. These come together, for example, in PA relations, where moral hazard, asymmetric information and divergent objectives increase the likelihood that outcomes might not be in the interest of the principal. By recognizing these issues, however, a policymaker can improve the incentives facing private forest companies so that outcomes are more in line with the desires of the public landowner and, thereby, socially efficient.

Over the past two decades, the NIE analytical framework has fruitfully been applied by forest economists to real-world problems. Studies in BC's context suggest that the choice of contractual forms has implications for opportunities to economize on transaction costs. Appropriate governance structures tend to align with transaction attributes and firm characteristics so that some costs of transacting can be minimized. In essence, the choice of governance mode should be dictated by the nature of the activities and transactions involved.

PA analysis has been applied to investigate responses to the MPB catastrophe in BC. Researchers and policymakers have come to recognize that desired outcomes in responding to the MPB epidemic depend not only on the structure of the incentives (mainly differentiated stumpage fees) that forest companies face, but also on the tenure arrangements made with the forest company. Using the PA theory, it is possible to target incentives at specific forest companies rather than attempt to regulate outcomes after setting incentives.

Finally, we find that in forestry there are too few studies that employ NIE methods to examine important problems related to forest tenures, trade in forest products (e.g. many issues in the ongoing Canada-US softwood lumber dispute might best be addressed within an NIE framework of analysis), carbon sequestration in forest ecosystems (van Kooten and de Vries, 2012) and forestry's role in economic development. As institutions and governance structures change over time (e.g. Clean Development Mechanism, REDD+), transaction attributes as well as the characteristics of economic agents are also subject to change. In this changing institutional environment, forest economists are unlikely to ever run out of problems to investigate.

Notes

1   The term 'common property' correctly refers to a situation where a community or nation owns a resource (e.g. forest, fishery), but lack of management leads to open-access exploitation. As common property, the relevant political jurisdiction has responsibility to allocate resource use to maximize net social benefits, but often this obligation is abrogated. Ostrom (1990, 1998, 2000) provides mechanisms in an institutional context to aid jurisdictions in resolving these sorts of social dilemmas. According to Ostrom, open-access resources (traditionally associated with the high-seas fishery, outer space, etc.) can simply be considered common property resources from the perspective of the global community.
2   Monetary values are in Canadian dollars, but the Canadian and U.S. exchange rates are currently equal, so $C = $US for all intents and purposes.

References

Alchian, A. A. and Demsetz, H. (1972). 'Production, information costs, and economic organization', *American Economic Review*, vol. 62, pp. 777–795.

Arrow, K. J. (1972). 'Gifts and exchange', Philosophy and Public Affairs, vol. 1, pp. 343–362.

Barzel, Y. (1989). *Economic Analysis of Property Rights*, Cambridge University Press, Cambridge, UK.

Bogle, T. N. and van Kooten, G. C. (2012). 'Why mountain pine beetle exacerbates a principal-agent relationship: Exploring strategic policy responses to beetle attack in a mixed species forest', *Canadian Journal of Forest Research*, vol. 42, pp. 621–630.

Bogle, T. N. and van Kooten, G. C. (2013). 'Options for maintaining forest productivity after natural disturbance: A principal-agent approach', *Forest Policy & Economics*, vol. 26, pp. 138–144.

Bogle, T. N. and van Kooten, G. C. (2014). 'Protecting timber supply on public land in response to catastrophic natural disturbance: A principal-agent problem', Manuscript submitted for publication.

Campbell, D. E. (2006). *Incentives: Motivation and the Economics of Information* (2nd ed.), Cambridge University Press, Cambridge, UK.

Coase, R. H. (1937). 'The nature of the firm', *Economica*, vol. 4, pp. 386–405.

Coase, R. H. (1984). 'The new institutional economics', *Journal of Institutional and Theoretical Economics*, vol. 140, pp. 229–232.

Coase, R. H. (1998). 'The new institutional economics', *American Economic Review*, vol. 88, no. 2, pp. 72–74.

Cumming, S. G. and Armstrong, G. W. (2001). 'Divided land base and overlapping forest tenure in Alberta, Canada: A simulation study exploring costs of forest policy', *The Forestry Chronicle*, vol. 77, pp. 501–508.

Demsetz, H. (1967). 'Toward a theory of property rights', *American Economic Review*, vol. 57, pp. 347–359.

Eggertsson, T. (1990). *Economic Behavior and Institutions*, Cambridge University Press, Cambridge, UK.

Fama, E. F. (1980). 'Agency problems and the theory of the firm', *Journal of Political Economy*, vol. 88, pp. 288–307.

Fukuyama, F. (1999). *The Great Disruption. Human Nature and the Reconstitution of Social Order*, The Free Press, New York, NY.

Fukuyama, F. (2002). 'Social capital and development: The coming agenda', *SAIS Review*, vol. XXII, no. 1, pp. 23–37.

Furubotn, E. and Pejovich, S. (1972). 'Property rights and economic theory: A survey of recent literature', *Journal of Economic Literature*, vol. 10, pp. 1137–1162.

Furubotn, E. and Richter, R. (1997). *Institutions and Economic Theory: The Contribution of the New Institutional Economics*, University of Michigan Press, Ann Arbor, MI.

Gelauff, G.M.M. (2003). 'Sociaal Kapitaal in De Economic', *Economisch Statistische Berichten*, vol. 88, no. 4398, pp. 3–5.

Hart, O. D. (1995). *Firms, Contracts, and Financial Structure*, Oxford Clarendon Press, Oxford, UK.

Hayek, F. A. (1937). 'Economics and knowledge', *Economica*, vol. 4, pp. 33–54.

Hayek, F. A. (1945). 'The use of knowledge in society', *American Economic Review*, vol. 35, pp. 519–530.

Holmström, B. (1979). 'Moral hazard and observability', *Bell Journal of Economics*, vol. 10, pp. 74–91.

Hyde, W. F. (2012). *The Global Economics of Forestry*, RFF Press, Washington, DC.

Jensen, M. C. and Meckling, W. (1976). 'Theory of the firm: Managerial behavior, agency costs, and ownership structure', *Journal of Financial Economics*, vol. 3, pp. 305–360.

Laffont, J. and Martimort, D. (2002). *The Theory of Incentives: The Principal-Agent Model*, Princeton University Press, Princeton, NJ.

La Porta, R., Lopez-de-Silanes, F., Shleifer, A. and Vishny, R. W. (1999). 'The quality of government', *Journal of Law, Economics and Organization*, vol. 15, no. 1, pp. 222–279.

Leffler, K. B. and Rucker, R. R. (1991). 'Transactions costs and the efficient organization of production: A study of timber-harvesting contracts', *Journal of Political Economy*, vol. 99, pp. 1060–1087.

Mathey, A. and Nelson, H. (2010). 'Assessing forest management strategies under a mountain pine beetle attack in Alberta: Exploring the impacts', *Canadian Journal of Forest Research*, vol. 40, pp. 597–610.

Nelson, H. (2007). 'Does a crisis matter? Forest policy responses to the mountain pine beetle epidemic in British Columbia, *Canadian Journal of Agricultural Economics*, vol. 55, pp. 459–470.

Niquidet, K. (2008). 'Revitalized? An event study of forest policy reform in British Columbia', *Journal of Forest Economics*, vol. 14, pp. 227–241.

North, D. C. (1990). *Institutions, Institutional Change and Economic Performance*, Cambridge University Press, Cambridge, UK.

North, D. C. (1991). 'Institutions', *Journal of Economic Perspectives*, vol. 5, pp. 97–112.

Ostrom, E. (1990). *Governing the Commons: The Evolution of Institutions for Collective Action*, Cambridge University Press, Cambridge, UK.

Ostrom, E. (1998). 'Scales, polycentricity, and incentives: Designing complexity to govern complexity', in L. D. Guruswamy and J. A. McNeely (Eds.), *Protection of Biodiversity: Converging Strategies*, Duke University Press, Durham, NC, pp. 149–168.

Ostrom, E. (2000). 'Collective action and the evolution of social norms', *Journal of Economic Perspectives*, vol. 14, no. 3, pp. 137–158.

Ostrom, E. (2005). *Understanding Institutional Diversity*, Princeton University Press, Princeton, NJ.

Ostrom, E. and Nagendra, H. (2006). Insights on linking forests, trees and people from the air, on the ground and in the laboratory, *Proceedings of the National Academy of Sciences of the United States of America*, vol. 104, no. 51, pp. 19224–19231.

Pejovich, S. (1995). *Economic Analysis of Institutions and Systems*, Kluwer, Dordrecht, The Netherlands.

Putnam, R. D. (2000). *Bowling Alone: The Collapse and Revival of American Community*, Simon & Schuster, New York, NY.

Ross, S. (1973). 'The economic theory of agency: The principal's problem', *American Economic Review*, vol. 63, pp. 134–139.

Rucker, R. R. and Leffler, K. B. (1988). 'To harvest or not to harvest? An analysis of cutting behavior on federal timber sales contracts', *Review of Economics & Statistics*, vol. 70, pp. 207–213.

Simon, H. A. (1957). *Administrative Behavior*, The Free Press, New York, NY.

Simon, H. A. (1962). 'New developments in the theory of the firm', *American Economic Review*, vol. 52, no. 2, pp. 1–15.

Simon, H. A. (1991). 'Organizations and markets', *Journal of Economic Perspectives*, vol. 5, pp. 25–44.

Sterner, T. (2003). *Policy Instruments for Environmental and Natural Resource Management*, Resources for the Future, Washington, DC.

Stigler, G. J. (1961). 'The economics of information', *Journal of Political Economy*, vol. 69, no. 3, pp. 213–225.

Stiglitz, J. E. (1974). 'Incentives and risk sharing in sharecropping', *The Review of Economic Studies*, vol. 41, pp. 219–255.

Stiglitz, J. E. (1975). 'Incentives, risk, and information: Notes towards a theory of hierarchy', *Bell Journal of Economics*, vol. 6, pp. 552–579.

van Kooten, G. C. (1993). *Land Resource Economics and Sustainable Development: Economic Policies and the Common Good*, UBC Press, Vancouver, British Columbia, Canada.

van Kooten, G. C. and Bulte, E. H. (2000). *The Economics of Nature: Managing Biological Assets*, Blackwell, Oxford, UK.

van Kooten, G. C. and de Vries, F. P. (2012). 'Carbon offsets', in J. Shogren (Ed.), *Encyclopedia of Energy, Natural Resource and Environmental Economics*, Elsevier, Oxford, UK, Chapter 165.

van Kooten, G. C. and Folmer, H. (2004). *Land and Forest Economics*, Edward Elgar, Cheltenham, UK.

Wang, S. and van Kooten, G. C. (2001). *Forestry and New Institutional Economics – An Application of Contract Theory to Forest Silvicultural Investment*, Ashgate Publishing Company, Aldershot, UK.

Williamson, O. E. (1975). *Markets and Hierarchies: Analysis and Antitrust Implications*, The Free Press, New York, NY.

Williamson, O. E. (1979). 'Transaction-cost economics: The governance of contractual relations', *Journal of Economic Behavior and Organization*, vol. 1, pp. 5–38.

Williamson, O. E. (1985). *The Economic Institutions of Capitalism*, The Free Press, New York, NY.

Williamson, O. E. (1998). 'Transaction cost economics: How it works; Where it's headed', *De Economist*, vol. 146, issue April, pp. 23–58.

Zhang, D. and Pearse, P. H. (1996). 'Differences in silvicultural investment under various types of forest tenure in British Columbia', *Forest Science*, vol. 42, pp. 442–449.

31

POLITICAL ECONOMY OF FORESTRY

Daowei Zhang

ALUMNI AND GEORGE W. PEAKE JR. PROFESSOR, FOREST ECONOMICS AND POLICY, SCHOOL OF
FORESTRY AND WILDLIFE SCIENCES, AUBURN UNIVERSITY, AUBURN, ALABAMA, USA. TEL.:
(334)844-1067; FAX: (334) 844-1084; ZHANGD1@AUBURN.EDU

Abstract

In this chapter, I first describe the origin and various theories of political economy. I then review
its study methods and present empirical studies of political economy in forestry in various countries in their domestic and international contexts. Finally, I propose a list of a few potentially
fruitful efforts in studying political economy of forestry.

Keywords

Political economy, rent seeking, rent extraction, rent seizing, public choice, neo-Marxism, interest group theory, forestry, environmental services

Introduction

Political economy of forestry deals with the interplay between individuals and the state regarding the production, use and conservation of forest resources. Individuals are defined broadly to
include individual persons, families, firms and groups such as producers and consumers of certain products, environmentalists and the general public. Similarly, the state covers national, sub-
national and super-national (such as European Union and the United Nations) governments
that have coercive and treaty powers and that function and serve as political institutions. Forest
resources include land and forests, which are ecosystems characterized by tree covers varying in
species composition, structure, age and associated process, and commonly including meadows,
streams, fishes and wildlife.

The phrase political economy (*économie politique* in French) first appeared in 1615 with the
book by Antoine de Montchrétien, *Traité de l'Economie Politique*. Originally, *political economy*
meant the study of the conditions under which production or consumption was organized in
the nation-states. The term emphasizes economics, which comes from the Greek *oikos* (meaning
'home') and *nomos* (meaning 'law' or 'order'). It thus was meant to express the laws of production of wealth at the state level, just as economics was the ordering of the home. As all things

at the state level involve law and politics, political economy is also seen as the interplay among economics, law and politics.

Adam Smith, David Ricardo, Karl Marx and Vilfredo Pareto were some exponents of political economy between the eighteenth and twentieth centuries. For them, it was impossible to understand politics without economics or economics without politics; the two systems were intertwined and each depended on the other for its existence. Adam Smith, for example, understood that while an efficient market would emerge from individual choices, those choices were framed by the political system in which they were made, just as the political system was shaped by economic realities. The followers of Adam Smith may believe in an autonomous economic sphere disengaged from politics, but Adam Smith was far more subtle. He called his great book *The Wealth of Nations*. This masterpiece was about wealth as well as nations, and was about political economy (Freeman, 2011).

Today, even though the term economics is more encompassing than political economy, understanding political economy is still very important. This is especially true for forest economists, who study the production and allocation of scarce forest resources which provide multiple beneficial outputs, such as timber, forage, water, recreation, habitats for species and carbon sequestration, that benefit various and often competing groups. The joint production process and increasing and overlapping demands on forest resources complicate their production and the problem of allocating them among alternative uses and combinations of uses. Often the state is involved in forest production and allocation, either directly through public ownership, law and regulations, taxation and subsidies, or indirectly by supporting market mechanism and other formal and informal institutions. This involvement or lack thereof impacts forest production, consumption and sustainability.

However, forest economists have not paid much attention to political economy of forestry. This is perhaps related to the facts that forest economics is basically an applied microeconomics and that most microeconomics textbooks are overwhelmed with analysis of economic market and include little or no discussion of redistributive activities. Nonetheless, the importance of political economy in forestry is well recognized by many, including international organizations and groups that want to promote sustainable forestry development, reduce deforestation and human poverty and protect global environment; and a small but increasing number of forestry-related political economy studies are conducted. I hope, in this chapter, to provide a brief overview of political economy theory and study methodology, to present the findings of a few empirical studies of political economy in forestry and to propose a list of potentially fruitful efforts in the study of political economy in forestry. The next section presents a literature review of political economy theory and methods, followed by political economy studies in forestry. The final section concludes and discusses future research topics.

Political economy theories

With a general agreement that political economy is about the interaction and interplay between state and individuals, classical political economists take various perspectives. For example, Adam Smith promoted a self-regulating economy within a market system while reserving three important roles for the state: to administer justice, to provide for the national defense and to maintain certain enterprises in the public interest that could never be profitable if undertaken privately (i.e. the 'public good' question). In the last century, not only has each of these functions weighed increasingly heavily on governments at every level, but it is widely recognized that markets rely on formal and informal institutions in which governments play important roles. Property rights, for example, are often defined, modified and distributed by governments. Karl Marx, on the other hand, started with production-determining relation and analyzed the class struggles between the capitalist elites and working classes and the role that states played in these struggles.

Two of the most popular political economy theories are the public choice theory and the neo-Marxian theory. The public choice theory is from the public choice school of Buchanan, Tullock and others that use economics to study politics, political process and institutions. In particular, public choice theory assumes that the political market is just like the market: Its equilibrium depends on the availability of rent, and it is often dominated by special interest groups against public interests (Rowley and Schneider, 2004). This special interest group theory is the opposite of the public interest theory of government. On the other hand, some public policies could be better explained by a mixture of special interest theory and public interest theory than by a single one of them. The neo-Marxism theory is sometimes called critical theory, meaning a special kind of social philosophy, and is loosely applied to any social theory or sociological analysis which draws on the ideas of Karl Marx and Friedrich Engels. It also gathers people who are severe critics of capitalism but believe that Marxism became too close to communism (Gunnoe, 2012).

Public choice: Special interest group theory

As we know, the market generally refers to supply of and demand for goods and services. Similarly, there are supply and demand in politics, political processes and institutions such as constitutions, elections, political parties, representative democracy, bureaucracies and interest groups. As these things are decided in the public arena, instead of in private transactions, they are matters of public choice – hence, the name *public choice school*.

The fundamental premise of public choice is that political decision making (voters, politicians, bureaucrats) and private decision making (consumers, producers) behave similarly: They follow the dictates of rational self-interest. Nonetheless, market behavior and political behavior differ in the constraints facing decision makers. One can distinguish the economic constraints in a market or proprietary setting and in a political or nonproprietary setting by analyzing the roles of agents and principals in each case. In the former, the agents are firms who respond to the principals – the consumers. In the latter, the agents are politicians who agree to perform a service for the principals, who are the voters. As the agents and the principals are both self-interested, the agents will not always act in the interest of the principals, particularly if the agents' behavior is costly to monitor.

Table 31.1 compares voting (a public choice) and purchasing of consumer goods (a private choice) and demonstrates the similarities and differences between economic and political markets. It shows that private and public choices differ in important ways, and political markets are more imperfect than private markets. As such, political principals (voters) face different incentives and have a more difficult task to control the behavior of the political agents (elected politicians and bureaucrats). As the interests of the principals and agents are often not aligned properly in political markets because of high transaction costs, political rent seeking is hard to cure. Thus, not only is government not perfect, it is often captured by special interest groups (thus the name *interest group theory* or *capture theory*).

Generally, economists recognize three types of political *rent seeking*: rent creation, rent extraction and rent seizing. *Rent creation* is a process in which firms or interest groups seek rents created by the state by bribing politicians and bureaucrats. Here, bribes include legally permitted rewards such as political campaign contributions as well as outright prosecutable bribery. *Rent extraction* is when politicians and bureaucrats seek rents held by firms by threatening firms with (and sometimes actually imposing) costly regulations or heavy taxations. Both of these types deal with

Table 31.1 Difference between economic and political markets.

Economic market	Political market
In a proprietary setting: Individuals directly bear the consequences of their decisions.	In a nonproprietary setting: Individuals do not always bear the consequences of their decisions.
Consumers/producers make marginal choices	Voters/politicians often make package deals.
Often one choice at a time, although package deals exist	Many choices (candidates/issues) at a time, adding complexity
Consumption is frequent and repetitive.	Infrequent and irregular
Consumers/producers have incentive to be informed.	Voters have little incentive to be informed.
Product or service is a private good.	Product or service is a common good or public good.
Information transferred is clear.	Information transferred to politicians is less clear.

Source: Adapted from Ekelund and Tollison (1994).

states (politicians) and individuals (firms, interest groups), and often imply a two-way, dynamic relationship between them. *Rent seizing*, on the other hand, occurs when competing state actors such as politicians and bureaucrats seek rents that are held by state institutions. These three types of rent seeking exist in varying degrees in all societies, whether they are democratic, capitalistic, authoritarian or socialistic, and a good reference on them is Rowley and Schneider (2004).

Rent creation

Rent creation starts with *interest groups* – collections of individuals with one or more common characteristics such as occupation, who seek collectively to affect legislation or government policy so as to benefit members of the group. To advance their interests, interest groups must overcome the *free-rider* problems and organize themselves, and then enter the political markets by helping elect those political candidates who share their views and supporting them while they are in public office. Their political actions may also be in the form of lobbying, pressure, political actions, campaign contributions and implicit threats.

In order for the interest groups to get organized and take actions, the issues at hand must benefit them greatly while the costs are spread widely on a per-capita basis. Schattschneider (1935, pp. 127–128) describes this condition succinctly, '[b]enefits are concentrated while costs are dispersed'. The higher the concentration in benefits, the easier for the interest groups to organize and get over the free rider problems. On the other hand, the wider the spread in costs, the more difficult for those who lose to find it worth their while to get organized. Interest groups are the *demanders* for certain legislation and public policy.

Not all interest groups are small. Many larger groups, such as farmers or trade associations, are effectively organized as interest groups. Interest groups may also become a *coalition* or form an alliance to advance common interests or causes. Farmers or commodity groups in the United States, for example, are now joined by groups concerned with nutrition, food safety, food processing and distribution, international trade, environmental quality, farm credit and the welfare of rural residents. The alliance may change over time, called *shifting alliance*, depending on the issues to be addressed. For example, the US timber industry and housing industry worked together to seek clarification and modification of the definition of 'harm' under the Endangered Species Act of the United States by filing legal challenges in US federal courts in the 1990s. Their purpose was to restrain federal logging regulations and reduce logging costs and lumber prices. Their opponents then were environmental groups. This alliance between the US timber industry and housing industry was broken up later in another issue: restricting Canadian lumber imports. In the latter case, the timber industry made an alliance with the environmental groups, which claimed that increasing Canadian lumber imports not only injured the US timber industry but also encouraged excessive logging in Canada and harmed the environment (Zhang, 2007).

The suppliers of public policies are legislators, while the public policies themselves are carried out by managers of political 'firms', or bureaucracies. Elected legislators and other politicians, including political appointees, make policy decisions based on their self-interests and their assessments of the benefits and costs of responding to the demand from various interest groups (or lack thereof). These politicians can be viewed as brokers because the real suppliers of policies are the individuals and groups that do not find it worthwhile to get politically organized and to effectively resist having their wealth taken away. This is an unusual concept of supply, but it is a supply function nonetheless.

Becker (1983) looked into the competition among political interest groups and politicians who rationally choose (that is, supply) policies in response to the competing interest group pressures to secure their rents. For politicians, rents may be in the form of election and re-election, campaign donations received, personal income and prestige, powers and influences and egos. Becker (1983) reasoned that political equilibrium depends on the efficiency of each group in producing pressure, the effect of additional pressure on their influence, the number of persons in each group and the deadweight loss. With competition among groups and the assumption that anything that benefits one group must either be financed directly through a tax or indirectly by charging higher prices to another group (including deadweight losses), Becker (1983) argued that resources allocated through the political process tend to be efficient (less deadweight loss) while favoring the politically influential. Gardner (1987) and Gisser (1993) found evidence that politicians favor efficient redistribution schemes that reduce the deadweight loss in American farm subsidy programs.

Once in office, legislators and other politicians may engage in *logrolling*, which is vote trading or vote exchange by legislators to gain sufficient support for a particular piece of legislation. Logrolling, starting with building a coalition of legislators on some issues that benefit certain districts or industries and ending with the redistribution of income toward them, are bad for state economy. Logrolling often leads to unnecessary and costly public works projects and legislation that protects an inefficient domestic industry in certain districts.

Sometimes interest groups, congressional committees responsible for drafting legislation and oversight, and federal agencies that are responsible for administering programs that benefit the interest groups may form a mutually beneficial relationship, referred to as an *iron triangle*. Iron triangles benefit the three parties involved by pursuing a favorable policy for the interest group at the expense of the constituencies that Congress and the federal bureaucracy are supposed to represent, namely the general public. Some of these iron triangles are stronger and last longer than others. But as society evolves, permanent iron triangles, or permanent political equilibriums, are rare. They are established and broken up, all because of rent seeking.

To get the attention of politicians, interest groups need to develop some political skills and connections with the governments. These skills and connections are part of their political capital (the other part includes campaign contributions and size of membership). To gain political capital quickly, some groups hire former staff members of legislators and top bureaucrats (the so-called *Washington insiders* in the United States) to work for them. Thus, we often see a *revolving door*, a term used to describe the movement of personnel between roles as legislators and regulators and the industries affected by the legislation and regulation. Sometimes the roles are performed sequentially, or in certain circumstances, they are performed simultaneously. This relationship, developed among the private sector, politicians and bureaucrats based on the granting of reciprocated privileges, often leads to regulatory capture.

Rent extraction

Whereas rent creation is based on the idea of private individuals benefiting themselves through political process, rent extraction points out that politicians are themselves players in the regulatory auction and have more than one strategy to benefit themselves. They gain not only when compensated by successful rent seekers, but also by threatening private individuals or groups with losses and then allowing themselves to be bought off rather than make good on the threats. Private wealth is extracted in the process.

Proposing onerous legislation and then – for a price – agreeing not to proceed or even withdrawing the legislation proposed is the essence of rent extraction. In this case, private individuals pay, not for special favors, but to avoid disfavor (McChesney, 1987, 1997). Unlike in rent creation, politicians are the demanders and private individuals become the suppliers in rent extraction. As in rent creation, rent extraction is detrimental because political entrepreneurs force private parties to spend scarcer resources to defend themselves.

Rent seizing

Rent seizing occurs when bureaucrats and politicians fight one another to control rents that are held by state institutions and that may rise due to rising demand or government monopoly (Ross, 2001). Rent seizing deals with the workings of the government, which is treated not as a single unit, but as a complicated network of individuals, each with an incentive to maximize his or her own interest. Rents in this case may be in the form of prestige and power, but also in the form of money and property being converted from public treasury to private bank accounts. The competition can, and does, take the form of destroying already-established institutions that might serve to limit the ability of individual bureaucrats to exploit the scarce resource. In other words, the politically powerful, in search of power and personal riches, could dismantle institutions that are set up to protect resources and ensure sustainability.

The often-mentioned 'turf war' among government agencies is usually rooted in rent seizing. Rent seizing could happen between the head of the state and his agencies, between agencies, between different branches of governments (executive vs. legislative), or between different levels of government (central vs. local government). The latter is particularly relevant to forestry as some countries are decentralizing their forest management. The worst turf wars could lead to constitutional crises and real civil wars.

Public interest theory

A contrary and longer-standing theory is the *public interest theory*, which assumes that legislators and other public officials make decisions that are in the 'public interest' or social welfare maximizing. These legislators and public officials are sometimes referred to as paternalists. Unlike 'careerists', whose major objective is to get re-elected, paternalistic politicians tend to act based on their ideologies and social welfare objectives.

This theory posits that governments are to correct externality and other market failures, and democratically elected politicians are the representative and responsive agents of, and make policy and decisions for, the public. Thus, politicians and government agencies are benevolent guardians of public interest, hampered perhaps only by innocent ignorance as they search for the best policies. This is the ideal of 'government of the people, by the people, and for the people'.

In practice, this theory and the ideal government assumption has hardly held up, at least not for all countries and in the long run. Thus, even in all countries that have representative democracy, there need to be mechanisms for check and balance. Most economists have by now abandoned the belief that the main purpose of government regulation is to correct for failures of private markets. At the minimum, it is increasingly difficult to fit all of the complexities and varieties of experiences into this traditional representative agent model of government in economic policymaking.

Mixed theory (private-public theory)

The public interest theory and private interest theory are not separate or mutually exclusive in all cases. For example, the special interest group theory is not divorced from market failures, because the presence of a market failure and a policy to correct it generate distributive consequences and opportunities for rent-seeking, although rent-seeking does not require the presence of a market failure. In other words, sometimes economically efficient choices may coincide with choices in the interest of one or more groups; hence, there is a need to disentangle economic and political influences. Accordingly, a hybrid theory which allows for the influence of politics and economic efficiency has been proposed (e.g. Joskow, 1972; Noll, 1989). This hybrid theory better explains some regulations that are actually public spirited and do resolve certain market failures. For example, trade tariffs have actually come down in the United States since the 1930s, not because of lack of demand for protectionism. The rise of exporter interest groups as a competing interest group to the protectionists is significant (Gilligan, 1997), but the increasing popularity of free trade ideology, after the public witnessed the devastation and destruction of escalating protectionism in the 1930s, may have something to do with US lawmakers' willingness to go with free trade legislations in the last 80 years.

This hybrid theory may be called private-public interest theory. A variation of it is the 'bootleggers and Baptists' paradigm. According to Yandle (1989), 'Baptists' are those who promote a public or private interest by attaching it to other issues that have broad public support (such as resource conservation and environment protection), whereas 'bootleggers' are those who support the same interest without attaching it to any public issue. Despite their different motives, by working together, Baptists and bootleggers can sometimes secure an economic regulation that would otherwise be unobtainable if they worked alone. In fact, many interest groups claim to represent and lobby for 'public interest'.

The political logic represented in these theories is applicable to politicians in various political systems. In the United States, it not only applies to individual members of Congress, but also to the President, even though the US president supposedly has a much broader constituency than individual members of Congress and would have less interest in favoring particular regions or industries. First, the President himself has self-interest. Further, those who help a presidential candidate get elected want and expect to get compensated in some form from him. Finally, the US Constitution, based on the separation of powers doctrine, limits presidential power, making the president listen to Congress on many matters.

Neo-Marxism theory

As stated earlier, neo-Marxism is a term loosely applied to any social theory that draws on the ideas of Karl Marx and Friedrich Engels and is critical of the capitalistic system. Marx and others used dialectics, which is a method of studying change and emphasizes how process and change result from the interaction or conflict of contradictory forces, to analyze economic and social evolution. In particular, Marx used this method to examine the dynamics of social change as they are rooted in class struggles taking place in the mode of production and between capital (the capitalistic elites) and wage labors. Marx saw capitalism as a process that requires ceaseless accumulation (or growth) in order to survive, and capitalists have to confront various barriers in their efforts to expand (Gunnoe, 2012). The study of these barriers and their manifestations is the focus of Marxian crisis theory and gives rise to Marxian political economy in class tension and struggles.

Political economy studies in forestry

Because forests produce multiple benefits, the political economy of forestry – whether it is growing trees, timber production (harvesting), forest products manufacturing, various recreational uses of forests or forest ecosystem services – may have many players (interest groups) and coalitions of groups, which sometimes include groups outside the country or jurisdiction. These groups demand certain forest policies, and the legislators make laws and direct bureaucracy to implement the policies.

Studies of political economy in forestry often use a combination of historical, institutional, analytical (process-oriented or procedural) and comparative approaches. Historical approaches review past events to describe the evolution of laws and regulations. Institutional approaches focus on the rules of the game and institutions that make the laws and regulations. Analytical approaches examine powers among groups and the process of policymaking. Comparative approaches are used to compare policies and institutions in various countries or regions or divide the policymaking process at various stages. Nonetheless, the general theoretical basis for these studies is public choice, public interest theory and neo-Marxism theory.

Public interest theory versus special interest theory: The case of the US Forest Service

The change in forest management regimes in the US national forests may illustrate the changing dynamic in political economy in forestry in the United States (Walker, 1994). Gifford Pinchot, the first chief of the US Forest Service, was sent to Europe to study forestry around 1890 because his father saw the massive economic and environmental devastation of large-scale timber harvesting. Once he became the chief of the US Forest Service, he saw that its mission was to achieve 'the greatest good to the greatest number for the longest run' and that the agency should use this ideal to manage US national forests and to adjudicate conflicts. As Gifford Pinchot is often regarded as a paternalist type of politician, perhaps this initiation phase – establishment of national forests, protecting forests from fires, and the practice of silviculture – can be best explained by the *public interest theory*.

The second phase, which could be called the 'timber phase', began after the Second World War when US population, personal income and housing demand grew drastically. So timber production was a priority for the US Forest Service, and the iron triangle among the forest industry (timber buyers), the US Forest Service and congressional subcommittees responsible for forestry legislation seemed to exist (Le Master, 1984). Thus, this phase may be better explained by the *special interest theory*.

In the later part of this phase, 'multiple uses' and environmental services of forests appeared, and the Forest Service encountered some opposition to the expanded timber program. The situation eventually became political enough to create the next phase: the environmental phase. Even when timber production started to yield to environmental concerns in the early 1980s, timber producers had considerable political influence; they lobbied and were able to get the federal government to spend $5 billion to bail them out of timber contracts they had signed when timber prices were high in previous years (Mattey, 1990). This bailout demonstrates how the political economy in forestry worked: It started with an economic cause (the collapse of the timber market and a demand from timber producers for a bailout) and a political action (the actual bailout that cost the country $5 billion).

Propelled by the National Environmental Policy Act in 1969, the environmental phase ushered in environmental assessments, environmental statements and formalized public participation in national forest policy. All these had the favorable effect of slowing the timber program. The environmental phase reached a new plateau when the spotted owl was listed as an endangered species in 1989 and more than 50 million acres of national forests were designated as 'working wilderness area' under the Clinton Administration. This, plus the more than 30 million acres of national forests that had already been designated as 'wilderness area', put nearly half of the 192 million acres of national forests out of timber production. Thus, the pendulum had swung from timber dominance to what in some cases amounted to eliminating wood production and a complete disintegration of the conventional timber iron triangle. This phase is perhaps best explained by the *mixed theory*.

Special interest theory: Rent creation, rent extraction and rent seizing

There are a number of political economy studies in forestry using the special interest theory, most of which focus on rent creation and rent seizing. In terms of forest issues covered, they include governance and tropical deforestation (e.g. Broad, 1995; Ross, 2001; Hapsari, 2011; Burgess, et al., 2011), subsidies (e.g. Repetto and Gillis, 1988; Mattey, 1990; Mehmood and Zhang, 2002), taxes (Tanger and Laband, 2010), community participation (e.g. Khan, 1998; Silva, 2004), biodiversity (e.g. Mehmood and Zhang, 2001; Thoyer and Tubiana, 2003), forest fires (e.g. Dauvergne, 1998; Sun, 2006), international trade (Zhang, 2007) and the role of forests in combating global warming (Hiraldo and Tanner, 2011; Chang, 2012). These studies cover many countries, but are mostly in North America and Asia.

Rent creation

Most of the earlier rent creation studies in forestry are descriptive and explanatory (e.g. Mattey, 1990; Broad, 1995; Dauvergne, 1998). One of the earliest quantitative studies with a theoretical foundation of rent creation was Mehmood and Zhang (2002), who studied state forestry subsidy programs in the United States. In particular, they hypothesized that existence and continuation of state forestry subsidy (cost-share) programs are results of interactions between demanders and suppliers of these programs in various states. The demanders for such programs are the forest industry, nonindustrial private forest (NIPF) landowners who actually receive the subsidy, and state forestry agencies. They used the importance of forest industry (GSP, which is the percent contribution of forestry to gross state products) to measure the power of the forest industry; the share of timber supply from NIPF (TIMBER), the total forest acreage under NIPF control (SIZE) and the total number of NIPF owners (NUMBER) to represent the power of NIPF landowners; and the total state expenditure (EXP) to represent the power of state forestry agencies. They expressed the demand for the cost-share program as

$$\text{Demand}_{csp} = f(P_{csp}, \text{GSP, TIMBER, SIZE, NUMBER, EXP}) \tag{1}$$

where P_{csp} is the price that a state has to pay by introducing a cost-share program. P_{csp} includes the amount of cost-share funds distributed to NIPF landowners, cost of administering the programs and deadweight loss as a result of wealth redistribution. As Mehmood and Zhang (2002) assumed that public willingness to pay for nonmarket environmental services associated with forestry is similar among the 50 states, this model does explain the responsiveness from policymakers on public goods (environmental services).

State lawmakers are on the supply side of these programs, and their decisions to subsidize a certain group are based upon two main factors. One is the cost; the other is the healthiness of the state economy. Because the subsidy comes from taxing the rest of the citizens, the cost of such programs (the deadweight loss) is a factor (Stigler, 1971; Peltzman, 1976; Becker, 1983). Therefore, the lower the cost, the more efficient is the transfer. In the context of the timber market, the subsidy to NIPF landowners can be measured as producers' gains in economic rents at the expense of taxpayers' incomes and consumer surplus, and the deadweight loss is affected by the timber supply and demand elasticity. The greater the elasticity of supply and the less the elasticity of demand, the less costly is the transfer (Gardner, 1987). Unfortunately, as state-specific estimates for timber supply and demand elasticities are not available, they focus only on the healthiness of state economy (DEBT and REVENUE) on the supply side:

$$\text{Supply}_{csp} = f(P_{csp}, \text{DEBT, REVENUE}) \tag{2}$$

P_{csp} and the outcome (CSP, if a cost-share program is present) are analogous to the equilibrium price and quantity in the usual supply and demand relations. Because P_{csp} and the outcome are simultaneously determined through interaction of the supply and demand equations, they used a reduced form equation in their econometric estimation. Deriving this equation requires some simple algebraic manipulation of the supply and demand equations. Because the outcome supplied equals the outcome demanded at the equilibrium, they set the equation equal:

$$D_{csp}(P_{csp}, \text{GSP, TIMBER, SIZE, NUMBER, EXP}) = S_{csp}(P_{csp}, \text{DEBT, REVENUE}) \tag{3}$$

They then solved for the price of cost-share program (P_{csp})

$$P_{csp} = f(\text{GSP, TIMBER, SIZE, NUMBER, EXP, DEBT, REVENUE}) \tag{4}$$

Equation (4) is the basic political economy model for forestry subsidy, although Mehmood and Zhang (2002) had to control for multicolinearity among variables and used the instrumental variable approach to control for simultaneity problems among expenditure, debt and revenue variables. Similar equations have been estimated for various roll call analyses of congressional votes in forestry-related matters (e.g. Mehmood and Zhang, 2001; Zhang and Laband, 2005; Tanger and Laband, 2010). The only difference is that, in the latter cases, opposing interest groups, the interaction between interest groups and politicians (in terms of campaign contributions) and the characteristics of politicians (party affiliations, voting records, ideology and key committee membership) are added.

Perhaps one of the most comprehensive political economy studies in forestry using special interest theory is in Zhang (2007), who analyzed the players, process, politics, laws and economic consequences of the large and long-lasting US-Canada softwood lumber trade dispute. This prolonged battle in the softwood lumber trade is rooted in the huge economic value of the bilateral lumber trade, the size of the lumber markets in both countries, the adverse impact of one country's policies on the other country's lumber market and especially interest group politics in both countries. By capturing the various intricacies of this complex, contentious and explosive topic, Zhang (2007) argued that well-organized US lumber interests have been able to successfully mount the equivalent of a full-court public-policy press by (1) appealing to their elected federal representatives, who in turn exerted enough power to influence the President, thereby (2) capturing the trade regulation bureaucracy, while (3) simultaneously accessing the federal courts and taking advantage of the fact that the chief opposing interest group – the home builders (lumber consumers) – does not have standing in trade dispute. This is a classic example of interest group politics, in which a small group with concentrated interests, financial resources and experienced leadership is able to outcompete a much larger group with diffuse interests.

Looking globally, Repetto and Gillis (1988) discussed government direct and indirect subsidies to forest industry. However, none of the case studies actually used a political economy model in general or rent creation in particular, although they concluded that subsidies have led to excessive forest clearance in six countries – Brazil, China, Indonesia, Malaysia, the Philippines and the United States. They state that in many countries, the policies have deliberately intended to reward special interest groups allied with or otherwise favored by those in power, and that the existence of large resource rents from harvesting mature timber has attracted politicians as well as businessmen to the opportunities for immediate gain (p. 388). Broad (1995) asserted that governments and major forest resource users create a powerful and destructive bond and resulted in deforestation in Indonesia and the Philippines. Dauvergne (1998) looked into the role that decades of 'reckless logging' in Indonesia, due to the collusion between President Suharto and his cronies and timber producers in Indonesia, had in the 1997 forest fire. Hapsari (2011) found that rent-seeking in forestry is still widespread after decentralization in post-Suharto Indonesia. Finally, Burgess et al. (2011) showed that increasing numbers of political jurisdictions lead to increased deforestation and demonstrated the existence of 'political logging cycles', in which illegal logging increases drastically in the years leading up to local elections in the tropics.

Rent extraction

Rent extraction is based on politicians as dishonest agents who broker policies in return for money to politicians who engage in a much more distasteful process: issuing threats to cause harm in order to be 'persuaded' not to implement said threats (McChesney 1987, 1997). Thus, if rent creation is bribes with carrots, rent extraction is extortion with sticks. For example, politicians and bureaucrats may attempt to use heightened environmental regulations or propose tax hikes to extort rents from industry. The most recent cases in forestry in the United States are whether some silvicultural activities and logging roads are exempt from the Clean Water Act.

Although money for nothing is a widespread phenomenon in political markets (McChesney, 1987), the concept of rent extraction has surfaced slowly and much more recently in economic literature than rent creation. It is also much more difficult to discern, and sometimes impossible to follow, not to mention the personal and political risks of documenting it in some countries. One might have to be an insider in order to know how politicians plot strategy and play the game of rent extraction from firms and industries.

Thus, even though rent extraction may be very common in countries where corruption in the political arena is high, I have not found any empirical study that has looked at politicians engineering threats to extort rents in forestry. The closest one that I can think of is a case described in Chapter 4 of Zhang (2007), which shows how the Vander Zalm government in the Province of British Columbia colluded with American lumber producers and then threatened and actually extracted rents from lumber producers in that province to fix the government budget woes in the mid-1980s.

Rent seizing

Ross (2001) is the first one who has studied rent seizing in forestry. He developed his analysis in the context of timber harvesting in four regions in Southeast Asia (the Philippines, two regions in Malaysia and Indonesia). His principal argument is that competition between incumbent politicians and bureaucrats for the right to control unexpected domestic quasi-rents created by an increase in export and high timber prices in international markets leads them to destroy political and economic institutions that implement the sustained yield policies and that may ordinarily promote economic growth. Such regions and countries end up worse off as a result of what ordinarily would be thought of as positive economic booms. He detailed the political process that dismantles forestry bureaucracies and professional forest management and described how politicians seize rents from booming timber exports from these regions. Hapsari (2011) showed that rent seizing takes place between local and central governments after decentralization in Indonesia.

Neo-Marxism

One of the most recent studies using the neo-Marxism theory to study forestry issues is a Ph.D. dissertation (Gunnoe, 2012). Starting with the concept of financialization – a gravitational shift from productive to financial forms of capital accumulation – Gunnoe (2012) conceptualized the relationship between productive and financial capitalists. He saw the rise of the shareholder value conception of managerial controls as an ideological manifestation of the shift from productive capitalists to financial capitalists. He used this framework to explain the historical development of the US forest products industry over the course of the second half of the twentieth century. He argued that both the decisions by managers to sell off their timberland holdings and the growth of institutional investors seeking to expand their investment portfolios are directly related to the process of financialization. He concluded that the financialization of the US forest industry led to favorable outcomes for financial interests, but has left the industry with higher levels of concentration, few employees, heightened risk and declining profits.

Another recent study in forestry using the neo-Marxism framework was Hudson (2011), who analyzed forest politics and the rise of megafires in the American West. He refutes that the rise of megafires in the American West is because of the US Forest Service's century-long fixation on fire suppression. Rather, he argues, the agency never has been the autonomous agent that its detractors and supporters have supposed, but is often caught in cross-fire and in a no-win situation. It has had to manage timber production and fire suppression on public lands without the authority to effectively do so at a landscape scale. Not only has this state of affairs come at a 'high cost to the healthy functioning of many forest ecosystems' (p.156), but it has also had a profoundly negative impact on the agency itself, reducing it to a manager of 'perpetual crisis' or US 'Fire Service'.

Hudson (2011) doubts that the US Forest Service will escape this trap as long as the nation-state remains subordinate to private capital, which is represented by the large timberland owners who lobby against the use of prescribed burning; the real-estate developers and rich individuals who build in the wild-urban interface, assuming that fire suppression is their rightful subsidy; and congressional representatives who use their investigative powers to haul agency leaders up to Capitol Hill to flail them for letting wildfires burn. 'This forest politic dynamics, plus climate change, constitutes a profound contradiction between the perpetuation of functioning, resilient, healthy ecosystems, . . . and the accumulative and the exploitative logic of capitalism' (p. 157). Subsequently, Hudson concluded that until the United States is able to counter the latter, it will have little chance at sustaining the former.

International political economy in forestry

Global forest policies focus on the allocation of global forest resources in economic development and environmental protection. Subsequently, the interaction between the state and individuals in the domestic political economy setting is often influenced by forces in the international arena. Studies of international political economy of forestry focus on how states and domestic interest groups respond to external ideas, conceptions, pressures and powers related to their forests and forest management; and how external forces influence domestic forestry political economy.

Based on per capita income as an indicator of economic development and per capita forest cover as an indicator of forest endowments, Maini (2003) proposed the four clusters of nations regarding their primary concerns of global forest management and the possible reasons that drive these concerns. Humphreys (1996) and Maini (2003) documented the positions that these countries took in the dialogues that led to international programs such as the Tropical Forest Action Plan and the Intergovernmental Panel (Forum) on Forests (IPF/IFF). Thoyer and Tubiana (2003) described the rising powers of developing countries which host a large share of global diversity in the international negotiation of biodiversity conservation.

Similarly, Tanner and Allouche (2011) paid attention to the way that ideas, power and resources are conceptualized, negotiated and implemented by different groups and countries related to climate change and development. Hiraldo and Tanner (2011) looked into the history, key issues and political economy of negotiations that shape the REDD+ (Reducing Emissions from Deforestation and Forest Degradation, enhancement of carbon stock and sustainable management of forests in developing countries initiative) at global level.

One of the most interesting international political economy studies is Chang (2012), which looked at the positions of various countries regarding reducing CO_2 emissions related to climate change. He found that, depending on their carbon balances and economic prosperity, countries either support or reject mandatory emission targets. For example, developing countries such as India and China see that carbon emission is necessary for economic development and thus a basic human right. They are not going to agree with any emission targets that may hinder their economic development. On the other hand, leaders in some developed countries have called on developing countries to do more to reduce their CO_2 emissions without committing their own countries to do more. This international political reality led to the failure of the 2009 Copenhagen Summit on Climate Change.

Studies on international political economy in forestry also cover some aspects of international forces (such as trade agreements and restrictions, market access, foreign aid) on domestic forest governance. Brown (2009) pointed out that paying attention to political economy of forestry improves the effectiveness of foreign aid in forest management in Ghana. Humphreys (1996) described how interest groups challenge the traditional government rule-making authorities and force forest industry and landowners in various countries to develop their own forest certification programs. Cashore, Auld and Newsome (2004) compared the politics of forest certification in five countries and reflected on why there are differences regionally, and assessed the ability of private forest certification to address global forest deterioration. Recently, forest certification/ market access has been translated into government trade and foreign aid policies in the United States and the European Union – as the United States has amended its laws and has started to ban forest products impacts without documentation of legal origins, and the United States and European Union have implemented the FLEGT (Forest Law Enforcement, Governance and Trade) initiative, which intends to alter the political economy dynamics, improve forest governance and enhance sustainable forest management in tropical timber exporting countries. The effectiveness and impacts of these regulations and initiatives remain to be seen.

Conclusions and discussion

In its simplest form, political economy is about the interplay between individuals and the state, and the political economy approach views all political choices – regulation or any other public policies – as demanded and supplied as a function of the interests of those who incur the distributive consequences of policy alternatives. Political economy of forestry deals with interaction and interplay between individuals and the state regarding the production, use and conservation of forest resources. Just as a smoothly functioning economic system is nurtured, fostered and supported by a well-functioning political system, a strong and healthy forest economy needs an efficient and effective government. The government, seen as a complicated nexus of politicians and bureaucrats, is supposed to work for the public interest, such as to correct externality and other market failures. But this is not often the case. Sometimes, they are captured by special interest groups and make decisions that are detrimental to the public interest. In still other cases, they may have an intention to serve the public interest but yield to the more politically powerful. In the worst case, politicians and bureaucrats are special interest groups and actively extort private individuals and firms to benefit themselves.

Because forests produce multiple goods and services that benefit various segments of society at the local, national, and global levels, many interest groups have a stake in forestry. These interest groups and their coalitions evolve over time; so do the relationships they build with politicians and bureaucrats.

Forest economists have not paid much attention to political economy. They are preoccupied with efficiency concerns such as land use, optimal rotation age, timber and nonpriced forest products supply and demand. Their analyses of forest policy are overwhelmingly from the perspective of efficiency and correction of market failure. Further, even though they are good at assessing economic trade-offs, few have assessed political trade-offs, or have studied forest policy from the perspectives and interplay of interest groups, politicians and bureaucrats. Yet many forest policies are not motivated solely by economic efficiency or correcting market failure or even by economic efficiency at all, but by distribution or redistribution of income. Not surprisingly, there are few studies of political economy in forestry, and most are done by economists, political economists and political scientists.

Future research in political economy of forestry could be done on various forest policy programs, such as forest property rights, regulations, taxes, subsidies, tariffs and quotas, forestry contracts in public forests and forest-based economic development. More interesting studies would be on the rising demand for various environmental services – which are often public goods – and the supply of these services by political means. For example, would rising urban populations translate to more stringent regulations on forest practices and land uses in rural areas, or higher taxes on forest lands?

Further research may also be devoted to comparative political economy studies. For example, what was the political economy of forest policies and their impacts in various Eastern European countries after the collapse of the Berlin Wall? Similarly, why did countries in Western Europe choose different ways to implement the 2001 EU Strategy for Reversing Biodiversity Loss? Finally, we forest economists need to devote some efforts to studying political economy in international settings, particularly in international forest policies, agreements, negotiations, implementation and disputes related to global commons such as global warming and biodiversity conservation. In some aspects, the recent negotiations and implementation of REDD+ and FLEGT projects in some countries could serve as good case studies of international political economy in forestry.

Before concluding this chapter, I must point out that, although applying a political economy approach to forestry is not a silver bullet, it does represent an improvement over the largely efficiency-oriented supply and demand (or benefit-cost) analyses of forest economy and management activities. At a minimum, understanding the political economy dimensions of forestry facilitates designing and implementing forestry projects and enhancing their effectiveness, efficiency and equity. To influence forest policymaking and changes in forestry practices, however, we need to understand not only the political economy contexts of forestry, but also how political changes occur and how various players in the forest sector of a jurisdiction secure their powers, play in a power game and respond to new ideas and internal and external demands. Breakthroughs on the supply side, such as breaking science and thoughts from visionaries (e.g. Rachel Carson, whose work fueled the rise of modern environmentalism) also influences public policy in general and political economy in particular.

References

Becker, G. S. (1983). 'A theory of competition among pressure groups for political influence', *Quarterly Journal of Economics*, vol. 98, pp. 371–400.

Broad, B. (1995). 'The political economy of natural resources: Case studies of the Indonesian and Philippine forest sector', *Journal of Developing Areas*, vol. 29, no. 3, pp. 317–340.

Brown, T. (2009). *Politics Matters: Political Economy and Aid Effectiveness*, The IDL Group, Bristol, UK.

Burgess, R., Hansen, M., Olken, B., Potapov, P. and Sieber, S. (2011). *The Political Economy of Deforestation in the Tropics*, Working paper, London School of Economics, London, UK.

Cashore, B., Auld, G. and Newsome, D. (2004). *Governing through Markets: Forest Certification and the Emergence of Non-state Authority*, Yale University Press, New Haven, CT.

Chang, S. J. (2012). *Solving the Problem of Carbon Dioxide Emission*, Paper presented in New Frontiers of Forest Economics, Zurich, Switzerland, June 26–30.

Dauvergne, P. (1998). 'The political economy of Indonesia's 1997 forest fires', *Australian Journal of International Affairs*, vol. 52, no. 1, pp. 13–17.

Ekelund, R. B., Jr. and Tollison, R. D. (1994). *Economics* (4th ed.), Harper Collins College Publishers, New York.

Freeman, G. (2011). 'Global economic downturn: A crisis of political economy'. Retrieved from www.stratfor.com/weekly/20110808-global-economic-downturn-crisis-political-economy

Gardner, B. L. (1987). 'Causes of U.S. farm commodity programs', *Journal of Political Economics*, vol. 95, pp. 290–310.

Gilligan, M. J. (1997). *Empowering Exporters: Reciprocity, Delegation, and Collective Action in American Trade Policy*, University of Michigan Press, Ann Arbor, MI.

Gisser, M. (1993). 'Price support, acreage controls and efficient redistribution', *Journal of Political Economy*, vol. 101, no. 4, pp. 584–611.

Gunnoe, A. A. (2012). 'Seeing the forest from the trees: Finance and managerial control in the U.S. forest products industry, 1945–2008', Ph.D. dissertation, University of Tennessee, Knoxville, TN.

Hapsari, M. (2011). 'The political economy of forest governance in post-Suharto Indonesia', in H. Kimura, Suharto, A. B. Javier, and A. Tangsupvattana (Eds.), *Limits of Good Governance in Developing Countries*, Gadjah Mada University Press, Yogyakarta, Indonesia.

Hiraldo, R. and Tanner, T. (2011). *The Global Political Economy of REDD+: Emerging Social Dimensions in the Emerging Green Economy*, Occasional Paper, United Nations Research Institute for Social Development. Geneva, Switzerland.

Hudson, M. (2011). *Fire Management in the American West: Forest Politics and the Rise of Megafires*, University Press of Colorado, Boulder, CO.

Humphreys, D. (1996). *Forest Politics: The Evolution of International Cooperation*, Earthscan, London.

Joskow, P. (1972). 'Determination of the allowed rate of return in a formal regulatory proceeding', *Bell Journal of Economics*, vol. 3, pp. 632–644.

Khan, N. A. (1998). *A Political Economy of Forest Resource Use: Case Studies of Social Forestry in Bangladesh*, Ashgate, Aldershot, UK.

Le Master, D. (1984). *Decades of Change*, Greenwood Press, Westport, CT.

Maini, J. (2003). 'International dialogue on forests: Impact on national policies and practices', in L. Teeter, B. Cashore and D. Zhang (Eds.), *Forest Policy for Private Forestry: Global and Regional Challenges*, CABI Publishing, Oxon, UK.

Mattey, J. P. (1990). *The Timber Bubble That Burst: Government Policy and the Bailout of 1984*, Oxford University Press, New York.

McChesney, F. (1997). *Money for Nothing: Politicians, Rent Extraction, and Political Extortion*, Harvard University Press, Cambridge, MA.

McChesney, F. S. (1987). 'Rent extraction and rent creation in the economic theory of regulation', *Journal of Legal Studies*, vol. 16, pp. 101–118.

Mehmood, S. and Zhang, D. (2001). 'A roll call analysis of Endangered Speies Act Amendments', *American Journal of Agricultural Economics*, vol. 83, no. 3, pp. 501–512.

Mehmood, S. and Zhang, D. (2002). 'Causes for continuation of state cost-share program for nonindustrial private forest landowners', *Forest Science*, vol. 48, no. 3, pp. 471–478.

Noll, R. (1989). 'Economic perspectives on the politics of regulation', in R. Schmalensee and R. Willing (Eds.), *Handbook of Industrial Organization: Vol. II*, North-Holland, New York.

Peltzman, S. (1976). 'Toward a more general theory of regulation', *Journal of Law and Economics*, vol. 19, pp. 211–240.

Repetto, R. and Gillis, M. (1988). *Public Policies and the Misuse of Forest Resources*, Cambridge University Press, New York.

Ross, M. L. (2001). *Timber Booms and Institutional Breakdown in Southeast Asia*, Cambridge University Press, Cambridge, UK.

Rowley, C. K. and Schneider, F. (2004). *The Encyclopedia of Public Choice*, Kluwer Academic Publisher, New York.

Schattschneider, E. E. (1935). *Politics, Pressure, and the Tariff*, Prentice-Hall, New York.

Silva, E. (2004). 'The political economy of forest policy in Mexico and Chile', *Singapore Journal of Tropical Geography*, vol. 25, pp. 261–280.

Stigler, G. (1971). 'The theory of economic regulations', *Bell Journal of Economics*, vol. 2, pp. 3–21.

Sun, C. (2006). 'A roll call analysis of the Healthy Forests Restoration Act and constituent interests in fire policy', *Forest Policy and Economics*, vol. 9, no. 2, pp. 126–138.

Tanger, S. and Laband, D. (2010). 'An empirical analysis of bill co-sponsorship in the US Senate: The TREE Act of 2007', *Forest Policy and Economics*, vol. 11, pp. 260–265.

Tanner, T., and J. Allouche. 2011. Towards a New Political Economy of Climate Change and Development. Institute of Development Studies Bulletin Volume 42, No. 3., Oxford, UK.

Thoyer, S. and Tubiana, L. (2003). 'Political economy of international negotiations on biodiversity: Players, institutions and global governance', *International Journal of Biotechnology*, vol. 4, no. 2–3, pp. 228–238.

Walker, T. (1994). *An Agency Gone Sour: A Brief History of the U.S. Forest Service and the Breakdown of Its Relationship with the North American Timber Industry*, Hatton-Brown Publishers, Montgomery, AL.

Yandle, B. (1989). *The Political Limits of Environmental Regulation: Tracking the Unicorn*, Quorum Books, New York.

Zhang, D. (2007). *The Softwood Lumber War: Politics, Economics and the Long U.S.-Canada Trade Dispute*, Resources for the Future Press, Washington, DC.

Zhang, D. and Laband, D. (2005). 'From senators to the President: Solving the lumber problem or else', *Public Choice*, vol. 123, pp. 393–410.

32

GAME THEORETIC MODELING IN FOREST ECONOMICS

Pradeep Kumar and Shashi Kant

FACULTY OF FORESTRY, UNIVERSITY OF TORONTO, 33 WILLCOCKS STREET, TORONTO, ONTARIO, M5S 3B3, CANADA. P.KUMAR@MAIL.UTORONTO.CA; SHASHI.KANT@UTORONTO.CA

Abstract

Game theory, a study of mathematical models of conflict and cooperation between rational agents, developed into a full-fledged discipline in the mid-1920s and then got impetus in the early 1950s with the work of John Nash. In the latter half of the twentieth century, game theory found application in many fields, including biology and economics, but the use of game theoretical models has been scarce in forest economics. Since the early 1990s, game theoretic models have been used to explain selected aspects of forest economics such as people's participation in co-management of forests, timber markets, interactions among stakeholders in the case of weak property rights and some other issues. These applications are reviewed and future directions of applications, such as applications to analyze nonmarket and multiple-stakeholder interactions, and inclusion of uncertainty, imperfect information, sequential game forms and evolutionary equilibrium concepts, are suggested.

Keywords

Game theory, forestry, noncooperative models, joint forest management, forest industry, property rights

Introduction

Game theory is the study of strategic decision making in which conflict and cooperation models among rational decision makers are developed and analyzed. These models have been used in varied fields, including biology, economics, political science and psychology (Osborne and Rubinstein, 1994). Forest management issues overlap with biology, economics and other social sciences. Though game theory has been applied extensively in these three disciplines, its application in forest economics has been limited.

In all game theoretic models, the basic entity is a player. The situation of cooperation and conflict arises when two or more players with different objectives interact for the same resource. With the help of game theoretic models, a player can find the optimal strategy in the face of an opponent who may have his or her own strategy (Limaei and Lohmander, 2008).

Some of the most often-used classes of games are cooperative and noncooperative games, strategic and extensive (sequential) games, perfect information and imperfect information games, symmetric and asymmetric games and zero-sum and non-zero-sum games. Depending on the real-life situation, a game may be modeled with any possible combination of the previous attributes, e.g. a game may be noncooperative, extensive, imperfect information, asymmetric and non-zero-sum.

In a strategic game, each player chooses a plan of action simultaneously and only once. In an extensive game, on the other hand, each player can choose an action from the possible sequence of actions, whenever it is his or her turn to make a move (Osborne and Rubinstein, 1994). In both strategic and extensive games, players may or may not be perfectly informed about other players' actions. The games in which players do have perfect information are referred to as games with perfect information, and others are referred to as games with imperfect information. There are also situations when players are perfectly informed about other players' actions but they are not sure about their payoffs. These models are called games with incomplete information.

A symmetric game is one in which the payoffs for playing a particular strategy depend only on the strategy profile (set of strategies of all the players), not on who is playing them. If players' identities can be changed without changing the payoff to the strategies employed by them, then a game is symmetric. Symmetric games are much easier to analyze than asymmetric games because the analysis can be performed by focusing on just one player.

In non-zero-sum games, the aggregate gains and losses between players need not be equal. In other words, one player receiving gains does not necessarily mean that the other players in the game must lose. In contrast, in a zero-sum game, each loss or gain is associated with a corresponding gain or loss to other players. In non-zero-sum games, the sum of the payoffs of all the players may increase or decrease as a result of the outcome of the game; in zero-sum games, it remains constant. In a two-player zero-sum game, each player knows exactly how the other player is affected, because each player knows how he or she is affected by a strategy profile.

The most common definition used to distinguish between cooperative and noncooperative games, according to Harsanyi (1966, p. 616), is: 'Cooperative game is a game where commitments (i.e. agreements, promises and threats) are fully binding and enforceable. In a noncooperative game, commitments have no binding force.'

There are many other ways to classify game theoretic models, but they are not relevant for the purpose of our discussion. It is expedient, however, to introduce evolutionary game theoretic models. Classical game theory requires that all players behave as rational economic agents, which means that each player has clear preferences, knows all alternatives, forms expectations about unknowns and uses economic optimization for his or her choices (Osborne and Rubinstein, 1994). In recent years many scholars have challenged the foundation of rational economic agent, and game theoretic models have started incorporating evolutionary aspects of human behavior. Evolutionary game theoretic models have shifted the focus of analysis to the learning and imitating trait of human behavior (Shahi and Kant, 2007). An evolutionary game is a dynamic process, which describes how players adapt their behavior over the repeated occurrence of a game. The dynamic process of the game provides a platform for an individual learning process (Samuelson, 1997).

In this chapter we briefly review the development of game theory and its applications in forest economics. The chapter is organized in five sections. In the next section we provide a short overview of the history of game theory, followed by four sections about the applications of game theory in forest economics. These sections are focused on the applications of game theory to analyze people's participation in forest management, wood markets, forest land property rights and other areas. In the concluding section, we suggest some potential areas of research on the application of game theory in forest economics.

Overview of game theory

Some game theoretic ideas can be traced back to the eighteenth century, but the major development of the theory began in the 1920s with the work of mathematician Emile Borel (1871–1956) and the polymath John von Neumann (1903–1957) (Osborne, 2004). Borel, between 1921 and 1927, gave the first modern formulation of a mixed strategy, along with finding the min-max solution for two-person games (Dimand and Dimand, 1992). von Neumann (1928) proved the min-max theorem. However, Cournot (1838), much earlier than Borel and von Neumann, had made a revolutionary contribution to the idea of noncooperative equilibrium in oligopoly setup, and Bertrand (1883/1992) had proposed an alternative model of oligopoly.

Zeuthen (1930) proposed a solution to the bargaining problem which Harsanyi (1956) showed is equivalent to Nash's bargaining solution. The year 1944 was decisive in the development of game theory because of the publication of the book *Theory of Games and Economic Behavior* (Osborne, 2004) by von Neumann and Morgenstern (1944), which laid the foundation of the field. The early 1950s also proved to be a defining period in the development of game theory. Kuhn (1950) proposed the formulation of extensive form games, which is being used now, and also gave some basic theorems on sequential games. In his seminal contributions to noncooperative game theory and bargaining theory, John F. Nash published four papers between 1950 and 1953. Nash (1950b, 1951) proved the existence of a strategic equilibrium for noncooperative games, now known as the Nash equilibrium. In his two papers on bargaining theory, Nash (1950a, 1953), he founded axiomatic bargaining theory and proved the existence of the Nash bargaining solution.

In a series of three papers, Harsanyi (1967–1968) constructed the theory of games of incomplete information. This laid the theoretical foundation for information economics that has become one of the major themes of economics and game theory.

Maynard Smith (1972) introduced the concept of evolutionary stable state (ESS), and the major impetus for the use of the ESS concept came from Maynard Smith and Price (1973). Selten (1975) introduced the concept of trembling hand perfect equilibrium that motivated many scholars, such as Kalai and Smorodinsky (1975), to refine the concept of the Nash equilibrium.

Kreps and Wilson (1982) extended the idea of subgame perfect equilibrium to the subgames that begin with information sets with imperfect information; they called it sequential equilibrium. Rubinstein (1982) considered an alternating offer game in which offers are made sequentially until one is accepted, and he showed that when each player's cost of delay is given by some discount factor, there is a unique subgame perfect equilibrium. Bernheim (1984) and Pearce (1984) discussed the issue of rationalizability. Tan and Werlang (1988) provided the theoretical foundations of iteratively undominated strategies and rationalizable strategic behavior. Fudenberg and Tirole (1991) introduced a formal definition of perfect Bayesian equilibrium (PBE) for multiperiod games with observed actions.

Game theoretic models to analyze people's participation in forest management

Forest management has been transforming during the last three decades from exclusion to inclusion of local peoples in forest management. The recognition and incorporation of forest values of local peoples in forest management is one of the key elements of sustainable forest management. Hence, there have been numerous efforts all around the globe to design new forest management programs aimed at active participation of local peoples in forest management. These programs are known by different names such as co-management, community-based management, participatory management and joint management. These programs have attracted the attention of forest economists, and cooperative, noncooperative and evolutionary game theoretic models have been used to analyze peoples' participation in forest management. Next, we discuss all three types of game theoretic models.

Cooperative game theoretic models

Cooperative game theory is not very popular, and in forestry there is only one application, by Kant and Nautiyal (1994). They applied the concepts of a cooperative game between a government agency and a community to determine the shares of two partners in timber harvest under the Joint Forest Management (JFM) program in India. The authors argued that both parties (government and community) in JFM make efforts to obtain cooperation from each other, and therefore cooperative, and not noncooperative, game theory should be used to analyze JFM. It seems that Harsanyi's (1966, p. 616) condition of cooperative games that 'commitments being fully binding and enforceable' is met to a reasonable extent in JFM; therefore, the application of cooperative game theory is appropriate.

In the case of JFM in India, according to the authors, forests are owned by the government; therefore, the government has a monopoly in supplying forest products to a community. In addition, most forest products are consumed by community members; therefore, the community enjoys a monopsony in demanding most forest products. Hence, the relationship between government and communities is of a bilateral monopoly; therefore, the authors used a bargaining model using bilateral gaming approach. Under the JFM program, community members have rights to collect and use all nonnationalized nontimber forest products (NTFPs), the collection of nationalized NTFPs is organized by the government and the final timber harvest is shared between a community and the government. The authors suggested a mechanism based on government and community fixed threat points to determine the shares of community and government in the final timber harvest.

The authors formulated a JFM situation as a two-player game (i.e. the government and the local community), and used Nash's (1950a, 1953) axiomatic bargaining model for its solution. According to the authors Kant and Nautiyal (1994, p. 256), 'A unique equilibrium solution for a bargaining or two-person fixed threat cooperative game, as proposed by Nash (1950a, 1953), is the set of utilities or payoffs which maximizes the product of the two players' utility gains measured from the status quo/fixed threat point or the product of their payoff increments.' The authors conceptualized that the owner's fixed threat payoff, or the payoff in the situation of negotiation failure, is equal to the revenue earned from the sale of forest products and the penalties charged to the offenders less the cost of management and policing. The fixed threat payoff of the community is equal to the value obtained from the illicit extraction of forest products minus penalties paid to the owner.

Suppose π_o and π_c are the fixed threat points per ha per year for the owner (government) and the community, respectively, and these are constants over time. A is the area of forest, R is the rotation age and r is the continuous discount rate. The per-unit area yield of timber products in the form of saw logs, pulpwood and so forth from the forest at rotation is Q_T units, and its market price is P per unit; the per-unit area yield of some composite NTFP is Q_{NT}, and its price P_{NT} is per unit. Assume that the community gets all the quantity of NTFP (which is the case under JFM), but its share as a fraction of the final timber product at rotation is α. Then the present values of the owner's payoff gain (μ_1) and the community's (μ_2) payoff gain are given by:

$$\mu_1 = \frac{(1-\alpha)PQ_T}{exp(rR)} - \frac{\pi_o A[exp(rR)-1]}{exp(rR)[exp(r)-1]} \tag{1}$$

$$\mu_2 = \frac{\alpha P Q_T}{exp(rR)} - \frac{(\pi_c - P_{NT}Q_{NT})A\left[exp(rR)-1\right]}{exp(rR)\left[exp(r)-1\right]} \tag{2}$$

The total payoff gain for both the owner and the community is $\mu = \mu_1 + \mu_2$.

Next, the authors analyzed the case of a single rotation because that is a realistic situation in JFM. They determined the value of forest area A and rotation age R which maximized the total gain μ. Suppose such values of area and rotation age are A^* and R^* and the corresponding values of μ, μ_1, μ_2 and Q_T are μ^*, μ_1^*, μ_2^* and Q_T^*, respectively. By maximizing the product μ_1^*, μ_2^* with respect to fraction α, the authors found:

$$\alpha = 0.5 - \frac{\left(\pi_o - \pi_c + P_{NT}Q_{NT}\right)A^*\left[exp(rR^*)-1\right]}{2PQ_T^*\left[exp(r)-1\right]} \tag{3}$$

The authors discussed the implications of this result for a simplified case in which the quantity of NTFP is negligible and there is a very low rate of time preference ($r \approx 0$). In that case, the community share of the timber harvest will depend upon the difference between the owner's and the community's threat payoffs and the relative value of the difference of total threat payoff (($\pi_o - \pi_c) A^* R^*$) from total area during rotation with respect to return from timber products. The community's share of the final timber product will be half, less than half and more than half when $\pi_o = \pi_c, \pi_o > \pi_c$ and $\pi_o < \pi_c$, respectively.

The authors suggested the use of their method for calculating the shares of different communities, but they also recognized the limitations of fixing different shares for different communities from the perspective of cooperation between communities and government, and suggested some alternative ways to address those limitations.

Noncooperative game theoretic models

The focus of Kant and Nautiyal (1994) was at the community level, and they treated a community as a homogenous unit. Lise (2001) extended the game theoretic analysis of peoples' participation in forest management to the level of an individual using a noncooperative game theoretic approach. Lise (2001) argued that economists generally study a system that is mapped in their minds and not necessarily in real life, and that is why they conduct analysis by assuming the type of game being played. The author argued that instead of assuming a particular type of game being played, the data should tell which game is being played by community members. Similarly, it is not known in advance whether the players are in equilibrium or not. In real life, community members may be playing different games in different communities, and they may not be in equilibrium. Hence, the main objective of Lise (2001) was to develop an algorithm to identify the type of noncooperative game being played by members of an organization such as a village-level organization of JFM.

Lise (2001) converted the n-person game into a two-person game by assuming that each player plays the game against all remaining members, and called it the *participation game*. In this game, each player has two choices – participate (P) or do not participate (D) – and depending on level of participation, they receive payoffs. Let us call player #1 the challenger and player #2 the contender. The payoff matrix of the participation game can be expressed as given in Table 32.1.

Table 32.1 The payoff matrix of the participation game.

		Contender	
	Action	P	D
Challenger	P	p, p	d, c
	D	c, d	q, q

The participation game can take any form of the two-person noncooperative game, depending upon the ordering of payoffs, and Lise (2001) identified six possible categories of this game – assurance, chicken, coordination, noncoordination, Pareto and prisoner's dilemma games.

The author suggested that on the basis of the level of participation and payoff of each member, the category of the game being played should be determined. The payoff (incremental income) will be the total sum of the imputed value of each product that accrues to a member due to his or her participation in the organizational activities. The choice of participation is voluntary, but the membership of the organization is not a good measure of participation, because some members may be participating very actively while others may be reaping the benefits only. Hence, the author suggested a multiattribute (e.g. involvement in decision making, involvement in monitoring and enforcement activities and sharing the benefits) measurement of participation, and the measurement of each attribute can be done using a five-point Lickert scale (ranging from 'not at all' to 'very much'). As far as the level of participation of the first person (challenger) is concerned, it can be determined either by a simple arithmetic sum of all participatory indicators or by using a factor analysis of the participation indicators.

To determine the participation level of the second person (contender), the author assumed that each member of an organization plays the game with all others as composite second player (contender). Hence, the level of participation of all other members (except the first player) needs to be transformed in the participation level of the second player, and for that the author used two measures: (1) the mean and (2) the variance of participation levels of all other members or players.

Once the participation levels of all combinations of the first and the second players are determined, all observations are divided into four groups to determine the values of four payoffs (p, q, c and d). The author suggested two methods of grouping: (1) Euclidian cluster method and (2) homogeneous grouping method. In the case of the Euclidian method, the Euclidian distance within clusters is minimized, while between clusters it is maximized. In the homogeneous grouping method, the observations are grouped into four groups in two steps. The first step separates participators and nonparticipators by using the median of participation as the separating threshold value. In the second step, each of the two subgroups (participators and nonparticipators) is further divided into two subgroups by treating the median of each subgroup as a dividing point.

Once all the participants are divided into four subgroups, the payoff values for different levels of participation are calculated by taking the average of payoffs in those subgroups. In the case of the mean, the grouping will be: large, large (P,P); large, small (P,D); small, large (D,P); and small, small (D,D), while in the case of variance the grouping will be large, large (P,D); large, small (P,P); small, large (D,P); and small, small (D,D).

Lise (2001) used this method to analyze JFM in the Haryana state of India. He calculated the payoffs and estimated regression equations for all four payoffs using a number of forest-related explanatory variables such as use of fuelwood, fodder and bamboo, forest dependence, number of animals and crop production. He found that the reduction in the total consumption of fuelwood from 15 quintals to 7.9 quintals per year will lead to a structural change in payoffs such that the Pareto game will become a reverse prisoner's dilemma game. Hence, with the Lise model, the point of transition from one type of game to another with change in socioeconomic variables can be determined, and this information can be used for policy intervention.

Lise (2005), in addition to estimating the games being played, modeled the choice situation as an infinitely repeated game and determined a trigger strategy or condition – a condition necessary for participation which, once violated, precludes participation forever to punish the deviant. The author used the discount factor as a measure of trigger point, and demonstrated that for the critical discount factor of zero, mutual participation can always be sustained, while for the discount factor of one, mutual participation cannot be sustained. The author used the model to analyze people's participation in JFM programs in three states of India – Haryana, Bihar and Uttar Pradesh – because there are structural differences in the JFM programs of these three states. (Note: The villages included in the study of the states of Bihar and Uttar Pradesh are now in the states of Jharkhand and Uttarakhand, respectively.) The author also derived a quantitative measure for the effectiveness and stability of an organization for forest management. The estimation results showed that the organizations studied for forest management in India varied in their effectiveness. Participation was difficult to achieve in Haryana, while in Bihar and Uttar Pradesh, the villagers were willing to participate. The organizations studied for forest management were more stable when villagers were sensitive to variation in the level of participation of other villagers. It was also found that mutual participation could be sustained in most of the cases; in 23 of the 32 studied villages the critical discount factor was zero, while in 7 villages the critical discount factor was one.

The author found that the willingness of people to participate in a JFM program differed across these three states, and there was partial participation in Haryana, conditional participation in Bihar and unconditional participation in Uttar Pradesh. The willingness to sustain participation was more favorable – mutual participation was sustainable in 28 of 32 villages studied.

Using the framework of Lise (2001, 2005) and Atmis, Dasdemir, Lise and Yildiran (2007) analyzed women's participation in forestry in the Bartin province of Turkey. The authors, using principal component analysis, found that the most important factors affecting women's participation were: (1) forest dependence, (2) quality of cooperatives, (3) quality of forest organization and (4) forest quality. Of these four factors affecting women's participation, three were used for the estimation of women's participation games. For forest dependence and the quality of cooperatives, the game was found to be a Pareto game (participation is a dominating strategy for both players and also leads to a Pareto optimal outcome), which implies that the payoffs are homogeneously distributed. However, for the quality of forest organization, the game was the chicken game, which means that the optimal strategy of the challenger is to choose the opposite of the strategy of the contender. In addition, mutual participation is Pareto optimal, but not stable. These results indicate that there is strong contrast in women's perceptions of the quality of the forest organization, and improvements in the forest organization may lead to improved women's participation.

Atmis, Günsen, Lise and Lise (2009) used the same model to analyze the participation of 71 forest cooperatives in the Kastamonu province of Turkey. The authors found that the most important factors affecting forest cooperatives' participation were: (1) member involvement, (2) forest ownership and administration and (3) harmony within cooperatives and between cooperatives and the state. Number of active members, sufficiency of work, use of published material and the availability of capital and credit are important variables to explain variations in participation. Game theoretic analysis confirmed that for all three factors, participants played the Pareto game, and there was a positive link between a cooperative's wood production (payoff) and levels of participation, indicating that a productive forest goes hand in hand with high levels of a forest cooperative's participation in forestry.

Evolutionary game theoretic models

Numerous spatial and temporal variations in the outcomes of people's participation in forest management programs have been reported in the literature. Shahi and Kant (2007) argued that these variations can be explained by evolutionary game theoretic models, because forest participatory games are repeated games in which players learn and imitate and follow successful strategies, while unsuccessful strategies disappear gradually. Shahi and Kant (2007) first formulated an *n*-person game for forest use by a community composed of law abiders (or cooperators, C) and lawbreakers (or defectors, D), under state (no people's participation) regime. The authors demonstrated that this game has a unique Nash equilibrium in which defectors earn a higher payoff, and they exploit the forest resource to its degradation and even extinction. Second, the authors formulated an *n*-person game for a community under a JFM regime composed of cooperators (C), defectors (D) and enforcers (E), and demonstrated that this game has many Nash equilibria, but has a unique subgame perfect defection equilibrium. Finally, the authors argued that in game situations in which the players learn and imitate successful strategies, the proportion of the three categories of players (C, D and E) will evolve with time, and used the concepts of evolutionary stable strategies (ESS) and asymptotically stable states to examine the evolutionary dynamics of the game for a JFM regime. The authors demonstrated that this game has four evolutionary strategy equilibriums: cooperators (C) equilibrium, defectors (D) equilibrium, defectors-enforcers (D-E) equilibrium and cooperators-enforcers (C-E) equilibrium. The authors asserted that generally, the E and C-D equilibria are not possible because there are always some, but only some and not all, enforcers under this regime. They demonstrated that only two equilibria, C-E and D-E, are asymptotically stable, and discussed the necessary conditions for their stability.

The authors asserted that the success of a JFM program depends upon policy prescriptions and their relevance to the conditions of forests and communities. In the short run, any of the outcomes are possible which explain spatial and temporal variability, but in the long run, only either the D-E equilibrium or C-E equilibrium will exist. Policymakers and forest managers have to ensure that the payoffs of enforcers are good enough to make sure that the D-E equilibrium is not stable. Similarly, policymakers and managers can enhance the chances of the stability of the C-E equilibrium by: (1) developing markets for NTFPs, (2) ensuring high prices for timber and NTFPs, (3) ensuring a substantial share of the final timber harvest and (4) providing wages to enforcers higher than their cost for protection and maintenance of the forest.

The second application of the evolutionary game theoretic model to JFM is also by the same authors, and in this model Shahi and Kant (2011) included forest managers as one category of players. Hence, they formulated an asymmetric bi-matrix game between a representative of the government and a representative of a community. Each player has two strategies, cooperate (C) and not cooperate (NC). The authors demonstrated that the JFM game has two pure-strategy Nash equilibria (C, NC) and (NC, C) and one mixed-strategy equilibrium in which players choose to cooperate with certain probabilities, and pure-strategy Nash equilibria are not evolutionarily stable, while a mixed-strategy equilibrium is evolutionarily stable. Authors argued that spatial variation in JFM outcomes across different communities and forest types can be explained for different values of these probabilities for different communities and forest types. In the long run, both government and local community representatives will learn from the outcome of the game being played by the other members of their respective population and try to imitate successful strategy. The authors showed through replicator dynamics that only the cooperation equilibrium, when both players decide to cooperate, is asymptotically stable. The authors observed that the outcome of this paper provides support to the JFM program, and also supports the argument that all communities and all government officials may not cooperate initially, because in realistic conditions, the communication between different communities and different government representatives may be limited within a subgroup of respective populations.

Game theoretic models for the forest industrial sector

Some forest economists have used game theoretic models to analyze timber markets, while others have used them to address some conflicts faced by forest industries, such as forest harvesting by forest owners and investment by the wood processing industry. Lohmander (1997) defined and analyzed a dynamic duopsony timber market game, a nonconstant sum game in which the players make use of local information and continuous observation of decision frequencies. The author investigated the trajectories of the combination of decision probabilities of two players. The author found that due to a large number of possible initial conditions, the combination of decision probabilities will follow a special form of attractor, and the probability that the Nash equilibrium will be the solution is almost zero. The concepts of Lohmander (1997) were further developed by Limaei and Lohmander (2008) and applied to analyze the timber market in northern Iran, and Limaei (2010) used the same approach to analyze the pulp market in northern Iran.

Limaei and Lohmander (2008) collected data from two sawmills in northern Iran, and found that these two mills bought more than 70% of the timber in the region; therefore, the situation was closer to a duopsony. The authors collected price data from these two mills and could not explain price variation using the auto-regressive (AR) method; therefore, they argued for the use of a dynamic game theory approach to interpret price variations. The authors assumed that the market was composed of two sawmills (X and Y), and forest owners, companies and local farmers sold timber to these two mills. In each transaction, each sawmill (player) gave a sealed bid. For simplicity of analysis, the authors used only two kinds of bids: low and high. It has been assumed that both mills used the mixed strategy in choosing low or high bid. Suppose mills X and Y assigned probability p and q, respectively, to choose a low bid. Using the available data, the authors found a mixed strategy Nash equilibrium; let us denote it by (p^*, q^*).

The authors argued that it was not likely that the managers of the mills had complete information about each other's mills. The costs and revenues of the competitor were not perfectly known, but each manager observed the mixed-strategy frequencies of the other; i.e. both managers were aware of how many times the other mill chose low bids. The expected marginal profits were calculated based on this information. It is assumed that the speed of adjustment (of p and q) was proportional to the expected marginal profits and that both mills, X and Y, had the same relation between speed of adjustment and expected marginal profit. That is, if expected marginal profits of mill X and mill Y at (mixed-strategy) Nash equilibrium are E_X and E_Y, respectively:

$$\dot{p} = c_1 \left(\frac{\delta E_X}{\delta p} \right) \tag{4}$$

$$\dot{q} = c_2 \left(\frac{\delta E_Y}{\delta q} \right) \tag{5}$$

Where c_1 and c_2 are speed of adjustment, which were assumed to be equal for both the firms. Using these two differential equations, the authors estimated the trajectories of $(p(t), q(t))$ for different mixed-strategy Nash equilibria and discussed their implications. The main findings of this paper, as well as of Limaei (2010), were similar to those of Lohmander (1997): (1) the decision probability combination follows a special form of attractor and (2) the probability that the Nash equilibrium will be reached is almost zero. The model explains why there is only slight variation in the prices of the two mills as shown by the real-life data.

A different game theoretic formulation of timber markets has been proposed by Koskela and Ollikainen (1998). The authors argued that the assumptions of the conventional demand and supply models of timber may not be a true representation of reality. The authors proposed a two-stage game theoretic model for price and quantity determination in the roundwood market. In the first stage, the forest owners' association and the industry decide on timber prices and capital stock, respectively. In the second stage, the industry determines the demand for timber, conditional on timber price and capital stock. The equilibrium concept is the subgame perfect Nash equilibrium, as in the first stage of the game, both the industry and the forest owners' association are aware that demand for timber is subsequently determined by the profit maximization strategy of the industry. The distinctiveness of the model is to incorporate investment decisions of firms into the model. Both players in the game know that new investments will have effects on the demand for timber and take this knowledge into account in formulation of their strategy. The authors used annual data for the Finnish pulp and paper industry over the period 1960–1992 to test their model, and the results were generally in conformity with the hypotheses implied by the game theoretic model.

In fact, the first application of the game theoretic model to the forest industrial sector was by Kivijärvi and Soismaa (1992), who formulated a finite horizon differential game model capturing dynamic interaction between forest owners and the wood processing industry in Finland. The main objective was to find the simultaneous solution for the investment by the wood processing industry and timber harvest by forest owners under three tax regimes – unit, lump-sum and yield taxes. In differential games, which are basically optimum control problems, there are two controls and two criteria, one for each player. Each player attempts to maximize its payoff, but the dynamic system responds to the input by both players. The authors used the Newton-Raphson algorithm to solve their differential game in which both players maximize their net incomes. The authors analyzed the cases of zero and positive salvage values of forest stock and wood-processing mill production capacity at the end of the planning horizon. In the case of zero value, the harvest rate remains constant under all three taxation methods, and the growing stock decreases if the harvest rate is higher than the forest growth rate. The wood processing industry manages its production capacity by substituting wood with capital at the equilibrium rate. In the case of positive salvage values, the harvest rate constantly decreases and forest stock increases. The wood processing industry makes the production capacity correspond to the supply of wood by the same process as in the case of zero salvage value.

In a somewhat different context, Erbas and Abler (2008) applied a game theoretic model to analyze regulatory mechanisms to reduce pollution and induce pollution prevention and abatement research and development (R&D) in the US pulp and paper industry. They formulated a two-stage (R&D and output) simultaneous Cournot duopoly game in which both stages are noncooperative. In the first stage, there is R&D competition in which each firm chooses pollution prevention R&D and pollution abatement R&D expenditures. In the second stage, there is output competition in which each firm chooses its output. Both firms have perfect knowledge of each other's characteristics. The results indicated different outcomes for pollution prevention R&D as compared to pollution abatement R&D, depending upon policy instruments and industry conditions.

Game theoretic models for property rights of forest lands

Property rights of forest lands have also attracted some attention from game theory modelers. Angelsen (2001) presented a game theoretic analysis to explain the impact of forest land appropriation on tropical deforestation. The author modeled the interaction between the state and the local community by applying noncooperative game theory. The focus is on two key questions: First, how does the structure of the game affect overall deforestation? Second, under what circumstances does state deforestation stimulate local deforestation?

The model characterizes the Nash equilibria in both – the Cournot game (static game with simultaneous moves) and the sequential leader-follower or Stackelberg game with either state or local community as leader. In a Stackelberg game, the follower observes the choice of the leader and chooses the optimal strategy in the same way as in a Cournot game; the leader anticipates the strategy of the follower and chooses its strategy accordingly by including the follower's response in the optimization problem.

The author discussed three specific cases. First, he considered interaction between state and local deforestation in the context of a poor and local community. The author assumed an imperfect market, and used a Cournot game. It was found that if the state wants to appropriate more land than before, the local community's response will be more forest clearing. State deforestation increases local deforestation in this case.

In the second case, the local community was assumed as an aggressive player that clears forest and pushes back the state. The Stackelberg game, with local community as leader, gave more overall deforestation than the Cournot game. Compared to the Cournot equilibrium, the community will clear more and the state will clear less forest. The community uses its strategic position as the leader (first mover) to squeeze the state and prevent it from converting as much forest as it would have done in the Cournot game.

For the third case, the author assumed that competition for forest land is strong and the local economy is well integrated with the regional and national economy. The game played is a Stackelberg game with the state as leader. The state now takes into account the negative effect of local forest clearance before deciding its own level of deforestation. The state uses its strategic position to squeeze the local community. Compared to the Cournot equilibrium, this Stackelberg game also results in more forest exploitation, just as in the second case.

In a slightly different context of the weak property rights of local communities over forestland, Engel, López and Palmer (2006) used a game theoretic model to illustrate the strategic interactions between local communities and logging companies in Indonesia. The game theoretic model consisted of three stages. In the first stage, the firm chooses the area it proposes to log. In the second stage, community and company bargain over a mutually agreed contract. The negotiation outcomes depend crucially on the bargaining positions of the company and community. A major influencing factor of bargaining position is the players' reservation utilities. The authors asserted that in a context of weak property rights, reservation utilities are not fixed, but depend on the outcome of a third-stage game in which each party attempts to attain de facto property rights. De facto property rights are modeled as the outcome of a 'war of attrition' between a logging company and a community. The model combines conflict and bargaining theory to analyze the interactions between communities and logging firms.

In the context of weak property rights, the community's ability to self-enforce its rights is critical for analyzing the negotiations of logging agreements in Indonesia. Conflict theory, in principle, models a situation where negotiations are absent and thus gives an indication of the enforcement ability of the community. The results of the conflict theory (third stage of the game) are then incorporated into bargaining theory to analyze why some communities receive higher payoffs than other communities. The conditions for successful negotiations are also predicted.

Building upon Engel et al. (2006), Engel and Palmer (2008) analyzed the potential of payments for environmental services (PES) to provide the communities an alternative to logging to promote forest conservation. The authors applied the same game theoretic model of community-firm with slight modification. They treated logging areas as exogenous and excluded the possibility of endogenous policy interventions and focused on community payoffs and the impacts of PES intervention on community-firm interactions.

Field observations showed that actual payoffs vary greatly among communities involved in the logging agreement. The authors argued that the game theoretic model explains the cause of variation. They analyzed the model's implications for effective and efficient PES design in the context of weak property rights. Contrary to the intuition that PES design should focus on communities with the lowest expected payments from logging, the authors showed from their model that these communities may not be able to enforce a PES agreement.

Game theoretic models in other areas of forest economics

In addition to the three areas discussed so far, game theoretic models have been applied to some other areas of forest economics, and we briefly discuss three applications.

First, Carlson and Wilson (2004) developed a game theoretic model to analyze the interactions between the US Forest Service (USFS) and environmentalists for national forest policy-making. The proposed model is an extensive game with perfect information, with USFS as the first mover to propose a policy change. Environmentalists respond to this proposal, and if they do not accept it, they put forth their proposal to which USFS responds, and so on. The authors in their analysis have shown that this game is not a zero-sum game. The model is used to account for outcomes associated with contemporary management policy and to examine changes in the rules of the game and how this may affect outcomes. The result indicated that some changes will have little or no effect while other changes may have a significant influence on the outcomes.

Second, Buckley and Haddad (2006) used a game theoretic model to identify three potential outcomes of restoration of ecosystem services and processes. The players are restorationists and farmers. The three outcomes correspond to nonstrategic equilibrium, noncooperative strategic equilibrium and cooperative bargaining solution. An important feature is that the authors analyzed the effect of imperfect information regarding expected payoffs to farmers and showed that it can lead to inefficient overshooting or undershooting the optimal scale, geographical positioning and form of restoration.

Finally, Busby and Albers (2010) developed a spatially explicit game theoretic model to examine the strategic interaction between public and private landowners with respect to their decision on wildfire hazard mitigation. The authors found that in areas where ownership is mixed, the private landowners do not do fuel treatment, as they get protected free of cost by the efforts of the state. The areas with public ownership remain underprotected because most of the funds are spent in the areas with mixed ownership.

Conclusion

The application of game theoretic models in forest economics is very limited compared to other similar fields such as environmental economics and economics of common pool resources. In addition, there are no efforts to enhance and refine the rare applications developed in the past, and only a few forest economists have shown some interest in game theoretic models. These applications are also very eclectic in terms of their origins, coming from forest economists, sociologists, economists and environmentalists.

There are many potential areas for enhancing game theoretic research in forest economics. First, forest management is moving from timber management to ecosystem management, and most ecosystem services are not traded in the market. In addition, the new forest management paradigm involves interactions among multiple stakeholders. Game theoretic models should be used to analyze these nonmarket and multiple-stakeholder interactions. Second, the same forest management programs result in different outcomes in different places. Evolutionary game theoretic models should be developed to analyze these diversities. Third, the game theoretic models used so far in the forestry sector by and large assume perfect and complete information, which is normally not the case in real-life situations. Fourth, most of the models are based on strategic decision making, which again may not fit real-life situations where normally decision making is sequential. Fifth, the work of Lise on the estimation of participation games needs to be extended to a real n-player game rather than a game between the first player and a composite second player representing $n - 1$ players. Sixth, cooperative game theory has escaped the real attention of forest economists thus far, while there may be many situations of cooperation of forest management specifically in the context of local communities, aboriginal and tribal peoples. In short, inclusion of uncertainty (incomplete information), imperfect information, sequential game forms and evolutionary equilibrium concepts is the way forward for game theoretic models in forest economics.

References

Angelsen, A. (2001). 'Playing games in the forest: State-local conflicts of land appropriation', *Land Economics*, vol. 77, no. 2, pp. 285–299.

Atmis, E., Dasdemir, I., Lise, W. and Yildiran, O. (2007). 'Factors affecting women's participation in forestry in Turkey', *Ecological Economics*, vol. 60, pp. 787–796.

Atmis, E., Günsen, H. B., Lise, B. B. and Lise, W. (2009). 'Factors affecting forest cooperative's participation in forestry in Turkey', *Forest Policy and Economics*, vol. 11, pp. 102–108.

Bernheim, B. D. (1984). 'Rationalizable strategic behavior', *Econometrica*, vol. 52, pp. 1007–1028.

Bertrand, J. (1992). 'Review by Joseph Bertrand of two books' (Trans. Margaret Chevaillier), *History of Political Economy*, vol. 24, pp. 646–653. (Original work published 1883).

Buckley, M. and Haddad, B. M. (2006). 'Socially strategic ecological restoration: A game-theoretic analysis shortened: Socially strategic restoration', *Environmental Management*, vol. 38, pp. 48–61.

Busby, G. and Albers, H. J. (2010). 'Wildfire risk management on a landscape with public and private ownership: Who pays for protection?' *Environmental Management*, vol. 45, pp. 296–310.

Carlson, L. J. and Wilson, P. I. (2004). 'Beyond zero-sum: Game theory and national forest management', *The Social Science Journal*, vol. 41, pp. 637–650.

Cournot, A. A. (1838). *Researches into the Mathematical Principles of Wealth* (English translation 1960), A. M. Kelly, New York.

Dimand, R. W. and Dimand, M. A. (1992). 'The early history of the theory of strategic games from Waldegrave to Borel', in *Toward a History of Game Theory* (Annual Supplement to Vol. 24, *History of Political Economy*).

Engel, S., López, R. and Palmer, C. (2006). 'Community-industry contracting over natural resource use in a context of weak property rights: The case of Indonesia', *Environmental and Resource Economics*, vol. 33, no. 1, pp. 73–93.

Engel, S. and Palmer, C. (2008). 'Payments for environmental services as an alternative to logging under weak property rights: The case of Indonesia', *Ecological Economics*, vol. 65, pp. 799–809.

Erbas, B. C. and Abler, D. G. (2008). 'Environmental policy with endogenous technology from a game theoretic perspective: The case of the US pulp and paper industry', *Environmental Resource Economics*, vol. 40, pp. 425–444.

Fudenberg, D. and Tirole, J. (1991). 'Perfect Bayesian equilibrium and sequential equilibrium', *Journal of Economic Theory*, vol. 53, pp. 236–260.

Harsanyi, J. C. (1956). 'Approaches to the bargaining problem before and after the theory of games: A critical discussion of Zeuthen's, Hicks', and Nash's theories', *Econometrica*, vol. 24, pp. 144–157.

Harsanyi, J. C. (1966). 'A general theory of rational behavior in game situations', *Econometrica*, vol. 34, pp. 613–634.

Harsanyi, J. C. (1967–1968). 'Games with incomplete information played by "Bayesian" players, parts I, II and III', *Management Science*, vol. 14, pp. 159–182, 320–334 and 486–502.

Kalai, E. and Smorodinsky, M. (1975). 'Other solutions to Nash's bargaining problem', *Econometrica*, vol. 43, pp. 513–518.

Kant, S. and Nautiyal, J. C. (1994). 'Sustainable joint forest management through bargaining: A bilateral monopoly gaming approach', *Forest Ecology and Management*, vol. 65, pp. 251–264.

Kivijärvi, H. and Soismaa, M. (1992). 'Investment and harvest strategies of the Finnish forest sector under different forest-tax policies: A differential game approach with computer-based decision aid', *European Journal of Operation Research*, vol. 56, pp. 192–209.

Koskela, E. and Ollikainen M. (1998). 'A game-theoretic model of timber prices with capital stock: An empirical application to the Finnish pulp and paper industry', *Canadian Journal of Forestry Research*, vol. 28, pp. 1481–1493.

Kreps, D. M. and Wilson, R. B. (1982). 'Sequential equilibria', *Econometrica*, vol. 50, pp. 863–894.

Kuhn, H. W. (1950). 'Extensive games', *Proceedings of the National Academy of Sciences of the United States of America*, vol. 36, pp. 570–576.

Limaei, S. M. (2010). 'Mixed strategy game theory, application in forest industry', *Forest Policy and Economics*, vol. 12, pp. 527–531.

Limaei, S. M. and Lohmander, P. (2008). 'A game theory approach to the Iranian forest industry raw material market', *Caspian Journal of Environmental Science*, vol. 6, no. 1, pp. 59–71.

Lise, W. (2001). 'Estimating a game theoretic model', *Computational Economics*, vol. 18, pp. 141–157.

Lise, W. (2005). 'A game model of people's participation in forest management in Northern India', *Environment and Development Economics*, vol. 10, pp. 217–240.

Lohmander, P. (1997). 'The constrained probability orbit of mixed strategy games with marginal adjustment: General theory and timber market application', *System Analysis – Modelling – Simulation*, vol. 29, pp. 27–55.

Maynard Smith, J. (1972). 'Game theory and the evolution of fighting', in J. Maynard Smith, *On Evolution*, Edinburgh University Press, Edinburgh, Scotland.

Maynard Smith, J. and Price, G. A. (1973). 'The logic of animal conflict', *Nature*, vol. 246, pp. 15–18.

Nash, J. F. (1950a). 'The bargaining problem', *Econometrica*, vol. 18, pp. 155–162.

Nash, J. F. (1950b). 'Equilibrium points in n-person games', *Proceedings of the National Academy of Sciences of the United States of America*, vol. 36, pp. 48–49.

Nash, J. F. (1951). 'Non-cooperative games', *Annals of Mathematics*, vol. 54, pp. 286–295.

Nash, J. F. (1953). 'Two person cooperative games', *Econometrica*, vol. 21, pp. 128–140.

Osborne, M. J. (2004). *An Introduction to Game Theory*, Oxford University Press, New York.

Osborne, M. J. and Rubinstein, A. (1994). *A Course in Game Theory*, MIT Press, Cambridge, MA.

Pearce, D. G. (1984). 'Rationalizable strategic behavior and the problem of perfection', *Econometrica*, vol. 52, pp. 1029–1050.

Rubinstein, A. (1982). 'Perfect equilibrium in a bargaining model', *Econometrica*, vol. 50, pp. 97–109.

Samuelson, L. (1997). *Evolutionary Games and Equilibrium Selection*, MIT Press, Cambridge, MA.

Selten, R. (1975). 'Reexamination of the perfectness concept for equilibrium points in extensive games', *International Journal of Game Theory*, vol. 4, pp. 25–55.

Shahi, C. and Kant, S. (2007). 'An evolutionary game-theoretic approach to the strategies of community members under Joint Forest Management regime', *Forest Policy and Economics*, vol. 9, pp. 763–775.

Shahi, C. and Kant, S. (2011). 'An evolutionary game theoretic approach to Joint Forest Management', *International Journal of Ecological Economics and Statistics*, vol. 23, pp. 37–55.

Tan, T. and Werlang, S. (1988). 'The Bayesian foundations of solution concepts of games', *Journal of Economic Theory*, vol. 45, pp. 370–391.

von Neumann, J. (1928). 'Zur theorie der gesellschaftsspiele', *Mathematische Annalen*, vol. 100, pp. 295–320.

von Neumann, J. and Morgenstern, O. (1944). *Theory of Games and Economic Behavior*, Princeton University Press, Princeton, NJ.

Zeuthen, F. (1930). *Problems of Monopoly and Economic Warfare*, George Routledge and Sons, London.

33

FOREST RESOURCES ACCOUNTING

Haripriya Gundimeda

ASSOCIATE PROFESSOR, DEPARTMENT OF HUMANITIES AND SOCIAL SCIENCES, INDIAN INSTITUTE OF
TECHNOLOGY BOMBAY, POWAI, MUMBAI – 400076, INDIA. HARIPRIYA.GUNDIMEDA@IITB.AC.IN

Abstract

The conventional system of national accounts (SNA) focuses on marketed activities and transactions. Hence, the measures derived from the SNA are not good indicators for measuring human well-being and sustainability of the economies. Though there was consensus on the need to have a better system of accounting, consensus on how to do it differed. This chapter reviews the arguments associated with the need for expanding forest accounts to better reflect their contribution in the national accounts. Two major approaches are reviewed: one based on income as return on wealth and the other based on the concept of income changes as an indicator of welfare change. A practical approach for expanding forest accounts is illustrated through a case study.

Keywords

System of national accounts, gross domestic product, net national product, sustainable development, SEEA, green accounting, India, depreciation, environmental accounting, well-being, green GDP, consumption, investment, asset accounts

Introduction

Since the creation of the system of national accounts (SNA) by Stone, the world gross domestic product (GDP) has increased manifold. However, this has come with a price. The world's forests, for example, have shrunk by 50%; the number of species lost have increased manifold, and air and water pollution have increased to levels of impending environment disaster. Thus, the costs of economic growth all over the world are the losses in natural resources and increasing environmental degradation. In the wake of the World Commission on Environment and Development (WCED, 1992), which popularized the term sustainable development, several researchers have asked the questions, 'Are current consumption patterns sustainable?' or 'What level of consumption would ensure that development is sustainable?' (Asheim and Weitzman, 2001; Dasgupta and Mäler, 2000; Hartwick, 1994; Weitzman and Löfgren, 1997).

In the context of this sustainability debate, the discussion has centered on the role of national accounting as an indicator of social well-being. The SNA considers natural resources within the production boundary but does not recognize their nonmarket contributions or the free services provided by various natural assets. The SNA disregards the impact of economic activity on natural assets and the subsequent decline in well-being due to their decline or degradation. In short, the traditional national accounting indicators fail to provide us with the answer to the effect of future scarcities of natural resources on societal well-being and economic sustainability. In the last three decades, green accounting measures, which account for all forms of capital – produced, natural, human and social – have received significant attention among researchers, governments and international organizations.

There are several reasons for green accounting measures. First, green accounting can be a yardstick to measure the sustainable use of natural resources. Second, the importance of ecosystems to human well-being and the lack of sensitivity of the traditional SNA to changes in ecosystems necessitate the need for a new accounting system. Third, green accounting measures reflect institutional and political commitments through *Agenda 21 Document – Our Common Future of 1992 Earth Summit* and *The Future We Want* accepted at Rio+20 in 2012. Recently, nations have agreed, among other commitments, to (1) explore alternatives to GDP as a measure of wealth, (2) recognize and capture the important environmental services provided by nature and (3) return ocean stocks to sustainable levels.

This chapter demonstrates the need to develop expanded forest accounts and illustrates practical implementation through a case study. It begins with a very simplified view of the standard SNA, followed by a section that ties some of the theoretical arguments linking sustainability, income and wealth with forest resources. The next section discusses how to extend theoretical principles to forest resources, followed by a discussion that looks more at the statistical applications of green accounting with emphasis on the system of environmental-economic accounting (SEEA). Next, a case study in India gives a practical illustration of expanded forest accounts, and then the chapter concludes.

The SNA

The SNA is an internationally accepted statistical standard first released by the United Nations (UN) Statistical Commission in 1953 using double-entry bookkeeping principles based on economic principles and statistical concepts. The basic unit for measuring economic activities and compiling national accounts is the 'institution'. The number is entered only once, but is interpreted differently (rows indicate receipts and columns indicate payments). The social accounting matrix (SAM) captures all the transactions and transfers between different economic agents in the system and represents national accounts in a single matrix. The SAM provides for cross-classification of output and value added by industry, institutions, functions of the government and individual consumption according to the purpose. From the supply and use tables of the SNA, one can derive the familiar input-output tables, which can provide the database for impact studies and productivity analysis. The system allows domestic and international comparisons of key economic variables such as rate of economic growth, changes in consumption, saving, investment, debts and wealth of the economy and its institutional sectors, due to its standardization (SNA, 2008).

The SNA views the relationship between the environment and economy purely from an economic perspective by covering only marketed goods and services. The national income accounts are grouped under three categories: current accounts, accumulation accounts and balance sheets. Current accounts deal with production, income and use of income (supply and use accounts). Accumulation accounts cover changes in assets and liabilities and changes in net worth. Balance sheets present stock of assets and liabilities and net worth.

The supply and use accounts compute income in three ways:

1. The production approach measures GDP by aggregating the output and deducting intermediate consumption across all industries of the economy. Thus,

    GDP = output + taxes − subsidies − intermediate consumption

2. The expenditure approach views GDP as the value of all goods and services available for domestic and final use (for consumption and investment) and exports. Thus,

    GDP = final consumption + gross capital formation + exports − imports

3. The income approach formulates GDP as the value of the incomes of inputs used in production. Thus,

    GDP = compensation of employees + taxes − subsidies + gross operating surplus/mixed income

These accounts reflect the three basic national accounts identities:

1. The supply-use identity:

    Production + imports = intermediate consumption + final consumption
    + gross capital formation + exports (1)

2. The value-added identity:

    Net value added = output − intermediate consumption
    − consumption of fixed capital (2)

3. The domestic product identity:

    GDP = final consumption + gross capital formation + exports − imports (3)

In addition to the supply and use accounts, the SNA also includes asset accounts. The 1993 SNA includes natural assets only if ownership rights exist and natural assets bestow economic benefits to their owners. Some examples of produced natural assets (referred as economic assets) include the value of livestock for breeding, orchards, private plantations, timber tracts and so forth. The economic assets are valued in the market either directly or indirectly. The asset balances for produced assets and nonproduced natural assets include the opening and closing stocks of produced assets, the elements explaining the change between the two (i.e. net capital formation, holding gains or losses of assets, other changes in volume of produced assets) and the closing stocks (equation 4).

For produced and nonproduced assets, the balances are identified as:

Closing stocks = opening stock + gross capital formation
− consumption of fixed capital + other changes in
volume of assets + holding gains/losses on assets (4)

The gross capital formation consists of (a) gross fixed capital formation and (b) changes in inventories in produced assets like building roads, machinery, stocks of commodities and so forth. Gross fixed capital formation also includes additions to the produced assets such as improvement of land, cost of transferring land and other nonproduced assets between owners. The value of capital formation is added to the value of nonproduced assets, but separately 'depreciated' as other changes in volume. Thus, the elements of the column related to nonproduced economic assets do not figure in the calculation of net domestic product (NDP), as all the changes in nonproduced natural assets between opening and closing stocks are explained in the SNA as holding gains or losses and other changes in volume of assets. Hence, the elements under other changes in volume are the most relevant items to be reclassified for analysis in the natural resources accounting.[1]

It is evident that the conventional economic accounts present an incomplete picture of human welfare. The net contribution to the national product from the forest sector is equal to the sum of value added (wages plus gross profits) in this sector. Nonmonetized forest stocks and flows are not reflected in the national accounts. The contribution of other forest services such as carbon, biodiversity, hydrological or cultural services, for example, is also left unaccounted and therefore, completely excluded. Hence, GDP underestimates the value of natural resources even when their use is monetized. In the forestry sector in India, for example, while timber and nontimber forest product (NTFP) uses contribute to GDP, there is significant underestimation of NTFP quantities, and the royalties may not reflect their true value. If countries increase their exploitation of forest resources, GDP increases. However, despite forests being a renewable resource, it is justifiable to treat their exploitation above the sustainable yield as capital consumption. In the case of deforestation, instead of accounting for benefits lost from forests, the expenditures incurred in forest harvesting are treated as gross capital formation. Thus, it is imperative to expand the forest accounts for appropriate economic, environmental and forest management decisions.

Green accounting in theory

The discussions on what constitutes income and how income relates to welfare are central to the green accounting framework. There are two distinct traditions in play. One is the macroeconomic concept of (sustainable) income as return on wealth as argued by Hicks, Lidhahl and others. Hicks (1946) defines sustainable income as 'the maximum amount of money which the individual can spend and still be able to spend the same amount in future' (Hicks, 1946, p. 174). 'If the present value of income received by an individual is equivalent to actual expected receipts, then that income is referred to as wealth equivalent income' (Hicks, 1946, p. 184). Later on, Weitzman (1976) conceptualized this by asking, 'At the current levels of existing resources, what is the maximum consumption that can be sustained to maintain a constant level of well-being?' Harwick (1976, p. 159) This is the basis for measures like green GDP. The welfare economics concept views income changes as an indicator of welfare change. This involves developing a single indicator of wealth that can measure whether our future well-being is increasing, decreasing or remaining constant. Nondeclining wealth per capita implies that we are starting with higher endowments. This notion is the basis for genuine savings and inclusive wealth indicators.

These two schools of thought raise the question of whether current income concepts like net national product (NNP) can be used as an indicator of sustainability, and if not, what modifications are required. Samuelson (1961) made the criticism that current income concepts like NNP cannot be used as meaningful welfare indices because they include investment in addition to consumption and emphasize the need for 'wealth-like-magnitude'. However, Weitzman (1976) argued that even though consumption is the ultimate end of economic activity, adding investment is essential to measure the economy's power to consume at a constant rate. In addition, he suggests accounting for capital consumption along with consumption to account for the decline (or growth) in natural resource stocks. In this case, the current national income does provide information concerning the welfare of present and future generations. He showed that, in a competitive economy, the current level of NNP is the same as the annuity equivalent of present discounted value of maximized consumption. For NNP to be construed as a measure of net welfare, Mäler (1991) suggests the following amendments: (a) deduct the flow of environmental damages, (b) add the net change in the stocks of all assets, (c) treat investments in enhancements of stocks of natural resources as intermediary products and (d) add the return on the total stock of assets to the existing wealth.

Extending the previous debate to forest resources involves 'deducting the rents (evaluated on a marginal unit of stock) on the physical amount of the natural resource that is "wasted", "run down" or "used up" in the accounting period. This will make NNP reflect economic depreciation of forest capital used in economic activity' (Hartwick, 1990, p. 301). The concept can be further extended to include all the ecological values of forests, as well. In the case of deforestation, where the land has an alternate use after harvests, Hartwick (1992) showed that the adjustments to the NNP depend on the use of forestland before and after conversion. As land is a capital asset, change in value of forested land should be treated as capital revaluation in the national accounts. In cases where perfect property rights exist, the appreciation in value of land should offset the depreciation in value.

Although the previous approach involves adjusting current NNP, Dasgupta and Mäler (2000) argued that the focus should be on changes in aggregate welfare (well-being). They showed that there exists a set of shadow (or accounting) prices of goods and services that can be used to obtain an index of welfare change. This index is the (net) national product, properly modified to include all nonmarket goods and services provided by different capital assets including natural, human and social capital. If the index increases, the aggregate well-being improves too. The accounting prices are the contributions an additional unit of asset would make to intergenerational well-being, other things being equal (Dasgupta and Mäler, 2000). The wealth measures are free from double-counting errors. In the case of a growing population, the net change in total wealth per capita is a measure of sustainability. Using this notion of sustainability for forest resources involves estimating the wealth of forests by taking into account the value contributed by forests to intergenerational well-being. Further theoretical exposition on sustainability measurements can be found in Vellinga and Withagen (1996), Aronsson and Löfgren (1995), Asheim and Weitzman (2001), Cairns (2000) and Heal and Kristrom (2005).

Forests and green accounting

Forest resources accounting requires two types of adjustments based on the two traditions in play. The first requires defining and valuing nonmarketed benefits provided by forests, and the second requires measuring and valuing forest stock changes.

Valuing benefits provided by forests

Defining and valuing nonmarketed environmental goods and services is the major challenge in green accounting and requires estimating the appropriate accounting prices. The concept of accounting prices has been widely used in project evaluation literature (Little and Mirrlees, 1990). The accounting prices of assets are defined as the contribution an additional unit of the asset would make to intergenerational well-being, which depends not only on its current use but also its future use and their current and future degree of substitutability with other assets (Dasgupta and Mäler, 2000). These are the same as market prices for all in the case of competitive economies, but they differ from market prices in the case of externalities, market failures and incomplete markets. The difference between the accounting and market prices is the social value of externality that the asset creates. There is a wealth of discussion in the environmental economics literature on how to value externalities and account for nonmarketed goods and services provided by environmental resources.

Where market prices for assets exist (e.g. timber, fuel wood, fish, etc.), it is possible to compute economic values. However, several ecosystem goods and services do not have observable market prices. Even where available, they are undervalued and distorted due to subsidies, price regulations and taxes, and hence require correction. If direct prices are not available, one can rely on indirect market prices such as replacement cost, shadow project approach, maintenance costs and so forth.

Replacement cost techniques look at the costs incurred in order to replace a damaged asset, which may be higher or lower than that of the actual good. A shadow project is based on costing alternate projects intended to substitute for lost environmental services. Maintenance costs observe the costs incurred to avoid a mishap (e.g. how much is being spent on maintenance of dykes to prevent typhoons causing destruction to the mainland). The production function approach estimates how a marginal change in one of the inputs impacts the marketed output of goods and services provided by the environmental resource under consideration (e.g. soil degradation affecting agricultural output leading to loss in productivity). Defensive or preventive expenditures refer to expenditures undertaken by individuals, firms and governments to avoid or reduce unwanted environmental effects.

In certain cases due to the absence of clearly defined markets, one may have to look at related markets (surrogate markets). The common techniques include revealed preference methods like hedonic price method and travel cost method. If revealed preference cannot be used, one can use stated preference techniques like contingent valuation method or choice experiments to evaluate people's preferences and choices to determine their willingness to pay. There are several good texts available on nonmarket valuation techniques (Freeman, 1993; Dixon, Scura, Carpenter and Sherman, 1994; Carson, 2004; Tietenberg and Lewis, 2008).

Valuing forest resource stocks and depreciation

On estimating accounting prices, the value of the assets and their depreciation can be computed using various methods. The value of forestland is estimated as the annual net return from the use of the resource over time, less a reasonable allowance for profit, which can be represented by:

$$V_t = \sum_{t=1}^{N} \frac{p_t q_t - c(q_t)}{(1+r)^t} \tag{5}$$

The depletion of the forest resource is given by $UC_t = V_t - V_{t-1}$

where p_t is the price of the asset, q_t is the quantity of the asset, $c(q_t)$ is the cost of harvesting, r is the discount rate and N is the life of the resource. The following assumptions have to be made to use the net present value approach:

1. Under the assumption of constant prices, any changes in asset values can only be attributed to changes in productivity of the resource and the time-value of money.
2. A lower discount rate favors future consumption, while a high discount rate gives preference to current consumption.

3. The cost of production is assumed to include the value of all material and labor inputs along with normal return on capital (i.e. profit).
4. Another important assumption concerns the variable N, the economic life of the resource.

Although theoretically this is the correct method, owing to the informational requirement inherent in the present value approach and associated difficulties in obtaining the information, alternate techniques to estimate the stock value and its changes have been proposed (Atkinson and Hamilton, 2007). Repetto (1989) proposed a total rent approach, which assumes an optimal extraction path. Using this method, the value of the resource at the beginning of the period t (V_t) is the volume of the resource R_t multiplied by the net price of the resource. Net price is the difference between the average market value per unit of the resource (P_t) and the per-unit average cost of extraction, development and exploration, including a normal return to the capital ($\bar{c}$). The depreciation (UC_t), if any, here would be the total rents

$$UC_t = (P_t - \bar{c})R_t \tag{6}$$

Ideally, this net price would be the Hotelling rent accruing to the owner of the resource, in which case the expected rate of growth of the unit rent would be equal to the discount rate. The approach is fairly easy to implement, but the main problem is that the marginal extraction is fixed, and for optimality, the total rent approach requires a rising price path to hold. A variant of this approach is to use marginal costs, in which case the depletion is given by:

$$UC_t = (P - c'(q))q \tag{7}$$

This approach assumes that the resource is extracted along an optimal path and the costs of extraction increase over time.

Applications of green accounting

The attempts to develop green accounts date back to the beginning of the 1970s. However, the approaches in developing environmental accounts differed. For instance, Nordhaus and Tobin (1972) provided a measure of economic welfare (MEW) by reclassifying certain items of expenditure, imputed for services of consumer capital, leisure and household work, and made a correction for the 'disamenities of urbanization'. These were further refined to include natural resources in the measures of well-being like the index of sustainable economic welfare (Daly and Cobb, 1994), genuine savings (Pearce, Hamilton and Atkinson, 1996), the genuine progress indicator (Anielski and Rowe, 1999) and genuine investment (Arrow et al., 2004). Another approach was to ensure consistency with the existing statistical system. The intention was to develop neither sustainable measure of income nor measure of welfare. Efforts made by different statistical agencies included extending the traditional indicators to account for the nonmarketed benefits associated with forests and adjusting NDP/NNP for the value of stock changes (Norwegian system of accounts, Netherlands [NAMEA], Philippines [ENRAP]).

The UN Statistical Division and the World Bank made efforts to develop a framework for the satellite system of integrated environmental and economic accounting system (SEEA) that further received an impetus with the *Interim Handbook on Environmental Accounting*, published by the UN in 1993. The main objective of these integrated accounts was to integrate environmental data sets with existing national accounts while keeping SNA concepts and principles as intact as possible. The handbook outlined various ways to revise the existing national account-ing indicators to compute indicators like green Green National Product (GNP) to reflect the costs of net resource depletion and environmental degradation. The handbook was revised further in 1998, 2003 and 2012.

The SEEA differs from the SNA in defining assets. The SNA considered two criteria – that they should be owned and should be able to provide economic benefits from using them. They do not include wild resources over which ownership rights do not exist. However, the SEEA expands the asset boundary by including all resources (terrestrial, aquatic, ecosystems, atmospheric and quality of assets) even if ownership rights cannot be exercised and economic benefits are not derived. The asset accounts are an extension of the accumulation accounts of the SNA. The second key feature in the SEEA is the reclassification of other changes in the volume of natural assets into (a) changes due to economic activity, (b) other accumulations, (c) other volume changes, (d) degradation in quality of assets and (e) holding gains and losses, similar to that of produced economic assets.

The latest version of the SEEA (2012) is composed of three parts: (1) SEEA central frame-work (formally accepted as an international statistical standard), (2) SEEA experimental ecosys-tem accounts and (3) SEEA extensions and applications. The SEEA central framework uses a systems approach to the accounting concepts, structures, rules and principles of the SNA and consists mainly of three parts:

1. Physical flows of material and energy within the economy and between the economy and the environment;
2. Stocks of environmental assets and changes in these stocks;
3. Economic activity and transactions related to the environment.

The SEEA experimental ecosystem accounts (which are currently under preparation) describe both the measurement of ecosystems in physical terms and the valuation of ecosystems consistent with market valuation principles. SEEA extensions and applications present various monitoring and analytical approaches that can be adopted. This set of accounts will describe ways in which SEEA data can be used to inform policy analysis such as resource efficiency, productivity, analysis of net wealth and depletion, sustainable production and consumption, structural input-output analysis and so forth.

In the central framework, the focus is on individual components of the environment that provide direct material benefits and space to all economic activities. All other benefits provided by environmental assets are considered in SEEA experimental ecosystem accounts (with special focus on ecosystems). Another feature in the SEEA central framework is recording of flows associated with economic activities related to the environment.

In contrast to the SNA, SEEA suggests constructing accounts in both physical and monetary terms for environmental assets. Environmental assets in the latest SEEA refer to the naturally occurring living and nonliving components of the earth, together comprising the biophysical environment that may provide benefits to humanity. Both the SEEA central framework and the SNA recognize the change in value of natural resources that is attributed to depletion. In the SNA, as already discussed, the differences in the stock of natural resources are treated as other changes in volume alongside flows such as catastrophic losses and uncompensated seizures. The SNA does not recognize this as cost against the income earned by the natural resource-extracting enterprise. However, the SEEA central framework considers the value of depletion as a cost against income; hence, it deducts depletion from the measures of value added, income and savings in the sequence of economic accounts. The adjustment for depletion is in addition to the already-deducted consumption of fixed capital (CFC) for the cost of using fixed assets from measures of value added, income and saving in the SNA. Table 33.4 illustrates the SEEA framework with forest resources as an example. The areas not in bold areas include the additional elements that are needed to supplement the SNA.

Practical applications of green accounting: Forest resources accounting for India

Based on the extended Hicksian contributions by Weitzman (1976), Hartwick (1990) and Mäler (1991), greening the national accounts involves adjusting the NNP for the drawdown of the natural capital in the current period. This approach involves taking stock of change in flows over a period. The welfare economic approach postulates 'an economy's development is sustainable over a period if and only if its wealth increases or remains constant'. The SEEA framework discussed in the earlier section gave a practical outline which in principle allows looking at stocks as well as flows. This section illustrates forest resources accounting for India based on earlier works (Gundimeda 2000, 2001; Gundimeda et al. 2005; Gundimeda, Sukhdev, Sinha and Sanyal, 2006, 2007; Atkinson and Gundimeda, 2006). The framework mentioned subsequently follows a structure similar to SEEA and can provide answers to both income-based and stock-based measures of sustainable development.

Forest resources accounting involves looking at forest asset stock changes along with their flows over a period, estimating accounting prices, monetizing the stock changes and integrating them with the national accounts. Developing forest accounts has several practical limitations. First is the choice of unit for analysis. For an environment minister, forest area accounts are appropriate; to a forester, timber volume matters; and biodiversity accounts are appropriate for a conservationist. Accordingly, forest accounts can be expressed in volume, weight, tons, hectares or number of species. Below we discuss the framework for area, volume and carbon accounts (Table 33.1).

Table 33.1 Framework for physical accounting of forest resources.

Area	Volume/carbon accounts
Opening stock	**Opening stock**
Changes in forest land	**Changes due to economic activities**
+ Natural expansion and afforestation	– Logging and logging damage
– Net transfer of forest land to nonforest uses (through deforestation or degradation)	– Forest encroachment and shifting cultivation
	+ Afforestation
– Loss of forest land due to shifting cultivation	– Loss due to livestock grazing
	Other accumulations
+ Net reclassification and other changes	**Changes due to natural causes**
= Closing stocks	+ Natural growth
	+ Natural regeneration
	Changes due to reclassification
	+ Net transfer of land
	Other volume changes
	– Stand mortality
	– Forest fires and pest damage
	= Closing stocks

Source: SEEA Framework.

Opening and closing stocks are the total standing volume of timber or area or tonnes of carbon present at the beginning and end of the accounting period. The volume of timber logged, above the long-term net growth of the forest, is nonsustainable and termed depletion. Forest clearances for nonforest use (e.g. forest encroachment, transfer to infrastructure), shifting cultivation and illegal logging are recorded as economic activity, while reforestation, natural growth and transfer of forest land to or from other economic activities are recorded as other accumulation. Other volume changes in environmental accounting include noneconomic factors like forest fires and stand mortality that contributes to changes in forest stock. Further, forests can be classified based on the economic function. If the use of the forest is for pure commercial purposes (plantations) and they have clearly enforced ownership rights, they are classified under produced economic assets, else classified under nonproduced economic asset or nonproduced natural asset.

Columns 3 and 4 of Table 33.2 give the accounts for timber and carbon. The opening stocks represent the growing stock of timber present in 2009 (assessment carried out during 2007–2008). To convert this estimate into units of carbon, the volume data has been converted into biomass and then converted to carbon by assuming 0.5 mg C per mg oven-dry biomass. The carbon estimates presented in Table 33.2 include the aggregate carbon content of biomass alone, but not the soil carbon stock. The closing stocks are equal to opening stocks less reductions plus additions.

Table 33.2 Physical accounts for forests for the years 2009–2011.

(1)		*Area (km²) (2)*	*Volume accounts (million m³) (3)*	*Carbon accounts (million tC) (4)*
Opening stock	(A)	6,92,394	4,498.66	6,813.16
Depletion/degradation due to economic activities (–)	(B)	–2,689	–17.47	–26.46
Afforestation (+)	(C)	4,972	32.10	48.92
Accumulations due to natural or artificial regeneration in forested land (+)	(D)	3,830	24.88	37.69
Transfer to nonforest purposes	(E)	–5,339	–34.69	–52.54
Other volume changes	(F)	–175	–0.21	–1.72
Changes in stock (A – G) = (C – E) for area accounts (A – G) = (C + D – B – E – F) for volume and carbon accounts		–367	–16.26	–9.335
Closing stock	(G)	692,027	11,877.50	6,822.49

Sources: Based on work done by the author for MOSPI (2013). Area accounts are based on SFR (2009, 2011). Growing stock and carbon per hectare is taken from Forest Survey of India. www.fsi.nic.in/fsi_projects/Biomass and Carbon in India's Forests.pdf. The rest of the figures are derived based on the assumptions specified in the text.

Note: B, D and F, though they have an impact on the volume, will not lead to additions or decreases in forested area because degradation can change dense to open forest or scrub, but the area officially still remains under the forest department. Hence, forest area would not change though tree cover might change. Hence, changes in forest area can only be due to additions in forest area and deductions in forest area. However, for volume and carbon stocks they have to be added or subtracted because disturbances can have an impact on the volume and carbon.

Changes in stock are the difference between opening and closing stocks. This should also equal the additions and reductions in the detailed accounts.

Negative signs indicate reductions, and positive signs indicate additions to the stock.

The monetary accounts are built on the physical data presented in Table 33.2 and require separation of value of land from the value of standing timber. This creates some complications as prices for forested land are seldom available, and, therefore, forests are best valued based on the total economic value of benefits they provide to the society. The values provided by forests can be categorized into the use values, nonuse values, existence values and option values. The use values arise from the use of forest goods such as timber, fuelwood and NTFPs, while nonuse can be from services like carbon sequestration, ground water recharge, soil erosion and flood prevention. Option values arise due to the possibility of possible future use that the asset provides. Some have value, not because of any use, but because of mere existence (spiritual, cultural, etc.). If the same piece of land provides all these benefits, then the benefits need to be aggregated in a meaningful way.

Different methods were used to estimate accounting prices for different components of forest wealth. Based on the previous discussion, where available, market prices were used as accounting prices and proxies were used where market prices were not available. Resource rent of timber was estimated as the average market price of roundwood and fuelwood minus the unit costs of extraction. Carbon was valued at $20/tC, based on the average marginal social damage costs (Atkinson and Gundimeda, 2006). Further, to avoid double counting, forests cannot be valued for both timber and carbon. Hence, protected forests are valued for timber and reserve forests are valued for carbon.

The accounting price of NTFPs, including fodder, was computed based on the fees received by collecting NTFPs per hectare (discounted value of NTFPs)[2] (Table 33.3). The accounting price of recreation is the consumer surplus derived for tourists. The ecotourism value can be captured through travel cost estimates or contingent valuation estimates. The benefit-transfer approach has been used to estimate the ecotourism values due to the difficulty in obtaining nationwide estimates. The approach involves adapting information from existing secondary studies on ecotourism to different protected parks in India. The biodiversity values of forests were estimated by assessing the value of marginal species, i.e. the contribution that one more species makes to the development of a new pharmaceutical product (optional value) for commercial medicinal use (see Gundimeda et al., 2006, for more details).

Table 33.3 Accounting prices and loss in wealth of forests for the year 2009.

Accounting price	Value in Rs
Timber (Rs/m³)	13,390
Fuelwood (Rs/m³)	1,943
NTFPs (Rs/ha)	6,289
Carbon (Rs/ton)	1,000
Recreation (Rs/ha)	96,355
Genetic material (Rs/ha)	33,471
Loss in value of timber, carbon and NTFPs (million Rs)	−19,859.95
Loss in ecotourism and genetic diversity (million Rs)	−4,765
Total loss in forest wealth (million Rs) for the period 2009–2011	−24,624.56

Sources: Based on work done by the author for MOSPI (2013). Timber, fuelwood and NTFPs based on national account statistics from MOSPI. Carbon, recreation and genetic material are from earlier studies but updated for the latest years.
Note: Loss in forest wealth: computed.

The net price and net present value approaches were used to estimate the asset value of forests (discussed previously). The value from NTFPs, fodder, ecotourism and bioprospecting are generated every year, contradictory to timber, which is lost forever unless replanted (hence can be valued only once). Hence, nontimber benefits were estimated using the present value approach, obtained by dividing the net price by the social discount rate of 4%.

An important final step in forest resources accounting is to integrate the estimates with the national accounts by adjusting three components of the national accounts:

1. Figures for the production of timber that adjust unreported production;
2. Capital accounts that expand capital formation to include accumulation in natural forests and depletion;
3. Consumption of capital to include the cost of depletion of natural forests, which decreases NDP.

Although the first adjustment is completely consistent with the SNA and simply represents a better estimate of conventional national income, the second set of adjustments is outside the SNA and represents revisions to the SNA proposed by SEEA.

The forest resource accounts integrated with the SNA in its simplified form are shown in Table 33.4. D_{np} (column 1, row vii) reflects the depletion of nonproduced economic assets ($-D_{np.ec}$) and the degradation of nonproduced environmental assets ($-D_{np.env}$). In SEEA, net capital formation is replaced with net accumulation. For produced assets, ($A_{p.ec}$), it is the net capital formation. For nonproduced assets, it reflects the net effects of addition and depletion/degradation of natural assets that are transferred to economic uses. Row ix, column 5 represents the transfer of natural assets to economic uses (such as transfer of land to economic uses, conversion of forests, etc.) as a change in the stock of nonproduced economic assets ($A_{np.ec}$) in physical/monetary terms. If environmental assets are transferred then it is reflected as ($A_{np.env}$).

Row x indicates other changes in the volume of produced ($Vol_{p.ec}$) and nonproduced ($Vol_{np.ec}$) assets, and row xi indicates the holding gains and losses of produced assets ($Rev_{p.ec}$) and nonproduced assets ($Rev_{np.ec}$) The closing stocks of both produced and nonproduced assets are given in row xii. Row viii, which is relevant only in the case of monetary environmental accounting, is included to derive environmentally adjusted net domestic product (EDP) and other environmentally adjusted measures.

The result of these adjustments shown in Table 33.4 is the EDP.

$$EDP = NDP + (A_{np} - Use_{np})$$

Depletion = other accumulations ± changes due to economic activities

where A_{np} represents the accumulation of nonproduced assets and Use_{np} is the use of forest capital. The accumulation and depletion come from the asset accounts.

The asset accounts are constructed as follows:

Closing stocks − opening stocks = changes due to economic activities ± other accumulations ± other volume changes ± omissions and errors

Table 33.4 Greening of national accounts: An illustration for the forest resources (units in rupees crores, 10 million) for the year 2009.

		Domestic production (1)	Rest of world (2)	Final consumption (3)	Produced assets (4)	Nonproduced assets (forests)[a] (5)	Other nonproduced natural assets (6)	Errors and omissions (7)
Opening stock of assets	(i)				$Ko_{p.ec}$	6,966,158		
Supply of goods and services	(ii)	13,102,022	1,639,872					
Economic use of goods and services	(iii)	6,551,751	1,298,371	4,567,456	2,016,186	17,420.58 ($Use_{np.ec}$)	$Use_{np.env}$	
GDP	(iv)	6,550,271	−341,501	4,567,456	2,389,213			−64,897
Use of fixed capital (capital consumption)	(v)	−655,673			−655,673			
Net value added/net domestic product (NDP)	(vi)	5,894,598	−341,501	4,567,456	1,733,540			−64,897
Use of nonproduced natural assets/natural capital consumption (depletion and degradation)	(vii)	−3,893.6 (− $D_{np.ec}$ + $D_{np.env}$)				−3,893.6 ($D_{n.ec}$)	$D_{np.env}$	
Environment adjusted domestic product (EDP)	(viii)	5,890,704.4	−341,501	4,567,456	1,733,540	−3,893.6		−64,897
Net capital formation/net accumulations of assets	(ix)				113,374	13,526.9 ($A_{np.ec}$)	$A_{np.env}$	
Changes in volume of assets	(x)				$Vol_{p.ec}$	160.97 ($Vol_{np.ec}$)	$Vol_{np.env}$	
Holding gains/losses	(xi)				214,619 ($Rev_{p.ec}$)	($Rev_{np.ec}$)	$Rev_{np.env}$	
Closing stock of assets	(xii)					6,962,465		
Errors and omissions	(xiii)				45,034			

Source: National Account statistics (2011) and based on the work done by the author for MOSPI (2013).
[a] The stock changes for forestland are for the assessment year 2007–2009, but the flows are annualized.

Conclusions and future direction

The SNA clearly suffers from limitations, as the boundary is restricted to marketed goods and services. This chapter has reviewed some ideas that led to the development of green accounts and the various traditions in play, and illustrated through a case study a practical way of expanding the forest resource accounts.

The practical application of green accounts across the world has been different and is still in progress. The UN Statistics Division, Eurostat and the London Group have worked and are continuing to refine the methodology to better account for natural resources in the traditional SNA. Several countries are testing the methodology provided by the latest SEEA manual revised in 2012. SEEA is firmly rooted and is consistent with the existing SNA. However, it does not completely respond to the criticism about the inability of national accounting measures to measure welfare and expand the production boundary to include nonmarketed benefits. SEEA does provide a valuable framework to organize the impact of economic activities on the environment and see whether the growth is sustainable.

The interest in developing alternate indicators to measure sustainable development is picking up in several directions. Attempts are being made to expand the notion of wealth to include human and natural capital as seen from the Inclusive Wealth Report (UNEP, 2012, UNU-IHDP). Researchers at the World Bank (Hamilton, 2006; Hamilton and Clemens, 1999) published adjusted net saving estimates for several countries. After the World Bank publications (World Bank, 1997, 2006, 2011) on measuring the wealth of nations, the effort is on to further strengthen the implementation of this approach through the Wealth Accounting for Valuation of Ecosystem Services (WAVES). In 2008, the President of the French Republic set up Stiglitz-Sen-Fitoussi, Commission on the Measurement of Economic Performance and Social Progress, with the aim of identifying the limits of GDP as an indicator of economic performance and social progress and assessing the feasibility of alternative measurement tools to measure sustainable economic progress. Developing countries like India have also taken initiatives to green their national accounts, as seen from the recently released report on *Green National Accounts in India: A Framework* (MOSPI, 2013).

International developments in the area of green accounting clearly show the importance of forest resource accounting, as forests are a valuable economic, ecological and cultural asset. Achieving well-being of future generations requires achieving forest sustainability. This means the integrity and diversity of forest resources needs to be preserved while providing a valuable economic resource. Though a great deal of work has been done on forest accounting, developing a complete set of forest accounts is quite challenging due to the complexity of diverse nonmarket values that forest provide to the society. The complexity comes from the poor quality of data, limited knowledge on the functioning of ecosystems and overlapping functions of forests. However, if the efforts are successful, forest accounts are useful to policy in a number of ways. They can provide a tool for analyzing the impacts of other sectors on forests, analyze the distributional aspects of the resource use, compensate the states that manage their forests better through additional fiscal transfers, design monetary payments for forest resource conservation and work out compensating mechanisms for degradation or transfer of forestlands. The next decade will see an increase in the number of green accounting applications, especially the forests.

Notes

1   Natural resource accounting deals with stocks and stock changes of natural resources usually in physical terms, while environmental accounting deals with the monetary value of these changes. The term 'green accounting' refers to accounting for the impact of the changes in the stocks of natural assets on the economy.

2   Like timber and fuelwood, the value of NTFPs is severely under-reported; the CSO estimates the value of unrecorded NTFP production as ten times the value recorded by the State Forest Departments, which are based on a nominal 'royalty' charged to forest users for collecting NTFPs, but which is largely unenforceable.

References

Anielski, M. and Rowe, J. (1999). 'The genuine progress indicator – 1998 update', in *Redefining Progress*, San Francisco, CA.

Aronsson, T. and Löfgren, K.-G. (1995). 'National product related welfare measure in the presence of technological change, externalities and uncertainty', *Environmental and Resource Economics*, vol. 5, pp. 321–332.

Arrow, K., Dasgupta, P., Goulder, L., Daily, G., Ehrlich, P., Heal, G., . . . Walker, B. (2004). 'Are we consuming too much?' *Journal of Economic Perspectives*, vol. 18, no. 3, pp. 147–172.

Asheim, G. B. and Weitzman, M. L. (2001). 'Does NNP growth indicate welfare improvement?' *Economics Letters*, vol. 73, no. 2, pp. 233–239.

Atkinson, G. and Gundimeda, H. (2006). 'Accounting for India's forest wealth', *Ecological Economics*, vol. 59, no. 4, pp. 462–476.

Atkinson, G. and Hamilton, K. (2007). 'Progress along the path: Evolving issues in the measurement of genuine saving', *Environmental and Resource Economics*, vol. 37, no. 1, pp. 43–61.

Cairns, R. D. (2000). 'Sustainability accounting and green accounting', *Environment and Development Economics*, vol. 5, no. 1, pp. 49–54.

Carson, R. (2004). *Contingent Valuation – A Comprehensive Bibliography and History*, Edward Elgar Publishing, Cheltenham, UK.

Daly, H. and Cobb, J. (1994). *For the Common Good: Redirecting the Economy toward Community, the Environment and a Sustainable Future*, Beacon Press, Boston, MA.

Dasgupta, P. and Mäler, K. G. (2000). ' Net national product, wealth, and social well-being', *Environment and Development Economics*, vol. 5, no. 1, pp. 69–93.

Dixon, J. A., Scura, L. F, Carpenter, R. A. and Sherman P. B (1994). *Economic Analysis of Environmental Impacts*, Earthscan Publications, London, UK.

Forest Survey of India (FSI). (2009). *State of Forest Report 2005*, Government of India, Ministry of Environment and Forests, Dehradun, India.

Forest Survey of India (FSI). (2011). *State of Forest Report 2009*, Government of India, Ministry of Environment and Forests, Dehradun, India.

Freeman, A. M. (1993). 'The measurement of environmental values and resources: Theory and methods', *Resources for the Future*, Washington DC.

Gundimeda, H. (2000). 'Integrating forest resources into the system of national accounts in Maharashtra, India', *Environment and Development Economics*, vol. 5, no. 1, pp. 143–156.

Gundimeda, H. (2001). 'Accounting for the forest resources in the national accounts in India', *Environmental and Resource Economics*, vol. 19, no. 1, pp. 73–95.

Gundimeda, H., S. Sanyal, R. Sinha, and P. Sukhdev (2005). '*The Value of Timber, Carbon, Fuelwood and Non-timber Forest Products in India's Forests*', Monograph 1, Green Accounting for Indian States Project, India.

Gundimeda, H., Sukhdev, P., Sinha, R. and Sanyal, S. (2006). *Value of India's Biodiversity*, Monograph 4, Green Accounting of Indian States Project, Green Indian States Trust, TERI Press, New Delhi: India.

Gundimeda, H., Sukhdev, P., Sinha, R. and Sanyal. S. (2007) 'Natural resources accounting for India: Illustration for Indian states', *Ecological Economics*, vol. 64, no. 4, pp. 635–644.

Hamilton, K. and Clemens, M. (1999). 'Genuine savings rates in developing countries', *The World Bank Economic Review*, vol. 13, no. 2, pp. 333–356.

Hartwick, J. M. (1990). 'Natural resources, national accounting and economic depreciation', *Journal of Public Economics*, vol. 43, no. 3, pp. 291–304.

Hartwick, J. M. (1992). 'Deforestation and national accounting', *Environmental and Resource Economics*, vol. 2, no. 5, pp. 513–521.

Hartwick, J. M. (1994). 'National wealth and net national product', *The Scandinavian Journal of Economics*, vol. 96, no. 2, pp. 253–256.

Heal, G. and Kristrom, B. (2005). 'National income and the environment', *Handbook of Environmental Economics*, vol. 3, pp. 1147–1217.

Hicks, J. R. (1946). *Value and Capital: Vol. 2*, Oxford University Press, Clarendon Press, Oxford, UK.

Little, I.M.D. and Mirrlees, J. A. (1990). 'Project appraisal and planning for developing countries', *Management of Indian Economy*, p. 233.

Little, I. M. D., & Mirrlees, J. A. (1990). Project appraisal and planning twenty years on. In *Proceedings of the World Bank Annual Conference of Development Economics 1990*. (pp. 351-382). World Bank.

Mäler, K. G. (1991). 'National accounts and environmental resources', *Environmental and Resources Economics*, vol. 1, no. 1, pp. 1–15.

Ministry of Statistics and Programme Implementation (MOSPI). (2013). *Green National Accounts in India: A Framework*, Report of the Expert Group on Green National Accounting, MOSPI, India. Retrieved from http://mospi.nic.in/mospi_new/upload/Green_National_Accouts_in_India_1may13.pdf

Nacions Unides. (2009). *System of national accounts 2008* (No. 2). Commission européenne (Ed.). United Nations Publications.

Nordhaus, W. D. and Tobin, J. (1972). 'Is growth obsolete?' in *Economic Research: Retrospect and Prospect: Vol. 5. Economic Growth*, National Bureau of Economic Research, Columbia University Press, New York.

Pearce, D. A., Hamilton, K. and Atkinson, G. (1996). 'Measuring sustainable development: Progress on indicators,' *Environment and Development Economics*, vol. 1, no. 1, pp. 85–101.

Repetto, R. C. (1989). *Wasting Assets: Natural Resources in the National Income Accounts*, World Resource Institute, Washington, DC.

Samuelson, P. A. (1961). 'The evaluation of "social income": Capital formation and wealth', in F. A. Lutz and D. C. Hague (Eds.), *The Theory of Capital*, St. Martin's Press, New York.

Tietenberg, T. H., & Lewis, L. (2008). *Environmental and Natural Resource Economics*, 8th edition, MA: Addison-Wesley.

Vellinga, N. and Withagen, C. (1996). 'On the concept of green national income', *Oxford Economic Papers*, vol. 48, no. 4, pp. 499–514.

UNU-IHDP, U. N. E. P. (2012). Inclusive wealth report 2012—measuring progress toward sustainability.

Weitzman, M. L. (1976). 'On the welfare significance of national product in a dynamic economy', *The Quarterly Journal of Economics*, vol. 90, no. 1, pp. 156–162.

Weitzman, M. L. and Löfgren, K. G. (1997). 'On the welfare significance of green accounting as taught by parable', *Journal of Environmental Economics and Management*, vol. 32, no. 2, pp. 139–153.

World Bank Group. (2011). *The Changing Wealth of Nations: Measuring Sustainable Development in the New Millennium*.

World Commission on Environment. (1992). *Our Common Future*. Centre for Our Common Future.

34

FOREST POLICY MODELLING IN AN ECONOMY-WIDE FRAMEWORK

Onil Banerjee[1] and Janaki R. R. Alavalapati[2]

[1]CSIRO ECOSYSTEM SCIENCES, 41 BOGGO ROAD, DUTTON PARK, QLD 4102, AUSTRALIA. ONIL@RMGEO.COM

[2]PROFESSOR AND HEAD, DEPARTMENT OF FOREST RESOURCES AND ENVIRONMENTAL CONSERVATION, COLLEGE OF NATURAL RESOURCES AND ENVIRONMENT, VIRGINIA TECH, BLACKSBURG, VA 24061, USA. JRRA@VT.EDU

Abstract

Computable general equilibrium (CGE) models are extensively used in agricultural economics and macroeconomic studies to assess the impacts of policy and programs. By providing a theoretically consistent mathematical representation of an economic system, these models effectively capture the linkages among sectors of the economy. This chapter provides an overview of CGE modelling in forest economics. As an illustration, a dynamic recursive CGE model is presented and applied to assess the regional economic impacts of Brazil's forest concessions policy in the Amazon. A comprehensive discussion on innovations in CGE analysis concludes the chapter. These innovations include institutional advancements, model validation techniques, systematic sensitivity analysis to address parameter uncertainty, and integration of ecosystem services.

Keywords

Computable general equilibrium model, dynamic model, input-output, land-use change, forest concessions, deforestation, Brazil

Introduction

The input-output (I-O) approach is by far the most commonly used technique in assessing economy-wide impacts of a policy or project. This approach rests on the idea that the economy is a system of interdependent industrial sectors and emphasizes incorporation of intersectoral linkages (Dixon, Parmenter, Powell and Wilcoxen, 1992; Shoven and Whalley, 1992). However, a closer look at the underlying assumptions of I-O models raises serious concerns about the validity of the information derived from these models (Alavalapati, Adamowicz and White, 1998). The I-O models' assumptions that prices of inputs and outputs are fixed, fixed amounts of inputs are required to produce a unit of output, there are no constraints on the supply of inputs and final demand for output is exogenous are likely to generate biased estimates (Alavalapati et al., 1998). These concerns have prompted economic modellers and policy analysts to develop an alternative interindustry, computable general equilibrium (CGE) approach, to assess economy-wide impacts of a policy, project or program.

Although CGE models have their foundation in the I-O framework developed by Wassily Leontief, they represent a theoretically consistent mathematical representation of an economic system. These models are formalized by a system of equations explaining demand for commodities, intermediate and factor inputs, equations relating prices and costs and market clearing equations for factors and commodities (Dixon et al., 1992). Both demand and supply equations describe the behavior of utility-maximizing consumers and cost-minimizing producers, respectively. This system of equations is simultaneously solved to determine the economic equilibrium (Bandara, 1991). CGE models are an improvement over I-O frameworks because they endogenize the price and demand system, enable substitutability of goods and services in production and demand, include a realistic treatment of factor scarcity, institutions and the macroeconomic environment and allow agent behavior optimization with producers competing for scarce resources and consumer expenditures.

At the core of a CGE model is the social accounting matrix (SAM), which empirically describes the structure of production and transactions between sectors, institutions and factors of production for a representative base year. The SAM serves to organize data transparently and provide the statistical basis for the model (King, 1985). SAMs are constructed based on I-O tables, national accounts data, government surveys such as household expenditure surveys and agricultural census data. A CGE model may be a static one-period or dynamic multiperiod model. Static models are used for estimating the order of magnitude and direction of effect of a policy shock, and depending on the model closure, they may be short-run or long-run. Dynamic CGE (DCGE) models shed light on the economic transition path resulting from a policy shock, including the short-term costs and longer-term gains (Cattaneo, 1999). DCGE models involve a deeper treatment of investment behavior and enable the modeller to update key growth parameters such as population, labor force, factor productivity, world prices and government consumption. As such, CGE models are considered flexible and powerful in assessing economy-wide and distributional impacts of a policy change by effectively incorporating intersectoral dynamics (Buetre, Rodriguez and Pant, 2003).

This chapter focuses on discussing the mechanics and application of a CGE model to provide an economy-wide assessment of a forest policy. The next section briefly reviews the literature regarding CGE applications to forestry issues. A stylized version of a CGE model is presented in the third section. In the fourth section, an application of a DCGE is presented through a case study of a recent forest policy in Brazil. The last section discusses innovations in CGE analysis.

Review of CGE applications to forestry

CGE models are frequently used to study international trade, taxes, economic policy packages and climate change issues (Stenberg, Mahinda and Siriwardana, 2005). In the last two decades they have been increasingly applied to the study of forest sector policies. For example, Dee (1991) developed a model to evaluate the impact of increasing the minimum harvest age of trees and variations in stumpage and discount rates in Indonesia. Wiebelt (1994) studied macroeconomic policy impacts on forestry in Brazil. Alavalapati, Percy and Luckert (1997) analyzed the distributional effects of an increase in the stumpage price in Canada. Thompson, Van Kooten and Vertinsky (1997) considered forest management options when nontimber values are accounted for in the modelling framework. Alavalapati et al. (1996) simulated land-use restrictions on a resource-dependent economy in Canada. Dufournaud, Jerrett, Quinn and Maclaren (2000) assessed the economics of an export ban and an increase in royalties and export taxes. Gan (2004) investigated potential impacts of trade liberalization on China's forestry sector, while Gan (2005) considered the socioeconomic impacts of forest certification. Stenberg and Siriwardana (2007) examined the economic effects of selective logging, stumpage taxes, set-aside areas and secure forest land tenure on the Philippine economy, integrating a CGE model and a forestry submodel. Ochuodho, Lantz, Lloyd-Smith and Benitez (2012) recently applied a CGE model to estimate economic impacts of climate change adaptation in Canadian forests.

A few modellers have addressed the interactions between land use and deforestation. Persson and Munasinghe (1995) assessed the impact of macroeconomic policy on deforestation and compared agent behavior under variable property rights arrangements. Cattaneo (2001, 2002) examined the relationship among economic growth, poverty and natural resource degradation in Brazil. Banerjee and Alavalapati (2010) considered both legal and illegal deforestation and forestry in their analysis of forest concessions in Brazil. Banerjee, Macpherson and Alavalapati (2012) evaluated the socioeconomic and land-use dynamics of ethanol expansion in Brazil. Huang, Alavalapati and Banerjee (2012) studied the economy-wide and welfare effects of bioenergy policies in the United States.

A stylized version of CGE

What follows is a stylized version of a static CGE model. Following Robinson, Yunez-Naude, Hinojosa-Ojeda, Lewis and Devarajan (1999), the model is organized into a series of building blocks, namely, price, quantity, income, expenditure and closure blocks.

Price equations

Prices in a CGE model may differ according to their origin and destination of use. The domestic import price P_i^m and the domestic export price P_i^e are a function of the world import price pw_i^m and the world export price PW_i^e in foreign currency, the exchange rate R and a tariff adjustment t_i^m on imports and a tariff adjustment t_i^e on exports (equations 1 and 2).[1] Interregional and international trade margins may also be included in the model specification. In single country or subnational models, the small country assumption is commonly made where world prices are taken as exogenous to reflect a country's often limited ability to affect world prices.

$$P_i^m = pw_i^m \left(1 + t_i^m\right) \cdot R \tag{1}$$

$$P_i^e = PW_i^e \left(1 + t_i^e\right) \cdot R \tag{2}$$

Equation 3 solves for the composite commodity price P_i^q, where composite goods supply Q_i is a CES aggregation of imported M_i and domestic D_i goods; P_i^d is the domestic goods price.

$$P_i^q = \frac{P_i^d \cdot D_i + P_i^m \cdot M_i}{Q_i} \tag{3}$$

Equation 4 describes output price P_i^x, which is a constant elasticity of transformation (CET) aggregation of goods supplied to both domestic and export markets.

$$P_i^x = \frac{P_i^d \cdot D_i + P_i^e \cdot E_i}{X_i} \qquad (4)$$

Equation 5 explains the value-added price P_i^v, which is the output price P_i^x less both indirect taxes t^x and the cost of intermediate inputs based on I-O coefficients described by the coefficient matrix a_{ij}. Sectoral value added is the product of the value-added price and sectoral output, represented by payments from sectors to factors in the SAM.

$$P_i^v = P_i^x \cdot \left(1 - t_i^x\right) - \sum_j P_j^q \cdot a_{ji} \qquad (5)$$

Capital used by different sectors is heterogeneous and therefore has a sector-specific price P_i^k as described in equation 6.

$$P_i^k = P_j^q \cdot b_{ji} \qquad (6)$$

The capital coefficient matrix b_{ji} is the composition of capital by its activity of origin. In static CGE models, capital stock is fixed; therefore, investment is treated as another type of demand with no supply-side effects. Accounting for capital heterogeneity is important in dynamic modelling where the composition of capital investment affects future economic growth pathways (Robinson et al., 1999).

Equation 7 defines the aggregate price indicator *PINDEX* as the *GDP* deflator, which is the quotient of nominal *GDP*, *GDPVA* at market prices and real *GDP*, *RGDP*.

$$PINDEX = \frac{GDPVA}{RGDP} \qquad (7)$$

This price index serves as the numeraire against which all changes in relative prices are measured. A numeraire is necessary because CGE models solve for changes in relative prices. Other indices or prices may be chosen as the numeraire, including the Consumer Price Index (CPI) or the exchange rate, respectively (Robinson et al., 1999).

Quantity equations

Equations 8 through 15 describe the supply side of the model, with equations 8 to 10 describing the production technology and factor demand. In CGE models, there are many ways to represent the structure of production. Options range from simple Leontief fixed shares specifications, to more complex configurations such as nested production structures, such as that presented in Figure 34.1.

At the top level, output is a fixed shares function of real value added and aggregate intermediate inputs. Real value added may be produced by a number of functional forms, including a Cobb-Douglas function of capital and labor or a constant elasticity of substitution (CES) function as in Figure 34.1. In other specifications, real value added may include land and other factors such as water. The level of domestic output X_i is given in equation 8.

$$X_i = a_i^D \prod_f FDSC_{if}^{\alpha_{if}} \qquad (8)$$

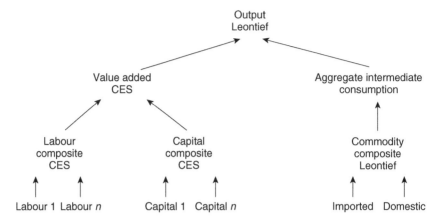

Figure 34.1 Structure of production.

where $FDSC_{if}$ is factor demand, a_{if} is a production function exponent and a_i^D is a production function shift parameter.

Firms are price takers and minimize costs subject to nested technological constraints. Sector output may be determined by combining value added with intermediate consumption through a Leontief production function. Labor and capital may be a CES aggregate of imperfectly substitutable types of labor and capital. Value added is a CES function of factors with firms employing factors until the value of the marginal factor product is equal to the factor price.

Factor demands are represented by equation 9. It is assumed that across all sectors, factors are paid the same average wage or rental rate WF_f. Factor market distortions may be simulated with the sector-specific *wfdist*$_{if}$ factor market distortion parameter. In the absence of distortions, this parameter is equal to one.

$$WF_f \cdot wfdist_{if} = P_i^v \cdot \alpha_{if} \cdot \frac{X_i}{FDSC_{if}} \tag{9}$$

Intermediate input demand INT_i in equation 10 is determined by fixed I-O coefficients, and each intermediate input is a CES aggregation of imported and domestic goods.

$$INT_i = \sum_j a_{ij} \cdot X_j \tag{10}$$

Similarly, imported goods M and domestic goods D are distinct from the composite good Q, with distinct sectoral prices. Any given sector may export and import a good simultaneously. The use of CET and CES functions insulates the domestic price system from large changes in world prices which could result in switching between exports and imports and vice versa. Equation 11 is a CET aggregate of exports and domestic sales where total domestic production X is either supplied to domestic markets D_i or foreign markets E_i and all have different prices.

$$X_i = a_i^T \left[\gamma_i E_i^{\rho_i^T} + \left(1 - \gamma_i\right) \cdot D_i^{\rho_i^T} \right]^{1/\rho_i^T} \tag{11}$$

where a_i^T is the CET function shift parameter, γ_i is the CET function share parameter and ρ_i^T is the CET function exponent.

Equation 12 solves for export supply, which is a function of the export price and domestic price ratio.

$$E_i = D_i \cdot \left[\frac{P_i^e \cdot (1-\gamma_i)}{P_i^d \cdot \gamma_i} \right]^{1/\rho_i^T} \tag{12}$$

Where certain sectors are able to exert market power, equation 13 represents a downward-sloping world export demand function, where $econ_i$ is the export demand shift parameter, $pwse_i$ is the world price of export substitutes and η_i is the export demand price elasticity.

$$E_i = econ_i \cdot \left[\frac{PW_i^e}{pwse_i} \right]^{\eta_i} \tag{13}$$

CES aggregation functions describe domestic and import demand for goods and the import demand function, which is determined by the ratio of domestic to import prices (equations 14 and 15).

$$Q_i = a_i^C \left[\delta_i M_i^{-\rho_i^C} + (1-\delta_i) \cdot D_i^{-\rho_i^C} \right]^{1/\rho_i^C} \tag{14}$$

where a_i^C is a CES function shift parameter and δ_i is a CES function share parameter.

$$M_i = D_i \cdot \left[\frac{P_i^d \cdot \delta_i)}{P^m \cdot (1-\delta_i)} \right]^{1/1+\rho_i^C} \tag{15}$$

Income equations

Income equations track the flow of income from value added to institutions to agents. Factor income Y_f^F, described by equation 16, is distributed to households as household income Y_h^H by equations 17 and 18, where *DEPREC* is the cost of depreciation.

$$Y_f^F = \sum_i WF_f \cdot FDSC_{if} \cdot wfdist_{if} \tag{16}$$

$$Y_{cap\in h}^H = Y_i^F - DEPREC \tag{17}$$

$$Y_{lab\in h}^H = \sum_{f \neq 1} Y_f^F \tag{18}$$

Government tariff revenue *TARIFF*, indirect tax revenue *INDTAX* and income tax revenue *HHTAX* are given by equations 19 through 21, export subsidies *EXPSUB* by equation 22 and total government revenue *GR* by equation 23, where t_h^h is the household income tax rate.

$$TARIFF = \sum_i pw_i^m \cdot M_i \cdot t_i^m \cdot R \tag{19}$$

$$INDTAX = \sum_i P_i^x \cdot X_i \cdot t_i^x \tag{20}$$

$$HHTAX = \sum_h Y_h^H \cdot t_h^h \, h = cap, lab \tag{21}$$

$$EXPSUB = \sum_i PW_i^e \cdot E_i \cdot t_i^e \cdot R \tag{22}$$

$$GR = TARIFF + INDTAX + HHTAX - EXPSUB \tag{23}$$

Equation 24 is financial depreciation *DEPREC* and equation 25 is household savings *HHSA* Equation 26 is government savings *GOVSAV* calculated as government revenues net of government expenditures. Finally, equation 27 represents total savings *SAVING*, which is the sum of household and government savings, less depreciation, plus foreign savings *FSAV* converted to local currency units, where $depr^i$ is the rate of depreciation, mps_h is household marginal propensity to save, and GD_i is government final demand.

$$DEPREC = \sum_i depr^i \cdot P_i^k \cdot FDSC_{il} \tag{24}$$

$$HHSAV = \sum_h Y_h^H \cdot \left(1 - t_h^H\right) \cdot mps_h \tag{25}$$

$$GOVSAV = GR - \sum_i P_i^q \cdot GD_i \tag{26}$$

$$SAVING = HHSAV + GOVSAV + DEPREC + FSAV \cdot R \tag{27}$$

Expenditure equations

Expenditure equations represent the demand for goods and services. Private consumption C_i is a function of household demands calculated by fixed expenditure shares β_{ih}^H in equation 28. Other options exist for representing household demand, such as CES or linear expenditure system (LES) functional forms. In an LES, households allocate their income to consume a minimum level of subsistence goods and services above which they purchase goods and services according to a linear relationship between income and consumption. LES functions differ from CES functions in that LES functions have nonunitary income elasticities for all pairs of goods (Annabi, Cockburn and Decaluwé, 2006) which enables flexible substitution possibilities in response to changes in relative prices (Decaluwé, Lemelin, Robichaud and Maisonnave, 2010).

Government demand *gdtot* in equation 29 is determined by fixed government expenditure shares β_i^G of real spending.

$$P_i^q \cdot C_i \cdot D_i = \sum_h \left[\beta_{ih}^H \cdot Y_h^H \cdot \left(1 - mps_h\right) \cdot \left(1 - t_h^H\right) \right] \tag{28}$$

$$GD_i = \beta_i^G \cdot gdtot \tag{29}$$

Changes in inventories DST_i are represented in equation 30 as fixed shares $dstr_i$ of sectoral production. Aggregate fixed nominal investment $FXDINV$ in equation 31 is total investment $INVEST$ less inventory accumulation. Fixed nominal shares $kshr_i$ in equation 32 are used to convert aggregate fixed investment to real sectoral investment DK_i. The capital composition matrix b_{ij} in equation 33 transforms investment by sector of destination into demand for capital goods by sector of origin ID_i.

$$DST_i = dstr_i \cdot X_i \tag{30}$$

$$FXDINV = INVEST - \sum_i P_i^q \cdot DST_i \tag{31}$$

$$P_i^k \cdot DK_i = kshr_i \cdot FXDINV \tag{32}$$

$$ID_i = \sum_j b_{ij} \cdot DK_j \tag{33}$$

Equations 34 and 35 solve for nominal and real GDP. Nominal GDP is the sum of nominal value added, indirect taxes and tariffs, net of export subsidies.

Real GDP is determined from the expenditure side, where imports are valued in world prices and exclude tariffs, while nominal GDP is calculated from the value-added side.

$$GDPVA = \sum_i P_i^v \cdot X_i + INDTAX + TARIFF - EXPSUB \tag{34}$$

$$RGDP = \sum_i (CD_i + GD_i + ID_i + DST_i + E_i - pw_i^m \cdot M_i \cdot R \tag{35}$$

Macroeconomic closure equations

Market clearing equations and macroeconomic closures represent the system constraints that the model must satisfy. In the competitive market economy system developed here, prices adjust to clear each market. Equation 36 defines the market clearing condition for product markets and equation 37 specifies the equilibrium condition for factor markets where factor supply fs_f is exogenous. Factors may be sector-specific or mobile.

$$Q_i = INT_i + CD_i + GD_i + ID_i + DST_i \tag{36}$$

$$\sum_i FDSC_{if} = fs_f \tag{37}$$

Equation 38 describes the balance of payments closure. To establish the balance of payments, foreign savings are the difference between total imports and total exports. Because foreign savings are defined exogenously, the exchange rate is the equilibrating variable and works through changes in the relative prices of tradable to nontradable goods. The model in effect solves for stability between the real exchange rate and the balance of trade. Alternative balance of payment closures are possible; for example, the exchange rate may be fixed with foreign savings acting as the equilibrating variable. Another closure option would be to fix the price index with the exchange rate, in which case foreign savings would adjust to balance the account.

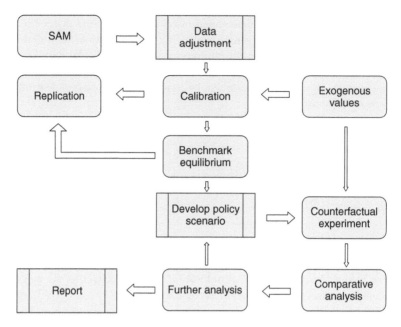

Figure 34.2   CGE workflow diagram with modifications.
Source: Shoven and Whalley (1992).

$$pw_i^m \cdot M_i = PW_i^e \cdot E_i + FSAV \tag{38}$$

Finally, equation 39 represents the aggregate savings *SAVING* and investment *INVEST* balance and is a neoclassical savings-driven closure where aggregate investment is the endogenous sum of the components of savings. Once again, alternative closures are possible such as the investment-driven closure, in which aggregate investment is fixed while a component of savings such as the marginal propensity to save or foreign savings is endogenous.

$$SAVING = INVEST \tag{39}$$

Once we have a system of equations in place, empirical analysis can proceed with a basic SAM (see Figure 34.2). The model is calibrated with base SAM data to estimate various model parameters and shares. Successful calibration to the base year SAM replicates the benchmark equilibrium. Next, counterfactual scenarios are developed and exogenous values are specified to introduce the policy shock. The counterfactual experiment is conducted and a counterfactual equilibrium is established. This counterfactual is compared with the benchmark equilibrium to evaluate the impact of the policy shock on the economy. Results of this analysis may provide insight into further policy counterfactuals and additional scenarios may be implemented.

An illustration of CGE analysis through a case study of Brazil's forest policy

In this section we explore the regional economic impacts of Brazil's implementation experience with forest concessions in the Amazon. In 2006, Brazil's Public Forest Management Law

(PFML; Law 11.284/2006) was passed to regulate the management of public forests for sustainable use and conservation. Prior to this law, Brazil lacked a mechanism to enable the legal harvest of timber on public land. Almost all timber harvesting was conducted on private land or illegally on public land. The law's principal mechanism for developing the natural forestry sector is through establishing forest concessions on public land (Banerjee, Macpherson and Alavalapati, 2009).

The DCGE model developed in Banerjee and Alavalapati (2010), based on the International Food Policy Research Institute's Standard CGE Model in GAMS (Lofgren, Harris, Robinson, Thomas and El-Said, 2002; Robinson and Thurlow, 2004), was applied in this analysis. This model has a regional dimension, representing Brazil's administrative areas, namely, the north, northeast, center west, south and southeast.[2] The underlying data source for the model and its calibration is based on the SAM developed in Banerjee and Alavalapati (2009, 2010). The SAM, with a base year of 2003, has 14 sectors/commodities and has significant forest sector detail, including natural forest management, forest plantations and both legal and illegal deforestation activities. There are three types of labor based on skill level and low, middle and high household income categories.

One salient feature of this model is its ability to track illegal and legal deforestation activities and land use. Although rates of deforestation have been on the decline in Brazil, deforestation remains an important feature of the rural economy, affecting all economic sectors, especially agriculture and forestry. The production structure of the deforestation sector and the customization of the SAM required to represent deforestation are described in detail in Banerjee and Alavalapati (2010). Key assumptions in the treatment of this sector are as follows:

1. Illegal deforestation produces less timber per unit area than the legal deforestation sector and pays less for access to forestland.
2. The illegal deforestation sector does not pay taxes; instead, it pays fines based on assumptions on probabilities of apprehension and rates of collection.
3. The difference between an illegal and legal sector's expenditure to produce a unit of forest product output is considered to be the level of above-normal profits earned for operating illegally. This profit is allocated to labor.

Finally, while the stylized model presented earlier was a static model, the model applied in this analysis is a recursive dynamic model enabling the accrual of investment and growth in factor stocks. In both the baseline and the policy shock scenarios, labor supply is updated based on the estimated labor force growth rate. Capital stocks are updated endogenously based on the previous period's allocation of investment and the rate of capital depreciation. Total factor productivity growth is updated based on OECD (2006) projections while the average capital to output ratio is derived from Morandi and Reis (2004).

The stock of agricultural land is also updated each year. Because the deforestation sectors clear forestland, the quantity of forestland cleared in 1 year is used to update the factor supply of agricultural land in the subsequent year and is thus made available to the agricultural and forest plantation sectors. Equation 40 demonstrates the updating of the agricultural land stock based on the previous period's level of deforestation using the stock of agricultural land in the north as an example.

$$QFS_{agriland\ north,a,t+1} = QFS_{agriland\ north,a,t} + QF_{forestland\ north,legal\ deforestation\ north,t}$$
$$+ QF_{forestland\ north,illegal\ deforestation\ north,t} \tag{40}$$

where:

$QFS_{f,a,t}$ = Quantity of factor f demanded by activity a in time t;
$QFS_{f,a,t}$ = Quantity of factor supply for activity a in time t.

Policy scenario design

The policy scenario developed here follows from the observed rate of growth in forest concessions. Since the PFML was passed, approximately 350,000 ha were made available for harvest as forest concessions in the Amazonian states of Rondônia and Pará. Between 2007 and 2012, forest concessions have been implemented at a rate of 70,000 ha per year.

The forecast baseline projects the Brazilian economy to the year 2030 in the absence of the PFML. In the policy scenario, it is assumed that forest concessions are established in the northern administrative region at a rate of 70,000 ha per year beginning in 2008 and implemented in the model as a forestland shock until 2030. The difference between the policy scenario and the forecast baseline scenario is the impact of the PFML on the Brazilian economy.

A balanced model closure is used in this analysis because it is the preferred closure for examining the probable economic impacts of policy shocks (Lofgren et al., 2002). A flexible real exchange rate is chosen for the balance of payments closure, while the government closure fixes direct tax rates enabling flexible government savings. The domestic price index is chosen as the numeraire. The model is implemented in the general algebraic modelling system (GAMS) and solved as a mixed complementarity problem using the PATH solver.

Results of CGE analysis

The results presented here are the deviation between the forecast baseline and policy scenario with regards to average annual growth rates (AAGR). All macroeconomic indicators grew faster as a result of the implementation of forest concessions, including gross domestic product (GDP), private and government consumption, fixed investment, exports and imports (difference in AAGR of 0.004%, 0.004%, 0.002%, 0.006%, 0.002%, 0.002% and 0.001%, respectively). Household consumption for low-, mid- and high-income households grew faster (difference in AAGR of 0.004%, 0.005% and 0.005%, respectively). Equation 41 summarizes the movement of these macroeconomic indicators, where C is consumption, I is investment, G is government consumption, X are exports and M are imports. Although the rate of growth of imports increased, GDP still grew faster as a result of the implementation of forest concessions.

$$GDP\uparrow = C\uparrow + I\uparrow + G\uparrow + X\uparrow - M\uparrow \tag{41}$$

All economic activities increased output in the baseline with the exception of plantations in the southeast and legal and illegal deforestation in the northeast. The policy shock resulted in slower growth in forestry in the northeast, southeast and south (−0.015%, −0.007% and −0.004%, respectively; Table 34.1). Forest plantation activity grew slower in the north, northeast, southeast, south and center west (−0.331%, −0.296%, −0.309%, −0.293% and −0.349%, respectively). Legal deforestation grew slower in the center west (−0.012%), while illegal deforestation contracted even further than that in the baseline. The policy shock resulted in slower though positive growth in the price of agriculture, forestry, deforestation, sawmilling and food processing (difference in

Table 34.1 Difference in AAGR of economic sectors (%).

Activity	Difference in AAGR	Activity	Difference in AAGR
Agriculture north	0.051	Plantations south	−0.293
Agriculture northeast	0.033	Plantations center west	−0.349
Agriculture southeast	0.005	Deforestation north	1.417
Agriculture south	0.022	Deforestation northeast	0.175
Agriculture center west	0.002	Deforestation center west	−0.012
Food processing	0.010	Illegal deforestation north	0.076
Forestry north	2.163	Illegal deforestation northeast	0.055
Forestry northeast	−0.015	Illegal deforestation center west	−0.099
Forestry southeast	−0.007	Industry	0.002
Forestry south	−0.004	Mining and petroleum	−0.002
Forestry center west	0.068	Utilities	0.003
Sawmilling	0.036	Construction	0.005
Pulp and paper	0.098	Communications	−0.005
Plantations north	−0.331	Transportation	0.002
Plantations northeast	−0.296	Services	0.002
Plantations southeast	−0.309	Public administration	0.001

AAGR of −0.0052%, −0.1430%, −0.0496%, −0.0364% and −0.0013%, respectively), while pulp and paper product prices fell faster (−0.0008%).

Figure 34.3 depicts the difference between the baseline and policy shock for output of domestically produced goods. Industrial commodities and services were the largest gainers, though beginning from an already large base. The increased growth rate of agriculture was tracked by that of the processed food sector. Construction and forest-sector commodity output also grew faster.

The implementation of forest concessions had a positive growth impact for the Brazilian economy, which was expected, because forest concessions were modelled as an increase in the land endowment, enabling increased forest sector output in the north. One of the strengths of the CGE approach highlighted in this case study is its ability to shed light on sectoral linkages and regional impacts. For example, while overall the impact of the policy was positive, forest concessions resulted in slower growth in Brazil's northeast, southeast and southern forestry sectors. Forest plantation activities also grew slower in all regions. This result is largely explained by the increase in output and greater competitiveness of the forest sector in the north as well as increased forest product output through deforestation in the north and northeast.

In the baseline scenario, legal forestry in the north and center west contracted. Legal forestry expanded in all other regions. With forest concessions implemented in 2008, the opposite became true, with a positive policy impact in the north and center west and negative impacts

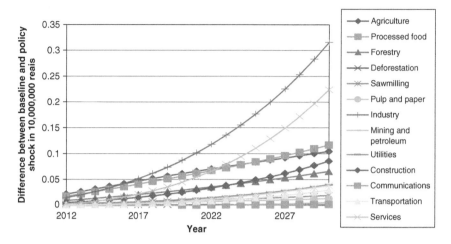

Figure 34.3 Difference between baseline and policy shock domestic commodity production.

in other regions. This is an indication of forestland scarcity, which was offset to a degree by an increase in the forestland endowment. Greater growth in the north and center west was accompanied by expansion in the sawmilling and pulp and paper sectors, which use forest products as intermediate inputs.

Illegal deforestation grew in the north and center west and contracted in the northeast in the baseline. Forest concessions caused a contraction in illegal deforestation in all regions. This contraction in illegal deforestation is a function of the increasing scarcity of forestland on which firms may operate illegally and the reduced returns to agricultural land. Greater growth of legal deforestation in the north and northeast produced agricultural land, allowing the sector to expand in these regions, while reduced growth in illegal deforestation in the center west slowed growth in this region.

Thus, in contrast to partial equilibrium approaches, the CGE approach highlights the interdependencies among economic sectors as well as regions. Constraints on resources (capital, labor and land) result in some sectors contracting while others expand. All sectors in the economy compete for scarce resources. Those sectors with comparative advantage are able to pull resources from less productive sectors. This also holds true for capital investment, in which sectors with above-average rates of return capture a greater share of investment. The dynamic among forestry, forest plantations, deforestation and agriculture is revealed in this analysis, with forestry in the north outcompeting forestry in most other regions. In the case of agricultural land, higher returns to agriculture draw land out of forest plantation production and into agriculture. A general equilibrium approach is required to shed light on this competitive dynamic.

Innovations in CGE analysis

Since Jones's (1965) seminal publication on 'The Structure of Simple General Equilibrium Models' in the *Journal of Political Economy*, several advancements have been made in CGE modelling. Specific institutions have been created to facilitate interaction among CGE modellers, policymakers and other end-users, to share experiences and innovations, and to provide modelling support and advice. The Global Trade Analysis Project (GTAP) at Purdue University is one of the largest such institutions. GTAP aims to improve quantitative analysis in an economy-wide framework by developing a common language for economic analysis through development of software and data, training and annual conference events. The Centre of Policy Studies (CoPS), based at Monash University in Australia, is another institution that does significant economic research and policy analysis for government and the private sector globally. The CoPS has developed a number of widely used models, including ORANI, MONASH and TERM, and offers training courses in CGE modelling worldwide. The CoPS has also developed GEMPACK, a CGE modelling software program, enabling accurate and efficient solution of CGE models, which provides powerful analytical tools to facilitate analysis and interpretation of results (Harrison and Pearson, 1996).

The Partnership for Economic Policy (PEP) was instituted in 2002 and now has over 8,000 members worldwide. PEP aims to ensure greater participation of local expertise in the analysis of issues related to poverty and socioeconomic development, using CGE models as one analytical approach. PEP's mandate is to produce evidence-based policy advice, develop analytical tools and build capacity in policy analysis. Finally, EcoMod is a global economic modelling network with over 3,000 members and offers training, support and advisory services to clients. EcoMod aims to promote advanced modelling and statistical techniques in economic policy and decision making through research, training, conferences and workshops.

Advances have also been made with regard to model validation, which is important to assure policymakers and other users that model results are credible. Validation refers to (1) verifying that solutions were computed correctly and (2) demonstrating that an explanation of results is an accurate reflection of the mechanics of the model. Validation may also be extended to demonstrating model consistency with history, as well as its effectiveness as a forecasting tool. Many procedures exist for validating a CGE (see Dixon and Rimmer, 2013); we present three of the more common approaches.

Testing model sensitivity to parameter inputs (e.g. shock and elasticity parameters) is one test of the robustness of a model and enables confidence intervals to be constructed around model results (e.g. welfare measures) and is referred to as systematic sensitivity analysis (SSA) (Alavalapati, Adamowicz and White, 1999). Monte Carlo analysis is one approach to SSA where a model is solved iteratively with the parameter of interest randomly selected from a sample distribution. For large models, however, this can make model solution cumbersome. The Guassian Quadrature approach provides a good alternative which estimates an approximate parameter distribution, thereby saving considerable computing time (Hertel, Hummels, Ivanic, and Keeney, 2004). Tools for conducting SSA have been built into the CGE modelling software, GEMPACK, facilitating their application significantly (Horridge and Pearson, 2011).

Demonstrating that results are consistent with the workings of the model may be achieved through telling a qualitative story, and quantitatively through back-of-the-envelope (BOTE) calculations. A qualitative story simply seeks to explain model results in broad descriptive terms in a way which demonstrates that model results are supported by assumptions on how the economy behaves (Dixon and Rimmer, 2013). BOTE calculations may then help explain a particular model feature that reproduces key aspects of the results. BOTE models may be as simple as containing only one domestically produced good which is consumed domestically and exported, an imported good and one type of capital and labor. The specification of the BOTE depends on the application (see Dixon and Rimmer, 2013, for examples).

The diversity of thematic areas to which CGE models are applied has increased considerably in the last decade. For example, there is a growing interest from policymakers and practitioners to consider policy impacts on ecosystem services – the benefits nature contributes to human well-being (Costanza et al., 1997; Daily, 1997; Millenium Ecosystem Assessment, 2005; Perrings, 2006; Kumar, 2010). Some ecosystem service values are traded in the market (e.g. carbon credits for climate change mitigation). Other ecosystem services are not traded in the marketplace and are therefore not captured in regular accounting frameworks. These services include regulating services such as erosion and natural disaster mitigation, the provision of habitat and gene pools and cultural and aesthetic values, to mention a few examples.

The System of Environmental and Economic Accounts (SEEA) was developed to hybridize economic data with environmental information in a common accounting framework such that the contribution of the environment to the economy as well as the impact of economic activity on the environment may be assessed in a quantitative way (Dube and Schmithusen, 2003). SEEA provides a promising framework for the accounting of ecosystem service supply and a platform for their integration with CGE models as it follows the United Nations' system of national accounts (European Commission et al., 2012). This framework is the first international statistical standard for environmental-economic accounting.

Significant advances have been made in regional CGE modelling, with increasing demand for evidence-based analysis to inform policies across regions. This demand has been driven in part by the transformative impacts of globalization and technical progress leading to regionally differentiated outcomes and a need to understand the underlying processes (Giesecke and Madden, 2013). Significant regional detail can be included in these models, from region-specific technologies, tastes and constraints, to regionally differentiated policies and policy shocks.

A regional model may be a single region which models an economy with little detail beyond what occurs beyond the regional border; a top-down model which disaggregates results by region through decomposition; a bottom-up model which models each region as a separate economy with a robust specification of interregional trade and factor flows subject to national and regional resource constraints; and finally, a hybrid top-down bottom-up model, such as the model presented in the Brazilian case study, which maintains a standard structure, though incorporates bottom-up detail as required by the nature of the inquiry. Bottom-up models enable deeper treatment of issues such as factor mobility, trade and transport, trans-boundary factor ownership, migration, interregional investment, transport margins, regionally differentiated treatment of taxes and the impact of regional and national government policy (Giesecke and Madden, 2013).

The development and application of DCGE has seen significant advances. Many CGE models are comparatively static and have been used to ask 'what if' questions (Dixon and Rimmer, 2009). Static models roughly distinguish between a short term, where capital is immobile between regions, and a long term, where capital stocks are interregionally mobile. Assumptions with regards to labor are made as appropriate to the particular region being modelled. Dynamic models are increasing their presence due to the additional economic insight they provide. Dynamic models portray the transition path of an economy, which permits a more robust specification of investment behavior, the potential to introduce lags such as sticky wages and interregional labor migration, as well as the ability to customize the policy or environmental shock through time for greater realism.

Understanding the past can help us better understand the future; that is, the relative importance of drivers of change can inform future policy development by highlighting those factors that were more important in driving economic outcomes (Giesecke and Madden, 2013). Historical modelling is one form of dynamic modelling and can shed light on how preferences and technologies as well as policies, trade and foreign demand have evolved over time (Dixon and Rimmer, 2009). This form of modelling occurs in two steps – first as a historical simulation and then a decomposition simulation. The power of this approach lies in its ability to provide information on the individual contributions of the historical movement of variables of interest, including those pertaining to policy and economic structure (Giesecke and Madden, 2013).

Economic forecasting is another emerging area for CGE modelling, providing forecasts for industries, employment and regional economies. Baseline forecasts use information derived from historical simulations; thus, the forecasts are consistent with historical trends, often supplemented by expert opinion (Dixon and Rimmer, 2009). Baseline forecasts also serve as the business-as-usual case used as a reference by which the impact of a policy change is evaluated. The net welfare impact of a policy or investment program depends heavily on the baseline forecast used (Wittwer and Banerjee, in review).

For the consideration of particular policy options, sometimes greater detail in agent behavior may be required, which has led to the integration of new theoretical constructs into the CGE model itself – for example, the TERM H_2O model, which incorporates a water accounting and trading framework within the CGE (Wittwer, 2012) or theories of lagged wage adjustment. Alternatively, CGE models may be linked with microsimulation models to enable more robust analysis of household level effects, particularly where understanding policy impacts on poverty and inequality are of interest (Ferreira Filho and Horridge, 2006; Agenor, Chen and Grimm, 2004; Savard, 2005; Bouet, Estrades and Laborde, 2013).

Notes

1   Notational convention: Upper case variables are exogenous; lower case variables are endogenous. The subscript i refers to commodities, and the subscript j refers to activities. The superscripts m and e are for imports and exports, respectively; d represents domestic, x is output, v is value added, q is composite commodity and k is for capital.
2   Brazil's administrative regions are north, northeast, southeast, south and center west. The northern region is composed of the states of Rondônia, Acre, Amazonas, Roraima, Pará, Amapá and Tocantins. The northeastern region includes Maranhão, Piauí, Ceará, Rio Grande do Norte, Paraíba, Pernambuco, Alagoas, Sergipe and Bahia. The southeast includes Minas Gerais, Espírito Santo, Rio de Janeiro and São Paulo. The south includes Paraná, Santa Catarina and Rio Grande do Sul. The center west includes Mato Grosso do Sul, Mato Grosso, Goiás and the Distrito Federal.

References

Agenor, P.-R., Chen, D.H.C. and Grimm, M. (2004). *Linking representative household models with household surveys for poverty analysis: A comparison of alternative methodologies*, World Bank, Washington, DC.

Alavalapati, J.R.R., Adamowicz, W. L. and White, W. A. (1998). 'A comparison of economic impact assessment methods: The case of forestry developments in Alberta', *Canadian Journal of Forest Research*, vol. 28, no. 5, pp. 711–719.

Alavalapati, J.R.R., Adamowicz, W. L. and White, W. A. (1999). 'Random variables in forest policy: A systematic sensitivity analysis using CGE models', *Journal of Forest Economics*, vol. 5, no. 2, pp. 321–335.

Alavalapati, J.R.R., Percy, M. B. and Luckert, M. K. (1997). 'A computable general equilibrium analysis of a stumpage price increase policy in British Columbia', *Journal of Forest Economics*, vol. 3, no. 2, pp. 143–169.

Alavalapati, J.R.R., White, W. A. Jagger, P. and Wellstead, A. (1996). 'Effect of land use restrictions on the economy of Alberta: A general equilibrium analysis', *Canadian Journal of Regional Science*, vol. 19, no. 3, pp. 349–365.

Annabi, N., Cockburn, J. and Decaluwé, B. (2006). *Functional forms and parameterization of CGE models*, MPIA Working Paper 2006–04, Poverty and Economic Policy Network, Quebec.

Bandara, J. S. (1991). 'Computable general equilibrium models for development policy analysis in LDCs', *Journal of Economic Surveys*, vol. 5, pp. 3–69.

Banerjee, O. and Alavalapati, J. (2009). 'A computable general equilibrium analysis of forest concessions in Brazil', *Forest Policy and Economics*, vol. 11, pp. 244–252.

Banerjee, O. and Alavalapati, J. (2010). 'Illicit exploitation of natural resources: The forest concessions in Brazil', *Journal of Policy Modeling*, vol. 32, pp. 488–504.

Banerjee, O., Macpherson, A. J. and Alavalapati, J. (2009). 'Toward a policy of sustainable forest management in Brazil: A historical analysis', *Journal of Environment & Development*, vol. 18, pp. 130–153.

Banerjee, O., Macpherson, A. J. and Alavalapati, J.R.R. (2012). 'Socioeconomic and land use trade-offs of ethanol expansion in Brazil', *Journal of Sustainable Forestry*, vol. 31, pp. 98–119.

Bouet, A., Estrades, C. and Laborde, D. (2013). *Household heterogeneity in a global CGE model: An illustration with the MIRAGE-HH (MIRAGE-HouseHolds) model*, LAREFI Working Paper No. 2013-01, Laboratoire d'Analyse et de Recherche en Economie et Finances Internationales (LAREFI), Pessac.

Buetre, B., Rodriguez, G. and Pant, H. (2003). *Data issues in general equilibrium modelling*, 47th Australian Bureau of Agricultural and Resource Economics Society, Freemantle, Western Australia.

Cattaneo, A. (1999). *Technology, migration and the last frontier: A general equilibrium analysis of environmental feedback effects on land use patterns in the Brazilian Amazon*, Johns Hopkins University, Baltimore, MD.

Cattaneo, A. (2001). Deforestation in the Brazilian Amazon: Comparing the impacts of macroeconomic shocks, land tenure, and technological change', *Land Economics*, vol. 77, no. 2, pp. 219–240.

Cattaneo, A. (2002). *Balancing agricultural development and deforestation in the Brazilian Amazon*, International Food Policy Research Institute Research Report 129, Washington, DC. Retrieved from www.ifpri.org/pubs/abstract/abstr129.htm

Costanza, R., D'arge, R., De Groot, R., Farber, S., Grasso, M., Hannon, B., . . .Van Den Belt, M. (1997). 'The value of the world's ecosystem services and natural capital', *Nature*, vol. 387, pp. 253–260.

Daily, G. C. (Ed.). (1997). *Nature's services: Societal dependence on natural ecosystems*, Island Press, Washington, DC.

Decaluwé, B., Lemelin, A., Robichaud, V. and Maisonnave, H. (2010). *The PEP standard computable general equilibrium model single-country, recursive dynamic version, PEP-1-t*, Partnership for Economic Policy, Université Laval, Quebec City, Canada.

Dee, P. S. (1991). *Modeling steady state forestry in computable general equilibrium context*, Working Paper No. 91/8, National Centre for Development Studies, Canberra, Australia.

Dixon, P. B., Parmenter, B. R., Powell, A. and Wilcoxen, P. J. (1992). *Notes and problems in applied general equilibrium economics*, North-Holland, New York.

Dixon, P. B. and Rimmer, M. T. (2009). *Forecasting with a CGE model: Does it work?* Centre of Policy Studies, Monash University, Clayton, Australia.

Dixon, P. B. and Rimmer, M. T. (2013). 'Validation in computable general equilibrium modeling', in P. B. Dixon and D. W. Jorgenson (Eds.), *Handbook of computable general equilibrium modeling*, North-Holland, New York.

Dube, Y. C. and Schmithusen, F. (2003). *Cross-sectoral policy impacts between forestry and other sectors*, Forestry Paper 142, Food and Agriculture Organization, Rome.

Dufournaud, C. M., Jerrett, M., Quinn, J. T. and Maclaren, V. (2000). 'Economy-wide effects of forest policies: A general equilibrium assessment From Vietnam', *Land Economics*, vol. 76, no. 1, pp. 15–27.

European Commission, Food And Agriculture Organization, International Monetary Fund, Organisation for Economic Cooperation and Development, United Nations and the World Bank. (2012). *System of environmental-economic accounting: Central Framework*.

Ferreira Filho, J. and Horridge, M. J. (2006). 'Economic integration, poverty and regional inequality in Brazil', *Revista Brasileira de Economia*, vol. 60, pp. 363–387.

Gan, J. B. (2004). 'Effects of China's WTO accession on global forest product trade', *Forest Policy and Economics*, vol. 6, no. 6, pp. 509–519.

Gan, J. B. (2005). 'Forest certification costs and global forest product markets and trade: A general equilibrium analysis', *Canadian Journal of Forest Research*, vol. 35, no. 7, pp. 1731–1743.

Giesecke, J. A. and Madden, J. R. (2013). 'Evidence-based regional economic policy analysis: The role of CGE modelling', *Cambridge Journal of Regions, Economy and Society*, vol. 6, no. 2, pp. 285–301.

Harrison, W. J. and Pearson, K. R. (1996). 'Computing solutions for large general equilibrium models using GEMPACK', *Computational Economics*, vol. 9, pp. 83–127.

Hertel, T., Hummels, D., Ivanic, M. and Keeney, R. (2004). *How confident can we be in CGE-based assessments of free trade agreements?* Global Trade Analysis Project, Working Paper No. 26, Purdue University, West Lafayette, IN.

Horridge, J. M. and Pearson, K. (2011). *Systematic sensitivity analysis with respect to correlated variations in parameters and shocks*, Global Trade Analysis Project, Purdue University, West Lafayette, IN.

Huang, M., Alavalapati, J.R.R. and Banerjee, O. (2012). 'Economy-wide impacts of forest bioenergy in Florida: A computable general equilibrium analysis', *Taiwan Journal of Forest Science*, vol. 27, no. 1, pp. 81–93.

Jones, R. W. (1965). 'The structure of simple general equilibrium models', *The Journal of Political Economy*, vol. 73, no. 6, pp. 557–572.

King, B. B. (1985). 'What is SAM?', in G. Pyatt and J. I. Round (Eds.), *Social accounting matrices: A basis for planning*, World Bank, Washington, DC.

Kumar, P. (Ed.). (2010). *The economics of ecosystems and biodiversity: Ecological and economic foundations*, Earthscan, London.

Lofgren, H., Harris, R. L., Robinson, S., Thomas, M. and El-Said, M. (2002). *A standard computable general equilibrium (CGE) model in GAMS*, International Food Policy Research Institute, Washington, DC.

Millenium Ecosystem Assessment. (2005). 'Ecosystems and human well-being: Wetlands and water synthesis'. Retrieved from www.millenniumassessment.org/documents/document.358.aspx.pdf

Morandi, L., and Reis, E. (2004). 'Estoque de capital fixo no Brasil, 1950–2002', *Anais do XXXII Encontro Nacional de Economia. ANPEC – Associação Nacional dos Centros de Pós-Graduação*.

Ochuodho, T. O., Lantz, V. A., Lloyd-Smith, P. and Benitez, P. (2012). 'Regional economic impacts of climate change and adaptation in Canadian forests: A CGE modeling analysis', *Forest Policy and Economics*, vol. 25, pp. 100–112.

OECD. (2006). *Brazil: OECD economic surveys*, Vol. 2006/18, OECD, Paris.

Perrings, C. (2006). 'Ecological economics after the millennium assessment', *International Journal of Ecological Economics and Statistics*, vol. 6, pp. 8–22.

Persson, A. and Munasinghe, M. (1995). 'Natural resource management and economywide policies in Costa Rica: A computable general equilibrium (CGE) modeling approach', *The World Bank Economic Review*, vol. 9, no. 2, pp. 259–285.

Robinson, S. and Thurlow, J. (2004). 'A recursive dynamic computable general equilibrium (CGE) model: Extending the standard IFPRI static model', Unpublished manuscript.

Robinson, S., Yunez-Naude, A., Hinojosa-Ojeda, R., Lewis, J. D. and Devarajan, S. (1999). 'From stylized to applied models: Building multisector CGE models for policy analysis', *North American Journal of Economics and Finance*, vol. 10, pp. 5–38.

Savard, L. (2005). 'Poverty and inequality analysis within a CGE framework: A comparative analysis of the representative agent and microsimulation approaches', *Development Policy Review*, vol. 23, pp. 313–331.

Shoven, J. and Whalley, J. (1992). *Applying general equilibrium*, Cambridge University Press, Cambridge, UK.

Stenberg, L.C. and Siriwardana, M. (2005). 'The appropriateness of CGE modelling in analysing the problem of deforestation', *Management of Environmental Quality*, vol. 16, no. 5, pp. 407–420.

Stenberg, L.C. and Siriwardana, M. (2007). 'Forest conservation in the Philippines: An economic assessment of selected policy responses using a computable general equilibrium model', *Forest Policy and Economics*, vol. 9, no. 6, pp. 671–693.

Thompson, W. A., Van Kooten, G. C. and Vertinsky, I. (1997). 'Assessing timber and non-timber values in forestry using a general equilibrium framework', *Critical Reviews in Environmental Science and Technology*, vol. 27, pp. 351–364.

Wiebelt, M. (1994). *Protecting Brazil's tropical forest: A CGE analysis of macroeconomic, sectoral and regional policies*, Kiel Working Paper No. 638, Kiel Institute for the World Economy, Kiel, Germany.

Wittwer, G. (Ed.). (2012). *Economic modeling of water, The Australian CGE experience*, Springer, Dordrecht, The Netherlands.

Wittwer, G. and Banerjee, O. (in revision). 'Investing in irrigation development in North West Queensland, Australia'. Submitted to Australian Journal of Agricultural and Resource Economics.

35

TWELVE UNRESOLVED QUESTIONS FOR FOREST ECONOMICS – AND TWO CRUCIAL RECOMMENDATIONS FOR FOREST POLICY ANALYSIS

William F. Hyde[1]

1930 SOUTH BROADWAY, GRAND JUNCTION, CO 81507, USA. WFHYDE@AOL.COM

Abstract

Faustmann showed us the basic economic formulation of the forestry problem 164 years ago. However, since then foresters and economists have learned of a number of situations in which Faustmann's marvelous contribution is either incomplete, inaccurate or inappropriate. Our recent experience addresses some of those – but we do not have fully accepted answers for all of them. This chapter identifies issues that remain unresolved within the discipline of economics as applied to forestry at the beginning of the twenty-first century, and then adds two more crucial concerns for policy applications.

Keywords

Faustmann, technological change, conservation, distribution, recreation supply, land tenure, collective management, environmental Kuznets's curve

Introduction

It has been 164 years since Faustmann's (1849/1968) famous formulation of the basic economic theory of the firm problem for forestry. In that time, numerous economists, many who are also trained foresters, have contributed to our understanding of the economics of forestry and the role of the forest sector in the aggregate economy. Nevertheless, it should be clear that some fundamental questions do still remain, and that those that remain have important implications for policy and management. (After all, wouldn't we direct our efforts elsewhere if the knowledge of forest economics was complete, if we found that little remained to challenge us and if our applications were infallible?)

What follows is one participant's organization of the outstanding unresolved questions. 'Unresolved', not 'unsolved', because the former term is more comprehensive. That is, for some of these twelve questions, the evidence, and the well-formed opinion too, is either incomplete or misunderstood. These do remain unsolved. For others, valid solutions too often go unrecognized or standard approaches are applied in inappropriate context. Those that remain unsolved need our attention. Those with solutions that go unrecognized or that are applied inappropriately need to be repeated and amplified until their comprehension is widespread.

Each of the twelve questions is important in varying degree wherever there are trees and forests. However, I would posit that many of both, the unsolved and the unresolved, are of special importance in the developing world. To be sure, issues such as the protection of diverse habitat, carbon sequestration to deter global change, and general environmental sustainability are of abiding interest for the people of the developed countries of Western Europe, North America and Japan, but they are especially relevant topics for tropical and developing country forestry. Important characteristics of the forest and important conditions of the political and market environments are different for many developing economies, different than the forest characteristics and political and market environments of the developed world where most forest economists of all nationalities were trained or the textbooks that underlie their training were written. Differences such as fewer forest plantations or more frequent occasions of political and economic instability cause some of the unresolved issues of forest economics to be much more important for the developing world, and, therefore, a developing economy focus is especially important for our attempts to examine many of the issues that I will outline. Finally, the greater diversity of developing country data, an asset for statistical analysis, may cause some developing country assessments to be more convincing than comparable assessments using forest data from more developed countries.

The twelve issues that follow are fundamental within the economics of forestry, but economics itself is a fundamental tool for understanding management and policy. Therefore, the applications of forest economics in general and the resolutions to these twelve problems can have their greater impacts on the global policy issues that are broader than forestry but which involve forestry at any moment in time. At this moment these are issues such as global change, protecting diverse habitat and environmental sustainability. I cannot anticipate the important policy issues that will include forestry in the rest of the twenty-first century. I can anticipate that, whatever these issues are, now or in the future, two cautions are critical for reliable applications of forest economics in general and for its applications for policy analysis in particular. I'll add these two cautions in the conclusion. Readers will agree that they are well known. Nevertheless, my own perspective is that way too many errors in modern policy analysis are due to their oversight.

Without further delay, let's look at the twelve questions – and suggest both some of the evidence and a few hypotheses.

The twelve questions

1. The Faustmann equation itself

The Faustmann equation is absolutely fundamental, as all forest economists know, but why is it that virtually no practical forest manager follows it? Shouldn't that signal something to our profession? Virtually no one anywhere harvests timber at an age corresponding to the simple Faustmann rule for financial maturity.

Consider each of the four general landowner categories: For vertically integrated industrial landowners, financial incentives are paramount. Their focus is not shortsighted, but it is not on their forests either. Their focus has to be on the return on their full investment; and their investments in manufactured capital, their mills, are usually very much greater than their investments in timber or forestland.[2] Therefore, integrated industrial landowners tend to hold forests as a guarantor of productive raw material input for the lifetime of their mills. The fundamental economic model for industrial forestry must account for this.

For nonindustrial private forest (NIPF) land managers, some have entirely different nontimber objectives for their forests; others combine income from timber with other objectives. For both, timber management and harvest decisions are more complex than simple variations on Faustmann and we shouldn't expect applications of Faustmann to predict the behavior of these landowners. For still others, timber management is primarily an income-generating activity, but market timing and immediate family needs, rather than applications of Faustmann, seem to be better predictors of their harvest decisions.

Governments are a third landowner category. In many countries they are the largest landowner. Government forestry agencies must respond to public values for nontimber forest services as well as for commercial timber, and also to different political motivations and budget constraints. But even when governments profess to follow economics in their commercial timber management, they tend to introduce their own version of 'sustainable yield' with little and generally convoluted reference to fundamental economics.

Furthermore, many public forests are mature and near or even beyond the boundary of successful commercial timber harvest activity. The Faustmann equation was designed for continuous commercial forest management. These mature forests do not fit the Faustmann description. Their harvests are better modeled as a one-time timber mining operation – perhaps with a share of the financial proceeds earmarked for reforestation to protect the local environment. Finally, forests beyond the frontier of commercial activity surely do not fit the Faustmann description – even if some public forestry agencies do find reason to finance timber harvests beyond this frontier.

Institutions are a newer, fourth, category of forest landowners, a category that has become important only in the last 30 years as comprehensive financial institutions have recognized the favorable risk-to-return relationship for forest ownership and the somewhat countercyclical nature of timber markets. For both reasons, timber contributes balance to broad investment portfolios. The new institutional landowners may follow Faustmann prescriptions more closely than landowners in the other three categories, but even the harvest behavior of these institutions may be tempered by exogenous demands for draws on their larger financial portfolios. Observations of the conduct of these institutions with respect to their forestlands in general and their timber harvesting in particular are probably too brief for us to draw confident conclusions as yet.

The problem becomes even more complex than the assessment of each specific landowner category when we attempt to explain the forest economics of a region that includes two or more of these landowner categories and, within them, both managed forests and mature natural forests at and beyond the boundary of viable commercial operation. There have been at least two attempts to explain forestry across this entire complex landscape (Angelson, 2000; Hyde, 2000, 2012) and one econometric application (Zhang et al., 2000), but one might question how widely these are known and accepted. As yet, no one has developed the full mathematical economic model explaining this complete landscape of multiple owner categories with multiple objectives possessing both managed and mature natural forests both within and beyond the frontier of economic activity.

2. The relationship between the macroeconomy and the forest sector

Do the effects of macroeconomic activity and macroeconomic policy dominate the effects of direct forest policy? Under what conditions? There is compelling evidence of the affirmative from isolated communities with undiversified forest sectors in the western United States (Burton and Berck, 1992; Burton, 1997). But what about somewhat less diverse communities with both timber and nontimber forest sectors of significance?[3]

And if the effects of macroeconomic activity and macroeconomic policy do dominate the effects of more direct forest policies like timber taxes, management and harvest regulations and incentive payments, then shouldn't this fact encourage a shift in the focus of our own professional analyses, causing us to direct relatively less attention to those narrower and more direct forest policies and relatively greater attention to the effects of macroeconomic variables on the forest sector and the forest itself? Shouldn't it encourage greater interaction between forest economists and those macroeconomists and macroeconomic policy institutions that also profess concern for the environment?

Stability, as one condition of the macroeconomic and political environment, deserves special attention. Economists know that stability, or at least predictability, is a necessary condition for any long-term investment – including long-term investments in forestry. Forest economists, most of whom have been educated in developed countries with relatively stable economic and political environments, tend to overlook this condition when we assess forest activity. Yet instability alone can dominate all other factors in the determination of a successful forest sector. Macroeconomic and/or political instability is often the most basic source of forest degradation and deforestation (e.g. Deacon, 1994), as well as the reason that forest management remains unattractive in some countries even where high prices for forest products and low costs for forest management might suggest greater opportunity.

These questions of macroeconomic market and policy impacts on forest management decision making deserve greater recognition. That will come only with further assessments by forest economists showing the effects of the macroeconomic environment on local and regional forests in a variety of situations. Many, but certainly not all, of these analyses will have to feature developing countries with significant forest sectors and contrasting recent periods of significant economic and political variability.

3. Technological change in forest products, forestry and forest conservation

Most of the analytical literature on technological change in forestry focuses on the wood processing industries, but some of this literature measures the economic effects of improved knowledge and improved technologies in forest management itself. The crucial conclusion to be taken from it is that the rate of technological improvement in wood processing exceeds both the natural growth rate for trees and the rate of technological improvement in forest management and timber production. This shouldn't be surprising. The potential physical gains from technological improvements in forest growth are well known – but few have been adopted widely even in the most technologically advanced regions of global forestry. Their adoption is likely to continue only slowly as long as inexpensive substitutes exist, particularly in the natural forests that compose one of forestry's extensive margins. (See Hyde, 2012, chapter 3 for a summary.)

The implications are twofold: First, if the beneficial effect of technological change in wood processing is relatively more important than that in timber management, then the secondary effect that extends from wood processing on to the forest may also be more beneficial to the conservation of forests and forestland than whatever improvements in on-the-ground forest production have been introduced to date – although how much more forest conservation has benefited from technological change in wood processing remains an interesting question. Furthermore, if this forest conserving impact is significant, then wouldn't it also be interesting to know whether research that leads to technological change in wood products could be designed for its favorable effect on particular forest environments or particular forest resource services?

Second, while the productive effects of improvements in on-the-ground timber management may not have progressed rapidly in the past, should we expect this experience to continue in the future? How great is the opportunity for substitution (suggested previously) and how long will it last? This too is worthy of inquiry. But might it be even more important to inquire about the potential for gains in the production of the increasingly important set of nonmarket forest services? Where (which resource services, at which locations, under what conditions) have these latter gains been greatest (whatever their source)? Where might research specifically designed for its direct effect on improving nontimber forest services be most productive in the future? These are, no doubt, difficult questions for analysis. The challenge of the inquiry, however, is no reason to disregard these potentially important questions of modern forest policy.

4. Forest extension and technology transfer

The forest industry rapidly adopts any new technology that either saves costs or increases production per unit of input. (See Hyde et al., 1992, for discussion and examples.) Forest land managers are a more dispersed group than the managers of industrial wood processing firms, and many land managers are more focused on alternative activities than on timber management. The latter may not be as aware of or as quick to adopt new technologies as their larger counterparts in the wood products industries. This creates targeted opportunity for forestry extension and for those international development agencies involved in the transfer of forest technologies. The question for them has to do with where their assistance can be most successful and under what conditions, with which new technologies and with which forest landowners and land managers.

Complete understanding of these related questions will require a diversity of case studies because the appropriate new technologies will differ across the many and varied local situations. A multitude of examples available to us now, however, make it clear that, despite numerous successes, there have also been many failures. Many extension agents have been unsuccessful in their attempts to introduce favorite new technologies, and the success of the many reforestation projects of numerous development agencies has also been less than desired.

Agricultural economists have some experience with their related technology transfer questions. There is an extensive literature that examines the conditions under which new agricultural technologies are most likely to be adopted and which farm landowners are the most likely adoption leaders. (See Ruttan, 1982, for an early summary or Rogers, 1983, for a noneconomic discussion; Feder et al., 1985, is the classic.) Might similar effort in forest economics provide improved guidance for forest extension agents and for those agencies that promote forestry projects?

5. (Income) distributive merit

Assessments of the distributive effects of public policy interventions in developed country forestry tend to focus on the sectoral and regional effects of forest activity. Some examine impacts on select groups of landowners. Negligible few, however, distinguish among rural income groups. Nevertheless, income redistribution remains a prime justification for public policy in general, and many continue to use distributive arguments for forestry interventions to assist women and the rural poor. These income distributive arguments may be justified in some cases, but there is precious little research literature whatsoever that examines (a) whether and how much distributive success has arisen from existing programs, (b) whether the target beneficiaries of public forestry programs are justified on distributive grounds or (c) whether there are other groups with greater distributive merit. Have income distributive programs been successful? Under what conditions? Or have they been misguided and less than successful despite their presumed distributive merit?

The Amacher et al., (1993) and Cooke (1998) assessments of household and gender dependence on forest resources in Nepal provide, to my mind, one useful approach for examining poor communities with varying degrees of integration into market economies. The assessments by Ruiz-Pérez et al. (Ruiz-Pérez et al., 1996; Ruiz-Pérez et al., 1999; Ruiz-Pérez and Belcher, 2001; Ruiz-Pérez et al., 2001a) of five counties in China provide another. All three provide convincing evidence that poor communities and households as well as particular minorities can benefit from available forest resources – but important economic conditions determine when they benefit most and, even then, other not-so-poor households often benefit even more.

The distributive examination needs to be broader, however, than just who benefits from forestry activity and the related (and seldom examined) question of whether these beneficiaries are among the more justified. It also needs to inquire about those in any community who do not benefit from the forestry activity (even poorer households? the landless?). And are these latter even more deserving of distributive assistance? Is there anything about the forest resource that would enable it to have potential benefit for them? None of the Ruiz-Pérez, Amacher or Cooke assessments continue their analyses to inquire of these questions – and very few other analyses do either.

6. The conduct and performance of public forest agencies

Public agencies, including forest ministries and domestic and international development agencies, have roles in each of the last three issues. The public forestry agencies conduct forest research and they and the development agencies provide technical assistance or fund forestry support programs. They also manage forestry programs which they justify on their distributive merit. In addition, public agencies manage the largest share of our global forest lands.

If these agencies were private institutions, they would come under almost continuous financial scrutiny from their investors. The public agencies certainly experience political oversight and some occasionally experience questions about their overall budget and personnel levels, but their detailed financial oversight is most uncommon. To my observation, support from their client groups for additional specialized funding is more common than any form of financial scrutiny. Financial and other managerial oversight, however, could be useful and it would be particularly timely in the current political climate questioning government deficits and the extent of government participation in general economic activity.

The means have always been available for assessing the financial viability of the public's commercial timber operations and there have been a few such assessments, at least in the United States. But what about the nonmarket activities of these agencies? Economists have developed two techniques, data envelopment analysis and stochastic frontier analysis, for assessing the performance of nonprofit organizations. They have applied these techniques in a wide variety of cases (e.g. hospitals, schools), including a very few public forestry examples (Kao and Yong, 1991; Vittala and Hanninen, 1998). One, for Poland (Siry and Newman, 2001), observes that twenty-six of forty timber management units of that country's state forestry agency operate more than 40% below the production frontier established by the agency's more efficient units. In particular, many units are inefficiently small and suffer from an excess of administrative personnel (but not an excess of forest workers). Imagine a 40% increase in efficiency simply by observing and adopting the best administrative structures and management practices of one's own colleagues. This is phenomenal.

Similar assessments of forestry and development programs and agencies around the world would tell us a great deal about the characteristics of successful, cost-effective, public forestry programs and the locations and levels of opportunity for the most beneficial adjustments in agency budgets and personnel levels.[4]

7. Recreation supply – together with demand

There is another, analytically easier, question having to do with public forest management but about which our knowledge is certainly inadequate. Public forests support huge and still rapidly increasing levels of outdoor recreation; for example, 270 million visitors each year in the United States (www.nps.gov; www.fs.fed.us) or increases as great as 25% over only 2 years (2000–2002) in China (Sayer and Sun, 2003). Economists have examined the demand for this forest resource service at a broad collection of locations.

However, economists are supposed to know that demand is only one-half of the information necessary for judgments about efficient allocation. Yet the assessment of the other half, supply for forest recreation, is limited to my knowledge to one single study (Daniels, 1986). This study, and its combined assessment of demand, found a remarkably efficient pattern of public agency decision making for improved campsites in a single valley in western Montana in the United States. Given the well-recognized costs of recreation facilities and the substantial personnel costs associated with the management of any recreation site, even remote wilderness sites, wouldn't additional assessments of recreation supply (together with demand) be useful – either to confirm the generality of the remarkable pattern of decisions in the Swan Valley in Montana, or to show public forestry agencies and consumers of outdoor recreation where improved resource allocation is possible, as well as what the gains from the improvement might be?

8. Distinctions between the various forest industries with respect to their demands on the forest resource

The modicum of inquiry into the structure, conduct and performance of the various wood processing industries is instructive about the forest industries' demands for each of their inputs: labor, manufactured capital, energy and natural resources. More information, however, would be better.

For our interests as forest economists, the different industrial demands for virgin wood fiber would be most instructive. Sawmills, for example, are highly dependent on this resource. Pulp mills are relatively much less so. The latter's investment in manufactured capital is much greater, both absolutely and relatively, and, unlike sawmills, pulp mills generally use fiber from other sources in greater volume than fiber directly processed from logs. For some pulp mills, water may be a more crucial resource than wood fiber. Nevertheless, wood may be the pulp mills' most important variable input. And what about the resource demands of the bioenergy market, or of the increasing number of producers of engineered wood products? How do they vary from the demands of either the pulp or sawmill industries and how and where do their industrial differences affect the forest itself?

The most important question for us may have to do with the anticipated pattern of growth for each of these industries and what this pattern suggests for future demands on the forest. To understand this question, we must anticipate growth in the demands for the processed products and then assess both the long-run and the short-run substitution between wood and other inputs for these industries, and between each industry's variable inputs in particular. And we must appreciate that our conclusions will not only be different for the different forest product industries, but these conclusions may vary even within one of these industries as the various establishments within that industry use somewhat different technologies, therefore different input mixes, to produce related products in regional and national economies characterized by different relative output prices and input costs.

9. NIPF landowners[5]

NIPF landowners are the subject of some of the best and most extensive forest economics research over the last 60 years. Much of that for the United States and northern Europe developed from an original interest in timber supply from the NIPF lands. The NIPF literature from developing countries has focused on land tenure and, particularly, the benefits of communal forest management.

The more recent research features distinctions within this class of landowners. NIPF landowners are no longer a homogeneous group. Perhaps they never were. Farm and nonfarm landowners within the broader group demonstrate different behaviors with respect to both their production and their consumption of various forest products and forest resource services. Kuuluvainen et al. (1996) finds even further distinctions – between landowners who manage their lands primarily (a) for timber revenues, (b) as an income safety net, (c) for family recreation or (d) for multiple objectives. Farm forest landowners in developing countries seem to manage their trees and forests differently depending on whether they obtain their primary employment and income on- or off-farm. In fact, can we determine that there really are substantial distinctions between the essential groups of NIPF landowners in either developed or developing countries?

Each category of NIPF landowners may sell timber sooner or later, but each also produces and consumes other forest resource services and, as Kuuluvainen et al. (1996) observe, each responds in a statistically significantly different manner to various market and policy signals. This is crucial. If their behaviors in response to market and policy signals are different, then we must examine them as independent, if sometimes related, groups. Assessments that lump them together will always be superficial, and often misleading.

In a related consideration, what are the policy questions having to do with NIPF landowners? As suggested earlier, these began with concern for long-run timber supply. Timber supply remains a motivating factor for some. For others, the production (often the protection) of environmental services from NIFP lands is a greater question.

These may be valid questions – although I suspect that the market will take care of timber supply, with rising prices inducing greater harvest levels and the additional management that leads to increasing forest stocks and, eventually, to further increases in harvest levels from the NIPF lands – as well as from other forestlands. The literature supports this argument, although it suggests that the responses of NIPF landowners are not especially price elastic (e.g. Binkley, 1981; Boyd, 1984; Kuuluvainen et al., 1996; among others).[6]

What about the performance and the financial burden of public programs that are designed to ensure a supply of either timber or environmental services from the NIPF lands? Some public programs provide financial assistance to NIPF landowners. The public bears the financial burden for these. Other public programs are prescriptive, requiring specific silvicultural treatments by the landowners. Where these treatments are additional to what the landowners would provide without the government regulation, as they must be in some cases, then surely the burden is on the landowners. But who are the beneficiaries? Just as surely, and in both cases, the additional future timber supply created by the public program has a market effect, reducing the price to the benefit of the purchasers, the mills.

Others, too, have suggested this sequence of effects for these public programs and the products and services that NIPF landowners provide selective consumers, but only in conversation (seldom in print, to my knowledge), and for the Nordic countries as well as the United States. Has anyone laid out the sequence clearly, and then estimated each component? Does the additional timber or improved environment induced by these public programs justify their costs, including the costs of program management and monitoring? Where and under what conditions? Can anyone provide a justification for programs that cost the public and small landowners and benefit, in particular, the wood processing industry – if, indeed, this is the case?

10. Land tenure and collective forest management

There has been extensive discussion, particularly by nongovernmental organizations (NGOs) and international development agencies, of the merits of collective forestland management in rural areas or by particular ethnic groups. There is, however, a body of cautionary, and even contrary, evidence. Various observers argue that local elites capture most of the benefits from collective properties. (Andersson and Agrawal, 2011, survey the literature.)

The comparable experience in agriculture suggests that farmers prefer to independently own and manage their lands, but that farmers often act collectively on activities that require large temporary outlays of capital or personnel, activities like harvesting or fire control, or activities for which individual farmers have no special personal advantage, activities like marketing. Why should our expectations be different for forestry? Indeed, evidence from China (Yin and Newman, 1997) shows that farm households grow larger forest stocks on more land when allowed to switch away from collective and toward individual household forest management.

In fact, we might expect the agricultural experience to duplicate itself in forestry – or, if we're not ready to accept the agriculture experience as fully transferable, then we might expect a spectrum of local situations such that it could be quite possible that collective action (formal or otherwise) works very well only where a unique set of conditions is present. The research literature increasingly recognizes these points, but until these conditions are clearly and more convincingly argued, I'm afraid that the single-minded enthusiasm of many NGOs and international development agencies for collective action will continue unabated – often to the detriment of their programs and to the local communities they intend to assist.

11. Is there an environmental Kuznets's curve for forestry?

Environmental Kuznets's curves (EKCs) are the graphic description for the argument that regional and national economic development begins in association with natural resource withdrawal and environmental depletion but, as development proceeds, the rate of depletion tapers off gradually until eventually, as growth and development proceed further yet, the environmental decline ceases. Even further along in the development process, above some critical level of welfare, environmental quality improvements and nonconsumptive public goods such as the amenity values received from natural forests become relatively more important than additional natural resource consumption, the economy invests in pollution control and other environmental improvements and overall environmental health improves along with improvements in human health and welfare.

The original statement of the EKC (Grossman and Krueger, 1993) refers to the condition of the full environment, and most developed countries seem to have followed the pattern of the EKC over the long course of their histories. Numerous empirical examinations have traced the pattern for specialized environmental conditions such as particular environmental pollutants.

The question for us is whether the pattern is accurate for forests. The existing evidence is most limited, and the question deserves further assessment.[7] (See Hyde, 2012, chapter 6, for a summary).

The question is important because, if the EKC is an accurate expectation for forests, then overall economic development is a positive response to our global concern with forest degradation and deforestation. Moreover, if the EKC is accurate for forests, then the turning point, the moment in the course of economic development at which forests begin to recover, is critical and the factors that affect this point should be of interest to all of us. If we can understand this turning point and the factors that could alter it, then we can have confidence regarding the fundamental means for correcting global deforestation.

The very limited current evidence is that the turning point for forests falls somewhere in the annual per capita income range of US$64 for one national case study for forestry (China) or seven times that for another (Vietnam) and US$5,000–8,000 for broader environmental assessments for numerous other countries. This is an exceedingly wide range. Are there better estimates? Is the evidence for either China or Vietnam a misestimate, or a special case? And regardless of the correct estimate, what market and policy factors are effective in reducing the turning point?

12. The wildland–urban interface[8]

As human populations have grown and as personal incomes have also grown, human intrusion into the forests of six continents, particularly at the forest edge, has increased – rapidly. This forest margin has become known as the wildland–urban interface. It is the location of many of twenty-first century forestry's most difficult problems and of professional forestry's greatest current concerns. In the United States, for only one example, eight million new homes, an increase of almost 70%, were built in the wildland–urban interface in the 1990s (Steelman, 2008).

The problems emerging with this development are of unexperienced magnitude for forest economists whose traditional foci have been on trees and the forest itself or on forest resource services, many of which tend to be dispersed (carbon, biodiversity) or off-site (watershed) and, perhaps, less immediate in their contribution to human value. The growing human activity in the wildland–urban interface, in contrast, is more concentrated and on-site and its demands are often immediate.

Wildfire is one increasingly important example, and one that promises to worsen further as global warming increases the risk of extreme weather conditions (Westerling et al., 2006; Hennessey et al., 2008). Growing human activity in the wildland–urban interface probably increases the occurrence of wildfire, and it certainly means that those fires that do occur are the source of greater risk and greater damage or loss. All activity in this region is of greater value and subject to greater management costs (and greater loss in the case of fire and other natural hazards) than when human activity in these lands was of smaller scale. The wildfire impact, however, is especially noteworthy. In six continents the number of human dwellings destroyed by wildfire is increasing annually. In the United States, wildfire management now consumes over 50% of the annual US Forest Service budget. The means for protecting and the requirements for insuring human communities within the wildland–urban interface have taken on new magnitudes of importance (e.g. Bradshaw and Lueck, 2012).

Our knowledge of the wildland–urban interface is in its early stages of development. Whatever economists – and other foresters – can learn about it is likely to be an important contribution.

Conclusion: Two further concerns for forest policy analysis

For each of these twelve unresolved research questions, and for others too, it is important that we rely on (1) empirical assessments of (2) incremental effects. This should go without saying. As foresters, one of our strengths should be our practical and empirical orientation. As economists, we should know to look for shifts at the margin. However, most any perusal of our literature will demonstrate the merit of this reminder. To my mind, too much of our literature is not empirical or, if it is empirical, it does not focus on marginal effects.

The empirical component of this reminder would argue, first of all, for getting out and mixing with landowners and forest users, learning their questions and directing our skills as forest economists toward the problems they identify. Some of us do this, but not often enough in my opinion.

Similarly, this argument would call for increased reliance on the observed behaviors of on-the-ground forest managers and consumers of forest-based resources and resource services. Readily available data alone are insufficient. Our models must mimic the observed behavior. That is, although both forest managers and consumers may be economically rational, their behaviors can be quite different than the researcher's supposition of economic rationality. In other cases, researchers and managers may agree on the narrow rationality of some particular behavior, but not on its relative importance. There are all too many examples of the latter. They can result in great expenditures of research energy on issues which managers understand to be less important and know to be dominated by other considerations. Only by putting our research models to the test of empirical data can we determine whether we are focusing on meaningful issues.

Economic empiricism also argues for reliance on economic rather than physical data. The former, economic data, can be a problem because most forest data report stocks measured to an arbitrary physical level. These stocks may be either less than or greater than an economic measure of the same resource, and even economic measures of the resource vary with the resource use in question. For example, all trees sequester carbon, including vast numbers of trees in home gardens, urban parks, backyards and other small plots, and along roadsides and fencerows that are excluded from the official measures of forests. For two examples, the official forest cover is absolutely minimal for New Jersey in the United States and Java in Indonesia. Yet when flying over either New Jersey or Java, one is impressed with the almost continuous forest cover. All of this 'unofficial' forest cover sequesters carbon – but it's seldom included in estimates of carbon sequestration. In contrast with the carbon example, potentially commercial timber is much less than many official measures of the forest stock, as these official measures include vast areas beyond the boundary of commercial access – in the interior of Alaska and northern Canada, in Siberia, in the Amazon and in Indonesian Kalimantan, for example. Of course, the forested regions of potential watershed or recreation value, or value for their threatened biodiversity, are different yet than both the carbon and the commercial timber examples. Valid economic assessments of each of these very different forest resource services must adjust for their unique differences from the official measures of forest stock.

The fair recognition of market and policy impacts on the forest requires still additional care in the accurate identification of marginal impacts. Consider most any modern forest policy question. Consider four widely different, currently important, issues: trade, certification, carbon offsets and controls on illegal harvest activity.

- Discussions of the effects of trade on the forest environment tend to feature the producing region, the region whose environment may be negatively affected by the expansion of trade and the resulting local increase in timber removal. The same discussions tend to overlook the counterpart decline in production and, therefore, environmental improvement in the consuming region. They may also overlook the (often longer-term) beneficial effects of trade-induced investment in the producing region.

- Discussions of certification tend to identify the gross area of forestland that has been certified. This is an overestimate of certification's effect. Those forests that are already managed, as are most of the forests of Western Europe, for example, are readily certified. Certification does not introduce a change in their management. A fair measure of the gains from certification includes only those lands whose managers perceive gains from formal recognition and, therefore, shift from an unsustainably to a sustainably managed category in order to qualify for the certification.

- Perhaps similarly, the true effect of carbon offsets is only on the frontier from which the forest would have been removed in absence of the offset. The remote forest beyond that frontier was not in danger of degradation or deforestation and, therefore, the carbon offset does not affect this remote region.[9]

  On the other hand, is there any market or policy activity which contributes significant adjustment to the volume of carbon stored in trees that are unaccounted in the standard physical measures of forests; for example, those trees previously mentioned as unaccounted but growing in large number in New Jersey in the United States or Java in Indonesia? Do these unaccounted trees, perhaps, grow and store even more carbon as local welfare improves? I suspect that, beyond some level of local economic development, they do. (Consider the EKC.) Yet no examination of trees and carbon, to my knowledge, has inquired whether the measure of these additional trees and this additional storage of carbon is significant.

- Finally, some illegal logging can be traced to corruption, as generally charged. Other occurs simply because forests are so dispersed that it is impossible to limit all trespass, and even the best designed program for monitoring and enforcement will not be fully effective. Therefore, total control is an unreasonable expectation at any price. In this case, the measure of effectiveness should compare the additional costs of any new controls on trespass and other illegal activity in the forest with the marginal reduction in illegal activity. Once more, the marginal economic effects are what we're after.

In sum, the point is to focus on the forest resource at risk, that which is either threatened or restored within our understanding of the particular resource question at hand and the prospective policy modification under discussion. This is usually a small, although crucial, share of the total forest, yet a different share for each forest resource use and for each prospective policy. Often, not always, it requires focus on three separate areas: the natural forest frontier plus, in some cases, the external margin of managed forest and any affected degraded forest. It means going beyond official measures of forest, often excluding vast regions beyond the frontier of commercial activity but also including large accumulations of generally uncounted trees in home gardens, in urban parks, in backyards and so forth.

Our challenge is to improve the understanding of the twelve questions in the body of this paper and to conduct better empirical assessments of market and policy effects similar to those identified in these last paragraphs. We don't have to provide definitive answers. The challenge is only to move toward improved answers with the hope that someday someone's improvement will be complete and definitive. Be bold. Tackle the important questions. If we cannot solve them, then let's sharpen our understanding of them and let's narrow the bounds of our uncertainty about them. This is the challenge that makes our job as researchers in the field of forest economics interesting.

Notes

1   Shashi Kant, Janaki Alavalapati and participants at the International Symposium of Forest Economics held at Beijing Forestry University in 2012 provided useful critique.
2   This is one reason that most large operators in the US wood processing industries have divested themselves of forestland over the course of the last 20–30 years. (See Hyde, 2012, chapter 8, for a review.) Nevertheless, many integrated wood processing firms in other countries continue to either own or hold under contract large areas of forest.
3   In a question related to these first two issues, Faustmann and macroeconomics, the Faustmann equation tells us that lower interest rates are associated with longer timber rotations. Therefore, as interest rates decline, we should expect landowners to delay harvests while their timber stocks increase with age. Yet we generally observe the opposite: Lower interest rates are followed by increases in timber harvests. Some of our distinguished predecessors identified this contradiction in conversation (but not in print, to my knowledge) at least 40 years ago. Does anyone have the answer? Is it just that lower interest rates encourage an expanding general economy and, therefore, an increase in the demand for wood products, and that macroeconomic effects do trump the microeconomics of forestry? What does this conclusion suggest for appropriate applications of Faustmann?
4   Of course, such assessments are less likely if the public agencies are also the funding sources for economic research in forestry – but that shouldn't make these questions any less meaningful for a few bold analysts.
5   The organization of this question benefits from discussion with Jari Kuuluvainen.
6   And related to this point, why do some farm landowners in both developed and developing countries continue to hold financially sub-marginal forestland? Some NIPF lands are, apparently, economically sub-marginal for any of the usual forest products or resource services. Are these simply lands in transition from one economic use to another? If not, then why do some landowners continue to hold them, and even continue to pay taxes on these unproductive lands? Is the total land area involved significant? Is it sufficiently large that disregarding its magnitude confuses our broader regional and national assessments of NIPF lands and landowners? (See Kuuluvainen, Karppinen and Ovaskainen, 1996; Hyde, 2012, chapter 9, for additional discussion.)
7   In order to understand the potential for an EKC for forests, we may have to examine underlying questions about the growth of nonmarket forest services. For example, what is the income elasticity of demand for environmental services like tourism, carbon sequestration, biodiversity or forested watersheds? Does it change with improving levels of per capita income?
8   Shashi Kant improved the definition of this twelfth question.
9   Modern spatial data permit better economic definition of the forest frontier – but only in local cases because, at this time, sorting national or global satellite data is so very time-consuming and expensive. There are now numerous applications of spatial data that trace economic forest use and forest land conversion near or at a local frontier. Two good recent examples are Pfaff et al. (2012) for Acre in Brazil's Amazon and Holland et al. (2012) for the Ecuadorean Amazon.

References

Amacher, G., Hyde, W. and Joshee, B. (1993). 'Joint production and consumption in traditional households: Fuelwood and agricultural residues in two districts of Nepal', *Journal of Development Studies*, vol. 30, no. 1, pp. 206–225.

Andersson, K. and Agrawal, A. (2011). 'Inequalities, institutions, and forest commons', *Global Environmental Change* (forthcoming).

Angelson, A. (2000). 'Faustmann meets von Thünen in the jungle: Combining forest rotation and spatial approaches in shifting cultivation', in S. Chang (Ed.), *Proceedings of the International Symposium: 150 Years of the Faustmann Formula*, Louisiana State University School for Forestry, Wildlife, and Fisheries, Baton Rouge, LA.

Binkley, C. (1981). 'Timber supply from private nonindustrial forests', Bulletin No. 92, Yale University School of Forestry and Environmental Studies, New Haven, CT.

Boyd R. (1984). 'Government support of nonindustrial production: The case of private forests', *Southern Economic Journal*, vol. 59, pp. 89–107.

Bradshaw, K. and Lueck, D. (Eds.). (2012). *Wildfire Policy: Law and Economics Perspectives*, RFF Press, New York.

Burton, D. (1997). 'An astructural analysis of national forest policy and employment', *American Journal of Agricultural Economics,* vol. 79, no. 3, pp. 964–974.

Burton, D. and Berck, P. (1992). 'Statistical causation and national forest policy in Oregon', *Forest Science,* vol. 42, no. 1, pp. 86–92.

Cooke, P. (1998). 'Intrahousehold labor allocation responses to environmental good scarcity: A case study from the hills of Nepal', *Economic Development and Cultural Change*, vol. 46, pp. 807–830.

Daniels, S. (1986). 'Marginal cost pricing and efficient resource allocation: The case of public campgrounds', PhD thesis, Duke University, Durham, NC.

Deacon, R. (1994). 'Deforestation and the rule of law in a cross section of countries', *Land Economics*, vol. 70, no. 4, pp. 414–430.

Faustmann, M. (1968). 'On the determination of the value which forest land an immature stands possess for forestry', in M. Gane (Ed.), Institute Paper 42, Commonwealth Forestry Institute, Oxford University, Oxford, pp. 27–55. (Original work published 1849)

Feder, G., Just, R. and Zilberman, D. (1985). 'Adoption of agricultural innovations in developing countries: A survey', *Economic Development and Cultural Change*, vol. 33, no. 2, pp. 255–298.

Grossman, G. and Krueger, A. (1993). 'Environmental impacts of the North American Free Trade Agreement', in P. Garber (Ed.), *US-Mexico Free Trade Agreement*, MIT Press, Cambridge, MA, pp. 13–56.

Hennessy, K., Fawcett, R., Kirono, D., Mpelasoka, F., Jones, D., Bathols, J., … Plummer, N. (2008). *An Assessment of the Impact of Climate Change on the Nature and Frequency of Exceptional Climatic Events*, CSIRO Marine and Atmospheric Research, Aspendale, Victoria, Australia.

Holland, M., de Koning, F., Morales, M., Naughton-Treves, L., Robinson, B. and Suárez, L. (2012). 'Complex tenure and deforestation: Implications for conservation incentives in the Ecuadorian Amazon', *World Development* (forthcoming).

Hyde, W. (2000). 'Applications of the Faustmann Model: Limitations and extensions', in S. Chang (Ed.), *Proceedings of the International Symposium: 150 Years of the Faustmann Formula*, Louisiana State University School for Forestry, Wildlife and Fisheries, Baton Rouge, LA, pp. 181–189.

Hyde, W. (2012). *The Global Economics of Forestry*, Taylor and Francis for RFF Press, New York.

Hyde, W., Newman, D. and Seldon, B. (1992). *The Economic Benefits of Forestry Research*, Iowa State University Press, Ames, IA.

Kao, C. and Yong, C. (1991). 'Measuring the efficiency of forest management', *Forest Science*, vol. 37, no. 5, pp. 1239–1252.

Kuuluvainen, J., Karppinen, H. and Ovaskainen, V. (1996). 'Landowner objectives and nonindustrial private timber supply,' *Forest Science*, vol. 42, pp. 300–309.

Pfaff, A., Robalino, J., Herrera, L., Lima, E. and Sandoval, C. (2012). 'The governance, locations and forest impacts of protected areas: Governance's development tradeoffs affect siting, outweighing its direct effect', *World Development* (forthcoming).

Rogers, E. (1983). *Diffusion of Innovations* (3rd ed.), The Free Press, New York.

Ruiz-Pérez, M., and Belcher, B. (2001). 'Comparison of bamboo production systems in six counties in China', in F. Maoyi, M. Ruiz Pérez and Y. Xiaosheng (Eds.), *Proceedings of the Workshop on China Social Economics: Marketing and Policy of the Bamboo Sector*, China Forestry Publishing House, Beijing, pp. 18–54.

Ruiz-Pérez, M., Fu, M., Xie, J., Belcher, B. and Zhong, M. (1996). 'Policy change in China: The effects on the bamboo sector in Anji County', *Journal of Forest Economics*, vol. 2, no. 2, pp. 149–176.

Ruiz-Pérez, M., Fu, M., Yang, X. and Belcher, B. (2001). 'Toward a more environmentally friendly bamboo forestry in China', *Journal of Forestry*, vol. 99, no. 7, pp. 14–20.

Ruiz-Pérez, M., Zhong, M., Belcher, B., Xie, C. and Fu, M. (1999). 'The role of bamboo plantations in rural development: The case of Anji County, Zhejiang, China', *World Development*, vol. 27, no. 1, pp. 101–104.

Ruttan, V. (1982). *Agricultural Research Policy*, University of Minnesota Press, Minneapolis.

Sayer, J. and Sun, C. (2003). 'Impacts of policy reforms on forest environments and biodiversity', in Hyde, W., Belcher, B. and Xu, J. (Eds.), *China's Forests: Global Lessons from Market Reforms*, Resources for the Future, Washington, DC, pp. 177–194.

Siry, J. and Newman, D. (2001). 'A stochastic production frontier analysis of Polish state forests', *Forest Science*, vol. 47, no. 4, pp. 526–533.

Steelman, T. (2008). 'Communities and wildfire policy', in Donoghue, E. and Sturtevant, V. (Eds.), *Forest Community Connections: Implications for Research, Management, and Governance*, RFF Press, Washington, DC.

Vittala, E. and Hanninen, H. (1998). 'Measuring the efficiency of nonprofit forestry organizations', *Forest Science*, vol. 44, no. 2, pp. 298–307.

Westerling, A., Hidalgo, H., Cayan, D. and Swetnam, T. (2006). 'Warming and earlier spring increase western U.S. wildfire activity', *Science*, vol. 313, pp. 940–943.

Yin, R. and Newman, D. (1997). 'The impact of rural reforms on China's forestry development', *Environment and Development Economics*, vol. 2, no. 3, pp. 289–303.

Zhang, Y., Uusivuori, J. and Kuuluvainen, J. (2000). 'Econometric analysis of the causes of forestland use/cover change in Hainan, China', *Canadian Journal of Forest Research*, vol. 30, pp. 1913–1921.

INDEX